D0086638

EXPLORING THE UNIVERSE WITH THE IUE SATELLITE

ASTROPHYSICS AND
SPACE SCIENCE LIBRARY

A SERIES OF BOOKS ON THE RECENT DEVELOPMENTS OF
SPACE SCIENCE AND OF GENERAL GEOPHYSICS AND ASTROPHYSICS
PUBLISHED IN CONNECTION WITH THE JOURNAL
SPACE SCIENCE REVIEWS

Editorial Board

R. L. F. BOYD, *University College, London, England*

W. B. BURTON, *Sterrewacht, Leiden, The Netherlands*

L. GOLDBERG, *Kitt Peak National Observatory, Tucson, Ariz., U.S.A.*

C. DE JAGER, *University of Utrecht, The Netherlands*

J. KLECZEK, *Czechoslovak Academy of Sciences, Ondřejov, Czechoslovakia*

Z. KOPAL, *University of Manchester, England*

R. LÜST, *European Space Agency, Paris, France*

L. I. SEDOV, *Academy of Sciences of the U.S.S.R., Moscow, U.S.S.R.*

Z. ŠVESTKA, *Laboratory for Space Research, Utrecht, The Netherlands*

VOLUME 129

EXPLORING THE UNIVERSE WITH THE IUE SATELLITE

Editor in Chief

Y. KONDO

NASA/Goddard Space Flight Center, Greenbelt, Maryland

Editors

W. WAMSTEKER

IUE Groundstation, Villafranca del Castillo

A. BOGGESS

Sciences Directorate/GSFC, Greenbelt, Maryland

M. GREWING

University of Tübingen, Tübingen, F.R.G.

C. DE JAGER

Sterrenwacht 'Sonnenborgh', Utrecht, The Netherlands

A. L. LANE

Jet Propulsion Lab., Pasadena, California

J. L. LINKSY

Joint Institute for Laboratory Astrophysics, Boulder, Colorado

R. WILSON

University College London, England

D. REIDEL PUBLISHING COMPANY

A MEMBER OF THE KLUWER ACADEMIC PUBLISHERS GROUP

DORDRECHT / BOSTON / LANCASTER / TOKYO

REF
QB
474
.E97
1987

Library of Congress Cataloging in Publication Data

Exploring the universe with the IUE satellite.

(Astrophysics and space science library; v. 129)
Bibliography: p.
Includes index.
1. Ultraviolet astronomy. 2. IUE (Artificial
satellite) I. Kondo, Yoji. II. Wamsteker, W.
(Willem). III. Series.
QB474.E97   1987   522'.68   87–9807
ISBN 90–277–2380–X

Published by D. Reidel Publishing Company,
P.O. Box 17, 3300 AA Dordrecht, Holland.

Sold and distributed in the U.S.A. and Canada
by Kluwer Academic Publishers,
101 Philip Drive, Assinippi Park, Norwell, MA 02061, U.S.A.

In all other countries, sold and distributed
by Kluwer Academic Publishers Group,
P.O. Box 322, 3300 AH Dordrecht, Holland.

All Rights Reserved
© 1987 by D. Reidel Publishing Company, Dordrecht, Holland
No part of the material protected by this copyright notice may be reproduced or
utilized in any form or by any means, electronic or mechanical
including photocopying, recording or by any information storage and
retrieval system, without written permission from the copyright owner

Printed in The Netherlands

TABLE OF CONTENTS

PREFACE

This book was conceived to commemorate the continuing success of the guest observer program for the International Ultraviolet Explorer (IUE) satellite observatory. It is also hoped that this volume will serve as a useful tutorial for those pursuing research in related fields with future space observatories. As the IUE has been the product of the three-way collaboration between the U.S. National Aeronautics and Space Administration (NASA), European Space Agency (ESA) and the British Engineering and Research Council (SERC), so is this book the fruit of the collaboration of the American and European participants in the IUE. As such, it is a testimony to timely international cooperation and sharing of resources that open up new possibilities.

The IUE spacecraft was launched on the 26th of January in 1978 into a geosynchronous orbit over the Atlantic Ocean. The scientific operations of the IUE are performed for 16 hours a day from Goddard Space Flight Center in Greenbelt, Maryland, U.S.A., and for 8 hours a day from ESA Villafranca Satellite Tracking Station near Madrid, Spain.

The opportunities for research with the IUE have been open to all astronomers of the world; indeed, scientists from all five continents have participated in research using this satellite observatory. Guest observers come to either of the two sites and obtain their observations in *real-time* assisted by resident astronomers and telescope operators. The total number of guest observers at Goddard should come to 755 individuals by the end of the 9th guest observer year ending on 31 May 1987, and that at Villafranca 720. This means that the IUE has been used for research by a very substantial fraction of all astronomers engaged in research, particularly in the U.S. and Europe.

The astronomical sources studied with the IUE range from solar system objects to external galaxies and quasars. The brightest object observed was Venus at −4th magnitude, and the faintest the central star of a planetary nebula at 20th magnitude. The results have had a profound influence on a large number of subdisciplines. The choice of themes for the chapters in this book was not altogether simple but the subjects chosen are, in the view of the editorial committee, representative of the outstanding research performed with the IUE.

Perhaps, the most important and difficult-to-judge criterion for a scientific program is the quality of the research performed. Some three dozen chapters in this book address that point eloquently. Another criterion is its productivity in terms of refereed papers published. At the end of 1985 the articles published in principal *refereed* journals using IUE observations had come to 1133. In this sense, as some guest observers have commented, the IUE has been the most productive telescope in the solar system over the past several years.

I should like to acknowledge the invaluable contributions, which have been made to the successful construction and operations of the IUE, by a large number of American and European scientists and engineers. Regrettably, it is not practical to cite all their names individually in these pages.

Y. Kondo (ed.), Scientific Accomplishments of the IUE, pp. vii.
© *1987 by D. Reidel Publishing Company.*

I wish to express my sincere thanks to my fellow editors, W. Wamsteker, A. Boggess, M. Grewing, C. de Jager, A. L. Lane, J. L. Linsky, and R. Wilson. Willem Wamsteker served as chief coordinator for our European colleagues and deserves a special recognition. Jeff Linsky's contribution was invaluable in starting this project.

YOJI KONDO
Editor in Chief

FOREWORD

The International Ultraviolet Explorer (IUE) has been observing astronomical spectra almost continually since its launch into a geosynchronous orbit on January 26, 1978, and has obtained 50 000 consecutive astronomical exposures by mid-1986. After nine years of successful operation of this satellite, a joint effort of the National Aeronautics and Space Administration (U.S.A.) and the European Space Agency, a survey of the scientific results obtained seems highly appropriate and is presented in this book.

The primary characteristic of the IUE which has made possible its prodigious scientific output is its capability for measuring a very wide region of the spectrum in a single exposure, with its two spectrographs covering the ranges 1150 to 1950 Å and 1900 to 3200 Å. With SEC Vidicons as the detectors, quantitative measures of the full spectrum within either of these regions are obtained either at high resolution (0.1 to 0.3 A) or low (6 to 7 A). The high observing efficiency possible in geosynchronous orbit also contributes to the high IUE data rate.

An area of scientific research which has been particularly well matched to the IUE capabilities is that of stellar atmospheres, particularly the high-excitation phenomena which can most readily be detected with ultraviolet observations. A number of factors contribute to this excellent match of IUE characteristics with the requirements of stellar research. Among these are the rich stellar spectra at ultraviolet wavelengths, the availability of numerous relatively bright stars with interesting and complex phenomena occurring in their atmospheres, and the adequacy for most stellar programs of IUE's spectroscopic resolving power of 10^4. Consequently, of the 31 articles in this book dealing with scientific results, 13 of these are devoted to stellar atmospheres, with many pioneering results obtained.

In the field of external galaxies and quasars the IUE has provided an exciting introduction to the results obtainable from ultraviolet studies. Since such systems have low radiation fluxes at the Earth, and the IUE clear aperture is only 0.45 m in diameter, most of the spectra observed in this research have been of low resolution with the echelle replaced by a mirror and the spectra produced by the predisperser alone. The 7 papers devoted to this topic describe a number of fascinating results, and whet one's appetite for the data anticipated from the Hubble Space Telescope, with its primary mirror 28 times larger in area.

Interstellar matter studies are also represented here, with 7 papers devoted to this topic. For observations of interstellar absorption lines the IUE is somewhat handicapped in that its resolving power and photometric accuracy are not quite high enough for ready observation of the weaker features, which tend to be unsaturated and hence simple to interpret. However, for emission-line studies and for a number of special absorption-line programs IUE results have made significant contributions to our understanding of the interstellar gas.

The volume is rounded out with three chapters of solar system studies and several others on the past and the future and on methods for using IUE data. In fact, use of the extensive data in the IUE archives is likely to support research for some years, having increased the number of IUE-based papers well above the 1200

Y. Kondo (ed.), Scientific Accomplishments of the IUE, pp. ix—x.
© *1987 by D. Reidel Publishing Company.*

published by mid-1986. It is clear that in the history of space astronomy the IUE will occupy a very important place.

LYMAN SPITZER, JR.

PART I

THE IUE PROJECT

Edited by R. WILSON

THE HISTORY OF IUE

ALBERT BOGGESS

Goddard Space Flight Center Greenbelt, MD, U.S.A.

and

ROBERT WILSON

Department of Physics and Astronomy, University College London, UK

with inputs from

PETER J. BARKER

Science and Engineering Research Council, London, UK

and

LESLIE M. MEREDITH

Goddard Space Flight Center, Greenbelt, MD, U.S.A.

1. Introduction

The International Ultraviolet Explorer, in the form of its acronym, IUE, has become a household name to astronomers throughout the world. Launched on January 26, 1978, it is still operational at the time of writing (end 1986, its ninth year) making it the longest-lived astronomical satellite ever. Although it is now showing some signs of age, its scientific performance is essentially unimpaired; for example, its photometric sensitivity has declined by only a few percent since launch.

IUE is a collaborative project of three agencies — the US National Aeronautics and Space Administration (NASA), the European Space Agency (ESA) and the UK Science and Engineering Research Council (SERC). Essentially, it is an astronomical space observatory consisting of a 45 cm Ritchey—Chrétien telescope which feeds either of two echelle spectrographs covering a total wavelength range of 1150—3200 Å with high ($\sim 10^4$) or low (~ 300) resolving powers. The basic detectors consist of solar-blind, uv-to-visible converters coupled to secondary electron conduction (SEC) television cameras. The sensitivity is such that, with several hours' exposure, blue objects can be observed down to about 12th visual magnitude in high resolution and down to about 17th visual magnitude in low resolution. Launched into an elliptical, geosynchronous orbit, it is controlled directly from one of two ground stations located at Goddard Space Flight Center near Washington and at Villafranca del Castillo (VILSPA) near Madrid. The continuous communication with the satellite allows an interactive and highly flexible operation in which the observer is present at the ground station and makes real-time decisions so as to maximise the quality and astronomical value of his/her data.

The division of responsibilities between the three agencies is as follows: SERC provided the on-board cameras with associated software and design inputs to

Y. Kondo (ed.), Scientific Accomplishments of the IUE, pp. 3—19.
© 1987 *by D. Reidel Publishing Company.*

the scientific instrument and sunshade; ESA provided the solar arrays for the spacecraft and the European ground observatory at Villafranca; NASA provided the rest of the scientific instrument and spacecraft, the launch and the US ground observatory at Goddard. Based on these contributions, NASA, the major partner, has responsibility for two-thirds of the observing time and exercises this by direct control of IUE from its Goddard ground station for sixteen hours every (sidereal) day. The remaining one-third is shared equally between the other two collaborating agencies who exercise this by direct control from the ESA ground station for eight hours every day. All three agencies operate a guest observer system and co-ordinate their time allocation procedures by issuing a common invitation for proposals which are accepted from anywhere in the world and judged entirely on scientific merit as assessed by peer review.

IUE is a very international project in its building and operation and even more so in its use. Those proposals which have been allocated time originated in thirty-one different countries and involved more than six hundred astronomers as principal or co-investigators. The data obtained resulted, by mid-1986, in more than one thousand papers appearing in the major refereed scientific journals (*Astrophysical Journal, Astronomy and Astrophysics, Monthly Notices, Nature*, etc.). Indeed, IUE became the basis for a higher publication rate than any other single astronomical facility and has inspired eight international symposia entirely devoted to its results, three in Washington, two in London and one each in Tübingen, Madrid and Rome; the proceedings of these conferences also contain papers numbering well in excess of one thousand. This prodigious output is not only a compliment to the foresight and achievements of the many people involved in the design, building, launching, commissioning, operation and use of IUE but a reflection on the astrophysical richness of the ultraviolet region of the spectrum which contains the resonance lines of many of the most cosmically abundant elements in the Universe.

IUE is clearly an outstanding success and has, through the efforts and ingenuity of its wide international user community, made a major impact on modern astronomy. This book is a testimony to that statement and demonstrates the great breadth of the IUE mission, covering as it does most areas of solar system, stellar, interstellar, galactic and extragalactic astronomy.

2. The Development of Ultraviolet Astronomy (pre-IUE)

The advent of the space age saw a rapid and remarkable development of cosmic ultraviolet astronomy in the United States. The pioneering observations were made by using the unstabilised Aerobee rocket to carry simple photometer systems which scanned the sky by the free motion of the vehicle. In this way, the first ultraviolet photometric data of non-solar objects were obtained in 1955—57 (Byram *et al.*, 1957; Kupperian *et al.*, 1958; Boggess and Dunkelman, 1959). Further advances in rocket-based UV instruments came rapidly, primarily from groups at the Goddard Space Flight Center, University of Wisconsin, Princeton University, the Naval Research Laboratory and Johns Hopkins University. The flight of a scanning, objective grating spectrometer in an Aerobee in 1961 resulted in the first stellar ultraviolet spectrophotometric observations (Stecher and Milligan, 1962) whereas the first ultraviolet stellar spectra (viz with sufficient resolution to detect individual spectral lines) were obtained in a rocket flight in 1965 (Morton and Spitzer, 1966) and were made possible by the development of a 3-axis star-pointing stabilisation system for the Aerobee rocket. These rocket programmes were remarkably successful in providing a foretaste of the research potential of UV astronomy. By the late 1960's they had produced photometric and spectroscopic observations of most of the bright O—B stars accessible from launch sites in the Northern Hemisphere, plus data on a few late-type stars and planets. Some of the major results were a revision of the temperature scale for hot stars, discovery of mass ejection in early supergiants, determination of the UV characteristics of interstellar scattering, and the first measurements of interstellar molecular hydrogen.

Since astronomy needs long time intervals for observations as much as large telescope apertures, the jump from the sounding rocket, with a typical operational time of minutes, to satellites, with a typical operational time of about one year, was inevitable. The first satellite instrument to provide UV data on stars was a photometer launched on a US Navy spacecraft in 1964 (Smith, 1967) and one of NASA's earliest, major, long-term programmes was the Orbiting Astronomical Observatories (OAO's). These were designed specifically for ultraviolet astronomy; the first OAO was planned to carry low-dispersion spectrometers covering the wavelength range 1200—4000 Å built by the University of Wisconsin plus UV cameras provided by the Smithsonian Astrophysical Observatory, the second OAO was to carry a medium-dispersion UV spectrometer from the Goddard Space Flight Center, and the third a high-dispersion spectrometer covering 900—3000 Å provided by Princeton University. The first of these payloads, OAO—A, failed after reaching orbit in 1966 but the spare parts were assembled into a second payload launched in 1968 which became the extremely successful OAO—2 (Code et al., 1970; Davis et al., 1972). Although the next in the series, OAO—B, was destroyed in a launch failure, the final one, OAO—C, launched in 1972 and designated Copernicus (Rogerson et al., 1973) was an outstanding success.

During this same period, astronomical observations were also being made from manned satellites. In fact, during his historic first orbital flight, John Glenn used a hand-held, objective-prism 35 mm camera, provided by a well-meaning bureaucracy, to record ultraviolet spectra of stars. Not until an astronomer was asked to interpret the developed film was it discovered that the window in Glenn's spacecraft was opaque to the ultraviolet! Despite this disappointing start, worthwhile data were obtained from cameras carried on Gemini flights and especially on Skylab (Henize et al., 1975) and also from the automated moon-based observatory set up by the astronauts on Apollo 16 (Carruthers, 1973). However, the most important and most extensive results came from the OAO's, whose data provided a major part of our knowledge of ultraviolet astronomy at the time IUE was launched. They had established accurate temperature scales and improved models for hot stars, revolutionized our ideas of the interstellar medium, and provided a start on many other topics to which IUE has contributed.

In Europe, a programme of ultraviolet astronomy was also being pursued and this also progressed from the use of sounding rockets to orbiting satellites. Although not at the same resource level as that in the United States nor quite as much at the forefront, it was highly complementary and made major contributions to the development of the subject. The first ultraviolet observations of stars in the Southern hemisphere were made in 1961 by a UCL group (Alexander *et al.*, 1964) with photometers carried in an unstabilised Skylark rocket launched from Woomera in Australia. The development of a 3-axis stabilisation system for the Skylark rocket allowed a British programme of ultraviolet and X-ray studies of the sun and stars to be carried out in the decade starting 1964. In the area of cosmic ultraviolet astronomy, a group at the Royal Observatory Edinburgh obtained objective prism spectra of some 70 stars in the region of Lupus with a launch in 1970, and the Astrophysics Research Unit (ARU) at Culham (now part of the Rutherford Appleton Laboratory) obtained good resolution (0.3 Å) spectra of γ Vel and ζ Pup over the wavelength range 900—2300 Å in a launch in 1973 (Burton *et al.*, 1973).

Some UV astronomy was also conducted from balloon platforms although atmospheric exinction limited observations to the wavelength range 2000—3000 Å. A Belfast/UCL team flew an objective grating spectrograph on a gondola with pointing capability and obtained high resolution spectra of several bright stars in a number of flights (Boksenberg *et al.*, 1972). Also, the Utrecht group, in association with US workers, flew a Balloon-borne Ultraviolet Stellar Spectrograph (BUSS) on a number of occasions and obtained high resolution spectra of a large number of stars (Kondo *et al.*, 1979). The instrumental concept had some similarity with IUE, a telescope feeding an echelle spectrograph using an SEC vidicon as a detector.

The first European Satellite to carry ultraviolet astronomy instrumentation was the European Space Research Organisation (ESRO) satellite TD—1, named, with some lack of imagination, after the acronym of the US Thor-Delta rocket which was to launch it in 1972. Of the seven experiments carried, two were devoted to cosmic ultraviolet astronomy. The first, which was the prime motivation for the mission and was carried out by groups in the UK (ROE and ARU) and in Belgium (Liège), conducted an all-sky ultraviolet spectrophotometric survey of objects down to visual magnitude of 9—10 (Boksenberg *et al.*, 1973). The second carried out stellar spectroscopy in regions between 2000 and 3000 Å, with equipment provided by the Utrecht group (Hoekstra *et al.*, 1972). Both were highly successful and enabled extensive studies to be made in stellar and interstellar astronomy. The former also made a contribution to the operation of IUE by allowing accurate estimates of the optimum exposure time for a large number of targets from the data in its all-sky catalogue (Thompson *et al.*, 1978).

The Astronomical Netherlands Satellite (ANS) also carried an ultraviolet telescope as its prime mission. Launched in 1974, it carried out broad band photometry of stars, clusters and galaxies and made a major contribution to ultraviolet astronomy.

3. The Origins of IUE

As is clear from the previous section, the early developments of ultraviolet astronomy in the United States and Western Europe were distinctly separate although, because the US programme was generally in the lead, the European programme was deliberately planned so as to be complementary rather than competitive, to the benefit of the subject as a whole. IUE was totally different in that it involved both the US and Europe in as close, intimate and successful a collaboration as possibly any other space project they have embarked upon.

To start at the beginning, we need to go back to the developments in Europe. The early programme of ESRO included plans for a major project and after a survey of possibilities, it was decided that this should be a Large Astronomical Satellite (LAS) devoted to ultraviolet spectroscopy of non-solar objects. Accordingly, in September 1964, an invitation was issued to member states to submit proposals for the LAS scientific package from groups suitable to undertake the responsibility of providing that package. Three proposals were received by the December 1964 deadline, one from the UK and the others from groups in Germany—Holland and in France—Belgium—Switzerland. All three proposing teams were contracted to carry out detailed design studies against a deadline of January 1966 and these were assessed by an extensive machinery specially set up for the purpose. This resulted in the UK design being adopted in May 1966 as the instrumental package for the LAS. A full project team was then assembled in which the UK group, drawn from Culham, UCL and Aldermaston and led by R. Wilson, formed the scientific component. ESRO persuaded NASA to release W. G. Stroud of Goddard to take charge of the LAS project and an intensive period of study followed which culminated in a detailed project development plan for consideration by ESRO. However, since the estimated cost was higher than expected, the Council of ESRO decided to abandon its major project in June 1967, nearly three years after it had initiated it.

During that time, the UK scientific team had been exposed to the wider problems of space technology and, from their central place in the LAS project study, had observed the development of a spacecraft design and operations system. The expensive areas were readily identifiable from the breakdown of the LAS estimates and it was believed that a complete re-examination of the system as a whole and the adoption of some recent technological developments might result in major savings at no loss in performance. Wilson was able to convince the relevant UK authorities on this point and the original team was reconvened and extended to include the Royal Aircraft Establishment (RAE) and British industry in the form of Elliott Bros. and EASAMS, thereby giving sufficient expertise to study the entire mission and its operation.

The resulting study (which ESRO later decided to fund) produced a design of a complete system called the Ultraviolet Astronomical Satellite (UVAS) which had a greater capability than the original LAS but was much simpler and therefore much cheaper. This was achieved by a radical re-design of the scientific instrument in order to relieve the problems in the rest of the system. The design of the scientific instrument was based on a 45-cm telescope feeding an échelle spectrometer with an SEC Vidicon as detector, giving a resolving power of about 10^4 over a spectral range 1150—3200 Å. The field of the telescope viewed by the acquisition system was increased so as to ensure the presence of the target star which was then located by a fine sensor and placed in the entrance slot of the spectrograph. The echelle grating was selected with a factor of ten higher resolving power than needed spectroscopically and this was used to relieve pointing, mechanical and thermal tolerances by that factor. The spectrograph consisted of a concave collimator, a plane echelle and a concave diffraction grating, dispersing ortho-gonally, to display the final spectrum in a compact, two-dimensional format ideally suited and matched to the SEC Vidicon. The imaging detector gave a great increase in efficiency over scanning photomultipliers, but additionally was used for data storage, thereby decreasing the on-board data-handling and computational requirements. These were further simplified by adopting an interactive operations system in which the spacecraft responded to commands which were mainly made directly from the ground with little onboard processing. Since the specified performance of the proposed launch vehicle, (the US Thor-Delta) at that time, could only place the spacecraft in a near-earth orbit, this meant that commands and data reception were restricted to those periods of ground station contact. The UVAS proposal (Wilson, 1968) was submitted to ESRO in November 1968 and, despite an exceptionally favourable report by a special assessment team (led by Stroud, again released by NASA for the purpose) it was not finally accepted.

This second disappointment was a much greater one for the UK team. Although they were involved in other space projects, they had nevertheless spent a significant part of their time over a period of four years on studies which now appeared to have been completely abortive. Convinced the UVAS was the right approach and that the considerable intellectual investment should not be wasted, Wilson sent the design report to L. Goldberg in his capacity as Chairman of NASA's Astronomy Missions Board, with the suggestion that it might be the right kind of project to build into the US space astronomy programme between the OAO's and the Space Telescope. With the commendation of his Board, Goldberg sent the study to NASA HQ, who in turn sent it to Goddard Space Flight Center with a request for a detailed evaluation. It arrived at a particularly appropriate time in that active consideration was being given in Goddard as to how best to follow the OAO accomplishments and how to broaden space astronomy participation by the development of a system 'friendly' to the use by experienced ground-based astronomers. The introduction of the UVAS concept into this process appeared to provide the means for efficiently accomplishing those objectives. Accordingly, the Goddard assessment team that was set up under L. Meredith was able to report particularly favourably and Goddard management agreed that Meredith should pursue the proposal in a NASA framework and seek the involvement of the UK team. As a result, Meredith invited Wilson to Goddard where they spent one week in discussions which included D. Krueger (later to become the first IUE project manager). That week, in May 1969, was one of the most important in the history of IUE. It saw the UVAS concept become the IUE concept we know today and it established a basis for collaboration and timesharing that has endured throughout the project.

The most important change in the UVAS concept resulted from the Goddard proposal to adopt a geosynchronous orbit in order to maintain continuous ground-contact from one ground station and to avoid the viewing constraints of a low earth orbit. This was feasible if plans proceeded on the basis of the *projected* rather than the *existing* capability of the Thor-Delta rocket and if payload weight control was made a key element of the management of the project. This greatly increased the scope of the interactive control and allowed the capability of the acquisition system to be extended to faint objects by using the observer to identify the star field by human pattern recognition and then to offset guide from the brightest field star using coordinates measured from ground-based surveys such as Palomar and the southern hemisphere programmes conducted by the European Southern Observatory (ESO) and the UK Schmidt Telescope. The move to an 'observatory' control philosophy, so well known to IUE users, had been made. In order to take advantage of the ability to acquire faint objects, it was also proposed to introduce a low dispersion mode into the spectrograph by the simple and elegant means of rotating a plane mirror in front of the echelle, thereby removing it from the optical train and leaving only the low dispersion spectrum of the cross-disperser.

The discussions on international collaboration during the meeting of May 1969 were made in the realistic framework of it being effectively a NASA project which involved external participation. It was agreed to proceed on the basis of the UK providing the onboard detectors and associated software as well as making design contributions to the sunshade and scientific instrument. Subsequently, Meredith introduced the idea of an additional ground station in Europe which could assume full control of the satellite. Given the provision of such a ground station and some additional contribution to the onboard hardware, it was agreed that a reasonable division of time sharing would be two-thirds to the United States and one-third to Europe. These could be assigned and exercised in a clean management sense by the separation of control allowed by the two ground stations.

Wilson returned to the UK and informed SRC and ESRO of the possible developments. After several discussions between those two agencies and NASA, SRC decided to support the UK contribution as proposed above and ESRO decided to build a European ground station and to provide the solar arrays for the spacecraft. These agreements were reached in 1970 and after Phase A studies were complete, substantive financial approvals for the mission were given by all three agencies in 1971. The approval process was not easy, requiring, as it did, the coordination of decision-taking in three agencies with quite different approval processes. That it was successful owes much to the enthusiastic support for the project by some key individuals in the headquarters of the different agencies, particularly, J. Mitchell, Director for Astrophysics and Physics, NASA HQ, J. F. Hosie, Director for Astronomy, Space and Radio, SRC, and H. Bondi, Director-General of ESRO.

Although the initiation of the discussion in 1969 on the concept that was to become IUE was prompted by European factors, it so happened that it was also a particularly opportune time for such an initiative from the point of view of the US Space Astronomy programme. OAO—2 was already a great success and astronomers were looking forward to obtaining increasingly high dispersion data with the remaining two OAO's. Although NASA had received several proposals for additional OAO satellites, it was reluctant to continue the series because the spacecraft design had been superseded by improved technology. The agency was conducting early feasibility studies on the Large Space Telescope (now named the Hubble Space Telescope) and was looking for a suitable mission to bridge the gap between the last OAO (Copernicus) and 'LST' (at that time thought to be only a few years!). In reviewing the OAO experience and considering the next step, several points seemed important:

(a) It would be essential to obtain both high and low dispersion spectra of objects much fainter than could be reached by the OAO's.

(b) Future satellite observatories should be available to a wide community of astronomers as research facilities, instead of benefitting primarily a Principle Investigator and his or her associates.

(c) If satellites were to be widely used by astronomers, their operations should be designed to emphasise the interaction between observer and scientific instrument and to minimize the highly specialised problems of commanding and orbital scheduling.

The distillation of the experience and lessons learned during the 1960's strongly influenced the design of IUE. The technical and organizational implications of these points were almost self-evident but they led directly to key features of the IUE mission.

On completion of formal approvals in 1971, the three agencies set up their project teams and vested overall control of IUE in a Management Team consisting of the project managers and project scientists of each agency with the NASA project manager in the chair. The resulting composition was: NASA — D. Krueger (later succeeded by G. Longeneckar) and A. B. Underhill (later succeeded by A. Boggess), ESA — M. G. Grensemann and F. Macchetto (who later took over the project manager role as well), SRC — P. J. Barker and R. Wilson. An Astronomy Working Group was also formed under the chairmanship of the NASA project scientist.

Technically, astronomers were well prepared to embark on IUE in the 1970's. Echelle spectrographs had been launched in rockets to observe the ultraviolet spectrum of the sun with high resolution by groups in the US (Purcell *et al.*, 1958) and the UK (Jones *et al.*, 1970). The television cameras flown on OAO—2 laid the technological basis for electronic imaging in the UV, and further development of the SEC vidicon had been vigorously pursued at Princeton University. The Star trackers were based on photometer designs developed at Wisconsin University and Goddard. Elaborate facilities for evaluating the calibrating astronomical instruments in the UV had already been developed at Goddard and Johns Hopkins University for the OAO and Apollo projects, and at Liège and Edinburgh for the TD—1 satellite. A lot of image analysis software had been developed at the Jet Propulsion Laboratory in support of imagers on the Mariner Mars missions and this could form the basis of the software needed to extract the spectra from the IUE camera data. All of this heritage, plus the extensive engineering and operational experience that had been gained with the ultraviolet astronomy programmes in the US and Europe, enabled the IUE scientists and engineers to take on a job that would have been difficult to contemplate only a few years before. This book is a testimony to how well they did that job.

4. The Design and Development of IUE

The design of the IUE scientific instrument and spacecraft were described in detail in the first of a series of articles published in Nature shortly after its launch (Boggess *et al.*, 1978a). Rather than repeat that information here, we have selected some of the important problems which arose during development and show how they affected the way in which IUE was built and how it performs. The reader should not be misled by the lack of names in this summary; the many sicientists and engineers who were involved in the design, construction, launch and operation of IUE are too numerous to mention but they are drawn from universities, research institutes, government laboratories and industry in many countries. Three groups participated in the development of IUE: the Goddard Space Flight Center (GSFC), a UK team formed from the Astrophysics Research Unit (ARU)*, the Appleton Laboratory (AL)* and University College London (UCL), and the European Space Research and Technology Centre (ESTeC). So far as the flight hardware was concerned, GSFC had responsibility for overall system design, design and construction of the telescope, spectrograph, pointing and attitude control system and spacecraft, pre-launch integration and testing, and the launch itself. The UK team was responsible for design and construction of the spectro-graph detectors and contributed to parts of the optical design and sunshade. ESTeC produced the solar arrays.

The operational lifetime of IUE was a major design issue. In 1971, most satellites had been designed for lifetimes of a few months to a year, but some were still healthy after five or more years of operation. Based on this experience, the scientists on the project argued that IUE should be designed to last for five years, but the engineers were reluctant to commit themselves to more than three. It was finally agreed that the design lifetime should be three years, but with a five year goal, and this had an important bearing on numerous project decisions throughout the development period.

Another major design issue arose from the selection of a geosynchronous orbit. This made payload weight the largest single problem during design and development and the resulting severe weight control management influenced the performance of many sub-systems. The telescope, spectrographs and every other major sub-system went through repeated weight-reduction exercises and weight was also one of the considerations in deciding that most of the spacecraft electronics would employ CMOS technology. However, after the CMOS chips were delivered it was discovered that they were more susceptible to radiation damage than had been expected, thus requiring the addition of extra shielding and thereby substantially reducing the hoped-for saving. Drastic weight reduction measures were taken at that time, including reduction of the battery capacity to a bare minimum. An additional step was to change the orbit from a 24-hr circular orbit to a 24-hr ellipse of eccentricity 0.24. This lower energy orbit increased the weight allowance substantially, but since perigee was now deeper into the radiation

---

* Both ARU and AL now form part of SERC's Rutherford Appleton Laboratory (RAL).

belts, some of that additional weight capability had to be used on still more radiation shielding. The three year design life played an important part in determining the amount of shielding required and, according to available test data, the thickness used would be sufficient to protect the electronics from damage for only three years. It was privately recognised, though, that the test conditions were more severe than in the IUE orbit and so three years would be a lower limit. In the event, as we now know, this has not been a limiting factor.

Another consequence of the changing radiation environment around the elliptical orbit is the existence of 'quiet' and 'noisy' observing shifts. The location of perigee was rather tightly constrained by the cumulative requirements of launch, of orbital insertion and ground station coverage. Since the portion of the orbit visible from VILSPA is constrained by the VILSPA horizon, it is somewhat fortuitous that the ESA shift is close to apogee and therefore 'quiet'. In fact, the exact tuning of the orbit and shift schedules have been re-negotiated on several occasions.

The mode of operation of IUE was very novel and greatly conditioned the design of the ground segment. The presence of a guest observer, awarded the time by competitive peer review, required data such as the telescope field and spectral image to be quickly and simply displayed. This allows decisions such as whether to repeat the observation if the data are inadequate or to slew to the next object, to be taken in near real time so as to maximise the astronomical return. Of course, such decisions are made against an already prepared observing plan and the observer is not allowed to press any control buttons! The operational staff and control system respond to the observer's requirements in a way which maintains the integrity and safety of the spacecraft. One function of the observer which is essential to the spacecraft operation is the identification of the telescope star field by pattern recognition against a previously prepared finder chart obtained from ground-based data; this allows the attitude coordinates to be updated after each slew.

The international aspect of the operation of IUE is one of its very important features, but it did cause some concern during the planning phase. It had been common to use ground stations distributed around the world to transmit commands to satellites, but there was always a single control centre responsible for operating the satellite and guaranteeing its safety. IUE introduced a completely novel idea; that actual decision-making control of the satellite, with all it implied, should routinely be passed back and forth between two stations on two different continents. This concept was questioned by many who believed that a strong centralised chain-of-command was essential to successful spacecraft operations. In fact, one of the particular successes of IUE has been the way the two Control Centres have been able to pursue independent programmes while maintaining common standards and criteria for operation of the satellite.

In the following paragraphs, some of the key IUE systems are considered:

THE TELESCOPE

The telescope primary was made of beryllium in order to save weight, while the secondary was made of quartz because it needed to be insensitive to the more extreme temperature changes it would experience. The specifications called for

1 arcsec image quality, and the unconstrained primary was polished to an excellent figure. Unfortunately, the support system did not work as well as intended and it was not possible to mount the mirror in an entirely stress-free condition, causing a degradation in image quality to about 3 arcsec. However, this was contained within the tolerances of the system as a whole since later tests in-orbit showed the spectral resolution to be optimum (see below). The choice of aluminium for the telescope structure, made to save cost, made it important to insulate the telescope tube well so as to minimize focus changes due to thermal expansion. The levels of in-orbital temperature changes predicted were such that a fine-focussing mechanism was considered necessary and this was provided by a stepping motor which moved the position of the secondary mirror along the telescope axis. However, after launch and the initial focussing procedure, the operations staff developed a much more efficient and reliable method for maintaining focus by adjusting the temperature of the telescope with on-board heaters. Another thermal problem which was exacerbated by the use of an aluminium structure is the bending of the telescope tube due to circumferential temperature gradients, which causes the image to drift as a new equilibrium is established after a large slew. The fine guidance system is designed to compensate for these drifts by performing closed-loop tracking on guide stars imaged very near the spectrograph apertures.

The adoption of a synchronous orbit required that observations had to be carried out in sunlight and this made the design and construction of an effective telescope baffle system critical to the success of the mission. The most demanding requirement was set by the attitude control sensor which viewed the telescope field in visible light and needed to detect and guide on stars down to 13th visual magnitude. This required a discrimination factor against sunlight of at least sixteen orders of magnitude and, at the time, it was difficult to persuade some critics that this was feasible. The baffle design was based on an elegant theoretical considera-tion of the initial diffraction and the subsequent trapping of the multiple scattered sunlight before it reached the focal plane of the telescope. In addition to baffles within the telescope tube, an angled sunshade enabled observations to be made to nearly 40° from the Sun, thus opening up about 80% of the sky to observation except for the presence of the bright Earth and Moon. The baffle has proven a complete success, no scattered sunlight having been detected by the telescope.

THE SPECTROGRAPHS

The entrance apertures to the spectrographs are located in an aperture plate placed in the telescope focal plane. The aperture plate is a surprisingly high-tech unit. Its front surface acts as a mirror which must reflect the telescope image into the fine guidance optics, so the plate is tilted 45 deg with respect to the optical axis. Since the reflection is at the focal plane, the mirror surface was given a special, low scattering polish to remove any blemishes that might confuse the image of the star field. Several small 'grain-of-wheat' lamps were embedded into the rear surface to act as fiducial false stars to calibrate the star tracker coor-dinates. A pattern of sharp lines was etched into part of the mirror as an aid in evaluating telescope focus, and a small low-reflectivity patch was deposited near the center as an aid in acquiring very bright stars.

The apertures themselves are four holes, a 100 micron circle and a 350 by 700 micron slot for each of the two spectrographs. The holes are drilled at 45 deg from the mirror normal, they open wide behind the mirror surface to accommodate the diverging $f/15$ optical beam, and the high-quality optical surface is maintained to the very edge of the drilled holes. Finally, a small 45 deg optical flat is mounted behind the two long-wavelength apertures and acts to separate the optical systems of the two spectrographs. After several attempts to machine these mirrors had failed, an instrument maker at the National Bureau of Standards was asked to show the project what he could do. His demonstration mirror was so good that he was then commissioned to machine more mirrors for flight. However, he never repeated his initial success and it is the demonstration mirror which is in orbit today. The 100 micron entrance apertures subtend 3 arcsec on the sky and the resolution of the spectrograph (optics plus detectors) was designed to match this aperture size. Consequently, the degraded telescope image does not materially affect the spectral resolution, but it does prevent the 3 arcsec aperture from being used for photometric purposes.

The gratings that were produced for IUE were all exceptionally good, the two echelles being among the best that have ever been ruled. The design of the cross-disperser gratings was of particular concern, because the geometry of their grooves is difficult to fabricate and still maintain low scattered light. A further difficulty was that the gratings had to be ordered long before the image quality of the combined optics/detector system was known. Since the separation of the echelle orders at the short wavelength ends of the two spectrographs was expected to be a problem, three different cross-dispersers were ruled for each spectrograph, corresponding to differing separations of the echelle orders. The gratings that were selected for flight gave the best compromise between order separation and complete wavelength coverage.

THE DETECTORS

An imaging detector is necessary to record the two-dimensional echelle format with reasonable efficiency. During the IUE design phase the only viable candidate as a low-light-level imager was the SEC Vidicon, and it had numerous problems. Vidicon lifetimes were typically shorter than the IUE goals, and so the cameras were specifically designed and operated in ways to maximise tube life. Developed vidicons were commercially obtainable in two versions: an electrostatically focussed tube which had a faceplate that would not transmit ultraviolet radiation and which also had limited resolution, and a magnetically focussed tube that avoided these problems but required a massive magnet. Special prototype tubes of both types were developed for IUE, but the magnetic tube was abandoned because of its cost and insurmountable weight problem.

The electrostatic tube's lack of UV sensitivity was solved by attaching a proximity-focussed image converter to the front face. The development of these converters was a major effort of its own. The choice of photocathodes was a critical issue; the original intent had been to use caesium telluride cathodes in the long wavelength spectrograph and to use caesium iodide cathodes in the short wavelength spectrograph to help discriminate against long-wavelength scattered light. It was finally decided that producing and testing a single type of converter-vidicon camera would be difficult enough, and that an attempt to develop and fabricate two different types would constitute an unacceptable programme risk. Consequently, caesium telluride cathodes were used for both spectrographs. In retrospect this decision still seems wise, but every observer of planets and cool stars at short wavelengths has had to cope with its consequences. Nevertheless, the use of a solar-blind uv-to-visible converter was of crucial importance to IUE and its successful development on a short timescale, was due largely to the skill and devotion of the Electro-Optical Products Division of ITT, Indiana.

The addition of the UV converter further complicated the resolution problem. The system resolution requirement was achieved by carefully optimizing the designs of the tubes and their control electronics. Establishing the control parameters, calibrations, and the investigation of photometric properties became a major research effort for each flight camera. The development of these cameras for the IUE mission was a major technical achievement and involved not only the UK team referred to above, but also its prime contractor Marconi Space Defense Systems (MSDS) of Portsmouth.

As mentioned earlier, it was recognized that trapped radiation in the Van Allen belts would have a deleterious effect on data quality. The principal concern was the production of Cerenkov radiation in the magnesium fluoride faceplates of the image converters, and an extensive test programme was undertaken to measure the seriousness of this problem. The analysis of these data showed that 4 mm of aluminium shielding around each of the cameras would allow them to be used, at least for short exposures, even at perigee. Great pains were taken during installation to ensure that the shielding covered the entire 4π steradians, including the optical path, but with minimum weight. The development of the cameras also necessitated development of a large data handling software system that could be used in a production mode, with minimum operator intervention. There were two major components: first to produce a photometrically correct camera image containing approximately half a million pixels, and second to extract from the echelle format in this image a calibrated listing of intensity versus wavelength. Two groups, one at AL and one at GSFC, took on these two jobs. Many refinements have been added over the years, but the basic software worked well and has been remarkably durable.

THE ATTITUDE CONTROL SYSTEM

The Attitude Control System consists of sensors (gyroscopes, star trackers, and a sun sensor); brains (a computer); and muscles (reaction wheels and gas jets). In order to make the control as tight as possible, the gyro package is mounted directly on the strong ring that holds the telescope. High quality gas-bearing gyros were selected for their slew accuracy and low noise. However, analysis of the failure rates of these gyros showed that a large number of spares would have to be mounted in each of the three control axes in order to achieve a lifetime goal of five years. This problem was minimised by devising a non-orthogonal mounting scheme that would allow any combination of three gyros to provide the necessary control information. It was calculated that only six gyros would be required initially in this configuration to yield a three-sigma probability of having three working at the end of five years.

At the end of a slew, a star tracker is commanded to scan the field and measure the positions of the operator-designated target and guide stars, so that the target can be manoeuvred into the correct spectrograph aperture. The tracker is then commanded to maintain the guide star at a fixed location. Originally a separate television camera was intended to provide images for target recognition, but this was eliminated to save both weight and cost. It was also planned that the trackers would have filter sets to provide visible and UV photometry of targets. These were eliminated as unwise reliability threats to the trackers which are essential to the mission. For similar reasons, the prime and spare trackers are made to share light by means of a semi-transparent pellicle instead of using a flip mirror to direct the entire beam to one or the other tracker. The limiting magnitudes of the trackers are thereby reduced, but a potential failure mechanism was eliminated.

IUE inherited its momentum wheels from a previous NIMBUS project. Although a faster slew rate would have been desirable, the availability of wheels which had already been space proven was too good an opportunity to miss. In order to improve reliability still further, IUE contains four wheels, one oriented in each of the three control axes plus a fourth canted at 45 deg to all the axes which can be used to control the spacecraft if any one of the other wheels fails.

The on-board computer is necessary to transform the ono-orthogonal error signals from the gyros into reaction wheel torquer commands. It is incidentally used for other functions, such as issuing command sequences to control camera exposures and read-outs. The computer is based on the controller flown on OAO—3 (Copernicus) and it served in turn as the prototype for computers later used on SMM, HST and others. Because it is critical to the mission, a spare CPU was incorporated into the IUE computer, but it has never been turned on in flight.

The hydrazine system was designed to fill several functions. It is used with fine jets to dump angular momentum from the spacecraft. The system can also employ larger jets for station-keeping in orbit and could have been used to adjust the orbit to compensate for errors in the launch sequence. In fact, the launch and orbit insertion were near perfect, and the hydrazine budgeted for that purpose is enough to take care of the momentum and station-keeping needs of IUE for several decades.

PRE-FLIGHT PREPARATIONS AND COMMISSIONING

Pre-launch testing was a complex and sometimes harrowing process. The large number of problems that were found and corrected was taken as a good sign; surely there was nothing left to go wrong in flight! The scientists and engineers on the project were quite confident as they prepared for the Launch Readiness Review, but more difficulties lay ahead. Because of concern about the effect of particle radiation on data quality, the Review Team commissioned an independent calculation of the effectiveness of the camera shielding. The results were several orders of magnitude worse than predictions made by the project staff — so much so that there was concern that physical damage to the cameras might result from a large solar event. It was decided at the last moment to install a radiation monitor which could alert operators to turn off critical components if radiation levels became too high. In fact, the Review Team's shielding calculations turned out to be incorrect, but the radiation monitor has been one of the most useful of all observing aids because it allows the operator to judge the rate of accumulation of background noise in the image.

There was no time to design and develop a radiation monitor specifically for IUE. Instead, spare parts from a previously flown monitor were found at the Johns Hopkins Applied Physics Laboratory. The monitor was assembled and tested over one weekend and then flown to the launch site for incorporation into the spacecraft. Disassembly of a satellite at the launch site is definitely not recommended practice and it was to happen not once, but three times for IUE! The first time was for the radiation monitor. Second was to replace wiring in the camera electronics that had been found to have damaged insulation. The final time was to adjust the spectrograph housing after scientists analysing test spectra inferred that the spectrograph structure must be in contact with the housing and was yielding to mechanical pressure.

The launch occurred on 1978 January 26, and the early operations were both tense and exciting. The initial check-out of the spacecraft hardware went remarkably smoothly, and the first spectrum was obtained on the third day of a calibration star, Eta Ursae Majoris. Although the telescope was obviously out of focus, it became clear as the spectrum appeared slowly on the monitor that IUE would be a success.

Over a year before launch, the IUE Science Working Group had established a small Commissioning Team of scientists expert in the principal instrumental and scientific areas relevant to IUE. It was their job to calibrate and evaluate the scientific performance of the instrumentation, obtain an initial set of scientific data, and develop the observing techniques that had to be in place before the observatory could be declared operational. It was expected that the commissioning activities would take about 60 days, and a strategy had been developed which called first for a crude calibration, then observations of a series of high priority targets whose images would comprise a worthwhile data set even if the IUE stopped working shortly afterwards. The next item of business was to do a careful in-flight optimisation of all the various parameters affecting camera performance. This required not only a great deal of spacecraft time, but also a major data analysis effort in order to derive the optimum control settings and install them quickly. Finally, the high-priority science targets were re-observed along with a sequence of photometric standards. Throughout this period, the operational staff were developing their observing procedures, while those looking at the initial data were characterizing the scientific instruments.

For a few days in the early stages of commissioning, it was thought that the project might have a serious deficiency. The spacecraft had been successfully activated with its solar arrays deployed and star acquisition established and commissioning had entered the camera switch-on phase. This went reasonably smoothly until the first image from the short wavelength prime (SWP) camera was displayed and showed noise and distortion at a level which made it quite useless as a detector. Since the SWP was the most important camera on board and, as such, had been selected for its high scientific performance, that first image induced an air of consternation and even crisis. A review meeting was held on a Saturday morning at which the camera experts claimed that the effect was microphonic noise caused by acoustic interference induced during read-out. The spacecraft engineers doubted whether anything on board could provide such an acoustic source but it was decided to embark on an investigative programme in which the SWP would be operated with successive spacecraft subsystems switched off. In planning that programme and before it was initiated, it was discovered that the panoramic attitude sensor (PAS) which scanned the sky in order to detect the Earth and which was only needed for initial attitude acquisition, was still in its scan mode. It was commanded to switch off during a read-out of an SWP image which showed the immediate disappearance of the interference. This induced sighs of relief all round.

One of the most critical moments of the commissioning period was caused by the onboard computer (OBC) which, among other things, played a crucial role in attitude control. Early in the mission, while in target control mode, it suddenly commanded the spacecraft to slew at maximum rate and, worse, towards the sun. Immediately after there was a break in communications for a short interval and the atmosphere in the control room became electric. Fortunately, it being early in the mission, most senior project staff were present, including the attitude control systems engineer. Retaining his composure, and after receiving his housekeeping data on resumption of communications, he was able to issue instructions which placed the spacecraft in a safe attitude. This 'schizophrenic' characteristic of the OBC became a matter of intense investigation. It was solved by ingenious reprogramming and the cause was traced to overheating. It should be stressed that the attitude control system of IUE was one of its many great achievements and its performance is far better than its specification.

A large amount of work was accomplished in the commissioning phase, allowing it to be completed in a very short time. Some members of the Commissioning Team spent most of their time working at the Goddard and VILSPA control stations. Others stayed at their own institutions, discovering bugs in the hardware and software through the time-honoured process of attempting to use the data for scientific reseach. Gradually the atmosphere changed from frenetic activity to cool professionalism. As an example, it took 24 hours from beginning to end to get the first spectrum of Eta UMa. By the end of the Commissioning period it was possible to obtain about one short exposure per hour. As soon as the operations staff were able to make a firm commitment, an observing schedule was drawn up calling for routine observations to start on April 1. When the first observer was contacted, he believed he was being made the butt of an elaborate April Fool's joke. Although suspicious to the end, he got good data.

The final product of the Commissioning Team was a series of papers in *Nature* which described the instrumentation (Boggess *et al.*, 1978a), the inflight performance (Boggess *et al.*, 1978b) and also gave surveys of data for each of the major areas to be studied by IUE: hot stars (Heap *et al.*, 1978), cool stars (Linsky *et al.*, 1978), the interstellar medium (Grewing *et al.*, 1978), X-ray sources (Dupree *et al.*, 1978), extragalactic objects (Boksenberg *et al.*, 1978), and solar system objects (Lane *et al.*, 1978). These papers demonstrated by example that the

IUE observatory was an exciting and major astronomical research facility. At the time of writing, as it enters its tenth year, it still is.

References

Alexander, J. D. H., Bowen, P. J., Gross, M. J., and Heddle, D. W. O.: 1964, *Proc. Roy. Soc.* A, **279**, 510.

Boggess, A. and Dunkelman, L.: 1959, *Astrophys. J.* **129**, 236.

Boggess, A. *et al.*: 1978a, *Nature* **275**, 372.

Boggess, A. *et al.*: 1978b, *Nature* **275**, 377.

Boksenberg, A., Evans, R. G., Fowler, R. G., Gardner, I. S. K., Houziaux, L., Humphries, C. M., Jamar, C., Macau, D., Malaise, D., Monfils, A., Nandy, K., Thompson, G. I., Wilson, R., and Wroe, H.: 1973, *Monthly Notices Roy Astron. Soc.* **163**, 291.

Boksenberg, A. *et al.*: 1978, *Nature* **275**, 404.

Boksenberg, A., Kirkham, B., Towlson, W. A., Venis, T. E., Bates, B., Courts, G. R., and Carson, P. P. D.: 1972, *Nature, Phys. Sci.* **240**, 127.

Burton, W. M., Evans, R. G., Griffin, W. G., Lewis, C., Paxton, H. J. B., Shenton, D. B., Macchetto, F., Boksenberg, A., and Wilson, R.: 1973, *Nature* **246**, 37.

Byram, E. T., Chubb, T. A., Friedman, H., Kupperian, J. E.: 1957, *The Threshold of Space*, Zelikoff, M. (ed.), Pergamon Press, London, p. 203.

Carruthers, G. R.: 1973, *App. Opt.* **12**, 2501.

Code, A. D., Houck, T. E., Mcnall, J. F., Bless, R. C., and Lillie, C. F.: 1970, **161**, 377.

Davis, R. J., Deutschman, W. A., Lundquist, C. A., Nozawa, Y., and Bass, S. D.: 1972, in *The Scientific Results from the Orbiting Astronomical Observatory*, (OAO—2), Code, A. D. (ed.), NASA SP-310, 1.

Dupree, A. K. *et al.*: 1978, *Nature* **275**, 400.

Grewing, M. *et al.*: 1978, *Nature* **275**, 394.

Heap, S. R. *et al.*: 1978, *Nature* **275**, 385.

Henize, K. G., Wray, J. D., Parsons, S. B., Benedict G. F., Bruhweiler, F. C., Rybski, P. M., and O'Callaghan, F. G.: 1975, **199**, L119.

Hoekstra, R., van der Hucht, K. A., de Jager, C., Kamperman, T. M., Lamers, H. J., Hammerschlag, A., and Werner, W.: 1972, *Nature* **236**, 121.

Jones, B. B., Boland, B. C., Wilson, R., and Engstrom, S. F. T.: 1970, in Houziaux, L. and Butler, H. E. (eds.), 'Ultraviolet Stellar Spectra and Related Ground-Based Observations', *IAU Symp.* **36**, 271.

Kondo, Y., de Jager, C., Hoekstra, R., van der Hucht, K. A., Kamperman, T., Lamers, H. J. G. L. M., Modisette, J. L., and Morgan, T. H.: 1979, *Astrophys. J.* **230**, 526.

Kupperian, J. E., Boggess, A., and Milligan, J. E.: 1958, *Astrophys. J.* **128**, 453.

Lane, A. L. *et al.*: 1978, *Nature* **275**, 414.

Linsky, J. L. *et al.*: 1978, *Nature* **275**, 389.

Morton, D. C. and Spitzer, L.: 1966, *Astrophys. J.* **144**, 1.

Purcell, J. D., Boggess, A., and Tousey, R.: 1958, *Trans. IAU* **10**.

Rogerson, J. B., Spitzer, L., Drake, J. F., Dressler, K., Jenkins, E. B., Morton, D. C., and York, D. G.: 1973, *Astrophys. J.* **181**, L97.

Smith, A. M.: 1967, *Astrophys. J.* **147**, 158.

Stecher, T. P. and Milligan, J. E.: 1962, *Astrophys. J.* **136**, 1.

Thompson, G. E., Nandy, K., Jamar, C., Monfils, A., Houziaux, L., Carnochan, D. J., and Wilson, R.: 1978, *Catalogue of Stellar Ultraviolet Fluxes*, SERC.

Wilson, R.: 1968, *Ultraviolet Astronomical Satellite*, Final Report, Vols. 1, 2, and 3, UK Atomic Energy Authority, Culham Laboratory.

OPERATION OF A MULTI-YEAR, MULTI-AGENCY PROJECT

JÜRGEN FÄLKER

European Space Operations Centre, Darmstadt, F.R.G.

FREDERICK GORDON

Goddard Space Flight Center, Greenbelt, MD

and

MICHAEL C. W. SANDFORD

Science and Engineering Research Council, London, U.K.

1. Operational Concepts

The IUE mission is based on real time control of the satellite and instrument and real time acquisition of the scientific data. Thus both the spacecraft operations staff and the science users must support IUE operations 24 hr per day, every day. Operations are shared between two very similar ground systems, one of which is located in the USA and operated by NASA. The other, provided and operated by ESA, is located in Spain and is used by European observers. According to the Memorandum Of Understanding between the three agencies, NASA has use of IUE for two thirds of the time, and ESA and SERC share equally the remaining third. For the purpose of operations it was decided to divide each day into three shifts of eight hours duration, and it was agreed that the NASA ground system would control the satellite for two shifts, and the ESA ground system for one shift.

The receiving site for NASA's ground system was, until April 1986, located at the Goddard Space Flight Center (GSFC), Greenbelt, Maryland. Subsequently it has been moved to NASA's facility at Wallops Island, Virginia (WPS). NASA's spacecraft and science operations and the data processing facilities have always been located at GSFC. Commands and received data are now transmitted between GSFC and WPS by a commercial communications satellite link.

NASA has three principal areas of responsibility in operations: firstly, overall responsibility for monitoring and maintaining the health of the spacecraft; secondly, providing a backup system for the purpose of spacecraft safety during the shift controlled by ESA's ground station; and thirdly, operating IUE for two eight-hour shifts each day, designated US1 and US2. The spacecraft-related tasks are carried out on a 24 hr per day basis by the GSFC IUE Operations Control Center (IUEOCC). The scientific operations are carried out by the IUE Science Operations Center (IUESOC), also at GSFC, in conjunction with the IUEOCC. The IUESOC comprises: a Telescope Operations Centre (TOC) from which astronomical observations are controlled during the US1 and US2 operational shifts; and an Image Processing Center (IPC) which carries out the standard processing of the IUE data, mainly during the third (European) shift.

All the elements of ESA's ground segment are located at Villafranca del Castillo near Madrid (VILSPA). ESA provides spacecraft control and science

Y. Kondo (ed.), Scientific Accomplishments of the IUE, pp. 21—42.
© 1987 *by D. Reidel Publishing Company.*

operations for one shift, designated VILSPA, and in a second shift the standard data processing is carried out. Because the ESA station has limited backup facilities, e.g. only one computer capable of controlling IUE, GSFC maintains readiness to take control and keep the spacecraft safe in the event of a failure in the VILSPA station.

The operational concept of sharing, internationally, the responsibility for operations was made possible by the choice of a geosynchronous orbit for IUE. Constraints on the orbital parameters arose since continuous viewing of IUE is required from the NASA receiving/transmission site and for at least one shift per day IUE must be in view from VILSPA. To ensure reliable communications, IUE must be at least 10° above the local horizon at VILSPA for more than about 10 hr per day which is the time required for an 8-hr shift, a shift handover of about 30 min, and a monthly adjustment by 2 hr in the shift starting time. The latter arises since the orbit is fixed in sidereal time but for the convenience of the operations staff the shift start remains at a fixed Universal Time during each month.

An elliptical geosynchronous orbit was chosen with a period of 23 hr 56 min, eccentricity 0.2, inclined at 28° to the equatorial plane and with the ground track initially centred over the Atlantic at about 70° west longitude. With this positioning the VILSPA viewing time usually averages 12 hr. As the orbital plane precesses westward due to natural perturbations the viewing time falls and when the 10 hr minimum is reached an orbit adjustment must be made. A maximum of 15 hr may be achieved at its most easterly position. Partly in consequence of these orbit adjustments, and partly as a result of other natural perturbations the ground track of the orbit has gradually evolved during life of the mission as shown in Figure 1.

An early project objective was that the IUE system would be an international research facility, available to a wide community of astronomers, and be organised much in the way that many ground based observatories are operated. The operations management plan and the ground system were specifically oriented towards achieving this. The degree to which the three agencies have succeeded in achieving this objective can be judged from the results described in this book. IUE has been operational since its launch, 26 January, 1978, for over nine years at the time of writing. It continues to provide excellent facilities to the scientific community. It can be unequivocally stated that all scientific objectives of the mission have been met or exceeded. The mission's success was made possible not only by the outstanding performance of the spacecraft and its scientific instrumentation, but also by the excellent cooperation of the technical and scientific staff of the participating agencies. These achievements have been widely acknowledged in the general scientific community as being unique in space exploration.

2. The IUE System

The operation of IUE, like any complex scientific satellite, involves many systems. The space segment includes the spacecraft and the payload systems. The ground segment includes: the ground station which transmits commands and receives telemetry; the operations control centre, housing the ground control computers used by the controllers to analyse the telemetry and issue appropriate commands;

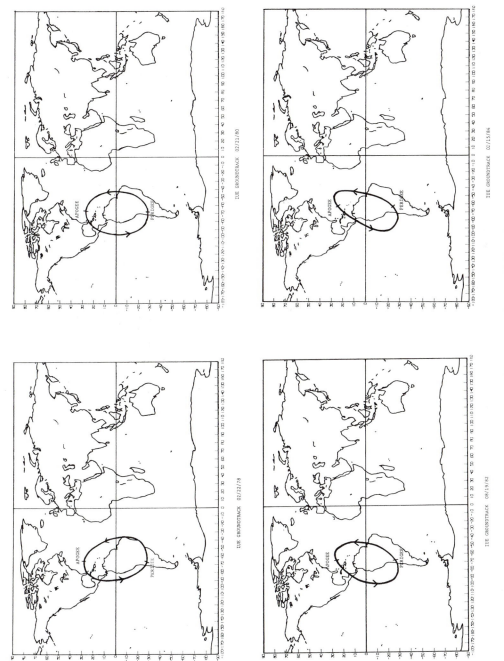

Fig. 1.   The Ground Track of IUE (1978–1984).

the science operations center comprising an instrument operations system and the data reduction system; and a communications system to link the parts together. Descriptions of some aspects of the system may be found in Boggess *et al.*, 1978, and article by Boggess and Wilson, pp. 3—19, Section 4. To prepare the reader for the discussion of operations that follows, this Section summarises the main characteristics of these systems and details are given of relevant systems not adequately covered elsewhere.

2.1. THE SPACECRAFT

The sole purpose of the spacecraft is to support the Scientific Instrument (S/I) in achieving the scientific objectives for the mission. These objectives give rise to the basic performance requirements for the spacecraft. First the satellite should be able to point the telescope anywhere within the celestial sphere except within 45° of the Sun. Secondly the Attitude Control System (ACS) should on command move the telescope to a new target with a slew rate of 4.5° min^{-1} axis^{-1} and guarantee that, at the end of the maneuver, the desired target star falls into the 16 arc min field of view of the Fine Error Sensor (FES), a unit which performs the dual functions of star field mapping and guide star tracking.

Three-axis stabilisation of IUE is accomplished by an Inertial Reference Assembly (IRA) composed of 6 gas bearing single-degree-of-freedom gyroscopes operated with pulse rebalance electronics. The ACS was designed to hold a 1 arc sec diameter star image within a 3 arc sec entrance aperture to permit an integrating exposure of at least 1 hr by the spectrograph camera. Now, under the two-gyro/fine sun sensor (FSS) control system, the FSS is also used as a sensor for spacecraft control. The ACS uses the outputs of the attitude control sensors (gyros, FES, and/or FSS) as inputs to the control program in the On-Board Computer (OBC). The OBC then controls the spacecraft attitude and slews by changing the rotation speed of the pitch, yaw, and roll momentum wheels.

The primary power for IUE is derived from the two solar cell arrays which are shown in the general layout, Figure 2. The power output is a function of the angle between the sun and the satellite pointing direction ($+x$ axis). The supplement of this angle is designated β. In normal operations the satellite is rolled about the x-axis to keep the sun close to the $x - z$ plane. Thus the single angle β is useful to define the approximate orientation of the satellite with respect to the sun, and will be so used here.

Central to the spacecraft control are the command system and the data multiplexer unit (DMU). The Command Decoder receives the commands, checks for errors, and then routes valid commands or OBC data block loads to the correct subsystem. The DMU samples the performance data from the spacecraft subsystems and generates two telemetry streams: one is dedicated to the OBC and its control of the spacecraft; the other is the ground telemetry stream, which includes both the science data and the spacecraft 'housekeeping' data. The OBC also uses the ground telemetry stream.

The final key spacecraft element is the communications system, which warrants description in some detail. Two wavebands, S-band and VHF, are used for communication between IUE and the ground. The normal data telemetry link

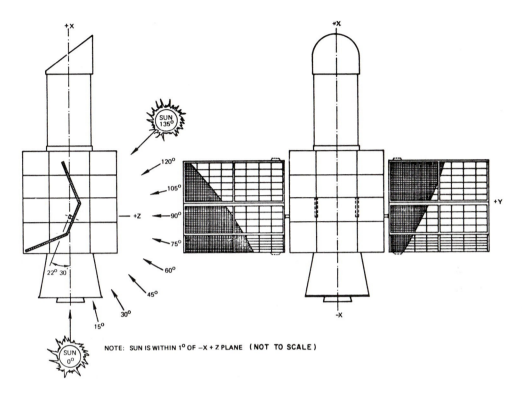

Fig. 2. IUE Configuration and Axes with the Solar Array Deployed. The angle β is measured between the sun and the $-x$ axis. The sun is normally within 1° of the $x - z$ plane. Roll, pitch and yaw are rotations about the x, y, and z axes respectively.

from IUE to the ground receiving site is by S-band (2249.8 mHz). This telemetry can be transmitted at several different bit rates: 40, 20, 10, 5, 2.5 and 1.25 kbs^{-1}. There is a convolved, half rate, data mode used to increase accuracy. The 40 kbs^{-1} telemetry rate is not used operationally because there is either a timing conflict in the generation by the DMU of the separate 40 kbs^{-1} ground and 40 kbs^{-1} OBC telemetry data streams, or the OBC has a timing conflict in reading the two 40 kbs^{-1} data streams. There are two telemetry streams, one special for the OBC and the other a ground telemetry stream that is also read and used by the OBC. The dedicated OBC telemetry rate is normally held at 20 kbs^{-1} for smooth spacecraft control. The telemetry rates are selected by the operations controller. Both spacecraft housekeeping data and science data come down in this S-band telemetry. There are four S-band power amplifier/antenna combinations distributed around the spacecraft, only one of which can be switched on at a time. Two are at the bottom on the sun- and the anti-sun-side respectively, and two are on the satellite upper body in similar locations. As a result, no matter what the attitude orientation relative to a ground station, there is one S-band antenna that can be used. There are two VHF transmitters (for redundancy) that operate at 136.86 mHz for the transmission of satellite data. A VHF transmitter is used during range

and range-rate operations to determine IUE's location and also under conditions
when S-band is not available, such as: spacecraft to ground S-band data link
problems; during eclipses because it uses less power than the S-band; and during
spacecraft emergency operations. The maximum data rate of the VHF system is 5
kbs^{-1}. There are two VHF command receivers (148.98 mHz) and decoders for
redundancy on board IUE. The command bit rate is 800 bs^{-1}. All operations and
control are achieved through this command system.

2.2. THE GROUND SYSTEM

The IUE ground system serves two general functions: controlling (commanding)
the spacecraft operations and payload utilisation; and receiving and processing
telemetry, both science data and spacecraft housekeeping data. Science data
processing is also carried out by the IUE computers and various ancilliary systems.
On account of the separation of the control center at GSFC from the command
and receiving site at WPS, a complex communications network is used when
operating IUE from GSFC. The ESA facility at VILSPA is very similar functionally but simpler partly because it is all situated at one site.

At GSFC the command of IUE is initated either in the TOC, or in the
IUEOCC. VILSPA also divides control between a main control room and the
observatory room. Operations of an observatory-type satellite are very complex
and are achieved for IUE by calling up a set of pre-coded and extensively tested

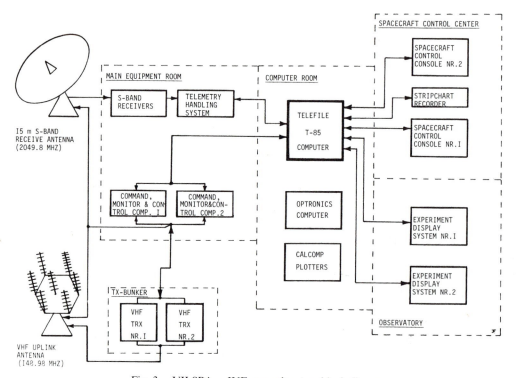

Fig. 3.   VILSPA — IUE ground system block diagram.

procedures, each carrying out a particular sequence of operations, such as camera preparation, exposure and reading. The procedures are run on the control centres' computers and can check the status of the satellite via the telemetry before selecting the appropriate command and issuing it at the appropriate time. A high level and user-friendly language known as "Control Center Interactive Language" (CCIL) was developed which is readily used by the operations staff to develop new procedures to meet the requirements of the guest observers (GOs).

The overall ground system is designed in such a way as to resemble functionally the operations of a modern ground-based telescope. A Resident Astronomer (RA) provides the necessary support to the GO. The Experiment Display System (EDS), which consists of an interactive control keyboard and display terminal, is operated by a Telescope Operator (TO), usually working in the TOC. These personnel, the RA and the TO, possess the required knowledge of spacecraft maneuvering, target acquisition, and S/I operations needed to advise the GO how to carry out critical operations in an efficient way. They also actually carry out the operations, since the GO is not permitted to use the control functions of the EDS console. The EDS provides the observer with all the information needed to plan maneuvers, identify the target, and verify the quality of the spectral image and carry out a 'quick-look' analysis of it.

3. Normal Operations

Normal operations are described below in an ordered sequence, commencing with proposal selection and the preparations made prior to the observing shift by a guest observer who has been allocated observing time, and followed by shift handover and normal spacecraft operations. Next, instrument operations are described in the form of the sequence for a typical observation. Finally, calibration observations and routine data reduction are considered.

3.1. PROPOSAL SELECTION AND OBSERVATION PLANNING

The proposals received by the deadline towards the end of each year are reviewed by panels of peers. In the USA, the panels report to NASA Headquarters which makes the final approval of shift allocations and funding levels. In Europe, ESA and SERC at first allocated their shares separately, but following an agreement reached in 1981, a single joint allocation committee (of peers) selects the proposals and determines the shift allocation. For the Europeans, funding is the responsibility of the observers' own national agencies. Allocations for collaborative proposals which require both US and VILSPA time are negotiated between the three agencies. An annual schedule commencing in April is prepared from the selected programmes, and these form IUE's observing years or observing episodes as they are known in the U.S.A.

Once a GO is granted observing time with IUE, usually in the form of a number of 8 hr shifts, all targets specified in his or her proposal will be checked against Sun, Earth, Moon, spacecraft power, and thermal constraints in order to investigate target availability throughout the year. This information, combined with any time-dependent requirements specified by the GO, is used by the two ground stations to construct the schedules that assign GOs to specific shifts. In due time the GOs are called to the ground station to carry out their observations. Meanwhile a preplanned observation tape (POT) is prepared by the observatory, containing the programme identification, the target identifications and their corresponding positions. The targets are run through a program which checks the coordinates provided by the GO against star catalogues. On arrival for his or her observing shifts the GO makes final preparations with the aid of an observatory RA, who checks the overall plan for the shift, in order to confirm the feasibility of the proposed observations, prepare any special procedures required, and establish whether suitable finder charts are available for star identification, etc.

3.2. SHIFT HANDOVER AND SPACECRAFT OPERATIONS

In the case of a transatlantic handover, the station assuming control will already be monitoring the spacecraft's telemetry and displaying its status when, about 30 min before handover is due, verbal contact is made between the control centers and details are passed to enable detailed planning of operations to commence. Handover itself is marked by the relinquishing station sending a command to change a telemetry bit that then indicates the other station has control.

Routine spacecraft operations include: maintaining communications with the spacecraft, principally by switching S-band antennas and sometimes adjusting the data rate; monitoring all spacecraft housekeeping data against given safety or operational limits for each system, carrying out corrective action when necessary, e.g. slewing to a power positive attitude to prevent excessive battery discharge; planning and execution of maneuvers ensuring pointing constraints will not be violated; correction of gyro drift using information from the fine error sensor; dumping of angular momentum by means of the hydrazine jets when the speed of a momentum wheel is too high; and recording processed telemetry data and its analysis for spacecraft trends. In addition, many of the problems described in Section 5, and the testing of new operating procedures give rise to new routine operational requirements. Finally, the operations team must always be prepared, in the event of an emergency, to take immediate action to ensure spacecraft safety pending a more detailed analysis by the relevant experts.

3.3. A TYPICAL OBSERVATION

The sequence of operations for a typical observation is as follows. The coordinates of the desired target are entered into the ground computer's maneuver generator which offers the operator the choice of several different maneuver sequences to reach the target. The operator selects one, configures the spacecraft appropriately and sends the maneuver command. On completion of the slew to the new target, the field of the FES is transmitted to the ground and displayed on the EDS, the target is identified by the observer, and IUE is moved to place the target in the required spectrograph entrance aperture. Next it is confirmed that the S/I is in the required observing mode. Then, assuming that the camera has already been prepared, an exposure may be started. During the exposure a camera in the other spectrograph can have its image read out, transmitted to ground and it can then be prepared for the next observation. At the end of the exposure the observer can choose to read out the image immediately, or, alternatively, he or she can commence the next operation without seeing the image and arrange for it to be read out later during an exposure with the other spectrograph or even during the slew to the next target. The observer, with the help of a skilled RA to advise, can get the maximum from an 8 hr shift by planning carefully the sequence of targets in order to minimise slewing time, and also by performing camera preparation and read-out during other operations if possible. This is especially important when the observer requires short exposures on several targets. Although a shift is planned in advance, many decisions have to be made on the spot. For example, there is in general no attempt made to coordinate the attitude at handover from one station (or observer) to the next, so the observer's choice of the first target in a shift is often made only 30 min before shift handover when the expected spacecraft position becomes known.

The commands to generate these operations are transmitted from the ground by running procedures in the ground station's computer with appropriate parameters. The options available are described in more detail below.

3.3.1. *Target Identification and Acquisition*

Located at the focal plane of the telescope, the FES, in its mapping mode, provides an image of a 16 arc min field of stars down to $V = 14$ mag. This has a dual role. Firstly it serves a spacecraft function; identification of a star with known celestial position which can be used to update the ACS to remove the errors in position introduced by a slew. Secondly it serves the astronomical function of a finder field in which, if bright enough, the target star can be identified and moved to the appropriate entrance aperture of the spectrograph. For fainter targets a blind offset can be made from a brighter nearby object using offset coordinates prepared in advance from Schmidt sky survey or astrographic plates. The FES has an additional function during exposure since it can be used in a tracking mode to follow a field star and provide a fine guidance signal. This is essential for long exposures to avoid movement of the target star from the spectrograph aperture as a result of gyro drift or thermally induced flexure of the telescope tube. The operations of the FES are controlled by using the appropriate procedure. All this is done by the observatory staff. The only related responsibility of the observer, but an essential one, is to provide a finder map at an appropriate scale, usually reproduced from Schmidt sky survey plates, and to identify the target.

3.3.2. *Telescope Focus*

The temperature of the structure of the telescope tube determines the separation of the primary and secondary mirrors and thus affects the focus of the telescope. It is possible to use a mechanism to refocus but this was only used for the initial adjustment in orbit. Thereafter the mechanism has not been used because it is not redundant and any failure in an out-of-focus position would reduce the quality of the data. Instead, the operators keep the telescope in focus by thermal control, achieved by switching heaters on and off at the back of the primary mirror and on the camera deck, following procedures established during the commissioning phase.

3.3.3. *Spectrograph Modes*

The spectrograph is configured for the observation mainly by operating mechanisms, of which the principal ones, duplicated in each spectrograph, are the high/low dispersion selection mirror, the shutter for the large entrance aperture, and the prime/redundant camera select mirror. Special procedures are used for taking wavelength calibration images, which operate the sun shutter to move a prism reflecting the calibration lamp into the apertures, and for taking flat field images from the UV flood lamps, which require special sequences to ensure they strike and warm up reliably.

3.3.4. *Camera Operations*

Camera operations are normally made very simple through use of CCIL procedures which carry out complex sequences involving 8 electrode voltages, scanning sequences, and checking for anomalous conditions that could lead to loss of image data or, worse, damage to a camera. For example, a single procedure call can carry out a timed exposure. Another carries out a read of the image, transmitting it to the ground computer, then automatically prepares the TV tube's SEC target by erasing the residuals from the previous image and establishing the correct bias voltage on the target for optimum performance on the next astronomical exposure.

As the target is read out the data are transmitted immediately to the ground and, shortly after completion of the whole read, a copy of the resulting image is transmitted from the control centre computer to the EDS in the TOC. There it can be examined by the observer using simple image processing facilities. The most important aspect of such examination is usually an assessment of the level of the exposure. Due to the restricted dynamic range of the SEC vidicon and the wide range of intensities present in many spectra, a repeat observation with a different exposure time may sometimes be necessary. Within 15 min of the termination of an exposure the observer can make a decision on the subsequent programme based on a quick evaluation of the image. This time is not always wasted, since an exposure in another mode may have already been initiated on the same target, or, when it is clear that a repeat is unlikely, a new target slewed to and another observation started. In camera operations, as in planning of slewing, and efficient observer can optimise the scientific return from an 8 hr shift.

3.4. CALIBRATION OBSERVATIONS

The RAs perform routine observations for the calibration of IUE. Regular images of the wavelength calibration lamp are used in the data reduction process to define the wavelength scale. The camera flat fields are monitored and occasionally some shifts are devoted to acquiring a set of images of the UV flood lamp in order to construct a new Intensity Transfer Function (ITF). The observation of a set of standard stars is regularly carried out to provide data for the absolute calibration of the fluxes measured by IUE and to monitor changes in the overall system sensitivity (see article by Harris and Sonneborn, pp. 729—749). Other calibrations, e.g. FES photometric calibration, can be carried out using the data acquired during the GO programmes. These astronomical calibrations, spacecraft engineering calibrations, and testing of new procedures are carried out during specific maintenance shifts which amount to about 8% of the all available shifts.

3.5. DATA REDUCTION

The IUE Spectral Image processing System (IUESIPS) is used to reduce the data acquired into products that are usable by the GO and are suitable for archiving for future use by other interested astronomers. The input to IUESIPS are the raw image data that result from the initial stage of image reconstitution performed on the incoming telemetry by the computer supporting operations. The prime scientific purpose of IUESIPS is to produce data that are as free as possible from instrumental effects. This is accomplished by resident image processing staff using a Sigma-9 computer, or equivalent, and associated peripherals. The processing operations include: an implicit compensation for geometric distortion of the camera system; photometric correction of the images using pixel-by-pixel ITFs; fitting of the spectral orders to a spectral format template and extraction of the spectrum; application of intensity calibrations derived from observations of standard stars; and various optional procedures. The principal data product that is delivered to the GO and transmitted to the archives, is a magnetic tape containing the raw data image and the various stages of reduction to a calibrated spectrum. In addition, other products available on request are photographic hardcopies of raw and processed images, plots of intensity against wavelength, and printouts of data values.

Each ground station has to keep up with the continuing daily flow of raw data images without incurring an ever-increasing backlog. The goal is to give the GO a data tape within 24 hr of the observations, prior to departure for his or her home institution. At GSFC IUESIPS runs on a Sigma-9 computer that shares a disk with the operational computer through which the images are transferred. At VILSPA there is only one computer for both operations and data reduction, so the latter is carried out during the US shifts following the VILSPA observing shift.

4. Spacecraft Operational Constraints

Constraints have to be placed on IUE operations for a number of reasons. Many of these result in restrictions, often contradictory, on pointing. Since these generally have a significant impact on the user, they are explained in detail below. Numerous others exist mostly directed towards maximising the useful life of IUE's subsystems, either by minimising wear, or by avoiding risky situations. Some constraints are programmed into the ground computers so that they cannot be inadvertently violated by ground station personnel.

4.1. SUN, EARTH AND MOON CONSTRAINTS

The only celestial body that IUE is absolutely forbidden to view is the sun, which would damage the detectors and the mirror coatings, and heat the focal plane. There is a sun shutter which should automatically close if sunlight enters the telescope tube, but it is only intended as a 'last ditch' safety device in the event of a failure elsewhere. In fact, IUE is not designed to operate closer than 45° to the sun

($\beta = 135°$), beyond which sunlight would enter directly into the sun shield. Prior to the failure of the fourth gyro (see Section 5.1) it was permissible to go to $\beta = 0°$, i.e. the anti-sun direction, but with the two-gyro/FSS system, $\beta > 15°$ must be maintained so that the FSS can play its role in controlling the satellite.

IUE can be, and has been, pointed at the earth and the moon, but as far as normal operations are concerned these bodies are considered to form constrained zones as they obscure a part of the celestial sphere. Furthermore they produce a high background in the FES due to scattered light within about 10° of the sunlit limb. Unlike the sun constraint these are not hard coded into the maneuver generator and so may be overridden.

4.2. ECLIPSES

One unavoidable constraint arising from the choice of orbit is that IUE experiences twice yearly seasons of solar eclipses caused by the earth. This season of inter-mittent earth shadow lasts from some 23 to 26 days, and during each season a daily eclipse occurs that varies from a few minutes partial eclipse up to 75 min total eclipse. During total eclipses the solar cells deliver no power and so IUE must operate from its batteries alone. To prevent excessive discharge of the batteries, operational constraints are imposed which restrict the collection of scientific data. There are adjustments to the shift times to share the burden of lost time between GSFC and VILSPA and to minimise the overall inconvenience. The annual shadow seasons are designated Spring and Fall, although they actually occur February/March and August/September respectively. The Fall eclipses are longer and deeper. While the first objective during eclipse is to maintain command and housekeeping telemetry contact, it has to be done in the least power-consuming manner possible. There are a set of flight operations directives (FODs) for use during eclipses which define various satellite configurations that progres-sively consume less power. These are employed as required to ensure that the maximum planned depth of discharge of the batteries is not exceeded, but always keeping in view the goal of minimising attitude disorientation, so that recovery time for science operations at the end of the eclipse is minimised. One power conservation measure that is adopted is to switch the telemetry downlink to VHF during deep shadow since VHF uses less power than S-band. Plans for shadow operations are made using predictions provided to the IUEOCC by the Flight Dynamics Group at GSFC. In addition to solar eclipses caused by the earth, the moon also eclipses the sun. However these tend to be brief, partial eclipses and early in the mission did not require spacecraft reconfiguration. The two-gyro/FSS uses the sun as one input to the control system, so all eclipses now require special spacecraft control configurations.

4.3. SPACECRAFT POWER

Figure 4 shows how the solar array power output has changed with time due to the degradation of the solar cells in their radiation environment; such degradation was, of course, anticipated and its effects planned for. To ensure long battery life it is essential that IUE is at a power positive attitude for nearly all the time, except for the unavoidable eclipses described above. At launch, the IUE power reqirements

FEBRUARY POWER HISTORY

1978-1986

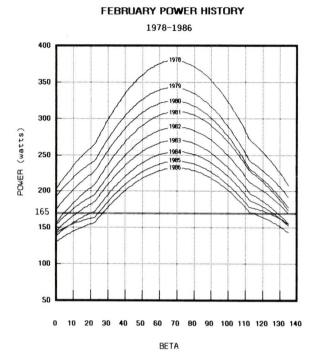

Fig. 4.   Solar Array Output as a Function of β for (February) 1978—1986. 165 W is the power consumed in the current operational configuration.

were about 180 W, but at the time of writing with the 2-gyro/FSS mode of operation in use (see Section 5.1), the requirement is 165 W. In February 1986, the power positive region was in the range $28° < \beta < 113°$. The region is narrower in August when the earth is at aphelion.

Although the batteries can be used to supplement the solar cell power and support operations over a wider range of β, this is minimised to prolong battery life and conserve their capacity for the more critical eclipse seasons. A FOD limits the number of times IUE may be used in a power-negative or power-neutral situation in any one year and the duration of any one such session. There are also constraints on the magnitude of current drain and other related battery operational parameters.

4.4. RADIATION

The principal source of radiation affecting IUE is the earth's out trapped electron belt. Compared to a circular geosynchronous orbit which would lie on the upper side of this belt, IUE's elliptical orbit dips into a region of higher electron flux at perigee. Towards apogee, however, there is a large portion of the orbit where the radiation is substantially less. Based on calculations of the expected radiation, the sensitive elements of the IUE systems were provided with shielding to the maximum extent possible within weight and design constraints (see article by Boggess and Wilson, pp. 3—19, Section 4). At a late stage, because of concern about radiation levels, a particle detector was added to IUE. The output voltage is continually monitored on the ground and at the predefined safety limit of 3.6 V the operating cameras are switched to the standby mode. This limit is only occasionally reached, and at this radiation intensity the background induced in the camera precludes useful exposures, even short ones. The orbit is such that the radiation principally affects the US2 shift which is consequently mainly scheduled for programs with short exposures and for maintenance. In addition to camera background, the radiation was expected to adversely effect some of the electronics and possibly the magnesium fluoride camera faceplates by the end of the 3 yr design life. In fact, other than the expected radiation-induced decay of the solar cell output, very little operational impact has been encountered.

4.5. TEMPERATURE

It is essential to keep the temperature of the various IUE components and subsystems controlled within certain limits. To monitor these temperatures, there are thermistors located in critical areas; their output is delivered to the telemetry via a subcommutator. In the IUEOCC, the thermistor outputs are converted to temperatures for display on the monitors. Preset limits are built into the computer system, and these cause the temperature displays to flash or blink when the limit is reached. The limits are set so that the operating analysts are warned that the parameter is outside its normal range. Usually no immediate action is required at a blink limit; however, if the 'red-line' limit, an absolute value set by the design engineers, is exceeded then corrective action must be taken. The OBC is a case in point. Since the temperature of the OBC is affected by the sun angle, β, and also by the Earth-Sun distance, there is a thermal constraint on β that is most restricting at perihelion. Presently there is no limit on science operations if the OBC is below 55.8 °C, but if the temperature stabilises at 55.8 °C, then normal operations may be performed only if β is outside the constrained range set for the month in question, as listed in the pertinent FOD. If the OBC temperature is greater than 55.8 °C, then as soon as possible IUE must be maneuvered outside the range $40° < \beta < 110°$ in order to cool off. There are also special circumstances, e.g. prior to performing an orbit adjustment, when it is expeditious to cool the OBC before taking a particular action, even though it may be below the blink limit. There are some 90 telemetered temperature points that are subject to blink or redline constraints covering such components as batteries, gyros, electronics, propulsion systems, solar arrays, etc.

4.6. MOMENTUM WHEEL SPEED

An FOD covers the control of the momentum wheels to provide smooth spacecraft control and to prevent excessive wear of the bearings. At the start of the mission the wheel speeds were normally kept between 250 and 1000 rpm in either direction. The wheel speeds are permitted to run through zero or to saturate during maneuvers. When the combination of gyros 1, 3, and 5 was no longer available for emergency attitude recovery, the momentum stored in the wheels had to be reduced so that they could always bring IUE to a safe hold attitude using the analog Sunbath mode. At that time the wheel speed for normal operations was changed to 200 to 500 rpm for roll and yaw and 200 to 1000 rpm for pitch. To date there have been no mechanical problems with the momentum wheels, and even if one of the orthogonally oriented ones should fail there is a spare that could contribute to any of the three axes.

5. Problem Areas

The problems that have occurred with IUE can generally be divided into two categories, those arising in spacecraft systems and those in ground facilities. With regard to the latter, most are related to an aging hardware system and operational software shortcomings which continue to be found throughout the lifetime of the programme. However, the ground system problems, hardware or software, can be and are dealt with, although often at the expense of considerable effort. We will not dwell here on these ground system problems, for they are for the most part those common to any complex computer system; particularly one that, understandably, now has a touch of age creeping in to it. Although retrieval and repair of failed spacecraft has been demonstrated, this is very costly and can at present only be done for a low earth orbit case. Until the development of robot repair missions, any anomaly occurring aboard the IUE spacecraft must be dealt with by changes to the commands transmitted to the satellite or by adopting work-around procedures or by switching to redundant subsystems.

In planning the operations of IUE a very high priority has been given to the safe preservation of the satellite and conservation of resources within the constraints of providing a scientifically effective and productive mission. Even with this approach anomalies can and have occasionally occurred in several systems, but those in two systems, the gyros and the OBC, have caused most concern and are described below. Other systems have degraded in an expected (non-anomalous) fashion, e.g. the solar cells which result in β constraints as described in Section 4.3. Problems have also occured in the S/I. Those occuring in the camera system have been potentially the most serious and these are also described below.

5.1. Gyros

The most troublesome operational system (from the point of near catastrophe) has been the gyros, which are essential to the attitude control of IUE. Thanks both to careful and original design concepts and also to subsequent foresight and ingenuity by a number of people associated with the program, it has been possible to maintain the performance of the ACS. As a result, satisfactory scientific data collection has continued for over 9 yr from launch to the time of writing. The original planning foresight was to build a six gyro package (see Figure 5) that was designed to maintain three-axis control as long as any three gyros were operational. This mode was used for 7.5 yr during which three of the gyros failed.

The first three gyro failures occured as follows. During the third eclipse season, in the Spring of 1979, three of the then six operating gyros (Gyros 2, 4 and 6) were turned off to conserve power. At the end of the eclipse season Gyro 6 failed to restart, although a number of attempts were made. (Hope has not been

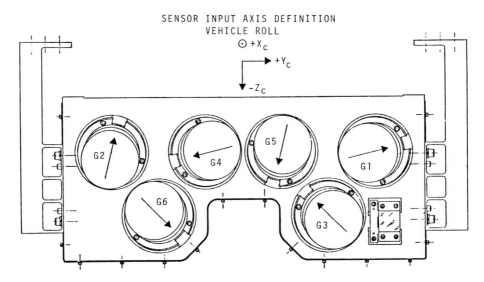

Fig. 5. Inertial Reference Assembly. The arrows show the gyro input axes orientation.

permanently abandoned for Gyro 6, as will be seen below.) In the middle of 1981, maneuver accuracy decreased and telemetry analysis indicated that Gyro 1 was suspect. It was taken off-line and a further analysis finally led to the conclusion that the feedback loop was open and the gyro was not recoverable. Gyro 1 was designated a permanent loss in March 1982. In July of the same year IUE began to slowly drift. Gyro 2 telemetry showed a rapid increase in current indicating that the gyro had stalled. At this point Gyro 2 was written off and the control software changed so that the spacecraft would be operating with the remaining Gyros 3, 4, and 5, at which point IUE had no spare gyros left. Seemingly, one more failure would mean the end of scientific observations. The IUE mission managers had anticipated this possibility even before the loss of the third gyro and had asked that the attitude control experts should prepare contingency plans.

They concluded that a method of attitude control could be worked out that used two active gyros and the fine-sun-sensor (FSS). A program was planned and teams formed to develop, build and test such a system. The testing was done on simulators, as tests on the spacecraft were considered an unacceptable risk. By Spring 1983 the two-gyro/FSS control system had been carried as far as it could go, short of spacecraft usage. As it turned out, for three years (from July 1982 until August 1985) the satellite operated satisfactorily on the three remaining gyros. Then on 17 August, 1985 the critical situation was eventually reached: Gyro 3 failed. The spacecraft gyro body angles began to drift in a situation when they should have remained fixed and the Gyro 3 current fell from a normal operating level of 65 mA to 2 mA. The duty IUEOCC Operations Director immediately recognised the situation and took the action necessary to place the satellite in the safe, SUNBATH, mode. The SUNBATH mode uses analog information from the sun sensors to orient the satellite to $\beta = 67°$ which gives the maximum output from the solar array. However the satellite slowly rotates about the yaw axis (z-axis) thus eventually requiring an attitude recovery exercise. After due consideration, it was decided to try to restart Gyro 3, but to no avail. A conference was called of all experts who could contribute to the situation, and it was decided to try to restart Gyro 6 (see above), but not so vigorously as to cause further damage in this already critical situation. If this failed, then the trusted but untried two-gyro/FSS system would be brought into play. Gyro 6 did not restart and the new system was now the only hope. It worked! Not smoothly at first, but with careful nurturing, it evolved in a few weeks into a system that enabled IUE to gather scientific data essentially as well as in the old days of the three-gyro mode. At the time of writing it continues to do so, always with new improvements being evolved. This was a very great and ingenious achievment by the staff involved.

Even this two-gyro system is not seen to be the terminal control system, for considerable progress has been made in the development of a promising one-gyro/FSS system, and it is possible that, in the event of another gyro failure a workable system could be operational within a few months. Finally, even a no gyro operation may be possible using the FES and FSS and making only small slews from star to star but science operations would be greatly restricted.

5.2. ON-BOARD COMPUTER (OBC)

The OBC is the other spacecraft component identified above as prone to operational anomalies. Indeed, prelaunch, the OBC was the cause of major concern due to malfunctions when running 'hot'. If the ACS can be considered the nerve and muscle of IUE, then the OBC is certainly its brain, for its function is to tell the ACS and other systems what to do and when to do it. It does this through a number of installed software routines called 'workers' which are in turn controlled by the computer executive system program. These workers are algorithms designed to perform specific spacecraft tasks. When there is a problem involving the OBC, it usually seems to involve the malfunctioning of the executive system hardware, with its subsequent effect on the performance of at least one of the workers. Over the lifetime of the satellite there have been numerous OBC malfunctions. While not always understood (mainly due to limited monitoring telemetry), most have not had any major impact on operations; i.e. the effect of the malfunction was easily and quickly corrected, usually by the simple retransmission of a command. However, there have been cases when attitude control was lost and recovery required a great deal of effort on the part of the RAs and the IUEOCC personnel. During the first three years of operation, 1978—80, the OBC was prone to crashes that caused loss of attitude. The OBC software analysts, OBC hardware engineers and operations personnel identified the conditions under which these crashes occurred. The software analysts then inserted code to detect the failures when they occurred, and restart the OBC control. As various failure modes occured with time, the OBC code was modified until OBC crashes were no longer a significant operational problem. This worked so well that continuing failures, 'hits', would have been undetected so a counter was inserted to record them. Many times there were several hundred "hits" per day from which recovery was automatically made without impact. The 'hit' have decreased to only a few per day in recent years.

Another class of anomaly in the OBC was concerned with camera control. A VILSPA study of this phenomenon in June 1980 suggested, but not conclusively, that these problems might be associated with passage of IUE through transitional regions of the radiation belts. These OBC-related camera anomalies still occur on occasion, but are usually caught quickly enough to minimise the effect on the science.

One type of anomaly that has occured twice, each time causing major concern, is the failure of the OBC to carry through a successful orbital adjustment (DELTA-V) to keep IUE on station. The first time that this occurred was on 12 January, 1984. Prior to this DELTA-V there had been eight without incident. The magnitude of the DELTA-V is controlled by the duration of the firing of IUE's high thrust jets. The durations are calculated and the values are inserted into the Worker 19 command sequence which controls the operation. In January 1984 it was planned to fire the appropriate jets for 8.2 s but the program aborted after 1.64 s, leaving IUE in a slow spin. It took over four hours to restabilise the spacecraft and several hours more to determine its attitude. After analysing the orbital data a new DELTA-V was planned and successfully carried out on 14 February 1984 without doing anything different except cooling the jets by orienting them away from the sun prior to firing. The next DELTA-V in November 1984 also proceeded without incident but July 1985 brought a repeat of the fault of January 1984. After collective consultation with the experts it was concluded that differences between successful attempts and failure lay in the yaw phase data. The operational plan was modified so that the next try would emulate the successful DELTA-Vs, which had a negative yaw angle start. On 9 August the exercise was repeated and was successful. It is not known for certain whether the problem had been correctly identified since only eight days later the failure of the fourth gyro occurred and Worker 19 had to be extensively modified for use in the two-gyro/FSS mode. The two DELTA-Vs carried out since have both been successful.

5.3. FINE SUN SENSOR (FSS)

Since the FSS is now a key part of the ACS, used to provide both roll and pitch information, its anomalies have assumed greater importance than before. It will play an even greater role in the attitude control if the one-gyro/FSS has to be installed. The FSS contains dual systems, each of which consists of an electronics package and dual detector heads. The two systems each cover an angular range of $+/- 32°$ in pitch and roll centered on $\beta = 45°$ and $\beta = 105°$ respectively. There is an anomaly in the low-β FSS that prohibits operations below $\beta = 20°$ since the two-gyro/FSS control system was put into operation.

5.4. CAMERAS

Four problems have occurred with the cameras. The first two, microphonic noise and scan control logic failures, have been largely overcome by changed operating procedures. The third has been the almost certain loss of use of the back-up or 'redundant' camera in the short wavelength spectrograph (SWR). Fourthly, a high voltage discharge developed in the redundant long wavelength camera (LWR) and it can now only be operated at reduced gain.

The IUE TV camera tubes are very sensitive to mechanical vibrations particularly at frequencies in the range 500 Hz — 10 kHz. This sensitivity arises because the support structure of the SEC target is a 2 cm diameter film of aluminium oxide 50 nm thick, which has high-Q resonances in this frequency range. The target, connected to a very sensitive amplifier, acts as a condenser microphone. During the commissioning of the S/I, severe microphonics frequently occurred. This was tracked down to the panoramic attitude sensor (PAS) which was used to sense the earth limb in order to determine IUE's attitude during the initial attitude acquisition process. After use, the PAS had been left switched on with its scanning prism rotating but, on discovery, was switched off.

Occasionally important data were lost due to a microphonic 'ping' lasting 10 sec that affected the LWR images by producing a band of interference across the image. It was eventually discovered that this 'ping' was induced by the warm up of the heater in the readout gun of the LWR tube. By increasing the delay between turning up the heater voltage and the commencement of the readout scan, it was possible to ensure the 'ping' occurred before the scan started.

The second problem occurred in the setting up of the digital logic circuits that peform the read-out scan of the LWP camera. When the circuits have remained unused for some hours, e.g. during a long exposure, the serially loaded set-up coordinates fail to pass a specific bit in the line scan register. When the scan is initiated the register then counts correctly but from the wrong starting point so that only part of the SEC target is read out. The solution to this, which took quite some time to develop into a reliable form, was to include in the operating procedure the initiation of a dummy scan to ensure the logic was unlatched prior to commencing the readout scan.

The third problem has been the intermittent operation of the SWR camera tube arising from a loss of the G1-on voltage, which controls the electron read-out beam current. The SWR camera was selected at the outset to be set up and calibrated in orbit for routine use in the short wavelength spectrograph. In the middle of this commissioning period the intermittent operation first occurred. At that time the decision was made to use the SWP camera routinely and this has operated flawlessly to date. At intervals, attempts were made to operate the SWR camera and it was soon observed that successful operation was much more likely when the camera electronics module was cold. As time has gone by successful reads have become less frequent even at low temperatures, and it seems unlikely that the SWR camera would be of any use in the event of the failure of the SWP. One possible cause of the failure is the loss of the clock pulse input to the G1 modulator. Such problems had occurred on the ground due to a poorly mating connector.

The fourth major problem has been the high voltage discharge or 'flare' that has developed in the LWR camera. A proximity focussed intensifier, whose output is coupled by fibre-optic windows to the TV tube, acts as a UV-to-visible wavelength converter. The high electric field (3.8 kV mm^{-1}) is near the maximum that the materials used for the photocathode and the anode can stand without a discharge developing. Any enhancement of the electric field caused, for instance, by a sharp point on the photocathode or contamination by a low workfunction electron emitter can result in an electron (or ion) discharge that gives rise to a point of light in the output image of the converter. The flare intensity increases rapidly as the voltage across the converter is raised, the image is enlarged due to scattering and halation, and further flares with a higher threshold voltage may appear. This phenomenon caused very considerable problems during the development of the IUE cameras. It was only after considerable labour by the project, and the eventually-successful converter manufacturer, ITT, that acceptable converters were made. The development was very successful. The high photocathode efficiency and very low background count rate (in the absence of radiation), which were better than specification, directly contributed to the performance of IUE. A difficult compromise had to be made between tube gain, safety margin from flares and reasonable production yields. The converters were formally rated at a maximum of 6 KV and some testing for flares and background was carried out at that voltage, but for all other testing and operations a voltage of 5 KV was selected.

In 1983 a weak flare was discovered in long exposures on the LWR camera. The centre of the discharge was just outside the field of view of the TV camera but well within the working area of the converter, so only a crescent of light was picked up by the TV tube. Subsequently the flare steadily increased in intensity and therefore affected shorter exposures. The size of the light patch has also greatly increased. Fortunately the LWP had been calibrated and was made available to observers since it offered improved sensitivity in some parts of the spectrum. In October 1983 the decision was made to switch to the LWP for all normal observations. At first, use of the LWR was still permitted at 5 KV to complete series of observations of variable objects, but from April 1985 operation has been restricted to 4.5 KV which is at present below the flare threshold. This has reduced the gain by 27%.

On one occasion in 1976, during laboratory acceptance tests on the LWR converter before coupling to its TV tube, an extremely weak pinpoint of light was detected when making a 2 hr exposure at 6 KV but this did not recur on subsequent tests. The position of that flare is exactly coincident with the deduced centre of the flare now seen through the TV camera. It is assumed that some aging process has lowered the flare threshold voltage. The extreme difficulty in pro-ducing completely flare-free converters forced the project to take the risk of including this one in the flight equipment. With hindsight this gamble has been well justified by five years of trouble-free operation by the LWR. Obviously such aging effects as have been seen in the LWR now raise the question of whether flares may in time appear in other cameras. These were not tested above 6 KV so there is no information as to whether they may have had incipient flare points which could appear at higher voltages.

6. Operational Performance

Despite the problems described above, IUE is still operating at high efficiency nearly nine years into a mission which was originally designed for three years with consumables sized for five years. The ability to reconfigure IUE and achieve the mission objectives in quite different ways has been one of the important factors contributing to its extended life.

The scientific efficiency of IUE was not high in the first month or two after launch. The operational procedures and the ground computer software needed some debugging and extensive tuning to reduce unproductive overhead time when IUE was essentially waiting for the next command to be uplinked. As the operations staff gained confidence in the use of the improving software IUE soon achieved an average efficiency (defined as the fraction of time spent collecting astronomical photons) of 60—70%. It was expected that the increased constraints on pointing in recent years would tend to reduce efficiency but this has been compensated by increased attention to efficient scheduling. At some point in the future it may be necessary to employ integrated operations so as to maintain efficiency by forming an optimum observing sequence from all the target lists of several GOs in one time interval.

At the time of writing, IUE is still in very good shape, with its scientific performance essentially unimpaired since launch over 9 years ago in January 1978. The targets it had observed by March 1986 included:

- 90 Solar System objects (planets, moons, asteroids and comets).
- 5330 Hot stars (O, B, A, S, WD, and WR).
- 2060 Cool stars (F, G, K, and M).
- 1340 Variable stars (novae, Cephids, T Tauri, and binaries).
- 910 Nebulae (planetary, HII, and SNR).
- 1200 Extragalactic objects (galaxies, Seyferts, and QSOs).

These observations had resulted in the production of 56 960 spectral images, collected on behalf of more than 600 astronomers, and have resulted in over 1000 papers being published in refereed journals.

The IUE archive has assumed increasing importance (see article Benvenuti *et al.*, pp. 759—769); any image can be obtained after expiry of the 6 months exclusive data rights held by the original observer. A total of over 40 000 spectra have been requested and the rate of requests now exceeds the rate of acquiring new spectra.

7. Abbreviations

It is convenient and customary amongst the IUE community to use abbreviations when referring to the various complex systems. We have restricted their use here to the more common ones. A list follows of all those used in this paper both those specific to IUE and those in more general use.

ACS	Attitude Control System
CCIL	Contol Center Interactive Language
DELTA-V	Orbit Adjustment Maneuver
DMU	Data Multiplexer Unit
EDS	Experiment Display System
ESA	European Space Agency
ESOC	European Space Operations Centre

FES	Fine Error Sensor
FOD	Flight Operations Directive
FSS	Fine Sun Sensor
G1	Grid 1 (of the vidicon)
GO	Guest Observer
GSFC	Goddard Space Flight Center
IPC	Image Processing Center
IRA	Inertial Reference Assembly
ITF	Intensity Transfer Function
IUE	International Ultraviolet Explorer
IUEOCC	IUE Operations Control Center
IUESIPS	IUE Spectral Image Processing System
LWP	Long Wavelength Prime (camera)
LWR	Long Wavelength Redundant (camera)
NASA	National Aeronautics and Space Administration
OBC	On Board Computer
PAS	Panoramic Attitude Sensor
POT	Preplanned Observation Tape
RA	Resident Astronomer
S/I	Scientific Instrument
SEC	Secondary Electron Conduction (vidicon tube)
SERC	Science and Engineering Research Council
SWP	Short Wavelength Prime (camera)
SWR	Short Wavelength Redundant (camera)
TO	Telescope Operator
TOC	Telescope Operations Center
US1	First US Shift
US2	Second US Shift
UV	Ultraviolet
VHF	Very High Frequency
VILSPA	Villafranca del Castillo, Madrid, Spain
WPS	Wallops Island Facility

Reference

Boggess, A. *et al.*: 1978, *Nature* **275**, 372.

PART II

SOLAR SYSTEM

Edited by A. L. LANE

PLANETARY ATMOSPHERES AND AURORAE

H. W. MOOS

Department of Physics and Astronomy, Johns Hopkins University, Baltimore, MA, U.S.A.

and

TH. ENCRENAZ

Observatoire de Paris — Meudon, Meudon, France

1. Overview

The observation of planetary atmospheres in the ultraviolet range provides specific advantages. A number of strong emissions from atoms and diatomic molecules appear in this spectral region; also a number of molecules have both continuum-like and narrow-band absorptions. In addition, for most atmospheres, the UV radiation does not penetrate to as deep levels as those probed in the visible, infrared and radio ranges because the UV penetration is limited by continuum molecular absorption, by absorption and scattering by small particles, and by rayleigh scattering. (The latter is particularly important for the thick hydrogen atmospheres of the outer planets.) Thus, the UV spectra are formed in the upper portions of a planet's atmosphere and the IUE is an excellent tool for studying the upper atmospheres of planets, their aeronomy and their photochemistry. It is possible to divide the ultraviolet spectra of the planets into two types. Those associated with the region above the homopause (at roughly densities of 10^{12} cm^{-2}) are of emissions from atoms and simple diatomic molecules in diffusive equilibrium above the more complex UV-absorbing molecules which are heavier and hence have smaller scale heights. The situation is illustrated in Figure 1 which shows altitude profiles of hydrocarbon mixing ratios for Jupiter computed by Gladstone (1982). H_2 is the dominant species with a mixing ratio near 1. Above 3×10^{-5} mbar, CH_4 falls off rapidly, He with a lower molecular weight more slowly, and H increases until at altitudes higher than those shown it becomes the dominant species. These emissions from the top of the atmosphere, which are produced by resonance reradiation, photochemical processes, photoelectron excitation and precipitated magnetospheric charged particles, permit determinations of the abundances of the observed species and the energy input for that region. At wavelengths above very roughly 1600 Å, there is a sharp decrease in absorption by atmospheric species (e.g. Mount and Moos, 1978) which permits the penetration of ultraviolet radiation to lower altitudes where the atmosphere is more homogenously mixed by turbulence. (Note in Figure 1 that H_2, He, and CH_4 approach constant values with decreasing altitude.) This is coupled with a sharp increase in the solar continuum giving a strong signal from rayleigh scattered radiation which shows characteristic absorptions from the molecules in the upper part of the homogenously mixed atmosphere. Due to the rapid changes in the

Y. Kondo (ed.), Scientific Accomplishments of the IUE, pp. 45–65.
© 1987 *by D. Reidel Publishing Company.*

rayleigh scattering cross section $(-\lambda^{-4})$ and changes in the molecular absorption cross sections, the radiation reflected at different wavelengths is associated with solar radiation which has penetrated to different depths in the atmosphere. In the case of the major planets, it is possible to measure the abundance of complex moleculars down into the stratosphere.

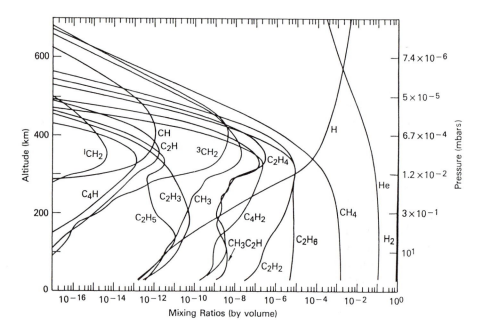

Fig. 1. Jovian altitude profiles of the hydrocarbon mixing ratios computed by Gladstone (1982). The pressure scale has been added by Strobel (1983). Calculations for NH_3, not shown, give a value for the mixing ratio of 10^{-5} at 500 mbar which decreases rapidly due to photochemical destruction to a value of 10^{-10} near 30 mbar, the bottom of the above figure (Strobel, 1983).

The spectra of Jupiter provide a good example. On Jupiter, the ultraviolet radiation above 1600 Å probes the region where NH_3 is largely photodissociated and where the species resulting from the hydrocarbon photochemistry are found (Figure 1). If only rayleigh scattering need be considered, radiation at 1600 Å would have an optical depth of one at a pressure of about 50 mbar; at 2100 Å, about 200 mbar. However, the effect of the absorbing species is to push the penetration depth to higher altitudes. Thus, most of the scattered signal comes from above the troposphere. Due to the large strengths of the UV electronic transitions, molecular species present even in very small concentrations can produce an absorption signature in the spectrum. Below 1600 Å, emissions due to higher altitude species appear, the Lyman and Werner bands of molecular hydrogen and the strong Ly α emission of atomic hydrogen at 1216 Å dominate the spectrum. These emissions are signatures of aurorae and other only partially understood excitation processes in the uppermost portions of the Jovian atmosphere.

A number of UV studies of the planets had been performed prior to IUE using rockets and satellites. As an example, Table I lists the studies of Jupiter up until the commissioning of IUE in 1978. Many of the Jovian results showed discrepancies due to the varying characteristics of the instrumentation, differences in absolute photometric calibration and possibly time dependent changes on the planet itself. (For example see the Comparison by Duysinx and Hendrist of the jovian albedo obtained by TD1A with those of Anderson et al., 1963; Stecher, 1965; Kondo, 1971; and Wallace et al., 1972. See also Giles et al., 1976). However, the studies showed reasonable agreement on the global shape of the geometrical albedo, on the presence of strong absorption below 2100 Å, the presence of strong Ly α emissions, and the presence of emission bands 1200—1600 Å, probably due to H_2. The absorption was attributed to saturated NH_3 vapor (Greenspan and Owen, 1967), but Tomasko (1974) demonstrated that, according to the UV data, ammonia had to be even more depleted than allowed by the saturation curve. In the case of Saturn, also observed with OAO—2 (Owen and Sagan, 1972; Wallace et al., 1972) the reflectivity curve was maximum around 2500 Å as it was for Jupiter, however a large uncertainty existed due to the unknown contribution of Saturn's rings. In the cases of Uranus and Neptune, the albedos between 2000 and 4000 Å were measured from OAO—2 and found to be lower than expected for the case of a pure rayleigh-raman scattering atmosphere (Savage and Caldwell, 1974). Scattergood and Owen (1977) proposed a variety of particles produced by photodissociation of H_2 and CH_4 in order to explain the low reflectivity observed on the giant planets and on Titan. H I Ly α emissions from Jupiter and Saturn were observed using rockets, the OAO, and the Pioneer 10 and 11 UV photometers (Table I). It was not certain if the wide range of H I brightness values measured for Jupiter were due to temporal variations because differences in the calibration and other characteristics of the instrumentation could not be

TABLE I

Ultraviolet observations of Jupiter prior to IUE

Type of experiment	Spectral range	Reference
Rocket	2700 Å	Boggess and Dunkelman, 1959
Rocket	1700—4000 Å	Stecher, 1965
Rocket	2300—3700 Å	Evans, 1967
Rocket	2400—3000 Å	Jenkins, 1969
		Jenkins et al., 1969
Rocket	1800—2600 Å	Anderson et al., 1969
Rocket	2100—3600 Å	Kondo, 1971
OAO—2	2000—3600 Å	Owen and Sagan, 1972
TD 1	1700—2500 Å	Duysink and Henrist, 1974
Rocket	1200—1800 Å	Moos et al., 1969
Rocket	1180—1900 Å	Rottman et al., 1973
Rocket	1200—1900 Å	Giles et al., 1976
Pioneer 10	500—1300 Å	Carlson and Judge, 1974

completely eliminated as possibilities. Weiser *et al.*, 1977 showed that there was a cloud of circumplanetary atomic hydrogen near Saturn. Finally, the Jovian H_2 emission bands were recognized as a signature of auroral processes. (Rottman *et al.*, 1973; Giles *et al.*, 1976). In summary, while the UV analyses performed before 1978 for the giant planets and in particular for Jupiter indicated the primary spectral features, they also indicated the need for systematic measurements by a single well calibrated instrument such as the IUE.

As mentioned by Lane *et al.* (1978), observations of planets with IUE present several problems. Solar system objects move on the celestial sphere and it is necessary to correct for this motion. These objects are often both unusually bright and extended which affects the ability of the IUE to track directly. Sometimes a planetary satellite can be used to get around this problem, but this technique requires accurate information on the time-dependent relative positions of the satellite and planet. Another problem comes from the steep slope of the spectrum for all solar system objects: the solar flux at 2000 Å is about 40 times stronger than at 1600 Å, so that due to the limited dynamic range of the IUE, several spectra of different exposures have to be recorded to cover the whole spectral range. Also, the photocathodes for the short wavelength cameras are not extreme-solar blind. As a consequence, the emission spectrum at short wavelengths is contaminated by grating scatter of the intense long wavelength radiation, a severe limitation to the sensitivity of IUE. Finally, serious problem is the acquisition of a solar comparison spectrum for comparison with the rayleigh scattered signal from the planet. Since the spectrum of the Sun shows significant temporal variations below 1800 Å, the best solution would be to record a solar spectrum simultaneously, but this is not possible with IUE. Thus the IUE planetary data are compared to solar spectra recorded at other times and with other instruments, presenting an important source of uncertainty.

We have divided the review into two areas: (1) the composition and structure of the upper parts of the atmospheres below the homopause and (2) the upper atmosphere above the homopause and the effects of the magnetosphere, including the Io torus. We have restricted this review almost exclusively to the outer planets although IUE has also been used for several investigations of Mars and Venus (e.g., Conway *et al.*, 1979; Feldman *et al.*, 1979; Durrance, 1981; Conway *et al.*, 1981).

2. Composition and Structure of Upper Atmospheres Below the Homopause

As mentioned above, the UV spectra of the major planets decrease very rapidly towards short wavelengths, following approximately the slope of the solar spectrum itself. At wavelengths above 2200 Å the reduction is relatively simple, because the solar spectrum exhibits no short period temporal variation, and because independent photometric measurements can be used for calibration. In this spectral range, the planetary spectra show to narrow absorption or emission features due to atmospheric species. At wavelengths shorter than 2200 Å, the situation is more complex. Some clear absorption features appear: C_2H_2 around 1700 Å on Jupiter, Saturn, and Uranus, NH_3 between 1800 and 2200 Å on

Jupiter. Some other broad absorptions, more difficult to interpret, are also present in the spectra of Jupiter and Saturn. A difficulty in reducing the interpreting these data comes from the choice of a solar reference spectrum for comparison. Solar spectra used in the different studies are those of Brinkmann *et al.* (1966), Broadfoot (1972), Kjeldseth-Moe *et al.* (1976), Mount *et al.* (1980), and Mount and Rottman (1981, 1983a, b). These various spectra show significant differences between one another, both in the line intensities and in the general shapes; these differences are sometimes due to temporal variations but may be attributed in some cases to possible instrumental effects. Wagener *et al.* (1985) have proposed the use of a corrected solar spectrum, adapted from Mount and Rottman (1981). However, the problem of the choice of the solar reference spectrum remains one of the major sources of uncertainty for the interpretation of planetary spectra below 1800 Å.

2.1. JUPITER

The long wavelength part of the UV spectrum ($\lambda > 2300$ Å) has been studied by Wagener *et al.* (1985) who have calculated the Jovian reflectivity between 2300 and 3150 Å, using a multi-layer scattering model. As shown on Figure 2, the best fit is obtained for a single scattering albedo of 0.42; the geometric albedo is about 0.25.

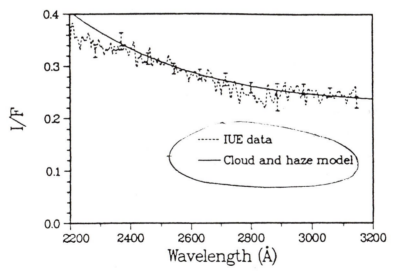

Fig. 2.   Model spectrum (line) for cloud and haze structure of Jupiter (see text) with single scattering albedo of 0.42. Dashed line: IUE data (Wagener *et al.*, 1985).

The 1900–2300 Å region is the domain of NH_3 absorption. As mentioned above, a broad feature had been tentatively identified around 2200 Å from previous rocket observations, but the individual bands had not been detected. A first identification of the NH_3 bands around 1900 Å was suggested by Owen *et al.* (1980), from low resolution SWP IUE spectra of Jupiter. More bands were identified (Figure 3) on LWR spectra recorded at high resolution and degraded to

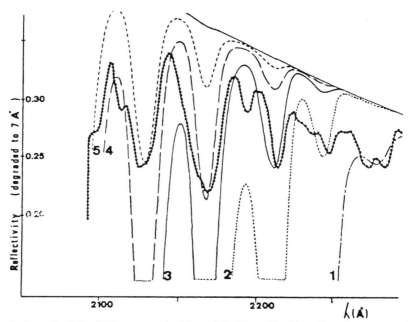

Fig. 3. Jovian reflectivity (I/F) compared with models (1) to (5) with various but constant NH_3/H_2 mixing ratios corresponding to NH_3 column density of: (1) 1×10^{21} cm^{-2}; (2): 3×10^{19} cm^{-2}; (3) 4×10^{18} cm^{-2}; (4) 1×10^{18} cm^{-2}; (5) 3×10^{17} cm^{-2} (Combes *et al.*, 1981).

3 Å to improve the signal to noise ratio (Combes *et al.*, 1981). A comparison with synthetic spectra corresponding to various NH_3 distributions shows that the NH_3/H_2 ratio must strongly decrease with pressure in the upper jovian atmosphere as an effect of photodissociation (Fricke *et al.*, 1982). The recent analysis of Wagener *et al.* (1985) leads to a mixing ratio NH_3/H_2 of 2.5×10^{-8} at the tropopause ($P \sim 100$ mbar).

The absorption spectrum of C_2H_2 is found between 1600 and 1900 Å. This signature was first identified for the case of Jupiter by Owen *et al.* (1980), who derived a mixing ratio of 2×10^{-8} (Figure 4). A recent study, based on a simple reflecting model, confirms this result (Encrenaz *et al.*, 1986), also in agreement with Wagener *et al.* (1985). Clarke *et al.* (1982) obtained a mixing ratio of 1.2×10^{-7} from the narrow C_2H_2 absorption bandsin the IUE spectra below 1750 Å. Gladstone and Yung (1983) analyzed the same data with a more detailed model and obtained $(1.0 \pm 0.1) \times 10^{-7}$. However, Wagener *et al.* (1985) used the 1600—1900 Å range and found that the best fit corresponds to a value of 3×10^{-8}. All of these authors conclude that an additional continuum absorber is required shortward of 1750 Å. As possible candidates, Wagener *et al.* (1985) suggest allene (C_3H_4) and cyclopropane (C_3H_6) with mixing ratios of 7×10^{-10} and 8×10^{-9} respectively.

2.2. SATURN

As in the case of Jupiter, the long wavelength part of the UV spectrum is

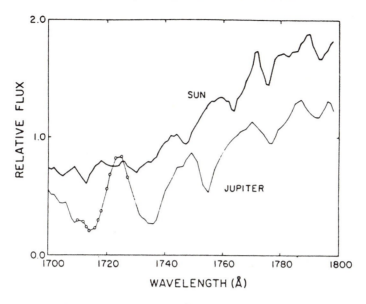

Fig. 4.   Jovian spectrum between 1700 and 1800 Å compared with the solar spectrum of Kjeldseth-Moe *et al.* (1976) showing the presence of C_2H_2 features at 1715, 1735, 1755, and 1775 Å (Owen *et al.*, 1980).

dominated by rayleigh scattering. Above 2000 Å, the reflectivity curve of Saturn (Figure 5) is a smooth curve which is well fitted with an atmospheric model consisting of a reflecting layer with a lambertian reflectivity of 0.20 with an

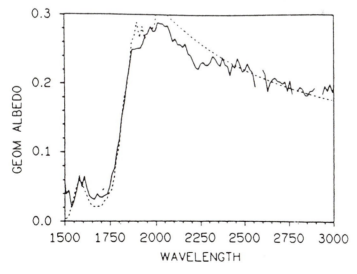

Fig. 5.   Saturn model containing C_2H_2, C_2H_6, and C_4H_2 (dotted line) compared to the geometric albedo of Saturn (solid line) (Winkelstein *et al.*, 1983). Both curves have been smoothed to 14 Å bins.

atmosphere of H_2 above, corresponding to a column-density of 3 km — Amagat above the reflecting layer (Winkelstein *et al.*, 1983).

Below 1900 Å, the spectrum of Saturn exhibits a series of strong narrow-absorption bands, first identified as C_2H_2 by Moos and Clarke (1979) (Figure 6). Clarke *et al.* (1982) reported evidence for a continuum absorber but were not able to fit the measured albedo. In a more recent analysis, Winkelstein *et al.* (1983) concluded that continuum absorbers had to be taken into account to fit the spectrum below 2000 Å. After considering a large number of possible candidates, they found a good agreement with a model including C_2H_2 and C_2H_6 with mixing ratios of 10^{-7} and 6×10^{-6} respectively, with, in addition, a layer of water with a column-density of 6×10^{-3} cm — Amagat at the top of the atmosphere. While the presence of C_2H_2 and C_2H_6 is quite understandable in the upper atmosphere of Saturn, the presence of H_2O is, in contrast, rather unexpected. If the identification is confirmed by further observations, its origin would need to be explained.

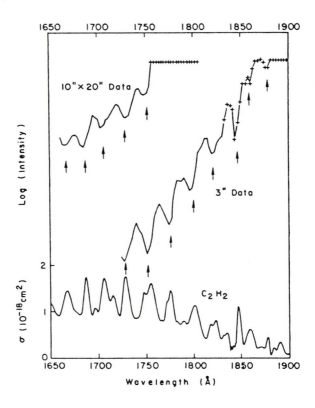

Fig. 6.   IUE spectra of Saturn, and acetylene cross sections (Moos and Clarke, 1979). The log of the intensity, in arbitrary units for each spectrum, is plotted versus wavelength for two spectra of Saturn, one obtained with a 10×20 arcsec entrance aperture and the other with a 3 arcsec diameter entrance aperture. The hatched line refers to data contaminated by saturated pixels. The vertical arrows indicate the wavelength position of maxima in the acetylene absorption cross sections (Nakayama and Watanabe, 1964).

2.3. URANUS AND NEPTUNE

Although they might look similar in view of their general physical properties, these two planets show significant differences. First, Neptune exhibits a strong internal energy source, while Uranus does not; second, the temperature of Neptune, above the tropopause, is much higher than the Uranus temperature. Finally, the emissions of C_2H_6 at 12 μ and CH_4 at 8 μ are much less for Uranus than for Neptune (Orton *et al.*, 1983). This third difference has two possible explanations: (1) CH_4 and C_2H_6 are depleted in Uranus, as compared to Neptune; (2) the upper temperature of Uranus is too low for the emissions to be observable.

UV observations of Uranus and Neptune are an important tool for solving this question. Indeed, the existence of a C_2H_2 signature in the UV does not depend upon the temperature. If CH_4 and C_2H_6 were significantly depleted on Uranus, we would expect C_2H_2 to be depleted also. We can thus use the C_2H_2 UV signature as a possible test of the abundance of methane and its derivatives in the upper atmosphere.

Between 2000 and 3000 Å, the spectra of Uranus and Neptune were analyzed by Caldwell *et al.* (1981, 1984). The authors compared the observations to a raman-rayleigh scattering curve and derived for both Uranus and Neptune a geometric albedo close to 0.5, i.e. two times higher than for Jupiter and Saturn.

At wavelengths lower than 2000 Å, observations of Uranus and Neptune with IUE are very difficult, and require long integration times. From 1980 to 1984, double shifts were used to accumulate integrating times of 10 to 15 hrs. In the case of Neptune, the signal to noise was too low and no information could be derived (Encrenaz, 1982). In contrast, the spectrum of Uranus showed an absorption band, interpreted as C_2H_2 by Caldwell *et al.* (1984, 1986) and Encrenaz *et al.* (1986) (Figure 7). From the 1980 data, the amount of C_2H_2 is about 3×10^{-8}. According to Encrenaz *et al.* (1986), this abundance implies implies an eddy diffusion coefficient between 10^5 and 10^6 cm^2 s^{-1}; the observed C_2H_2 is controlled by photochemistry and not by saturation, which takes place at lower latitudes (Figure 8). These results have been recently confirmed by the Voyager UVS experiment.

3. The Upper Atmosphere: Above the Homopause and the Effects of the Magnetosphere

This section is concerned with IUE studies at shorter wavelengths, where the IUE spectra provide information on the atmosphere above the homopause. Emissions at these wavelengths also provide extensive information on the magnetosphere of a planet. After discussing the important excitation processes and then the general scientific goals of studies of this type, three examples drawn from the giant planet are given: Jupiter, Io torus, and Uranus. The IUE has become an important tool for planetary research, and we can only present a portion of the results published to date as illustrative examples. A review of IUE studies of the Jovian magnetosphere and the Io torus has also been given by Feldman (1986).

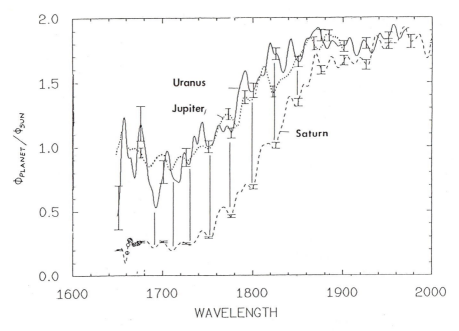

Fig. 7. IUE spectra of (a) Uranus (1980) and (b) Saturn between 1650 and 1900 Å. The data have been divided by the solar spectrum of Kjeldseth-Moe *et al.* (1976).

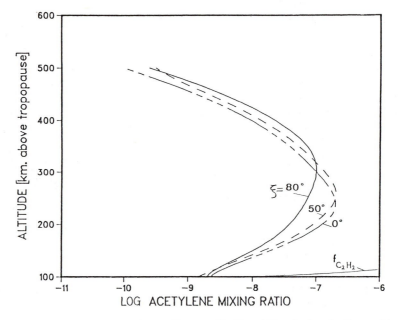

Fig. 8. Photochemical profiles of C_2H_2 at Uranus with $K = 10^5$ cm²s⁻¹ and solar zenith angles of 80°, 50°, and 0° (Encrenaz *et al.*, 1986). Also shown is the curve for saturated vapor mixing ratio of C_2H_2 ($f_{C_2H_2}$).

3.1. EXCITATION PROCESSES

Above the homopause, almost all of the emitted radiation is due to line emissions of atoms, ions and molecules. (Molecular dissociative continua may also be present). Where there are matches between strong solar lines and the upward transition energy, resonance reradiation (e.g., the pervasive Ly α) or fluoresence occurs. Electron excitation is also significant and often dominates. Sources of the electrons include photoelectrons due to EUV solar radiation and auroral secondaries caused by ionization of the atmospheric gas by higher energy primary precipitating from the magnetosphere. The electroglow discussed below is probably produced by a different as yet undetermined source of electrons (or electron acceleration). Excitation by chemical reactions is less common but is a possible source.

3.2. FUNDAMENTAL QUESTIONS AND SCIENTIFIC GOALS

The first questions answered by any spectroscopic study are the identities of the species and their abundances. Note te quantitative values for the abundances depend on knowledge of the excitation processes and the accuracy of the models employed to analyze the spectral data. These models in turn require good spatial resolution for the determination of quantities such as scale heights for their formulation. This requirement could limit the accuracy of the abundance determinations by observatories such as the IUE. Fortunately, observations by flybys and orbiters have provided spatial information and fairly accurate models for many of the planets. As a result, by using these models, reliable abundances as well as changes in the abundances and excitation processes can be determined from the near Earth astronomical data.

Secondly, the data provides a determination of the energy deposition rate in the upper atmosphere of the planet. This deposited energy is emitted as radiation used to heat the atmosphere or converted to chemical energy by the creation of atoms and ions. Some of these atoms in turn are transported to lower altitudes where they play a role in chemical processes in the lower mesosphere and stratosphere (see Section 2). Both the Sun and the magnetosphere serve as significant sources of energy with characteristically different spectral signatures. In addition, other processes such as the presently poorly understood 'electroglow' process discussed below must be considered as significant energy sources for the upper atmosphere.

Planetary magnetospheres may be studied by using ultraviolet spectroscopy. The trapping, heating and transport of plasma in these magnetospheres involve a set of fairly complex physical processes — of interest not only for planetary studies but for a broad range of astrophysical phenomena. The trapped magnetospheric plasma can be observed when it precipitates, exciting atmospheric emissions — i.e. aurorae — or it can be observed directly as circumplanetary gas — e.g. the Io torus. Ultraviolet astronomical observations of these emissions can be used to examine magnetospheric processes remotely and it appears that remote observations over an extended length of time with photometrically stable instruments may be the best way available for monitoring changes in the magnetospheric activity of the outer planets.

3.3. JUPITER

Rocket measurements (Rottman *et al.*, 1973; Giles *et al.*, 1976) demonstrated the presence of intense molecular hydrogen emissions implying signifigant amounts of auroral activity. The auroral zones were mapped and a large number of spectra were obtained by the Voyager UVS experiment (Broadfoot *et al.*, 1979). The IUE obtains spectra of these bright emissions with good signal to noise and a spectral resolution of approximately 11 Å, about three times better than that of the Voyager UVS (Durrance *et al.*, 1982) (see Figure 9.). The intensity of emissions varies both in time and with the location in magnetic longitude. As shown in Figure 9, by aligning the ten arcsec wide spectrograph aperture in an approximately north-south direction and placing it on the northern (or southern) portions of the 40 arcsec disc, it is possible to map the longitudinal distribution of auroral emissions. As the planet rotates over its 10 hr period, the changing signals reflect both changes in the observing geometry and the effects of the longitudinal distribution. Figure 10 shows the results of eight rotational studies compared with

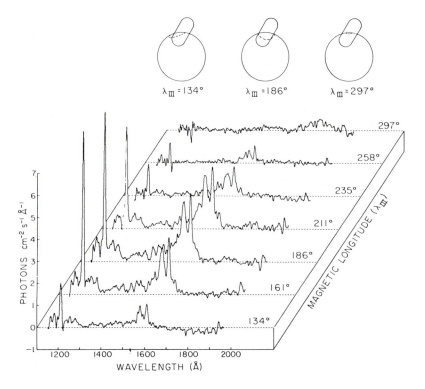

Fig. 9. Spectra of the Jovian aurora as a function of magnetic longitude (λ_{III}) (Durrance *et al.*, 1982). The orientation of the entrance aperture in relation to the disk of Jupiter is shown for four of the exposures above the spectra. The dashed line indicates the position of the auroral zone, taken to be a mapping of magnetic field lines that intersect the Io torus. Seven exposures of 15 min each were separated by ~45 min. The central meridian at the midpoint of each exposure is given. Only the auroral emissions are included in the spectra shown; contributions due to non-auroral planetary light and non-Jovian Lyman-α emission have been subtracted.

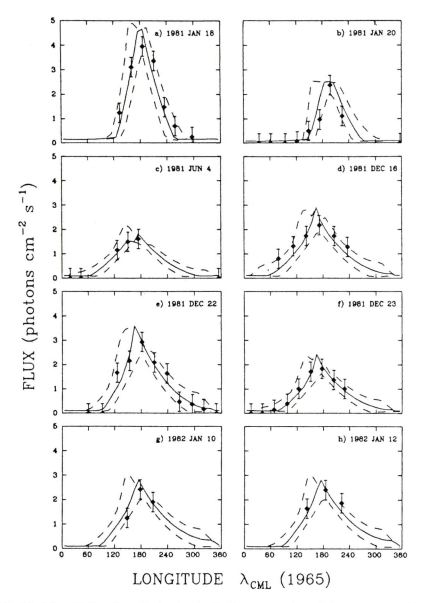

Fig. 10.   Model predictions for a Jovian longitudinally asymmetric emission concentrated between magnetic longitudes of 120° and 240° compared with the integrated IUE flux 1300—1650 Å for individual observations as a function of central meridian longitude (Skinner *et al.*, 1984).

a particular model in which the emissions were restricted to a range of latitudes corresponding to the location of magnetic field lines coming from the Io torus and central meridian magnetic longitudes between 120° and 250° (Skinner *et al.*, 1984). The agreement is quite good. Models which assumed the same emission intensity at all longitudes or emission at higher latitudes (corresponding to the magnetotail, the normal location of terrestrial aurorae) do not fit. Studies of the southern auroral zone also show a longitudinal assymmetry centered near 0° (Skinner and Moos, 1984). The total radiated auroral power is estimated to be $1-3 \times 10^{12}$ W.

In the equatorial region, Voyager data and a rocket observation shows that a region of enhanced and variable H I Lyα brightness exists near 100° magnetic longitude (Sandel *et al.*, 1980; Clarke *et al.*, 1980). The cause of this enhancement, often referred to as the hydrogen or Lyman-alpha bulge, is poorly understood. It is clearly related to the magnetosphere as it is tied to the magnetic longitude and the IUE observations show that it continues to persist near the same magnetic longitudes after a number of years (Clarke *et al.*, 1981; Skinner *et al.*, 1983).

Away from the bulge, the equatorial Lyα emission brightness is almost constant as a function of magnetic longitude and time. The source of this emission was thought to be due to resonance scattering of the solar line and thus it could be used to determine the atomic hydrogen column density. This analysis becomes more complex if charged particles are responsible for a significant portion of the excitation. The stable photometric calibration of the IUE has proved to be a significant advantage in tracking the brightness over a large portion of a solar cycle and the IUE measurements have shown that the brightness has in fact slowly decreased with the decrease in the solar Lyα (Deland, 1985; Skinner, 1986). However, this result is not conclusive as the solar EUV would also decrease in a similar fashion. While the solar EUV production of photoelectrons is not sufficient to provide the observed intensities, Shemansky and Smith (1986) have argued that a mechanism similar to the electroglow mechanism which is so important for Uranus is also significant for the equatorial region of Jupiter. What we have then are some of the pieces of a puzzle. It is known that the electroglow excitation disappears on the dark side of a planet implying that solar radiation serves as a stimulator and/or control. If in fact this mechanism also dominates the equatorial emissions for Jupiter, the IUE results imply that the solar EUV serves as a quasilinear regulator of the much stronger electroglow excitation rates. Determining whether the definite physical cause of these emissions can be obtained by using the capabilities of the IUE or whether this advance must await more refined observations is a present challenge to workers in this area.

3.4. IO TORUS

The concept of a circumplanetary nebula or torus of atoms of satellite origin was first proposed by McDonough and Brice (1973) for Titan. However, ground based observations first showed the presence of an such a nebulae in the orbital path of Io about Jupiter. Sodium optical emission from Io's vicinity was first discovered by Brown (1974). Kupo *et al.* (1976) discovered the existence of optical emission in the red S II lines. The Voyager encounters showed that the torus was the dominant source of the magnetospheric plasma. The Voyager UVS instruments showed strong signals, particularly below 1200 Å, from the ions of sulfur and oxygen. (Broadfoot *et al.*, 1979). The ultimate source of the sulfur and oxygen is volcanic activity on Io driven by the orbital motion of the satellite in the Jovian gravitational field. However, the mechanisms which introduce the material into the plasma torus are not well understood. Likewise, te detailed mechanisms for plasma heating, transport and loss are the subject of considerable discussion. Observations by IUE at the time of the Voyager 1 encounter showed that the IUE could detect emissions from the nebulae with exposure times of about eight hours and a spectral resolution of about 11 Å (Moos and Clarke, 1981). The observed features are S II 1256, S III 1199 and S IV 1406. S III 1199 tends to be contaminated by the intense geocoronal Ly α; however, it has been shown that a feature at 1729 is due to a previously undetected intercombination transition (^{5}S—^{3}P) which is not quenched at the low electron densities in the torus (Moos et al., 1983). O III 1664 has not been detected with certainty. (There is a blemish on the camera near 1664 Å). It is unfortunate that there is no feature due to O II in this spectral region as most of the oxygen is probably in this ionization state and the ratio of sulfur to oxygen provides insight into the nature of the processes whereby material from Io is injected into the plasma torus. Because the IUE spectra are relatively clean with just a few isolated features, it is also possible to set upper limits for a number of neutral and ionic species with emissions in the 1200—1900 Å region (Moos and Clarke, 1981).

The ion and electron concentrations can be obtained from the IUE spectra by using a model of the torus (Smyth and Shemansky, 1983) based on the Voyager spectroscopic and in situ measurements (Moos et al., 1985). The major uncertainty is the percentage composition of O$^+$ which requires measurements at wavelengths below the IUE range. The photometric stability of the IUE instrumentation has permitted addressing the question of whether there are long term changes in the Io torus. Figure 11 shows concentrations obtained from IUE measurements at very roughly six month intervals over a five year period from the Voyager 1 encounter to March 1984. Over this five year interval the torus has remained remarkably stable.

The most desirable way to obtain electron temperatures would be to measure the relative population of two levels with quite different excitation energies above the ground state (and with respect to kT). This is not possible for the Io torus in the IUE spectral range although such measurements are possible at shorter wavelengths. (S III 1199/1729 shows a slight sensitivity, but λ1199 is often contaminated by Ly α.) However, it is possible to estimate the electron temperature by comparing the *ionization balance* with the predictions of models. Figure 12 compares the S II/S III and S III/S IV abundance ratios with the predictions of a simple zero dimensional model which incorporated the transport properties into a characteristic confinement time for the plasma. The temperature appears to have remained remarkably stable with a value near 55000 K. (The one discordant point may be due to an injection event or other time dependent phenomena, whereas the model assumes steady state.) Although the exact value of the temperature might change slightly in a more refined model, the ionization rates change rapidly with temperature and as a consequence the implied variability in the temperature would remain small. Thus, on the basis of the IUE data, it appears that the Io torus plasma has a natural thermostat mechanism which maintains the plasma near this temperature.

It is also possible to use the IUE to study the atmosphere of Io itself. A recent observation by Ballester et al. (1986) with Io centered in the large aperture of the

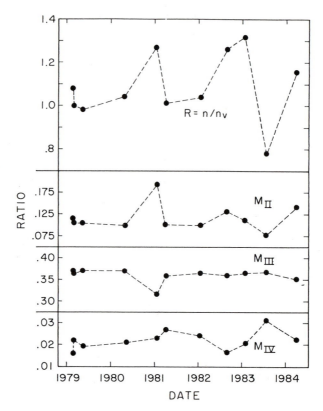

Fig. 11.   Ratio of electron density to the Voyager 1 case (R) and mixing ratios for S^+, S^{++}, and S^{+3} (M II, M III, and M IV) determined from the modeling of IUE data obtained between 1979 March 1 and 1984 March 21. (Moos *et al.*, 1985).

short wavelength spectrograph has detected emission lines of atomic sulfur and oxygen near Io with a spatial extent along the slit close to that expected from the spatial resolution limit of the IUE (about 6 satellite radii). Observations of this type will provide information on the interaction of the atmosphere of Io with the plasma torus.

3.5. URANUS

Early work by Darius and Fricke (1981) using the small aperture showed that the value of the Ly α brightness was surprisingly high, 1.9 kR. This value was uncertain because it was not certain how much of the 3.8 arcsec planetary disc passed through the nominal 3 arcsec diameter aperture and because the ratio of the solid angles of the large and small apertures — used to subtract the terrestrial airglow from the total signal — was uncertain. Durrance and Moos (1982) repeated this measurement using an imaging technique. The planetary disc was placed in the center of the large aperture and the monochromatic image at 1216 Å was used to determine the baseline value of the geocorona. Figure 13 illustrates the technique showing geocorona only, geocorona plus Uranus and the difference.

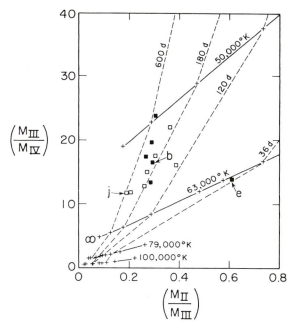

Fig. 12. Comparison of the measured ratios of the mixing ratios presented in Figure 11 with the predictions of a zero-dimensional model which included the effects of ionization, recombination and plasma diffusion expressed as a confinement time in days. M_{II} is the mixing ratio for S^+ etc. The measured values are marked with filled squares for values obtained prior to 1981.7 and open squares for later observations. The points marked with plusses were computed for plasma confinement times of $^\infty d$, 600d, 180d, 120d, 60d, 48d, 36d, 24d, and 12d. Solid lines connect points with the same electron temperature; dashed lines connect points with the same confinement time. Observation (b) was obtained at the time of the Voyager 1 ecounter, (e) near the time ofan intense aurora, and (j) at an elongation of 7.0 R_J (all others were at $\leqq 6\ R_J$).

The value obtained, 1.6 ± 0.4 kR, was quite high if scattering of solar radiation was the only source. Precipitation of charged particles trapped in a magnetic field, analogous to magnetospheric processes on the Earth, Jupiter, and Saturn was suggested. At about the same time Clarke (1982) showed that the Ly α was not only bright but variable implying that at the least, the variable portion of the signal was due to a process such as electron precipitation. Thus, both of these studies concluded that Uranus had a significant magnetic field. Repeated IUE observations, displayed in Figure 14, show that the phenomenon of a bright variable Ly α signal has persisted over a three and a half year period (Clarke *et al.*, 1986).

Although Voyager has shown that in fact Uranus does have a significant magnetic field, it has also shown that another excitation mechanism in addition to aurorae must be considered in discussing the Ly α and the H_2 Werner and Lyman emissions. By modeling the H_2 spectra obtained by the Voyager UVS, Shemansky and Smith (1986) showed that the energy spectrum of the exciting electron fell off much more sharply with energy than expected for secondary electrons produced by precipitating auroral primaries. This process, which is also important for the equatorial regions of Jupiter and Saturn, requires solar radiation. It is not seen on

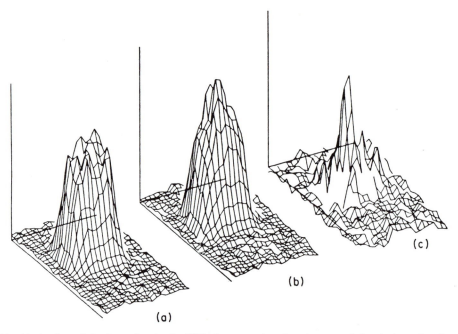

Fig. 13.   Projection of the Ly α flux in the IUE short wavelength spectrograph focal plane showing the Ly α emission from Uranus. (a) Sum of three exposures with geocorona only normalized to (b); (b) sum of two 2-hr exposures with Uranus at the centre of the large entrance aperture; (c) the uranian Ly α emission obtained by subtracting the geocoronal contribution for (b) and multiplying the result by 3. The long dimension of the aperture is approximately parallel to the axes pointing upward and to the right. The dispersion direction is approximately parallel to the axes pointing downward and to the right (Durrance and Moos, 1982).

the dark side of the planet (Broadfoot *et al.*, 1986) — and is greater than two orders of magnitude above that expected from solar EUV. It appears to be a new type of excitation process for planetary atmospheres which differs qualitatively from either auroral or the usual airglow processes. To emphasize the difference, it has been dubbed the 'electroglow'. At present, the basic mechanism for heating the electrons and the roles of the Sun and the magnetic field in the heating process are not known. The nature of the electroglow and its relative importance for different planets are important questions in the understanding of planetary atmospheres.

4. Summary

A remarkable number of important molecules, atoms and ions important for the study of planetary atmospheres have strong absorption and emission lines in this spectral region. As a consequence, the ultraviolet is a powerful technique for planetary astronomy. The power of the IUE observatory is underlined when one notes that the spacecraft was not optimized for planetary studies due to technical compromises of the type that always must be made in a cost constrained project: the short wavelength camera does not have an extreme solar blind cathode;

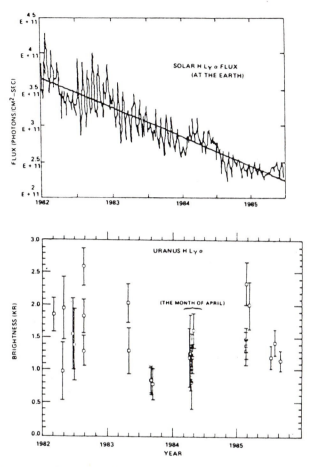

Fig. 14. The disk-averaged H Ly α brightness of Uranus for all measurements performed with the IUE (bottom) from 3 March, 1982 through 24 September, 1985 (Clarke *et al.*, 1986). Plotted for comparison is the solar H Ly α flux at the Earth over the same period, as measured by the SME spacecraft (top).

precision pointing at planets requires some care; and solar power and thermal considerations constrain the available areas of the sky. Despite this, as this survey shows, the IUE has become an important tool for the study of the planets. This success is a compliment to the original design concept, to those who implemented the concept, and to the observatory staff who have found ingenious ways to optimize the spacecraft performance for planetary studies.

Acknowledgement

The preparation of this review was partially supported by NASA under grant NSG-5393 to the Johns Hopkins University.

References

Anderson, R. C., Pipes, J. G., Broadfoot, A. L., and Wallace, L.: 1969, *J. Atmospheric Sci.* **26**, 874.

Ballester, G. E., Feldman, P. D., Moos, H. W., and Skinner, T. E.: 1985, *Bul. Am. Astron. Soc.* **17**, 697.

Ballester, G. E., Moos, H. W., Strobel, D. F., Summers, M. E., Feldman, P. D., Bertaux, J.-L., Festou, M. C., Skinner, T. E., and Lieske, J.: 1986, 'Detection of Neutral Atomic Cloud Near Io Using IUE', Second Neil Brice Memorial Symposium Magnetospheres of the Outer Planets.

Boggess, A. and Dunkelman, L. 1959, *Astrophys. J.* **129**, 236.

Brinkman, R. T., Green, A. E. S., and Barth, C. A.: 1966, JPL Technical Report No. 32—951.

Broadfoot, A. L.: 1972, *Astrophys. J.* **152**, 682.

Broadfoot, A. L., Belton, M. J. S., Takacs, P. Z., Sandel, B. R., Shemansky, D. E., Holberg, J. B., Ajello, J. M., Atreya, S. K., Donahue, T. M., Moos, H. W., Bertaux, J. L., Blamont, J. E., Strobel, D. F., McConnell, J. C., Dalgarno, A., Goody, R., and McElroy, M. B.: 1979, *Science* **204**, 979.

Broadfoot, A. L., Herbert, F., Holberg, J. B., Hunten, D. M., Kumar, S., Sandel, B. R., Shemansky, D. E., Smith, G. R., Yelle, R. V., Strobel, D. F., Moos, H. W., Donahue, T. M., Atreya, S. K., Bertaux, J. L., Blamont, J. E., McConnell, J. C., Dessler, A. J., Linick, S., and Springer, R.: 1986, *Science* **233**, 74.

Brown, R. A.: 1974, in A. Woszczyk and C. Iwaniszewska (eds.), 'Exploration of the Planetary System', *IAU Symp.* **65**, 527.

Caldwell, J., Owen, T., Rivolo, A. R., Moore, V., Hunt, G. E., and Butterworth, P. S.: 1981, *Astron. J.* **86**, 298.

Caldwell, J., Wagener, R., Owen, T., Combes, M., and Encrenez, T.: 1984, *Uranus and Neptune*, J. T. Bergstrahl (eds.), NASA CP 2330, p. 157.

Caldwell *et al.*, 1986

Carlson, R. W. and Judge, D. L. 1974, *J. Geophys. Res.* **79**, 3623.

Clarke, J. T.: 1982, *Astrophys. J. Letters* **263**, L105.

Clarke, J. T., Moos, H. W., and Feldman, P. D.: 1981, *Astrophys. J. Letters* **245**, L127.

Clarke, J. T., Moos, H. W., and Feldman, P. D.: 1982, *Astrophys. J.* **255**, 806.

Clarke, J. T., Weaver, H. A., Feldman, P. D., Moos, H. W., Fastie, W. G., and Opal, C. B.: 1980, *Astrophys. J.* **240**, 696.

Clarke, J. T., Moos, H. W., Atreya, S. K., and Lane, A. L.: 1981, *Nature* **290**, 226.

Clarke, J. T., Durrance, S. T., Atreya, A., Barnes, A., Belcher, J., Festou, M., Mihalov, J., Moos, H. W., Murthy, J., Padhan, A., and Skinner, T. E.: 1986, *J. Geophys. Res.*

Combes, M., Courtini, R., Caldwell, J., Encrenaz, Th., Fricke, K. H., Moore, V., Owen, T., and Butterworth, P. S.: 1981, *Adv. Space Res.* **1**, 169.

Conway, R. R., McCoy, R. P., Barth, C. A., and Lane, A. L.: 1979, *Geophys. Res. Letters* **6**, 629.

Conway, R. R., Durrance, S. T., Barth, C. A., and Lane, A. L.: 1981, *The Universe at Ultraviolet Wavelengths*: The First Two Years of IUE, (NASA CP-2171), Washington, D.C. p. 33.

Darius, J. and Fricke, K. H.: 1981, *The Universe at Ultraviolet Wavelengths: The First Two Years of IUE*, (NASA CP-2171), Washington, D.C. p. 85.

Deland, M. T.: 1985, M. A. Essay, The Johns Hopkins University, Baltimore, MD.

Durrance, S. T. 1981, *J. Geophys. Res.* **86**, 9115.

Durrance, S. T. and Moos, H. W.: 1982, *Nature* **299**, 428.

Durrance, S. T., Feldman, P. D., and Moos, H. W.: 1982, *Geophys. Res. Letters* **9**, 652.

Duysinx, R. and Henrist, M.: 1974, in A. Woszczyk and C. Iwaniszewska (eds.), 'Exploration of the Planetary System'.

Encrenaz, T.: 1982, *Proceedings of the Third European IUE Conference*, (ESA SP 176).

Encrenaz, T., Combes, M., Atreya, S. K., Romani, P. N., Fricke, K., Moore, V., Hunt, G., Wagener, R., Caldwell, J., Owen, T., and Butterworth, P.: 1986, *Astron. Astrophys.* (in press).

Evans, D. C.: 1966, NASA GSFC Report X-613-66-172.

Feldman, P. D.: 1986, 'The Io Torus and the Jovian Magnetosphere', *New Insights in Astrophysics, 8 Years of UV Astronomy with IUE*, 14—16 July, 1986, London, UK.

Feldman, P. D., Moos, H. W., Clarke, J. T., and Lane, A. L.: 1979, *Nature* **279**, 221.

Fricke *et al.*, 1982.

Giles, G. J., Moos, H. W., and McKinney, W. R.: 1976, *G. Geophys. Res.* **81**, 5797.

Gladstone, G. R.: 1982, Ph.D. Thesis, California Institute of Technology, Pasadena, CA.

Gladstone, G. R. and Yung, Y. L.: 1983, *Astrophys. J.* **266**, 415.

Greenspan, J. A. and Owen, T. 1967, *Science* **156**, 1489.

Jenkins, E. B., Morton, D. C., and Bohlin, R. C.: 1969, *Astrophys. J.* **154**, 661.

Kjeldseth-Moe, O., Van Hoosier, M. E., Bartoe, J. D. F., and Brueckner, G. E.: 1976, USNRL Report 8057.

Kupo, T., Mekler, Y., and Evitar, A.: 1976, *Astrophys. J. Letters* **205**, L51.

Lane, A. L., Boggess, A., Evans, D. C., Gull, T. R., Owen, T. C., Moos, H. W., Tomasko, M. G., Gehrels, T., Hunt, G. E., Wilson, R., Conway, R., Barth, C. A., Schiffer, F., Turnrose, B., Perry, P., Holm, A., Macchetto, F., and Hamrick, E.: 1978, *Nature* **275**, 414.

McDonough, T. R. and Brice, N. M.: 1973, *Nature* **242**, 513.

Moos, H. W. and Clarke, J. T.: 1979, *Astrophys. J. Letters* **229**, L107.

Moos, H. W. and Clarke, J. T. 1981, *Astrophys. J.* **247**, 354.

Moos, H. W., Fastie, W. G., and Bottema, M.: 1969, *Astrophys. J.* **155**, 887.

Moos, H. W., Durrance, S. T., Skinner, T. E., and Feldman, P. D.: 1983, *Astrophys. J. Letters* **275**, L19.

Moos, H. W., Skinner, T. E., Durrance, S. T., Feldman, P. D., Festou, M. C., and Bertaux, J-L.: 1985, *Astrophys. J.* **294**, 369.

Mount, G. H. and Moos, H. W.: 1978, *Astrophys. J. Letters* **224**, L35.

Mount, G. H. and Rottman, G. J.: 1981, *J. Geophys. Res.* **86**, 9193.

Mount, G. H. and Rottman, G. J.: 1983a, *J. Geophys. Res.* **88**, 5403.

Mount, G. H. and Rottman, G. J.: 1983b, *J. Geophys. Res.* **88**, 6807.

Mount, G. H., Rottman, G. J., and Timothy, J. G.: 1980, *J. Geophys. Res.* **85**, 4271.

Nakayama, T. and Watanabe, K.: 1964, *J. Chem. Phys.* **40**, 558.

Orton, G. S., Tokunaga, A. T., and Caldwell, J.: 1983, *Icarus* **56**, 147.

Owen, T. and Sagan, C.: 1972, *Icarus* **16**, 557.

Owen, T., Caldwell, J., Rivolo, A. R., Moore, V., Lane, A. L., Sagan, C., Hunt, G., and Ponnamperuma, C.: 1980, *Astrophys. J. Letters* **236**, L39.

Rottman, G. J., Moos, H. W., and Freer, C. S.: 1973, *Astrophys. J. Letters* **184**, L89.

Sandel, B. R., Broadfoot, A. L., and Strobel, D. F.: 1980, *Geophys. Res. Letters* **7**, 5.

Savage, B. D. and Caldwell, J. J.: 1974, *Astrophys. J.* **187**, 197.

Scattergood an Owen 1977.

Shemansky, D. E. and Smith, G. R.: 1986, *Geophys. Res. Letters* **13**, 2.

Skinner, T. E.: 1986, Private Communication.

Skinner, T. E. and Moos, H. W.: 1984, *Geophys. Res. Letters* **11**, 1107.

Skinner, T. E., Durrance, S. T., Feldman, P. D., and Moos, H. W.: 1983, *Astrophys. J.* **265**, L23.

Skinner, T. E., Durrance, S. T., Feldman, P. D., and, Moos, H. W.: 1984, *Astrophys. J.* **278**, 441.

Smyth, W. H. and Shemansky, D. E.: 1983, *Astrophys. J.* **271**, 865.

Stecher, T. P.: 1965, *Astrophys. J.* **142**, 1186.

Strobel, D. F.: 1983, *International Reviews in Physical Chemistry* **3**, 145.

Tomasko, M. G.: 1974, *Astrophys. J.* **187**, 641.

Wagener, R., Caldwell, J., Owen, T., Kim, S. J., Encrenaz, T., and Combes, M. 1985, *Icarus* **63**, 222.

Wallace, L., Caldwell, J. J., and Savage, B. D.: 1972, *Astrophys. J.* **172**, 755.

Weiser, H., Vitz, R. C., and Moos, H. W.: 1977, *Science* **197**, 755.

Winkelstein, P., Caldwell, J., Kim, S. J., Combes, M., Hunt, G. E., and Moore, V.: 1983, *Icarus* **54**, 309.

PLANETARY SATELLITES

ROBERT M. NELSON and ARTHUR L. LANE

Jet Propulsion Laboratory, Calfornia Institute of Technology, Pasadena, CA, U.S.A.

1. Introduction

Spectrophotometry of planetary satellites at wavelengths less than 3000 Å has long been desired by the planetary science community in order to augment the rapidly expanding database that has become available in recent decades as a result of the many successful planetary probe missions to the outer solar system. From the perspective of a planetary geologist, the spatial information about planetary satellites observed by the imaging systems on these probes is critical to developing an understanding of the physical processes which determine the surface morphology of airless solar system bodies. Therefore, a primary goal of such imaging systems is to obtain the highest possible spatial resolution. However, the attainment of high spatial resolution from a spacecraft imaging system comes, in part, at the expense of spectral resolution, and the latter is critical to making inferences regarding the chemical and mineralogical composition of a planetary surface. Furthermore, because of the limited spectral sensitivity of the past spacecraft imaging systems at wavelengths less than 3500 Å, the spectral region attainable by IUE remains essentially unexplored by the previous planetary probes to the Jovian and Saturnian systems. Thus, the information returned by planetary satellite observations with IUE forms a unique data set, in spite of the obviously limited spatial resolution (whole disk only) that constrains an earth orbiting spacecraft with a telescope of IUE's 45 cm aperture.

By 1989, planetary space probes will have returned high spatial resolution images of all the satellites of Mars, Jupiter, Saturn, Uranus and Neptune. Of this set of objects, only the brightest ones are observable with IUE. These include the Galilean satellites of Jupiter (Io, Europa, Ganymede, and Callisto) and some of the large satellites of Saturn that are also of sufficient distance from the primary body so as to prevent scattered light from Saturn from entering the spectrograph. During the lifetime of IUE, an extensive set of spectra of the Galilean satellites has been obtained, as well as a limited set of Saturnian satellite observations. These observations have had significant impact on discussions regarding the composition and distribution of chemical species present on the satellite surfaces, the effect of the modification of the surfaces by bombardment from the changed particles in the magnetosphere of the primary bodies, and physical texture of the small scale particles comprising a satellite's surface regolith.

Even though IUE only observes these satellites as point sources (they are all ~ 1 arcsec diameter objects), limited spatial resolution of the spectral data has been obtained. This is done by making observations at times which consider the differences in illumination geometry that are apparent when observing a superior solar system object during the course of an observing year. Figure 1 shows the

Y. Kondo (ed.), Scientific Accomplishments of the IUE, pp. 67–99.
© 1987 *by D. Reidel Publishing Company.*

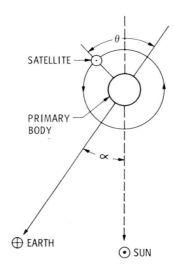

Fig. 1.   Definition of the terms orbital phase angle, θ, and solar phase angle, α, to describe the
observational aspects of a solar system satellite observed from earth or near-earth orbit.

position of an earth based (or IUE based) observer when studying an outer solar
system planetary satellite. The solar phase angle, α, is the earth-object-sun angle
and the angle θ is the rotational (or orbital) phase angle. For objects in phase
locked orbits (synchronous rotation), θ at the time of observation is equal to the
longitude of the sub-earth point on the satellite. Thus, the spectral variation as a
function of longitude on a planetary satellite is obtained by making observations at
many orbital phase angles.

A primary goal of a solar system astronomical observing program is to
determine the Bond albedo of the object under study. The Bond albedo is defined
as the total radiative flux reflected in all directions divided by the total incident
flux. Following Russell (1916), if S is the solar energy flux at the earth's orbit, then
the amount of energy impinging on a solar system object of radius r at a distance
R (in astronomical units) from the Sun is:

$S\pi r^2/R^2$.

If an observer were exactly on the sun-object line ($\alpha = 0$) a fully illuminated
disk would be observed. The intensity measured in this case is defined to be $C(0)$.
For most observations, $\alpha \neq 0$, and therefore a different intensity $C(\alpha)$ will be
observed. The phase function, $B(\alpha)$ is defined as the ratio of the intensity at zero
phase angle to the intensity at phase angle α. In this more general case, the rate at
which the object reflects radiation in all directions is,

$$2\pi d^2 C(0) \int_0^\pi B(\alpha) \sin \alpha \, d\alpha$$

where d is the distance between the observer and the object. The bond albedo, A, is then

$$A = \frac{2d^2R^2C(0)}{r^2S} \int_0^\pi B(\alpha)\sin\alpha\,d\alpha.$$

The term on the right in the above expression is often divided into two parts, the phase integral, q, and the geometric albedo, p, where:

$$q = 2\int_0^\pi B(\alpha)\sin\alpha\,d\alpha \qquad \text{and} \qquad p = d^2R^2C(0)/r^2S.$$

The geometric albedo is the ratio of the brightness of the object to the brightness of a perfectly diffusing disk of the same radius at the same distance from the Sun. For the outer planets and their satellites, earthbased (and IUE) observations can only be made over a narrow range of solar phase angles. Thus, the phase integral, q, cannot be determined and therefore, neither can the Bond albedo. However, the geometric albedo as a function of wavelength (i.e. the spectral geometric albedo, $P(\lambda)$ is obtainable and proves to be of great use in making comparisons to laboratory spectra of hypothesized surface materials in order to attempt a compositional identification. $P(\lambda)$ is of greater use than a normalized relative reflection spectrum because it is possible for a laboratory spectrum of a particular set of materials to have a good relative spectral match to $P(\lambda)$ and yet the materials have different albedos. In such cases the likelihood of such a material being abundant on a planetary surface is quite small.

Because all of the planetary satellites studied by IUE are in synchronous rotation (the same face always is oriented toward the primary body), an observing program can be developed to determine the variation in the spectral geometric albedo as a function of orbital phase angle and thus the variation in abundance of spectrally active absorbers across the surface of the object can be mapped. By ratioing the spectra of an object at one longitude to those at other longitudes, mineralogical mapping in one spatial dimension (longitude) can be accomplished by only using point source spectra.

Textural information on the small scale properties of a planetary regolith can be obtained by studying the limited portion of the phase curve (the variation in brightness as a function of solar phase angle) that is observable from IUE observations. Previous observational and theoretical studies have shown that the phase curve is determined by the object's surface texture (Hapke, 1963; Veverka, 1977). The most important parameters are: the single scattering albedo, the single particle phase function, the macroscopic roughness, and the compaction state of the particles which comprise the regolith (Buratti, 1984). It is the effect of mutual shadowing between regolith particles that contributes to the brightness increase as phase angle decreases. High albedo objects have smaller phase coefficients because multiple scattering tends to dilute or fill in shadows. An increasing amount of large scale roughness causes the phase coefficient to increase significantly but only for phase angles greater than 30—40° (Hapke, 1981; Buratti and Veverka, 1983).

Thus, the solar phase curve data for the planetary satellites studied with IUE (small phase angle observations only) can be used to infer the amount of compaction of the surface regolith particles.

In light of the above discussion, the goals of the IUE observational program for atmosphereless planetary satellites are to:

(1) Obtain the spectral geometric albedo of each object and compare this to laboratory data to identify the surface components of the satellites,

(2) Use the technique of ratioing spectra taken at different orbital phase angles to map the variation in longitudinal distribution of the already identified surface components across the satellites' surfaces,

(3) Obtain the largest possible coverage of the object's phase curve in order to make inferences regarding the nature of the surface texture.

The results of eight years of IUE observations are discussed below along with the interpretations that developed during the reduction process.

2. The Galilean Satellites

Background
Usable quantitative (spectro)photometric data on the Galilean satellites extends over 6 decades. Sixty years ago, groundbased photoelectric photometric observations at many orbital phase angles established the synchronous rotation of the Galilean satellites (Stebbins, 1927; Stebbins and Jacobson, 1928). Three color photometry by Harris (1961) found that all four satellites were more absorbing in the near-UV at ~ 3500 Å than in the visual (~ 5000 Å) and that Io has the strongest UV absorption of the four satellites. Higher resolution spectrophotometry detected broad absorption features on Io and Europa at ~ 5600 Å and established clearly the spectral reflectivity decrease shortward of 5000 Å on all the satellites (Johnson and McCord, 1970; Johnson, 1971). Io and Europa were found to have very high visual geometric albedos P_v ~ 0.6 for both) (Morrison and Cruikshank, 1974; Johnson and Pilcher, 1977).

Observations of the Galilean satellites at wavelengths shorter than 3000 Å were done from space by Caldwell (1975) using the OAO—2 spacecraft. He reported that Io had a very low ultraviolet geometric albedo (P_{uv} < 3% at 2590 Å). Continued groundbased observations at higher spectral resolution (~ 40 Å) established the spectral geometric albedos of all the satellites in the range 3200 to 8600 Å (Nelson and Hapke, 1978). This work identified an absorption feature in Io's spectrum shortward of 3300 Å in addition to a previously identified absorption feature at 4000—5000 Å. The 3300 Å feature was consistent with the very low albedo for Io that had been reported by Caldwell. Further measurements in the visible and near UV have confirmed these results (McFadden *et al.*, 1980).

Infrared observations (Moroz, 1966, 1968; Fink *et al.*, 1973; Lebofsfy, 1977; Cruikshank *et al.*, 1978; Pollack *et al.*, 1978; Clark, 1980) identified frozen water ice in great abundance on the surfaces of Europa, Ganymede, and Callisto and lead to the identification of condensed sulfur dioxide on Io (Fanale *et al.*, 1979; Nash and Nelson, 1979; Smythe *et al.*, 1979; Hapke, 1979).

The Voyager encounters with Jupiter and its satellites provided a wealth of information regarding the nature of the satellites and their environment. However, the absence of high spectral resolution information in the Voyager images severely limits attempts at mineralogical mapping. The extension of the available spectral range to shorter wavelengths with IUE allows for the detection of more adsorption features thereby providing further constraints on proposed compositional models. The observations obtained with IUE and (hopefully) the Hubble Space Telescope will form the major set of ultraviolet specroscopic data on the planetary satellites for many years to come.

The Observations

More than 300 observations of the Galilean satellites were observed with IUE during the Jupiter apparitions from 1978 through 1985 and have been reported in the literature (Nelson *et al.*, 1987). All the results discussed here were obtained using the large ($10'' \times 20''$) aperture of the long wavelength spectrograph 1800— 3300 Å at low resolution (~ 11 Å). Data from wavelengths shorter than 2400 Å are not reported because of the very low signal-to-noise in the UV due to the decreased solar luminosity at short wavelengths, the low surface reflectivity of the objects observed and the added complications of correction for scattered light inside the spectrograph.

One of us (A.L.L.) developed, with the assistance of the IUE staff astronomers, a method for taking two long wavelength exposures of the same object in the large aperture. These observations appear in the IUE observatory records as one spectrum but in the case of the Galilean satellite Io they are treated as two separate observations. This is because Io moves a significant amount in orbital phase angle in the time required for two separate spectra to be recorded. In the case of the other satellites observed, the two spectra were co-added and the result was treated as one spectrum.

2.1. THE BROADBAND GEOMETRIC ALBEDO

To calculate the geometric albedo for a solar system object, an absolute solar spectrum is required in the spectral range under consideration. The solar flux can be divided into the raw spectrum of an object in order to give an absolute reflected flux from the source thus allowing the geometric albedo to be determined as outlined above. Absolute, whole integrated disk UV solar spectra were obtained from rockets and from the Solar Mesospheric Explorer (SME) spacecraft (Rottman, 1983). At the UV wavelengths > 2400 Å, the solar flux is not appreciably variable (Lean, 1982a, b). These data were updated regularly and generously provided by the University of Colorado Solar Studies Team (Mount, private communication). The solar spectrum was multiplied by the IUE spectral response function provided by the IUE spacecraft project. A typical IUE spectrum of Io is shown in Figure 2a prior to the removal of the solar spectral features by ratioing to a solar spectrum. Figure 2b is the ratio of the spectrum in Figure 2a to the product of the solar spectrum and the IUE response function. The spectral information in Figure 2b is unique to Io. To calculate a geometric albedo, the 11 Å resolution IUE data were averaged over three wide bandpasses to reduce noise. These bandpasses are hereafter referred to as Bands 1, 2, and 3. The wavelength

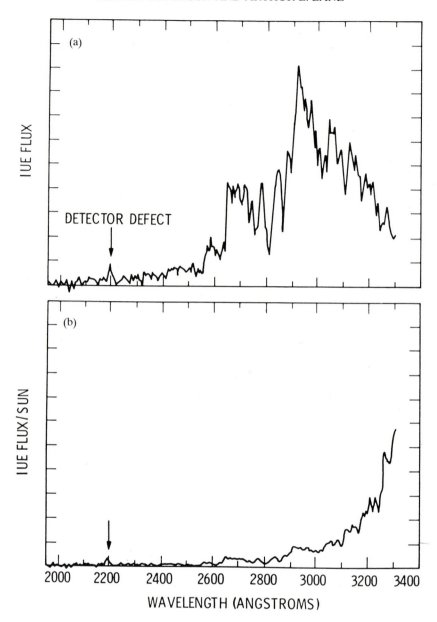

Fig. 2(a). Typical IUE raw spectrum of Io. The solar spectrum has not been removed. (b). Same spectrum as in Figure 2a, now ratioed to a solar spectrum. The features in this spectrum are unique to the object being observed, in this case, Io. The fine structure is due to slight wavelength differences in constructing the ratio and the rapid variation of solar flux intensity over short wavelength intervals especially near the Frauenhofer line regions. The apparent emission feature near ~ 2200 Å is a detector defect and does not represent valid data.

ranges for these bandpasses are 2400 < Band #1 < 2700 Å, 2800 < Band #2 < 3000 Å, and 3000 < Band #3 < 3200 Å. This bandpass definition allows data from IUE to be easily compared with groundbased data in the region of spectral overlap between IUE's longest wavelength range (Band #3) with the shortest wavelengths attainable from groundbased spectral observations. If the other observational parameters are the same, then the IUE Band 3 data (3000—3200 Å) and the groundbased data at 3200 Å should be in close agreement unless the spectrum of the object is rapidly changing within those wavelengths.

Because the geometric albedos of the Galilean satellites change as a function of the satellite's orbital phase angle at the time of observation, the albedos derived from the three bandpasses have been grouped according to the specific orbital phase angles at which the observations were made. The term 'leading side' and 'trailing side' to refer to orbital phase angles at or near 90° and 270° respectively (see Figure 1). The leading side includes observations between orbital phase 45° and 135° and the trailing side includes observations between 235° to 315°. Only data from solar phase angles between 7° and 12° are included for this albedo/orbital phase variation determination in order to minimize the contribution of the opposition effect — the non-linear increase in brightness which is observed for solar system bodies at very low solar phase angles. The geometric albedos of the leading and trailing sides in each IUE UV photometric bandpass were determined by deriving the linear phase coefficients using the data from solar phase angles 7° to 12° and extrapolating these data to 0°. The UV geometric albedos of the satellites are shown in Table I. The groundbased data of Nelson and Hapke (1978) are also shown for comparison. The errors shown are the standard deviations of the mean value.

Because Io and Europa's trailing sides have absorption features between 2800 and 3200 Å, there is disagreement between the groundbased data ($\lambda > 3200$ Å) and the IUE Band #3 data. However, the data in Table I show that there is reasonable agreement between IUE and groundbased albedo results for Ganymede and Callisto at ~ 3200 Å. This serves as an important confirmation of the quality of the reduction procedure employed in the analysis of the IUE data.

The data in Table I also show that at UV wavelengths, Io's trailing side has a higher albedo than it's leading side; just the opposite of what is observed at longer wavelengths. This effect is also confirmed by the orbital phase curves and the spectra that are shown later. This reversal in orbital phase behavior is more pronounced for Io than for any other object in the solar system and proves to be important in efforts to determine the surface composition variation in longitude across Io's surface.

2.2. THE ORBITAL PHASE CURVES

The orbital phase curves (the variation of the disk integrated brightness as a function of orbital phase angle) of the Galilean satellites haev been quantitatively studied in groundbased observations for more than five decades. Recent review of the groundbased data reduced to the uvby wavelengths has found that the albedos of Io, Europa and Ganymede all are higher on the leading hemispheres than on the trailing hemispheres (Morrison et al., 1974). At the same wavelengths Callisto's trailing side albedo is higher than it's leading side albedo. Figures 3—6 show the albedo variations with respect to orbital phase angle at the IUE bandpasses defined above.

TABLE I

Galilean Satellites Ultraviolet Geometric Albedos (%)

	B1	B2	B3	Groundbased (N & H)
Io (L)	1.5 + −0.1 (N = 34)	1.7 + −0.1 (N = 35)	4.2 + −0.1 (N = 27)	7
Io (T)	2.8 + −0.2 (N = 28)	3.0 + −0.5 (N = 24)	3.8 + −0.3 (N = 27)	6
Europa (L)	18.0 + −0.4 (N = 22)	26.0 + −1.0 (N = 18)	37.0 + −2.0 (N = 24)	36
Europa (T)	9.6 + −0.2 (N = 31)	12.9 + −0.4 (N = 33)	17.1 + −0.6 (N = 28)	24
Ganymede (L)	12.8 + −0.7 (N = 26)	16.8 + −0.9 (N = 20)	20.0 + −0.1 (N = 20)	14
Ganymede (T)	7.0 + −0.3 (N = 16)	7.5 + −0.04 (N = 16)	10.5 + −0.8 (N = 17)	10
Callisto (L)	4.0 + −0.8 (N = 15)	4.9 + −0.1 (N = 13)	6.6 + −0.02 (N = 14)	no data
Callisto (T)	5.6 + −0.2 (N = 26)	6.4 + −0.02 (N = 21)	10.5 + −0.8 (N = 26)	8

N & H = Nelson and Hapke (1978); L = Leading side ($45° < \theta < 135°$).

T = Trailing side ($235° < \theta < 315°$); N = Number of observations.

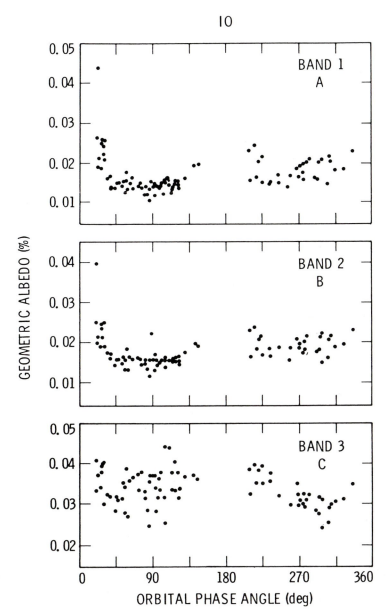

Fig. 3. The variation in geometric albedo as a function of orbital phase angle is an important indicator of the change in abundance with longitude of any spectrally active surface components on a satellite's surface. (a) Geometric albedo variation of Io as a function of orbital phase angle in the wavelength range 2400–2700 Å. (b) Same as (a) except for the wavelength range 2800–3000 Å. (c) Same as (a) except for the wavelength range 3000–3200 Å.

For Io, the data in Figure 3 show more clearly the UV hemispherical albedo asymmetry apparent in Table I. Shortward of the IUE bandpass #3 (~ 3200 Å) Io's leading hemisphere becomes less reflective than its trailing hemisphere. This

EUROPA

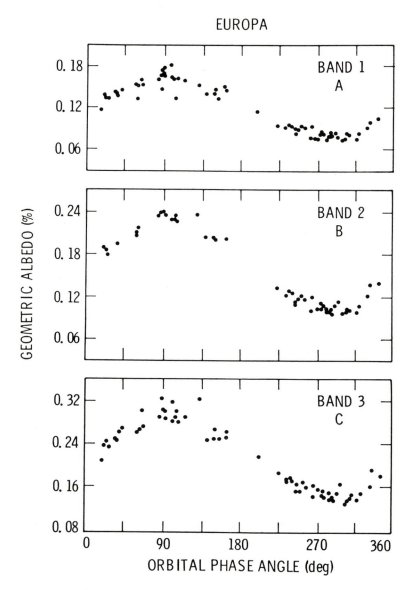

Fig. 4(a). Geometric albedo variation of Europa as a function of orbital phase angle for the wavelength range 2400–2700 Å. (b) Same as (a) except for the wavelength range 2800–3000 Å. (c) Same as (a) except for the wavelength range 3000–3200 Å.

absorption on Io's leading side is somewhat stronger in band #2 (~2900 Å) and very strong at IUE band #1 (~2500 Å). It can be directly inferred from the Io data that there is a longitudinally asymmetric distribution of a spectrally active surface component on Io's surface. The material is strongly absorbing shortward of ~3200 Å and is strongly reflecting longward of that wavelength. It is in greatest abundance on the leading hemisphere of Io and it is in least abundance on the

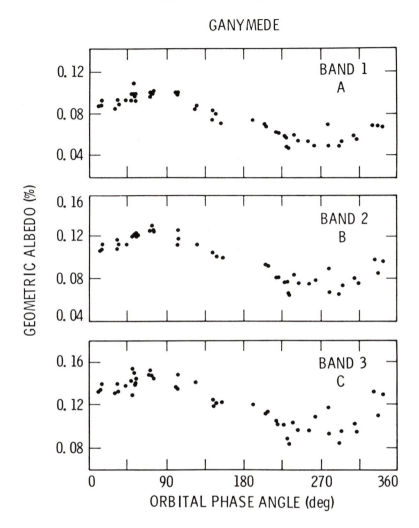

Fig. 5(a). Geometric albedo variation for Ganymede as a function of orbital phase angle for the wavelength range 2400–2700 Å. (b) Same as (a) except for the wavelength range 2800–3000 Å. (c) Same as (a) except for the wavelength range 3000–3200 Å.

trailing hemisphere. The Band #3 data show a large amount of variation in the albedo of Io's leading hemisphere when compared to it's trailing hemisphere and to the other satellites. Because a spectral absorption feature has been identified in the spectral region under consideration, this variability may be due to redistribution of surface material on Io's surface.

For Europa and Ganymede (Figures 4 and 5) the variation in UV brightness with longitude is in the same sense as the variation at the uvby wavelengths; at all wavelengths these objects are brighter on their leading sides than on their trailing sides. A gradual decrease in brightness toward shorter wavelengths occurs on both hemispheres of both objects.

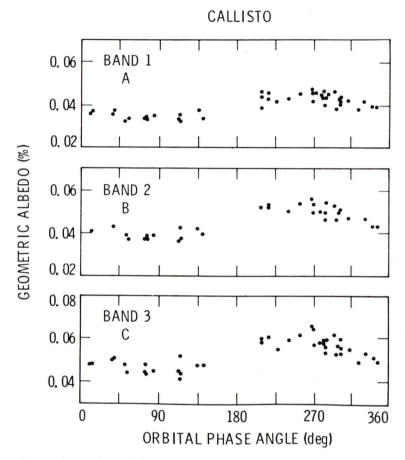

Fig. 6(a). Geometric albedo variation of Callisto as a function of orbital phase angle in the wavelength range 2400–2700 Å. (b) Same as (a) except for the wavelength range 2800–3000 Å. (c) Same as (a) except for the wavelength range 3000–3200 Å.

Groundbased observations of Callisto have found that the trailing side has a higher albedo than the leading side. This is also true in the three IUE bands (Figure 6). The albedo of Callisto decreases shortward of 5500 Å and its albedo at all wavelengths is lower than the albedo of Europa and Ganymede.

2.3. HEMISPHERICAL SPECTRAL ASYMMETRIES

Broadband orbital phase curves of the type shown above indicate that UV albedo changes are apparent on all four objects. These albedo differences between individual hemispheres imply that differences in chemical composition exist across the surfaces of each object. The UV spectral signature of different absorbing materials on a satellite's surface becomes apparent when the spectra of the opposite hemispheres of each object are ratioed. In order to improve signal-to-noise, the individual spectra taken near one hemisphere of an object have been co-added and ratioed to similarly co-added spectra of the opposite hemisphere. The resulting opposite hemisphere ratio spectra are shown in Figure 7 for the four satellites. All the ratio spectra have been normalized to unity at 2700 Å. For Io and Callisto, the leading/trailing side ratio is shown. Europa and Ganymede have the trailing/leading side spectral ratio is shown. This was done in order to display all spectrally active absorbers in the same sense as is commonly seen in catalogues of reflection spectra of possible planetary surface constituents.

The leading/trailing side UV spectral ratio for Io is shown in Figure 7a. It is immediately obvious that Io's UV reflectance is not only characterized by a hemispherical albedo asymmetry (as Figure 3 has previously shown) but it also has a UV hemispheric spectral asymmetry. This is consistent with the above suggestion that a strong, spectrally active, UV absorber is asymmetrically distributed across Io's surface. The material is characterized by a strong absorption shortward of 3300 Å, and is distributed in greater abundance on Io's leading side. The strength of this feature has not changed over eight years indicating that there has been no large scale redistribution of the spectral absorber during that time.

The trailing/leading side UV spectral ratio for Europa is in Figure 7b. This ratio confirms the previous report of Lane *et al.* (1981) that Europa's opposite hemispheric UV ratio spectrum is characterized by a broad weak absorption centered at approximately 2800 Å. The spectral absorption is relatively stronger on Europa's trailing hemisphere compared to its leading hemisphere. Ockert *et al.* (1987) have searched for variations in intensity of this absorption and they report no detectable change at a ~ 20% discrimination level over an eight year period.

Figure 7c shows the trailing/leading side ratio spectrum of ·Ganymede The opposite hemispheric UV ratio spectrum of Ganymede is characterized by a weak spectral absorption beginning at ~ 3200 Å which gradually increases in absorption down to 2600 Å. The absorption is more pronounced on Ganymede's trailing hemisphere than on it's leading hemisphere. The ratio spectrum longward of 2800 Å is strikingly similar to the ratio spectrum of the opposite hemispheres of Europa (Figure 7b). However, shortward of 2800 Å the Europa ratio spectrum rises while Ganymede's continues to fall. The absorber has not been positively identified but several possibilities will be discussed later.

The leading/trailing side ratio spectrum of Callisto is Figure 7d. This hemispheric UV ratio spectrum is flat and virtually featureless at the present data noise level which indicates that Callisto has no UV spectral asymmetry. It should be noted that Figure 5 indicates a variation in the strength of a UV absorber across the surface of Callisto; however, the absence of a spectral signature in Figure 7d indicates that the absorbing material is spectrally featureless in the 2400—3400 Å spectral range.

2.4. SPECTRAL ORBITAL PHASE VARIATIONS

While the orbital phase curves shown in Figures 3—6 are helpful in determining gross color characterizations of the Galilean satellites, improved spectral resolution is required in order to attempt compositional identification of chemical species distributed on the satellite's surfaces. In order to offset the lower signal-to-noise that results from increased spectral resolution, the spectra from each object

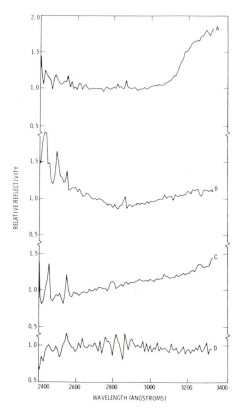

Fig. 7. Opposite hemispheric spectral ratios of the Galilean satellites. By co-adding many spectra of a planetary satellite at or near a common orbital phase angle and ratioing the result to a set of co-added sepctra at or near another orbital phase angle, the variation in hemispherical abundance of spectrally active absorbers can be seen on the surface of the satellite. In this case the opposite hemispheres of each object are ratioed. All ratio spectra have been normalized at 2700 Å. (a) Ratio of the sum of 55 spectra of Io taken at orbital phase angle $45° < \theta < 135°$ divided by the sum of 27 spectra of Io taken at orbital phase angle $225° < \theta < 315°$. The increase of albedo longward of 3000 Å is consistent with the laboratory spectrum of sulfur dioxide frost. The significance of the change in the ratio spectrum shows that there is more sulfur dioxide frost on Io's leading hemisphere than on it's trailing hemisphere. (b) Ratio of the sum of 47 spectra of Europa taken at orbital phase angle $225° < \theta < 325°$ to the sum of 30 spectra taken at orbital phase $45° < \theta < 135°$. The absorption feature centered at ~ 2800 Å is consistent with the spectrum that would be expected from water ice that had been bombarded with sulfur ions. Because the hypothesized source of the sulfur ions is the Jovian magnetosphere, which sweeps past Europa as the satellite orbits Jupiter, the effect of this surface modification process would be greatest on Europa's trailing hemisphere. (c) Ratio of the sum of 27 spectra of Ganymede taken at orbital phase angle $225° < \theta < 315°$ to the sum of 25 spectra taken at orbital phase angle $45° < \theta < 135°$. The weak spectral decrease toward shorter wavelengths may indicate a magnetospheric ion implantation process similar to the sulfur ion implantation process that is occurring on Europa. However, the shape of the spectral absorption feature seen on Ganymede is not the same. Therefore, another ion may be responsible. The spectral absorption may be a combined effect of implanted sulfur and oxygen ions (principal Jovian torus components) which would cause the matrix isolation spectrum to resemble the sum of gaseous sulfur dioxide and ozone spectra. (d) Ratio of the sum of 22 spectra of Callisto taken at orbital longitude $45° < \theta < 135°$ to the sum of 31 spectra taken at orbital phase angle $225° < \theta < 315°$. The flat spectral response indicates that there are no asymmetric distributions of UV spectrally active absorbers on Callisto's surface.

have been co-added into 30° orbital phase segments and these segments are
presented for each object in Figures 8—11. The spectrum for each longitude bin is
normalized to unity at 2700 Å. Each spectrum has been smoothed to a resolution
of ~ 50 Å. Each satellite spectrum was divided by a solar spectrum in order to
remove the solar UV spectral features from the reflection spectrum of the object.
Great care has been taken to examine the individual spectra within each bin to test
for time variability and the spectral geometric albedo for each object at all phase
angles has been unchanged over eight years.

Io

The co-added, solar corrected, normalized spectra for Io are shown in Figure 8a, b.
Io's UV spectra reflectivity is characterized at all longitudes by a slight decrease in
spectral albedo from 2400—2900 Å. Between 2900 and 3200 Å Io's spectral
reflectivity increases greatly, the amount of increase is dependent on longitude.
The increase is greatest between longitudes 90°—240° and least between longitudes
300° and 30°. These results are consistent with the best available groundbased data
(Nelson and Hapke, 1978).

Europa

The co-added, albedo normalized, UV spectra of Europa are shown in Figure 9a, b.
When ratioed to the solar spectrum Europa's spectral reflectivity is characterized
at all longitudes by a featureless spectrum throughout the range 2400—3300 Å
with gradually decreasing reflectance toward shorter wavelengths. The ratio
spectrum from the region near ~ 270° has a slightly smaller slope between 2500—
2800 Å than the spectrum at ~ 90°. This is in agreement with the albedo
variations shown in Figure 4 and the hemispherical spectral asymmetries shown in
Figure 7b which clearly demonstrate that Europa has a non-uniform distribution
of ultraviolet absorbing material. The absorber is present in greater abundance at
longitudes near 270° relative to longitude near 90°.

Ganymede

The co-added, albedo normalized, spectra of Ganymede are shown in Figure
10a, b. These spectra indicate that Ganymede is characterized by a featureless
spectrum throughout the 2400—3200 Å range. Nevertheless, the albedo variation
as a function of orbital phase angle (Figure 5) shows that Ganymede has a
longitudinally asymmetric distribution of absorbing material. The material is
present in greater areal abundance at longitudes near 270° than at longitudes near
90°, and it is spectrally active in the ultraviolet. The ratio spectrum of Ganymede is
similar to the ratio spectrum of Europa longward of 2800 Å. However, shortward
of 2800 Å the ratio spectrum of Ganymede decreases while the ratio spectrum of
Europa increases. This indicates that the same absorbing material cannot explain
the spectral behavior of both objects. Another absorbing material must be present
on Ganymede's surface.

Callisto

The co-added albedo normalized spectra of Callisto are shown in Figure 11a, b. The
resultant ratio spectra indicate that the UV spectrum of Callisto is characterized

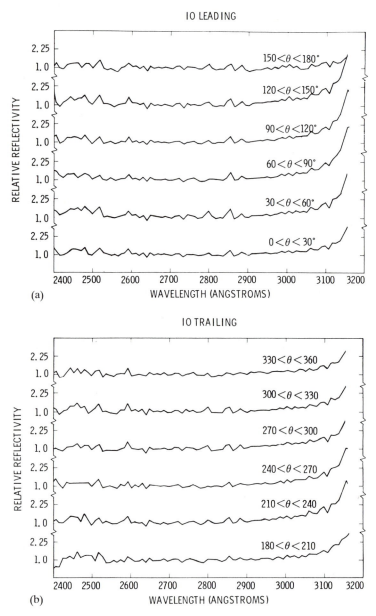

(a)

(b)

Fig. 8.   Higher spatial resolution longitudinal mapping of spectrally active surface constituents is possible (at the expense of an increase in noise) by subdividing the spectra into longitude bins of smaller size. In this case the spectra of the Galilean satellites are shown as co-additions into 30° longitude bins as ratioed to the solar spectrum. Thus the changing character of the spectral absorptions can be observed in 30° longitude increments. The data have been normalized at 2700 Å. (a) The sum of spectra of Io's leading side as co-additions into 30° bins. The increase in reflectivity longward of 3000 Å is due to the presence of sulfur dioxide frost on Io's surface. The distribution of the frost is highest between longitudes 90° and 240°. (b) Same as (a) except for the trailing side. The sulfur dioxide frost is east abundant at longitudes between 300° and 30°.

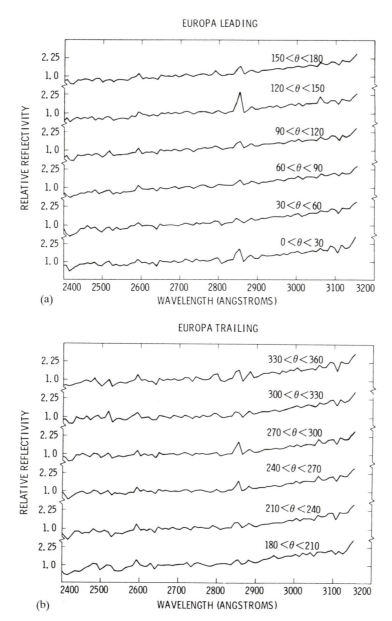

Fig. 9(a). The sum of the spectra of Europa's leading side in 30° bins. (b) Same as (a) except for the trailing side.

by a featureless spectral reflectivity throughout the wavelength range 2400—3300 Å decreasing towards shorter wavelengths. Callisto differs from Europa and Ganymede because it's trailing side in the UV is less reflective than it's leading side. Furthermore, while the three other satellites have UV spectral signatures which are apparent in the solar corrected ratio spectra, and in the ratio of opposite

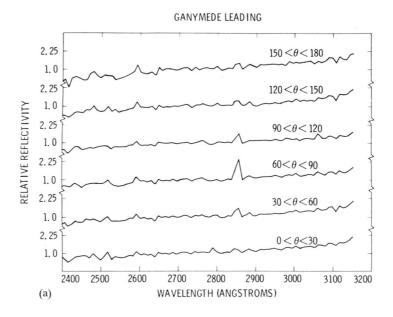

(a)

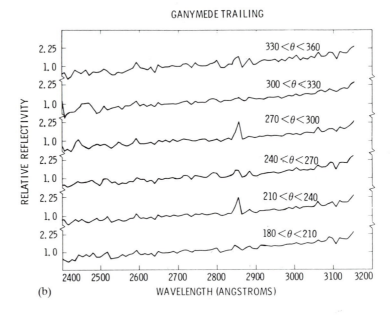

(b)

Fig. 10(a). The sum of the spectra of Ganymede's leading side in 30° bins. (b) Same as (a) except for the trailing side.

hemispheres, Callisto's ratio spectrum shows no UV spectral asymmetries. Therefore, while the albedo variation with orbital phase angle indicates that there is an asymmetrical distribution of a spectral absorbing material on Callisto, the ratio spectra indicate that this absorber is featureless in the ultraviolet.

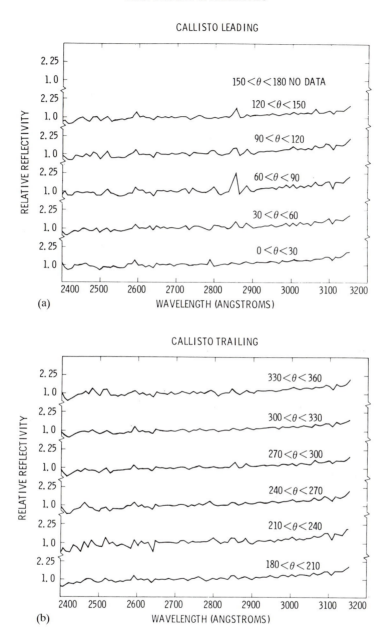

Fig. 11(a). The sum of the spectra of Callisto's leading side in 30° bins. (b) same as (a) except for the trailing side.

2.5. SOLAR PHASE ANGLE VARIATIONS

Observations at low solar phase angles are difficult with IUE because the configuration of the solar photovoltaic panels on the spacecraft is such that the spacecraft batteries discharge when the telescope is pointed at or near the anti-sun direction. In addition, observations at low solar phase angle create increased difficulties in spacecraft attitude control. Therefore only a few observations were obtained.

Because all the satellites show albedo differences in longitude due to large scale compositional differences, the solar phase angle data have been grouped separately into leading side ($45° < \theta < 135°$) and trailing side ($235° < \theta < 315°$) classes and these have been treated separately. These data were corrected for longitude variations using the orbital phase/albedo curves shown in Figures 3—6. Therefore, the solar phase coefficients for the opposite hemispheres of each object are presented separately. The linear phase coefficients have been calculated using the data for solar phase angle $7° < \alpha < 11°$, where the solar phase variation is not influenced by the non-linear opposition surge. These are presented in Table II.

TABLE II

IUE Phase Coefficients (mags/degree)

	Band # 1	Band # 2	Band # 3	Previous work (visual)
Io (L)	0.017 + −0.002	0.020 + −0.001	0.034 + −0.004	0.023 (S & V)
Io (T)	0.043 + −0.007		0.025 + −0.006	0.021 (S & V)
Europa (L)	0.012 + −0.001	0.013 + −0.006	0.024 + −0.009	0.154 (B & V)
Europa (T)	0.025 + −0.003	0.028 + −0.004	0.023 + −0.004	
Ganymede (L)	0.032 + −0.006	0.035 + −0.005	0.035 + −0.006	0.026 (S)
Ganymede (T)	0.033 + −0.005	0.010 + −0.008	0.02 + −0.01	
Callisto (L)	0.021 + −0.003	0.029 + −0.003	0.02 + −0.01	0.027 (S)
Callisto (T)	0.028 + −0.004	0.026 + −0.005	0.02 + −0.01	

L = Leading side ($45° < \theta < 135°$).
T = Trailing side ($235° < \theta < 315°$).
S & V —Simonelli and Veverka, 1985 ($6°—29°$, Voyager imaging).
B & V — Buratti and Veverka, 1983 ($3°—32°$, Voyager iamging).
S-Soderblom et al., 1983 ($5°—16°$, Voyager imaging).

The low albedo of the Galilean satellites coupled with the restricted exclusion to phase angles to less than 12° means that the observations are sensitive primarily to the compaction state of the surface particles which comprise the upper regolith. The data in Table II indicate that Io, Ganymede and Callisto all have phase coefficients in the range 0.02—0.03 mags/deg which is typical of dark airless bodies. Europa's leading side has a markedly smaller value of ~ 0.012 mags/deg which is consistent with Europa's leading side having a more compacted regolith than the other Galilean satellites or its own trailing side.

In several cases observations were made at sufficiently low solar phase angle so as to measure the size of the opposition effect. At visual wavelengths the geometric albedos of Io and Europa are very high ($P_v = ~60\%$, Morrison et al., 1974) and multiple scattering may diminish the size of the opposition surge at visual wavelengths. At ultraviolet wavelengths, however, the albedos of the Galilean satellites are all sufficiently low (Table I) that the opposition surges will not be influenced by multiple scattering between regolith particles. These results are presented in Table III.

TABLE III

Opposition surges
Fraction relative to extrapolated linear phase curve

	Band # 1	Band # 2	Band # 3
Io (L)	1.37 + −0.17	1.29 + −0.16	1.18 + −0.39
Io (T)			1.27 + −0.18
Europa (L)	1.13 + −0.14	1.02 + −0.16	
Europa (T)	1.14 + −0.06	1.13 + −0.06	1.15 + −0.13
Ganymede (L)	1.09 + −0.10		
Ganymede (T)	1.04 + −0.05		
Callisto (L)	1.45 + −0.15		
Callisto (T)	1.27 + −0.20	1.12 + −0.20	

L = Leading side (45° < θ < 135°); T = Trailing side (235° < θ < 315°).

Inspection of Table III shows that there are significant differences in the size of the opposition surges of the Galilean satellites. The large surge reported for Io (compared to Europa and Ganymede) indicates that Io's regolith is more loosely packed. The leading side of Callisto exhibits a more pronounced opposition surge which is consistent with groundbased observations at longer wavelengths (Morrison *et al.*, 1974).

3. Discussion

The observational data that have been presented in the preceding sections for the Galilean satellites are important for developing new models (and test existing ones) of the satellite surfaces. With the exception of Io, infrared observations of large airless satellites in the solar system have reported that a principal surface component is water ice (Moroz, 1966, 1968; Lebofsky, 1977; Clark, 1980; Brown and Cruikshank, 1983). However, the icy Galilean satellites are distinguished from the icy satellites of Saturn and Uranus by their very low geometric albedo at UV wavelengths and the very low UV/IR color ratio (see Lane *et al.*, 1986). Yet pure water ice is characterized by high reflectivity at IUE wavelengths (Hapke *et al.*, 1981) and therefore the low UV albedos reported for the Galilean satellites in Table I require the presence of additional absorbers on the surfaces of Europa, Ganymede and Callisto. The most likely darkening agents are elemental sulfur and sulfur bearing compounds originating from the very young and active surface of Io which are transported as ions and neutrals outward from Io's orbit by Jovian magnetospheric processes. These energetic ions and neutrals interact with the icy surfaces of Europa, Ganymede and Callisto and cause the ices to become darkened at UV wavelengths. This process competes with other processes of surface modification such as infall of interplanetary debris.

Io

Io's UV/near-IR color ratio is the lowest of all the large satellites in the solar system. The areal coverage by UV absorbing material(s) is very large. For example, given the leading side UV albedo (Band #1) that has been determined for Io with IUE then, if most of Io's surface were covered by material of ~ 1% albedo, at best only ~ 2% of the surface could have an albedo as high as ~ 50%. Groundbased spectrophotometry has shown that Io has a strong absorption feature in the 4000—5000 Å range and this is consistent with elemental sulfur being a major surface constituent (Wamsteker *et al.*, 1974). Mixtures of sulfur and various evaporite salts provide improved spectral agreement in the 4000—5000 Å spectral region (Fanale *et al.*, 1974; Nash and Fanale, 1977). Allotropic forms of elemental sulfur other than S_8, the most common allotrope, provide improved agreement between astronomical observation and laboratory spectral reflectance measurements (Nelson and Hapke, 1978). The presence of these allotropes might indicate an interesting color-morphology relationship for selected Io surface features seen in the Voyager images (Sagan, 1979). Photometric studies of flow features in the Voyager images are consistent with this hypothesis (Pieri *et al.*, 1984). While there is some evidence to suggest that elemental sulfur may not be present or is limited in extent (Young, 1984; Hammel, 1985), the UV spectrum of elemental sulfur in any of the suggested allotropic forms is consistent with the very low UV albedo that has been measured by IUE (Nelson *et al.*, 1983). Evidence for the presence of sulfur bearing compounds on Io surface is quite strong and this is consistent with the 4000—5000 Å absorption feature and with the low UV albedo reported in this study (see e.g. Nash and Nelson, 1979).

Observations of Io in the infrared have identified a strong absorption band at 4.08 μm (Cruikshank *et al.*, 1978) and on the basis of laboratory spectral reflectance measurements this feature has been attributed to sulfur dioxide frost (Smythe *et al.*, 1979; Fanale *et al.*, 1979; Hapke, 1979). The source of the frost is most probably a condensate of sulfur dioxide gas which has been detected in the Io volcanic plumes (Pearl *et al.*, 1979).

The spectral absorption shortward of 3300 Å observed in groundbased observations (Nelson and Hapke, 1978) is confirmed by the IUE observations (Table I, Figures 7a, 8a, b). The IUE data show that the strength of this feature varies as a function of longitude across Io's surface. Laboratory spectral reflectance measurements of sulfur dioxide frost have found that it is characterized by an absorption feature at this wavelength (Nash *et al.*, 1979; Hapke *et al.*, 1981). Hapke *et al.* also report that the laboratory spectral reflectance of sulfur dioxide frost increases slightly toward shorter wavelengths between ~ 2900 and 2500 Å. This change is also evident in the Io spectra shown in Figure 8a, b. The spectra shown in Figure 8a, b and the orbital phase curves shown in Figure 3 demonstrate that the strength of the UV absorption is variable. This confirms the conclusion based on preliminary IUE data (Nelson *et al.*, 1980) that this absorption feature is 50—75% stronger at longitudes ~ 70°—~ 150° than it is at longitudes ~ 250°— ~ 330°. Therefore the sulfur dioxide condensate is more abundant on the leading side than on the trailing side. Infrared observations of the variability of the 4.08 μm sulfur dioxide absorption feature also confirm this distribution asymmetry

(Howell, 1984). The short term variability of the geometric albedo on Io's leading side in Band # 3 suggests short timescale redistribution of sulfur dioxide frost on Io's surface. This is because the laboratory reflection spectrum of the frost has a strong increase in reflectance within the Band # 3 wavelength range and therefore a small change in the amount of surface coverage of sulfur dioxide frost will be detected by a change in the size of the Band # 3 albedo. The source of the frost is vapor condensation of the sulfur dioxide gas from the Io volcanos which may have short term variable output.

It is difficult to make a quantitative assessment of the absolute abundance of spectral absorbing material on the surface of a solar system object without a full understanding of the object's phase function (Veverka *et al.*, 1978a, b; Veverka *et al.*, 1979; Gradie and Veverka, 1982). However, simple linear ratioing of the albedo on either side of the absorption feature shows that the surface distribution of sulfur dioxide on Io's leading side is ~ 20%, which is consistent with the areal coverage of the 'white material' in the Voyager images. In the Voyager color filter bandpasses sulfur dioxide frost would appear white. At the IUE wavelengths sulfur dioxide frost would appear black. However, using the same spectral feature ratioing technique as was used with the IUE data, the infrared observations yield a much higher surface coverage. This may be due to the more than order of magnitude difference in wavelength and hence the difference between what appears to be an optically active regolith to the IUE observer and the groundbased IR observer. This is particularly complicated by the possibility that elemental sulfur is also present as a component of the regolith. At IR wavelengths sulfur is transparent, while at UV wavelengths, sulfur is opaque and therefore the IR observations contain information from much deeper in the regolith than do the UV. Furthermore, the fluxes measured in the IR may also include a thermal emission component (e.g. Morrison and Telesco, 1980) which must be considered.

The large opposition surge found in the IUE observations of Io is consistent with a loosely packed, highly porous, surface regolith. Comparison of these data with theoretical models and laboratory studies suggest that the surface is comprised of material which has ~ 80% void space (Buratti *et al.*, 1985) as would be expected from a surface that is being continually renewed from precipitates.

Europa
The UV geometric albedo of Europa is the highest of all the Galilean satellites (see Table I). While IR observations report that water ice is a major surface component of Europa's surface, the relative depths of the water bands suggest that another absorber may be present (Clark, 1980). The UV albedo of this object is too low on either hemisphere for large areas of the surface to be explained by water ice alone. The normalized spectra for each 30° longitude bin for Europa (Figure 9a, b) show a very slight variation with longitude. At longitudes near ~ 270° the spectrum is flatter in the 2500—2800 Å range than at longitudes near ~ 90°. This longitudinal difference is most obvious in the ratio spectrum shown in Figure 7b. This spectrum shows that Europa's trailing side relative to its leading side is characterized by a broad spectral absorption feature with a minimum at ~ 2800 Å. Lane *et al.* (1981) have suggested that this absorption is due to the presence of

isolated sulfur atoms in a matrix of water ice which comprises most of Europa's surface. The sulfur atoms are implanted in the ice as a result of their motion as ions in the Jovian magnetosphere. Io is the most probable source of the sulfur.

Ganymede

The UV geometric albedo of Ganymede (Table I) is lower than that of Europa. If IR spectroscopy measurements indicate that water ice is a major surface component then Ganymede's low albedo implies that other spectral absorbers must be present. Examination of the broadband geometric albedos at the three IUE bandpasses (Table I) indicates that the slope of Ganymede's UV spectral reflectance is much flatter on its trailing side than on it's leading side. Inspection of the trailing to leading side ratio spectrum (Figure 7c) as a function of longitude (Figure 10a, b), indicates that at longitude $\sim 120°-180°$ Ganymede is more absorbing shortward of 2800 Å than at $\sim 300°-300°$. Although Europa has a similar spectral asymmetry between it's leading and trailing hemispheres, the spectral character of Europa's hemispheric ratio (Figure 7b) differs from Ganymede's (Figure 7c). Europa's ratio spectrum increases shortward of 2800 Å while the comparable ratio spectrum of Ganymede's opposite hemispheres decreases. This implies that there is a different (or additional) UV absorber on Ganymede than on Europa. Assuming that the water ice on Ganymede's surface is being subjected to magnetospheric ion implantation processes that are similar to those on Europa (although at a reduced flux), then sulfur ions embedded in water ice cannot alone explain Ganymede's UV spectrum. If an additional spectral absorber were present which had a broad absorption feature centered near 2500 Å, then the combined effect of the sulfur ion absorption at 2800 Å with the hypothesized absorber at 2500 Å would adequately explain the features seen in Ganymede's ratio spectrum. Logical candidates for this spectral absorption have been examined and one possible spectral match might be magnetospheric oxygen ions. Oxygen atoms suspended in a water ice matrix could stimulate a gaseous ozone spectrum. This, combined with the spectrum of gaseous sulfur dioxide caused by the matrix isolated sulfur might explain the Ganymede UV spectrum.

Callisto

The UV geometric albedo of Callisto is the lowest of all the Galilean satellites. This implies that water ice resurfacing processes may be occurring at a lower rate than on Europa and Ganymede or that the flux of UV absorbing material falling onto the surface of Callisto is higher than on the other icy Galilean satellites. The leading and trailing hemispheres of Callisto have a different UV albedo (Table I). However, unlike Europa and Ganymede, it is Callisto's trailing hemisphere that has a higher albedo than it's leading hemisphere. This is also true at visual wavelengths (Morrison *et al.*, 1974). Inspection of Callisto's opposite hemisphere UV ratio spectrum (Figure 7d) indicates that there are no spectrally active UV absorbers on Callisto. Given that the hemispheric UV albedo differences on Callisto are in the same sense as the visual albedo hemispheric differences then one material may be responsible for both absorptions. Water ice contaminated by trace amounts of elemental sulfur is not inconsistent with this observation although a host of other possibilities exist.

4. The Saturnian Satellites

Background

Quantitative spectrophotometric studies of the major Saturnian satellites using groundbased telescopes with standard UBV or uvby filter sets have only been undertaken during the last decade. Until then, the only Saturnian satellites that had been subjected to extensive observation were the two most unusual planetary satellites, Iapetus and Titan. Iapetus has a pronounced leading to trailing hemisphere albedo asymmetry that is the largest of any satellite in the solar system and first was reported by Cassini, the satellite's discoverer, in 1671 (see Cruikshank *et al.*, 1981). Titan is also unusual and was known to have a thick methane atmosphere (Kuiper, 1944). Because of its thick atmosphere, Titan is unique among planetary satellites for its surface has not been seen. The other Saturnian satellites have been studied at wavelengths accessible to groundbased observers and the brighter satellites (Tethys, Dione, Rhea and Iapetus) have had limited study by IUE.

The groundbased observations confirm that all these satellites are in synchronous rotation (Blanco and Catalano, 1971: McCord *et al.*, 1971; Millis, 1973; Franklin and Cook, 1974; Blair and Owen, 1974; Franz and Millis, 1975). At visual wavelengths, Tethys, Dione and Rhea all have leading side albedos that are 10—30% higher than their trailing sides, which is indicative of longitudinal differences in chemical/mineralogical abundance and/or composition in the optically active regoliths.

The hemispheric albedo asymmetry of Iapetus at visual wavelengths is extremely large (the trailing side is brighter by a factor of 5) and the cycle to cycle repeatability is variable which has been interpreted as being caused by the effect of different scattering properties of the optically active regoliths of the two very different hemispheres (Millis, 1973). Morrison *et al.* (1975) constructed a global albedo map in which the leading hemisphere was covered mostly by dark material ($P_v = \sim 0.07$) and the trailing hemisphere and the south polar cap was covered mostly by bright material ($P_v = 0.35$) (Morrison *et al.*, 1975). This albedo map, developed entirely from disk integrated observations, provided a strikingly good approximation to the images of Iapetus returned by the Voyager spacecraft (Smith *et al.*, 1982).

Infrared observations of the large satellites of Saturn have identified water ice as the principal absorbing specie comprising the optically active surface of Tethys, Dione, and Rhea and the trailing (bright) hemisphere of Iapetus (Johnson *et al.*, 1975; Morrison *et al.*, 1976; Fink *et al.*, 1976; Lebofsky and Feierberg, 1985). The leading (dark) hemisphere of Iapetus does not show spectral features consistent with water ice and has an infrared spectrum that is featureless (Soifer *et al.*, 1979; Cruikshank *et al.*, 1981). The albedo of water ice alone is too high for the surfaces of the satellites to be covered only by this material. Other material(s) must be present in varying amounts in order to explain the albedos of all the

Saturnian satellites. In the case of the dark hemisphere of Iapetus, the darkening material is most probably the dominant specie on the surface.

The IUE Observations

A limited number of observations of the Saturnian satellites with IUE were made during the 1981—1985 apparitions with the long wavelength spectrograph. The number of usable observations (N = 33) is considerably less than the number of Galilean satellite observations and the uncertainties in the reduced data are correspondingly greater. Because Saturn is always near the aperture, absolute photometric observations of the type required for geometric albedo determination of planetary satellites are limited to orbital phase positions that are at or near elongation in the satellite's orbit about Saturn. Therefore orbital phase curves and 30° binned ratio spectra are not available. Also, the variation of brightness as a function of orbital phase has not been studied because of the limited number of observations that comprise the Saturnian satellite data set and the geometry constraints.

4.1. BROADBAND GEOMETRIC ALBEDO

The ultraviolet geometric albedos for the Saturnian satellites were calculated for the three IUE wavelength bandpasses using the same method as was used for the Galilean satellites. In the case of the Saturnian satellites, the shortest wavelength attainable from the groundbased observations and from Voyager photometry does not overlap the IUE longest wavelength bandpass. However, given that the reduction technique for the Saturnian satellites is identical as that used for the Galilean satellites (which were directly comparable to groundbased spectrophotometry at or very near the IUE long wavelength bandpass) the absolute albedos of the Saturnian satellites reported in Table IV should be valid.

TABLE IV

Saturnian Satellites
Ultraviolet Geometric Albedos

	Band # 1		Band # 2		Band # 3	
Tethys (LE)	0.61	(N = 1)	0.61	(N = 1)	0.62	(N = 1)
Dione (LE)	0.27 + −0.04	(N = 5)	0.27 + −0.03	(N = 5)	0.29 + −0.04	(N = 5)
Dione (TE)	0.22 + −0.03	(N = 4)	0.27 + −0.02	(N = 2)	0.26 + −0.04	(N = 4)
Rhea (LE)	0.26 + −0.05	(N = 10)	0.27 + −0.05	(N = 8)	0.30 + −0.06	(N = 9)
Rhea (TE)	0.16 + −0.09	(N = 4)	0.19 + −0.09	(N = 4)	0.22 + −0.12	(N = 4)
Iapetus (LE)	0.03 + −0.009	(N = 5)	0.03 + −0.001	(N = 2)	0.03 + −0.002	(N = 5)
Iapetus (TE)	0.21 + −0.01	(N = 4)	0.24 + −0.01	(N = 2)	0.25 + −0.02	(N = 4)

L = Leading Side (70° < θ < 110°).
T = Trailing Side (240° < θ < 300°).
LE = Leading Side Elongation (80° < θ < 100°).
TE = Trailing Side Elongation (260° < θ < 280°).
N = Number of observations.

The UV albedo of Tethys (P_{uv} = ~60%) is the highest of the Saturnian satellites and this comparable to the high visual albedo reported by Voyager and groundbased visual filter observations. The leading side of Dione is ~10% brighter than its trailing side which is somewhat less than the ~30% hemispheric albedo brightness variation reported from groundbased V wavelength observations (Franz and Millis, 1975). The leading side of Rhea is ~40% brighter than its trailing side at the IUE wavelengths. This is more than the ~20% observed from the ground at V wavelengths (Blair and Owen, 1974) and this is consistent with the increase in albedo difference seen in the B − V and U − B color ratios observed from the ground indicating an increase in hemispherical contrast toward shorter wavelengths. The UV albedo of Iapetus is consistent with the albedos reported at longer wavelengths from groundbased observations and the Voyager spacecraft. In the UV, as in the visual, the leading side of Iapetus is very absorbing, and the trailing side is comparable to the trailing side albedos of other Saturnian satellites. The leading side albedo is at least 7 times less than the trailing side at the IUE wavelengths. This is greater than the 5 times darker reported at visual wavelengths. The implies that the spectral absorber which darkens the leading hemisphere of Iapetus is more absorbing toward shorter wavelengths.

4.2. HEMISPHERICAL SPECTRAL ASYMMETRIES

The broadband UV albedos reported by IUE observations of the Saturnian satellites confirm the suggested differences in chemical/mineralogical composition on Dione, Rhea and Iapetus that the longer wavelength observations imply. In the case of Dione these observations indicate that there are no strong UV absorptions in the unidentified materials on the satellite's surface. For Rhea and Iapetus, the UV absorption becomes greater toward shorter wavelengths. This may be a gradual decrease in reflectance or may be the effect of an absorption band. If such absorptions occur at the IUE wavelengths they should be detectable in the opposite hemispheric spectral ratios as occurs in the case of the Galilean satellites most notably on Io. The individual spectra for each object are few in number and quite noisy when compared to spectra of the much brighter Galilean satellites. For the sake of comparison, the spectral ratios (normalized at 2700 Å) are presented (unsmoothed) in Figure 12a, b, c for Dione, Rhea and Iapetus respectively. Figure 12a is the ratio spectrum of Dione's leading side to trailing side. The ratio spectrum suggests the presence of a very slight absorption feature centered at ~2900 Å and perhaps a second feature shortward of ~2600 Å. Neither of these absorptions are significantly above the noise to permit a positive confirmation of their existence. Further work with IUE and other orbital UV instruments may allow for confirmation.

The spectral ratio of Rhea's leading side to trailing side is shown in Figure 12b. The noise level in the data is unusually high given that Rhea is brighter than Dione and the number of observations is greater. Given the high level of noise, no absorption features can be inferred, in spite of the fact that the albedo asymmetries reported for Rhea at the IUE wavelengths are different than those reported at visual wavelengths which is suggestive of an absorption feature being present somewhere in the 2600−5600 Å range.

The ratio spectrum of the leading to trailing hemispheres of Iapetus is shown in Figure 12c. Although there is a very great difference in albedo between the two hemispheres (a factor of 7 at the IUE wavelengths), the differences in spectral

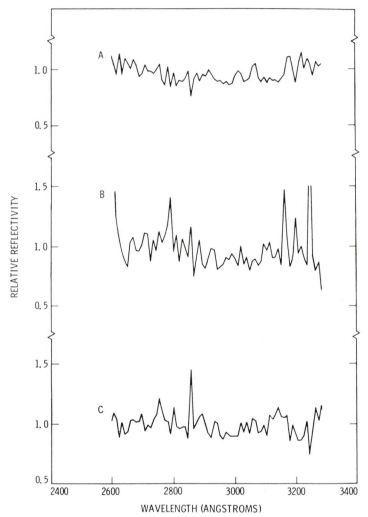

Fig. 12. Opposite hemisphere ratio spectra of the Saturnian satellites. For the Saturnian satellites, the spectra are much noisier than the Galilean satellites due to the fewer number of total observations and the faintness of the sources. All ratio spectra have been normalized at 2700 Å. (a) Sum of 5 co-added spectra of Dione's leading side divided by the sum of 4 spectra of Dione's trailing side. Given the noise in the data positive identification of spectral features is not possible, however there is suggestion of a very broad absorption feature centered at ~ 2900 Å. No spectra absorber has been associated with this feature. (b) The ratio of the sum of 10 co-added spectra of Rhea's leading side divided by the sum of 4 co-added spectra of Rhea's trailing side. Given the noise in the data no spectral features can be identified. (c) The ratio of 5 co-added spectra of the leading hemisphere of Iapetus to the sum of 4 co-added spectra of the trailing side. In spite of the great difference in geometric albedo between the hemispheres of this object, no sharp spectral features are discernible in the ratio spectrum.

reflectance are very slight. However since the hemispheric albedo ratio does decrease from 7 at the IUE wavelengths compared to 5 at visual wavelengths, the possibility of an absorption feature somewhere between 2400 and 5600 Å can be inferred.

5. Summary Discussion

The results of IUE observations of planetary satellites reported above can now be integrated with the results of groundbased and spacecraft observations of the families of large planetary satellites in the solar system to gather some information regarding comparative planetology. The data set comprises observations of the Galilean satellites and the larger Saturnian and Uranian satellites and each set of objects has been subjected to different processes of surface evolution.

The geometric albedos of the Saturnian satellites indicate that throughout the Saturnian system there is wide variation in UV geometric albedo, ranging from a high of ~0.6 for Tethys to a low of 0.03 for Iapetus. The icy Galilean satellites Europa, Ganymede and Callisto all have low UV geometric albedos. The Voyager Photopolarimeter (PPS) experiment reported low UV albedos for the Uranian satellites. However, the PPS experiment reported that the Uranian satellite near IR albedos are also quite low. The Galilean satellites, with similar low UV albedos have high IR albedos. These photometric relationships are displayed in Figure 13 which is a plot of UV albedo vs. UV/IR color ratios for the Galilean, Saturnian, and Uranian satellites. It is obvious from Figure 13 that for the Galilean and Uranian satellites, each set of objects falls in a very close range on the UV–UV/IR albedo plot. This is not true for the Saturnian satellites. This indicates that the surfaces of the Galilean and Uranian satellites are each modified in a manner that is unique with respect to the primary body. The Saturnian satellites apparently have no common surface modification process or perhaps that process has been overwhelmed by other ones which make each satellite in the Saturnian system a class unto itself.

While water ice is common to all these objects (except Io), the additional spectral absorbers which are present on the surfaces of the icy Galilean satellites are different from those which are on the surfaces of the Saturnian and Uranian satellites. In the case of the Galilean satellites, the darkening agent is most likely sulfur or sulfur bearing compounds which originate from the highly active surface of Io and spiral outward as ions trapped in Jupiter's magnetic field and ultimately impact the surfaces of the other satellites. In the case of the Uranian satellites, the common surface modification process may be due to hydrocarbons which are formed on the surface over time from the primordial condensation of methane-water ice mixtures. No positive spectral identification of any particular specie has been made.

The direction for future study in satellite surface spectroscopy is obvious. Higher resolution and greater wavelength range is required to maximize the probability of identification of spectrally active absorbers. Because many of these future observations will be done from spacecraft such as the Hubble Space Telescope, great care should be given to allow such earth orbiting observing

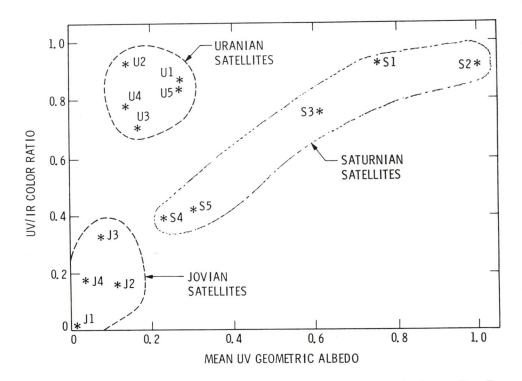

Fig. 13. Comparisons of geometric albedos of the Galilean, Saturnian and Uranian satellites. For the Galilean satellites and for Dione and Rhea the UV albedos are from IUE. For Mimas and Enceladus the UV albedos are from Voyager images (Buratti, 1984) and are of longer wavelength (3500 Å) than those from IUE. The IR albedos of the Galilean and Saturnian satellites are from groundbased spectrophotometry (Nelson and Hapke, 1978; Noland *et al.*, 1974). All the Uranian satellite data are from the Voyager Photo-polarimeter (Lane *et al.*, 1986). With the exception of the slight difference in UV wavelength noted above for Mimas and Enceladus all the wavelength ranges are similar. Jovian and Uranian satellites each have distinct color ratios which distinguish the two groups of satellites from each other. The Saturnian satellites have litte albedo similarity among themselves. This is consistent with the hypothesis that the surfaces of all the icy Galilean satellites are being modified by a common process. Likewise, the Uranian satellite surfaces may also have a common process of surface modification that is different from that in the Jovian system. The Saturnian satellite system may not have a common process of surface modification or the system may have been disturbed and the albedos of the satellites have been altered.

platforms to observe moving targets with variable track rates and to ensure that scattered light rejection is maximized so that faint objects such as planetary satellites can be observed while near their very bright primary bodies. For at least the next 15 yr there wil be few spacecraft encountering the outer planets to make spectroscopic measurements on the satellites. The earth orbital telescope will become the mainstay of this type of research — pioneered mostly by the International Ultraviolet Explorer Satellite.

Acknowledgements

The authors would like to thank their JPL co-investigators on this project, D. L.

Matson, B. J. Buratti, G. J. Veeder and E. F. Tedesco who participated in many of the observations. Particular thanks are due for the patience and understanding of the two IUE observatory directors whose tenure spanned the length of this program, A. Boggess and Y. Kondo. The resident astronomers and telescope operators of the IUE project devoted many hours of general support to this effort and demonstrated important extra consideration in accommodating the unique needs of a solar system observational program. The support of E. Hamrick, F. Motteler, M. E. Ockert and M. D. Morrison and L. Watson in software development and data processing is gratefully appreciated.

This work represents one phase of research carried out at the Jet Propulsion Laboratory, California Institute of Technology under contract with NASA.

References

Blair, G. N. and Owen, F. N.: 1974, *Icarus* **22**, 224.

Blanco, C. and Catalano, S.: 1971, *Astron. and Astrophys.* **14**, 43.

Boggess, A., Carr, F. A., Evans, D. C., Fischel, D., Freeman, H. R., Fuechsel, C. F., Klinglesmith, D. A., Krueger, V. L., Longanecker, G. W., Moore, J. V., Pyle, E. J., Rebar, F., Sizemore, K. O., Sparks, W., Underhill, A. B., Vitagliano, H. D., West, D. K., Macchetto, F., Fitton, B., Barker, P. J., Dunford, E., Gondhaklekar, P. M., Hall, J. E., Harrison, V. A. W., Oliver, M. B., Sanford, M. C. W., Vaughan, P. A., Ward, A. K., Anderson, B. E., Boksenberg, A., Coleman, C. I., Snijders, M. A. J., and Wilson, R.: 1978, *Nature* **275**, 372.

Buratti, B.: 1984, *Icarus* **59**, 392.

Buratti, B. and Veverka, J.: 1983, *Icarus* **55**, 93.

Buratti, B. J., Smythe, W. D., Nelson, R. M., Gharakhani, V., and Hapke, B. W.: 1985, *Bull. Am. Ast. Soc.* **17**, 918.

Brown, R. H. and Cruikshank, D. P.: 1983, *Icarus* **55**, 83.

Caldwell, J.: 1975, *Icarus* **25**, 384.

Clark, R. N.: 1980, *Icarus* **44**, 388.

Cruikshank, D. P., Jones, T. J., and Pilcher, C. B.: 1978, *Astrophys. J. Let.* **25**, L89.

Cruikshank, D. P.: 1980, *Icarus* **41**, 246.

Cruikshank, D. P., Bell, J. F., Gaffey, M. J., Brown, R. H., Howell, R., Beerman, C., and Roganstad, M.: 1981, *The Dark Side of Iapetus*, University of Hawaii Planetary Geosciences Division Pub. No. 333.

Fanale, F. P., Johnson, T. V., and Matson, D. L.: 1974, *Science* **186**, 922.

Fanale, F. P., Brown, R. H., Cruikshank, D. P., and Clark, R. N.: 1978, *Nature* **280**, 760.

Fink, U., Deckkers, N. H., and Larson, H. P.: 1973, *Astrophys. J. Let.* **179**, L155.

Fink, U., Larson, H. P., Gautier, T. N., and Treffers, R. R.: 1976, *Astrophys. J. Let.* **207**, 163.

Franz, O. G. and Millis, R. L.: 1975, *Icarus* **24**, 433.

Franklin, F. A. and Cook, A. F.: 1974, *Icarus* **23**, 355.

Gradie, J. and Veverka, J.: 1982, *Icarus* **49**, 109.

Hammel, H. B., Goguen, J. D., Sinton, W. M., and Cruikshank, D. P.: 1985, *Icarus* **64**, 125.

Hapke, B. W.: 1963, *J. Geophys. Res.* **68**, 4571.

Hapke, B. W.: 1979, *Geophys. Res. Let.* **6**, 799.

Hapke, B. W.: 1981, *J. Geophys. Res.* **86**, 3039.

Hapke, B. W., Wells, E., Wagner, J., and Partelo, W.: 1981, *Icarus* **47**, 361.

Harris, D. L.: 1961, Photometry and Colorimetry of Planets and Satelites', in G. P. Kiuper and B. P. Middlehurst (eds.), *The Solar System, Vol. III: Planets and Satellite*, University of Chicago Press, pp. 232—342.

Howell, R. R., Cruikshank, D. P., and Fanale, F. P.: 1984, *Icarus* **57**, 83—92.

Irvine, W. M.: 1966, *J. Geophys. Res.* **71**, 2971.

Johnson, T. V. and McCord, T. B.: 1970, *Icarus* **13**, 37.

Johnson, T. V.: 1971, *Icarus* **14**, 94.

Johnson, T. V., Veeder, G. J., and Matson, D. L.: 1975, *Icarus* **24**, 428.

Johnson, T. V. and Pilcher, C.: 1977, 'Satellite Spectrophotometry and Surface Compositions', in J. Burns (ed.), *Planetary Satellites*, pp. 232—269. U. of Arizona Press, Tucson,

Kuiper, G. P.: 1944, *Astrophys. J.* **100**, 378.

Lane, A. L., Nelson, R.M., and Matson, D. L.: 1981, *Nature* **292**, 38.

Lane, A. L., Hord, C. W., West, R. A., Esposito, L. W., Simmons, K. E., Nelson, R. M., Wallis, B. D., Buratti, B. J., Horn, L. J., Graps, A. L., and Pryor, W. R.: 1986, *Science* **233**, 65.

Lean, J. L., White, O. R., Livingston, W. C., Heath, D. F., Donnelly, R. F., and Skumanich, A.: 1982a, *J. Geophys. Res.* **87**, 10307.

Lean, J. L., Donnelly, R. F., and Heath, D. F.: 1982b, *Calculated Solar Flux Variability 200—300 nm.* Paper SSS41A-11, Fall Meeting Amer. Geophys. U. 1982.

Lebofsky, L. A.: 1977, *Nature* **269**, 785.

Lebofsky, L. A. and Feierberg, M. A.: 1985, *Icarus* **63**, 237.

McCord, T. B., Johnson, T. V., and Elias, J. H.: 1971, *Astrophys. J.* **165**, 413.

McFadden, L. A., Bell, J. F., and McCord, T. B.: 1980, *Icarus* **44**, 410.

Millis, R. L.: 1973, *Icarus* **18**, 247.

Moroz, V. I.: 1966, *Soviet Astronomy—A. J.* **9**, 999.

Moroz, V. I.: 1968, *Physics of Planets*, NASA TT F-515, Clearinghouse for Federal Scientific and Technical Information, Springfield VA, pp. 389.

Morrison, D. and Cruikshank, D. P.: 1974, *Space Sci. Rev.* **15**, 641.

Morrison, D., Morrison, N. D., and Lazarewicz, A. R.: 1974, *Icarus* **23**, 399.

Morrison, D., Jones, T. J., Cruikshank, D. P., and Murphy, R. E.: 1975, *Icarus* **24**, 157.

Morrison, D., and Telesco, C. M.: 1980, *Icarus* **44**, 226.

Nash, D. B. and Nelson, R. M.: 1979, *Nature* **280**, 763.

Nash, D. B., Fanale, F. P., and Nelson, R. M.: 1980, *Geophys. Res. Let.* **7**, 665.

Nelson, R. M., and Hapke, B. W.: 1968, *Icarus* **36**, 304.

Nelson, R. M., Matson, D. L., Lane, A. L., Motteler, F. C., and Ockert, M. E.: 1980a, *Bull. Am. Astron. Soc.* **12**, 713.

Nelson, R. M., Lane, A. L., Matson, D. L., Fanale, F. P., Nash, D. B., and Johnson, T. V.: 1980b, *Science* **210**, 784.

Nelson, R. M., Pieri, D. C., Baloga, S. M., Nash, D. B., and Sagan, C.: 1983, *Icarus* **56**, 409.

Nelson, R. M., Lane, A. L., Matson, D. L., Veeder, G. J., Buratti, B. J., and Tedesco, E. F.: 1987, 'Spectral Geometric Albedos of the Galilean Satellites from 0.24—0.34 Micron: Observations with the International Ultraviolet Explorer Spacecraft', submitted to *Icarus*.

Ockert, M. E., Nelson, R. M., Lane, A. L., and Matson, D. L.: 1987, 'Europa's Ultraviolet Absorption Band (2400 to 3200 Å): Temporal and Spatial Evidence from IUE', submitted to *Icarus*.

Pieri, D. C., Baloga, S. M., Nelson, R. M., and Sagan, C.: 1984, *Icarus* **60**, 685.

Pearl, J., Hanel, R., Kunde, V., Maguire, W., Fox, D., Gupta, S., Ponnaperuma, C., and Rawlin, F.: 1979, *Nature* **280**, 755.

Pilcher, C. B., Ridgeway, S. T., and McCord, T. B.: 1972, *Science* **178**, 1087—1089.

Pollack, J. B., Witteborn, F. C., Erickson, E. F., Strecker, D. W., Baldwin, B. J., and Bunch, T. E.: 1978, *Icarus* **36**, 271.

Rottman, G. J.: 1983, *Planet. Space Sci.* **31**, 1001.

Russell, H. N.: 1916, *Astrophys. J.* **43**, 173.

Sagan, C.: 1979, *Nature* **280**, 750.

Simonelli, D. and Veverka, J.: 1984, *Icarus* **59**, 406.

Smith, B. A., Soderblom, L. A., Batson, R., Bridges, P., Inge, J., Masursky, H., Shoemaker, E., Beebe, R., Boyce, J., Briggs, G., Bunker, A., Collins, S. A., Hansen, C. J., Johnson, T. BV., Mitchell, J. L., Terrille, R. J., Cook, A. F., Cuzzi, J., Pollack, J. B., Danielson, G. E., Ingersoll, A. P., Davies, M. E., Hunt, G. E., Morrison, D., Owen, T., Sagan, C., Veverks, J., Strom, R., and Suomi, V. E.: 1982, *Science* **215**, 504.

Smythe, W. D., Nelson, R. M. and Nash, D. B.: 1979, *Nature* **280,** 766.

Soderblom, L. A., Mosher, J. A., Danielson, E. G., Cook, A. F., and Kupferman, P.: 1984, *Geophys. Res.* **88**, 5789.

Soifer, B. T., Neugebauer, G., and Gatley, I.: 1979, *Astron. J.* **84**, 1644.

Stebbins, J.: 1927, *Lick Obs. Bull.* No. 385, **13**, 1.

Stebbins, J. and Jacobson, T. S.: 1928, *Lick Obs. Bull.* No. 401, **13**, 180.

Veverka, J.: 1977, 'Photometry of Satellite Surfaces', in J. Burns (ed.), *Planetary Satellites*, U. of Arizona Press, Tucson, pp. 171—210.

Veverka, J., Goguen, J., Yang, S., and Elliot, J.: 1979, *Icarus* **37**, 249.

Veverka, J., Goguen, J., Yang, S., and Elliot, J. L.: 1978a, *Icarus* **34**, 63.

Veverka, J., Goguen, J., Yang, S., and Elliot, J. L.: 1978b, *Icarus* **33**, 368.

Wamsteker, W.: 1972, 'Narrowband Photometry of the Galilean Satellites', *Comm. of the Lun. and Planet. Lab.* No. 167, **9**, 171.

Wamsteker, W., Kroes, R. L., and Fountain, J. A.: 1974, *Icarus* **23**, 417.

Young, A. T.: 1984, *Icarus* **58**, 197.

COMETS

M. C. FESTOU

Observatoire de Besançon, Besançon, France

and

P. D. FELDMAN

Department of Physics and Astronomy, Johns Hopkins University, Baltimore, MD, U.S.A.

1. Introduction

Observations of comets in the vacuum ultraviolet since 1970 have contributed to significant progress in our understanding of cometary comae and, indirectly, of the cometary nucleus itself. The first ultraviolet observations, of comets Tago—Sato—Kosaka (1969 IX) and Bennett (1970 II), made in 1970 by both OAO—2 and OGO—5, demonstrated the existence of a hydrogen envelope that extended for millions of kilometers from the comet's nucleus (Bertaux *et al.*, 1973; Code *et al.*, 1972). Analysis of this H I Lyman-α envelope and the accompanying strong emission from OH at 3085 Å (until then seen only weakly in ground-based spectra) was taken to provide strong confirmation of Whipple's icy conglomerate model which had been proposed two decades earlier (Whipple, 1950, 1951) on the basis of the non-central force perturbations of cometary orbits. The observed emissions could be accounted for by the photodissociation by sunlight of H_2O evaporated from the surface of the 'dirty snowball' nucleus and the derived H_2O production rate, typically of the order of 10^{29}—10^{30} molecules s^{-1}, was exactly the magnitude predicted by Whipple's model (Bertaux *et al.*, 1973; Keller and Lillie, 1974). Comet Tago—Sato—Kosaka was also observed in Ly-α by a rocket experiment (Jenkins and Wingert, 1972).

Prior to IUE, opportunities for vacuum ultraviolet observations of comets were very limited. Comet Kohoutek (1973 XII) was discovered 10 months prior to its perihelion and its promise led to an extensive campaign of coordinated space and ground-based observations. Atomic oxygen and carbon were discovered in the UV spectra obtained by two sounding rocket experiments (Feldman *et al.*, 1974; Opal and Carruthers, 1977) and direct Ly-α images of the hydrogen envelope were obtained with rocket (Opal *et al.*, 1974) and Skylab (Carruthers *et al.*, 1974) ultraviolet cameras. Two years later, at the apparition of comet West (1976 VI), the rocket instrumentation developed for the Kohoutek observations was used to obtain comprehensive ultraviolet spectra of a comet revealing for the first time the emissions of CO, CS, CO_2^+, and S (Feldman and Brune, 1976; Smith *et al.*, 1980). The OAO—3 (*Copernicus*) observatory was used to obtain very high resolution line profiles of the H I Ly-α emission from Comet West and several other comets (Festou *et al.*, 1979) during this time period.

Since January 1978, the International Ultraviolet Explorer (IUE) satellite observatory has been available for cometary observations and to date has

Y. Kondo (ed.), Scientific Accomplishments of the IUE, pp. 101—118.
© 1987 *by D. Reidel Publishing Company.*

observed 26 different comets (although one, periodic Comet Encke, was observed on two successive apparitions). The long lifetime of IUE, coupled with its stable photometric calibration over this period (due in large measure to its geosynchronous orbit), makes possible quantitative, comparative studies of these comets. This chapter will present a representative sample of cometary spectra and summarize the significant contributions of IUE to our understanding of the physics and chemistry of the coma. A number of outstanding problems remain unsolved and are identified for future UV observations with the next generation of orbiting observatories (e.g., ASTRO, HST, and LYMAN). Also, some preliminary results from the extensive IUE program on Comet Halley are given. Earlier reviews of the ultraviolet spectroscopy of comets have been given by Feldman (1982, 1983), and of IUE observations by Festou *et al.* (1982) and Feldman (1984).

2. Survey of IUE Observations

2.1. LOW RESOLUTION ULTRAVIOLET SPECTROSCOPY OF COMETARY COMAE

Table I lists the 26 comets observed with IUE, almost all of which produced positive spectra containing at least the strong OH(0, 0) band at 3085 Å. Most of them were observed when close to the Sun (i.e. 0.7 AU $< r <$ 1.7 AU) except for two which had distant perihelia, comets Bowell and Cernis, and comet Halley which was observed between 2.64 AU pre-perihelion and 2.57 AU post-perihelion. An attempt to detect P/Halley at 4.4. AU in April 1985 yielded only an upper limit to the OH emission (Festou *et al.*, 1986a). Of these comets, eleven are short period comets from Jupiter's family; four are periodic comets of slightly longer periods, among them P/Halley; seven have moderately long periods ranging from 300 to 65 000 yr; the four remaining are very long period comets, some of which could have arrived directly from the Oort cloud (e.g. comet Bowell). The comets observed with IUE are thus dynamically very different, and it is possible to compare the composition as reflected in the UV spectra of comets which have made different numbers of revolutions around the Sun. Evidently, direct comparisons are valid only as long as the comet spectra which are compared are obtained at similar heliocentric distances and heliocentric velocities and in objects of similar nucleus output. Otherwise, it is necessary to model the physical processes taking place in the coma to distinguish between real differences and effects due to geometrical factors or Doppler shifts (Festou, 1981a).

One of the first striking discoveries made with IUE is that, at first sight, all comets have a similar spectrum. As an example, the spectrum of comet Bradfield, a moderate period 'gassy' comet, is shown in Figure 1, and that of comet Encke, whose period is the shortest known, is shown in Figure 2. Aside from relative intensity differences, which result from differences in orbital parameters, and the poorer signal-to-noise ratio in the case of the fainter comet, these spectra indicate a fundamentally constant composition of the cometary ice from comet to comet. Weaver *et al.* (1981b) reached this conclusion from a model independent comparative study of seven comets. From our knowledge of the characteristics of subsequently obtained IUE spectra of additional objects, we can now state that

TABLE I

Comets observed with IUE

Comet	Observing period	Orbital period (yr)[a]	Observing site[b]
Seargent (1978 xv)	Oct—Nov 1978	8960	G
Bradfield (1979 x)	Jan—Mar 1980	291	G, V
Encke (1980 xi)	Oct—Nov 1980	3.30	G
Stephan—Oterma (1980 x)	Nov—Dec 1980	37.7	G
Meier (1980 xii)	Dec 1980	4800	G
Tuttle (1980 xiii)	Dec 1980	13.7	G
Panther (1981 ii)	Mar 1981	64920	G
Borrelly (1981 iv)	Mar 1981	6.77	G
Bowell (1982 i)	Apr—Nov 1982	—	G, V
Grigg—Skjellerup (1982 iv)	Apr 1982	5.09	G
Austin (1982 vi)	Jul—Nov 1982	35500	G, V
d'Arrest (1982 vii)	Sep—Oct 1982	6.38	G
Churyumov—Gerasimenko (1982 viii)	Nov 1982	6.61	V
IRAS—Araki—Alcock (1983 vii)	May 1983	1000	G, V
Sugano—Saigusa—Fujikawa (1983 v)	Jun 1983	—	G
Tempel 2 (1983 x)	Jul 1983	5.29	V
Cernis (1983 xii)	Sep 1983	—	G
Kopff (1983 xiii)	Jul—Sep 1983	6.44	V
Crommelin (1984 iv)	Feb—Apr 1984	27.4	V
Encke (1984 vi)	Apr—May 1984	3.31	G
Takamizawa (1984 vii)	Aug 1984	7.24	G
Arend—Rigaux (1984 xxi)	Dec 1984	6.84	V
Schaumasse (1984 xxii)	Dec 1984	8.26	V
Levy—Rudenko (1984 xxiii)	Dec 1984	42800	V
Halley (1982 i)	Apr 1985—Jul 1986	76.0	G, V
Giacobini—Zinner (1984 e)	Jun—Oct 1985	6.59	G, V
Hartley—Good (1985 ℓ)	Nov 1985	—	V

[a] Osculating period after Marsden (1986).
[b] V = Vilspa; G = GSFC.

this general property holds for all comets observed with IUE, with two exceptions. First, the continuum strength (usually measured by the flux in the 2900—3000 Å interval) can vary by orders of magnitude relative to the gas emission. While comet Bradfield had an almost dust-free coma (A'Hearn et al., 1981), comets Bowell (A'Hearn et al., 1984) (see Figure 3) and Cernis (Feldman and A'Hearn, 1985) possess a well developed solid grain coma despite the fact they were observed at large heliocentric distances. As Figure 4 indicates, comet Halley also had a very strong continuum which tended to mask weak gas emission such as that of CO_2^+. The variation in gas-to-dust ratio between individual comets is a well known property of these objects. IUE, however, has made it possible to simultaneously observe the dust and the *principal* gas emission (OH), which often is not successfully modelled on the basis of the visible gas emissions which arise from less than 1% of the total content of the gaseous coma (A'Hearn et al., 1985). For several

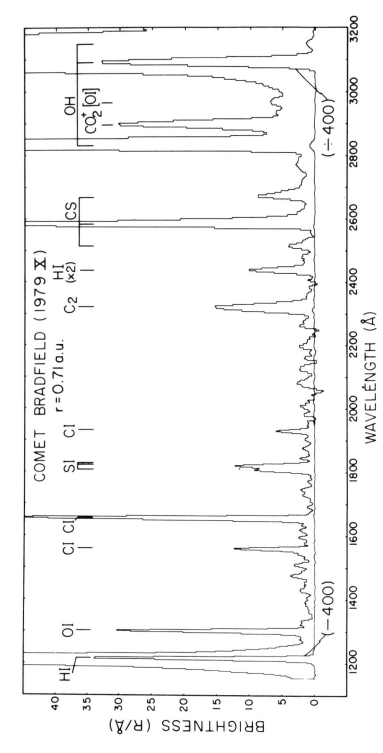

Fig. 1.   Composite spectrum of comet Bradfield (1979 X) taken from four IUE low dispersion images obtained on 10—11 January 1980. The heliocentric distance of the comet was 0.71 AU. The brightness is an average over the central 10″ × 15″ rectangular portion of the large aperture. The principal emission features are indicated.

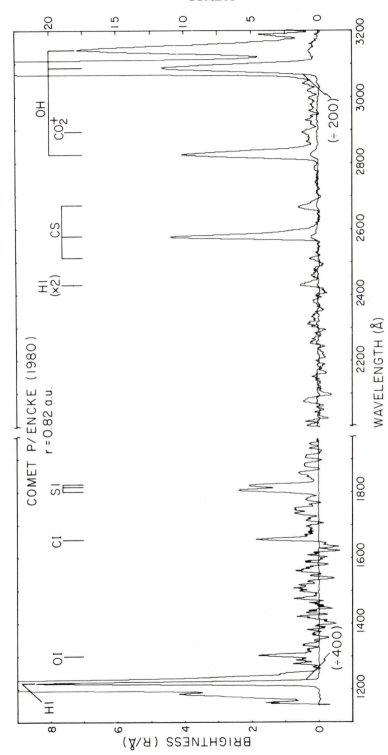

Fig. 2.   Same as Figure 1 for comet P/Encke recorded on 4 November, 1980. The heliocentric distance was 0.82 AU.

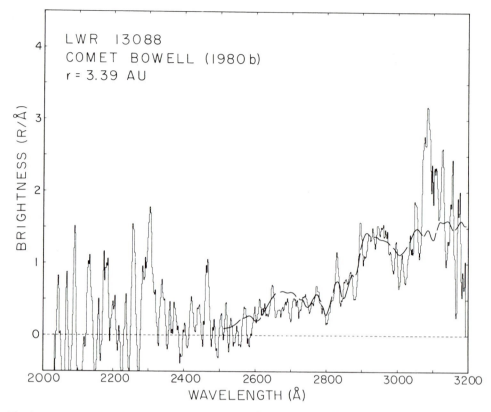

Fig. 3. Long wavelength spectrum of comet Bowell recorded on 27 April, 1982 when the comet was 3.39 AU from the Sun. A solar spectrum at comparable resolution is shown for comparison as the dashed line.

comets observed over an extended period by IUE, the apparent gas-to-dust ratio also appeared to vary with time (or heliocentric distance), though not in a predictable fashion (Feldman and A'Hearn, 1985).

The second noticeable difference between the UV spectra of various comets is the variable intensities of the C I multiplets and of the bands of the CO 4th positive system. The UV emission of CO is difficult to record because of the weakness of the solar flux around 1500 Å. CO emission in comet West was detected by sounding rocket experiments in 1976 and, prior to Halley (Figure 5), was detected by the IUE only in comet Bradfield at a level just above the sensitivity of the SWP camera. In comparing these two observations, Feldman (1986) concluded that the spatial distribution suggested that CO was a parent molecule directly evaporated from the nucleus, but with CO 25 times more abundant, relative to water, in comet West, than in comet Bradfield. Weaver *et al.* (1981a) and Festou (1984), in analyzing the atomic carbon emissions, which appear in many of the comets observed by IUE, could not reconcile the strength of the C I emissions with the absence of detectable CO emission and Festou (1984) postulated that the source of the carbon may be a relatively abundant, to date unidentified carbon-bearing

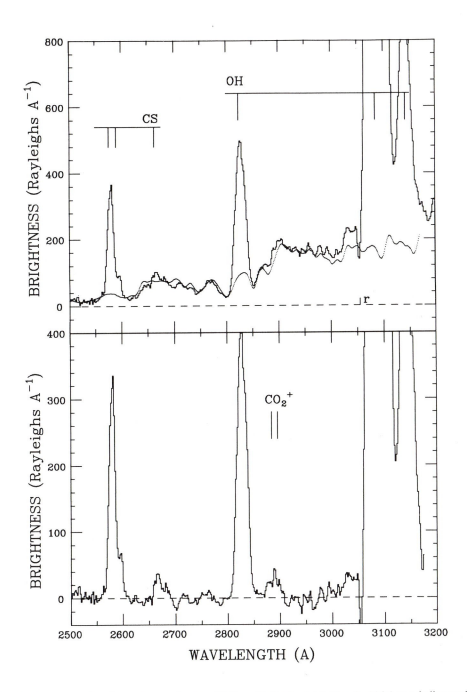

Fig. 4. Long wavelength spectrum of comet Halley obtained on 18 March, 1986 at a heliocentric distance of 0.98 AU. The top panel shows the large contribution of dust reflected solar radiation. The lower panel is the same spectrum but with a normalized, reddened solar spectrum subtracted. The CO_2^+ emission is now apparent.

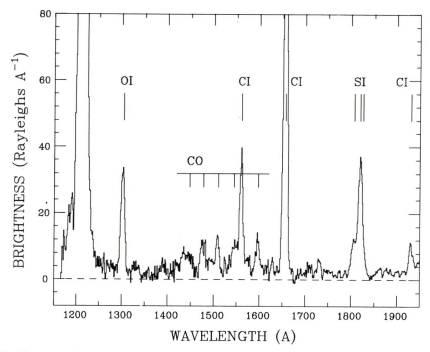

Fig. 5. Short wavelength spectrum of comet Halley from 9 March, 1986, taken shortly after the encounter of Halley by the *Vega 2* spacecraft.

molecule. This problem has been highlighted by a similar report of an over-abundance of atomic carbon in comet Halley by the neutral mass spectrometer experiment on the *Giotto* spacecraft (Krankowsky *et al.*, 1986). Woods *et al.* (1986), on the basis of long-slit UV spectra of Halley obtained from a sounding rocket experiment, have suggested that the C I and CO observations of Halley could be reconciled if an additional collisional source of CO dissociation were present in the inner coma to enhance the production of carbon atoms over that expected on the basis of CO and CO_2 photodissociation alone. This mechanism may be applicable to the C I emission observed in the majority of comets by IUE to date. However, a definitive solution of this problem probably awaits the next generation of more sensitive ultraviolet spectrographs in space.

With the exceptions discussed above, the spectroscopic evidence to date leans toward a unique composition of comets. Variations in relative abundance and inhomogeneties in the surface distribution, ascribed to evolutionary effects or ageing, have been inferred from observed temporal and spatial variations as described below. However, it is important to note here that there is no evidence from IUE to suggest that any of the comets observed to date is not dominated by water ice or that CO_2, NH_3 or CH_4 are ever present in abundances comparable to that of H_2O.

2.2. DETECTION OF NEW SPECIES

The detection of new emitting species in comet spectra has always been a major

goal of cometary spectroscopists. For a long time, the identity of the main constituents of comet heads was not known because the principal emissions were
either in the ultraviolet or the infrared wavelength ranges. At optical wavelengths,
identification of weak features requires a lot of care because of the contamination
of comet spectra by the continuum emission and by extended series of C_2 and NH_2
bands. In the UV, the problem is less severe.

Almost all of the known UV cometary emissions were first identified in the
rocket spectra of comet West (1976 VI) by Feldman and Brune (1976) and Smith,
Stecher and Casswell (1980). The first extensive study of a comet with IUE was
that of comet Bradfield (Figure 1). It was first thought that the intensity of the CO^+
first negative bands, seen previously in comet West, was abnormal (Feldman *et al.*,
1980) until it was realized that CO^+ had not actually been detected. In fact, the
$\Delta v = 0$ sequence of the Mulliken band system of the C_2 molecule had been
observed instead as well as H I Lyman-α in second order (A'Hearn and Feldman,
1980). It is still puzzling that the CO^+ first negative system, so prominent in the
spectrum of comet West, has still not been seen in any comet observed to date by
IUE. The same spectrum showed the presence of the $^1S-^3P$ O I transition at
2972 Å (Festou and Feldman, 1981). This transition, difficult to observe because
of its intrinsic weakness and because of the presence of an underlying continuum
having the same spatial distribution, is of interest as being a tracer of the water
production if no other molecule contributes to the production of ^1S oxygen atoms.
The analogous $^1S-^1D$ (5577 Å) and $^1D-^3P$ (6300, 6364 Å) transitions have been
used as a ground-based monitor of water production in a number of comets
(Spinrad, 1982). The strong dust continuum emission in comet Halley (Figure 4)
prevented the detection of this line in that comet.

Perhaps the most significant discovery made with IUE is that of the S_2
molecule, never previously observed in a celestial object (A'Hearn *et al.*, 1983).
The observation was made possible by the close approach to Earth (0.032 AU) of
comet IRAS–Araki–Alcock on 11 May, 1983. Figure 6 shows two spectra of this
comet, the lower one showing the S_2 bands was taken with the spectrograph
aperture centered on the comet while the upper spectrum had the slit displaced
15″ parallel to the x-axis of the IUE Fine Error Sensor (FES). The short lifetime
of the S_2 molecule ($\sim$500 s) which limits its spatial extent also makes it an ideal
tracer of short-term cometary activity (Feldman *et al.*, 1984), but it has not been
detected in any other comet to date and it may require the much higher spatial
resolution of the Hubble Space Telescope to detect S_2 in future comets. The
implications of the presence of S_2 in cometary ice are not yet fully understood
(A'Hearn and Feldman, 1985), but there is no doubt that the presence of such a
compound in cometary ice is a key for understanding the formation of comets in
the early solar system.

Many IUE cometary spectra show apparent emission features at or slightly
above the non-statistical camera noise background. Some of these features are
probably real, but many are artifacts of the IUE detectors, and it is not possible to
separate the two except by taking another spectrum with two to three times the
exposure time. Unfortunately, this is not usually practical for comet observations.
Low frequency variations in camera background can also distort the relative

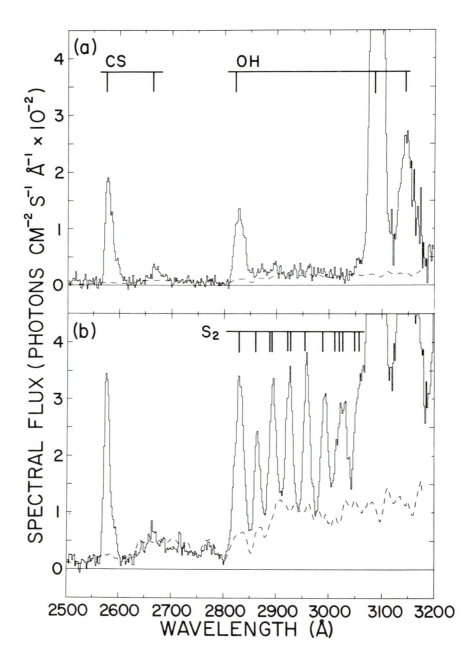

Fig. 6.   Two long wavelength spectra of comet IRAS—Araki—Alcock recorded on 11 May, 1983 near the time of the comet's closest approach to Earth. The lower panel is a 30 min exposure with the comet centered on the aperture and shows the appearance of S_2 emission bands near the cometary nucleus. The upper panel is also a 30 min exposure, but offset 15″ from the center in a direction parallel to the long dimension of the slit. The $10″ \times 15″$ aperture corresponds to 240×350 km^2 projected at the comet.

intensities of weak emission features such as the CO fourth positive bands seen in Figure 5, when compared with observations obtained nearly simultaneously with a photon-counting spectrograph (Woods *et al.*, 1986). It is also interesting to note that the sensitivity to minor constituents in the ultraviolet has decreased since 1980 due to the decrease in solar extreme ultraviolet flux since the maximum of the solar cycle in that year. The principal effect is a decrease in the photodissocia-tion rate, leaving fewer radicals and atoms in the inner coma. There is also an effect due to lower solar H I Lyman-α and O I $\lambda 1302$ fluxes so that the cometary resonance scattering of these lines yields H I and O I emissions relative to OH at 3085 Å which is considerably lower for comet Halley in 1986 than for comet Bradfield in 1980.

2.3. HIGH DISPERSION SPECTROSCOPY

The use of the high dispersion mode of the IUE spectrographs for cometary investigations has concentrated mainly on two species, OH and CS. Schleicher and A'Hearn (1982) have modelled the fluorescence scattering of sunlight by OH radicals as a function of the relative velocity between the Sun and comet. The presence of strong Fraunhofer absorption bands in the solar spectrum which shift in and out of coincidence with individual OH lines produces a fluorescence efficiency for the entire band (see below) which can vary by a factor of five over a complete cometary orbit near perihelion (the well known 'Swings effect'). The fluorescent pumping also produces the population inversion in the ground rota-tional state which gives rise to the Λ-doubled 18 cm transitions routinely observed by radio astronomers, and one of the goals of the high-dispersion analysis is to reconcile the long-standing discrepancy between UV and radio determinations of the OH production rate in a number of comets (Despois *et al.*, 1981). Extensive observations of Halley at different positions in the coma were made to determine the extent of collisional quenching of the upper level.

In addition to OH, A'Hearn *et al.* (1985) have calculated the fluorescence spectrum of OD as a function of heliocentric velocity and have set an upper limit to [OD]/[OH] of 0.004 based on IUE high dispersion spectra of the most favorable case to date, comet Austin. Again, observations of Halley were made at an optimum heliocentric velocity to enhance the OD emission relative to OH and can be expected to lead to an order of magnitude improvement over the value from comet Austin. High dispersion studies of CS, also vigorously pursued with Halley, have previously been used to establish the rotational temperature of the CS molecules in the coma (Jackson *et al.*, 1982).

3. Physical Studies of the Coma

3.1. SPATIAL MAPPING OF THE EMISSIONS

A comet, as seen from the Earth, extends over a few arc minutes at most wavelengths so that the comparatively small size of the IUE slits requires that the spatial distributions of the various emissions can be obtained only by moving the entrance aperture of the IUE spectrographs on the image of the comet in the focal plane of the IUE telescope. Such studies are mandatory if one wants to derive coma abundances from integrated brightness measurements (Festou, 1981a). A detailed study of the water dissociation product distributions in comet Bradfield was performed by Weaver *et al.* (1981a) and confirmed that that comet was dominantly composed of water. The nature of the probable parent of CS, carbon disulfide, was inferred by Jackson *et al.* (1982) using a similar technique, and the results were improved by A'Hearn *et al.* (1983) when comet IRAS—Araki—Alcock passed within 0.032 AU of the Earth in May 1983. The coma of comet Encke was mapped in 1980 (Feldman *et al.*, 1984). It was found that its OH emission presented an asymmetric emission surprisingly mimicking the appearence of the visible fan shaped coma due to the emission of C_2 radicals, and consequently, it was concluded that the outgassing of the nucleus of comet Encke was non-uniform with the main emissive area situated on the pre-perihelion sunlit hemisphere of the nucleus (Whipple and Sekanina, 1979). Finally, by offsetting the IUE slit in the tailward direction in the coma of comet Bradfield, features of OH^+ and CO_2^+ were identified by Festou *et al.* (1982) as being clearly due to ionic species.

Another source of information on the spatial distribution of the UV emissions is obtained by using the spatial information contained in the IUE spectral images made with the $10'' \times 20''$ aperture. The IUE spatial resolution of ≈ 5 arc seconds corresponds to a resolution of a few thousand kilometers at the comet in most cases (with the obvious exceptions of the Earth grazing comets of 1983). It was found that the emissions of CS (Jackson *et al.*, 1982), CO (Feldman, 1986), and S_2 (A'Hearn *et al.*, 1983) as well as the continuum light were sharply peaked at the nucleus, indicating that those species were either escaping directly from the nucleus or were dissociation products of a short lived parent. This is illustrated in the photowrite image reproduced in Figure 7. Emissions of O I (Festou *et al.*,

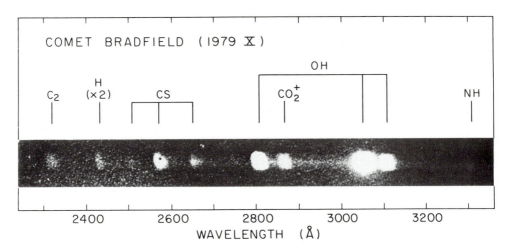

Fig. 7. Photowrite image of the long LWR exposure of comet Bradfield shown in Figure 1. Note that the CS bands and continuum near 2950 Å do not completely fill the $10'' \times 20''$ aperture. The OH bands are saturated in this exposure.

1982), H I, C I, and OH (Weaver *et al.*, 1981a) present much flatter profiles, typical of daughter products.

3.2. Abundance Determinations

If the nature of the complete production chain of a coma species is known, one can derive the production rate of that species. This is why the early observations made with IUE attempted to monitor the evolution of the coma output over the largest possible range of heliocentric distances. Such studies were completed in comets Bradfield (Weaver *et al.*, 1981a), Encke (Feldman *et al.*, 1984), Austin (Feldman *et al.*, 1984), Crommelin (Festou *et al.*, 1985), Giacobini—Zinner (McFadden *et al.*, 1986) and Halley (Festou *et al.*, 1986b). In the case of the first three of these comets, it was found that the gaseous output varied with heliocentric distance approximately as $r^{-3.5}$, contrary to the prediction of the vaporization theory of water (Delsemme, 1982). This contradictory result has not yet been fully explained. The evolution of the OH production with heliocentric distance in comets Bradfield and Halley is shown in Figures 8 and 9. OH production rates are evaluated by the method discussed by Festou (1981b, c). As for the other comets, it was found that the gaseous output of the nucleus did not follow a simple power law, a strong indication that the outgassing of the nucleus was asymmetric with respect to perihelion.

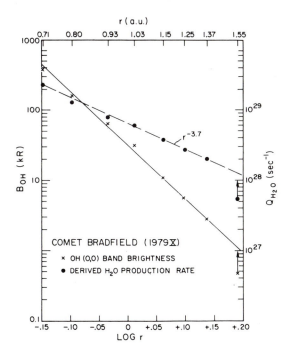

Fig. 8. Brightness of the OH(0, 0) band in comet Bradfield (1979 x) as a function of heliocentric distance. The water production rate, derived using a radial outflow model, is also shown.

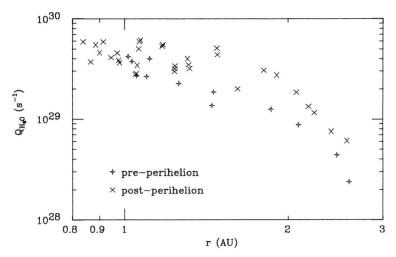

Fig. 9. Water production rate for comet Halley derived from observations made between 12 September, 1985 and 8 July, 1986. In this case, the vectorial model of Festou (1981a) was used to derive the production rate.

Crucial to these studies is the conversion of the observed surface brightness of a given emission feature, B_i, into the average column density of the emitting species along the line-of-sight, $\bar{N}_i$. If B_i is measured in Rayleighs, then

$$B_i = 10^{-6} g_i \bar{N}_i$$

where g_i, the fluorescence efficiency, is a function of both heliocentric velocity and distance. The derived values of $\bar{N}_i$ can then be compared with the various models of the density distribution to derive the production rate of the species. There are several caveats to this procedure. The models largely assume an isotropic, steady-state vaporization rate, a condition which is hardly valid for a number of recently observed comets, most importantly Halley. However, to the extent that the photodissociative lifetimes of most of the species are of the order of one day or longer, the observation along a column path effectively integrates the production of the species over a lifetime and the derived production rate thus represents a time-averaged value. Festou (1981b) has shown that for light daughter species the radial outflow model is inappropriate and that the proper spatial distribution requires application of the dissociation kinematics in the rest frame of the parent molecule. Finally, optical depth effects, both for the absorbed solar radiation and the emitted fluorescent photons, if not accounted for can lead to underestimates of the column densities.

Below 1700 Å, the solar spectrum consists mainly of emission lines. As in the case of OH fluorescence at 3085 Å, the resonance scattering of O I $\lambda 1302$ and C I $\lambda\lambda 1561$, 1657 will depend strongly on the heliocentric velocity of the comet. Feldman *et al.* (1976) first analyzed the 'Swings effect' in the emission g-factors for these lines and a summary of values adopted for the early IUE work was given by Feldman (1982). Additional work on the fluorescence of carbon has been presented by Festou (1984), on sulfur by Azoulay and Festou (1986) and on CS by Butterworth *et al.* (1986). Once again, we note the additional variable of the solar far ultraviolet fluxes as a function of time within the 11-yr solar cycle.

3.3. SHORT-TERM VARIABILITY

The appearance and observation by IUE of comet IRAS—Araki—Alcock at a distance of 0.032 AU from Earth on 11 May, 1983 has been discussed above. The acquisition of the spectra of this comet demonstrated the exceptional skill of the IUE observatory staff as the comet's motion on the sky reached $2°\ hr^{-1}$ (or 2 arc seconds s^{-1}). In order to update the tracking rate of the IUE telescope, every five minutes during the observation the comet image in the focal plane was removed from the spectrograph aperture and placed at the FES reference point where the gyro drift rates were trimmed. Each such operation also automatically recorded the 'FES count rate'. When on 12 May it was found that the UV emissions had all decreased and the newly discovered S_2 all but disappeared, it was realized that the FES count data could be used to construct a visible light curve and that there were strong variations in this curve on the scale of a few hours. In fact, the discovery of S_2 on 11 May occurred at the time of an apparent outburst (Feldman *et al.*, 1984).

The FES count rate measures the amount of light which is diffused by C_2 radicals and dust particles in an area whose size depends on the FES mode being used, typically $\sim 18''$ square. Since this area is small, the response of the FES count rate for the observations made in March 1986. As seen in Figure 11, the OH provided an unexpected opportunity to record nucleus activity changes over timescales on the order of a few hours. Figure 10 shows the variation in FES count rate for the observations made in March 1986. As seen in Fig. 11, the OH emission and UV continuum flux exhibit parallel behavior. This behavior was in agreement with what was observed from the ground at large heliocentric distances (Festou *et al.*, 1986c). During the 18 March, 1986 outburst, detailed in Figure 12, it was possible to record a series of spectra showing that the CO_2^+ coma content

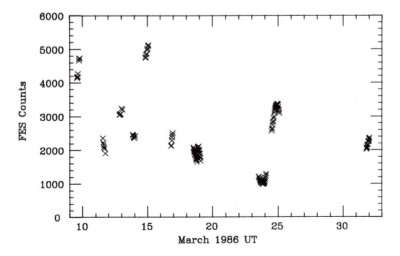

Fig. 10.   FES photometry of comet Halley during March 1986.

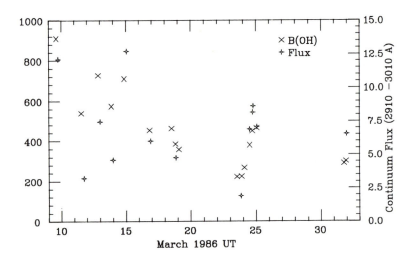

Fig. 11.   Variation of UV fluxes of OH and continuum for the same interval as Figure 10. The UV fluxes are found to follow the FES count rate very closely. The continuum flux is in units of photon cm^{-2} s^{-1} in an effective 10″ × 15″ aperture.

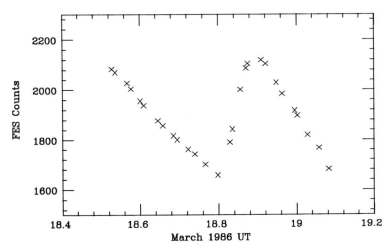

Fig. 12.   Detail of Figure 10, showing the FES light curve during the outburst of 18−19 March, 1986.

was directly related to the occurrence of this outburst. Thus, it was possible to establish from remote observations that the composition of a comet nucleus was not uniform on a scale of the order of that revealed by the *Giotto* images (Keller *et al.*, 1986).

4. Summary

Spectroscopic observations of comets made between 1940 and 1970 were mostly intended to study the emission mechanisms of the then-known cometary radicals. Few quantitative studies had been conducted to derive coma abundances because of the difficulty of obtaining well calibrated measurements on moving objects, almost systematically poorly placed in the sky. The discovery of a huge hydrogen cloud surrounding the nucleus of comet Bennett in April 1970 gave a new impetus to cometary spectroscopy. Rocket observations of comets Kohoutek and West confirmed and extended the results obtained by the OAO—2 and OGO—5 spacecraft and produced the first comprehensive UV spectra of the coma. The availability of IUE made it possible to develop an extensive program of UV observations of comets of various orbital characteristics. A wealth of new and often unexpected results was collected by comparing these different objects. One can deduce from these observations that most comets, if not all, are water dominated and have a similar chemical composition. However, their apparent instantaneous composition may differ from object to object as a result of their evolution under the influence of the solar radiation which continuously exposes new material as the comet traverses the inner part of the solar system. One knows that exotic species such as S_2 exist in their nuclei, molecules which are probably the remnants of the interstellar matter from which comets were formed. The legacy of IUE is to have opened a new means of remote study of the most primitive bodies in our solar system, one which eagerly awaits the next generation of orbiting ultraviolet observatories.

Acknowledgement

The preparation of this review was partially supported by NASA under grant NSG-5393 to The Johns Hopkins University.

References

A'Hearn, M. F. and Feldman, P. D.: 1980, *Astrophys. J. Letters* **242**, L187—L190.

A'Hearn, M. F. and Feldman, P. D.: 1985, 'S$_2$: A Clue to the Origin of Cometary Ice?', in J. Klinger, D. Benest, A. Dollfus, and R. Smoluchowski, (eds.), *Ices in the Solar System*, D. Reidel Publ. Co., Dordrecht, Holland, pp. 463—471.

A'Hearn, M. F., Birch, P. V., Feldman, P. D., and Millis, R. L.: 1985, *Icarus* **64**, 1.

A'Hearn, M. F., Feldman, P. D., and Schleicher, D. G.: 1983, *Astrophys. J. Letters* **274**, L99.

A'Hearn, M. F., Millis, R. L., and Birch, P. V.: 1981, *Astron. J.* **86**, 1559.

A'Hearn, M. F., Schleicher, D. G., Feldman, P. D., Millis, R. L., and Thompson, D. T.: 1984, *Astron. J.* **89**, 579.

A'Hearn, M. F., Schleicher, D. G., and West, R. A.: 1985, *Astrophys. J.* **297**, 826.

Azoulay, G. and Festou, M. C.: 1986, in C. I. Lagerkvist, B. A. Lindblad, H. Lundsted, and H. Rickman (eds.), *Asteroids, Comets, Meteors II*, Uppsala University, pp. 237—277.

Bertaux, J. L., Blamont, J. E., and Festou, M.: 1973, *Astron. Astrophys.* **25**, 415.

Butterworth, P. S., Russell, J. and Jackson, W. M. 1986, in C. I. Lagerkvist, B. A. Lindblad, H. Lundsted, and H. Rickman (eds.), *Asteroids, Comets, Meteors II*, Uppsala University, pp. 269—272.

Carruthers, G. R., Opal, C. B., Page, T. L., Meier, R. R., and Prinz, D. K.: 1974, *Icarus* **23**, 526.

Code, A. D., Houck, T. E., and Lillie, C. F.: 1972, in A. D. Code (ed.), *The Scientific Results from Orbiting Astronomical Observatory (OAO—2)*, NASA SP-310, pp. 109—114.

Delsemme, A. H.: 1982, in L. L. Wilkening, (ed.), *Comets*, University of Arizona Press, Tucson, pp. 85—130.

Despois, D., Gerard, E., Crovisier, J., and Kazès, I.: 1981, *Astron. Astrophys.* **99**, 320.

Feldman, P. D.: 1982, in L. L. Wilkening (ed.), *Comets*, University of Arizona Press, Tucson, pp. 461—479.

Feldman, P. D.: 1983, *Science* **219**, 347.

Feldman, P. D.: 1984, *Adv. Space Res.* **4**, 177.

Feldman, P. D.: 1986, in C. I. Lagerkvist, B. A. Lindblad, H. Lundsted and H. Rickman (eds.), *Asteroids, Comets, Meteors II*, Uppsala University, pp. 263—267.

Feldman, P. D. and Brune, W. H.: 1976, *Astrophys. J. Letters* **209**, L145.

Feldman, P. D. and A'Hearn, M. F.: 1985, 'Ultraviolet Albedo of Cometary Grains', in J. Klinger, D. Benest, A. Dollfus, and R. Smoluchowski (eds.), *Ices in the Solar System*, D. Reidel Publ. Co., Dordrecht, Holland, pp. 453—461.

Feldman, P. D. *et al.*: 1980, *Nature* **286**, 132.

Feldman, P. D., Takacs, P. Z., Fastie, W. G., and Donn, B.: 1974, *Science* **185**, 705.

Feldman, P. D., Opal, C. B., Meier, R. R., and Nicolas, K. R.: 1976, in B. Donn *et al.* (eds.), *The Study of Comets*, NASA SP-393, pp. 773—795.

Feldman, P. D., A'Hearn, M. F., and Millis, R. L.: 1984, *Astrophys. J.* **282**, 799.

Feldman, P. D., A'Hearn, M. F., Schleicher, D. G., Festou, M. C., Wallis, M. K., Burton, W. M., Keller, H. U., and Benvenuti, P.: 1984, *Astron. Astrophys.* **131**, 394.

Feldman, P. D., Weaver, H. A., and Festou, M. C.: 1984, *Icarus* **60**, 455.

Festou, M. C.: 1981a, *Proc. of the Alpach Summer School*, ESA SP-164, pp. 213—219.

Festou, M. C.: 1981b, *Astron Astrophys.* **95**, 69.

Festou, M. C.: 1981c, *Astron Astrophys.* **96**, 52.

Festou, M. C.: 1984, *Adv. Space Res* **4**, 165.

Festou, M. C. and Feldman, P. D.: 1981, *Astron Astrophys.* **103**, 154.

Festou, M. C. *et al.*: 1986a, *Astron Astrophys.* **155**, L17.

Festou, M. C. *et al.*: 1986b, *Nature* **321**, 361.

Festou, M. C. *et al.*: 1986c, *Astron Astrophys.* **169**, 336.

Festou, M. C., Carey, W. C., Evans, A., Wallis, M. K., and Keller, H. U.: 1985, *Astron Astrophys.* **152**, 170.

Festou, M. C., Feldman, P. D., and Weaver, H. A.: 1982, *Astrophys. J.* **256**, 331.

Festou, M. C., Feldman, P. D., Weaver, H. A., and Keller, H. U.: 1982, *Proc. Third IUE Conf.*, ESA SP-176, pp. 445—449.

Festou, M., Jenkins, E. B., Keller, H. U., Barker, E. S., Bertaux, J. L., Drake, J. F., and Upson, W. L.: 1979, *Astrophys. J.* **232**, 318.

Jackson, W. M., Halpern, J., Feldman, P. D., and Rahe, J.: 1982, *Astron Astrophys.* **107**, 385.

Jenkins, E. B. and Wingert, D. W.: 1972, *Astrophys. J.* **174**, 697.

Keller, H. U. and Lillie, C. F.: 1974, *Astron Astrophys.* **34**, 187.

Keller, H. U. *et al.*: 1986, *Nature* **321**, 320.

Krankowsky, D. *et al.*: 1986, *Nature* **321**, 326.

Marsden, B. G.: 1986, Catalogue of Cometary orbits, 5th Edition, Smithsonian Astrophysical Observatory, Cambridge, MA.

McFadden, L. A. *et al.*: 1986, submitted to *Icarus*.

Opal, C. B., Carruthers, G. R., Prinz, D. K., and Meier, R. R.: 1974, *Science* **185**, 702.

Opal, C. B. and Carruthers, G. R.: 1977, *Astrophys. J.* **211**, 294.

Schleicher, D. G. and A'Hearn, M. F.: 1982, *Astrophys. J.* **258**, 864.

Smith, A. M., Stecher, T. P. and Casswell, L.: 1980, *Astrophys. J.* **242**, 402.

Spinrad, H.: 1982, *Pub. A. S. P.* **94**, 1008.

Weaver, H. A., Feldman, P. D., Festou, M. C., and A'Hearn, M. F.: 1981a, *Astrophys. J.* **251**, 809.

Weaver, H. A., Feldman, P. D., Festou, M. C., A'Hearn, M. F., and Keller, H. U.: 1981b, *Icarus* **47**, 449.

Whipple, F. L.: 1950, *Astrophys. J.* **111**, 375.

Whipple, F. L.: 1951, *Astrophys. J.* **113**, 464.

Whipple, F. L. and Sekanina, Z.: 1979, *Astron. J.* **84**, 1894.

Woods, T. N., Feldman, P. D., Dymond, K. F., and Sahnow, D. J.: 1986, *Adv. Space Res.* **5**, 289.

PART III

STARS

Edited by C. DE JAGER and J. L. LINSKY

UV STELLAR SPECTRAL CLASSIFICATION

A. HECK

C.D.S., Observatoire Astronomique, Strasbourg, France

1. Introduction

Stellar observers in the ultraviolet tend to use the MK classification system as a spectral reference also in this spectral range, implying that it characterizes the whole spectrum, even if it is defined only from the visible range. However, an analysis of about two thousand spectra collected by the S2/68 experiment on board the TD1 satellite showed that stars which are spectrally normal in the visible range do not always behave normally in the ultraviolet range (see Cucchiaro *et al.*, 1976, 1977, 1978a, b, 1979, and 1980).

About 10% of the survey stars were normal in the visible and abnormal in the UV or vice versa as illustrated by Figure 1. An optical programme conducted to

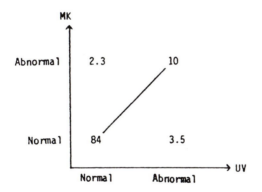

Fig. 1.   Percentages of coincidences between MK classifications and TD1 spectral types for 1904 stars (from Jaschek and Jaschek, 1984).

re-classify some of the stars (Jaschek and Jaschek, 1980) confirmed significant discrepancies and thus the lack of full coincidence between the UV and visible ranges. Jaschek and Jaschek (1982) showed also that this was not resulting from the use of a different spectral resolution. About 75% of the standards they examined agreed in both systems, with about 10% of large discrepancies. Examples will be given in Section 3 once introduced the UV classification symbolism.

Lamers *et al.* (1979) studied at a much higher resolution (S59 experiment on board the TD1 satellite) a hundred stars classified as normal in the visible, and found that about 15% were abnormal in the UV. Consequently, MK spectral classifications defined from the visible range cannot be extrapolated to the UV range without precaution.

Y. Kondo (ed.), Scientific Accomplishments of the IUE, pp. 121–137.
© 1987 *by D. Reidel Publishing Company.*

Other pre-IUE works related to UV spectral classification were reported by Henize *et al.* (1975) and by Panek and Savage (1976), respectively based on Skylab S—019 experiment spectra and on OAO—2 data. Both groups recognized the variation with temperature and luminosity of the Si IV (1400 Å) and C IV (1550 Å) resonance lines in OB stars. Henize *et al.* (1981) proposed a classification system for O—B2 stars by using the log (Si IV/C IV) versus C IV diagram.

Stellar spectral classifications are more than taxonomical exercises aiming just at labelling stars and putting them into boxes by comparison with standards. They are used for describing fundamental physical parameters in the outer stellar atmospheres, for discriminating peculiar objects, and for other subsidiary applications like distance determinations, interstellar extinction and population synthesis studies.

It is important to bear in mind that the classification systems are built independently of the stellar physics in the sense that they are defined completely by spectral features in a given wavelength range in selected standards (see e.g., Jaschek, 1979; Morgan, 1984). If the schemes are based on a sufficiently large number of objects, it appears easily that they are intimately linked with the physics, but not necessarily of the same stellar layers if they refer to different wavelength ranges. Consequently, the discrepancies reported between the MK system and the UV frames are not too surprising.

2. IUE Versus TD1

In view of these discrepancies, a UV stellar classification programme has been initiated in 1978 with the specific goal to define, from IUE low-resolution spectra, smooth spectral sequences proper to the UV and describing the stellar behavior in the UV while staying as far as possible in accordance with the MK scheme in the visible.

This undertaking was strongly supported by the participants in the VILSPA workshop on UV Stellar Classification (proceedings published as ESA SP-182, 1982) who stressed in their final resolution its importance not only for stellar and extragalactic astronomy, but also for the preparation of spectroscopic programmes for future space missions.

Compared with TD1, IUE provides a significant improvement because it offers a better resolution (even at low dispersion: ≈ 7 Å against ≈ 36 Å) and a larger spectral range (≈ 1150—3200 Å against ≈ 1250—2550 Å). It covers a larger variety of spectral types and reaches fainter magnitudes (see, for instance, Heck, 1982, 1984).

TD1, however, carried out a survey and the S2/68 data have the advantage to represent a magnitude-limited unbiased sample of bright stars, whereas IUE is pointed only to preselected targets from accepted observing proposals, thereby providing a highly biased sample. The number of stars with abnormal spectra is therefore about twice as large as the number of stars with normal spectra. Moreover, the normal stars constitute also an heterogeneous sample because they were observed mainly for comparison purposes. In spite of complementary observations carried out to bridge the gaps, there remained some bias, essentially as an overrepresentation of hot stars compare to cooler ones.

3. The IUE Classification System

The first volume of an IUE Low-Dispersion Reference Atlas has been produced (Heck *et al.*, 1984a), together with reference sequences and standard stars. The considerable underlying classification work has been carried out following a classical morphological approach (Jaschek and Jaschek, 1984) and it essentially confirmed that there is no one-to-one correspondence between the UV and visible ranges.

To avoid confusion with the MK system, a specific set of designations has been introduced (Table I), but it has been chosen similar to the HD—MK notation to

TABLE I

Luminosity designations used for the UV spectral types (from Heck *et al.*, 1984a)

Designation	Name	Rough MK equivalent
d	dwarf	V
g^-	subgiant	IV
g	giant	III
g^+	bright giant	II
s^-	supergiant	Ib
s^+	bright supergiant	Ia

render comparisons easier. The finer subdivisions ($^+$ and $^-$ signs) were mostly used for B-type stars since their number in the sample at hand was large enough to allow for a finer classification. Jaschek and Jaschek (1984) showed that spectral types and luminosities can be assigned from UV criteria alone. The selected standard stars are gathered in Table II, while the lines used for classification of O- and B-type stars are listed in Table III to V. The corresponding spectral sequences are displayed in Figures 2 and 3, while Figures 4 and 5 illustrate luminosity effects in B0 and B2-type stars.

As for A- and F-type stars, the most important spectral-class criterion is the slope of the continuum towards shorter wavelengths. Isolated features that are also spectral-type dependent are located at 1850, 1933, 2670, 2800 (2795 + 2803), and 2855 Å. Luminosity criteria are more difficult to apply in this range. One sensitive feature is at 1850 Å.

Table VI collects a few illustrations of the discrepancies mentioned earlier. If the MK system were fully applicable to the UV range, one should be able to transit smoothly from one UV spectrum of an MK standard to the next, as it can be done in the visible range. However, a comparison of the IUE spectrum of HD 46769 (B8 Ib) with those of HD 125888 (B6 Ib) and HD 202850 (B9 Iab),

A. HECK

TABLE II

Standard stars (HD numbers; from Heck et al., 1984a)

	s	g	d
O3			303308
4			46223
5	210839		46150
7	152248		165052
8			46056
9	123008	209481	214680
B0	122879	75821	36512
0.5	152667		55857
1	2905	173502	144470
2	41117	51283	148605
2.5		207330	
3	53138		32630
5	167838	22928	83754
6		23302	90994
7	183143		
8	166937	23850	23324
9	21291	149212	108767
A0	87737		103287
1			29646
2	197345		48250
3	210221		17138
4			5448
5	59612	159561	116842
7	148743	28319	87696
F0	36673	12311	27176
2	74180	89025	40136
3		99028	128167
5	20902		
6			173667
8	8890		
9			102870
G1.5			10307
5			20630

reveals a lack of continuity in the behavior of the features, noticeably around 1900 Å (see, for instance, the representations in the atlas by Wu *et al.*, 1983).

Another example is furnished by the stars HD 90994, 29335, and 46769 classified respectively as B6 V, B7 V and B8 Ib in the 3500—4800 Å region and whose ultraviolet spectra look very much alike (*d* B6). Other examples are furnished by HD 147547 and 12311 (respectively A9 III and F0 V in the visible, and *g* F0 in the UV), and by HD 6961 and HD 102647 (respectively A7 V and A3 V in the visible, and *d* A5 in the UV).

TABLE III

Lines characteristic of O-type stars (from Heck *et al.*, 1984a)

Wavelength (Å)	Identif.	Comments
1175	C III	Increases from O4 toward B1, where it has its maximum.
1255		Decreases from O3 to O7.
1300	Si III	Decreases from O3 to O9.
1333		Decreases from O3 to O9, where it disappears.
1371	O V	Decreases from O3 to O7, where it disappears.
1394 / 1403	Si IV	Becomes well visible from O7 onwards. Has strong positive luminosity effect.
1428	C III	Increases from O4 toward B1, where it has its maximum.
1453		Has its maximum at O4. Disappears at B0.
1548	C IV	Decreases from O3 toward B-type. Disappears at B2. Has a positive luminosity effect.
1718	N IV	Has a maximum at O4. At O7, is blended with 1721 Al III present in B-type stars.

It is important to stress that such discrepancies do not arise because of flaws in the MK classification system itself. They illustrate the point — not always remembered — that, although the ordering of the spectra obtainable from different wavelength regions coincide in a large number of cases, there are some definite exceptions.

It is however not feasible to provide, as with TD1, statistical figures on the match between the MK and IUE classification systems. Indeed, IUE did not carry out a survey and thus the sample at hand is not exhaustive. Any attempt to carry out statistics of this kind from the IUE Low-Dispersion Reference Atlas would be even more hazardous since the contents of any atlas result from selecting procedures.

4. The IUE Low-Dispersion Spectra Reference Atlas and Catalogue

The first volume of this work has already been published, providing the astronomical community in general, and the users of IUE images in particular, with reference sequences of ultraviolet low-resolution spectra.

TABLE IV

Lines characteristic of B-type stars (from Heck *et al.*, 1984a)

Wavelength (Å)	Identif.	Comments
1175	C III	Has its maximum at B1. Disappears at B6 in the L_{alpha} wing.
1216	L_{alpha}	When not affected by interstellar or circumstellar effects, has a half width increasing monotically from 10 Å at O9 to 100 Å at B8.
1265	Si II	Visible from B1 on. Increases toward B9.
1300	Si	Blend of Si II and Si III. In later types, Si II predominates. Grows between B2 and B8, where its usefulness disappears because of L_{alpha}.
1336	C II	Has its maximum at B8.
1400	Si IV	Blend of 1394 and 1403. Has a maximum at B1. Disappears at about B8.
1465		Appears at about B2. Increases toward later types.
1548	C IV	Strong in O-type stars. Disappears at B2 in dwarfs.

The body of the Atlas consists of the presentation, on the left-hand pages, of 5 Å-step flux tables for 229 selected spectra, essentially of stars exhibiting a normal behavior in the ultraviolet. A few peculiar objects have however been included for illustration of typical abnormalities. The right-hand pages of the Atlas display the corresponding composite graphs for the full IUE range. These graphs have 2 Å steps.

The flux (y) scale has been optimised to give the best display of the spectra. In some cases, in order to present fully the important details at short wavelengths, the spectrum has been truncated on the right-hand side, especially when the gradient was too steep towards the long-wavelength end. In each case however, the flux tables are complete and no valuable information has been omitted from the graphs.

Adjacent to the graphs are given the identifications of the IUE images used, as well as basic astronomical data retrieved from the SIMBAD data base at the Strasbourg Data Centre (Figure 6). Finally a set of transparencies has been provided for the most representative stars, to be used as overlays for direct comparison with the spectra and easy illustration of the spectral sequences.

TABLE V

Lines (Å) showing a luminosity effect in B-type stars (from Heck *et al.*, 1984a)

1400	Blend of 1394 and 1403 of Si IV. Has a pronounced positive luminosity effect. Can be used for supergiants up to B5.
1548	C IV. Has a pronounced positive luminosity effect. Disappears in dwarfs at B2. Thus its appearance in middle B-type stars indicates a supergiant.
1608 1622 1629 1640	Fe II, M. 8. Strongest line is 1608. All are enhanced in B-type supergiants and show little variation with temperature.
1723	Al II, M. 6 (blend of three lines). Has a positive luminosity effect.
1855	Al III, M. 1. Blended with 1862 Al II + Al III.
1891	Fe III, M. 52 + 53.
1926	Fe III, M. 57 + 34.
1967	Fe III. Blend of several lines.

The catalogue (Heck *et al.*, 1984b) gathers the same spectra on magnetic tape (available from the Strasbourg Data Center) and lists the fluxes by 2 Å-steps for the interval 1150–3200 Å. For each object, the first record gives the identification of the star and a value recommended as maximum flux for graphically representing the spectrum.

A statistical confirmation of the morphological classification procedure is described in the following section.

5. Statistical Confirmation of the IUE Classification System

For the reasons explained earlier, the only way to confirm independently the correctness of the UV classification frame introduced in the Atlas was to use data from the same wavelength range with a methodology, independent of any bias that might arise a priori from the visible and the UV ranges.

Therefore statistical algorithms working in a multidimensional parametric space were applied to variables expressing, as objectively as possible, the information contained in the continuum and the spectral features (Heck *et al.*, 1986a). This was done through, on the one hand, an asymmetry coefficient describing the continuum shape and empirically corrected for the interstellar reddening, and, on the other hand, the intensities of sixty objectively selected lines (including all the lines retained as discriminators in the atlas).

TABLE VI

A few examples of discrepancies between MK and IUE spectral classifications (from Heck *et al.*, 1984b)

Identifier		IUE	Sp.	MK Sp.
HD	6619	*d*	A5	A1 V, Am
HD	12311	*g*	F0	F0 V
HD	23480	*d*	B6	B6 IV (e)
HD	23753	*g*	B8	B8 V n
HD	27290	*d*	F1	F4 III
HD	36879		O7 p	O7.5 III
HD	40932	*d*	A4	Am SB, VB
HD	46769	*d*	B6	B8 Ib
HD	47755	*d*	B3	B9
HD	50138		B9 pec	B6 III
HD	89025	*g*	F2	F0 III
HD	109387	*d*+	B7	B6 III p
HD	150898	*g*+	B0:	B0.5 Ia
HD	156208	*g*	A1	A2 V
HD	161817	*d*:	A3 p	A2 VI
HD	163181	*s*	B1	B1 Ia pe
HD	165024	*g*+	B1	B2 Ib
HD	170153	*d*	F6	F8 Vb, vw
HD	202444	*d*:	F3	F1 IV
HD	210424	*d*	B7	B7 III
HD	212571	*d*	B1	B1 III—IV e
HD	216701	*d*	A5	A7 III
BD	+60°2522	*s*	O7	O6.5 III ef

These line intensities were weighted in a way that has been called the 'Variable Procrustean Bed (VPB) method' because, contrary to a standard weighting where a given variable is weighted in the same way for all the individuals of a sample, they were weighted according to the asymmetry coefficient which varies with the star at hand.

The stellar material was essentially the same as that of the Atlas. The SPAD algorithm applied to the set of the variables consisted of a Principal Component Analysis and a Cluster Analysis (Lebart and Morineau, 1982; Lebart *et al.*, 1982). The UV classification symbolism introduced in the Atlas served as a reference to study the composition of the groups of stars resulting from the Cluster Analysis.

The results obtained were excellent for the relatively small size of the sample used as compared with the number of variables needed to fully describe the information contained in the spectra. The correctness of the morphological classification has been confirmed in the sense that the homogeneity obtained in the groups was quite satisfactory, as was the discrimination for spectral types and luminosity classes. This was especially true for the early spectral types which were well represented in the sample used for this study.

Under its generalized form, the asymmetry coefficient turned out to be an excellent discriminating factor (corresponding to a sorting on the continuum

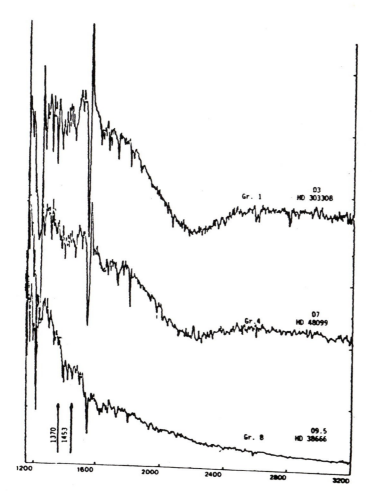

Fig. 2.   Illustration of the main IUE sequence in O-type stars (from Jaschek and Jaschek, 1984). The abcissa gives the wavelengths in Å and the ordinate, the fluxes in arbitrary units. Characteristic features are arrowed.

shape) for both reddened and unreddened normal stars. The individual classifications resulting from the morphological approach used for the atlas were confirmed, and ipso facto the discrepancies with the MK classifications in the visible range. The standard stars could be found in the neighbourhood of the barycenters of the groups.

6. Peculiar Stars

The preparation of the second volume of the IUE Low-Resolution Spectra Reference Atlas, gathering stars with anomalous spectra (see i.a. Egret *et al.*, 1985), is raising a number of difficulties implying that it is impossible to translate

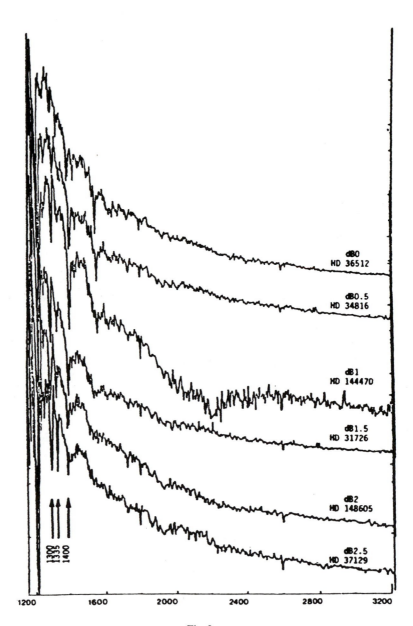

Fig. 3a.

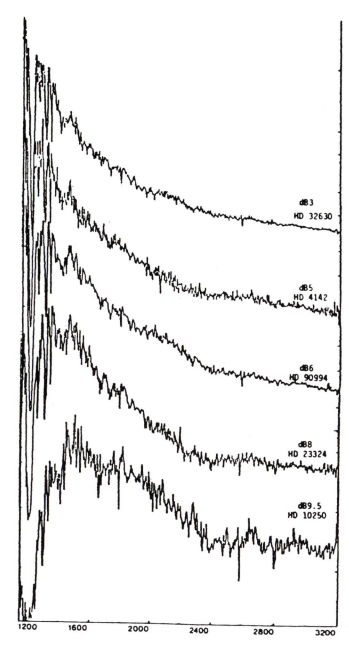

Fig. 3b.

Fig. 3a, b.   Illustration of the main IUE sequence in B-type stars (from Jaschek and Jaschek, 1984).
Comments as in Figure 2.

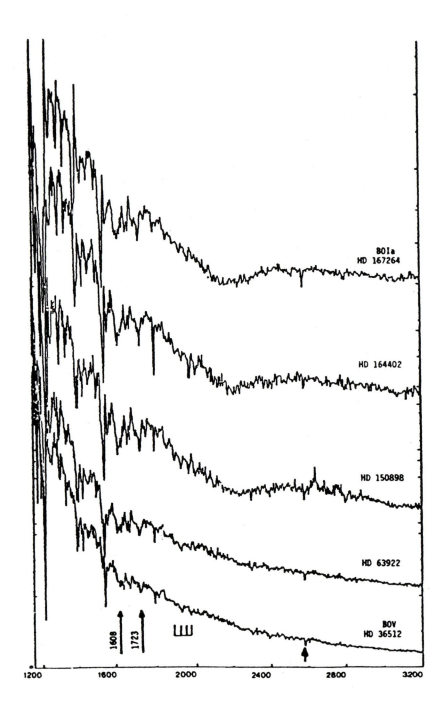

Fig. 4. Illustration of luminosity effects in B0-type stars (from Jaschek and Jaschek, 1984).
Comments as in Figure 2.

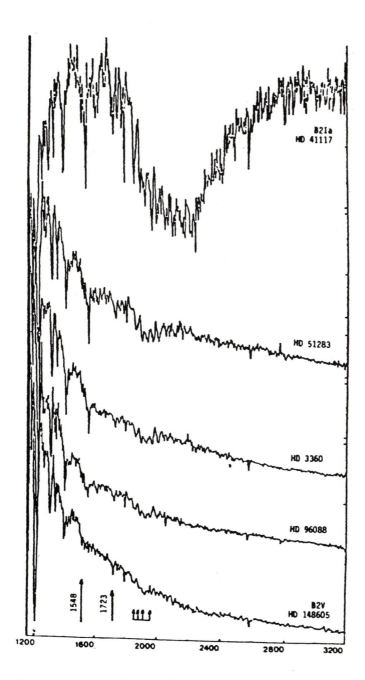

Fig. 5. Illustration of luminosity effects in B2-type stars (from Jaschek and Jaschek, 1984).
Comments as in Figure 2.

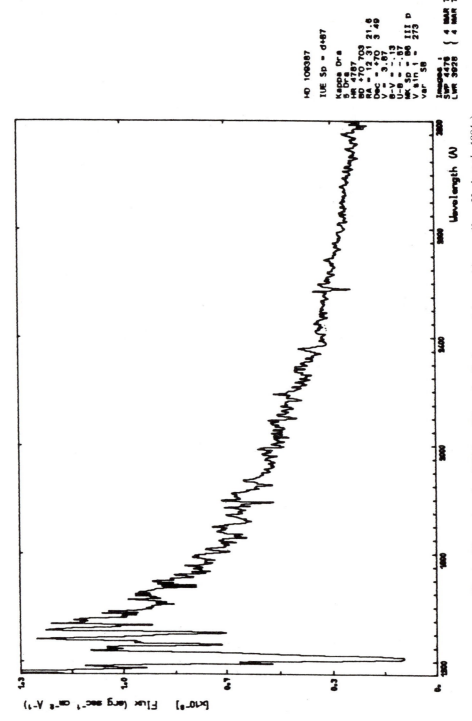

Fig. 6.   Illustration of an IUE Low-Dispersion Reference Atlas graphic page (from Heck *et al.*, 1984a).

all peculiarity groups of the 3500—4800 Å region into similarly structured peculiarity groups in the 1200—3200 Å region (Heck *et al.*, 1986b).

Thus the Hg—Mn subgroup of Ap stars is characterized by strong Ga lines in the visual region. The converse is however not true and the group of ultraviolet Gallium stars comprises a wide variety of objects besides Hg—Mn stars (Jaschek and Jaschek, 1987).

The group of Horizontal-Branch stars is heterogeneous in the ultraviolet (Jaschek *et al.*, 1985). Besides the metal weakness, which is shared by all members of the group, half of the spectra exhibit a depression around 1600 Å, whereas the others do not exhibit it: This discontinuity exists however for the lambda Boo group (Baschek *et al.*, 1985).

The group of early-type hydrogen-deficient stars is also heterogeneous in the ultraviolet: some stars exhibit strong C II lines whereas others do not. On the other hand, some stars called normal in the visual region do show enhanced C II.

In conclusion, peculiar stars bring up the same phenomenon as normal stars: classifications based upon one spectral region have no strict equivalence in another spectral region.

7. Other Work from IUE Data

Wu *et al.* (1981, 1983) carried out a programme to obtain IUE low-dispersion spectra of stars with well-determined MK spectral classifications (and thus of many primary MK standards) covering spectral types from O3 to M4 and luminosity classes from V to Ia. The stars had also to be sufficiently faint and as unreddened as possible to obtain high-quality trailed spectra. The corresponding tape is available from NSSDC at Goddard Space Flight Center.

Nandy (1985) investigated the measurements of major stellar features in the UV over a range of luminosity and spectral types for a large number of B stars in our galaxy and the Magellanic Clouds. He found that an accurate stellar classification from the strengths of these features is severely limited by large observational scatter and possible abundance effects as seen from the LMC and SMC stars. It is possible that the parameters such as stellar wind and metallicity are important.

Walborn and Panek (1984a, b) drew attention to Si IV as a luminosity criterion for O-type stars from a sample of IUE high-resolution spectra. The resulting frame allows also to identify some abnormalities in the CNO features among the ON and OC stars. Walborn *et al.* (1985) produced an IUE high-resolution atlas of O-type spectra (1200—1900 Å range) based on these discriminative features.

Rountree *et al.* (1985) announced a program of spectral classification of B stars, using high resolution IUE spectra, which will both test the applicability of the MK standard sequences and establish new classification criteria in the ultraviolet. See also Rountree (1985).

Finally, a number of more specific works have been produced. Cacciari (1985) studied the UV fluxes of Population II stars from IUE low-resolution spectra. Baschek *et al.* (1984) offered a discrimination of lambda Bootis stars based on strong identified (C I 1657 Å and C I 1931 Å) and unidentified (1600 Å and

3040 Å) absorption features. Jaschek *et al.* (1985) described similar features in Horizontal—Branch stars.

Acknowledgement

Valuable comments by C. Jaschek, M. Jaschek and C. C. Wu duly acknowledged.

References

Baschek, B., Heck, A., Jaschek, C., Jaschek, M., Köppen, J., Scholz, M., and Wehrse, R.: 1984, *Astron. Astrophys.* **131**, 378.

Cacciari, C.: 1985, *Astron. Astrophys. Suppl.* **61**, 407.

Cucchiaro, A., Jaschek, M., Jaschek, C., and Macau-Hercot, D.: 1976, *Astron. Astrophys. Suppl.* **26**, 241.

Cucchiaro, A., Jaschek, M., Jaschek, C., and Macau-Hercot, D.: 1977, *Astron Astrophys. Suppl.* **30**, 71.

Cucchiaro, A., Jaschek, M., Jaschek, C., and Macau-Hercot, D.: 1978a, *Astron. Astrophys. Suppl.* **33**, 15.

Cucchiaro, A., Jaschek, M., and Jaschek, C.: 1978b, *An Atlas of Ultraviolet Stellar Spectra*, Liège and Strasbourg.

Cucchiaro, A., Jaschek, M., Jaschek, C., and Macau-Hercot, D.: 1979, *Astron. Astrophys. Suppl.* **35**, 75.

Cucchiaro, A., Jaschek, M., Jaschek, C., and Macau-Hercot, D.: 1980, *Astron. Astrophys. Suppl.* **40**, 207.

Egret, D., Hassall, B. J. M., Heck, A., Jaschek, C., Jaschek, M., and Talavera, A.: 1985, in M. Jaschek and P. C. Keenan (eds.), *Cool Stars with Excesses of Heavy Elements*, D. Reidel Publ. Co., Dordrecht, Holland, p. 47.

Heck, A.: 1982, in A. Hech and B. Battrick (eds.), *Ultraviolet Stellar Classification*, ESA SP-182, p. 55.

Heck, A.: 1984, *IUE NASA Newsl.* **24**, 215.

Heck, A., Egret, D., Jaschek, M., and Jaschek, C.: 1984a, *IUE Low-Resolution Spectra Reference Atlas: Part 1. Normal Stars*, ESA SP-1052.

Heck, A., Egret, D., Jaschek, M., and Jaschek, C.: 1984b, *Astron. Astrophys. Suppl.* **57**, 213.

Heck, A., Egret, D., Nobelis, Ph., and Turlot, J. C.: 1986a, *Astrophys. Space Sci.* **120**, 223.

Heck, A., Egret, D., Hassall, B. J. M., Jaschek, C., Jaschek, M., and Talavera, A.: b, in E. Rolfe (ed.), *New Insights in Astrophysics — Eight Years of UV Astronomy with IUE*, ESA SP-263, 661.

Henize, K. G., Wray, J. D., Parsons, S. B., Benedict, G. F., Bruhweiler, F. C., Rybsky, P. M., and O'Callaghan, F. G.: 1975, *Astrophys. J.* **199**, L119.

Henize, K. G., Wray, J. D., and Parsons, S. B.: 1981, *Astron. J.* **86**, 1658.

Jaschek, C.: 1979, in D. Ballereau (ed.), *Classification Spectrale*, Ecole de Goutelas, Obs. Meudon.

Jaschek, M. and Jaschek, C.: 1980, *Astron. Astrophys. Suppl.* **42**, 115.

Jaschek, M. and Jaschek, C.: 1982, in A. Heck and B. Battrick (eds.), *Ultraviolet Stellar Classification*, ESA SP-182, p. 9.

Jaschek, M. and Jaschek, C.: 1984, in R. F. Garrison (ed.), *The MK Process and Stellar Classification*, David Dunlap Obs., p. 290.

Jaschek, M. and Jaschek, C.: 1987, *Astron. Astrophys.* **171**, 380.

Jaschek, M., Baschek, B., Jaschek, C., and Heck, A.: 1985, *Astron. Astrophys.* **152**, 439.

Lamers, H. J. G. L. M., Faraggiana, R., and Burger, M.: 1979, *Astron. Astrophys.* **79**, 230.

Lebart, L. and Morineau, A.: 1982, *Système portable pour l'analyse de données*, C.E.S.I.A., Paris.

Lebart, L., Morineau, A., and Fénelon, J. P.: 1982, *Traitement des données statistiques*, Dunod, Paris.

Morgan, W. W.: 1984, in R. F. Garrison (ed.), *The MK Process and Stellar Classification*, David Dunlap Obs., p. 18.

Nandy, K.: 1985, Meeting on 'Difficulties of Extrapolating the MK System to Other Spectral Ranges', *IAU Comm.* **45**, New Delhi.

Panek, R. J. and Savage, B. D.: 1976, *Astrophys. J.* **206**, 167.

Rountree, J. C.: 1985, Meeting on 'Difficulties of Extrapolating the MK System to Other Spectral Ranges', *IAU Comm.* **45**, New Delhi.

Rountree, J. C., Sonneborn, G., and Panek, R. J.: 1985, in D. S. Hayes, L. E. Pasinetti, and A. G. D. Philip (eds.), *Calibration of Fundamental Stellar Quantities*, D. Reidel Publ. Co., Dordrecht, Holland, p. 411.

Walborn, N. R. and Panek, R. J.: 1984a, in R. F. Garrison (ed.), *The MK Process and Stellar Classification*, David Dunlap Obs., p. 305.

Walborn, N. R. and Panek, R. J.: 1984b, *Astrophys. J.* **280**, L27.

Walborn, N. R., Nichols-Bohlin, J., and Panek, R. J.: 1985, *International Ultraviolet Explorer Atlas of O-type Spectra from 1200 to 1900 Å*, NASA Ref. Publ. 1155.

Wu, C. C., Boggess, A., Holm, A. V., Schiffer III, F. H., and Turnrose, B. E.: 1981, *NASA IUE Newsl.* **14**, 2.

Wu, C. C., Ake, T. B., Boggess, A., Bohlin, R. C., Imhoff, C. L., Holm, A. V., Levay, Z. G., Panek, R. J., Schiffer III, and F. H., Turnrose, B. E.: 1983, *NASA IUE Newsl.* **22**.

WINDS FROM HOT YOUNG STARS

J. P. CASSINELLI

Department of Astronomy and Washburn Observatory,
University of Wisconsin, Madison, WI, U.S.A.

and

H. J. G. L. M. LAMERS

SRON Laboratory for Space Research and
Sonnenborgh Observatory, Utrecht, The Netherlands

1. Introduction

Early observations of the UV spectra of luminous early-type stars showed, totally unexpectedly, that the stars have rapidly expanding atmospheres and undergo severe mass loss. Later observations with the Copernicus satellite revealed a second mystery: the winds are superionized. These two facts completely upset earlier ideas that these stars would not be very interesting as far as atmospheric dynamical effects are concerned. They also showed that our concept about the evolution of massive stars, with its assumption of conservation of mass, was drastically wrong.

The IUE observations have extended our knowledge to a much larger number of stars. In some cases this has led to a development of explanations, such as our better understanding of narrow absorption features, while in other cases the observations have produced more riddles, for example, the differences in UV spectra among some stars with very similar optical spectra. Ultraviolet observations have turned the study of early-type stars into a very lively subject. Prior to the ultraviolet era the stars were primarily of interest because of spectroscopic non-LTE effects. Now the stars are at the focus of theoretical research on stellar evolution, gas dynamics, and ´of observational studies from X-ray through radio spectral regions. It is also a field with many controversies as alternative explanations and models are proposed, criticized, and rejected or improved.

In this paper we present the status of topics of particular interest to us, dealing primarily with stellar winds and mass loss. Hundreds of early-type stars have been observed with IUE. Because of this great increase in available data, it has become possible to recognize general properties of mass loss in the HR diagram. We will not be able to confine ourselves to ultraviolet results alone. During the IUE years, crucial new data have become available also from ground-based observatories at radio wavelength and satellite observations from the infrared through the high energies. Radio observations from the Very Large Array (VLA) have been especially valuable for deriving mass loss rates and for discovering non-thermal emission processes. X-ray observations, especially those made with the Einstein satellite, have led to some understanding of the super-ionization seen in UV spectra.

Y. Kondo (ed.), Scientific Accomplishments of the IUE, pp. 139—155.
© 1987 *by D. Reidel Publishing Company.*

2. Boundaries and Limits in the HR Diagram for Hot Stars

Over the past decade, several zones and limits in the HR diagram for early-type stars have been determined from observational studies. The recognition of discrete zones is of great importance for the development and testing of theoretical models. For a model to be acceptable, it must not only be able to explain the presence of some anomaly at one point in the HR diagram, but it must also explain the reason why the anomaly persists up to the limits of the zone and no farther.

Figure 1 shows a theoretical HR diagram with recently recognized zones and limits identified. The Figure is also useful for defining the topics to be covered in this review as opposed to topics covered in other chapters in this book. Early-type stars are usually considered to be those with effective temperatures between roughly 10^5 and 8000 K, or spectral type A and earlier. There is a wide variety of stars in this effective temperature range. The central stars of planetary nebulae are a reasonably well-defined class of low mass stars that often show strong emission

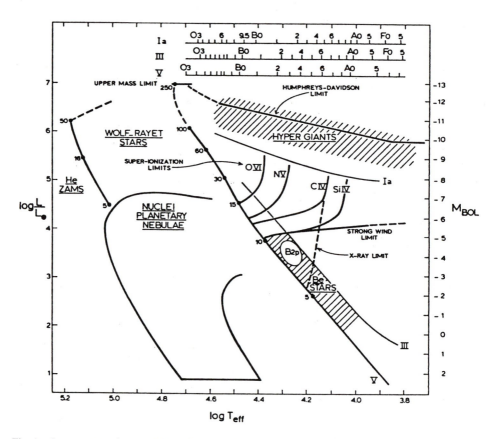

Fig. 1. Important regions and boundaries for hot stars in the HR diagram. The spectral type scales for different luminosity classes are indicated in the top part. The ZAMS for normal stars and for helium stars are shown with the corresponding masses. The regions and boundaries are discussed in Section 2.

lines in their optical and ultraviolet spectra. The Wolf—Rayet class consists of highly evolved stars that are noted for their extremely massive winds. Closer to the hydrogen burning main sequence are the Be stars, which have emission lines often attributed to effects of their rapid rotation speeds. These three classes of stars — the central stars of planetary nebulae, the Wolf—Rayet stars, and the Be stars — are discussed elsewhere in this volume. The rest of the objects in Figure 1 will be discussed to a greater or lesser degree in this paper.

Several interesting zones and limits can still be recognized in our 'sample'. Starting at the highest luminosities and proceeding toward fainter and cooler stars, we see the following zones and limits.

(A) There is a now well-determined observational upper limit in the HR diagram. A portion of this, the Humphreys—Davidson limit (Humphreys and Davidson, 1979) is of particular interest because it lies at a luminosity that is somewhat lower than the luminosity of the most massive stars on the main sequence. Stellar evolutionary tracks of the very massive main sequence stars computed as if the stars had no mass loss, would take the stars well above the Humphreys—Davidson limit because the tracks are nearly iso-luminosity lines. The presence of the limit can be considered observational evidence that mass loss occurs at a rapid enough rate to significantly affect stellar evolution (e.g., Chiosi and Maeder, 1986).

The luminosity (or mass) of the stars at the top of the main sequence has been a topic of great interest in the past few years because of the arguments that R136a, the central object of the 30 Doradus Nebula in the Large Magellanic Cloud, contains a star with a mass of the order 1000 $M_\odot$ (Feitzinger et al., 1980; Cassinelli et al., 1981). Recently Weigelt and Baier (1985), using speckle inter-ferometry techniques, have been able to resolve R136a into several components, the brightest of which would have a mass of about 250 $M_\odot$.

(B) There is a very interesting class of stars associated with the Humphreys—Davidson limit, called the hypergiants or Luminous Blue Variables (LBV). The class extends across much of the early-type star region of the HR diagram and beyond to the cooler parts of the diagram. Some examples of this type of stars are the well known objects P Cygni, η Car, AG Car, and S Doradus. The stars have winds that are quite peculiar as compared with somewhat less luminous stars of the same effective temperature (Lamers, 1986b; Wolf, 1986). The mass loss rates are extremely large, $\sim 10^{-4.5}$ $M_\odot$ yr^{-1} The velocities of the winds are rather slow, $\sim$ a few hundred km s^{-1}. The stars are thought to be at the limit of stability because of tubulent pressure gradients (de Jager, 1984), or radiation pressure (Lamers, 1986a).

(C) Among the lower luminosity stars there is a sequence of ionization limits in the HR diagram. One of the most interesting discoveries in ultraviolet studies of early-type stars was the finding of strong P Cygni lines of anomalously high ionization stages, such as the O VI λ 1030 lines. These are called 'superionization' stages. Surveys have shown that there are rather well-defined limits in the HR diagram where the high ionization stages are present (Lamers and Snow, 1978; Morton, 1979). The superionization stages are those that should not be present in the wind of the stars if the temperature were determined by radiative equilibrium conditions. The lines of O VI, for example, continue to be present in the spectra of supergiants as late as B1, while the lines 'should' be absent in stars later than about O3. Similarly, lines of N V, C IV, and Si IV persist to later spectral types than expected from a radiative equilibrium model. The fact that O VI, N V, and C IV are from ion stages that are two stages higher than the predicted dominant stages, has led to the suggestion that the superionization stages are produced by K-shell ionization by X-rays present in the wind, the second electron being removed by the Auger ionization process (Cassinelli and Olson, 1979). The Einstein X-ray satellite gave support to this idea in that it found that essentially all O stars are X-ray sources, as are some B stars (Harnden *et al.*, 1979).

(D) Strong wind lines are not seen for stars all along the main sequence. There is an interesting transition from easily detected winds to undetected winds for stars near spectral class B0 V. This 'strong wind limit', as we refer to it in Figure 1, extends to higher luminosity classes along a line at roughly $M_{\text{bol}} \simeq -6$. The stars near the limit have observationally interesting properties in that one star will have a clear wind profile in its UV spectra while another of the same spectral class will show no wind (Walborn and Panek, 1984).

Near this limit and below, stars can have strong winds, that is with mass loss rates greater than about 10^{-9} $M_\odot$ yr^{-1}, if there is some effect that can initiate an outflow. Examples of effects that appear to be important are (a) the rapid rotation of the Be stars, discussed in another chapter, (b) enhanced metal abundances that allow radiation to initiate winds for lower luminosity stars (Massa *et al.*, 1984), (c) pulsation, either radial (Burger *et al.*, 1982) or non-radial (Vogt and Penrod, 1983; Abbott *et al.*, 1986), and (d) open magnetic field regions of the chemically peculiar stars.

(E) For the chemically peculiar stars, that tend to occur at spectral classes B2 V and B6 V, the UV spectra shows a variation in the mass ejection that is related to the periodicity of the kilogauss magnetic fields in the stars (Brown *et al.*, 1984). The outflows have been discussed in the context of oblique-rotator models. Later we briefly discuss the possibility that magnetic fields influence the outer atmospheres of other early-type stars.

(F) The last of the limits that we note in Figure 1 is the X-ray limit derived from the observations of the Einstein Observatory. The early-type stars have an X-ray luminosity that tends to follow the simple relation $L_x \simeq 10^{-7}$ L_{bol} (Seward and Chlebowski, 1982). There is then a void in the X-ray emission through the A V spectral classes to about F5 V where the X-ray emission commenses again, presumably because of mechanical energy generated in the convective envelopes of the stars of later spectral type. For the early-type stars the X-ray emission has not been explained in a satisfactory way as yet. X-rays can be formed in the shocks that are expected to exist in the winds of stars with strong mass outflow (Lucy and White, 1980). However, the X-rays appear to come also from stars with winds that are too weak to form strong shocks, for instance, τ Sco B0 V. This suggests that gas at coronal temperatures exists elsewhere in the atmospheres, not only in hot shocks embedded in the winds (Cassinelli and Swank, 1983). The zone in the HR diagram where X-ray emission occurs in Figure 1 is somewhat uncertain because of the sensitivity limits of the Einstein satellite. There are a few main sequence stars as late as B5 with X-ray emission, but it is not clear whether all of the earlier main sequence stars are emitting X-rays. Also in the supergiant region it is not certain whether all of the B stars up to B5 emit X-rays, nor whether there is a limit at B5. Odegard and Cassinelli (1982) have used the superionization stages observed with the IUE satellite to estimate the X-ray flux of B supergiants and found that the expected fluxes are near or below the Einstein detection limits.

There are certainly other zones and limits in the HR diagram for early-type stars than the ones identified here. Their presence is not so clearly defined as the ones that we have noted. Of particular interest are the displaced narrow absorption features that often occur in the broad shortward shifted wind absorption lines of early-type stars. At the present time these narrow features do not seem to be confined to any particular part of the HR diagram other than the broad zone that we have described as having 'strong winds'. These narrow components will be discussed in a later section. Recent observations have also shown that early-type stars at nearly all spectral classes show evidence for non-radial pulsation in their photospheric lines (Baade, 1984; Smith, 1986). Although we will not be discussing this topic in our review, it is likely to be of central importance in the eventual explanation of several of the boundaries that we have noted in Figure 1.

3. Mass Loss in the Upper Part of the HR Diagram

3.1. THE LINES OBSERVED WITH IUE

The UV spectra of luminous early-type stars clearly show the presence of mass loss by high velocity winds: strong violet-shifted resonance lines often accompanied by redshifted emissions.

The lines which are the best indicators of mass loss from OB stars in the spectral range accessible to IUE are the resonance transitions of $N V$ (1239, 1243), $C VI$ (1548, 1551), and $Si IV$ (1394, 1403). Fortunately, these ions have a lithium-like structure, otherwise their resonance transitions would be in the inaccessible wavelength region below the Lyman limit. Unfortunately, however, the quantitative interpretation of these lines for the determination of mass loss is usually difficult and uncertain: the lines are often saturated so that only lower limits of their column densities can be derived, or when they are not saturated, they represent minor ionization fractions of these elements so that large corrections for the unobservable major ionization stages are necessary.

In the hottest stars of spectral type early-O, the lines from excited levels of $N IV$ (1719), $O IV$ (1339, 1343), and $O V$ (1371) are formed in the wind and can be used for mass loss determination. However, in this case the excitation fraction of the relevant atomic levels, which is controlled by the photospheric flux in the inaccessible regions below 200 Å, is uncertain. So, again, the resulting mass loss rates depend strongly on uncertain factors.

For late-B and A-type stars the situation is not much better. The strongest UV resonance lines are those of $Mg II$ (2796, 2803), which are usually saturated. The many resonance lines of $Fe II$, $Ni II$, and $Cr II$ are usually not saturated, but they represent a minor ionization stage in the winds of late-B stars. For these stars the

Al III (1855) line can be used for mass loss studies. For the A-supergiants the resonance lines of singly ionized metals are the best indicators of mass loss.

The problems with the mass loss studies of hot stars from UV spectra are intrinsic to the stars: the mass loss rates are so large, 10^{-8} to 10^{-5} $M_\odot$ yr^{-1}, that the accessible UV lines are either saturated or represent minor ionization stages. There are only two ways out of this problem. (A) Extend the observed λ-range to wavelengths below 1200 Å. The Copernicus satellite has shown that important ions can be observed between 900 and 1200 Å (e.g., C III, O VI, P V, S IV and VI). Future satellites, such as the proposed Lyman satellite, will make this λ-range accessible for many stars. (B) Detailed and reliable calculations of the ionization fractions of minor ionization stages. The calculations by Olson and Castor (1981) and the empirical study by Lamers *et al.* (1980) represent only the first steps in this direction.

3.2. THE VELOCITIES OF THE WINDS

One of the easily derived properties of a hot star wind from UV observations is the wind terminal velocity. Measurements of the violet edge of the absorption features yield wind speeds ranging from 300 km s^{-1} for the A supergiants to 3500 km s^{-1} for the main sequence O3 stars. There is clear evidence that this velocity, which reflects the maximum speed that can be reached in the winds, increases with increasing temperature and increasing escape velocity. Figure 2 shows the dependence of the ratio between the terminal wind velocity, V_∞, and the escape velocity at the stellar surface, V_{esc}, and T_{eff} (from Abbott, 1982). The dashed line is our best-fit through the data:

$$V_\infty/V_{esc} = \quad 24.0 - 4.6 \log T_{eff} \quad \text{for} \quad T_{eff} \geq 25\,000 \text{ K}$$
$$V_\infty/V_{esc} = -25.3 + 6.6 \log T_{eff} \quad \text{for} \quad 10\,000 \geq T_{eff} < 25\,000 \text{ K}$$

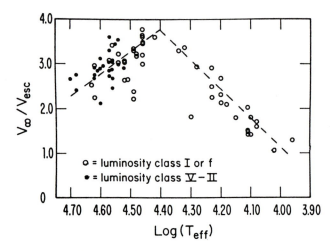

Fig. 2.   The relation between the terminal velocity of the wind normalized to the escape velocity in the photosphere, V_∞/V_{esc}, plotted versus T_{eff} (from Abbott, 1982). The dashed line is our best fit through the data.

This trend agrees reasonably well with the one predicted for the radiation driven wind theory (Abbott, 1982), especially when the finite size of the star is properly accounted for in the calculation of the radiation pressure (see Pauldrach *et al.*, 1986; Friend and Abott, 1986).

3.3. SUPERIONIZATION

To derive mass loss rates from ultraviolet line profiles it is necessary to know the fractional abundance of the ion that produces the line. This is especially uncertain because of the presence of the superionization stages that appear in the Copernicus and IUE spectra. As has been discussed by Cassinelli *et al.* (1978) the superionization stages that are seen can be produced either by electron collisions in a wind with temperatures greater than $\sim 10^5$ K or by the Auger ionization process in which two electrons are removed from C, N, and O following the K shell absorption of X-rays. Lamers and Snow (1978) derived the observed limits for the superionization assuming the wind is at an elevated temperature (i.e., the warm wind model of Lamers and Morton 1976). Cassinelli and Olson (1979) derived the limits from the assumption that X-rays from a corona are incident on the wind. Now that X-rays have been observed with the Einstein satellite, it appears that the Auger ionization process is an important one. Olson and Castor (1981) have estimated mass loss rates for several O and B supergiants under the assumption that a thin corona with a temperature of 4×10^6 K is present. Hamann (1981) however has found that it is necessary to use warm wind temperatures to explain the line profiles of superionization stages in τ Sco.

There is still a great deal of uncertainty concerning the ionization structure in the winds of hot stars. The location of the X-ray emitting material is not yet understood. The Einstein spectra show the presence of X-rays just above the K-shell absorption edge of oxygen. This means that X-ray emission is not occurring just at the base of the winds, say in a thin corona, but is occurring also from shocks of hot zones embedded in the wind. However, the shock models have difficulty explaining the X-ray emission at energies higher than 2 keV which indicate the presence of gas at temperatures ranging from 5 to 30 million deg (Cassinelli and Swank, 1983).

3.4. MASS LOSS RATES

The data in Figure 3 show a set of stars with accurately determined mass loss rates. The rates are from two different sources: (a) *from radio observations* (Abbott *et al.*, 1980, 1981, 1984). These rates are considered the most accurate ones, since they depend only weakly on the adopted model for the stellar wind. The stars Cyg OB 2 Nr 9 and 9 Sgr are omitted since their radio flux has a non-thermal origin and cannot be used for mass loss determinations. This limits the sample to only 11 stars of $\log L > 5.4$ and $\dot{M} \gtrsim 2.5 \times 10^{-6}\ M_\odot$ yr^{-1}. These stars are shown by hatched symbols in Figure 3. The three brightest B-stars in this sample, P Cygni, ζ^1 Sco and Cyg OB 2 Nr 12 are all hypergiants of class Ia$^+$. The spectral type, luminosity, and temperature of the latter star is uncertain, and has been classified between B3 Ia$^+$ or B8 Ia$^+$. (b) *From UV observations* (Gathier *et al.*, 1981; Olson and Castor 1981; Garmany *et al.*, 1981; Garmany and Conti,

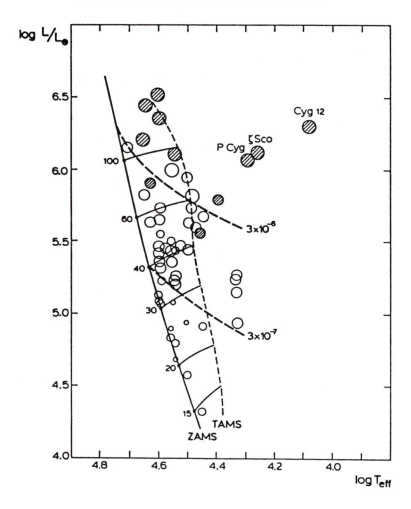

Fig. 3.   The stars with the most reliable mass loss rates are plotted in the HR diagram. The size of the symbols indicates the mass loss rate in four intervals: $\log \dot{M} > -5.5$ (largest symbols); $-6.5 < \log \dot{M} \leqslant -5.5$; $-7.5 < \log \dot{M} \leqslant -6.5$; $-8.5 < \log \dot{M} \leqslant -8.5$ (smaller symbols). Two dividing lines for $\log \dot{M} = -5.5$ and -6.5 are shown. They clearly slope down, which indicates that the mass loss rate increases as the star evolves away from the main sequence.

1984). In these studies the UV line profiles were fitted with theoretical P Cygni profiles and the column densities of different ions were determined. In the studies of Olson and Castor and those of Garmany, the column densities are converted to mass loss rates by assuming an ionization equilibrium produced by a wind of $T_e \simeq 35\,000$ K and Auger ionization by X-rays from a thin corona of $T \simeq 4 \times 10^6$ K. In the study of Gathier *et al.* (1981) the stars with known mass loss rates from radio data were used to calibrate a relation between the total column density N_H in the wind and the observed column densities of the different ions. This calibration was used to derive $\dot{M}$ for all the stars. Since the calibration relation was derived for the most luminous stars, the extrapolation to lower luminosity might produce errors in the mass loss determinations. For instance, a study of τ Sco (B0 V) by Lamers and Rogerson (1978), and Hamann (1981) in which a warm wind model was assumed for the interpretation of the ionization equilibrium results in a ten times smaller rate than derived by Gathier *et al.* (1981).

In principle, it is possible to determine mass loss rates from the IR excess that is produced by bound-free and free-free emission of the wind (e.g., Barlow and Cohen, 1976; Berthout *et al.*, 1985). However, recent studies of the IR excess by Castor and Simon (1983) and Abbott *et al.* (1984) have shown that the rates derived from the IR excess are poorly correlated with those derived from radio and UV data from the same stars. The reason for this discrepancy is our poor knowledge of the structure of the lower parts of the stellar wind. Since the IR excess depends on n_e^2 it is particularly sensitive to the density distribution in the layers between the photosphere and the wind. The stellar wind models that have been calculated up to now do not provide reliable results for this region. In addition, density inhomogeneities such as clumping, prominence-like structures or plumes, or temperature inhomogeneities may affect the IR excess drastically. A similar problem applies to the determination of mass loss from the $H\alpha$ emission from winds, which is sensitive to the same effects.

The stars with reliable mass loss rates are plotted in Figure 3. The most reliable ones, i.e., those derived from radio data, are shown by filled symbols.

Almost all early-type stars with reliable mass loss rates are of spectral type B1 or earlier. In the HR diagram they cluster in a narrow strip in between the ZAMS and the TAMS, which indicates that they are in the phase of core hydrogen burning. This is due to the fact that up to now most mass loss rates have been determined from the UV lines in the spectra of O to B1 stars. The reason is simply that no systematic study of the ionization in the winds of late-B and A-type stars has been made yet and so, the mass loss estimates based on the UV lines for those stars are very uncertain.

The data in Figure 3 show clearly that the most luminous stars have the highest mass loss rates and that the rates decrease toward lower luminosity. The dividing lines of $\dot{M} \simeq 3 \times 10^{-6}$ and 3×10^{-7} $M_\odot$ yr^{-1} occur on the ZAMS at $\log L \simeq 6.3$ and 5.3, respectively, and on the TAMS at $\log L \simeq 5.8$ and 5.0, respectively. The dividing lines clearly show a negative slope in the HR diagram. This implies that during the evolution of the stars with nearly constant luminosity the mass loss increases when the temperature and the gravity decrease. Based on this consideration, we have tried to express the mass loss rates of the stars in Figure 3 in terms of a simple relation:

$$\log \dot{M} = a + b \log(L/10^6) + c \log(T/T_{ZAMS}).$$

In this expression, the first two terms describe the mass loss on the ZAMS and the third term describes the change in mass loss when the star evolves away from the main sequence at almost constant luminosity. We have only used the stars with $\log L \geqslant 5.0$ since the lower luminosity stars are only at the ZAMS. We also omitted the star Cyg OB 2 nr 12 because of its uncertain temperature. The best fit is found if

$$a = -5.5$$
$$b = +1.6$$
$$c = -1.0.$$

This relation is valid in the range of $5.0 < \log L < 6.5$ and $4.2 < \log T < 4.7$. The measured mass loss rates scatter around this relation by 0.4 dex (standard deviation). This relation has about the same dependence of L as found in various other studies (see e.g., Lamers, 1981; Garmany and Conti, 1984), and agrees with the prediction of Friend and Castor (1986). The dependence on T/T_{ZAMS} shows that $\dot{M} \sim (R/R_{ZAMS})^{0.5}$ when the star evolves at constant L.

4. Narrow Absorption Components

One of the topics in which IUE has contributed enormously is the study of the blue shifted narrow absorption components (NAC), also called 'discrete absorption lines' or 'shifted narrow components'. These terms refer to narrow absorption components of UV resonance lines, which are blue shifted by about -300 to -3000 km s^{-1}, and which often appear superimposed on the P Cygni profiles of C IV, N V, and Si IV.

After the discovery of these components by Morton (1976) and Snow and Morton (1976) in Copernicus spectra, and after the first systematic quantitative investigation of their properties by Lamers *et al.* (1982) many studies with IUE have been devoted to the explanation of this phenomenon. Especially the capability of IUE to observe the variability of these NAC and discover their time-history has provided considerable new information. We will summarize the major results.

Reviews on this topic have been written by Lamers *et al.* (1982), Henrichs (1984) and by Prinja and Howarth (1986). A very extensive review has been written recently by Henrichs (1986), who also discusses the comparison of the observations with predictions based on proposed mechanisms.

4.1. GENERAL PROPERTIES OF THE NAC AND THEIR IMPLICATIONS

(i) The NAC have been observed with Copernicus and IUE in all the resonance lines that indicate mass loss in hot stars, i.e., the resonance lines of C III, C IV, N V, O VI, Si III, and Si IV. The studies by Snow (1977) and Lamers *et al.* (1982) of 26 stars, Prinja and Howarth (1986) of 22 stars and Henrichs and Wakker (1986) of 241 stars from the IUE data bank have shown that at least 65% of the stars with $M_{bol} < -7$ show NAC at one time or another. Since the NAC are known to be variable and not always present, we suspect that NAC can occur in all stars that show mass loss effects in their UV spectrum.

(ii) If present, the NAC occur in both the superionized atoms (O VI, N V) and in the normal atoms simultaneously. For instance, Lamers *et al.* (1982) found these components in several stars in Si III, C III, N V and O VI. This indicates that the same effects that determine the stage of ionization in the wind also occur in the material of the NAC. Lamers *et al.* (1982) and Prinja and Howarth (1986) found that the ratio between the column densities of the superions and the normal ions, e.g., N(N V)/N(Si IV) is higher in the NAC than in the underlying P Cygni profile. This suggests that the mechanism responsible for the superionization, probably Auger-ionization by X-rays (see III—C), is more efficient in the NAC material. If this is true, it implies the production of X-rays by shocks with temperatures of 10^6 to 10^7 K that are associated with the material of the NAC.

(iii) The velocities of the NAC are in the range of -300 to -3000 km s^{-1}, with typical values of about -2000 km s^{-1} for O-stars and -1000 km s^{-1} for B-supergiants. These velocities scale with the terminal velocity, v_∞, of the wind: the NAC have velocities of about 0.5 v_∞ to 0.9 v_∞. (The rare occurrence at much smaller velocities may represent an early phase in the development of the NAC.) This shows that the NAC material is usually accelerated to a lower velocity than the wind. (The possibility that the NAC material does reach velocities higher than v_∞, but that it can only be observed at $v < v_\infty$ can be excluded on the basis of the variation of the NAC with time; see IV—B.)

When the NAC are seen in the lines of different ions, their velocity is the same for all ions within the accuracy of the measurements, i.e., within about 20 km s^{-1}. This excludes the interpretation of the NAC in terms of a distance-dependent ionization in the wind.

(iv) Sometimes multiple NAC at different velocities are observed, with a maximum of four. Good examples of these are found in the O-subdwarf HD 128220 (Bruhweiler and Dean, 1983) and in the hypergiant ζ^1 Sco (Burki *et al.*, 1982). This excludes the interpretation of the NAC in terms of a plateau in the velocity law of the wind. Such a plateau would produce a large column density at the velocity of the plateau. Multiple components would require multiple velocity plateaus, for which there is no explanation.

(v) The column densities of the NAC are of the order of 10^{14} to 10^{15} cm^{-2} for the observed ions (Lamers *et al.*, 1982; Prinja and Howarth, 1986). In order to derive the total hydrogen column density of the NAC, corrections have to be made for the abundance and the ionization. Assuming approximately the same degree of ionization as in the wind, the values of N_H turn out to be of the order of 10^{19} to 10^{22} cm^{-2}. These values are uncertain, however, as they depend strongly on the ionization in the NAC material.

4.2. VARIABILITY OF THE NAC

The NAC are variable in strength and in velocity. Numerous studies with IUE have provided a wealth of observations of these variations. Most of these observations, however, are scattered in time and can only be used to detect variability, but not to study the nature of these variations. (For an extensive list of references, see Henrichs, 1986). A few series of sequential observations that show the history of a NAC exist. These are very important for our understanding of the nature and origin of the NAC.

The best time-resolved series of observations is by Prinja *et al.* (1984, 1986) of the star ξ Per 07.5 III((f)). See Figure 4. The spectra obtained between October 18 and 20 1984 show the appearance and development of a NAC. On October 18

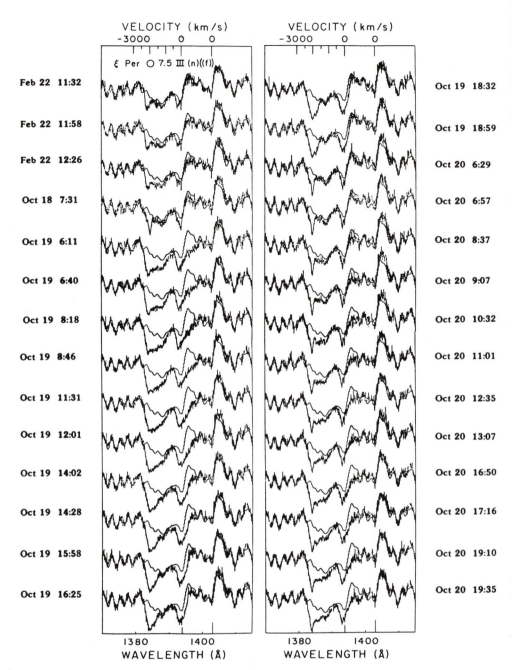

Fig. 4. Variations in the Narrow Absorption Components of the Si IV resonance lines of ξ Per 07.5 III ((f)) (from Prinja *et al.*, 1986). The long tick marks at the horizontal axis indicate the laboratory wavelengths of the two components. The two short tick marks indicate the wavelengths of 1380 and 1400 Å. The shape of the underlying P Cygni profile is shown for comparison in each observation. Notice the appearance of a strong absorption at $v \simeq -1000$ km s^{-1} on Oct 19 6:11 and its gradual shift to shorter wavelength ($v \simeq -2100$ km s^{-1}) and narrowing within one day. On Oct. 20 8:37, a new component appears at $v \simeq -1000$ km s^{-1}.

the Si IV resonance lines show the presence of a NAC at $v \simeq -2100$ km s^{-1}. On October 19 6^h:11^m a broad absorption feature appears at $v = -1000$ km s^{-1}. As time progresses, this feature shifts to shorter wavelengths and becomes narrower. On October 20 6:57 it is reduced to a very narrow feature at $v = -2150$ km s^{-1}. This is one day after its first appearance. From then onward the velocity remains approximately constant, but the feature reduces in strength, i.e., the column density decreases. On October 20 8:37 a new broad absorption component appears near -1000 km s^{-1} and within one day, the absorption gets narrower and shifts to shorter velocity in a way very similar to that of the previous NAC.

This sequence of observations shows that NAC can develop in timescales of an hour or less and remain visible for days up to weeks.

One might be tempted to conclude from these observations that the NAC are at about -1000 km s^{-1} ($\sim 0.5\ v_\infty$) when they are formed. This, however, is not necessarily the case. The presence of strong photospheric components and the saturation of the P Cygni profiles at low velocity may make it difficult to detect the NAC at low velocity. (See the predictions of the NAC as a function of time due to shell ejection; Prinja and Howarth, 1985).

In some stars NAC have been observed at low velocity. For instance, Gry *et al.* (1984) observed variable NAC at $v \simeq -100$ km s^{-1} in the Lyman lines of several early-type stars. *If* these low velocity components are the precursors of the features that are observed later at higher velocities in the UV lines, they would indicate that the NAC material is ejected from the photosphere or from the base of the wind. *If*, on the other hand, the observed NAC in the UV are *not* preceded by low velocity components, they have to be created at a considerable distance of about 1 R_* in the wind, possibly in the form of shocks or blobs. Therefore, it is important to look for the NAC almost simultaneously in the UV resonance lines and in unsaturated density-sensitive lines such as Hα.

4.3. POSSIBLE EXPLANATIONS OF THE NAC

A large number of explanations have been proposed for the NAC. As more and better observations become available, largely due to IUE, most of these explanations had to be rejected. (For reviews of suggested explanations and reasons for rejection, see Lamers *et al.*, 1982; Prinja and Howarth, 1986; and Henrichs, 1986). The two models that seem to agree best with the observations are the Corotating Interaction Region (CIR) model by Mullan (1984, 1986) and the shell ejection model by Henrichs *et al.* (1980, 1983).

The CIR model is based upon on analogy with the solar wind, which in the equatorial plane consists of fast and slow moving regions, due to the magnetic configuration. If a similar system of fast and slow streams exist in the stellar wind of a rotating star, the fast streams will collide with the slow streams and produce shocked interaction regions with an enhanced density. These higher density shock regions (CIR) may produce the NAC. Mullan (1986) and Prinja and Howarth (1986) have argued that this mechanism can explain the fact that NAC are usually seen at high velocity $v(\text{NAC})/v_\infty \gtrsim 0.5$, that some stars with a high ratio of $v \sin i/v_\infty$ have the smallest ratio $v(\text{NAC})/v_\infty$, and that the NAC with the highest column density have the lowest velocity.

The shell ejection model or episodic mass loss model explains the NAC as being due to shells or puffs that are ejected at a density higher than in the ambient wind. This model can explain very well the time series of the NAC that have been observed in, e.g., ξ Per (Figure 4) and in γ Cas (Henrichs *et al.*, 1980). It also explains the fact that most NAC are seen at $0.5\ v_\infty \lesssim v \lesssim 0.9\ v_\infty$: the shells are rapidly accelerated at low velocity, so the probability of observing this phase is small, and since the density in the shells is higher than in the ambient wind, the shell will be accelerated to a velocity lower than v_∞. The fast variations observed at low velocity in Hα and in the Lyman lines of some stars may also support this model.

It is clear that both mechanisms described here can explain some of the features of the NAC. The more fundamental questions, however, such as "why does a hot star have fast and slow streams" and "why does a hot star suffer frequent shell ejections" are not answered. There is growing evidence that stellar rotation plays an important role. For instance, in the Be stars that are in general more rapidly rotators than normal B stars, the mass ejection is concentrated in the equatorial plane and stars with a higher $v \sin i$ show more variability in the NAC. The way in which rotation affects the winds and the NAC's is not fully understood, but seems to operate via the non-radial pulsations. There are some good examples of a correlation between mode-changes of the non-radial pulsations and enhanced mass loss (see Henrichs, 1984, 1986).

The qualitative picture that is suggested by these observations is the following: Many (or most?) early-type stars turn out to be non-radial pulsators. In these stars mode-switching seems to be a common phenomenon. The release of energy during mode-switching may trigger enhanced mass loss, which produces the NAC. Since rapid rotation splits the mode-frequency spectrum into a wider range the excitation of the modes might be easier. If more modes can be excited in rapidly rotating stars, mode-switching is more like to occur. This may explain why rapid rotators are generally more 'active' in terms of wind variability and NAC.

5. Magnetic Field Effects

IUE observations have shown that for the chemically peculiar stars the outflowing wind can be controlled by the presence of a strong magnetic field. This has been illustrated most clearly by Brown, Shore, and Sonneborn (1985) for the CP2 star HD 21699 (B6 V) and in the helium strong star HD 184927 (B2 V) by Barker *et al.* (1982). HB 21699 has a kilogauss bipolar magnetic field that varies with the stellar rotation period. The UV spectra show large variations in the C IV λ1550 resonance doublet. The lines show strong shortward absorption extending to -600 km s^{-1} at phases in which there is the largest line-of-sight magnetic field. The line is absent, both in shortward absorption and in longward emission at other phases. The observations can be understood as caused by mass loss directed along a magnetic polar jet or plume. The profiles are well matched by the theoretical profiles of Kunasz (1984) for flow along a cylinder and as seen at different aspect angles. The C IV doublet is from a superionization state and indicates that some sort of non-radiative heating of the flow is occurring. The kilogauss magnetic field is strong enough to enforce co-rotation of the outflowing material, as has been shown in the oblique rotator model of Shore and Bolton (1986). HD 21699 provides the best evidence for magnetic control of a radiative stellar wind.

In a series of papers, Underhill (1983, 1984) has argued that magnetic fields are important in many classes of early-type star, either by restricting the source region from which winds can originate, or by geometrically confining the flow to plumes. Underhill (1984) argues that there are major differences in the UV line profiles in stars with essentially identical classical stellar parameters, and attributes this stellar individuality to the stellar magnetic fields. Other suggestive evidence is presented in the review of Cassinelli (1985). For example, in ζ Orionis 09.5Ia and in τ Sco B0 V there is rather hard X-ray emission or KeV emission lines in Einstein satellite spectra that indicates the presence of gas at temperatures of 15 million deg or more. This exceeds the escape temperature of the stars and may require magnetic field confinement like that usually postulated to explain the hot X-ray emission in cool giants.

If loops or closed field regions exist, then by analogy with the sun, there should be plumes of flow occurring. Possible evidence for this is the variability of the optical polarized light from OB supergiants, in which there are changes not only in the magnitude of the polarized flux but also in the angle of polarization on the plane of the sky (Lupie and Nordsieck, 1987). Also the concentration of flows to specific zones of the stellar disk could lead to the 'stream-stream' interaction regions postulated by Mullan (1984) to explain the narrow absorption features seen in the cores of UV wind lines.

Epilogue

In this review we have concentrated on the contribution of IUE observations to our knowledge about the envelopes of early-type stars. There are several other aspects of the study of hot star to which IUE has contributed.

— Classification studies that show the existence of a correlation between visual characteristics and UV characteristics, and the deviating stars.

— Effective temperature determinations from integrated energy distributions for stars in our galaxy and in the Magellanic Clouds.

— Studies of the surface abundances, especially of He, C, N, and O that change in time as the stars evolve and lose mass.

— Stellar winds in stars outside our galaxy for which a different metal abundance, Z, may result in mass loss rates and evolution different from the galactic stars.

These topics were not discussed in this review. We felt that the topics that were described are exciting enough to illustrate the importance if IUE for the study of winds from hot stars.

Future observations with ST will make it possible to study the winds and their variability in stars of other galaxies where different chemical compositions, different environments, and different star formation may produce unexpected effects. In the more distant future, the exciting possibility to observe the winds of stars in our Galaxy and in other galaxies in the extremely rich wavelength range of $\lambda < 1200$ Å with the proposed Lyman mission, will enormously increase our knowledge of the physical processes in stellar winds.

Acknowledgement

This review was written while J. P. C. held a Fulbright Fellowship at the University of Utrecht.

References

Abbott, D. C.: 1982, *Astrophys. J.* **259**, 282.
Abbott, D. C., Bieging, J. H., and Churchwell, E.: 1981, *Astrophys. J.* **250**, 645.
Abbott, D. C., Bieging, J. H., and Churchwell, E.: 1984, *Astrophys. J.* **280**, 671.
Abbott, D. C., Bieging, J. H., Churchwell, E., and Cassinelli, J. P.: 1980, *Astrophys. J.* **238**, 196.
Abbott, D. C., Telesco, C. M., and Wolff, S. C.: 1984, *Astrophys. J.* **279**, 225.
Abbott, D. C., Garmany, C. D., Hansen, C. J., Henrichs, H. F., and Pesnell, W. D.: 1986, *Publ. Astr. Soc. Pacific* **98**, 29.
Baade, D.: 1984, *Proceedings 25th Liege Int. Astroph. Coll.*, A. Noels and M. Gabriel (eds.), Université de Liege, p. 115.
Barker, P. K., Brown, D. N., Bolton, C. T., and Landstreet, J. D.: 1982, in Y. Kondo and J. Mead (eds.), *Advances in UV Astronomy: The Fourth Year of IUE*, NASA CP-2238, p. 589.
Barlow, M. J. and Cohen, M.: 1977, *Astrophys. J.* **213**, 737.
Bertout, C., Leitherer, C., Stahl, O., and Wolf, B.: 1985, *Astron. Astrophys.* **144**, 87.
Brown, D. N., Shore, S. N., Bolton, C. T., Hulbert, S. J., and Sonneborn, G.: 1984, in J. Mead, R. Chapman, and Y. Kondo (eds.), *Future of UV Astronomy Based on Six Years of IUE Research*, NASA CP-2349, p. 483.
Brown, D. N., Shore, S. N., and Sonneborn, G.: 1985, *Astron. J.* **90**, 1354.
Bruhweiler, F. C. and Dean, Ch. A.: 1983, *Astrophys. J.* **274**, L87.
Burger, M., de Jager, C., van den Oord, G. H. J., and Sato, N.: 1982, *Astron. Astrophys.* **107**, 320.
Burki, G., Heck, A., Bianchi, L., and Cassatella, A.: 1982, *Astron. Astrophys.* **107**, 205.
Cassinelli, J. P.: 1985, in A. B. Underhill and A. G. Michalitsianos (eds.), *The Origin of Nonradiative Heating/Momentum in Hot Stars*, NASA CP-2358, p. 2.
Cassinelli, J. P. and Olson, G. L.: 1979, *Astrophys. J.* **229**, 304.
Cassinelli, J. P. and Swank, J. H.: 1983, *Astrophys. J.* **271**, 681.
Cassinelli, J. P., Castor, J. I., and Lamers, H. J. G. L. M.: 1978, *Publ. Astr. Soc. Pacific* **90**, 477.
Cassinelli, J. P., Mathis, J. S., and Savage, B. D.: 1981, *Science* **212**, 1497.
Castor, J. I. and Simon, T.: 1983, *Astrophys. J.* **265**, 304.
Chiosi, C. and Maeder, A.: 1986, *Ann. Rev. Astron. Astrophys.* **24**, 329.
de Jager, C.: 1984, *Astron, Astrophys.* **138**, 246.
Feitzinger, J. V., Schlosser, W., Smidt-Kaler, T., and Winkler, C.: 1980, *Astron. Astrophys.* **84**, 50.
Friend, D. B. and Abbott, D. C.: 1986, *Astrophys. J.* **311**, 701.
Garmany, C. D. and Conti, P. S.: 1984, *Astrophys. J.* **284**, 705.
Garmany, C. D., Olson, G. L., Conti, P. S., and van Steenberg, M. E.: 1981, *Astrophys. J.* **250**, 660.
Gathier, R., Lamers, H. J. G. L. M., and Snow, T. P.: 1981, *Astrophys. J.* **247**, 173.
Gry, C., Lamers, H. J. G. L. M., and Vidal-Madjar, A.: 1984, *Astron. Astrophys.* **137**, 29.
Hamann, W. R.: 1981, *Astron. Astrophys.* **93**, 353.
Harnden, F. R., Branduardi, G., Elvis, M., Gorenstein, P., and Grindlay, J. E.: 1979, *Astrophys. J.* **234**, L51.
Henrichs, H. F.: 1984, in E. Rolfe and B. Battrick (eds.), *Proc. Fourth European IUE Conference*, ESA SP-218, p. 43.
Henrichs, H. F.: 1986, in P. S. Conti and A. B. Underhill (eds.), *O, Of and Wolf-Rayet Stars*, NASA/CNRS Monograph (in press).
Henrichs, H. F., Hammerschlag-Hensberge, G., and Lamers, H. J. G. L. M.: 1980, in B. Battrick and J. Mort, (eds.), *Proc. Second European IUE Conference*, ESA SP-157, 4, p. 147.
Henrichs, H. F., Hammerschlag-Hensberge, G., Howarth, I. D., and Barr, P.: 1983, *Astrophys. J.* **268**, 807.
Henrichs, H. F. and Wakker, B. P.: 1986 (preprint).

Humphreys, R. M. and Davidson, K.: 1979, *Astrophys. Letters* **232**, 409.

Kunasz, P. B.: 1984, *Astrophys. J.* **276**, 677.

Lamers, H. J. G. L. M.: 1981, *Astrophys. J.* **245**, 593.

Lamers, H. J. G. L. M.: 1986a, *Astron. Astrophys.* **159**, 90.

Lamers, H. J. G. L. M.: 1986b, 'P Cygni Type Stars: Evolution and Physical Processes', in C. de Loore, A. Willis, and P. Laskarides (eds.), *Luminous Stars and Associations in Galaxies*, D. Reidel Publ. Co. Dordrecht, Holland, pp. 157—182.

Lamers, H. J. G. L. M. and Morton, D. C.: 1976, *Astrophys. J. Suppl.* **32**, 715.

Lamers, H. J. G. L. M. and Rogerson, J. B.: 1978, *Astron. Astrophys.* **66**, 417.

Lamers, H. J. G. L. M. and Snow, T. P.: 1978, *Astrophys. J.* **219**, 504.

Lamers, H. J. G. L. M., Gathier, R., and Snow, T. P.: 1980, *Astrophys. J. Letters* **242**, L33.

Lamers, H. J. G. L. M., Gathier, R., and Snow, T. P.: 1982, *Astrophys. J.* **258**, 186.

Lucy, L. B. and White, R. L.: 1980, *Astrophys. J.* **241**, 300.

Lupie, O. L. and Nordsieck, K. H.: 1987, *Astron. J.* **92**, 214.

Massa, D., Savage, B. D., and Cassinelli, J. P.: 1984, *Astrophys. J.* **287**, 814.

Morton, D. C.: 1976, *Astrophys. J.* **203**, 386.

Morton, D. C.: 1979, *Monthly Notices Roy. Astron. Soc.* **189**, 57.

Mullan, D. J.: 1984, *Astrophys. J.* **283**, 303.

Mullan, D. J.: 1986, *Astron. Astrophys.* **165**, 157.

Odegard, N. and Cassinelli, J. P.: 1982, *Astrophys. J.* **256**, 568.

Olson, G. L. and Castor, J. I.: 1981, *Astrophys. J.* **244**, 179.

Pauldrach, A., Puls, J., and Kudritzki, R. P.: 1986, *Astron. Astrophys.* **164**, 86.

Prinja, R. K. and Howarth, I. D.: 1984, *Astron. Astrophys.* **133**, 110.

Prinja, R. K. and Howarth, I. D.: 1985, *Astron. Astrophys.* **149**, 73.

Prinja, R. K. and Howarth, I. D.: 1986, *Astrophys. J. Suppl.* (in press).

Prinja, R. K., Henrichs, H. F., Howarth, I. D., and van der Klis, M.: 1984, in E. Rolfe and B. Battrick, (eds.), *Proc. Fourth European IUE Conference*, ESA SP-218, p. 319.

Prinja, R. K. *et al.*: 1986 (in preparation).

Seward, F. D. and Chlebowski, T.: 1982, *Astrophys. J.* **256**, 530.

Shore, S. N. and Bolton, C. T.: 1986 (preprint).

Smith, M. A.: 1986, *Publ. Astron. Soc. Pacific* **98**, 33.

Snow, T. P.: 1977, *Astrophys. J.* **217**, 760.

Snow, T. P. and Morton, D. C.: 1976, *Astrophys. J. Suppl.* **32**, 429.

Vogt, S. S. and Penrod, G. D.: 1983, *Astrophys. J.* **275**, 661.

Underhill, A. B.: 1983, *Astrophys. J. Letters* **268**, L127.

Underhill, A. B.: 1984, *Astrophys. J. Letters* **287**, L874.

Walborn, N. R. and Panek, R. J.: 1984, *Astrophys. J.* **286**, 718.

Weigelt, G. and Baier, G.: 1985, *Astron. Astrophys.* **150**, L18.

Wolf, B. 1986, 'S Doradus Type; Hubble-Sandage Variables', in C. de Loore, A. Willis, and P. Laskarides (eds.), *Luminous Stars and Associations in Galaxies* D. Reidel Publ. Co., Dordrecht, Holland, pp. 151—156.

WOLF—RAYET STARS

ALLAN J. WILLIS

*Department of Physics and Astronomy, University College London,
London, UK*

and

CATHARINE D. GARMANY

Joint Institute for Laboratory Astrophysics, Boulder, CO, U.S.A.

1. Introduction

Our perception of Wolf—Rayet stars has undergone a remarkable transformation over the past decade largely through the advent of new observations spanning a wide range of the electromagnetic spectrum, from X-rays to the radio, coupled to improvements in theoretical atmospheric and evolution models and the discovery and study of WR populations in external galaxies. It is no coincidence that this decade of change in perception has embraced the IUE era since UV spectroscopy has provided key data for advancing knowledge of this stellar class, and herein we discuss results from a broad range of IUE studies of WR stars that have contributed to the recent advances. A more general summary of the field can be found in *IAU Symposium* No. 99 — 'WR Stars; Observations, Physics, Evolution', (de Loore and Willis, 1982) and *IAU Symposium* No. 116 — 'Luminous Stars and Associations in Galaxies', (de Loore *et al.*, 1986).

It is now generally recognised that WR stars represent an important and integral stage in the evolution of massive stars; that this stellar class are signposts to the IMF's and SFR's in galactic environments, and may well be the precursors of Type II supernovae. We consider WR stars are the helium burning cores of stars whose initial masses were greater than about 30—40 $M_\odot$, which have lost their outer hydrogen-rich layers either through mass loss or by Roche lobe overflow to a companion. This scenario was first proposed by Smith (1973) and has been expanded and elaborated upon by Van den Heuvel (1976), Conti (1976), Schild and Maeder (1984) and many others.

Evidence for this picture includes atmospheric abundance analyses of the heavy elements, the H/He ratio and, somewhat less conclusively the bolometric magnitudes and effective temperatures, and statistics concerning the frequency of WR stars and their progenitor O-stars.

One expects the ultraviolet to be particularly important for WR studies since these stars are hot and emit the bulk of their radiation shortward of the visible, and exhibit substantial mass loss and stellar winds which can be readily probed using UV P-Cygni profiles. Additionally, the UV provides numerous important line diagnostics for quantitative atmospheric analyses principally through the availability of resonance and low excitation transitions in common ions. Ultraviolet observations of WR stars prior to IUE have been reviewed by Willis (1980) and,

Y. Kondo (ed.), Scientific Accomplishments of the IUE, pp. 157—182.
© 1987 *by D. Reidel Publishing Company.*

mainly, consisted of broad-band photometric (OAO—2, ANS) or low resolution spectrophotometric (OAO—2, S2/68, SKYLAB) observations of a relatively small sample of galactic objects. High resolution UV spectra were available essentially only for Gamma—2 Velorum — the brightest WR star in the sky.

2. IUE Observations of WR Stars

IUE has brought a large number of WR stars in both the Galaxy and in the Magellanic Clouds under detailed scrutiny in both its high- and low-resolution mode, spectroscopically placing this class on the same footing in the UV as has been available in the optical. These IUE studies have led to improved measurements of the continuum energy distributions (Section 3), stellar wind and mass loss properties (Section 4), binary system interactions and intrinsic variability (Section 5), atmospheric chemical abundances (Section 6) and in overall terms contributed greatly to the development of current ideas concerning their evolutionary status (Section 7).

Initially most IUE—WR programmes were aimed at spectral surveys embracing individual subtypes in the WN and WC sequences. The most comprehensive HIRES data set has been given by Willis et al. (1986a) who present a spectral atlas for 14 galactic WN and WC stars covering the full SWP and LWR wavelength ranges and include for each star a comprehensive line identification list. Low resolution spectrophotometry of 15 galactic WR stars has been published in both numerical flux and spectral plot form by Nussbaumer et al. (1982) and similarly for 9 WR stars in the LMC by Smith and Willis (1983). Tabulations of emission line strengths in a wide sample of galactic and Magellanic WN stars is given by Conti et al. (1983) and for WC stars by Torres et al. (1986).

The general characteristics of the IUE spectra of WR stars at LORES have been discussed by Nussbaumer et al. (1982). As expected the UV spectra are dominated by numerous, strong emission lines in He II, N III–V in the WN stars and by C II–IV, He II, O III–V in WC spectra. The behaviour of relative line intensities within a given ion along both the WN and WC sequences is found to be consistent with that expected from their optical classification and excitation class. The HIRES UV spectra given by Willis et al. (1986a) confirm these broad trends and also highlight the numerous emission lines attributable to metallic ions like Fe IV and Fe V which abound in the vacuum ultraviolet. Examples of such high resolution WR spectra are shown in Figure 1.

Although carbon lines are definitely present in WN spectra (e.g., C IV 1550 is seen throughout the WN sequence and C III 1175, 1247, and 2297 are seen in WN7 and WN8 spectra), Willis et al. (1986a) find no umbiguous evidence for any nitrogen lines in WC spectra. In particular N IV 1486, 1718 identified in the pre-IUE era are not confirmed in the new, higher resolution IUE spectra of WC stars. A tentative identification by Willis (1982a, b) of N V 1240 P-Cygni absorption in the WC8 star HD 192103 superimposed on that of C III 1247 has not been confirmed in further spectra secured by Schmutz et al. (1984), Schmutz and Nussbaumer (1986).

Fitzpatrick (1982) and Fitzpatrick et al. (1982) have reported narrow absorption lines of Fe V and O V in high resolution IUE spectra of HD 93162 (WN7 abs) and HD 193793 (WC7) which, however, are not seen in the IUE spectra of other WR stars discussed by Willis *et al.* (1986a). In the case of HD 193793 there is evidence for an O-star companion which may give rise to the observed absorption spectrum. In the spectra discussed by Willis *et al.* (1986a) the only obvious absorption features seen are those associated with P-Cygni profiles.

3. UV Energy Distributions and WR Temperatures

The low resolution IUE spectrophotometry now available for many WR stars can be combined with ground-based optical data to study WR continuum energy distributions over a large wavelength baseline. Using such combined data several attempts have been made to derive effective temperatures for WR stars from colour temperatures derived from a comparison of the UV-optical continua and models. Using this method Nussbaumer *et al.* (1982) derived black-body colour temperatures for 15 galactic WN and WC stars in the range 20 000—40 000 K with the earlier subtypes giving the higher values. They also computed Zanstra temperatures from the observed strengths of optically thin He II emission lines which gave similar results and thus concluded that the true effective temperatures of the WR stars are close to these values. Deviations from a black body are observed for some WR stars, probably due to an extended continuum emitting region. For these stars they calculated a Zanstra temperature and pointed out that their assumption predicted continuum absorption at 2050 Å, which was subsequently not observed by IUE. Schmutz (1982) suggested that the Zanstra analysis is not valid for WR stars and he performed model calculations assuming a black body core and spherically symmetric wind to deduce an effective temperature for HD 50896 (WN5) of 43 000 K.

Underhill (1980, 1981), also using IUE and optical continuum measurements compared to Kurucz (1979) models, concludes that the effective temperatures for all WR stars are close to 30 000 K. For 9 WR stars in the LMC, Smith and Willis (1983) combine IUE LORES spectrophotometry with complementary IPCS optical data and again deduce WR colour temperatures lying in a similar range, although with some evidence that the LMC WN7 and WN8 stars may be hotter by about 5000 K than their galactic counterparts.

Barlow *et al.* (1981), in their study of the mass loss rates of 21 galactic WR stars, also derived colour temperatures based on IUE-optical data, extrapolated into the unobserved far-UV and also computed the observed flux contribution in the emission lines to deduce Bolometric Luminosities. Fitzpatrick *et al.* (1982) found that a black body of 43 000 K gave a best fit to HD 193793 (WC7) from the UV to the near-IR region.

A summary of these various temperature estimates is given in Table I, where it is seen that values in the range 25 000—50 000 K seem appropriate. As pointed out by Barlow *et al.* (1981), such temperatures are too low to produce radiative luminosities sufficient to drive the observed WR mass loss rates (about $2 \times 10^{-5} M_\odot$ yr^{-1}, see below) with usual radiation driven wind theory. It is

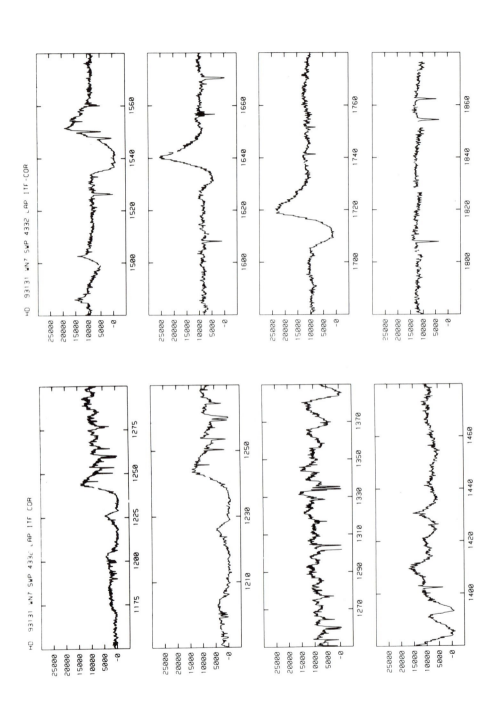

Fig. 1(a).   The SWP HIRES spectrum of HD 93131 (WN7) taken from Willis *et al.* (1986a), showing many, strong emission lines and P-Cygni profiles in N III, IV, V, He II and C IV lines.

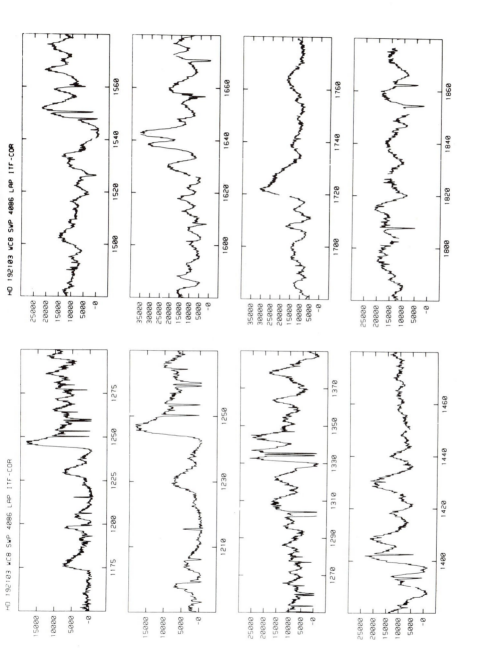

Fig. 1(b). The SWP HIRES spectrum of HD 192103 (WC8) showing emission lines and P-Cygni profiles in CII, III, IV, OIII, IV, HeII, SiII, III, IV, and FeIV, V transitions.

TABLE I

Effective temperature and luminosity estimates for WR stars based on IUE data

Star	Sp	E_{B-V} [a]	$T_{\rm eff}$ [1]	$\log L/L_\odot$ [1]	$T_{\rm eff}$ [2]	$\log L/L_\odot$ [2]	$\log L/L_\odot$ [3]	$T_{\rm eff}$ [4]	$\log L/L_\odot$ [4]
HD 9973	WN3	0.35	39 000	5.03			5.18		
HD 16523	WC6	0.55	41 000	4.90			5.08		
HD 50896	WN5	0.00	41 000	4.86	29 900	4.73	5.68		
HD 86161	WN8	0.58	30 000	5.42			5.68		
HD 92740	WN7abs	0.28	30 000	5.60	25 100	5.63	5.65		
HD 93131	WN7abs	0.16	30 000	5.41	29 700	5.76	5.65		
HD 96548	WN8	0.44	32 000	5.50	24 900	5.38	5.68		
HD 151932	WN7	0.45	24 000	5.33	25 000	5.63	5.65		
HD 156385	WC7	0.25	30 000	4.71			5.18		
HD 164270	WC9	0.47	22 000	4.39					
HD 165763	WC5	0.26	41 000	4.90	30 000	4.77	5.44		
HD 184738	WC9	0.20*			18 000	2.38			
HD 187282	WN4	0.20	36 000	4.70			4.71		
HD 191765	WN6	0.42	41 000	5.06	24 800	4.65	5.47		
HD 192103	WC8	0.38	26 000	4.72	25 200	4.80	5.05		
HD 192163	WN6	0.52	34 000	5.00	39 900	5.24	5.47		
FD 13	WN4	0.05						24 000	4.80
FD 23	WN7	0.05						36 000	5.27
FD 24	WN7	0.05						38 000	5.55
FD 70	WN7	0.10						29 000	5.93
FD 71	WN7	0.30						36 000	6.13
FD 12	WN8	0.05						34 000	5.22
FD 5	WC5	0.05						38 000	4.65
FD 27	WC5	0.05						37 000	4.99
FD 46	WC5	0.05						30 000	4.90

For V444 Cyg (WN5 + 05) Cherepaschuk *et al.* (1984) derive $T_{\rm eff}$ = 90 000 K, $\log L/L_\odot$ = 5.69 for the WN5 star.

[a] E_{B-V} from Nussbaumer *et al.* (1982) except for HD 184738 which is from Underhill (1983b).
[1] Values from Nussbaumer *et al.* (1982).
[2] Values from Underhill (1980, 1981, 1983b).
[3] Values from Barlow *et al.* (1981).

therefore pertinent to consider whether such measured colour temperatures (albeit based on extended baseline data) really do translate into reliable values of T_{eff}.

The first source of uncertainty arises from the interstellar reddening corrections that must be applied to the observations, particularly important in the UV where the extinction is large and where the bulk of the radiation is emitted. The requirement is twofold — accurate values of E_{B-V} are needed and one must be confident of the applicability of the extinction law used. Nussbaumer *et al.* (1982) address these problems and conclude that their stellar sample followed the usual galactic extinction law and that accurate colour excesses could be derived from the observed strengths of the 2200 Å bands. They checked the individual colour excesses by independently computing the level of extinction from the observed relative intensities of optically thin He II recombination lines. In the case of the LMC stars analysed by Smith and Willis (1983) the reddening in all cases was small and probably does not introduce significant uncertainties in the intrinsic energy distributions that are modelled.

However, it is now known that in many sight lines the UV extinction does not follow the usual galactic law (most notably in a global sense in the LMC and SMC) and Garmany *et al.* (1984) have found suprious results from several galactic WR stars based on UV-optical data, suggestive in many cases of additional, possibly circumstellar, extinction. Anomalous far-UV extinction has also been found for HD 147419 (WN6) by Willis and Stickland (1981) and for CQ Cep (WN7 + O) by Stickland *et al.* (1984). Thus the interstellar extinction corrections need to be carefully assessed for individual objects in order to derive meaningful intrinsic energy distributions. However, current uncertainties in values of T_{eff} for WR stars from this cause are probably not too great — there are sufficient cases where one is confident of the colour excess and extinction law which still produce what may be regarded as uncomfortably low temperatures.

The larger question is the validity of deriving T_{eff} from continuum observations of WR stars. Abbott and Hummer (1985) have argued that the continuum method is unreliable primarily because the stellar flux distribution is as sensitive to gravity as to T_{eff}. They show that the dispersion in the colour temperature for models with the same T_{eff} but different values of log g is larger than for models with the same log g but different T_{eff}. Even if the gravity is known, which is not the case for WR stars, the dependence of colour temperature on T_{eff} is weak.

Following earlier work, based on ANS data, by Van der Hucht *et al.* (1979), Schmutz and Smith (1980) have assessed the confidence level which can be applied in transforming from colour to effective temperatures from UV and optical measurements. Both papers agree that it may not be possible to assign a unique value of T_{eff} since a combination of different temperatures and atmospheric extensions can produce very similar continuum shapes. Equally, as Garmany *et al.* (1984) stress, even down to 1200 Å one is still only measuring the low energy tail of the energy distribution for stars hotter than about 20 000 K, and it may not be ruled out that the true WR effective temperatures are considerably higher than the 30 000—50 000 K range implied by the observed continua.

Support for much higher values of T_{eff} (which could alleviate the mass loss rate problem) has recently come from a detailed study of the WR eclipsing binary system V444 Cyg by Cherepaschuk *et al.* (1984) who deduce from an analysis of the UV, optical and near-IR light curves a small radius of 3 $R_\odot$ and high T_{eff} of about 90 000 K for the WN5 component. Support for these values has been given by Pauldrach *et al.* (1985) who show that the mass loss rate theoretically expected from such a star is similar to that deduced from radio data. If such high values of T_{eff} (and correspondingly increased radiative luminosities) are indeed correct and the norm for WR stars, the problem of reconciling observed mass loss rates with radiation pressure models may well be resolved. However, Barlow *et al.* (1981) point out that the unobserved WR energy distributions shortward of the IUE range would require equivalent temperatures of $> 10^5$ K to produce bolometric luminosities sufficient to radiatively drive the observed mass outflow, which may not be consistent with the observed level of spectral ionisation. In particular it seems unlikely that such high values of T_{eff} would be appropriate for WNL and late-type WC stars which exhibit He I and other low ionisation species in their UV and optical spectra. Values of T_{eff} close to 30 000 K may well be the case for these classes. Very recent non-LTE atmosphere models for WR stars produced by Schmutz and Hamman (1986) support this view.

4. Mass Loss Rates and Wind Velocity Characteristics

Although there is little doubt now that the primary cause of the predominant emission line spectra of WR stars is the presence of very dense supersonic stellar winds, it is only comparatively recently that accurate mass loss rates for this class has been determined. Currently the most reliable method of determining mass loss rates for hot stars uses a combination of UV and radio (or IR) data; the former providing the wind velocities and the latter a measure of the density in the outermost regions (taken to be flowing at a terminal velocity) where the emission arises from free-free processes. Then one can use the simple formulation of Wright and Barlow (1975) for the mass loss rate.

Measurements of v_∞ are now available for a substantial number of WR stars from HIRES IUE spectra, and a plot of v_∞ vs subtype is shown in Figure 2 for both WN and WC stars, adapted from Willis (1982a, b) using additional data given by Willis *et al.* (1986a). In general measured values of v_∞ from UV P-Cygni profiles are significantly larger than previous estimates based on line widths in optical spectra (see below). In the case of WC stars, there is a clear correlation between v_∞ and subtype, analogous to the corresponding correlation of optical line width with subclass found by Smith (1968) and often used as a classification criterion. For the WN stars, the spread within any subclass is greater, although a rough trend is apparent, with the higher excitation subclasses showing the higher speed winds.

Using such values of v_∞ and measurements of IR 10 μm flux excesses, Barlow *et al.* (1981) deduced mass loss rates for 21 galactic WR stars. The results showed a range of only a factor of four, with the mean values being dM/dt(WC) $=$ $4.1 \times 10^{-5} M_\odot$ yr^{-1} and dM/dt(WN) $= 2.7 \times 10^{-5} M_\odot$ yr^{-1}. Subsequent measurements of the radio free-free emission from a substantial sample of WR stars, where one is more sure that the emitting region is moving at a constant, terminal speed,

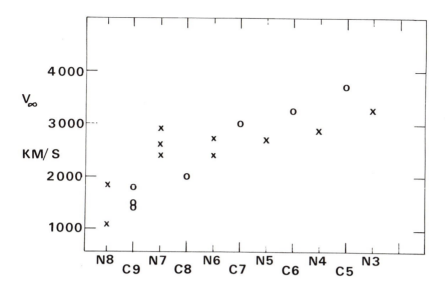

Fig. 2.   Terminal velocities vs. WR subtype derived from IUE HIRES spectra of UV resonance line P-Cygni profiles, adapted from Willis (1982a).

has given similar results (e.g., Abbott *et al.*, 1982; Abbott, 1986; Hogg, 1982). A compilation of such results based on IR and radio data has been given by Willis (1983) and Abbott (1986), and these mass loss rates are shown as a function of subclass in Figure 3a. The main conclusions from these recent studies are that WR mass loss rates are (i) very high (ii) show no difference between WN and WC stars, to within a factor of two or so, (iii) no correlation within the WN and WC subtypes (despite gross spectral differences that are seen), (iv) no differences between single WR's and those in binary systems. If the true effective temperatures of WR stars are 30 000—50 000 K, the derived mass loss rates are too large by factors of 4—50 than can be accomodated by usual radiation pressure driven wind models with single scattering (cf., Figure 3b) and would require an alternative(s) mass loss mechanism. On the other hand it may be possible to reconcile radiation pressure theory and observation if the WR effective temperatures are 10^5 K, although in that case there may be additional problems in reconciling the observed level of spectral ionisation in the winds.

    IUE spectra, particularly in the HIRES mode, have given additional insight into the velocity and ionisation characteristics of the WR winds. Willis (1981, 1982a) has carried out a semi-quantitative analysis of P-Cygni profiles in the IUE spectra of 10 galactic WR stars, complemented with optical measurements, embracing a wide range of I.P. and E.P. of the relevant ionic species and transitions. For each star the velocity associated with the central part of the violet-displaced absorption component in each line (denoted v_0) is found to correlate well with the E.P. of the transition lower level, demonstrating lower degrees of spectral excitation at higher wind velocities. An example of such a correlation is shown in Figure 4a for HD 192103 (WC8) showing conclusively that the WR lines arise in an accelerating

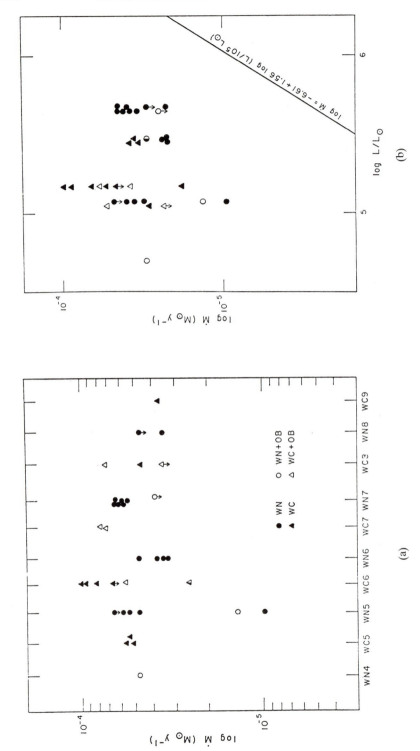

Fig. 3(a). Mass loss rates for WR stars as a function of subtype derived from a combination of IUE measurements of v_{inf} and radio or IR measurements of the free-free wind emission (from Willis, 1983). (b) WR mass loss rates plotted against luminosity adopting values of T_{eff} deduced from WR UV-optical continua. The full line is the correlation found by Abbott et al. (1981) for OB stars, accomodated by usual radiation pressure stellar wind theory.

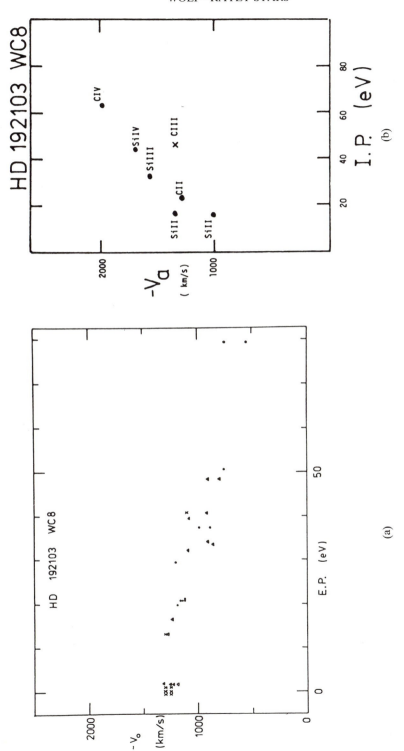

Fig. 4(a). The correlation between velocity of the centre of violet-shifted P-Cygni absorption components versus E.P. of the transition lower level, measured from IUE and optical spectra of HD 192103 (WC8), taken from Willis (1982a). (b) The edge velocity of the UV resonance line P-Cygni profiles versus I.P. of the relevant ion measured from the IUE spectrum of HD 192103 (taken from Willis 1982a).

wind irrespective of whether the level populations are determined by photo-excitation or collisional excitation.

In the case of WC and some WN stars, there is also a marked correlation between displaced absorption edge velocity (denoted v_a) with I.P. for the UV resonance lines, with the higher I.P. species exhibiting the larger values (cf., Figure 4b). This is in marked contrast to previous conclusions based on visible spectra alone (cf., Kuhi, 1973) where higher I.P. emission lines showed narrower profiles interpreted as evidence for a temperature decrease in the outflowing winds. Willis (1983) is able to show that by combining UV and optical measurements, thus spanning the full range of E.P. and I.P., that the earlier, optical, correlations were misleading. Taking both data sets, a good correlation is apparent in emission line width (or v_0) with E.P. but not with I.P. To a first approximation this result suggests that the wind ionisation is roughly constant (or frozen) with radius (at least in the bulk of the emission line region) with the different behaviour of the various transitions reflecting excitation (primarily controlled by density) effects alone. Support for this view has recently been given by Koenigsberger and Auer (1985) in their analysis of the variability of WR emission lines in WN binary systems.

From IUE LORES spectra, Smith and Willis (1983) find similar velocity characteristics for the stellar winds of WR stars in the LMC, suggesting that any global metallicity differences between the Galaxy and the LMC is not reflected in the gross properties of their WR stellar winds.

The extensive mass loss of the WR stars produces important interaction effects with the ambient interstellar medium, most obviously manifested by the occurrence of pronounced ring nebulae surrounding many WR stars (for a review of such ring nebulae see Chu, 1981).

High resolution IUE spectra of HD 192163 (WN6), surrounded by the ring nebula NGC 6888 are reported by Huber et $al.$ (1979) which show strong interstellar lines in the resonance transitions of Si IV, C IV, and Al III with in each case a blue shifted, weak absorption component, displaced by about $90\,\mathrm{km^{-1}\,s}$ — comparable to the expansion velocity of the nebula as measured from optical data. Huber et $al.$ (1979) thus infer that these shifted absorptions are produced in NGC 6888, providing the first detection of absorption lines arising in a nebula associated with a hot, early-type star. From simple absorption line spectroscopy arguments they conclude that these highly ionised absorptions arise in the shock heated inner boundary of the nebula at an ionisation temperature of about $60\,000\,\mathrm{K}$. Further studies of the interstellar Si IV and C IV resonance absorptions seen in the IUE high resolution spectra of 10 WR stars are reported by Smith et $al.$ (1980) who conclude that in most cases the observed line strengths are consistent with those expected to arise in the surrounding H II regions for stellar temperatures $> 30\,000\,\mathrm{K}$.

The star HD 96548 (WN8) is surrounded by the ring nebula RCW 58 which consists of stellar ejecta enclosed in a wind-blown bubble. Smith et $al.$ (1984) report important new results for this nebula based on a detailed analysis of high quality IUE high resolution and optical spectra. These data show blue-shifted absorption components (cf. Figure 5a) in a number of ionic transitions, and it is

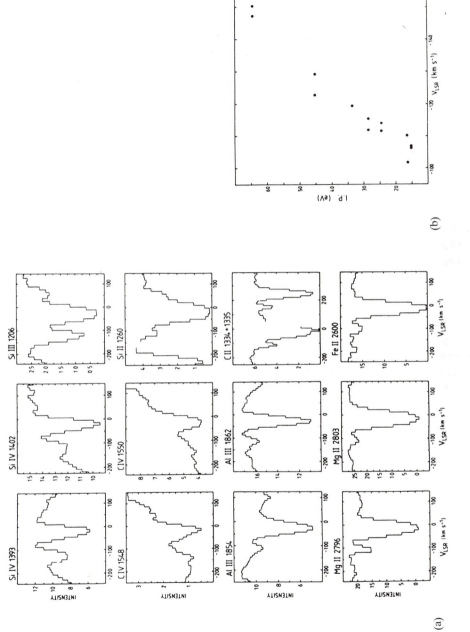

Fig. 5(a). Examples of the UV interstellar lines observed in the IUE spectrum of HD 96548 (WN8) showing the violet shifted components arising in the surrounding ring nebula RCW 58 (taken from Smith et al., 1984). (b) The correlation between velocity and I.P. of the violet shifted components of the UV interstellar lines in HD 96548 (from Smith et al., 1984).

found that the displaced velocities of the components are very well correlated with I.P. in the sense that C IV has the highest velocity (-150 km s^{-1}) and Fe II the lowest (-102 km s^{-1}) as shown in Figure 5b. Smith *et al.* (1984) discuss the significance of this correlation, which is the first such example known, and argue that the absorption components arise behind the inwards-facing shock within the wind blown bubble. However, the results show that the velocity/I.P. slope is too high for a simple post-shock flow requiring significant modifications to usual standard bubble models. They propose that a complex flow occurs in which shocked stellar wind material mixes with cold stellar ejecta and cools as it flows outwards and decelerates. This could explain the apparent lack of energy conservation in WR ring nebulae.

Van der Hucht *et al.* (1981, 1982) report the discovery of displaced narrow absorptions due to Fe III (uv34) transitions arising from a metastable lower level in two WC9 stars, HD 164270, HD 136488. HD 157541 (WC9) also shows this effect. In all 3 stars these features share the common characteristic of being displaced by the same amount relative to the wind terminal velocity, viz. 0.6 v_∞ and are believed to be formed in the decelerating outer region of the WC9 stellar winds. Van der Hucht *et al.* (1982) conclude that these Fe III absorptions are formed at large stellar distances, comparable to those deduced for the substantial dust shells discovered around WC9 stars (typically about 100 AU) and thus infer that some mechanism must be operating in the WC9 winds at this distance to reduce the velocity to about 0.6 its maximum value, accompanied by the formation of dust. It should be noted that Hackwell *et al.* (1979) have given evidence that the dust content around HD 193793 (WC7) is mainly iron, which may therefore also be the case for WC9 dust shells.

5. WR Binaries

Views on WR binaries have altered greatly since the 1960's when it was often assumed that the WR phenomenon was linked to the star's duplicity. It is now recognised that the fraction of WR stars in binary systems is similar to that of normal massive stars (Massey, 1981), although WR binaries as desribed theoretically by Paczynski (1967) may play an important role.

The long-lifetime and observatory nature of IUE has greatly facilitated UV studies of WR + OB binary systems and WR stars with suspected neutron star companions (collapsars). For reviews of such systems based on optical data see Massey (1982) and Moffat (1982) respectively.

5.1. WR + OB SYSTEMS

Stickland *et al.* (1984) have carried out an extensive analysis of the 1.6 day eclipsing binary system CQ Cep (WN7 + O) based on 85 IUE spectra covering essentially all binary phases complemented by new optical and IR photometry and re-measurements of archival optical radial velocities. They find a colour excess of $E_{B-V} = 0.75$ with evidence for an anomalously high level of far-UV extinction. Modelling the observed continuum light curves in 15 wavebands between 0.13 and 3.6 μ indicates that the stellar components are in or near to contact, with typically half the amplitude of the brightness variations arising from geometrical eclipses and half from ellipticity effects. The radial velocity curves from several optical emission and absorption lines are highly disparate and none are found to be a good indicator of the secondary's motion, but rather indicate substantial wind stratification and gas streaming. However, an analysis of the overall excitation-velocity characteristics of the UV P-Cygni profiles indicates little modification of the WN7 wind by its enigmatic companion. With a mass function $f_m = 4.5$, the derived minimum allowed masses are about 20 $M_\odot$ for each star. However in that case inconsistency is found in the luminosity ratio from the available photometry with allowable radii and separation values. It is deduced that to obtain overall consistency the WN7 temperature might need to be greater than about 55 000 K. Another interpretation is offered by Leung *et al.* (1983).

The two WC8 + O systems, Gamma—2 Velorum (WC8 + O9I) and CV Ser (WC8 + O9III − V) have been extensively observed with IUE by Willis *et al.* (1979) and Howarth *et al.* (1982) respectively. Gamma—2 Velorum does not exhibit optical eclipses or substantial line variability, whilst CV Ser was known to be eclipsing in the 1950—1960's but apparently stopped eclipsing by 1970 (cf., Kuhi, 1973) — a puzzling circumstance. For both systems substantial variability in the UV emission lines and P-Cygni profiles are evident in the IUE data, which are interpreted in terms of selective line eclipses of the O-star by the WC8 winds first proposed by Willis and Wilson (1976), with the eclipse effects being confined to resonance and low excitation transitions. For CV Ser Howarth *et al.* (1982) place an upper limit to UV continuum variability of less than 5% which implies an upper limit for the WC8 mass loss rate of $8 \times 10^{-5}\ M_\odot\ yr^{-1}$. Both Willis *et al.* (1979) and Howarth *et al.* (1982) make use of the eclipses seen in the C III 1909 inter-combination transition to derive mass loss rates of $6.3 \times 10^{-5}\ M_\odot\ yr^{-1}$ for CV Ser (which coupled with the above result gives dM/dt (CV Ser) $= 7 \pm 1 \times 10^{-5}$) and for Gamma—2 Velorum of $1.1 \times 10^{-4}\ M_\odot\ yr^{-1}$. In addition for Gamma—2 Vel, Willis *et al.* (1979) deduce a Boltzmann excitation temperature of about 10^4 K for the eclipsing region of the WC8 wind from the derived column densities of C III 1909 and C III 2297. A further IUE study of the CV Ser system has been published by Eaton *et al.* (1985) who suggest that the observed UV variability implies a photospheric flux distribution for the WC8 component below the Lyman Limit characterised by a temperature much larger than 30 000 K, analogous to results secured for V444 Cyg (see below).

The WC6 + O binary Theta Mus does not show the substantial UV line variability exhibited by Gamma—2 Vel and CV Ser (Beekmans *et al.*, 1982). Howarth *et al.* (1982) discuss this apparent anomaly and conclude that the system may be triple with an unseen secondary in close orbit with the WC6 star and the O star sufficiently remote to preclude eclipse effects. Optical RV studies of Theta Mus by Moffat and Seggewiss (1977) provide support for this assertion.

A survey of LORES IUE variability in six WN + OB systems has been carried out by Koenigsberger and Auer (1985) who conclude that (as for the WC systems discussed above) the observed UV changes as a function of binary phase are consistent with selective spectral line atmospheric eclipses of the O companion. Based on this interpretation applied to N IV 1718 data they deduce an optical depth distribution in the WN winds of the form $\tau \sim r^{-1}$ for $16 < r < 66\ R$. They also find supporting evidence for the approximately constant ionisation structure of WR emission line regions proposed by Willis (1982a).

The eclipsing binary V444 Cyg (WN5 + O6) is one of the best studied systems in the optical and UV and has recently become the focus of much attention. Cherpaschuk *et al.* (1984) and Eaton *et al.* (1982) discuss the UV variability of the system based on respectively OAO–2 and IUE data coupled with eclipse measurements at optical and IR wavelengths. From an analysis of the multi-waveband eclipse curves from 0.246 to 3.5 μ they deduce a small core radius of about $3\ R_\odot$ and a high effective temperature of about 90 000 K for the WN5 component. The velocity law in the wind is also derived from these eclipse data. Pauldrach *et al.* (1985) show that such stellar parameters are consistent with a mass loss rate of about $10^{-5}\ M_\odot\ \mathrm{yr}^{-1}$ using their improved model codes for moving atmosphere calculations. If such high values of T_{eff} (and hence radiative luminosities) turn out to be the norm for WR stars, then the apparent discrepancy between deduced mass loss rates and radiative pressure theory for mass loss may become reconciled.

5.2. WR + COLLAPSARS

Van den Heuvel (1976) has reviewed the theoretically expected sequential stages of close binary evolution in massive systems: observationally they correspond to (a) O + O systems, (b) WR + O stage, (c) collapsed object + O (viz. massive X-ray) binaries, and (d) collapsed object + WR systems. All stages but (d) have been unambiguously observed.

Moffat (1982) has reviewed the available optical evidence for the existance of systems involving a WR primary and a collapsed (viz. neutron star) secondary. However, the observed levels of X-ray emission (typically $< 10^{33}\ \mathrm{erg\ s}^{-1}$, e.g., Sanders *et al.*, 1982) are comparable to single O and WR stars and much lower than values of $> 10^{36}\ \mathrm{erg\ s}^{-1}$ appropriate for OB X-ray binaries. To explain this it is usually assumed that the postulated n.s. companion is buried in a dense WR wind which severely attenuates the intrinsic X-ray emission. A very common characteristic of the UV spectra of known massive X-ray binary systems is the pronounced phase dependent variations seen in the P-Cygni profiles of UV resonance lines. This is explained as arising from aspect linked anisotropic ionisation caused by the X-ray emission coming from the orbiting neutron star — the Hatchett and McCray (1977) effect and the observation of this in WR + c candidates would thus provide strong support for their reality. To date several searches for such effects have been made.

Smith *et al.* (1985) do find considerable UV variability in their HIRES IUE spectra of the WR + c candidate system HD 96548 (WN8) as shown in Figure 6, but conclude that the data can only be explained in terms of a compact companion if the X-ray ionisation were to modify the dynamics of the entire wind — (at variance with the requirement to invoke localised and heavy attanuation). Moreover the UV variability is not consistent with the derived photometric optical periodicity. Rather they suggest that the observed line profile changes result from

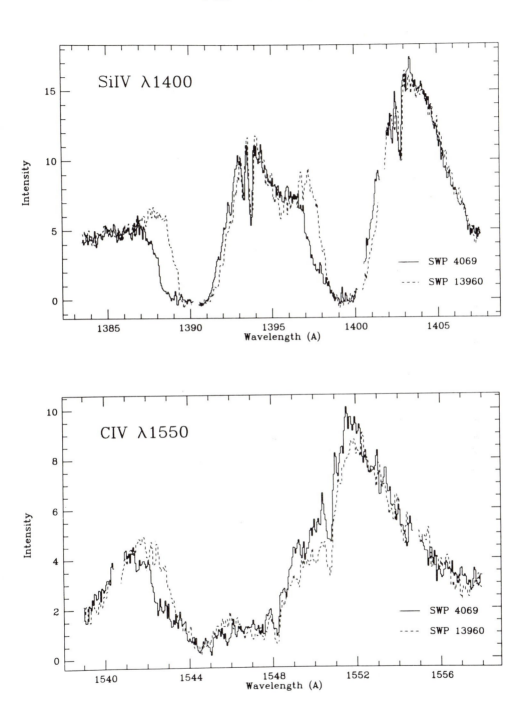

Fig. 6.   Examples of the dramatic profile variations in the Si IV and C IV P-Cygni profiles in two IUE
spectra of HD 96548 (WN8) secured about two years apart (from Smith *et al.*, 1985).

intrinsic stellar wind variability caused by fluctuations in the continuum radiative flux possibly linked to some type of stellar pulsation.

Willis *et al.* (1986b) report initial results of extensive IUE HIRES monitoring of the candidate collapsar system HD 50896 (WN5) based on over 100 IUE spectra collected over 1978—83 and, in particular, continuous monitoring over a 6 day period in 1983. They again find substantial UV profile variations (cf. Figure 7) in

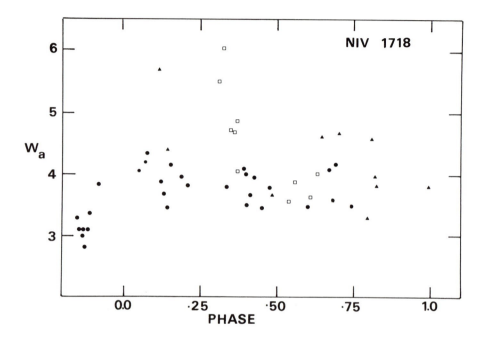

Fig. 7. Measured equivalent widths of the N IV 1718 P-Cygni absorption component in a large number of IUE spectra of HD 50896 (WN5) plotted against phase in the 3.7 day purported binary period. (i) data secured in September 1983 showing relatively small, short-time scale changes at all phases, (ii) data obtained over consecutive days in 1980 showing much larger, secular changes, (iii) data obtained at sporadic epochs during 1978—82 **(taken from Willis *et al.*, 1986b).**

N V, C IV, He II, and N IV transitions but with no phase dependence in the 3.7 day purported binary period which can be associated with a Hatchett and McCray effect. The data provide strong evidence for variability intrinsic to the WN5 star on both the short timescale (viz. hours) and secular, large scale, epoch changes in the velocity and/or ionisation structure of the WR mass outflow. A similar effect has been noted in the X-ray emission from HD 50896 by White and Long (1986), who find changes by a factor of two in only 30 min. A possibility that might explain the observed line profile variations is atmospheric pulsation. WR stars are close to the theoretical limit for vibrational instability (Maeder, 1984) and Vreux (1985) has argued that the observed optical variations are most likely due to non-radial oscillations of a single star rather than linked to any collapsar companion.

Such UV profile variations are not confined to Pop I WR stars; large variations are seen in the IUE spectra of the low mass qWR star HD 45166 as reported by Willis and Stickland (1983).

Thus, to date, the IUE data have been unable to provide any supporting evidence for the existance of neutron star companions to WR stars, but have discovered substantial intrinsic WR wind variability whose origin has yet to be understood.

6. The Chemical Nature of WR Stars

A central question concerning the WR stars is whether their spectral dichotomy into the WN and WC sequences is the result of grossly different chemical abundances or highly peculiar and selective physical effects of excitation and ionisation. Qualitatively from the spectral appearance one expects WN stars to be C-poor (since they show mainly He and N lines) and WC stars to be N-poor (since this sequence shows mainly He, C, and O transitions). In both sequences, optical measurements of the He II Pickering decrement is taken to imply low, and in many cases effectively zero hydrogen content (cf. Smith, 1973, Perry and Conti, 1982). Quantitative measurements of WR abundances have long been plagued by the inherent difficulties in developing suitable models to interpret the observed spectra in which the highly complex treatment of line formation in a rapidly expanding, dense atmosphere involving full non-LTE effects is required. Most attempts make use of the Sobolev approximation and utilise transitions whose relevant level populations are determined by bound-bound processes alone, and thus the continuum transfer can be suitably approximated. For many C and N ions such a treatment is best suited to the analysis of lines arising between resonance and low excitation transitions, which occur in the vacuum ultraviolet. Thus UV observations have made a significant impact on the determination of the chemical nature of WR stars.

Willis (1982b) reviewed knowledge of the chemical composition of WR atmospheres including the first results based on analyses of IUE spectra of both WN and WC stars coupled to deductions (mainly about the H/He ratio) gained from optical data. Garmany and Conti (1982) discuss the IUE and optical evidence for abundance anomalies in WN stars. Preliminary results for the C/N ratio in galactic WR stars is given by Smith and Willis (1982a) and a fuller treatment, including the background theory, for 6 WN and 4 WC stars is given by Smith and Willis (1982b). The latter paper finds a C/N ratio lying in the range 2×10^{-2} to 6×10^{-2} by mass for WNE stars; 6×10^{-3} to 4×10^{-2} for WNL stars and > 60 for WC stars. The lower limit for the WC stars is based on non-detections of nitrogen lines in WC spectra (see below). These Sobolev analyses have been extended to IUE and optical spectophotometry of nine WR stars in the LMC (6 WN and 3 WC) by Smith and Willis (1983) who find that derived C/N, C/He, and N/He ratios exhibit no systematic differences with those deduced for galactic objects. Similar considerations apply to the global mass loss properties of the LMC stars, and Smith and Willis conclude that global metallicity differences between the LMC and the Galaxy are not reflected in the physical and chemical properties of their respective WR populations. Figure 8 shows the C/N

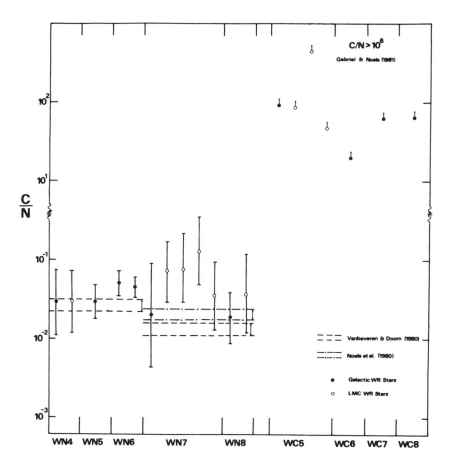

Fig. 8. The C/N abundance ratio for WN and WC stars in the Galaxy and the LMC derived from IUE spectra by Smith and Willis (1983) compared to recent stellar evolutionary calculations. C/N > 10^6 Gabriel and Noels (1981). (— — —) Van Beveren and Doom (1980); (— · — · —) Noels *et al.* (1980); (•) Galactic WR Stars; (○) LMC WR Stars.

ratios deduced for the galactic and LMC stars alluded to above, taken from Smith and Willis (1983). Also shown are results from stellar evolutionary models in which interior nuclear processing products have been exposed either through mass exchange in binary systems or through extensive mass loss for single stars at the end of core H-burning (to be associated roughly with the WN stars) and some way into He-burning (WC stars). The agreement, particularly for the WN stars, is very good and provides strong support for the scenario that WR stars are chemically evolved objects at different stages in the H-burning and He-burning phases of massive hot stars.

With the development of more sophisticated treatments of spectral line formation and continuum transfer in dense stellar winds, it is becoming possible to test and refine these results based on simplified Sobolev modelling. Such developments are still at an early stage, but recent work appears to confirm the earlier results.

Hillier (1986) has carried out a very detailed analysis of the optical and IUE spectrum of the WN5 star HD 50896 and found the following abundance ratios for this star (by number): $C/He = 3 \times 10^{-5}$ and $N/He = 10^{-3}$, very similar to the results for this star given by Smith and Willis (1982b).

A high carbon abundance has been derived for the (rare) WC11 star CPD −56 8032 based on IUE LORES spectra by Houziaux and Heck (1982) who find $\log[C] = 9.39$ compared to the solar value of 8.55. However, the status of this object is somewhat uncertain, being denoted as the central star of a compact planetary nebula by Van der Hucht *et al.* (1981).

For WC stars, available results based on IUE spectra point only to a lower limit for the C/N ratio of > 60 (Smith and Willis, 1983), a result which is constrained by the lack of any unambiguous nitrogen-line identifications in WC spectra, as discussed above. Theoretically from evolutionary models (cf., Maeder, 1984) it is expected that at the onset of helium burning, any N present (e.g. from CNO-burning in the WN stage) will be very rapidly destroyed whilst the carbon is quickly replenished by triple-alpha processing, so that the expected N-abundance in WC stars should be effectively zero. Whilst current results do not conflict with this expectation, it may be that the true N-abundance is not zero for WC stars. Stickland and Willis (1982) present an analysis of the IUE spectra of the WN6—C4 star HD 62910 (presumed to be a single) which does show both C and N lines suitable for Sobolev modelling. Their abundance analysis gives a C/N ratio of about 0.5 by mass (intermediate between WN and WC results) whilst the N/He ratio derived of 7×10^{-2} is close to that for normal WN stars. Clearly the carbon (very deficient in WN stars) has been quickly replenished (presumably by the triple-alpha process) without destroying much of the nitrogen in this object. Of course, the galactic population of known WN—C stars (3) is very low and there may not be a major problem in reconciling this result with evolutionary expectations when lifetime considerations are taken into account.

7. The Evolutionary Status of WR Stars

The evolutionary status of the WR stars has been, and still is, the subject of much diverse opinion, reflecting the considerable uncertainties that have existed in our knowledge of their basic parameters of mass, T_{eff}, luminosity and chemical composition. Because it is extremely difficult to quantitatively and unambiguously analyse WR spectra, progress has relied upon piecing together evidence from a wide range of observational techniques. There is still much uncertainty but over the past decade considerable consensus has been reached that WR stars represent evolved stars which are the core He-burning descendents of more massive early-type progenitors in which outer atmospheric stripping by mass loss (and/or mass transfer in binary systems) has played a major role. Smith (1973) gave the first comprehensive discussion of the possibility that WR stars are core He-burning objects whilst Conti (1976) first suggested that once mass loss has stripped the hydrogen-rich outer layers from an O-star to the point where CNO-burning products are revealed, the star would be identified as WN. These views have generally been confirmed and amplified in recent years according to the following evidence:

(a) the atmospheric chemical analyses point to an abnormally low H-content (Willis, 1982b). For the WN stars, direct measurements of the observed He II Pickering decrement yield H/He ratios (by number) between 0 and 0.8, with about 1/3 of all WN stars showing evidence for some H and in a few cases relatively high values of < 5 (Conti *et al.* 1983). For the WC stars there is no evidence for any hydrogen (Willis, 1981; Torres *et al.*, 1986). The C/N ratio, about 0.01 by mass, deduced from UV spectra (Section 6) for WN stars is found to agree well with values expected for the CNO-burnng equilibrium products and is transformed to much higher values of > 60 for WC stars consistent with expectations for He-burning products. In both cases the results imply the exposition of interior nuclear processed material which requires extensive mixing and/or mass loss.

(b) we now know that the WR mass loss rates are very large, 10^{-5} to $10^{-4} M_\odot \mathrm{yr}^{-1}$ (Section 4) and moreoever that mass loss during the H-burning lifetime of an O-star can radically effect its evolution and can peel off sufficient of the outer atmosphere to reveal nuclear processed material. Indeed an enhanced He abundance has been found for the O4f star Zeta Pup (Hamman, 1985; Kudritzki and Hummer, 1986) which illustrates this, and some Of stars are known to show spectroscopic characteristics in the UV and optical intermediate between Of and WN spectra (Willis and Stickland, 1980; Nandy *et al.*, 1980). From a study of WR + O double lined spectroscopic binaries Massey (1981) has deduced WR masses in the range $10-25\ M_\odot$. If the current O-star companion's mass is shifted along its evolutionary track to the ZAMS, and assuming an initial mass ratio of unity, Massey concludes that the WR component has lost at least 40% of its initial mass.

(c) Although the T_{eff}'s for WR stars are still uncertain (Section 3) there is no doubt that we are dealing with high radiative luminosities. If T_{eff} is in the range 30 000—50 000 K (as deduced from the UV-optical continua) then B.C.'s of 3—4 mag are implied which suggests $M_{\mathrm{Bol}} < -9$. Theoretical evolution models would place such stars either on, or near the He-burning main-sequence for stellar masses of 8—14 $M_\odot$ or as H-burning stars with masses in the range 30—45 $M_\odot$. Since WR masses in binaries lie in the range $10-25\ M_\odot$ (Massey *et al.*, 1981) the second possibility is the less likely. WR T_{eff}'s of > 10^5 K would reinforce this conclusion.

Large Bolometric Corrections are also inferred, independent of temperature considerations, from the visual magnitudes of galactic WR stars in clusters. Although a somewhat circular argument, if one assumes that WR stars have evolved from stars at least as massive as the earliest O-star in a cluster the difference between M_{Bol} for the O-star and M_V for the WR stars gives a minimum B.C. for the latter. Using this method Humphreys *et al.* (1984) and Schild and Maeder (1984) have deduced B.C.'s for WR stars of about 3 mag.

(d) Smith (1973) discussed the statistics of WR and O stars and showed that the former are not particularly rare if they represent He-burning descendents of more massive progenitors of mass > 10 $M_\odot$. Garmany, Conti and Chiosi (1982) from a census of galactic O stars showed that the most massive O stars and the WR stars are preferentially found within the Sagitarrius-Carina arm, implying a lower limit to the initial mass of the WR progenitor of $> 30\ M_\odot$. Althugh Conti *et al.* (1983) have stressed that the asymmetric galactic distribution of WR stars could be explained by different IMF's in different parts of the Galaxy for the massive O stars, such possible differences in IMF are not crucial; there is no question that there are more stars of initial mass greater than about 40 $M_\odot$ in the galactic arm that also contains the majority of WR stars.

We have emphasised the current consensus that WR stars represent evolved objects and reviewed the supporting evidence for this view. A completely different scenario is given by Underhill (1983, and references therein) who considers WR stars as unevolved, yet abnormal, late O and early B stars which possess extensive outer 'mantles' in which high magnetic fields and plumes control the plasma in which the observed emission line spectra are formed.

Clearly there is still much to learn about the WR stars and the WR phenomenon. We need to unambiguously locate them on the HR diagram, requiring accurate measurements of effective temperatures and luminosities and further work on their chemistry is needed. Do the different subclasses in the WN and WC sequences represent evolution sequences and do all WN stars evolve into WC stars? What is the driving mechanism(s) for the WR mass loss and what are the velocity and temperature laws in their winds? What is the cause of the observed WR wind variability? Do WR collapsar systems really exist? Finally, what is the reason for the gross differences in the WR populations in different galaxies and is this linked to different IMF's and SFR's in these different environments?

Clearly these, and other questions will require extensive further observations across the electromagnetic spectrum for their solution coupled to improved theoretical models. Future UV observations with the Hubble Space Telescope and, in particular, with missions emphasising the largely unexplored region below 1200 Å will continue to play a major role in furthering this field.

Acknowledgements

We are grateful to Dr David Abbott for several helpful comments on this manuscript.

References

Abbott, D. C.: 1985, in R. Hjellming and D. Gibson (eds.), *Radio Stars*, D. Reidel Publ. Co., Dordrecht, Holland, p. 61.

Abbott, D. C. and Hummer, D. G.: 1985, *Astrophys. J.* **294**, 286.

Abbott, D. C., Bieging, J. H., Churchwell, E. B., and Torres, A. V.: 1986, *Astrophys. J.* **303**, 239.

Abbott, D. C., Bieging, J. H., and Churchwell, E.: 1982, in C. de Loore and A. J. Willis (eds.), 'Radio Continuum Measurements of Mars Loss from Wolf-Rayet Stars', *IAU Symp* **99**, 215.

Barlow, M. J., Smith, L. J., and Willis, A. J.: 1981, *Monthly Notices Roy. Astron. Soc.* **196**, 101.

Beeckmans, F., Grady, C. A., Macchetto, F., and van der Hucht, K. A.: 1982, in C. de Loore and A. J. Willis (eds.), 'Spectral Variations of Theta Muscae (WC + 09.5I) in the Ultraviolet', *IAU Symp.* **99**, 311.

Cherepaschuk, A. M., Eaton, J. A., and Khalliullin, Kh. F.: 1984, *Astrophys. J.* **281**, 774.

Chu, Y. H.: 1981, *Astrophys. J.* **249**, 195.

Conti, P. S.: 1976, *Mem. Soc. Roy. Sci. Liege* **9**, 193.

Conti, P. S., Garmany, C. D., de Loore, C., and Vanbeveren, D.: 1983, *Astrophys. J.* **274**, 302.

Conti, P. S., Leep, E. M., and Perry, D. N.: 1983, *Astrophys. J.* **268**, 228.

Eaton, J. A., Cherepaschuk, A. M., Khaliullin, Kh. F.: 1982, in Y. Kondo, J. M. Mead, and R. D. Chapman (eds.), *Advances in UV Astronomy*, NASA CP-2238, p. 542.

Eaton, J. A., Cherepaschuk, A. M., and Khaliullin, Kh. F.: 1985, *Astrophys. J.* **296**, 222.

Fitzpatrick, E. D.: 1982, *Astrophys. J. Letters* **261**, L91.

Fitzpatrick, E. D., Savage, B. D., and Sitko, M. L.: 1982, *Astrophys. J.* **256**, 578.

Gabriel, M., Noels, A., 1981, *Astron. Astrophys.* **94**, < 1.

Garmany, C. D. and Conti, P. S.: 1982, in C. Loore and A. J. Willis (eds.), 'Chemical Composition of WR Stars: Abundant Evidence for Anomalies', *IAU Symp.* **99**, 105.

Garmany, C. D., Conti, P. S., and Chiosi, C.: 1982, *Astrophys. J.* **263**, 777.

Garmany, C. D., Massey, P. M., and Conti, P. S.: 1984, *Astrophys. J.* **278**, 333.

Hackwell, J. A., Gehrz, R. D., and Smith, J. R.: 1974, *Astrophys. J.* **192**, 383.

Hamman, W. R.: 1985, *Astron. Astrophys.* **145**, 443.

Hatchett, S. P. and McCray, R.: 1977, *Astrophys. J.* **211**, 552.

Hillier, D.: 1986, in C. de Loore, A. J. Willis, and P. Laskarides (eds.), 'The Formation of Nitrogen and Carbon Lines in HD 50896 (WNS)', *IAU Symp.* **116**, 261.

Hogg, D. E.: 1982, in C. de Loore and A. J. Willis (eds.), 'Radio Emission from WR Stars', *IAU Symp.* **99**, 221.

Howarth, I. D., Willis, A. J., and Stickland, D. J.: 1982, *Proc. 2nd European IUE Conf.*, B. Battrick (ed.), ESA Sp-176, 331.

Houziaux, L. and Heck, A.: 1982, in C. de Loore and A. J. Willis (eds.), 'Carbon Abundance in the WC 11 Star CPD −56°8032', *IAU Symp.* **99**, 139.

Huber, M. C. E., Nussbaumer, H., Smith, L. J., Willis, A. J., and Wilson, R.: 1979, *Nature* **27**, 697.

Huber, M. C. E., Nussbamer, H., Smith, L. J., Willis, A. J., and Wilson, R.: 1979, *Nature* **278**, 697.

Humphreys, R. M., Crampton, D., Cowley, A. P., and Thompson, I. B.: 1984, *Publ. Astron. Soc. Pacific* **96**, 811.

Koenigsberger, G. and Auer, L. H.: 1985, *Astrophys. J.* **294**, 255.

Kudritzki, R. P. and Hummer, D. G.: 1986, in C. de Loore, A. J. Willis, and P. Laskarides (eds.), 'Intrinsic Parameters of Hot Blue Stars', *IAU Symp.* **116**, 3.

Kuhi, L. V.: 1973, in M. Bappu and J. Sahade (eds.), 'Wolf—Rayet Binaries and Atmospheric Stratification', *IAU Symp.* **49**, 205.

Kurucz, R. L.: 1979, *Astrophys. J. Suppl.* **40**, 1.

Leung, K. C., Moffat, A. F. J., and Seggewiss, W.: 1983, *Astrophys. J.* **265**, 961.

de Loore, C. W. H. and Willis, A. J. (eds.): 1982, 'Wolf—Rayet Stars: Observations, Physics and Evolution', *IAU Symp.* **99**, 636 pp.

de Loore, C. W. H., Willis, A. J., and Laskarides, P. (eds.): 1986, 'Luminous Stars and Associations in Galaxies', *IAU Symp.* **116**, 534 pp.

Maeder, A.: 1984, in A. Maeder and A. Renzini (eds.), 'Evolution of Massive Stars, Supergiants and Wolf-Rayet Stars', *IAU Symp.* **105**, 299.

Massey, P. M.: 1981, *Astrophys. J.* **246**, 153.

Massey, P. M.: 1982, in C. de Loore and A. J. Willis (eds.), 'WR Stars with Massive Comparisons', *IAU Symp.* **99**, 251.

Massey, P. M., Conti, P. S., and Niemela, V.: 1981, *Astrophys. J.* **246**, 145.

Moffat, A. F. J.: 1982, in C. de Loore and A. J. Willis (eds.), 'WR Stars with Compact Comparisons', *IAU Symp.* **99**, 263.

Moffat, A. F. J. and Seggewiss, W.: 1977, *Astron. Astrophys.* **54**, 607.

Nandy, K., Morgan, D. H., Gondhalekar, P. M., Willis, A. J.: 1980, *Monthly Notices Roy. Astron. Soc.* **193**, 43.

Noels, A., Conti, P. S., Gabriel, M., Vieux, J. M., 1980, *Astron. Astrophys.* **92**, 242.

Nussbaumer, H. and Schmutz, W.: 1986, *Astron. Astrophys.* **154**, 100.

Nussbaumer, H., Schmutz, W., Smith, L. J., and Willis, A. J.: 1982, *Astron. Astrophys. Suppl.* **47**, 257.

Paczynski, B.: 1967, *Acta Astron.* **20**, 47.

Pauldrach, A., Puls, J., Hummer, D. G., and Kudritzki, R. P.: 1985, *Astron. Astrophys.* **148**, L1.

Perry, D. and Conti, P. S.: 1982, in C. de Loore and A. J. Willis (eds.), 'H/He Ratios for WN Stars in the LMC and the Galaxy', *IAU Symp.* **99**, 109.

Sanders, W. T., Cassinelli, J. P., and van der Hucht, K. A.: 1982, in C. de Loore and A. J. Willis (eds.), 'X-Rays from Wolf-Rayet Stars Observed by the Einstein Observatory', *IAU Symp.* **99**, 589.

Schild, H. and Maeder, A.: 1984, *Astron. Astrophys.* **136**, 237.

Schmutz, W.: 1982, in C. de Loore and A. J. Willis (eds.), 'The Effective Temperatures of Early WR Stars', *IAU Symp.* **99**, 23.

Schmutz, W. and Smith, L. J.: 1980, *Proc. Symp. 2nd European IUE Conf.*, B. Battrick (ed.), ESA Sp-157, p. 249.

Schmutz, W. and Hamann, W. R.: 1986, *Astron. Astrophys.* **166**, L11.

Schmutz, W., Morossi, C., Ramella, M.: 1984, *Proc. Symp. 4th European IUE Conf.*, E. Rolfe and B. Battrick (eds.), ESA Sp-218, p. 325.

Smith, L. F.: 1968, *Monthly Notices Roy. Astron. Soc.* **138**, 109.

Smith, L. F.: 1973, in M. Bappu and J. Sahade (eds.), 'Classification and Distribution of WR Stars and an Interpretation of the WN Sequence', *IAU Symp.* **49**, 15.

Smith, L. J. and Willis, A. J.: 1982a, in C. de Loore and A. J. Willis (eds.), 'The C/N Ratio in WN and WC Stars', *IAU Symp.* **99**, 113.

Smith, L. J. and Willis, A. J.: 1982b, *Monthly Notices Roy Astron. Soc.* **201**, 451.

Smith, L. J. and Willis, A. J.: 1983, *Astron. Astrophys. Suppl.* **54**, 229.

Smith, L. J., Willis, A. J., and Wilson, R.: 1980, *Monthly Notices Roy. Astron. Soc.* **191**, 331.

Smith, L. J., Pettini, M., Dyson, J. E., and Hartquist, T. W.: 1984, *Monthly Notices Roy. Astron. Soc.* **211**, 679.

Smith, L. J., Lloyd C., and Walker, E. N.: 1985, *Astron. Astrophys.* **146**, 307.

Stickland, D. J. and Willis, A. J.: 1982, in C. de Loore and A. J. Willis (eds.), 'IUE Observations of the WN–C Star MD 62910', *IAU Symp.* **99**, 491.

Stickland, D. J., Bromage, G. E., Budding, E., Burton, W. M., Howarth, I. D., Jameson, R., Sherrington, M. R., and Willis, A. J.: 1984, *Astron. Astrophys.* **134**, 35.

Torres, A., Conti, P. S., and Massey, P. M.: 1986, *Astrophys. J.* **300**, 379.

Underhill, A. B.: 1980, *Astrophys. J.* **239**, 220.

Underhill, A. B.: 1981, *Astrophys. J.* **244**, 963.

Underhill, A. B.: 1983, *Astrophys. J.* **265**, 933.

Vanbeveren, D., Doom, C., 1980, *Astron. Astrophys.* **88**, 230.

Van den Heuvel, E. P. J.: 1976, in P. Eggleton, S. Mitton, and J. Whelan (eds.), 'Late Stages of Close Binary Systems', *IAU Symp.* **73**, 35.

Van der Hucht, K. A., Cassinelli, J. P., Wesselius, P. R., and Wu, C. C.: 1979, *Astron. Astrophys. Suppl.* **38**, 279.

Van der Hucht, K. A., Conti, P. S., Lundstrom, I., Stenholm, B.: 1981, *Space Sci. Rev.* **28**, 227.

Van der Hucht, K. A. and Conti, P. S.: 1981, 'The Iron Curtain of the WC9 Star HD 164270', in C. Chiosi and R. Stalio (eds.), *Effects of Mass Loss on Stellar Evolution*, D. Reidel Publ. Co., Dordrecht, Holland, pp. 35–37.

Van der Hucht, K. A., Conti, P. S., and Willis, A. J.: 1982, in C. de Loore and A. J. Willis (eds.), 'The Iron Curtain of WC9 Stars', *IAU Symp.* **99**, 277.

Vreux, J. M.: 1985, *Publ. Astron. Soc. Pacific* **97**, 274.

White, R. L. and Long, K. S.: 1986, *Astrophys. J.* **310**, 832.

Willis, A. J.: 1980, in *Proc 2nd European IUE Conf.*, B. Battrick (ed.), ESA Sp-157, p. il.

Willis, A. J.: 1981, 'Effects of Mars Loss or Stellar Evolution', in C. Chiosi and R. Stalio (eds.), *The Velocity Characteristics of WR Stellar Winds*, D. Reidel Publ. Co., Dordrecht, Holland, pp. 27–33.

Willis, A. J.: 1982a, *Monthly Notices Roy. Astron. Soc.* **19**, 897.

Willis, A. J.: 1982b, in C. de Loore and A. J. Willis (eds.), 'The Chemical Composition of the WR-Stars', *IAU Symp.* **99**, 87.

Willis, A. J.: 1983, In *Proc. 1st Trieste Workshop on Stellar Atoms. Velocity Fields*, R. Stalio (ed.), Trieste, p. 111.

Willis, A. J. and Wilson, R.: 1976, *Astron. Astrophys.* **47**, 429.

Willis, A. J. and Stickland, D. J.: 1980, *Monthly Notices Roy. Astron. Soc* **190**, 27.

Willis, A. J. and Stickland, D. J.: 1981, *Monthly Notices Roy. Astron. Soc.* **197**, 1.

Willis, A. J. and Stickland, D. J.: 1983, *Monthly Notices Roy. Astron. Soc.* **203**, 619.

Willis, A. J., Wilson, R., Macchetto, F., Beekmans, F., Stickland, D. J., and van der Hucht, K. A.: 1979, in *Proc Symp: 1st year of IUE*, A. J. Willis (ed.), Univ. Coll. London., p. 394.

Willis, A. J., van der Hucht, K. A., Conti, P. S., and Garmany, C. D.: 1986a, *Astron. Astrophys. Suppl.* **63**, 417.

Willis, A. J., Howarth, I. D., Conti, P. S., and Garmany, C. D.: 1986b, in C. de Loore and A. J. Willis (eds.), *IAU Symp.* **99**, 257.

Wright, A. E. and Barlow, M. J.: 1975, *Monthly Notices Roy. Astron. Soc.* **170**, 41.

BE STAR PHENOMENA

THEODORE P. SNOW

Center for Astrophysics and Space Astronomy, University of Colorado, Boulder, CO, U.S.A.

and

ROBERTO STALIO

Department of Astronomy, University of Trieste, Italy

1. Introduction

In order to discuss the impact of IUE observations on studies of Be star phenomena, we must first address the question of what a Be star is. The fact that this is required says something significant about the field: Be stars represent an assortment of complex phenomena for which there is still no comprehensive theoretical understanding. If there were, we might have less trouble saying just what a Be star is.

Classically, Be stars have been defined as B stars on or near the main sequence that show emission in the Balmer lines. It was soon discovered that many stars show emission at some times but not at others, so the definition extended to include stars that have *ever* shown emission in the Balmer lines. Then B stars were found with circumstellar absorption lines that seemed to indicate the presence of an equatorial disk, and the 'shell' stars found themselves joined with Be stars in a general class of B stars with circumstellar gas. Soon there were discovered the so-called 'P Cygni' stars, which clearly show Balmer emission, but just as clearly have very properties from the classical Be stars. Add to the list the "peculiar" Be stars, which show forbidden emission lines and are generally associated with nebulosity; the 'Herbig Ae and Be stars', which appear to be pre-main sequence objects; and finally the Oe stars, main sequence O stars that may either be viewed as extensions of the more luminous Of star class, or as simply hotter versions of classical Be stars. Every conference devoted to the topic of Be stars brings together people with diverse views as to what a Be star is, and every time, the first session is devoted to wrangling with the issue.

This review paper is no exception, but there are only two of us, and we have the uncontested power to define Be stars as we see fit for our purposes. The reader may or may not agree, and will be correspondingly satisfied or disappointed in our coverage.

We choose to define Be stars as stars on or near the main sequence which have optical emission or absorption lines produced in *intrinsic* circumstellar material; that is, circumstellar material that is not acquired from a companion or a surrounding nebulosity, but which instead apparently is produced by the star itself. Thus, we include the 'classical' Be stars and shell stars as well as their extensions into the late O and early A main sequence, but we exclude the P Cygni stars (which are supergiants), the peculiar Be stars (which are embedded in nebulosity),

Y. Kondo (ed.), Scientific Accomplishments of the IUE, pp. 183–201.
© 1987 *by D. Reidel Publishing Company.*

and the Herbig Ae and Be stars (also associated with nebulosity, although some have been interpreted as Be stars with unusually high-density intrinsic circumstellar material; de Freitas Pacheco and Gilra, 1982; Selvelli and De Araujo, 1984). Some readers may be very unhappy that we have also excluded B stars that have acquired circumstellar material via mass transfer from a binary companion, because it can be argued that some Be stars are Be stars because of mass transfer. Here we expose a bias: we believe that there is a large class of Be stars that are not in mass-exchange binary systems, and these are the stars we have assigned ourselves to discuss.

The Be stars as we have defined them lie near or on the main sequence, in the spectral type range from about O9 to B9, and are usually assigned to luminosity classes IV or V. As already mentioned, one of the chief defining characteristics of Be stars is the presence of Balmer line emission, sometimes limited to only H-alpha, and sometimes extending to high members of the Balmer series. Linear polarization is often present in visible wavelengths. Historically, it has been noticed that Be stars often have high projected rotational velocities, and the idea that these stars are generally rapid intrinsic rotators has been a persistent feature of many models.

The impact of IUE data on Be star research has been enormous. As early-type stars, Be stars have strong ultraviolet continua, so the ability to study the properties of the stars themselves is enhanced by the availability of ultraviolet spectra. Furthermore, ultraviolet observations have revealed wholly new phenomena associated with Be stars, particularly high-velocity stellar winds, that are not detected by optical techniques. Many ultraviolet observations of Be stars were made with earlier instruments such as *S2/68* experiment aboard *TD−1, ANS*, and *OAO 2*, but the spectral resolution in these cases was limited, and the primary results were studies of flux distributions and the measurements of very strong (usually photospheric) lines. Moderate- to high-resolution ultraviolet spectroscopy of Be stars has been carried out for limited numbers of objects by such experiments as sounding rocket spectrographs (e.g., Heap, 1975), the *S59* spectrometer aboard the *TD1* satellite, the Dutch-U.S. *BUSS* project (e.g., Morgan *et al.*, 1977; Bruhweiler *et al.*, 1978, 1979), and especially *Copernicus*, which had the advantage of an eight-year lifetime but the disadvantage of being a scanning instrument, requiring several minutes or hours to observe separate spectral features in a single star. The IUE has profoundly extended the ultraviolet spectroscopic study of Be stars, in at least two ways: (1) the IUE is more sensitive than *Copernicus* (and the other spectroscopic instruments mentioned above), so the number of objects for study has increased dramatically; and (2) IUE provides full spectra simultaneously, so that studies of time variability of Be phenomena are much more meaningful. A further major advance made by IUE in this field has been the long lifetime of the satellite and the resulting opportunity to study Be stars for variability over many timescales.

With the IUE, it has proven feasible in many cases to make either coordinated or simultaneous observations of objects in different wavelength bands. This kind of approach has been especially fruitful for Be star studies, because there are very important questions about the relationship of, for example, the ultraviolet indicators of stellar winds and the visible-wavelength emission and shell lines.

Simply measuring the flux distribution over broad wavelength bands such as has been done by Sitko and Savage (1980) is very useful, but even more important is to make such measurements at different times, to see how the ultraviolet, visible, and perhaps infrared diagnostics of circumstellar material compare as the star varies. Some long-term studies of this type have been carried out (e.g. Doazan *et al.*, 1985; Grady *et al.*, 1984), and will be discussed in a later section of this review.

There was a comprehensive review of ultraviolet observations of Be stars by Marlborough (1982), which included a complete summary of IUE Be star work up to that time, so we will not refer in detail to results covered in that review. Otherwise, we are attempting to provide complete coverage, but cannot guarantee that we have not overlooked some results (particularly those still in preprint form at the time of this writing).

2. Properties of the Stars

The IUE has been used more for studies of circumstellar material around Be stars than for studies of the stars themselves. Most of the work on basic stellar parameters such as flux distributions and spectral classification using UV lines was done using data from earlier ultraviolet instruments. Therefore, we will say little here, except to briefly review some of the chief results of those earlier studies in order to set the context for the IUE research to be described later.

It was demonstrated long ago (see Hack and Struve, 1970, and references cited therein) that Be stars have photospheric properties very similar to normal B stars of the same spectral type (with the exception that Be stars tend to be more rapid rotators). Ultraviolet data appear to bear this out, with some differences. The photospheric spectrum is very similar for Be and normal B stars, for example, but the continuous flux distribution is different: Be stars, particularly the early types, have a UV flux deficiency that has been attributed primarily to hydrogen bound-free absorption in the circumstellar shell (Marlborough, 1982, and references cited therein). The evidence on this point has been quite contradictory, however, with some reports of UV flux *excesses* in some stars at times when strong shell absorption lines were present, just the opposite of what would be expected from shell line blanketing (Barylak and Doazan, 1986). In yet other cases, no difference between the flux distributions in Be and normal B stars was seen (e.g., Polidan *et al.*, 1986).

Spectroscopic observations of Be stars have shown some subtle contrasts between B and Be stars, most of which may actually be attributable to circumstellar material. For example, it was found that certain photospheric lines, such as Fe III lines (e.g., Heap, 1975) are generally enhanced in Be stars, but it has been found that the enhanced lines are due to circumstellar material. Some of the enhanced Fe III lines are attributed to circumstellar shell absorption (e.g., Beeckmans, 1975) and others, first observed at low resolution, have been found due to C IV (e.g., Barbier and Swings, 1979), an ion not initially expected to be present in B-star spectra. It now appears safe to conclude that the underlying photospheric line spectra of Be stars are indistinguishable from normal B stars.

One interesting trend, first noticed in Be stars (Heap, 1975), but evidently linked to rotation, rather than the Be character of a star *per se*, is that the photospheric lines often appear narrower in the ultraviolet than in the visible. The apparent reason for this, first suggested by Morton *et al.* (1972) and later analyzed quantitatively by Hutchings (1976) and Sonneborn and Collins (1977), is that in a rapidly rotating star the equatorial region has a lower effective temperature than the polar latitudes, so that a disproportionate fraction of the observed UV flux comes from the poles, which have a lower projected rotational velocity than the equator. One of the most useful aspects of this is that the ratio of line widths between the visible and ultraviolet spectra of a given star could in principle be used to deduce the inclination angle of the star and therefore the true rotational velocity, but in practice this is difficult because of complications due to rotational distortion of the star.

Another approach to the problem of determining the inclination angles of Be stars has been used by Polidan *et al.* (1986), who have compared the observed far-UV flux distribution (from the *Voyager* spacecraft) with model results. These authors find that the true rotational velocity is always less than about 85% of 'breakup' velocity.

3. The Extended Atmosphere

The most important contributions to Be star research from ultraviolet observations in general and from IUE in particular have come in the study of circumstellar material. Here it becomes difficult to organize a review such as this, because so many seemingly different phenomena are doubtless intimately related. Nevertheless, in an attempt to make some organized sense of it, we discuss the phenomena in the following order: (1) the so-called 'cool' circumstellar material; (2) the 'superionized' material; (3) the high-velocity stellar winds characteristic of most, if not all, Be stars; and (4) the time variability that occurs in *all* aspects of circumstellar material. We will describe the observational results in the following subsections, and then briefly discuss some attempts at tying the diverse phenomena together into models.

3.1. MATERIAL OF NORMAL IONIZATION

By 'normal ionization', we mean 'cool' material that is no more ionized than the underlying photosphere, and which therefore could be in radiative equilibrium with it. This therefore includes the gas that produces shell absorption lines due to such species as Fe II and Fe III, but not the C IV lines that are often seen. In the visible, the cool circumstellar material produces hydrogen shell absorption lines, is probably the origin of the infrared excesses seen in many Be stars (and usually attributed to free-free emission), and is usually considered responsible for the linear polarization that characterizes many Be stars. The ultraviolet signatures of this material have proven more elusive.

One of the major reasons for analyzing circumstellar absorption lines is to derive the geometry of the extended envelope in which the lines arise. Visible-wavelength data have long been interpreted as showing that the circumstellar material that produces shell lines is equatorially concentrated, because of the observed correlation between projected rotational velocity and the presence of shell lines (e.g., Hack and Struve, 1970), but some early ultraviolet spectroscopy with *Copernicus* seemed to indicate that low-ionization UV shell lines (such as Fe II and Fe III lines) were seen even in stars that are viewed pole-on (Snow *et al.*, 1979). This has been disputed more recently, however, (Oegerle and Polidan, 1984), and the question remains open. Clearly it is of great importance for Be star models to know whether the cool circumstellar material is in the form of a disk or is more globally distributed. Unfortunately, the IUE does not cover the region (near 1130 A) where most of the Fe II and Fe III lines observed with *Copernicus* arise, so no definitive study of these low-ionization UV shell lines has been carried out using IUE.

Some IUE studies of other features arising in low-ionization circumstellar material have been carried out, however. In a study of three objects by Ringuelet *et al.* (1981), evidence for two distinct zones was found. One, characterized by relatively narrow absorption lines, rotates with the star, while the other, represented by broad, shallow lines, apparently does not. The three stars (27 CMa, π Aqr, and 48 Lib) all have high projected rotational velocities and are therefore seen more or less equator-on, so no information was derived on the distribution of the observed material as a function of latitude.

Some evidence for *very* cool circumstellar material in Be stars has been reported by Tarafdar (1983) and Tarafdar and Krishna Swamy (1981), who report finding carbon monoxide absorption in the spectra of Be stars. Arguments are made that the CO must be circumstellar, but it remains to be explained how the molecules could survive the intense ultraviolet radiation field so near a B star. If the identification of CO is correct, the possibility that it is due to foreground interstellar material needs to be reexamined.

3.2. SUPERIONIZED CIRCUMSTELLAR MATERIAL

It has been found in all classes of O and B stars that species can be present showing a higher degree of ionization than can be explained by radiative equilibrium with the underlying star. A summary of this phenomenon based on *Copernicus* data was published by Lamers and Snow (1978), but few Be stars were included in their list (furthermore, *Copernicus* had low sensitivity at the wavelengths of two of the most important indicators of superionization in B stars, the C IV doublet near 1550 Å, and the Si IV doublet at 1400 Å). Some early evidence for species with anomalously high ionization was found in relatively hot Be stars, based on *Copernicus* data (e.g., Morton, 1976; Snow and Marlborough, 1976; Marlborough, 1977), but by far the most extensive work has been done using IUE data.

As noted above, the best indicators of superionization for B stars are C IV and Si IV, although N V and even O VI have been reported in the hottest cases. Several important surveys of superionization in Be stars have been carried out using IUE data (Marlborough and Peters 1982; Barker *et al.*, 1984; Grady *et al.*, 1986), with the general result that C IV and Si IV both extend to cooler spectral types in Be stars than in normal B stars. No C IV absorption is seen in normal B stars beyond

B2, whereas this species is seen in Be stars as late as B9. Similar results are found for Si IV, which can be present in Be stars as cool as A0 or A1 (Slettebak and Carpenter, 1983).

The presence of anomalously high ionization in Be stars and its absence in normal B stars indicates that the Be stars have a source of ionization not available in normal B stars. It suggests, furthermore, that this source of ionization is somehow linked to the cause of Be properties in B stars. The source of the excess ionization is not established, but for the more luminous O and B stars, ionization by soft X-rays produced in a stellar corona (Cassinelli *et al.*, 1978; Cassinelli and Olson, 1979) has been suggested. Unfortunately, few data exist on possible soft X-ray emission from Be stars, clearly an important test of the hypothesis that superionization in these objects may have the same cause as in more luminous early-type stars. The fact that X-ray luminosity in OB stars scales with total luminosity suggests that Be stars will be weak X-ray sources, and for the most part they would lie below the detection threshold for past and present X-ray instruments such as the *Einstein Observatory* and *EXOSAT* (except, of course, for Be stars such as X Per which have compact companions). We may expect the study of Be stars with *ROSAT* to provide important information on this question.

3.3. STELLAR WINDS AND MASS LOSS

Considerable effort was made using *Copernicus* data to discover whether stellar winds occur in Be stars, and if so, whether the winds differ in any significant way from those in the more luminous O and B stars. The early results, based mostly on measurements of Si IV in bright Be stars, were that winds are nearly always present in Be stars, but often present also in normal B stars (e.g., Snow, 1981, 1982a). It was not clear whether winds were more common in the Be stars, or whether they extended to later spectral types in Be than normal B stars. IUE data have been used extensively in attempts to answer these questions.

High-velocity winds in Be stars produce at least two different signatures in ultraviolet line profiles. One is the appearance of extended short-wavelength wings in resonance line profiles, most commonly Si IV. Another manifestation of winds in Be stars is the occasional presence of relatively narrow absorption features, displaced toward short wavelengths from the rest wavelength, sometimes superposed upon a broad absorption line and sometimes not (see Figure 1). The narrow components are most often seen in C IV, and are particularly prone to variability with time (see the next section). Both types of wind features are similar to those seen in the more luminous O and B stars, but different in degree. The more luminous stars often have fully developed P Cygni profiles, whereas in the Be stars the emission components of the resonance lines are usually too weak to do more than fill in the absorption component. Narrow, shifted absorption components are common in the high-luminosity O and B stars, and may in fact *always* be present (Gathier *et al.*, 1982; Prinja and Howarth, 1986).

The asymmetric line profiles in Be stars are generally interpreted as due to an accelerating outflow, and using the theory of radiatively-driven winds (e.g., Castor and Lamers, 1979), it is possible to derive mass-loss rates from the profiles. This was done by Snow (1981) for a number of stars observed with *Copernicus*, with

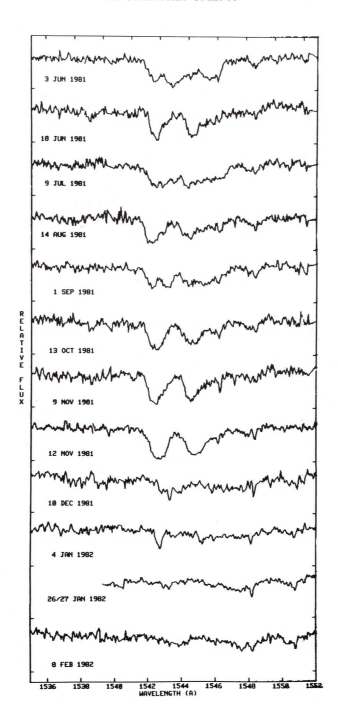

Fig. 1.   Variations in the C IV profile in 59 Cyg. This time sequence, covering a nine-month period, illustrates many of the modes of variation seen in Be star ultraviolet line profiles.

the resulting mass-loss rates ranging generally between 10^{-11} and 10^{-9} $M_\odot$ yr^{-1}. No clear dependence of mass-loss rate on stellar parameters was found among the Be stars, but the general level of mass loss is consistent with an extension of the OB star winds to lower luminosities (Snow, 1982a). There is a lot of scatter in the correlation of mass-loss rate with luminosity, however, indicating that factors other than luminosity must affect the rate of mass loss in Be stars (certainly time variability contributes to this scatter as well).

Numerous IUE studies have resulted in the determination of mass-loss rates for individual Be stars (e.g., Hammerschlage-Hensberge *et al.*, 1980; Codina *et al.*, 1984; de Freitas Pacheco, 1982), with the derived rates usually falling within the range mentioned above (studies carried out with other instruments have yielded similar results; e.g., Bruhweiler *et al.*, 1978, 1979). Larger samples of stars have been analyzed for mass-loss rates by Slettebak and Carpenter (1983) and by Grady *et al.* (1986), with similar results. Interestingly, much higher mass-loss rates are often found from the analysis of infrared continuum data (e.g., Waters, 1986), which may be reconciled with the ultraviolet results if the infrared emission and the ultraviolet wind indicators form in different physical regimes, as discussed below.

An important but elusive question regarding Be star winds is whether and how they depend on stellar characteristics. Closely related is the question of the relationship between winds in Be stars and those in normal B stars of similar spectral type. Both questions are made difficult by the variability of the wind indicators, and especially by the weakness of the asymmetric absorption lines in many cases. Whereas it may be easy to see displaced narrow absorption components, it is often difficult to determine when a strong resonance line such as Si IV is weakly asymmetric. Thus, it is not easy to say with certainty where in the H-R diagram winds are present and where they are absent, and no clear correlation of wind presence with stellar parameters has been established.

Despite these difficulties, it can be said that, among B stars hotter than B5 or so, Be stars seem always to have winds, while normal B stars do often but not always (e.g., Snow, 1981, 1982b; Slettebak and Carpenter, 1983). There may be a tendency for the mass-loss rates to be higher among the Be stars (Grady *et al.*, 1986). For the cooler Be and B stars, it becomes most difficult to assess the presence or absence of winds, and it is not certain whether the winds extend to later spectral types in Be than in normal B stars, although this is suggested by the fact (discussed earlier) that superionization extends to later spectral types among the Be stars.

The geometry of Be star winds is an important issue, because it places constraints on both the nature of the winds themselves and on the relationship of the winds to the rest of the circumstellar environment. Whereas optical shell-line data suggest that the cool circumstellar material is equatorially concentrated, and recent work on infrared excesses from Be stars indicates the same (Waters, 1986; Bjorkman, 1986), the ultraviolet data on high-velocity winds suggests that the winds are global, with little or no concentration in the equatorial plane. The latter point is supported by the lack of clear dependence of mass loss on projected rotational velocity (Snow, 1981), and, perhaps more convincingly, by the presence of displaced narrow components in some 'pole-on' Be stars (Peters, 1982a, b; Dachs and Hanuschik, 1984). This latter result is, however, contradicted by a survey of a very large number (62) of Be stars, in which there was a clear tendency for stars with high values of $v \sin i$ to be more likely to show displaced narrow components (Grady *et al.*, 1986). This may be evidence for equatorial concentration or, perhaps equally likely, that intrinsically rapid rotators are more prone to the formation of the narrow components. These components are nearly always variable, and will therefore be further discussed later in this review.

3.4. VARIABILITY

We begin the discussion of variability by remarking that, if one attempted to define Be stars entirely on the basis of ultraviolet spectra, they would hardly stand out as a distinct class. They would not be identified as emission-line stars, but might be classified together because of their enhanced superionization with respect to normal B stars, and their enhanced variability in the ultraviolet resonance lines that indicate mass loss in all B stars. These points make it clear that the UV data refer to a part of the atmosphere of a Be star whose presence is not inferred from optical emission lines. This emphasizes the importance of systematically integrating UV data with observations in other wavelength regions such as the optical, in order to construct a global picture of the phenomenology of Be stars.

A few isolated reports of variability in the UV spectra of Be stars appeared before IUE data became available. For example, Snow and Marlborough (1976) found dramatic short-term variations in the N v lines in 59 Cyg (in *Copernicus* scans), and Snow, Oegerle, and Polidan (1980) were the first to show (in a *Copernicus* study of δ Cen) that variations in UV profiles probably indicate true variations in the rate of mass loss.

The IUE, although not specifically designed to be used as a monitoring telescope, and not optimized for studying UV resonance lines because an important number of these lines (e.g., C III, N II and N III, O VI, P IV, and P V, S III, S IV, and S VI) are inaccessible to observations, is providing the best survey of a few of the high ionization stellar wind profiles in Be stars and the best time coverage for variability of such profiles in selected objects. These results have been obtained thanks to access to archive data and, most important, thanks to coordinated efforts of several groups of European and NASA users, which have organized regular observations since the first year of IUE. A few of these groups have also organized simultaneous ground based observations (in the spectral lines and continua) with the declared goal of studying the interaction of the variable stellar wind with the optically identified intrinsic circumstellar environment.

The most important stellar wind lines which are accessible to IUE (in the spectral regions of better instrumental sensitivity) are the resonance doublets of N V, C IV and Si IV. Their profiles display a variety of shapes and structures not only in different Be stars but often in the same star at different epochs. As stated earlier, these shapes can be ascribed to asymmetric absorption profiles with violet extended wings indicating outflowing material. In many Be stars the wings are so strong and extensive that the components of the N V and C IV doublet merge together to form a single, broad absorption feature which may extend from zero velocity to more than 1000 km s^{-1}. In other cases the profiles are characterized by complex structures which are superposed on the violet wings or are the only signatures of high velocity outflow. These structures (in absorption) are often considered to have similar origin as the shifted narrow absorption components observed to be superposed on the P Cygni profiles in O stars and B supergiants (see for example Barker and Marlborough, 1985; Henrichs, 1986). In Be stars several narrow components may coexist at different velocities and vary in strength and velocity with time.

From survey studies of individual Be stars we derive information on the existence of different modes of UV line profile changes, how these modes vary with time in a given Be star and how they are related to the variability characteristics of the optical spectrum.

Modes of profile changes are summarized in Barker and Marlborough (1985) and Doazan *et al.* (1985). Both papers emphasize C IV which, indeed, seems to be the clearest indicator of stellar wind variability for many Be stars. The variability modes which have been found are (1) the appearance and disappearance of broad, displaced absorptions, and (2) the changes in presence, number and velocity of the shifted narrow components.

A well-studied example of the former mode is presented by 59 Cyg (B1.5 Ve; example C IV profiles are displayed in Figure 1) where the equivalent widths of the C IV ions have been measured to vary from 1.1 to 5.4 A (Doazan *et al.*, 1985). This star, which has been regularly monitored since 1978, displays changes in shape (e.g., in the position of maximum absorption) and edge velocities during epochs of comparable C IV strengths and an apparently cyclic behavior of the equivalent widths with a period of approximately two years. Equivalent width changes of the magnitude that has been observed in 59 Cyg cannot be explained without assuming a variable mass loss rate from the star and a variable ionization structure of the stellar wind. This variability mode is generally observed on long time scales, from several months to a few years.

The shifted narrow components of resonance lines have FWHM ranging from about 100 to a few hundreds of km s^{-1} and are characterized by variability in strength and velocity. The properties of these components have recently been reviewed by Henrichs (1986) who underlines the fact that their variability is enhanced in Be stars relative to OB stars. Most of the types of variability seen in the narrow components have been observed in 59 Cygni. This star displays at different epochs irregular appearance and disappearance of single shifted narrow components, changes in strength on time scales of a few days, changes in number from one to many (as many as five of these components were measured in the N V and C IV profiles on May 26, 1980), and merging of several of them to form a broad absorption feature.

Besides 59 Cygni, several other Be stars display variable shifted narrow components. A few of them have been studied with some detail as, for example, ζ Oph, 09.5 Ve (Snow 1977; Howarth, Prinja, and Willis 1984), ω Ori, B2 IIIe (Peters 1982, Sonneborn *et al.*, 1986), λ Eri, B2 III(e)p, 6 Cep, B2.5 Ve, 105 Tau, B3 Ve and HR 7739, B3 Ve (Barker and Marlborough 1985). A few additional cases are discussed below in order to illustrate this short time scale (from a few hours to days) mode of variability.

γ Cas (B0.5 IVe) is a very well studied Be star whose UV wind lines display one narrow component at a velocity of about 1400 km s^{-1}. Variability is strong in the equivalent widths; Henrichs *et al.* (1983) have calculated logarithmic column densities decaying from approximately 14.5 to 14.0 in N V and C IV and from 13.9 to 13.3 in Si IV over a period of 19 days. A small but apparently significant velocity shift accompanies these changes in strength. By combining data taken at different epochs Henrichs *et al.* (1983) found an inverse linear relation between the logarithm of the ionic column densities and the velocity of the shifted narrow components.

Observations of 66 Oph (B2 IV—V) from 1979 to 1981 (Barker and Marlborough, 1985) show a broad, variable and displaced C IV absorption which seems to be formed by several shifted narrow components as the only constituents. Some components disappeared slowly in about three months starting at the end of 1981 to leave, in February 1982, a relatively strong component at about -240 km s^{-1} and a weak one at -840 km s^{-1}. The fact that no C IV absorption close to zero velocity was observed is indicative, according to Barker and Marlborough, that these ions are not formed in the wind until some distance from the photosphere.

The original interpretation of the shifted narrow components has been in terms of 'puffs' of high velocity material moving through the outflowing gas (Lamers *et al.*, 1978, Henrichs *et al.*, 1983; this model is demonstrated in Figure 2). Several other different explanations have been proposed (see e.g., Lamers *et al.*, 1982); in particular for γ Cas, Doazan *et al.* (1980) have suggested that the observed variability of these components is the result of density variations at a location in the outflow where there is a strong temperature gradient, so that the observed velocity of a given ion is very sensitive to the temperature structure of the stellar wind.

Recently it has been suggested that the origin of the shifted narrow components is related to the existence of variable, nonradial pulsation in early-type stars (Henrichs, 1986; observational evidence for nonradial pulsations in Be stars is found in Baade 1982; Bolton 1982; Smith 1985). The possible role of nonradial pulsations in the Be phenomenon is discussed more fully in the next section.

In θ CrB (B6 V), another well-studied case, narrow C IV absorption lines, not necessarily related to narrow shifted components, persisted at nearly constant strength and low (~ 30 km s^{-1}) velocities from February 1980 to March 1982 (Doazan *et al.*, 1986). Subsequently the star has exhibited a series of oscillations in the C IV strength from a relatively strong line to a weak or undetectable one; these oscillations persisted from May 1982 to June 1984 with a time scale of about three months, leaving at the end a C IV absorption that was steadily weak or absent until at least July 1985. A similar general behavior was observed in the resonance lines of Si IV and A1 III and in some exited lines of Fe III although these lines always retained their individuality and displayed specific differences among them.

Underhill's (1985) analysis of IUE archive data, using a subset of the material collected by Doazan and collaborators from the beginning of their UV — optical

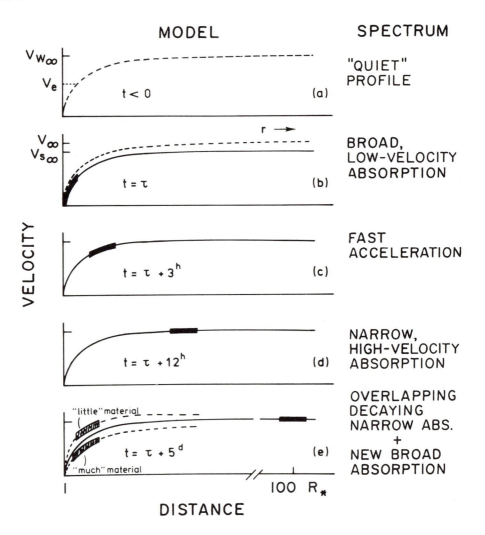

Fig. 2. One interpretation of variable narrow shifted components. The scenario depicted here was developed by Henrichs and co-workers, showing schematically how the observed development and eventual dissipation of a shifted narrow component may be interpreted as due to the ejection of a 'puff' of gas which then flows out with the wind, spending most of its time at or near the wind terminal velocity.

campaign on θ CrB until August 1983, concluded that the highly ionized material is suspended above the stellar photosphere and is moving at velocities of less than 100 km s^{-1} in regions close to the star. In Underhill's view there is no measurable mass loss and the changes in the highly ionized lines are explained in terms of different column densities of magnetically-confined material along the line of sight. The claim that there is no mass loss in this star is difficult to reconcile with the occasional appearance of higher-velocity C IV components (there is one spectrum in the set considered by Underhill that shows a component at -330 km s^{-1}; Stalio, 1985). Observational programs currently underway will probably clarify the question of mass loss from θ CrB, which is a test for models.

The question of a relationship between the stellar wind and the low velocity envelope is very important, and the subject of disagreement. Statistical studies of large numbers of stars appear to show little or no relationship (Grady *et al.*, 1986), as do some studies of individual stars (e.g., Sonneborn *et al.*, 1986). On the other hand, long-term observational studies of some stars indicate a definite relationship. Some examples are described in the following paragraphs.

In 59 Cyg, during the initial period of a new Be phase (which started in 1978), the optical spectrum exhibited an initial irregular increase of Hα emission strength to a maximum value, followed by an irregular decrease to an almost stationary low value lasting from about 1981 to 1984. The ratio of blue- and red-shifted emission peaks (the *V/R* ratio) displayed a well defined cyclic variation from positive to negative *V/R* with a period of about two years. At the same epoch the C IV resonance doublet was strongly correlated with the *V/R* curve in its absorption equivalent widths (Doazan *et al.*, 1985).

A similar correlation between the occurrence of a high velocity narrow component observed in N V, C IV, and Si IV and *V/R* changes at Hα is suggested for γ Cas during a well established Be phase of that star (Doazan *et al.*, 1984).

In 66 Oph the strengthening of the C IV profiles was accompanied by a decrease in the far-UV continuum flux measured from *Voyager* 1 and 2 far-UV spectrometers (912 to 1600 Å: Polidan and Peters, 1986). A similar rapid far-UV flux decrease, corresponding to a change in effective temperature of about 700 K, was observed in α Eri, B5 IVe, by Polidan *et al.* (1986) but no simultaneous observations in the UV lines nor in the optical were obtained at that time.

At the beginning of a B-normal phase for θ CrB (Doazan *et al.*, 1986), after the end of a weak shell phase which lasted at least through March 1979, the C IV profiles showed a progressive narrowing and low velocities (approximately -30 km s^{-1}) before undergoing the series of oscillations in strength which were described above; this second epoch is characterized by larger shortward displacements (-110 km s^{-1}) of the C IV cores when they were present. A similar narrowing of the absorption cores shell lines at Hα occurred from 1979 to 1980, followed by a shift towards negative velocities; subsequently the shell spectrum disappeared, leaving the hydrogen lines always in absorption. In this case there was a phase lag between similar phenomena occurring in the optical and the UV.

In 88 Her (B7 Ve), the transition from a quasi-normal B phase to a B-shell phase (Barylak and Doazan, 1986) was accompanied by a luminosity decrease at all observed wavelengths (unfortunately not including the infrared) and by a luminosity increase as the B-shell spectrum developed. These observations seem to contradict what is usually accepted on the basis of the traditional disk model for a Be star, which attributes the observed luminosity variations to changes in the physical conditions of the Hα-forming region.

Clearly there are very important questions concerning the relationship between the superionized winds and the less ionized circumstellar material in Be stars. The examples just cited, in which correlated behavior is reported, contradict some other studies of other stars, as noted above. The importance for Be star models of the presence or absence of such a correlation cannot be overemphasized.

4. Models

The recent Be star modeling history has three important dates which signify the evolution of the field: 1982, 1984, 1986.

In 1982, at the Munich Symposium on Be stars, Poeckert (1982) summarized what was then the current theoretical and empirical-theoretical understanding of the atmospheric structure of Be stars and the origin of the Be phenomenon. His explanatory diagrams, shown in Figure 3, were essentially oriented towards considering equatorial mass loss produced by critical rotation as playing a major role in producing the disk geometry for the cool part of the atmosphere; hot winds, imposed by UV observations, were ejected radially from polar latitudes or from a latitudinally expanding disk. Variability was mentioned *en passant* as something that has to be incorporated in some way in the proposed scenarios.

An alternative model was proposed by Doazan *et al.* (1981; we abstract from them the schematic γ Cas atmosphere presented in Figure 4) and Doazan and Thomas (1982). This model is based on observations of variability, and describes the outer atmosphere of a Be star as a sequence of atmospheric regions as shown in Figure 4 for γ Cas. These regions are defined according to their physical conditions and are physically connected. Variability in the mass loss and in the non-radiative energy flux are the necessary conditions for the production of the Be phenomenon because it somehow creates the condition for the formation of the Hα-emitting envelope, for its filling up, and for its dispersion in the nearby environment. Challenges to this quasi-spherically-symmetric model are presented by observations of linear polarization and by various observational arguments favoring the existence of equatorial disks in Be stars (e.g. a reported correlation of Hα with vsini by Andrillat, 1983; and statistical studies of UV wind indicators, such as those of Oegerle and Polidan, 1984 and Grady *et al.*, 1986; it is also noteworthy that analyses of infrared flux distributions of Be stars, such as those by Waters, 1986; Bjorkman, 1986; Lamers and Waters, 1984, are consistent with the assumption of equatorial concentration of the circumstellar envelope).

In 1984, at the Trieste-Sac Peak Workshop on 'Relations between Chromospheric-Coronal Heating and Mass Loss in Stars' (Stalio and Zirker, 1985), there was a definitive recognition that the critical rotation mechanism for the production of the disk geometry by equatorial mass outflow is dead. There was general agreement that the rate of mass loss varies, but discordant opinions remained on the geometry, on the basis of observational arguments just cited here.

Theoretical arguments favoring an equatorially-concentrated circumstellar envelope have been advanced by Poe and Friend (1986). They have modified a previously-developed stellar wind model in which magnetic forces on the out-flowing material are treated analytically (Friend and MacGregor, 1984). When rapid stellar rotation is incorporated into the model, the mass loss rate, hence the wind density, becomes strongly enhanced in the equatorial zone.

All of the models cited so far attempt to explain the Be phenomena in terms of

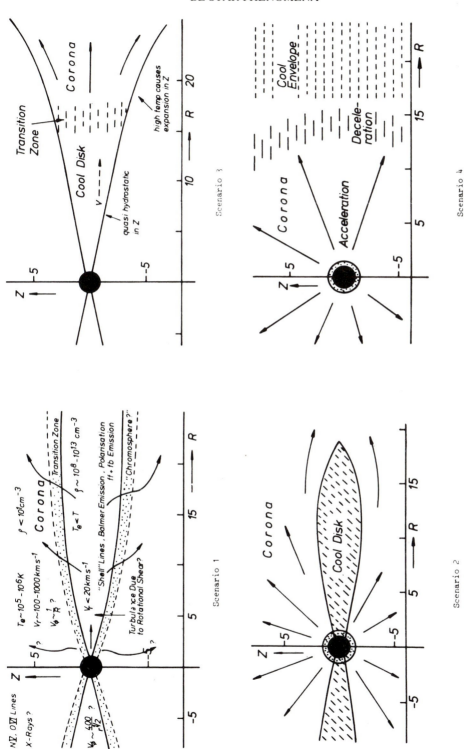

Fig. 3.   Schematic models of equatorially-concentrated circumstellar envelopes. These four general pictures are from the review of Poeckert (1982). Three of the four invoke a 'cool' disk and a global coronal region where the high-velocity wind flows.

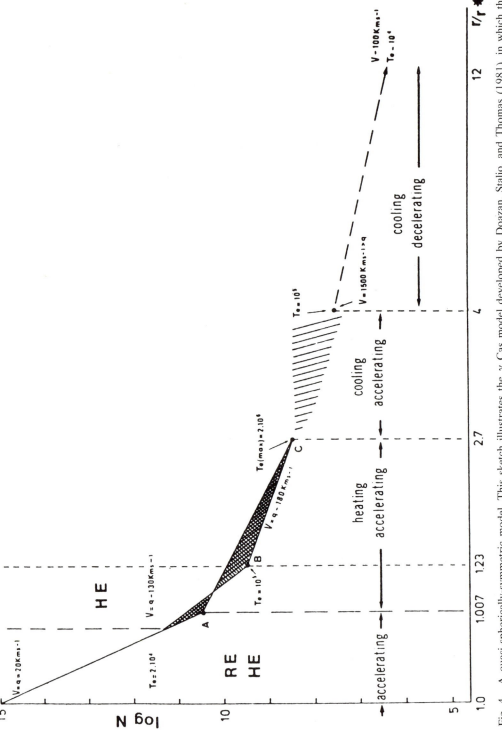

Fig. 4. A quasi-spherically-symmetric model. This sketch illustrates the γ Cas model developed by Doazan, Stalio, and Thomas (1981), in which the

outflowing material, but none treats the root *cause* of the outflow. Now a promising new line of evidence may do just that. Variability in nonradial pulsations has been, for the first time, invoked as mechanism able to initialize enhanced mass loss near the stellar equator Smith (1985) and Penrod and Smith (1985). The variable character of nonradial pulsations seems to be stronger in Be stars because they are often pulsating at both short period, $l = 2$, modes and at long period, high l, modes. The $l = 2$ modes penetrate more deeply into the star and contain more energy, which is transferred into the superficial, high l modes. It is suggested that, because the high-l modes have an intrinsically low energy content but large wave amplitudes, these modes can progress into shocks and eject a portion of a star's atmosphere (Smith and Penrod, 1985). A challenge to the theory that nonradial pulsations are responsible for the Be phenomenon will be to show why there might be a connection between rapid rotation and the existence of the appropriate pulsation modes.

We expect that at the next IAU Symposium on Be stars, to be held in Boulder in August 1986, there will be further significant developments in modeling. Presently all interested authors are jealously keeping secret their new observations, correlation diagrams and interpretations.

5. Future Research

There are several promising lines of attack on the problem of understanding the Be stars. More work needs to be done on nonradial pulsations, both theoretically and observationally. The organization by some groups of a long term IUE and optical monitoring programs of several individual Be stars has the potential of firmly establishing the relationship between UV-optical and line-continuum variability patterns, which are important tests for models. Due to the long-term nature of these programs, some results are appearing only now and need a more extensive observational basis to be confirmed in general.

References

Andrillat, Y.: 1983, *Astron. Astrophys. Suppl.* **53**, 319.

Baade, D.: 1982, in M. Jaschek and H.-G. Groth (eds.), 'An Unusually Stable and Short Spectroscopic Period of the Be Star 20 CMA', *IAU Symp.* **98**, 167.

Barbier, R. and Swings, J. P.: 1979, *Astron. Astrophys.* **72**, 374.

Barker, P. K. and Marlborough, J. M.: 1985, *Astrophys. J.* **288**, 329.

Barker, P. K., Marlborough, J. D., and Landstreet, J. D.: 1984, *Future of Ultraviolet Astronomy Based on Six Years of IUE Research, NASA-CP 2349*, J. L. Mead, R. D. Chapman, and Y. Kondo (eds.), Washington: NASA, p. 219.

Barylak, M. and Doazan, V.: 1986, *Astron. Astrophys.* (submitted).

Beeckmans, F.: 1975, *Astron. Astrophys.* **45**, 177.

Bjorkman, K. S.: 1986 (in prep.).

Bolton, C. T.: 1982, in M. Jaschek and H.-G. Groth (eds.), 'A Preliminary Report on Simultaneous Ultraviolet and Optical Observations of Lambda Evidence', *IAU Symp.* **98**, 181.

Bruhweiler, F. C., Morgan, T. H., and van der Hucht, K. A.: 1978, *Astrophys. J. Letters* **225**, L71.

Bruhweiler, F. C., Morgan, T. H., and van der Hucht, K. A.: 1979, *Astrophys. J.* **262**, 675.

Cassinelli, J. P. and Olson, G. L.: 1979, *Astrophys. J.* **229**, 304.

Cassinelli, J. P., Olson, G. L., and Stalio, R.: 1978, *Astrophys. J.* **220**, 573.

Castor, J. I. and Lamers, H. J. G. L. M.: 1979, *Astrophys. J. Supp.* **39**, 481.

Codina, S. J., de Freitas Pacheco, J. A., Lopes, D. F., and Gilra, D.: 1984, *Astron. Astrophys. Suppl.* **57**, 239.

Dachs, J. and Hanuschik, R.: 1984, *Astron. Astrophys.* **138**, 140.

Doazan, V., Grady, C. A., Snow, T. P., Peters, G. J., Marlborough, J. D., Barker, P. K., Bolton, C. T., Bourdonneau, B., Kuhi, L. V., Lyons, R. W., Polidan, R. S., Stalio, R., and Thomas, R. M.: 1985, *Astron. Astrophys.* **152**, 182.

Doazan, V., Marlborough, J. M., Morossi, C., Peters, G. J., Rusaconi, L., Sedmak, G., Stalio, R., Thomas, R. N., and Willis, A.: 1986, *Astron. Astrophys.* **158**, 1.

Doazan, V., Sedmak, G., Stalio, R., and Thomas, R. N.: 1984, *Future of Ultraviolet Astronomy Based on Six Years of IUE Research, NASA-CP 2349*, J. L. Mead, R. D. Chapman, and Y. Kondo (eds.), Washington: NASA, p. 212.

Doazan, V., Stalio, R., and Thomas, R. N.: 1981, *The Universe at Ultraviolet Wavelengths, NASA CP-2171*, R. D. Chapman (ed.), Washington: NASA, p. 149.

Doazan, V. and Thomas, R. N.: 1982, in *B Stars With and Without Emission, NASA/CNRS Monograph Series, NASA-SP 456*, A. B. Underhill and V. Doazan (eds.), Washington: NASA, Chapter 13.

de Freitas Pacheco, J. A.: 1982, in M. Jaschek and H.-G. Groth (eds.), 'Mass Loss from π Aquarri', *IAU Symp.* **98**, 391.

de Freitas Pacheco, J. A. and Gilra, D.: 1982, in M. Jaschek and H.-G. Groth (eds.), 'The Peculiar Be Star MD 87643', *IAU Symp.* **98**, 399.

Friend, D. B. and MacGregor, K. B.: 1984, *Astrophys. J.* **282**, 591.

Gathier, R., Lamers, H. J. G. L. M., and Snow, T. P.: 1981, *Astrophys. J.* **247**, 173.

Grady, C. A.: 1984, *The Origin of Nonradiative Heating/Momentum in Hot Stars, NASA-CP 2358*, A. B. Underhill and A. G. Michalitsianos (eds.), Washington: NASA, p. 57.

Grady, C. A., Bjorkman, K., and Snow, T. P.: 1986, *Astrophys. J.* (in press).

Hack, M. and Struve, O.: 1970, *Stellar Spectroscopy: Peculiar Stars*, Trieste: Osservatorio Astronomico di Trieste.

Hammerschlage-Hensberge, G., van den Heuvel, E. P. J., Lamers, H. J. G. L. M., Burger, M., de Loore, C., Glencross, W., Howarth, I., Willis, A. J., Wilson, R., Menzies, J., Whitelock, P. A., van Dessel, E. L., and Sanford, P.: 1980, *Astron. Astrophys.* **85**, 119.

Heap, S. R.: 1975, *Phil. Trans. Roy. Soc. London A* **279**, 371.

Henrichs, H. F.: 1986, *O, Of, and Wolf-Rayet Stars, NASA/CNRS Monograph Series*, P. S. Conti and A. B. Underhill (eds.) (in press).

Henrichs, H. F., Hammerschlag-Hensberge, G., Howarth, I. D., and Barr, P.: 1983, *Astrophys. J.* **268**, 807.

Howarth, I. D., Prinja, R. K., and Willis, A. J.: 1984, *Monthly Notices Roy. Astron. Soc.* **208**, 525.

Hutchings, J. B.: 1976, *Publ. Astron. Soc. Pacific* **88**, 5.

Lamers, H. J. G. L. M., Gathier, R., and Snow, T. P.: 1982, *Astrophys. J.* **258**, 186.

Lamers, H. J. G. L. M. and Snow, T. P.: 1978, *Astrophys. J.* **219**, 504.

Lamers, H. J. G. L. M., Stalio, R., and Kondo, Y.: 1978, *Astrophys. J.* **223**, 207.

Marlborough, J. M.: 1977, *Astrophys. J.* **216**, 446.

Marlborough, J. M.: 1982, in M. Jaschek and H.-G. Groth (eds.), 'Ultraviolet Observations, Stellar Winds, and Mass Loss for Be Stars', *IAU Symp.* **98**, 361.

Marlborough, J. M. and Peters, G. J.: 1982, in M. Jaschek and H.-G. Groth (eds.), 'Variation of Anomalous Stages of Iomization with Spectral Type for Be Stars', *IAU Symp.* **98**, 387.

Morgan, T. H., Kondo, Y., and Modisette, J. L.: 1977, *Astrophys. J.* **216**, 457.

Morton, D. C.: 1976, *Astrophys. J.* **203**, 386.

Morton, D. C., Jenkins, E. B., Matilsky, T. A., and York, D. G.: 1972, *Astrophys. J.* **177**, 219.

Oegerle, W. R. and Polidan, R. S.: 1984, *Astrophys. J.* **285**, 648.

Peters, G. J.: 1982a, *Astrophys. J. Letters* **253**, L33.

Peters, G. J.: 1982b, in M. Jaschek and H.-G. Groth (eds.), 'Evidence for Mass Loss at Polar Latitudes in ω Ori and 66 Oph', *IAU Symp.* **98**, 401.

Poe, C. H. and Friend, D. B.: 1986 (preprint).

Poeckert, R.: 1982, in M. Jaschek and H.-G. Groth (eds.), 'Model Atmospheres of Be Stars', *IAU Symp.* **98**, 453.

Polidan, R. S. and Peters, G. J.: 1986, private communication.

Polidan, R. S., Stalio, R., and Peters, G. J.: 1986, *Astrophys. J.* (submitted).

Prinja, R. K. and Howarth, I. D.: 1986 (preprint).

Ringuelet, A. E., Fontenla, J. M., and Rovira, M.: 1981, *Astron. Astrophys.* **100**, 79.

Selvelli, P. L. and De Araujo, F. K.: 1984, *Proc. Fourth IUE Conference, ESA Spec. Pub. SP-218*, E. Rolfe and E. Battrick (eds.), p. 291.

Sitko, M. L. and Savage, B. D.: 1980, *Astrophys. J.* **237**, 82.

Slettebak, A. and Carpenter, K. G.: 1983, *Astrophys. J. Suppl.* **53**, 869.

Smith, M.: 1985, *Relations between Chromospheric-Coronal Heating and Mass Loss in Stars*, R. Stalio and J. B. Zirker (eds.), Triste: Tabographys-TS, p. 342.

Smith, M. and Penrod, D.: 1985, *Relations between Chromospheric-Coronal Heating and Mass Loss in Stars*, R. Stalio and J. B. Zirker (eds.), Trieste: Tabographys-TS, p. 394.

Snow, T. P.: 1977, *Astrophys. J.* **217**, 760.

Snow, T. P.: 1981, *Astrophys. J.* **251**, 139.

Snow, T. P.: 1982a, *Astrophys. J. Letters* **253**, L39.

Snow, T. P.: 1982b, *Advances in Ultraviolet Astronomy, NASA-CP 2238*, Y. Kondo, J. L. Mead, and R. D. Chapman (eds.), Washington: NASA, p. 61.

Snow, T. P. and Marlborough, J. M.: 1976, *Astrophys. J. Letters* **203**, L87.

Snow, T. P., Peters, G. J., and Mathieu, R.: 1979, *Astrophys. J. Suppl.* **39**, 359.

Snow, T. P., Oegerle, W. R., and Polidan, R. S.: 1980, *Astrophys. J.* **242**, 1077.

Sonneborn, G. and Collins, G.: 1977, *Astrophys. J.* **213**, 787.

Sonneborn, G., Grady, C. A., Wu, C.-C., Hayes, D. P., Guinan, E. F., Barker, P. K., and Henrichs, H. F.: 1986 (preprint).

Stalio, R.: 1985, *Relations between Chromospheric-Cornoal Heating and Mass Loss in Stars*, eds. R. Stalio and J. B. Zirker (Trieste; Tabographys-TS), p. 29.

Stalio, R. and Zirker, J. B. (eds.): 1985, *Relations between Chromospheric-Coronal Heating and Mass Loss in Stars*, Trieste: Tabographys-TS.

Tarafdar, S. P.: 1983, *Monthly Notices Roy. Astron. Soc.* **204**, 1081.

Tarafdar, S. P. and Krishna Swamy, K. S.: 1981, *Monthly Notices Roy. Astron. Soc.* **196**, 67.

Underhill, A. B.: 1985, *Astron. Astrophys.* **148**, 431.

Waters, L. B. F. M.: 1986 (preprint).

SYMBIOTIC STARS

HARRY NUSSBAUMER

Institute of Astronomy, ETH, Zürich, Switzerland

and

ROBERT E. STENCEL

Center for Astrophysics and Space Astronomy, University of Colorado, Boulder, CO, U.S.A.

1. Symbiotic Stars as Historically Understood from Purely Optical Observations

What are symbiotic stars? They form a small class of astronomical objects whose nature has long been controversial. The traditional definition is that of a starlike object with a spectrum that simultaneously contains cool star features, like TiO molecular bands, and nebular emission lines, like [O III] or more highly ionized features. This suggests the presence of two components: a cool giant star plus a nebula of moderate ionization. The light output of such stars is variable on timescales of months and years. Symbiotic stars also show abrupt changes, increases in visual brightness by several magnitudes on timescales of days and months. A catalogue of 145 symbiotic stars was compiled by Allen (1984).

The mystery of symbiotic stars is due to our traditional, optical view of their spectra. The visual domain is the dividing line between the two sources presumed present in these systems: infrared emission from the cool, highly evolved star, and ultraviolet emission from a nebula ionized by a source of approximately 100 000 K or hotter. Variability of the two sources is not necessarily correlated.

Why should the physicist be interested in such exotica of the astronomical zoo? There are several reasons. First: as with all binary stars, it is possible to derive fundamental properties of the stars involved, using essentially Kepler's laws. Knowledge of masses and radii of stars is essential to understanding their evolutionary state. Second: binary stars interact. They can transfer mass across their gravitational potentials and alter each others evolutionary development. Third: the symbiotic stars are among the binaries with the longest known periods. This provides the volume required for each star to potentially fulfill its evolutionary destiny before interaction with its companion begins. This way, with a sequence of symbiotics of increasing separation, we can map out developmental stages in the evolutionary history of asymptotic red giants and supergiants, independently assess masses and radii, monitor changes in the outer atmospheric structure of such stars and define the parameters associated with the onset of rapid mass loss stellar winds and the formation of circumstellar dust. Today, with the combined data taking capabilities in the X-ray, ultraviolet, optical, infrared and radio regions, the mysteries of symbiotic stars are being replaced with fiducial information relevant to understanding the last stages of evolution of stars.

The binary nature of all symbiotic stars is not a foregone conclusion. Only one system was thought to exhibit optical eclipses (AR Pav). Thackeray warned that the presence of TiO in the spectrum of RR Tel did not by itself prove the presence

Y. Kondo (ed.), Scientific Accomplishments of the IUE, pp. 203—222.
© 1987 *by D. Reidel Publishing Company.*

of a cool companion star. Single star models were prevalent during the 1950s and 1960s. Boyarchuk (1967) postulated the duplicity of AG Peg on the basis of periodic emission line velocity displacements. Penston and Allen (1985), however, argue that the emission line radial velocities are not repeatable and no detailed orbital solution is possible. Cowley and Stencel (1973) established the elements of orbital motion by measuring radial velocity changes in the M giant's photospheric absorption line spectrum. An 830 day period was derived. For CI Cyg, SY Mus, and EG And, binarity has been claimed on the basis of periodic luminosity variations (cf. Stencel, 1984), but only recently have additional studies of periodic wavelength shifts been completed (Oliversen *et al.*, 1985; Kenyon *et al.*, 1984; Kenyon and Garcia, 1985). Although not observationally proven in every case, the presence of the red giant star has been less controversial than the origin of the optical emission line spectrum.

Symbiotic stars divide into two major classes according to their infrared properties: the '*s*' (stellar) type and the '*d*' (dusty) type. The *s* or stellar types are relatively dustfree, with the stellar flux dominating in the 2.3μ K-band photometry. The *d* or dusty types exhibit thermal radiation from dust at the K band. This major distinction is likely to hold a key to an overall understanding of the symbiotic phenomenon. At present, the distinction seems to be one of orbital separation: whether the cool star has been permitted to evolve to a state with substantial dust production in its envelope. In addition to this, much more insight is needed to determine the relationship between symbiotic stars and other classes of active binary stars, such as recurrent novae.

2. The Ultraviolet Observations and Discoveries

Access to the ultraviolet has finally given us a clearer look at the inner environments near the hot object in symbiotic binaries. The remarkable contributions of the IUE Observatory are due in part to its longevity. With the multi-year characteristic period of symbiotics, only recently have we obtained sufficient phase coverage to discuss detailed structure of the binaries. This has lead to possible comparisons with well studied short period cataclysmic variables and dwarf novae with much better defined disk and stream structures.

In addition, IUE's nine years of contribution (thus far!) has also made it possible to follow the onset and evolution of outbursts in several objects. This type of data allows us to discriminate among intrinsic variability of the cool object and episodic accretion disk events and accretion-driven thermonuclear ignition at the surface of the hot star.

The first UV spectra of symbiotic stars were taken already in the first few months of IUE during 1978: CH Cyg on April 22, Z And on May 5, and V 1016 Cyg on May 30. The first published ultraviolet spectrum of a symbiotic star was that of V 1016 Cyg. The authors (Flower *et al.*, 1979) considered it to be a candidate for a young planetary nebula. However, IUE was not the first artificial satellite to observe symbiotic stars. On May 11, 1970 the Orbiting Astronomical Observatory, OAO−2, had secured a broadband spectrum of AG Peg. This observation was however not published until May 1979 (Gallagher *et al.*, 1979).

That this data could languish for so long, compared to the rapid pace of publications from IUE, reflects positively on the scientific policy established for IUE: observations are considered to belong to the whole astronomical community, the observer simply has a headstart of 6 months. We are convinced that this open policy has done much to foster such tremendous interest not only in observing with IUE, but also in using its databank, leading to scientific advances.

The symbiotic stars observed with IUE up to the end of 1985 are shown in Table I. We do not want to give an exhaustive description of all the symbiotic stars observed with IUE. We shall give the characteristics of a few which we consider to give a representative collection of ultraviolet spectra of symbiotic stars.

Another uniqueness of IUE's contribution comes with the high resolution ultraviolet spectra of symbiotic stars as well. P Cygni profiles, indicative of high speed winds, have been detected in ultraviolet emission lines in several of the dust-free symbiotic stars. Multi-component profiles have also been found, with broad and narrow line features combined. Much of the current research focuses on interpreting phase dependent changes in these complex profiles. IUE has helped demonstrate binarity in several cases: EG And, HBV 475 = V1329 Cyg, SY Mus.

2.1. S-TYPE SYMBIOTIC STARS

AG Peg

AG Pegasi may have the longest observational history of any symbiotic star. Lundmark (1921) reviewed the light curve including the 1855—1871 rise in light from 9th to 6th magnitude, followed by the 100 yr long decline to the present brightness (8.6). Its optical spectrum has been intensively studied (cf. Merrill, 1959). The OAO—2 spectrum of AG Peg had too low a resolution to allow identification of emission lines, but at least it showed that AG Peg had a higher luminosity in the UV than in the red and infrared. For their analysis, Gallagher *et al.* (1979) could also consult the extensive set of observations of Wolf Rayet stars obtained by Willis and Wilson (1978) with the ESRO satellite TD—1. Their model fitting procedures suggested to them a temperature for the hot component in the range of an O8 or WN5 star. With the advent of IUE the brightest and least reddened of the symbiotic stars was observed very early and a first brief report was given by Keyes and Plavec (1980) and by Slovak and Lambert (1982). A much more detailed investigation has been given by Penston and Allen (1985). Making full use of the high resolution facility on IUE they present a complex picture. But concerning a satisfactory understanding of AG Peg, the two authors admit defeat, they stress the necessity for long term observations to disentangle the orbital and evolutionary effects.

AR Pav

AR Pav shows optical eclipses with a period of 604.6 days (Thackeray and Hutchings, 1974), but there is no clear cut case for orbital motion from line shifts. Hutchings *et al.* (1983) have observed that system with IUE. They looked for eclipse in the stronger short wave emission lines. For most of the lines, including He II $\lambda 1640$, the observed variations are not obviously phase related. A similar

TABLE I

List of symbiotic stars observed with IUE up to 1985. The objects chosen are those listed in Allen (1984), whose coordinates have been retained.

Object	α			δ			First observed
	h	m	s	°	′	″	
EG And	00	41	53	+40	24	22	30/12/1978
AX Per	01	33	06	+54	00	07	31/12/1978
Sanduleak	05	46	03	−71	17	13	21/3/1982
BX Mon	07	22	53	−03	29	51	6/1/1979
RX Pup	08	12	28	−41	33	18	19/7/1979
SY Mus	11	29	55	−65	08	36	20/9/1980
BI Cru	12	20	40	−62	21	39	26/3/1981
RW Hya	13	31	32	−25	07	29	2/1/1979
HE2−106	14	10	23	−63	11	45	26/7/1984
Hen 1092	15	42	30	−66	19	58	1/3/1980
T CrB	15	57	25	+26	03	39	5/1/1979
AG Dra	16	01	23	+66	56	25	29/6/1979
He2−171	16	30	47	−34	59	12	1/3/1980
Hen 1242	16	40	00	−62	31	40	1/3/1980
HK Sco	16	51	30	−30	18	17	26/3/1981
CL Sco	16	51	40	−30	32	30	19/7/1980
RT Ser	17	37	04	−11	55	04	29/8/1982
AE Ara	17	37	20	−47	01	50	8/10/1979
H1−36	17	46	24	−37	00	36	10/1/1979
RS Oph	17	47	32	−06	41	40	15/7/1979
V2416 Sgr	17	54	16	−21	41	10	20/2/1982
AS 289	18	09	35	−11	40	55	12/8/1980
Y CrA	18	10	47	−42	51	27	10/10/1979
YY Her	18	12	26	+20	58	20	25/5/1980
AS 296	18	12	33	−00	19	53	12/8/1980
AS 295B	18	12	52	−30	52	16	3/8/1979
AR Pav	18	15	25	−66	06	07	17/7/1979
V443 Her	18	20	03	+23	25	47	25/5/1980
BF Cyg	19	21	55	+29	34	34	5/1/1979
CH Cyg	19	23	14	+50	08	31	22/4/1978
HM Sge	19	39	41	+16	37	33	6/6/1978
CI Cyg	19	48	21	+35	33	23	5/1/1979
V 1016 Cyg	19	55	20	+39	41	30	30/5/1978
RR Tel	20	00	20	−55	52	04	20/6/1978
HBV 475	20	49	03	+35	23	37	25/6/1979
CD−43.14304	20	56	49	−42	50	34	9/10/1979
V407 Cyg	21	00	24	+45	34	41	29/8/1982
AG Peg	21	48	36	+12	23	27	27/7/1978
Z And	23	31	15	+48	32	31	5/5/1978
R Aqr	23	41	14	−15	33	43	3/1/1979

observation was also reported by Stencel *et al.* (1982) for CI Cyg. This is in contrast to the visual lines where for example He II $\lambda4686$ reportedly disappears during eclipse. No wavelength shifts were established in the AR Pav IUE spectra. As expected, the eclipse is clearly seen in the IUE FES counts, which represents essentially V-band photometry. No satisfactory model has yet been proposed for AR Pav, although Kenyon and Webbink (1984) advocate that the hot component is a main sequence star surrounded by an accretion disk.

Z And

Z And is occasionally described as prototype of symbiotic stars. It is doubtful whether such a title should be attributed to any object, given the diverse properties of symbiotic stars. Notwithstanding this reservation, Z And is one of the best documented symbiotic stars. Pickering (1901) cites it in a list of 'Sixty-four new variable stars' and mentions W. P. Fleming as its discoverer. Z And shows phases of intense activity alternating with extended quiescent periods. The active phases show recurrent nova type variations. In a preliminary report Sahade *et al.* (1981) compared spectra taken in February 1979 with others taken in July 1979 and found spectral variations on a short timescale. Altamore *et al.* (1981) discussed a series of coordinated ground based and IUE observations of the quiet phase of Z And. They favour a binary star model, in which the emission lines are formed in a solar type transition region around the cool star. Based on long term monitoring with IUE, Fernandez-Castro *et al.* (1986) conclude that the system is composed of an M 6.5 giant and a hot, $T_{eff} \approx 150\,000$ K, radiation source, ionizing a bow-shaped nebula where the variable emission lines and the Balmer continuum are formed. The variability is ascribed to orbital motion of the nebula, with a period of 760 days.

HBV 475

IUE helped to establish binarity beyond doubt for HBV 475 (= V1329 Cyg). This object has become one of the best observed symbiotic systems. The periodic shifts in the line positions (e.g. Figure 5 in Mueller *et al.*, 1986) can best be explained with a double star model. In the absence of well calibrated spectra in the visual domain, the finer error sensor (FES) of IUE has provided very valuable measurements of the broadband emission extending into the near infrared (see Figure 4 of Mueller *et al.*, 1986). This is one of many examples, were the FES counts have become an integral part of IUE observations. Periodic wavelength shifts of HBV 475 have been observed in the visible by Grygar *et al.* (1979). If the periodic wavelength-shifts are interpreted as being due to an elliptical motion of the emitting region, assumed to be placed around the hot star, we can deduce a mass function. From visual observations Grygar *et al.* (1979) found $f(M) = 23\,M_{\odot}$. The uncomfortably high masses led Iijima *et al.* (1981) to abandon the orbital explanation of the wavelength shifts, they suggest instead a moving ionization front in a nova-like event. Nussbaumer and Schmutz (1983) as well as Nussbaumer *et al.* (1986) advocate a model where a hot source in elliptical orbit around a mass losing giant, ionizes a fraction of the red giants wind. Model calculations show, that the ionized region can be very close to the cool star.

AG Dra

To date AG Draconis is the strongest X-ray source among symbiotic stars, the first observations are described by Anderson *et al.* (1981). This X-ray emission was ob-

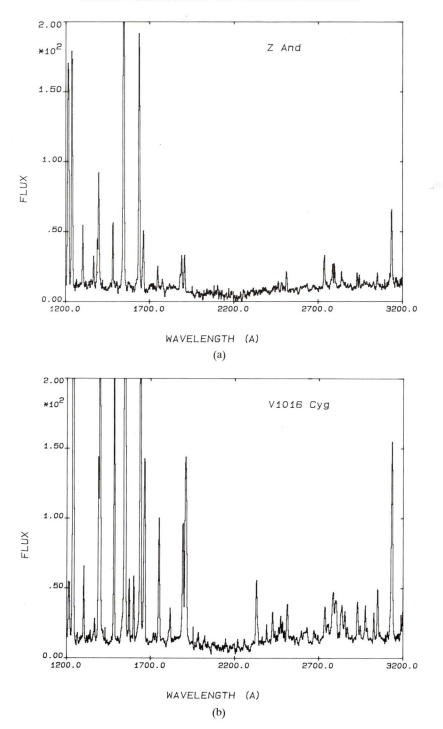

Fig. 1. Observed IUE spectra of symbiotic stars. (a) The s-type symbiotic star Z And. (b) The d-type symbiotic star V 1016 Cyg. The flux is given in units of 10^{-14} erg/(cm^2 s Å).

served 25 years after the outburst in 1955, but 6 months before the outburst at the end of 1980. IUE spectra were taken before and after the 1980 outburst. A study of this object has been given by Viotti *et al.* (1983, 1984) and Slovak *et al.* (1985). They assume that the UV emission line spectrum observed by IUE emanates from a nebula radiatively ionized by a white dwarf, and that the nebula itself is the result of a massive wind from a K giant star.

CH Cyg

In Allen's (1984) catalogue CH Cyg is listed as 'possible symbiotic star'. In its quiet phase it shows a normal M 6 III spectrum. Bursts of activity appear at irregular intervals, separated by several years. An active phase lasted from September 1963 until August 1965, the next one lasted from June 1967 to December 1970, in August 1977 another one began (Persic *et al.*, 1984) and it is assumed to die away during 1986. During the activity phase the visual magnitude does not change appreciably, whereas the blue magnitude increases and emission lines of He I, Fe II up to [O III] appear (Hack *et al.*, 1982). The IUE observations show emission of O I], Si III] and C III], the strong continuum, rather uncharacteristic of symbiotic stars, shows a large number of absorption lines, such as C I, N I, Fe II, Ni II, but also Si IV and C IV. After its drop in brightness and its radio outburst in July 1984 (IAU circular 4145), CH Cyg entered its nebular phase in March 1985. In the final phase of activity there is a dramatic change in the EUV spectrum (Selvelli and Hack, 1985). The continuum becomes flat, emission lines dominate, former absorption lines, like C IV or Si IV turn into emission, C III λ1907 appears besides λ1909, indicating a lowering in density.

Except for the rapid flickering observed by Slovak and Africano (1978), which they compare to cataclysmic variables, there is no compelling observational evidence for CH Cyg to be classified as a double star system.

2.2. D-TYPE SYMBIOTIC STARS

R Aqr

R Aqr is different from other symbiotic stars in that it possesses an extended nebulosity seen on direct photographs (e.g. Wallerstein and Greenstein, 1980). This apparent uniqueness however, could be due to its relative proximity. R Aqr is at a distance of only $\approx$ 200 pc, in contrast to other symbiotic stars observed with IUE, which have distances $\gtrsim$ 1000 pc. Taking into account the Mira light variation, Willson *et al.*, 1984 (see Garnavich and Mattei, 1981) proposed that R Aqr is an eclipsing binary with a 44 yr period.

The first IUE observations of the jet of R Aqr were reported by Kafatos and Michalitsianos (1982). Since then, the jet has become a regular IUE target. Hollis and Michalitsianos (1983) and Hollis *et al.* (1985) have observed the jet of R Aqr also at radio wavelengths. This feature is explained as being ejected near periastron as the two stars approach each other and being illuminated by a high energy beam arising from the inner region of an accretion disc. As mentioned before, the R Aqr jet may not be unique among symbiotic stars, asymetric features have been observed in several of them either in the radio or the optical emission.

R Aqr has also been detected as an SiO maser source (Zuckerman, 1979), and it was detected in the soft X-ray region with EXOSAT (Baratta *et al.*, 1985).

V 1016 Cyg

V1016 Cygni underwent a nova-like outburst in 1965. Prior to the eruption it was classified as a long period variable with Hα emission. Post-outburst, spectral similarities to planetary nebula were discussed (cf. Boyarchuk, 1968; Baratta *et al.*, 1974). Ultraviolet aspects of the nebular spectrum have been discussed by Flower *et al.* (1979), Nussbaumer and Schild (1981), Feibelman (1982), Deuel and Nussbaumer (1983), Feibelman and Fahey (1985). Despite the infrared variations observed by Taranova and Yudin (1983) or Lorenzetti *et al.* (1985), there are no clear indications of marked variations in the ultraviolet fluxes observed by IUE. In that respect V 1016 Cyg resembles RR Tel.

Although single star protoplanetary nebula models were proposed, a nova-like model with an M giant supplying matter via accretion to a subdwarf or a white dwarf matches the energetics required rather well (Kenyon and Truran, 1983). High spatial resolution VLA maps reveal a bipolar appearance to the expanding nebula (Hjellming and Bignell, 1982).

RR Tel

Discovered in 1908 as a variable and extensively analysed by Thackeray (1977), our knowledge of RR Tel benefitted from extensive IUE observations. The main characteristics of its UV spectrum have been described by Ponz *et al.* (1982) and Penston *et al.* (1983). From the spectrum of RR Tel Penston *et al.* (1983) established and identified a list of over 400 emission lines — this list has helped many IUE observers with their own IUE observations of emission line objects. RR Tel shares with other symbiotic stars the fate of having been called different names at different times and by different observers (see Penston *et al.*, 1983): slow nova, symbiotic star, Mira type variable; none of them is wrong. The observations of Feast *et al.* (1983) have established beyond doubt a periodic variability in the infrared with a period of 387 days. But Heck and Manfroid (1985) show that it is unlikely that the visual spectrum is due to the Mira type star. Hayes and Nussbaumer (1986) collected all the IUE observations from 1978 to 1984 to see whether such a periodicity can also be established in the UV. The IUE FES data (Fine Error Sensor: a broad band detector in the 4000—7000 Å range) show variations which are compatible with the period and phase found in the infrared. Yet there is no evidence for periodic variations in the 1200—3200 Å line emission. However, they found a systematic decrease in the line fluxes at those wavelengths.

Hayes and Nussbaumer (1986) established an electron density in the range of $5 < \log N_e[\text{cm}^{-3}] < 9$. They also tried to derive an electron temperature, T_e, by relying upon collisionally excited Si III $\lambda\lambda 1206$, 1892. They also compared collisionally excited and dielectronic recombination lines and derived $T_e <$ 20 000 K. Collisional ionization is clearly ruled out by this result.

Hayes and Nussbaumer (1986) attempted a determination of the stellar radiation field. They only give a lower limit of $T^* \gtrsim 150\,000$ K. They point out that with the present observational capabilities (including IUE) the Stoy and Zanstra methods are not well suited for accurate T^* determinations if $T^* \gtrsim$ 150 000 K.

2.3. SYMBIOTIC STARS IN OTHER GALAXIES

To the surprise of many, it has proven possible to observe symbiotic stars in external galaxies with the IUE. Kafatos *et al.* (1983) reported the first low dispersion spectra of LMC S63 and Sanduleak's star, and found overabundance of nitrogen in both objects. This overabundance was correlated with CNO processing such as is seen in η Car. Both objects contain very hot and very luminous companions. Further observations with the Hubble Space Telescope are anticipated.

Kenyon and Gallagher (1985) present new observations of Hubble-Sandage variables in M31 and M33 and argue that, given their location outside of active star forming regions, several of their observed nova-like properties may be explained by massive binaries undergoing rapid mass exchange. While not strictly symbiotic stars, H-S variables may be undergoing an evolutionary transformation analogous to the symbiotic stars and suggest the ubiquity of this phenomenon in other galaxies.

3. Ultraviolet Spectral Diagnostics

3.1. GENERAL REMARKS

Through its spectroscopic diagnostic possibilities IUE opened a new continent to astronomy. IUE was not the first far UV observatory, it was not in every respect better than what had existed before, but it had the right facilities for answering crucial questions in many domains of astronomy.

For research into the properties of symbiotic stars, the diagnostic possibilities offered by IUE are tremenduous, yet at the same time frustrating. At 1100 Å $<$ $\lambda < 3200$ Å we not only find strong resonance and intercombination lines from ions already observed in the visual, but also from ions which do not pocess low lying metastable levels, in particular C^{2+}, C^{3+}, N^{2+}, N^{3+}, O^{3+}, Si^{3+}. If symbiotic stars contain indeed a hot radiative source, then the far UV is the place to investigate. And if the 'nebular' spectrum originates in a collisionally ionized transition region then we look again to the far UV for information and confirmation. The far UV spectrum of symbiotic stars consists of emission lines and a continuum. The emission lines have very large equivalent widths. As example we might mention V 1016 Cyg where typical line widths are ≈ 0.4 Å. The well defined lines have fluxes between 1 and 50 times 10^{-12} erg cm^{-2} s^{-1}, the underlying continuum is $\approx 1 \times 10^{-13}$ cm^{-2} Å^{-1} s^{-1}. The dynamical possibilities of IUE, with its theoretical range of 1 to 256, are therefore not well suited for observing simultaneously emission lines and underlying continuum. The same problem hampers the simultaneous observation of strong and feeble lines. Despite this, the high resolution capability and longevity of IUE have permitted unique observations at different orbital and activity phases of symbiotic systems.

3.2. THE CONTINUUM

As mentioned above, in IUE observations of symbiotic stars a well exposed line spectrum is very often hopelessly underexposed for a reliable measurement of the continuum. Where continuum observations are possible, often the slope implies color temperatures in excess of 20 000 K and up to 150 000 K. However, these color temperatures must be taken with caution as recent far-UV observations of cataclysmic variables indicate a turnover of an analogous continuum rise short-ward of Ly α, suggesting the source is in an accretion disk rather than the photosphere of the hot companion.

As mentioned above, in IUE observations of symbiotic stars a well exposed line spectrum is very often hopelessly underexposed for a reliable measurement of the continuum. For those cases where the continuum is observed, Cassatella *et al.* (1986) succeed in most cases when fitting the observations to a superposition of a hot and a cool stellar continuum in addition to a nebular recombination continuum (see Fernandez-Castro *et al.* (1986) for the case of Z And). Kenyon and Webbink (1984), however, distinguish between hot stellar sources and main sequence accretors as source of the hot continuum.

3.3. EMISSION LINES

3.3.1. *The Well Known Diagnostics*

There are several multiplets in the IUE range which contain information about densities and temperatures (see for example Nussbaumer, 1982). For N_e deter-minations we think in particular of C III] $\lambda\lambda 1907$, 1909 (Nussbaumer and Schild, 1979), N IV $\lambda\lambda 1483$, 1487 (Nussbaumer and Schild, 1981) and Si III] $\lambda\lambda 1883$, 1892 (Nussbaumer, 1986). The two components of these multiplets lie close together and thus present no problem of reddening, the ratios are practically insensitive to the electron temperature T_e. In symbiotic stars we always find the magnetic quadrupole transition $^3P^0_2 \rightarrow {}^1S_0$, which lies at the shorter wavelength, to be much feebler than the electric dipole $^3P^0_1 \rightarrow {}^1S_0$ transition at the longer wavelength. This immediately indicates $N_e \gtrsim 10^5 \text{ cm}^{-3}$. N III] $\lambda 1749$ (Nussbaumer and Storey, 1979) and O IV] $\lambda\lambda 1401$ (Nussbaumer and Storey, 1982) are two multiplets, the components of which could in principle be employed for deter-mining N_e if densities fall in the range $10^8 < N_e[\text{cm}^{-3}] < 10^{11}$. Those multiplets are usually weak and they have problems of blends and, as mentioned by Altamore *et al.* (1981), the radiation field cannot be neglected a priori. However, when they were consulted, the results usually indicate $N_e < 10^{10} \text{ cm}^{-3}$. The N III] and O IV] line ratios only change by a factor ≈ 2 in the region of interest, this is little when considering the quality of the observational data in those multiplets.

The line ratios mentioned up to now have the advantage of being practically independent of the electron temperature and of reddening. In contrast the lines [Ne IV] $\lambda\lambda 1601$, 2423, corresponding to the multiplets $^2P^0 \rightarrow {}^4S^0$ and $^2D^0 \rightarrow {}^4S^0$ have to be corrected for reddening, they also depend on T_e. However, that ratio has the advantage of changing by a factor of ≈ 1000 from $N_e \approx 10^4 \text{ cm}^{-3}$ to $N_e \approx 10^9 \text{ cm}^{-3}$ (see Nussbaumer, 1982). It may indeed have its highest sensitivity at those densities of greatest interest to symbiotic stars. Mueller and Nussbaumer (1985) derive from these multiplets for HM Sge a density of $6.5 \times 10^5 < N_e$ [cm^{-3}] $< 7.5 \times 10^8$. But the observation of these multiplets needs an IUE long- and shortwave spectrum, and in general their flux is rather low.

3.3.2. *The Case of* Si III] $\lambda 1892$/C III] $\lambda 1909$

The flux ratio of the two lines Si III] $\lambda 1892$ and C III] $\lambda 1909$ represents a tempting solution for a N_e determination. This ratio was already used in solar work, and also in the interpretation of IUE observations of outer stellar atmospheres (Doschek *et al.*, 1978). However, severe assumptions have to be made in order to employ those two lines. Let us designate $\Lambda(\lambda 1892/\lambda 1909)$ as the ratio of the observed fluxes

$$\Lambda(\lambda 1892/\lambda 1909) = F(\text{Si III}] \lambda 1892)/F(\text{C III}] \lambda 1909). \tag{1}$$

We assume further that the observed emission has its origin in a homogeneous gas. In that case the observed ratio equals the ratio of emissivities. (As the lines are close in wavelength we disregard reddening effects.) Thus

$$\Lambda(\lambda 1892/\lambda 1909) = \frac{\varepsilon(\lambda 1892)}{\varepsilon(\lambda 1909)}, \tag{2}$$

the emissivity, ε, being defined as

$$\varepsilon(\lambda) = N(X_3^{2+}) h \frac{c}{\lambda} A_\lambda. \tag{3}$$

$N(X_3^{2+})$ stands for the density by number of the twice ionised atom of C or Si in the level $2s2p$ $^3P_1^0$ or $3s3p$ $^3P_1^0$ respectively, which is the third energy level. With the expansion

$$N(X_3^{2+}) = N(X) \frac{N(X^{2+})}{N(X)} \frac{N(X_3^{2+})}{N(X^{2+})} \tag{4}$$

we obtain

$$\Lambda(\lambda 1892/\lambda 1909) = \frac{N(\text{Si})}{N(\text{C})} \frac{N(\text{Si}^{2+})/N(\text{Si})}{N(\text{C}^{2+})/N(\text{C})} \rho(\lambda 1892/\lambda 1909), \tag{5}$$

where ρ represents the ratio of relative emissivities:

$$\rho(\lambda 1892/\lambda 1909) = \frac{\varepsilon(\lambda 1892)/N(\text{Si}^{2+})}{\varepsilon(\lambda 1909)/N(\text{C}^{2+})}. \tag{6}$$

At low densities the only processes of importance for determining the relative population, $N(\text{Si}_3^{2+})/N(\text{Si}^{2+})$ and $N(\text{C}_3^{2+})/N(\text{C}^{2+})$, which enter explicitly in expression (6), are collisional excitation from the ground term ^1S and radiative decay of $^3P_1^0$ to the ground term. The rate of collisional excitation, q_{lu}, is given by

$$q_{lu} = \frac{8.63 \times 10^{-6}}{g_l \sqrt{T_e}} \mathrm{T}_{lu} \exp(-\Delta E_{lu}/kT_e) \text{ [cm}^3 \text{ s}^{-1}], \tag{7}$$

g_l is the statistical weight of the ground level ($= 1$ in our case), T is the collision strength averaged over a Maxwellian electron distribution of temperature T_e, and ΔE_{lu} is the energy difference between levels 1 and 3. For our particular case ΔE has, to better than 1%, the same value for Si^{2+} and C^{2+}. We might therefore assume that $q(\lambda 1892)$ has practically the same energy dependence as $q(\lambda 1909)$. To a first approximation this is indeed the case. However, the collision strengths for $^1S - {^3P^0}$ have different T_e dependences for Si^{2+} and C^{2+}. The reason for this difference is mainly the position and strength of the resonances in the cross-sections, these resonances being very important for the total value of T. In Figure 2 we show $\rho(\lambda 1892/\lambda 1909)$. The rapid increase in ρ for $N_e > 10^9\,\mathrm{cm}^{-3}$ should provide at least a reliable indication whether the dominating densities are higher or lower than $10^{10}\,\mathrm{cm}^{-3}$.

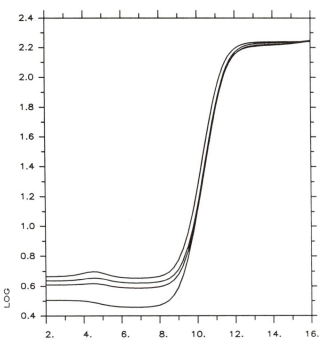

Fig. 2.   Ratio of relative emissivities in Si III $\lambda 1892$ and C III $\lambda 1909$ as given in relation (6). The ratio is given on a logarithmic scale as a function of $\log N_e$ [cm^{-3}]. The bottom line corresponds to $T_e = 5000$ K, the upper curves to 15 000, 20 000, and 50 000 K.

Expression (5) contains additional uncertainties, in particular the relative abundance

$$a(Si/C) = N(Si)/N(C),\qquad\qquad (8)$$

and the fractional abundances of the doubly ionized elements

$$f = \frac{N(Si^{2+})/N(Si)}{N(C^{2+})/N(C)}.\qquad\qquad (9)$$

The function f depends on physical conditions. In Table II we give f for a

TABLE II

The function f for different combinations of T^*, T_e, and N_e. The second line to each radiation temperature gives the mean T_e of the Si III and C III emission region.

T^* [K]	T_e [K]	N_e [cm^{-3}]			
		10^3	10^6	10^9	10^{11}
40 000		1.20	1.14	0.85	0.70
	model	7820/7800	10 500/10 300	12 900/13 200	14 700/15 500
60 000		0.65	0.50	0.19	0.08
	model	8520/8570	11 200/11 200	14 800/14 800	17 100/17 300
100 000		0.52	0.37	0.16	0.10
	model	10 100/10 200	13 300/13 100	16 200/16 000	17 000/16 800
160 000		0.53	0.39	0.23	0.18
	model	12 000/11 800	14 400/14 200	16 400/16 700	16 900/17 300
$T^* = 0$	40 000	2.6	2.6	2.6	2.6
$T^* = 0$	60 000	0.69	0.69	0.69	0.69

radiatively ionized gas with several different radiation temperatures T^*, and for a collisionally ionized gas at $T_e = 40\,000$ and $60\,000$ K, they were calculated with the program described in Nussbaumer and Schild (1981). Model calculations of radiatively ionized emission regions of symbiotic stars usually require $T^* \gtrsim 10^5$ K. This results in electron temperatures of the Si III, C III regions of $10\,000$ K $< T_e <$ $20\,000$ K, thus $0.1 < f < 0.5$. With the function defined above, expression (2) takes the form

$$\Lambda(\lambda 1892/\lambda 1909) = \frac{a(\text{Si/C})}{a_\odot(\text{Si/C})} a_\odot(\text{Si/C}) f \rho. \tag{10}$$

Assuming solar abundance ($a_\odot(\text{Si/C}) = 0.1$), the Λ-curves were calculated for $T_e = 15\,000$ K and $f = 0.1$ and $f = 0.5$ respectively. The results are given in Figure 3.

From Table III we see that in symbiotic stars the observed flux ratio is $F(\lambda 1892/\lambda 1909) \lesssim 1$. We can therefore state that $N_e \lesssim 10^{11}$ cm^{-3}. Together with the evidence from C III $\lambda\lambda 1907, 1909$, N IV $\lambda\lambda 1483, 1487$, Si III $\lambda\lambda 1883, 1892$ we find for symbiotic stars $10^6 \lesssim N_e$ [cm^{-3}] $\lesssim 10^{11}$. It would of course be of particular interest to investigate the exception, AG Dra, more closely in other density sensitive lines, such as mentioned in Section 3.3.1. This conclusion depends on the assumption $a(\text{Si/C}) \approx a_\odot(\text{Si/C})$. Lower Si abundances would lead to higher upper limits for the densities. It also depends on the assumption that the $^1S - {}^3P^0$ intercombination line is not fed by dielectronic recombination. This assumption has to be checked in each case, particularly if C IV $\lambda 1550$ or Si IV $\lambda 1400$ are strongly present.

There are several symbiotic stars with an observed $F(\lambda 1892/\lambda 1909) \gtrsim 0.4$. According to Figure 3 this might indicate $10^9 \lesssim N_e$ [cm^{-3}] $\lesssim 10^{11}$. However, a value of ≈ 1 seems to be the upper limit to the observed ratio. In Table II we also show the typical T_e of the Si III and C III emitting regions. An increase in N_e above

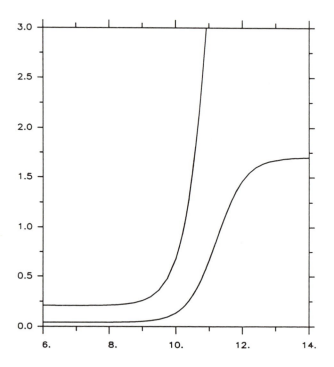

Fig. 3. Ratio of emissivities in Si III λ1892 and C III λ1909 as given in relation (10). The upper curve corresponds to $f = 0.5$, the lower to $f = 0.1$. Λ was calculated with a solar abundance of $a(Si/C) = a_\odot(Si/C) = 0.1$ and $T_e = 15\,000$ K. The ratio is given on a linear scale as a function of $\log N_e$ [cm^{-3}].

10^9 cm^{-3} causes an increase in collisional de-excitation of C^{+2} 2s2p ^{3}P^0 and leads therefore to a decrease in cooling through the C III λ1909 transition. At slightly higher densities the same process weakens the cooling effect of O III] $\lambda\lambda$1661, 1666. This may have far reaching consequences, as the inhibition of radiative cooling for $N_e \gtrsim 10^9$ cm^{-3} leads to an increase T_e and thus to a rapid increase in the electron pressure, $p_e \sim N_e T_e$. We suspect that this increase is then compensated by expansion. This self-regulating mechanism may be the reason for 10^{10} to 10^{11} cm^{-3} to be an upper limit to the electron density in symbiotic stars.

3.3.3. *Abundance Determinations and Electron Temperatures*

It may be astonishing to see temperature and abundance determinations grouped together. The emissivity in a collisionally excited resonance line of an m-times ionized element X is given by

$$\varepsilon(\lambda) = q_{lu} N_e N(X^{+m}) \times h \frac{c}{\lambda} \ [\text{erg cm}^{-3} \text{ s}^{-1}], \qquad (11)$$

q has been defined in expression (7). The emissivity of a recombination line in the spectrum of X^{+m} is due to recombination of the next higher ionization stage and

TABLE III

Observed Si III λ1892/C III λ1909 flux ratios. If variations are small, a mean value is given.

Object	$\Lambda(\lambda1892/\lambda1909)$	Source of observational flux
HM Sge	0.15	Mueller and Nussbaumer (1985)
V 1016 Cyg	0.2	Nussbaumer and Schild (1981)
RR Tel	0.23	Hayes and Nussbaumer (1986)
R Aqr	0.33	Michalitsianos and Kafatos (1982)
AR Pav	0.38	Hutchings *et al.* (1983)
HBV 475	0.18—0.49	Nussbaumer and Schmutz (1983)
RX Pup	0.23—0.45	Kafatos *et al.* (1985)
CL Sco	0.48	Michalitsianos *et al.* (1982)
YY Her	0.57	Michalitsianos *et al.* (1982)
Z And	0.67	Altamore *et al.* (1981)
AG Peg	0.7	Penston and Allen (1985)
RW Hya	0.74	Kafatos *et al.* (1980)
SY Mus	0.71, 1.5	Michalitsianos and Kafatos (1984)
CH Cyg	0.54—0.92	Selvelli and Hack (1985)
BX Mon	1.0	Michalitsianos *et al.* (1982)
Mira B	1.1	Cassatella *et al.* (1985)
AG Dra	2.0	Viotti *et al.* (1983)

can be written as

$$\varepsilon(\lambda) = N_e N(X^{+m+1})\alpha_{\mathrm{eff}}(\lambda)h\,\frac{c}{\lambda}\;[\mathrm{erg\,cm^{-3}\,s^{-1}}]. \tag{12}$$

Abundance determinations should preferentially be done with lines which are insensitive to changes in N_e and T_e. Recombination transitions come fairly close to fulfilling the second condition. If $T_e \lesssim 20\,000$ K, as seems to be the case in the line emission region of symbiotic stars, then dielectronic recombination dominates. Nussbaumer and Storey (1984, 1986) have given α_{eff} for all the important dielectronic recombination lines of up to three times ionized C, N, O, Mg, Al, Si. The low dynamic range of IUE and the weakness of most of the IUE spectra of symbiotic stars prevent a full exploitation of the diagnostic possibilities of these lines. Those most readily observable are: C III λ2297, 1577, 1247, N IV λ1719, O IV λ1341, O V λ1371. Thus from the flux in λ2296 or λ1719 the abundance of C^{+3} or N^{+4} can be derived. If, on the other hand, the recombination line N IV λ1719 is compared to λ1240 (the strongly T_e sensitive resonance multiplet of N V) we should be able to derive T_e of the N^{+4} region. — These possibilities exist in principle for IUE, and they have been employed. Thus Hayes and Nussbaumer (1986) find $T_e < 20\,000$ K for the C^{+2}, C^{+3}, N^{+4} regions of RR Tel. However, most of the dielectronic recombination lines are either too feeble for IUE or are marred by instrumental problems such as the reseau mark in λ2297.

Because of the difficulties mentioned above, the exceedingly important problem of the T_e determination is usually tackled by a comparison of two collisionally excited lines. Alas, some of the lines which might be best suited for such determination, lie outside or just at the limit of the IUE wavelength range. This is the case for the C III pair of $\lambda 1176/1909$ or Si III $\lambda 1299/1892$. This latter pair was employed by Hayes and Nussbaumer (1986) to derive an upper limit for T_e in RR Tel, they found $T_e \lesssim 25\,000$ K. Considering the even lower values found from dielectronic recombination, we do not hesitate to claim that the line emission region in at least some of the symbiotic stars is radiatively ionized. This conclusion is of primary importance for modelling symbiotic stars.

3.3.4. *Doppler Shifts and Doppler Broadening and Line Profiles*

Friedjung *et al.* (1983) measured line positions in several symbiotic stars (Z And, EG And, CI Cyg, RW Hya, AG Peg, RR Tel, AG Dra). In three of them (Z And, CI Cyg, RW Hya) they found a significant redshift of the N V, C IV, and Si IV resonance emission lines with respect to the C III, N III, N IV, O III, O IV, O V, Si III, and Si IV intercombination lines. This result, if verified with higher resolution instruments, suggests systematic flows between different line forming regions.

Cassatella *et al.* (1985) discuss the line widths of Mira Ceti obtained from a long (570 min) high resolution exposure in the short wavelength range observed on July 7, 1983. They intended to follow up on IUE observations of Stickland *et al.* (1981) who distinguished narrow and wide emission lines, attributing them to an extended atmosphere of Mira A and a disc around Mira B respectively. However, on their higher quality observation Cassatella *et al.* (1985) did not find the narrow lines. The only lines they can detect unambiguously are Lα C II $\lambda 1335$, Si IV, and O IV] at $\lambda 1400$, C IV $\lambda 1550$, Si II $\lambda 1808{-}16$, Si III] $\lambda 1892$, C III] $\lambda 1909$. Their half widths at half maximum and half widths at base are between 200 and 500 km s^{-1}, with Lα between 400 and 1000 km s^{-1}. Reimers and Cassatella (1985) interpret these emission lines as originating from a gas of $N_e = 10^{10}$ cm^{-3} and $T_e = 2 - 3 \times 10^4$ K in the case of intercombination lines like C III] and Si III], or of $N_e \lesssim 10^{12}$ cm^{-3} and $T_e = 1 \times 10^5$ K for the allowed lines of C IV, Si IV, N V. They also find that the continuum shortward of 2600 Å is due to a hydrogen recombination continuum, formed at $T = 11\,000$ K in a disc around Mira B. The gaseous disc and cloud around Mira B is supposed to be formed by accretion from the wind of Mira A. Reimers and Cassatella (1985) conclude that this disc is completely replaced within approximately 5 hr and that the dwarf companion of Mira has $T^* \lesssim 14\,000$ K.

4. Multispectral Information and Theory

The analysis of symbiotic stars is most crucially dependent on a wide spectral coverage. As mentioned above, more rapid progress in understanding symbiotic stars can be made if we seek spectral information from those wavelengths that best characterize the hot object (UV continuum and X-rays), the cool object (IR) and the nebular material in the system (radio continuum and UV emission lines).

In addition to the ultraviolet spectroscopy made possible with IUE, the HEAO−2 and EXOSAT instruments have performed surveys and individual observations of several symbiotic stars. Allen (1981) reported several detections among 19 objects observed with HEAO—2. He noted that only the nearby, unreddened or recent activity objects were detected. All the sources showed extremely soft spectra. The decrease in X-ray flux with age in the sequence HM Sge — V1016 Cyg — RR Tel could be explained by cooling of the white dwarf following an accretion powered outburst. Anderson *et al.* (1981) discussed HEAO—2 observations of AG Dra. Willson *et al.* (1984) discussed HEAO—2 observations of HM Sge. Cordova (1985 IAU Circular No. 4049) reported intense X-ray emission from RS Oph. Viotti (IAU Circular No. 4083 and personal communication) reported detection of R Aqr (twice), AG Dra (thrice), CH Cyg (by Leahy and collaborators), and RS Oph in the soft X-ray region with EXOSAT. Further reports from EXOSAT are anticipated.

A miniature UV spectrometer on the Voyager spacecraft, now in the outer reaches of our solar system, has surprising potential for making far-UV observations of symbiotic stars. Polidan and Keyes (1986) reported detection on September 13, 1983 of O VI emission at 1036 Å in AG Peg, with an integrated flux of 1.2×10^{-10} erg cm^{-2} s^{-1}. Additional prominent emission features were found in the 20 Å resolution data at 920—960 Å, 990 Å, and 1175 Å. The continuum is matched by a 35 000 K model atmosphere, with $E_{(b-v)}$ of 0.12. Additional observations are underway.

Catalogues of calibrated optical spectra are now being published, following criticism of the lack of same. Blair *et al.* (1983) and Allen (1984) both contain flux calibrated spectra obtained with modern detectors with large dynamic ranges. Archives of this type will prove very useful for long term studies of variations of symbiotic stars.

As infrared detector technology advances, improved observations and characterisation of the red giant components in symbiotic stars become possible. An important early report was given by Kenyon and Gallagher (1983), where an IR color index was used to measure the strength of molecular bands in the red giant continuum and to assess their luminosity class. With the new far infrared data from the successful infrared astronomical satellite (IRAS), Kenyon, Fernandez-Castro and Stencel (1986) found strong segregation of symbiotics into s and d types according to far IR colours. They also concluded that the longest period systems are preferentially type d. The larger separation permits the red giant to reach advanced evolutionary stages, which include significant levels of dust formation. The dust greatly obscures the red giant star continuum light. Among such systems the lack of optical and IR TiO bands, the odd shape of the IR continuum and uncertain distances could all be explained in terms of enhanced dust effects.

Radio observations give at present the highest spatial resolution of astronomical objects. Some symbiotic objects have benefitted from high spatial resolution radio observations, V 1016 Cyg is one of them (Hjellming and Bignell, 1982), the jet feature of R Aqr (Hollis *et al.*, 1985), and AG Peg (Hjellming, 1985) are others. If this spatial structure is a consequence of present day dynamical activity, then it may also be observable in the line profiles. An interpretation along these lines is given by Feibelman and Fahey (1985) to the observed IUE line profiles of V 1016 Cyg, or to RX Pup by Kafatos *et al.* (1985). With the advent of the Very Large Array (VLA), Seaquist *et al.* (1984) reported detection of 17 symbiotic stars out of a sample of 59 surveyed at 4.9 and 1.5 GHz. A positive correlation between radio luminosity and red giant spectral type was noted. For the objects with multifrequency detection, they advocate that the positive spectral index argues for thermal Bremsstrahlung due to ionized gas in an optically thick asymmetric mass outflow. A simple model of a hot star photoionizing part of the neutral stellar wind from the red giant accounts for the observations. Further radio studies, correlated with multi-spectral observations, are in progress. Several symbiotic stars seem to have been detected as maser sources as well.

Finally, the existence of a class of low-moderate mass stars with long orbital periods presents a challenge to our understanding of binary star evolution. The usual assumptions of total mass and angular momentum conservation may not apply. Close pairs quickly evolve to their gravitational Roche limits and adiabatically transfer much of their mass (resulting in the Algol paradox: cf. Paczynski, 1967). Widely separated pairs have the physical space for expansion to the red giant and supergiant state, where dust formation and stellar wind driven mass loss can occur, resulting in substantial mass and angular momentum loss from the system. The courses of mass loss in long period systems also can lead to common envelope evolution. This can convert binaries into exotica, such as the FK Comae stars, cataclysmics or possibly supernovae (Webbink, 1976). The study of the matter flow between widely separated stars holds the key for further elucidation of this missing part of the theory of binary star evolution.

References

Allen, D. A.: 1981, *Monthly Notices Roy. Astron. Soc.* **197**, 739.

Allen, D. A.: 1984, *Proc. ASA* **5**, 369.

Altamore, A., Baratta, G. B., Cassatella, A., Friedjung, M., Giangrande, A., Ricciardi, O., and Viotti, R.: 1981, *Astrophys. J.* **245**, 630.

Anderson, C. M., Cassinelli, J. P., and Sanders, W. T.: 1981, *Astrophys. J.* **247**, L127.

Baratta, G. B., Cassatella, A., and Viotti, R.: 1974, *Astrophys. J.* **187**, 651.

Baratta, G. B., Piro, L., Viotti, R., Cassatella, A., Altamore, A., Ricciardi, O., and Friedjung, M.: 1985, ESA SP-239, p. 95.

Blair, W. P., Stencel, R. E., Feibelman, W. A., and Michalitsianos, A. G.: 1983, *Astrophys. J. Suppl. Ser.* **53**, 573.

Boyarchuk, A. A.: 1967, *Sovjet Astronomy* **11**, 8.

Boyarchuk, A. A.: 1968, *Astrofizika* **4**, 289.

Cassatella, A., Eiroa, C., and Fernandez-Castro, T.: 1986, *Astron. Astrophys.* (preprint).

Cassatella, A., Holm, A., Reimers, D., Ake, T., and Stickland, D. J.: 1985, *Monthly Notices Roy. Astron. Soc.* **217**, 589.

Cowley, A. P. and Stencel, R.: 1973, *Astrophys. J.* **184**, 687.

Deuel, W. and Nussbaumer, H.: 1983, *Astrophys. J.* **271**, L19.

Doschek, G. A., Feldman, U., Mariska, J. T., and Linsky, J. L.: 1978, *Astrophys. J.* **226**, L35.

Feast, M. W., Whitelock, P. A., Catchpole, R. M., Roberts, G., and Carter, B. S.: 1983, *Monthly Notices Roy. Astron. Soc.* **202**, 951.

Feibelman, W. A., 1982, *Astrophys. J.* **263**, L69.

Feibelman, W. A. and Fahey, R. P.: 1985, *Astrophys. J.* **292**, L15.

Fernandez-Castro, T., Gimenez, A., Friedjung, M., Cassatella, A., and Viotti, R.: 1985, *Proc. ESA Workshop: Recent Results on Cataclysmic Variables, Bamberg 1985*, ESA SP-236, 225.

Fernandez-Castro, T., Cassatella, A., Gimenez, A., and Viotti, R.: 1986, *Astrophys. J.* (submitted).

Flower, D. R., Nussbaumer, H., and Schild, H.: 1979, *Astron. Astrophys.* **72**, L1.

Friedjung, M., Stencel, R. E., and Viotti, R.: 1983, *Astron. Astrophys.* **126**, 407.

Gallagher, J. S., Holm, A. V., Anderson, C. M., and Webbink, R. F.: 1979, *Astrophys. J.* **229**, 994.

Garnavich, P. and Mattei, J.: 1981, IBVS, No. 1961.

Grygar, J., Hric, L., Chocol, D., and Mammano, A.: 1979, *Bull. Astron. Inst. Czech.* **30**, 308.

Hack, M., Rusconi, L., Sedmak, G., Engin, S., and Yilmaz, N.: 1982, *Astron. Astrophys.* **113**, 250.

Hayes, M. A. and Nussbaumer, H.: 1986, *Astron. Astrophys.* **161**, 287.

Heck, A. and Manfroid, J.: 1985, *Astron. Astrophys.* **142**, 341.

Hjellming, R. M. and Bignell, R. C.: 1982, *Science* **216**, 1279.

Hjellming, R. M.: 1985, 'Radio Stars', in R. Hjellming and D. Gibson (eds.), *The Radio-Emitting Wind, Jet and Nebular Shell of AG Pegasi*, D. Reidel Publ. Co., Dordrecht, Holland, pp. 97—100.

Hollis, J. M., Kafatos, M., Michalitsianos, A. G., and McAlister, H. A.: 1985, *Astrophys. J.* **289**, 765.

Hutchings, J. B., Cowley, A. P., Ake, T. B., and Imhoff, C. L.: 1983, *Astrophys. J.* **275**, 271.

Iijima, T., Mammano, A., and Margoni, R.: 1981, *Astrophys. Space Sci.* **75**, 237.

Kafatos, M., Hollis, J. M., and Michalitsianos, A. G.: 1983, *Astrophys. J.* **267**, L103.

Kafatos, M., Michalitsianos, A. G., Allen, D. A., and Stencel, R. E.: 1983, *Astrophys. J.* **275**, 584.

Kafatos, M., Michalitsianos, A. G., and Fahey, R. P.: 1985, *Astrophys. J. Suppl. Ser.* **59**, 785.

Kafatos, M., Michalitsianos, A. G., and Hobbs, R. W.: 1980, *Astrophys. J.* **240**, 114.

Kenyon, S. J., Fernandez-Castro, T., and Stencel, R. E.: 1986, *Astron. J.* **92**, 1118.

Kenyon, S. J. and Gallagher, J. S.: 1983, *Astron. J.* **88**, 666.

Kenyon, S. J. and Truran, J. W.: 1983, *Astrophys. J.* **273**, 280.

Kenyon, S. J. and Webbink, R. F.: 1984, *Astrophys. J.* **279**, 252.

Kenyon, S. J. and Gallagher, J. S.: 1985, *Astrophys. J.* **290**, 542.

Kenyon, S. J. and Garcia, M.: 1985, *Astron. J.* **90**, No. 12.

Kenyon, S. J., Michalitsianos, M., Lutz, J., and Kafatos, M.: 1984, *Publ. Astron. Soc. Pacific* **97**, 268.

Keyes, C. D. and Plavec, M.: 1980, *Int. Astr. Union Symp.* **88**, 535.

Lorenzetti, D., Saraceno, P., and Strafella, F.: 1985, *Astrophys. J.* **298**, 350.

Lundmark, K.: 1921, *Astron. Nachr.* **213**, 93.

Merrill, P.: 1959, *Astrophys. J.* **129**, 44.

Michalitsianos, A. G. and Kafatos, M.: 1982, *Astrophys. J.* **262**, L47.

Michalitsianos, A. G. and Kafatos, M.: 1984, *Monthly Notices Roy. Astron. Soc.* **207**, 575.

Michalitsianos, A. G., Kafatos, M., Feibelman, W. A., and Hobbs, R. W.: 1982, *Astrophys. J.* **253**, 735.

Mueller, B. E. A. and Nussbaumer, H.: 1985, *Astron. Astrophys.* **145**, 144.

Mueller, B. E. A., Nussbaumer, H., and Schmutz, W.: 1986, *Astron. Astrophys.* **154**, 313.

Nussbaumer, H.: 1982, 'The Nature of Symbiotic Stars', in M. Friedjung and R. Viotti (eds.), *UV Line Emission of Symbiotic Stars*, D. Reidel Publ. Co., Dordrecht, Holland, pp. 85—102.

Nussbaumer, H.: 1986, *Astron. Astrophys.* **155**, 205.

Nussbaumer, H. and Storey, P. J.: 1979, *Astron. Astrophys.* **71**, L5.

Nussbaumer, H. and Schild, H.: 1979, *Astron. Astrophys.* **75**, L17.

Nussbaumer, H. and Schild, H.: 1981, *Astron. Astrophys.* **101**, 118.

Nussbaumer, H. and Storey, P. J.: 1982, *Astron. Astrophys.* **115**, 205.

Nussbaumer, H. and Schmutz, W.: 1983, *Astron. Astrophys.* **126**, 59.

Nussbaumer, H. and Storey, P. J.: 1984, *Astron. Astrophys. Suppl. Ser.* **56**, 293.

Nussbaumer, H., Schmutz, W., and Vogel, M.: 1986, *Astron. Astrophys.* **169**, 154.

Nussbaumer, H. and Storey, P. J.: 1986, *Astron. Astrophys. Suppl. Ser.* **64**, 545.

Oliversen, N., Anderson, C., Stencel, R., and Slovak, M.: 1985, *Astrophys. J.* **295**, 620.

Paczynski, B.: 1967, *Acta Astr.* **17**, 193.

Penston, M. V. and Allen, D. A.: 1985, *Monthly Notices Roy. Astron. Soc.* **212**, 939.

Penston, M. V., Benvenuti, P., Cassatella, A., Heck, A., Selvelli, P., Macchetto, F., Ponz, D., Jordan, C., Cramer, N., Rufener, F., and Manfroid, J.: 1983, *Monthly Notices Roy. Astron. Soc.* **202**, 833.

Persic, M., Hack, M., and Selvelli, P. L.: 1984, *Astron. Astrophys.* **140**, 317.

Pickering, E. C.: 1901, *Astrophys. J.* **13**, 226.

Polidan, Keyes: 1986, *Bull. Amer. Astron. Soc.* **17**, 886.

Ponz, D., Cassatella, A., and Viotti, R.: 1982, 'The Nature of Symbiotic Stars', in M. Friedjung and R. Viotti (eds.), *Ultraviolet Observations of RR Telescopii*, D. Reidel Publ. Co., Dordrecht, Holland, pp. 217—218.

Reimers, D. and Cassatella, A.: 1985, *Astrophys. J.* **297**, 275.

Sahade, J., Brandi, E., and Fontenla, J. M.: *Rev. Mexicana Astron. Astrof.* **6**, 201.

Seaquist, E. R., Taylor, A. R., and Button, S.: 1984, *Astrophys. J.* **284**, 202.

Selvelli, P. L. and Hack, M.: 1985, *Astronomy Express* **1**, 115.

Slovak, M. H. and Africano, J.: 1978, *Monthly Notices Roy. Astron. Soc.* **185**, 591.

Slovak, M. H. and Lambert, D. L.: 1982, 'The Nature of Symbiotic Stars', in M. Friedjung and R. Viotti (eds.), *Ultraviolet Properties of the Symbiotic Stars*, D. Reidel Publ. Co., Dordrecht, Holland, pp. 103—113.

Slovak, M. H., Cassinelli, J. P., Anderson, C. M., and Lambert, D. L.: 1985 (preprint).

Stencel, R. E.: 1984, *Astrophys. J.* **281**, L75.

Stencel, R. E., Michalitsianos, A. G., Kafatos, M., and Boyarchuk, A.: 1982, *Astrophys. J.* **253**, L77.

Stickland, D. J., Cassatella, A., and Ponz, D.: 1981, *Monthly Notices Roy. Astron. Soc.* **199**, 1113.

Taranova, O. G. and Yudin, B. F.: 1983, *Astron. Astrophys.* **117**, 209.

Thackeray, A. D.: 1977, *Mem. Roy. Astr. Soc.* **83**, 1.

Thackeray, A. D. and Hutchings, J. B.: 1974, *Monthly Notices Roy. Astron. Soc.* **167**, 319.

Viotti, R., Ricciardi, O., Ponz, D., Giangrande, A., Friedjung, M., Cassatella, A., Baratta, G. B., and Altamore, A.: 1983, *Astron. Astrophys.* **119**, 285.

Viotti, R., Altamore, A., Baratta, G. B., Cassatella, A., and Friedjung, M.: 1984, *Astrophys. J.* **283**, 226.

Wallerstein, G. and Greenstein, J. L.: 1980, *Pub. Astr. Soc. Pacific* **92**, 275.

Webbink, R. F.: 1976, *Astrophys. J.* **209**, 829.

Willis, A. J. and Wilson, R.: 1978, *Monthly Notices Roy. Astron. Soc.* **182**, 559.

Willson, L. A., Wallerstein, G., Brugel, E. W., and Stencel, R. E.: 1984, *Astron. Astrophys.* **133**, 154.

Zuckerman, B.: 1979, *Astrophys. J.* **230**, 442.

INTRINSICALLY VARIABLE STARS

ERIKA BÖHM-VITENSE

Department of Astronomy, University of Washington, Seattle, WA, U.S.A.

and

MONIQUE QUERCI

Observatoire du Pic du Midi et de Toulouse, Toulouse, France

1. Introduction

The present paper, dealing with intrinsically variable stars, consists of two parts. In the first part the pulsating stars of medium spectral types are discussed (mainly the δ Cepheids, W Vir stars and related groups), while the second part deals with late-type variables, mainly the M, S, and C giants and supergiants, hence stars close to and on the Asymptotic Giant Branch (AGB) in the HR diagram. Few of these latter stars appear to be 'quasi-constant' and have been in priority analysed by IUE. Their discussion is included since they are a clue to the understanding of the real variables.

2. The Cepheids and Related Pulsating Stars

Among the pulsating stars the δ Cephei, the RR Lyrae and the W Virginis stars appear to be the most important kinds, we shall therefore concentrate our discussions on these groups.

The β Cephei stars are also very interesting objects, mainly because we do not yet understand their excitation mechanism and their mode of pulsation. But exactly for this reason they are at the moment not very useful for other studies, though they may turn out to be very important probes for the interior of massive stars during their early stages of evolution, when interior mixing above the convective core may be very important but is very little understood at present.

Delta Scuti stars have attracted little attention from IUE researchers. Long period variables and RV Tauri stars will be discussed in the second part of this chapter on intrinsically variable stars.

Observations of population I and population II Cepheids are of major importance because:

(1) The period luminosity (PCL) relation provides a rather simple way to determine distances to nearby extragalactic systems, especially those in the local group of galaxies.

(2) The Cepheids provide an excellent means to check our understanding of stellar structure and evolution.

In the following we shall discuss how IUE observations have helped in the calibration of the PCL relation and how IUE observations of δ Cephei stars have been useful to check stellar evolution calculations.

Y. Kondo (ed.), Scientific Accomplishments of the IUE, pp. 223–257.
© 1987 *by D. Reidel Publishing Company.*

3. Calibration of the Period — Luminosity Relation

3.1. PREVIOUS CALIBRATIONS AND PROBLEMS

Twenty years ago it appeared that mainly due to the work of Sandage and Tammann (1969) the distances to the galactic Cepheids were well established. The PCL relation calibration was based on Cepheids in galactic clusters whose distances had been determined by main sequence fitting to the Hyades main sequence. When the Hyades distance modulus was increased (Hodge and Wallerstein, 1967) the distances of the Cepheids were increased correspondingly. Crawford (1978, 1979) emphasized the importance of evolutionary effects on the upper parts of the cluster main sequences used for the main sequence fitting. He based his distance scale on trigonometric parallaxes of F stars with $\pi \geqslant 0''.04$, thereby avoiding the problem of the Hyades distance. He tried to take into account evolutionary effects by means of Strömgren photometry. He obtained generally smaller cluster distances than previous observers, but we realize that the trigonometric parallaxes of F stars show a large scatter and may therefore also be not very reliable.

Schmidt (1984) used Crawford's calibration to determine new distances to galactic clusters with Cepheids and obtained distance moduli which are smaller by 0.4 mag than those obtained with the Sandage Tammann relation, corrected for the larger Hyades distance.

Schommer *et al.* (1984) redetermined the distance to the LMC by measuring the apparent magnitudes of unevolved main sequence stars in the LMC clusters NGC 2162 and 2190 and fitting them to theoretical main sequences calculated by VandenBerg (1984). They also obtained a smaller distance for the LMC ($m_v - M_v = 18.2 \pm 0.2$) than derived so far. These authors did not determine possible interstellar extinction in the neighborhood of this cluster. They used an estimated metal abundance for the theoretical main sequence.

A new distance determination for the LMC clusters H4 by means of main sequence fitting (Hodge and Mateo, 1986) also derives a distance modulus which is even smaller than the one derived by Schommer *et al.* (1985) namely $m_v - M_v = 18.1 \pm 0.3$.

Clube and Dawe (1983) also found an even smaller distance to the LMC from a discussion of RR Lyrae stars, namely $m_v - M_v = 17.9 \pm 0.3$.

Cepheid distances could in principle be determined from visual binaries with one Cepheid component, but suitable visual binaries are not known. Cepheid distances can also be determined by means of the Baade-Wesselink method, but the results are somewhat uncertain for several reasons:

(1) This method needs a determination of the expansion and contraction velocities for the continuum forming layer while only radial velocities for metallic lines can be measured, averaged over the visual hemisphere. The relation between the measured radial velocities and the expansion velocity of the continuum forming layer is not at all clear.

(2) The method may lead to incorrect results if the Cepheid has a companion, which influences the colors. For a blue companion the radii come out too small and for a red companion too large radii are obtained (Balona, 1977).

3.2. Cepheid distance determinations by means of IUE observation

Cepheid distances can also be determined from spectroscopic parallaxes of Cepheid binaries if the absolute magnitudes of the companions can be determined from their spectral types, colors or energy distributions. This method works in the same way as the distance determination from Cepheids in clusters except that we do not have to worry about cluster membership if close binaries are observed (though it cannot be absolutely excluded that occasionally we may find an optical companion).

Also we are not limited to rather distant clusters with very faint main sequences. Several Cepheid binaries turned out to be closer than the clusters with Cepheids, fainter main sequence companions can therefore be observed, whose luminosities are not yet influenced by evolutionary effects. However, in the optical spectral region Cepheids are intrinsically very bright stars. More massive companions have evolved to become compact objects and less massive close companions are too faint to be observable in this spectral region. In the ultraviolet the rather cool Cepheids are intrinsically faint stars and blue main sequence companions can easily be detected for wavelengths $\lambda \leqslant 2500$ Å. By means of IUE observations 16 blue companions of Cepheids have been detected (Mariska *et al.* 1980; Schmidt and Parsons, 1982; Böhm-Vitense and Proffitt, 1985; Arellano Ferro and Madore, 1985). Five of these companions are main sequence A stars, whose energy distributions longward of 1250 A are very sensitive to small changes in T_{eff} and for which T_{eff} can therefore be determined very accurately. This means their absolute visual magnitudes M_V can also be determined well if we know the relation between M_V and T_{eff} for zero age main sequence (ZAMS) stars. If Cepheids are indeed as massive as their evolutionary masses indicate (see Cox, 1980) then main sequence A stars should have a much smaller mass than the Cepheids, they should be still close to the ZAMS. With the $E(B - V)$ values determined by Dean *et al.* (1978) and with the average galactic extinction law as given for instance by Savage and Mathis (1979) a reddening corrected ultraviolet energy distribution for the companion can be determined from the measured energy distribution, which can then be compared with model energy distributions (Kurucz, 1979) in order to determine T_{eff} and M_V. From the model ratio of far *uv* fluxes to visual fluxes the apparent visual magnitudes can be obtained and the distance modulus be determined (Böhm-Vitense, 1985). This comes out to be even smaller than determined by Schmidt (1984), a distance modulus smaller by 0.45 mag than the original Sandage Tamman relation is obtained. Assuming an uncertainty in T_{eff} by $\pm$ 200 K (an uncertainty of $\pm$ 100 K was estimated) and an uncertainty of $\pm$ 0.15 mag in the relation between the far *uv* fluxes and the visual fluxes in addition to an uncertainty of 0.1 mag in the $M_V(T_{\text{eff}})$ relation, the distance modulus derived by Schmidt would be just at the error limits of this determination. If the masses of the Cepheids are actually considerably smaller than given by the evolutionary masses (see below) then the luminosities of the A star companions may be increased due to evolutionary effects. We estimate that the ΔM_V may be between -0.1 and -0.25 depending on the mass of the Cepheid. This lead us to correct the distance moduli determined previously from the main sequence A star companions by $\Delta(m_v - M_v) = +0.2$.

The adopted P.-L. relation then corresponds to the original Sandage—Tammann relation but the M_V corrected by $+0.2$, i.e.

$$M_V = -3.425 \log P + 2.52 \langle B - V \rangle_0 - 2.16. \qquad (1a)$$

as compared to Schmidt's calibration

$$M_V = -3.8 \log P + 2.70 (\langle B \rangle_0 - \langle V \rangle_0) - 2.21. \qquad (1b)$$

In Table I we give both values of M_V for selected Cepheids. Our values are about 0.25 mag fainter in the average than those obtained by Schmidt's calibration.

4. Stellar Evolution Theory and IUE Studies of δ Cephei Stars

4.1. A CHECK OF STELLAR EVOLUTION THEORY BY MEANS OF GIANT COMPANIONS OF CEPHEIDS

The best way to check stellar evolution theory is to observe stars of equal age with nearly equal masses ($\Delta M/M \leqslant 0.05$). If such stars are at an advanced stage of evolution they outline an evolutionary track. Cepheids are in an advanced evolutionary state. Unfortunately they do not occur in clusters which are populous enough to find enough stars of nearly identical mass to outline such evolutionary tracks. If Cepheids can be found in populous clusters in the LMC we may have a chance. In binaries, generally pairs of nearly equal mass are preferred (Trimble, 1978), nevertheless, we have to be lucky to find a companion whose mass agrees to within 5% with that of a Cepheid. It is, however, to our advantage that giant companions are relatively brighter in the optical region than main sequence companions. Several of these giant companions could therefore be detected by their veiling effect on the Ca II K lines (Miller and Preston, 1964) or by their influence on the *UBV* colors (Madore, 1977).

For RW Cam, AW Per, SV Per, RY Nor and AX Cir blue giant companions have either been detected or confirmed by means of IUE. As we have only one companion for each Cepheid we cannot outline the whole evolutionary track but we can at least compare two points on the track.

From the far UV energy distributions of the giant companions, as measured on the IUE SWP spectra, the effective temperatures can be determined after correction for interstellar extinction. With distances obtained from the PCL relation for the Cepheids the luminosity of the giant companion can also be determined and we can check whether Cepheid and companion fit on nearly the same track. It was found consistently that the Cepheids are too luminous in comparison with their giant companions by about $\Delta \log L \approx 0.4$ (Borutzki *et al.*, 1984; Böhm-Vitense and Proffitt, 1985), see Figure 1. AX Cir is the only exception. The companion is still in an early phase of evolution. The companions for AW Per, SV Per, RY Nor and RW Cam fit on an evolutionary track for a star whose mass is about 25% less than the standard evolutionary mass of the Cepheid. Since the masses of the Cepheids can at most be 5% larger than those of the giant companions the blue loops of the evolutonary tracks must occur at higher luminosities than calculated.

TABLE I

Masses of Cepheids determined in different ways with distances according to Equation (1a)

Star	S Mus	V636 Sco	SU Cyg	δ Cep	β Dor	ζ Gem	ℓ Car	AW Per	SV Per	RW Cam
$\langle B \rangle_0 - \langle V \rangle_0$	0.55	0.70	0.45	0.57	0.71	0.66	(1.00)	0.62	0.66	0.86
$\langle B - V \rangle_0$ [a]	0.51	0.74	0.52	0.64	0.76	(0.71)	1.07	(0.67)	(0.71)	0.91
$\log P$	0.98	0.83	0.58	0.73	0.99	1.00	1.55	0.81	1.05	1.21
M_v (Equation 1a)	−4.33	−3.24	−2.94	−3.15	−3.77	−3.90	−4.87	−3.35	−4.07	−4.11
M_v (Equation 1b)	−4.45	−3.47	−3.20	−3.45	−4.06	−4.23	−5.40	−3.61	−4.42	−4.49
BC	0.02	0.11	0	0.03	0.13	0.09	0.40	0.06	0.09	0.25
$M_{\rm bol}$ (1a) [b]	−4.35	−3.35	−2.94	−3.18	−3.90	−3.99	−5.27	−3.41	−4.16	−4.36
$\log L/L_\odot$ [c]	3.640	3.240	3.076	3.172	3.460	3.496	4.008	3.264	3.564	3.644
$M/M_\odot$ (evol.)	5.25	4.29	3.95	4.15	4.79	4.88	6.32	4.34	5.05	5.26
$M/M_\odot$ (puls.) [e]	5.53	3.77	2.96	3.26	4.42	4.33	6.74	3.71	4.52	4.20
$M/M_\odot$ (Wess)	4.33	3.96	—	4.87	5.08	5.43	6.55	4.75	—	—
$M/M_\odot$ (dyn)	5.25 ± 1	4.8 ± 1	5.5 ± $^{0.5}_{2}$							
$M/M_\odot$ (giant comp.) [f]								3.5 ± 0.6	5.0 ± 0.6	4.6 ± 0.6
$\log T_{\rm eff}$ [d]	3.762	3.736	3.783	3.758	3.734	3.742	3.686	3.749	3.742	3.708

[a] Numbers in brackets are estimated.

[b] Using M_v (1a).

[c] Assuming $M_{\rm bol\,\odot} = 4.75$.

[d] Obtained from $\langle B \rangle_0 - \langle V \rangle_0$.

[e] With the M.-L. relation given in Equation (3).

[f] For the distance scale adopted here.

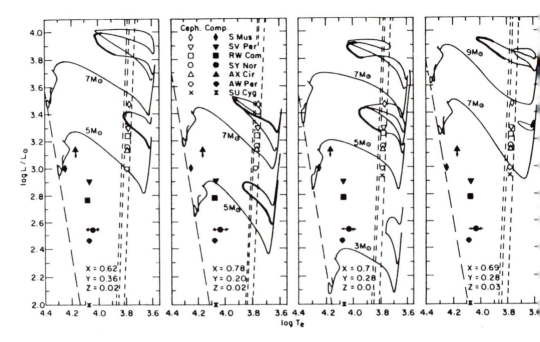

Fig. 1. The positions of the Cepheids and their giant companions are shown in the HR diagram. Also shown are evolutionary tracks for different masses and different chemical compositions according to Becker *et al.*, 1979. No matter which chemical composition is used, the luminosities of the Cepheids are too large in comparison with the luminosities of their giant companions. Copied from Böhm-Vitense and Proffitt (1985).

What could cause such a discrepancy? It appears that convective overshoot mixing leads to relatively higher luminosities of the Cepheids with respect to late B giants of equal mass (see for instance Becker and Cox, 1982; or Matraka *et al.*, 1982). In fact it seems that one can estimate the degree of mixing by comparing the luminosities of the Cepheids and the giant companions assuming that the Cepheid did not have measurable mass loss during its giant evolution. If indeed mass loss did occur it has reduced the Cepheid luminosity and we will underestimate the amount of mixing. *If* convective overshoot mixing is the explanation than the giant luminosity is also larger than calculated with 'standard' evolution theory, i.e., without taking into account convective overshoot mixing. If the relative luminosity of giant and Cepheid is changed by $\Delta \log L \approx 0.4$ then the giant luminosity is also increased, though by a smaller amount as we infer from Becker and Cox (1982). We can then estimate that for the Cepheid the increase in luminosity is probably between $\Delta \log L \approx 0.5$ and 0.7, as compared to the 'standard' evolution.

This has important consequences for the mass luminosity relation for Cepheids. For the standard evolution Becker *et al.* (1977) find for $Y = 0.28$ and $Z = 0.02$ that (see their equation 7)

$$\Delta \log M = 0.066 + 0.219 \, \Delta \log L \qquad (2)$$

with $\Delta \log M = \log M - 0.7$ and $\Delta \log L = \log L - 3.25$. Increasing $\log L$ by 0.6 (presumably due to mixing) leads to

$$\log L = \log M / 0.219 + 0.353 \qquad (3)$$

and

$$\log M = \log L \times 0.219 - 0.077. \qquad (4)$$

increasing $\log L$ by 0.7 leads to

$$\log M = \log L \times 0.219 - 0.099. \qquad (4a)$$

We *tentatively* adopt the corrected mass-luminosity relation (4) for Cepheids with periods of $16.5 > P > 6.5$ days, the period range to which the observations of giant companions refer. The evolutionary masses given in Table I were calculated using Equation (4).

4.2. THE CEPHEID MASSES

4.2.1. *The Problem*

It has long been realized (Christy, 1966) that the masses derived from the phase of the bumps in the velocity curves with periods around 7 days are much smaller than the evolutionary masses for the same Cepheids or, in other words, the observed luminosities of the Cepheids are much larger than expected from the phases of the bumps in the velocity curves.

With the new distance scale the masses derived from the pulsational periods are also much smaller than the evolutionary masses and tend to agree with the bump masses (Schmidt, 1984).

For the 'beat' Cepheids with periods around two days masses can be determined from the ratio of the beat periods. Masses around two solar masses are found, while the evolutionary masses are around 5 $M_\odot$. With the new distance scale the pulsational masses, determined for Cepheids with only slightly larger periods, come out closer to the beat masses (Schmidt, 1984).

Radii of Cepheids can be determined by means of the Baade–Wesselink method. Inserting these into the theoretical period density relation yields the so-called Wesselink masses, which are independent of the distance scale, but depend rather sensitively on the adopted relation between the velocity observed for the metallic lines and the true expansion velocity of the photosphere, which is not known very well. For periods less than 6 days the average Wesselink masses tend to agree with the evolutionary masses, for $P > 11$ days they are in the average much smaller (Cox, 1980).

4.2.2. *Determination of Dynamical Masses by Means of IUE Observations*

IUE has offered the possibility to determine dynamical masses for Cepheids with companions. Once the effective temperature of a main sequence companion has been determined (see Section 3) its mass is also known. The ratio of the orbital velocities of the two components then determines the mass ratio and thereby the mass of the Cepheid. The orbital velocity of the Cepheid and the system velocity can be determined by groundbased observations. It is then in principle sufficient to measure one radial velocity of the companion for a known orbital phase. These latter measurements have been made possible by IUE.

On IUE spectra it is difficult to measure absolute velocities but relative velocities of different sets of lines on one spectrum can be measured with an accuracy of about ± 2 km s^{-1} by means of cross-correlation and if many orders of the echelle spectra are used. On LWR images the spectrum of the companion as well as the one of the Cepheid can be seen. For $\lambda > 2700$ Å we see the light from the Cepheid while for $\lambda < 2500$ Å mainly the light of the companion is seen if it has a spectral type of A0 or earlier. The radial velocity of the Cepheid spectrum can be measured by cross-correlation with the spectrum of the G supergiant β Aqr, and the velocity of the companion by cross-correlation with a spectrum of α Lyrae. The velocity difference between the β Aqr spectrum and the one of α Lyrae can again be measured by cross-correlation. We can thus determine the velocity difference between the Cepheid and the companion. With the known pulsational phase of the Cepheid the pulsational radial velocity can be subtracted. The remaining velocity difference between the components gives the sum of the orbital amplitudes. If the orbital velocity of the Cepheid for the time of observation has been previously determined the ratio of the orbital velocities and the dynamical mass of the Cepheid is known. Due to the many intermediate steps the final result is only reasonably accurate if the orbital amplitudes are large at the time of observation.

So far measurements have been made for S Mus and V636 Sco (Böhm-Vitense, 1986). For these stars Lloyd Evans (1982) has determined the orbital parameters and phases. Unfortunately the orbital amplitudes were not very large during the times of the IUE observation, and the derived dynamical masses have rather large error bars ($\pm 25\%$). For S Mus with a period of 9.5 days the dynamical mass came out to be 5.25 $M_\odot$ and for V636 Sco with $P = 6.8$ days a dynamical mass of 4.8 $M_\odot$ was obtained.

Radial velocity measurements were also obtained for SU Cyg, for which Evans and Bolton determined the orbital parameters. Large short period velocity variations were observed for the companion indicating the binary nature of the companion itself (Evans et $al.$, 1985). The period of these rapid radial velocity variations is 4.7 days, while the pulsational period of the Cepheid is 3.8 days. The period of the close binary is short enough to treat it as one body when deriving the mass ratio of the Cepheid and the sum of the companions, which indicates a slightly larger mass for the Cepheid than for the sum of the companions, which is rather unusual for a triple system of this kind. For the companion seen in the SWP spectrum a spectral type B9 V was derived by Böhm-Vitense and Proffitt, leading to a mass of about 3 $M_\odot$ for this star. At this point the nature of the third star is still debated. If it is indeed an A0 V star as Evans and Bolton (1986) suggest then SU Cyg with a period of 3.8 days has a larger mass than obtained for S Mus and V636 Sco which have much longer periods. A similar trend is, however, indicated by the average Wesselink masses as was shown by Böhm-Vitense (1986). If the third star is indeed invisible and has a smaller mass then we can only say that the dynamical mass of SU Cyg is larger than say 3.5 $M_\odot$.

5. The Importance of the Hydrogen Convection Zone in Cepheids

5.1. THE PROBLEM

While it is suspected on theoretical grounds that the red boundary of the Cepheid instability strip is determined by the onset of efficient convective energy transport in the hydrogen convection zone the time dependent convection theory is very uncertain. We would therefore like to support this suspicion by observations. Ultimately such observations may help us to improve our convection theory.

We are also not sure how much the pulsations of the Cepheids might be influenced by convection. Since the Cepheids are right at the borderline in the HR diagram where convection has to stop rather abruptly because the convection zones become too thin, mixing length calculations for the convective energy transport have to be very careful in choosing the size of the mixing length, which cannot be larger than half the depth of the convection zone. While we do not expect any influence of the convective energy transport for stars on the blue side of the instability strip, convection may be of importance for Cepheids on the red side of the instability strip, especially during the low temperature phases of the pulsational cycle. During the high temperature phases convection will probably die out except perhaps for stars right at the red boundary line. It would be assuring if such an effect could really be observed. Could the phases of the bumps in the velocity curve perhaps be influenced by time dependent convection? Or could the beat periods be influenced by convection, thereby explaining the low beat masses?

IUE observations can and have shed some light on the importance of convective energy transport in Cepheids.

5.1.1. *The Cepheid Instability Strip and the Boundary Line for Chromospheric Emission*

If our assertion is correct that the red boundary of the Cepheid instability strip is due to the onset of efficient convective energy transport, which kills the excitation mechanism, then we should find stars with efficient convection on the cool side of the red boundary of the Cepheid instability strip and no convection close to the blue boundary line. Since we believe that the temperature rise in the solar type chromospheres and transition layers is due to mechanical energy input which has its origin in the hydrogen convection zones we expect to find chromospheric and transition layer emission to be visible only on the red and not on the blue side of the Cepheid instability strip. In an IUE study Böhm-Vitense and Dettmann (1980) confirmed the earlier conclusion by Böhm-Vitense and Nelson (1976), that this is indeed the case.

5.1.2. *Chromospheric and Transition Layer Emission in Cepheids*

IUE studies by Schmidt and Parsons (1982, 1984a, b) looked for chromospheric emission lines in δ Cephei stars during different phases of the pulsational cycle, see also Eichendorf *et al.* (1982). The first question is of course: if indeed chromospheric emission is observed is it due to a solar type chromosphere or is it due to the pulsational shock propagating through the outer atmospheric layers? If the shock is the origin we expect the emission to occur for all Cepheids during and shortly after maximum light when the shock reaches the high layers. If a true solar type chromosphere, due to heating by some mechanical flux, originating in a convection zone, is responsible we expect the emission to become visible during the coolest phases of the pulsational cycle. That start of the chromospheric emission must depend on the timescale for the development of full convection. We also expect then that Cepheids closer to the red boundary of the instability strip show chromospheric emission for larger fractions of the pulsational cycle than those on the blue side of the instability strip. For shockwave emission the position within the instability strip should be of no influence.

Schmidt and Parsons did indeed find chromospheric emission for several δ Cephei stars during part of their pulsational cycle. We have used their data to study the correlation with the positions in the instability strip. No emission was observed for δ Cephei for any phase of the pulsational cycle except right at the phase of maximum acceleration. For β Dor emission lines of O I were observed but only around phases 0.8 which means at phases of increasing light. The C II (1335 Å), Si IV (1400 Å) and C IV (1550 Å) lines were detected once at phase 0.96. For ζ Gem chromospheric emission was observed through the whole cycle though of varying intensity. No transition layer emission was seen except for one observation of the C II (1335 Å) line. ℓ Car was observed essentially only during a very narrow interval around phase 0 (maximum light), when it showed strong O I emission and also C IV emission which decreased rapidly with progressing phase. Schmidt and Parsons also find that the maximum chromospheric and transition layer emissions agree with the ones for non pulsating stars with the same $B - V$ as the Cepheid has during the phase of maximum emission. These authors emphasize the correlation of the chromospheric and transition layer emission with the length of the pulsational period, which leads them to believe that the chromospheric heating is due to pulsational energy. We arrive at a different conclusion. In Figure 2 we have plotted the positions of these Cepheids in the HR diagram. The luminosities were determined according to our Equation (1a). As a reference line we have also drawn the theoretical blue edge for fundamental mode pulsations with the equations given by Becker *et al.* (1977) and with our calibrations given in Equations (1a) and (3). Becker *et al.* (1977) estimate the width of the instability strip to be about $\Delta \log T_{\text{eff}} \approx 0.06$. We have drawn a line corresponding to $\Delta \log T_{\text{eff}} = 0.07$. If convective overshoot mixing is important for the evolution of massive stars then the position of the theoretical instability strip might perhaps also shift somewhat.

In any case it is quite obvious that the observed chromospheric emission is weakest for the stars closest to the blue edge and is strongest for the stars close to the red edge, which also happen to have the longest periods. It is very interesting that the reddest star ℓ Car is the only star which ever shows strong C IV emission. It is also interesting to note that β Dor shows more transition layer emission than does ζ Gem. The studies of chromospheric emission of δ Scuti stars by Fracassini

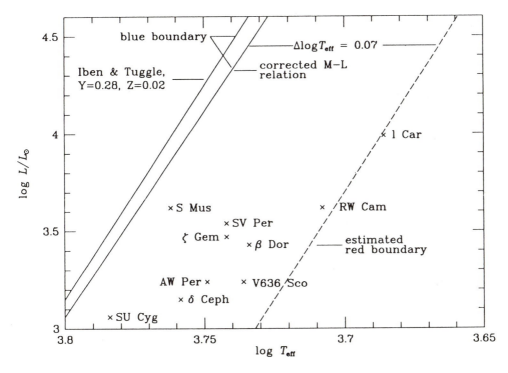

Fig. 2. The positions of the Cepheids studied by IUE and discussed here are shown in the T_{eff} — luminosity diagram. The stars which show chromospheric emissions lines during part of the pulsational cycle have positions closer to the red boundary line for pulsational instability than δ Ceph which does not show chromospheric emission lines.

and Pasinetti, 1982 also show that only the stars on the cool side of the instability strip show Mg II emission (γ Cyg, ρ Pup and β Cas) while those on the blue side do not show emission. Again the δ Scuti stars on the red side are also the ones with the longest periods. If, however the length of the period would be the decisive question, then it is hard to understand why δ Cephei with a period of 4 days has no emission while ρ Pup with a period of 0.14 days shows Mg II emission.

Studies of the Mg II lines for the beat Cepheid TU Cas by Henden *et al.*, 1984, may show some Mg II emission shortly before maximum light.

Even though the phases for the emission lines will have to be studied in more detail the positions of the stars in the instability strip generally confirm our expectations for true solar type chromospheric emission and therefore tell us when chromospheric heating takes place. This conclusion is supported by the fact that except for δ Cephei, which is apparently too blue to show emission, the maximum chromospheric and transition layer emission of the Cepheids agrees with that of non pulsating stars of the same color. This is natural if the maximum emission occurs for fully developed convection and chromospheres. This would be a peculiar accident if the emission would be due to pulsational energy. We probably could derive the efficiency of convective energy transport and wave flux generation as a function of time, if we follow the time evolution of the chromospheric emission.

6. Population II Pulsating Stars

6.1. THE PROBLEMS

RR Lyrae variables are also very important distance indicators for globular clusters and for galaxies in the local group. It has become clear during the last decades of research that their luminosities depend on the chemical abundances. Whether other parameters like age or cluster membership are also important is not clear.

Good abundance determinations are needed to get closer to the answer of these questions. These can only be obtained in combination with good temperature determinations of these stars.

There are two periods in the life of RR Lyrae and W Vir stars for which we have essentially no information. The first is the transition from the tip of the red giant branch to the blue horizontal branch (BHB) which must be connected with mass loss because only stars with $M \leqslant 0.6\ M_\odot$ can become BHB stars, while the red giants have masses between 0.8 and 1 $M_\odot$. The second dark period is the transition from the asymptotic red giant branch to the white dwarf stage. It will be quite interesting to see whether the white dwarfs in globular clusters also have masses around 0.6 $M_\odot$ like most of the white dwarfs in our neighborhood seem to have, or whether they have smaller masses as expected for remnants of the BHB.

It is generally accepted that population II Cepheids get into the instability strip during double shell source flashes (Schwarzschild and Härm, 1970), but this conclusion is mainly by default because it seems to be the only way to bring them into the instability strip. Maybe some W Vir stars get into the instability strip during their transition from the red giant tip to the BHB? Or they might get there during their transition from the asymptotic giant branch to the white dwarf sequence? In the latter cases we would probably expect rather efficient mass loss for the population II Cepheids. Willson (1985) asserts that mass loss takes place mainly during phases of pulsational instability. In metal poor clusters no long period variables are found. How then do red giants lose their mass? Mass loss of population II Cepheids may be one possibility.

The question how cluster morphology is related to chemical abundances and/or mass loss or perhaps age has been debated in many papers during the last decades.

6.2. TEMPERATURE AND ABUNDANCE DETERMINATIONS BY MEANS OF IUE OBSERVATIONS

Most abundance determinations for population II stars are only done by means of broad band colors and/or using the Ca II K line. W Vir is one of the few stars for which the abundances of the heavy elements have been determined from high resolution spectra (Barker *et al.*, 1971). The abundance deficiencies derived by these authors showed a dependence on the ionization potential of the elements indicating problems with the temperature determination. Later temperature determinations from broad band colors by Böhm-Vitense (1974) yielded higher temperatures.

IUE observations now offer the possibility to observe the ultraviolet energy distributions of these stars, which are much more temperature sensitive than the energy distributions in the optical region. Much more flux is received in the far UV for these metal poor stars than for otherwise similar population I stars. With IUE we can indeed get measurable fluxes for 10th magnitude stars with temperatures around 6000 K like W Vir, AL Vir and ST Pup.

The RR Lyrae stars have higher temperatures and can be observed even better. The rather strong discontinuity in the energy distribution around 1700 Å offers an excellent opportunity to study the average deficiencies of Si, Mg, Fe, if temperatures can be determined from the energy distribution at other wavelengths.

Bonnell and Bell (1985) made use of the far UV energy distributions of RR Lyrae and X Arietis measured with IUE in order to check earlier temperature and abundance determinations. They mainly confirmed the earlier results which had estimated $[A/H] = -1$ for RR Lyrae and $[A/H] = -2$ for X Arietis. For X Arietis very good agreement was found between calculated and observed fluxes, confirming that the UV fluxes are a good means to determine metal abundances and temperatures. For RR Lyrae the agreement was less good. Lower metal abundances seemed to be indicated by the weak lines in its spectrum, but the low UV fluxes seemed to require a higher metal abundance. It is suspected that these discrepancies may be related to the beat phenomenon and possibly missing line opacities in the computed spectra.

From the far UV energy distributions for ST Pup, W Vir and AL Vir, measured by means of IUE, Böhm-Vitense et al. (1984) determined metal abundances and T_{eff} for ST Pup and AL Vir and redetermined the metal abundances and temperatures for W Vir for phases near maximum light. The higher temperatures found by the earlier investigation of Böhm-Vitense were reconfirmed and a metal abundance of $[A/H] = -1.0 \pm 0.3$ was found from the UV energy distribution and from the discontinuity at 1700 A. For ST Pup a lower metal abundance of $[A/H] = -1.8 \pm 0.2$ was derived and a temperature of $T = 6500$ K for the phase two days after maximum light. For AL Vir the data are consistent with $[A/H] = -1$.

6.3. MASS LOSS IN POPULATION II CEPHEIDS

For W Virginis low resolution long wavelength IUE spectra were obtained by Parthasarathy and Parsons (1984) and by Böhm-Vitense et al. (1984). Mg II emission was observed at all phases with a surprisingly constant flux in the center of the lines, inspite of the large variations in temperature and radius of the star during the pulsational cycle, see Figure 3. This constancy of the Mg II emission line flux during the pulsational cycle makes us wonder whether these emission lines may not arise in a shell detached from the star rather than in a chromosphere. A surprise is seen, however, at minimum light when the Mg II lines become weaker and an additional line is observed at 2828 A, possibly an Fe I line due to fluorescence with the Mg II emission line which only works if there is a relative velocity of about 50 km s^{-1} between the Mg II emitting layer and the layer

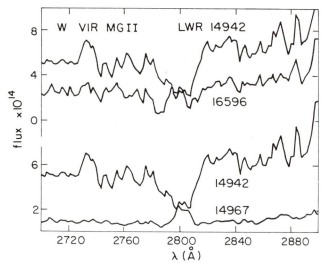

Fig. 3. The spectral energy distribution around the Mg II lines in W Vir is shown for different phases. In the top part we compare the line profile for the phase 0.9, with the profile at phase 0.16. At the bottom we compare the energy distribution at phase 0.02 with the same profile at the phase 0.9 shown in the top spectrum. It can be seen that the flux in the center of the line does not change only the flux in the neighborhood of the emission line changes with phase.

absorbing the Fe I line. It is hoped that high resolution spectra of the Mg II lines will clarify whether the Mg II emission lines take part in the pulsation or whether they remain stationary. In the latter case they cannot originate in the star proper and probably indicate important mass loss.

Unusually strong Mg II absorption components were observed in RR Lyrae by Bonnell and Bell at a velocity which agrees with that of the interstellar medium, they were therefore attributed to interstellar medium, but their abnormal strength was thought to indicate that this "interstellar" medium might have its origin in the star.

6.4. IMPORTANCE OF CONVECTIVE ENERGY TRANSPORT IN RR LYRAE AND W
     VIRGINIS STARS

It is well known that W Vir stars show hydrogen emission during phases of increasing light, which is attributed to an outward progressing shock. Bonnell and Bell found the Al III doublet at 1854 Å and at 1862 Å in emission in the RR Lyrae spectrum taken at phase 0.38 but not in the spectra taken at other phases. No other emission lines were seen. In their ST Pup spectrum Böhm-Vitense and Wallerstein saw some indication of possible emission in the C III lines at 1909 A a few days after maximum light. No spectra are available for other phases. While no firm conclusion can be drawn at this time it appears that observed emisison lines in RR Lyrae and W Virginis stars are attributable to an outward propagating shock rather than to chromospheric emission.

7. The Cold Giants and Supergiants

The number of M, S, and C star spectra recorded by IUE spacecraft is relatively low and addressed to the brighter stars. The reason is that, even at low resolution in the long-wavelength region, exposure times are prohibitive for systematic surveys of late M and late C (N type) stars, due to their faintness in the UV. The case of carbon stars, which have the largest $(B - V)$ index, is particularly striking. The first attempt of recording low-resolution LWR spectra of 2 Mira and 2 semiregular late N stars was unsuccessful: no light was got in 2.3 to 4.0 hr; this fact was used to set useful limits on their chromospheric emission (Querci *et al.*, 1982).

As discussed by Wing *et al.* (1983), since the Planckian photospheric component of the ultraviolet spectrum is very steep in any cold star, there is in each star a fairly well-defined wavelength (depending upon the stellar photospheric temperature) at which the spectrum changes over from an emission-line to an absorption-line spectrum. The hotter the star, the more violet shifted is the cross-over wavelength. In fact, absorption lines are only detected in the long-wavelength range.

7.1. SHORT-WAVELENGTH RANGE (SWP IMAGES)

The 1150—2000 Å region of giant and supergiant M star spectra at low resolution is illustrated by Figure 4. The emission lines are of *neutral* or *lower ionization state species*. No emission from species formed at temperature above ~ 2×10^4 K has been detected, as first pointed out by Wing (1978), showing that the cold giants and supergiants are non-coronal (i.e., clearly located at the right of the Linsky-Haisch dividing line in the HR diagram; Linsky and Haisch, 1979). The early M giant spectra are dominated by O I and S I $\lambda 1302$, and Si II and S I $\lambda\lambda 1807—1826$ emission features. Weaker lines such as C I (UV2) $\lambda 1656$, S I (UV3) $\lambda\lambda 1473$ and 1484, S I $\lambda\lambda 1899$ and 1914 give a contribution. H Ly α ($\lambda 1215$) is saturated in M giants, predominantly due to geocoronal contamination, but, as noted by Stickland and Sanner (1981), a strong stellar component is obvious to this line in most of the stars in their sample. Ly α is much stronger in γ Cru (M3.4 III) than in g Her (M6.6 III), two normal giants (Wing and Carpenter, 1978). It is not so strong in the Mira o Cet (Cassatella *et al.*, 1980a). In the spectrum of the irregular supergiant α Ori (M1—2 Iab) (Figure 4a), the most conspicuous feature is the Si II plus S I emission blend, whereas the feature at $\lambda 1302$ becomes very weak, or insignificant (e.g., Dupree, 1979; Basri *et al.*, 1981).

The UV continua has been recorded in SWP low resolution spectra of several bright M stars (Stickland and Sanner, 1981): they are rather flat and comparable to the flux in the visual region (the ratio is 2×10^{-5} for M0 III star, but increases with later type stars and also with luminosity to about 10^{-4} for M3 II stars).

High resolution spectra confirm the identification of emission lines blended at low resolution and specify what lines are the dominant contributors to the blends. Figure 5 shows that the single broad feature observed at low resolution between 1800 and 1830 Å is due primarily to the triplet of S I (UV2) rather than to the doublet of Si II which is generally preponderant in warmer stars (e.g., in α Tau (K5 III), Brown and Jordan, 1980). In the supergiant α Ori (Figure 5b), the S I lines

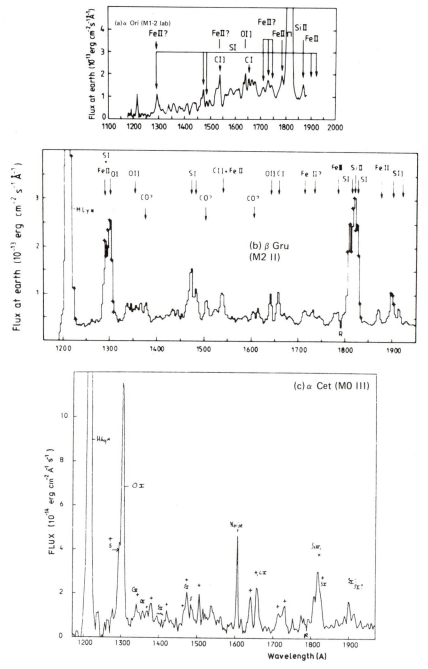

Fig. 4.   SWP low-resolution spectra of M stars. (a) α Ori: the composite spectrum is from Basri *et al.*, 1981; the identification of the features is from Johansson and Jordan, 1984; (b) β Gru (exposure time: 175 mn); reseau marks occur at ~ 1200 and 1800 Å; the spectrum is from Brown and Jordan, 1980, whereas the identification is from Johansson and Jordan, 1984; (c) α Cet (exposure time: 90 mn), from Brown *et al.*, 1979.

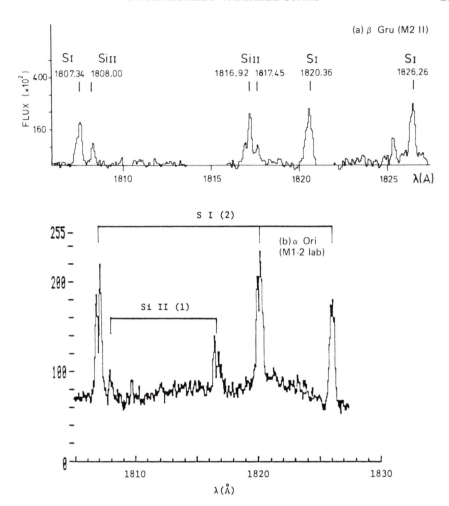

Fig. 5. SWP high resolution spectra. (a) β Gru (exposure time: 370 mn) showing the vicinity of SI (UV2) lines; the ordinate is IUE flux numbers (from Brown and Jordan, 1980); (b) α Ori: raw spectrum in the same region as β Gru (from Carpenter and Wing, 1979, and private communication).

appear doubly reversed, suggesting very high opacity (Stencel *et al.*, 1982). Line identifications in a SWP high resolution spectrum of β Gru (M2 II) are discussed by Brown and Jordan (1980) and by Johansson and Jordan (1984):

(1) The low resolution blend at around $\lambda 1304$ (Figure 4b) shows now three lines of SI (UV9) resolved from the OI triplet (UV1). Two other SI lines of this multiplet lie in the strong OI wings. There is evidence that the SI lines are pumped by two of the strong OI lines. The OI emission itself is likely excited through resonance fluorescence with H Ly β (Haisch *et al.*, 1977), which is anticipated to be as strong as H Ly α (Stickland and Sanner, 1981).

(2) Lines near $\lambda\lambda 1473$ and 1483 are confirmed to belong to SI (UV3). However, the relative intensity of the multiplet UV3 and UV5 (lines at $\lambda\lambda 1425$

and 1433) is really lower than in the solar spectrum; several causes are possible, particularly photo-ionization (or excitation) followed by recombinaison or cascade spectrum. In fact, photo-excitation by strong lines appears to be important in these stars (Brown *et al.*, 1981).

(3) A line at 1641 Å is O I (UV146) pumped by the O I resonance lines (UV2).

(4) Some lines are definitely attributed to Fe II. The source of excitation for Fe levels with excitation energy around 10 eV above the ground state is emission in the H Ly α line, as suggested by the presence of particular transitions and the absence of others. The feature around 1785 Å (mutilated by a reseau in Figure 4b) identified as Fe II (UV191) is an example. Also, the emission around 1870 Å is due to decays from a new discovered further Fe II level at $\sim$ 13 eV pumped by H Ly α emission. Other decays from levels which can be populated by H Ly α may contribute to the feature around 1539 Å.

(5) A C I] transition (an intersystem line pumped by C I (UV4)) may also contribute to the total low resolution flux at 1539 Å.

(6) Some other emission features are noted on Figure 4b as possible CO Fourth Positive System, pumped by the strong O I lines (Ayres *et al.*, 1981).

(7) Finally, in β Gru the broad H Ly α line plays an important role in exciting the observed multiplets of S I (Judge, 1984; Jordan and Judge, 1984).

Many of the features described above are recognized in the spectrum of α Ori (Figure 4a). The exception comes from the O I 1300 Å and S I 1295 Å fluorescent lines. They are indicated totally lacking in a high resolution spectrum recorded by Stencel *et al.* (1982) in August 1981, whereas we note a weak feature, however identifiable, in the low dispersion composite spectrum of Basri *et al.* (1981) recorded in 1978 (Figure 4a). Such a variation in strength should argue in favor of a variable circumstellar absorption or a variable activity in the α Ori extended chromosphere. It is known that α Ori has a close companion (Karovska *et al.*, 1986) influencing its chromospheric layers through tidal effects (Querci and Querci, 1986). Basri *et al.* (1981) suggest that a weak or absent O I λ1304 resonance triplet implies weak hydrogen Lyman emission (and thus no pumping by Ly β) or attenuation by interstellar oxygen. In fact, the H Ly α line must have sufficient available flux to excite the Fe II (UV191) multiplet near 1785 Å visible in Stencel *et al.*'s (1982) spectrum, while not observed in previous SWP spectrum of α Ori.

A summary of close blends in S I and O I lines resolved in SWP high resolution spectra of β Gru and α Ori are given by Linsky *et al.* (1984) with comments on line strength.

In the best low resolution SWP spectrum of an early R star (HD 156074) the chromospheric Si II λ1810 lines are detected at about the flux level of the quiet Sun (Eaton *et al.*, 1985). To our knowledge, no other SWP spectrum of carbon stars has been published. The SWP spectrum of an S star (a particular one discussed in Section 9) is to be mentioned (Johnson and Ake, 1984).

7.2. LONG-WAVELENGTH RANGE (LWP OR LWR IMAGES)

In the long-wavelength range (from $\sim$ 2000 to $\sim$ 3200 Å), spectra of some M giants and supergiants are easily obtained at high resolution (e.g., Wing *et al.*,

1983). Figure 6 shows an example of a normal M giant, γ Cru (M3 III), where the cross-over wavelength occurs at about 2900 Å: above this wavelength the photospheric absorption lines dominate, whereas the chromospheric emission lines are the conspicuous features below it (Wing, 1978; Carpenter and Wing, 1979). High-resolution spectra have been poorly recorded for N-type stars (in fact, the only published example is a weak high-resolution spectrum of the irregular TX Psc; Eriksson *et al.*, 1986). A high dispersion LWP spectrum of an early R-type star (HD 16115) is currently analyzed (Johnson and Baumert, 1986). For S stars, to our knowledge, only one high-resolution spectrum, has been published that of the S Mira χ Cyg (Cassatella *et al.*, 1980b). Nevertheless, we have a few low-resolution spectra of N stars (O'Brien and Johnson, 1982; Johnson and O'Brien, 1983; Querci and Querci, 1983, 1985a; Johnson and Luttermoser, 1986; Johnson *et al.*, 1986) , of S stars (Johnson and Ake, 1984, Johnson *et al.*, 1985a) and of early- and late-R stars (Eaton *et al.*, 1985), mainly used for comparative purposes on the chromospheric activity of the various type stars.

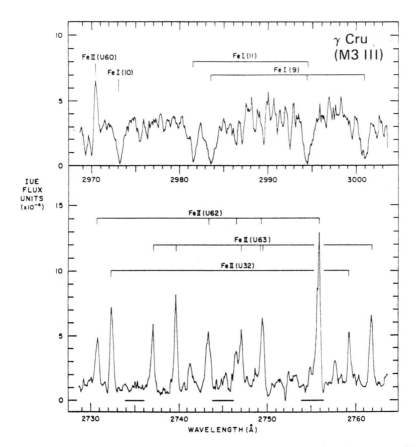

Fig. 6. LWR high resolution spectra of γ Cru illustrating in this giant the change around 2900 Å from a photospheric absorption line spectrum (above this wavelength) to a chromospheric emission line spectrum (from Carpenter and Wing, 1979, and private communication).

7.2.1. *Emission Lines and Chromospheric Activity*

As suggested by Figure 6, the full range of the spectra is crowded by lines of Fe II. Carpenter (1984a) remarks that in the spectra of γ Cru (M3 III) and α Ori (M2 Iab), over 80% of the emission features are due to Fe II lines; among them, the strongest transitions are from multiplets UV1—3, 32, 33, 35, 36, and 60—64. Fe II emission also contributes clearly to N spectra (e.g., Johnson *et al.*, 1986). Other outstanding bright features in these cool stars are the emission components of the Mg II h and k resonance lines (at $\lambda2795$ and $\lambda2802$, respectively), the $\lambda2669$ Al II (UV1) line and the C II (UV0.01) intercombination lines near 2325 Å. A list of emission lines identified in the 2500—3230 Å region of γ Cru is given in Wing *et al.* (1983), and suggested identifications of features in low resolution $\lambda\lambda2250$—3350 spectra of seven N stars are listed by Johnson and Luttermoser (1986). A fluorescence mechanism is often assumed in the line excitation; in particular Mg II emission is responsible for line pumping, such as for the Fe I (UV44) line. Interlocking between multiplets of a given element is possible, such as Fe II (UV32) pumped by Fe II (UV1) (Brown *et al.*, 1981), and emission in H Ly α is the source of emission for Fe II levels around 10 eV, such as for lines at 2506.48 and 2508.34 Å observed in β Gru (M3 II) and now identified as Fe II lines (Johansson and Jordan, 1984).

Among the dominant ultraviolet lines, *Fe II lines* can provide indicators and probes of the chromospheric dynamics in luminous cool stars. They have been carefully analysed in high-resolution IUE spectra of the M supergiant α Ori (Carpenter, 1984b). Comparison of Fe II profiles in α Ori and in the normal giant γ Cru shows that in α Ori many lines are self-reversed and quite broad (FWHM ~ 120 km s^{-1}), whereas in γ Cru the same lines are single-peaked and narrower (~ 40 km s^{-1}). Also, Fe II lines are often asymmetric in α Ori. Carpenter finds a correlation of Fe II line asymmetry with intrinsic line strengths: inside the Fe II line forming region exists a velocity field varying with radius (it is initially constant, then increases with radius to a maximum value and then decreases significantly before reaching an asymptotic flow in the upper chromospheric layers). However, if line asymmetries indicate velocity gradients, they do not allow us to give the flow direction. In fact, by combining the photospheric radial velocities measured by Carpenter from UV Fe I absorption lines (which provide a reference velocity to the Fe II emission lines) at the dates of the observations and an averaged absolute photospheric radial velocity known from blue optical lines, Querci (1986a) and Querci and Querci (1986) show that: (1) the mean velocity of Fe II emitting region indicates an infall during the observational period studied by Carpenter, and (2) this enters in a general temporal motion of respective infall and outfall of the α Ori high chromospheric layers, which is also verified by other chromospheric indicators (such as Mg II h and k lines) and likely caused by the close companion to α Ori (Karovska *et al.*, 1986), when crossing through its chromosphere.

Other IUE emission lines which play an important role in the knowledge of chromospheres of late-type stars, are *the Mg II resonance doublet*, h ($\lambda2803$) and k ($\lambda2796$) lines. Figure 7 shows examples of Mg II line profiles in M, S, and C stars.

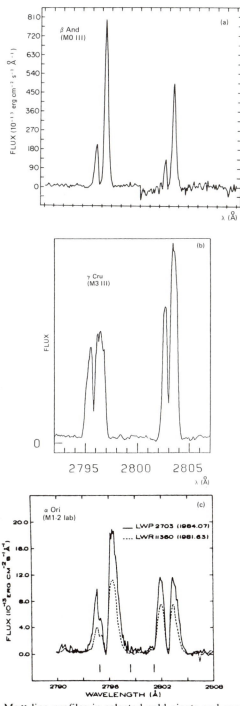

Fig. 7. High resolution Mg II line profiles in selected cold giants and supergiants showing different appearance in the asymmetry of the emission peaks as discussed in the text. (a) β And (from Hartmann *et al.*, 1982); (b) γ Cru (from Carpenter and Wing, 1979, and private communication); (c) α Ori: on 24 January, 1984 (solid line) and on 19 August, 1981 (dashed line); the vertical lines mark the position of circumstellar Mn I absorption lines (from Dupree *et al.*, 1984).

244

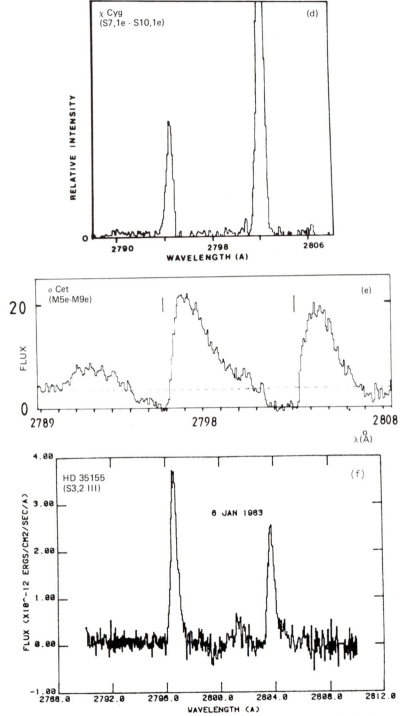

Fig. 7. High resolution Mg II line profiles in selected cold giants and supergiants showing different appearance in the asymmetry of the emission peaks as discussed in the text. (d) χ Cyg, an S Mira, at postmaximum phase 0.22 (Cassatella *et al.*, 1980b); (e) o Cet, the Mira prototype, recorded near the 1979 maximum. Vertical lines indicate the positions of the components shifted by the systemic velocity. The dashed lines indicate the probable continuum level. Ordinate is flux in units of 10^{-13} erg s^{-1} cm^{-2} Å^{-1} (from Stickland *et al.*, 1982); (f) HD 35155 (from Johnson and Ake, 1984).

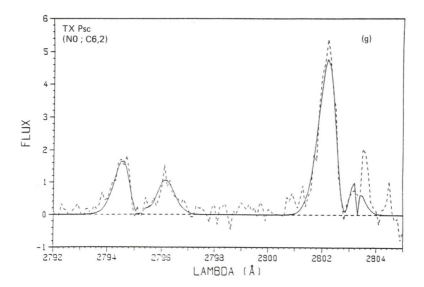

Fig. 7. High resolution Mg II line profiles in selected cold giants and supergiants showing different appearance in the asymmetry of the emission peaks as discussed in the text. (g) TX Psc: the dashed line is the observed IUE spectrum, the solid line is a synthetic spectrum from modeling self-absorption and heavy circumstellar absorption (from Eriksson *et al.*, 1986).

In M stars, generally speaking, the Mg II line widths show a positive luminosity effect (as the Ca II H and K lines in the visible): the M supergiants have much broader Mg II emission profiles than the M ordinary giants, as illustrated by the M supergiant α Ori (Figure 7c) in which the emission lines are ~ 4 Å wide, and by the giant γ Cru (Figure 7b) with its lines about 2 Å wide (Wing and Carpenter, 1978). The Mg II profiles contain deep *self-reversals* displaced slightly shortward of the emission line centers. This might indicate the presence of a cool circumstellar shell expanding rather moderately outward with respect to the chromosphere (e.g., Wing, 1978). However, the chromospheres of M giants and supergiants are now proved to be geometrically quite extended, and intrinsic self-absorption from Mg II in the chromosphere likely contributes either totally or partly or not at all to the reversals. In β And (Figure 7a) in which the mass-loss rate is low, circumstellar absorption is expected to be quite weak, but absorption by interstellar Mg II is to be considered (Baliunas, 1984). In α Ori which possesses an extensive CS shell, the circumstellar contribution to line cores might be preponderant, without ruling out the absorptions due to corotating interactions regions (CIRs) in stellar winds (Mullan, 1984).

Drake and Linsky (1984) measured the positions of the self-reversals in the Mg II lines in a sample of M giants and supergiants and find typical 'chromospheric' expansion velocities of 10 to 25 km s^{-1}. Only in the case of β And (M0 III) has the wind velocity significantly varied in time.

M giants and supergiants show *asymmetric k-line emission peaks.* Various available examples will show in the following that such an asymmetry may present a different appearance from star to star. Redward asymmetry (long-wavelength peak higher than the short-wavelength peak) is generally observed: it is taken as being indicative of differential expansion due to mass loss. In addition, the geometrical extent of the line-forming region is an important contributor to the red-dominance (Drake and Linsky, 1983).

The red-dominated k emission in the luminous M supergiant α Ori (Figure 7c) is an illustration of the geometrical extend of its chromosphere. Moreover, in this star, selective circumstellar absorptions, particularly in the k_2^- peak (by Fe I $\lambda 2795.01$ and Mn I $\lambda 2794.82$ lines; e.g., de Jager *et al.*, 1979) strengthens the asymmetry. As pointed out by Basri *et al.* (1981), it is surprising that, in this context, the h-line in α Ori shows an almost symmetric profile. However, an intrinsic asymmetry may be masked by the selective absorption due to the low excitation Fe I $\lambda 2803.16$ line (Bernat and Lambert, 1976).

In the Mira S star χ Cyg (S7, 1e—S10,1e; $P \sim 406.84$ days), the Mg II h and k lines observed at phase 0.22 (about two months after the 1979 light maximum) (Figure 7d) are striking (Cassatella *et al.*, 1980b): the red-peaks are absent; this is quite different from the profiles described above for M irregular giants or supergiants. Such an asymmetry suggests a decelerating flow. In addition, the 'fictituous' deep core is violet-shifted with respect to laboratory wavelength, especially for the k-line. This might indicate that the k-line is formed in an outstreaming, decelerating flow, in agreement with the excitation of such an emission line in the front of a shock wave running through the Mira photosphere and reaching the higher atmospheric layers where it begins to damp. The complete absence of the usual red-wings evokates very dense, overlying outer shells (Hinkle *et al.*, 1982) in which self-absorption from Mg II as well as other selective absorptions (e.g., by Mn I $\lambda 2797.19$) strongly act. In fact, near visual maximum light ($\varphi \sim 0.04$), low resolution spectra show a *Mg II broad absorption feature* and not yet emission; this fact implies that the Mg II line-forming region is located in the Mira high atmosphere reached by the shock-wave after maximum light (e.g., Querci, 1986a).

As for the irregular carbon star TX Psc (N0 or C6,2), Figure 7g shows the first high-resolution Mg II profile obtained for an N star (Eriksson *et al.*, 1986). Heavy circumstellar absorption and self-absorption are reflected in the Mg II profiles. Again, the blue peak in the line profiles is preponderant, but the red peak is here noticeable with respect to χ Cyg. Like the latter, an expanding shell, decelerating at the time of observation might be assumed for explaining the features. This might agree with the general picture of acoustic waves acting in cool stars of all type of variability, but turning into shocks and dissipating their energy at atmospheric levels that vary from star to star (e.g., according to the extension of its radiative damping zone), as stressed by Querci (1986a). Observational evidence leads this author to conclude that, in semiregular (SR types) or irregular (L types) stars, the shocks *do not* develop as soon as the low photosphere is reached, as is generally shown by Mira stars, but preferably at higher atmospheric levels — in other words, at the stellar chromosphere level. We deal with the shock-wave mechanism in the so-called short-period acoustic heating theory by Ulmschneider and Stein (1982). The chromosphere is very extended in these cold giants and supergiants (at least up to 2 R_* above the photosphere), and emission lines may be permanently seen, in contrast to their periodic presence in Mira photospheres as illustrated by the χ Cyg previous example. In brief, a *permanent chromosphere* (in the sense of the atmospheric region where the emission lines form) is plausible in SR and L stars, whereas the term 'chromosphere' is questionable for Miras. (A theoretical support to the last statement is given by Willson and Bowen, 1985). This means also that the identification of a given stellar feature as either emission or absorption line is more reliable in SR or L variables than in Miras, the observed phase in the cycle influencing generally only the line strength.

The giant (S3,2) star, HD 35155, a quasi-constant star, shows an inverse phenomenon with respect to the previous variable stars, χ Cyg or TX Psc (Figure 7f): only red peaks appear in the Mg II profiles, the entire blueward portion of the lines is obliterated by overlying, outflowing gas (Johnson and Ake, 1984). In this case of a normal giant, the suppression of the blue emission peak is likely linked in priority to the geometrical extension of the Mg II line-forming region. As emphasized by Linsky (1986), according to Drake and Linsky (1983): "with increasing flow speed and geometrical extension, the effect on Mg II line profiles is to suppress the blue emission peak, shift the central minimum (k_3) to the blue, and enhance the red emission peak . . .".

Finally, the case of Mira Ceti (Figure 7e) with its very broad Mg II emission lines which suffer heavy absorption in the blue side of the profile, is interesting (Cassatella *et al.*, 1980a; Stickland *et al.*, 1982). This star is a close visual double star and the characteristics of the Mg II lines are attributable to a large volume of gas around the hot companion.

Generally speaking, the Mg II emission lines are stronger in M stars than in N stars (Table II; see also Johnson and Luttermoser, 1986), making Johnson and O'Brien (1983) to suggest a substantially weaker chromosphere in N stars than in M stars. The heavy overlying absorption found in these lines on a high-dispersion spectrum of the N star, TX Psc (Eriksson *et al.*, 1986) might change that inference: the lines, and hence the chromospheres of N-type stars, are possibly as strong as in M stars, but there is more enormous overlying material. Johnson *et al.* (1985a) find that S stars show a wide variation in the strength of Mg II lines: warmer S stars are comparable to M giants in both the Mg II flux and the continuum emission, cool S stars show Mg II emission comparable to the warmer N stars.

The normalized Mg II flux, and hence the chromospheric activity, decreases rapidly with effective temperature, as demonstrated in a sample of M giants by Steiman-Cameron *et al.* (1985). In addition, these authors find that giants which are deficient in TiO relative to the mean abundance levels for stars of similar color, also have Mg II lines that are weaker than the mean, implying less chromospheric activity; a possible origin of this correlation may be related to aging, because this affects the stellar composition and the chromospheric activity. On the basis of a two-color diagram involving a Mg II index (measuring the ratio of the flux in a 30 Å band at 2800 Å to the flux at V) against an IUE color (the ratio of the flux between 2585 and 3200 Å (Mg II excluded) to the flux at V), Eaton *et*

TABLE II

Carbon II and Magnesium II emission-line fluxes[a]

Star	Spectral type	$\dfrac{10^8\,f(\text{Mg})}{f(\text{bol})}$	$\dfrac{10^8\,f(\text{C})}{f(\text{bol})}$	$\dfrac{f(\text{Mg})}{f(\text{C})}$
TX Psc	N	10	5	2
T Ind	N	22	6	3.5
BL Ori	N	5	8	0.7
HD 37212	R8	< 140	< 240	—
HD 52432	R5	< 210	< 175	—
α Her	M5 II	130	—	—
72 Leo	M3 III	670	—	—
β Peg	M2 II—III	360	—	—
α Ori	M2 Iab	320	20	28
α Tau	K5 III	280	19	14
α Boo	K2 III	950	19	51

[a] From O'Brien and Johnson (1982).

al. (1985) show the wide range of chromospheric emission encountered in cold stars and confirm the decrease in chromospheric flux with the effective temperature. The K and M giants show a great range in Mg II strength and are located up to 2 mag above the sequence formed by carbon stars which are separated into three groups. The early R stars fall among the G5—K2 III stars, the late R stars have colors similar to late K and M stars, and the N stars have colors overlapping the blue end of the late R stars and extending 2.5 mag farther to the red.

The Mg II, Al II, and Fe II emission line flux have been examined in luminous M stars with varying gas-to-dust ratios (Carpenter *et al.*, 1985b): the flux in these UV chromospheric indicators is substantially smaller (~ 10 X) in the stars with a high dust/gas index than in the less dusty stars.

In the high-resolution long-wavelength IUE spectra, *the C II $\lambda 2325$ intersystem multiplet* has been used to estimate the plasma properties of the M star chromospheres: densities, temperatures (involving also the C II (UV1) doublet near 1335 Å), as well as the geometrical extent of the C II emitting regions (Brown *et al.*, 1981; Stencel *et al.*, 1981; Stencel and Carpenter, 1982; Brown and Carpenter, 1984; Carpenter *et al.*, 1985c). However, Judge, as reported by Linsky (1986) points out inadequacies in the diagnostic technique, whereas Stencel (1986) notes that this simple analytic treatment provides only upper limits to the physical extent of the line-forming region. Stencel adds that the method used by Judge is an improvement, but probably reflects a lower limit to the size. (For details and for summary of more performant modeling techniques of the extended chromospheres, see Jordan and Linsky, pp. 259—293).

Table II shows the quite smaller ratio of $f(\text{Mg II})/f(\text{C II})$ in the N stars with respect to the M stars. The abundance effect or a line-variability effect may not be the only cause.

7.2.2. *Absorption Lines*

Several absorption features are common to M and N stars, such as the low excitation lines of Fe I, Ti I, Ni I. However, there are differences. The Mg I $\lambda 2852$ line is a strong absorption line in M stars as well as in R stars, whereas no feature appears at this wavelength in N stars. Other differences linked to differences in chemical composition (at first in the C/O ratio differentiating the oxygen-rich and carbon-rich stars) arise from the molecular bands: the $(0, 0)$ and $(1, 1)$ bandheads of the CH 3143 Å system are visible in N spectra, whereas the $\lambda\lambda 2811$ and 3064 OH bandheads are detected in M spectra. Also, the identification of CaCl, well known in the visible range of certain C and SC stars, is proposed at ~ 3000 Å by Bennett and Johnson (1985). In R stars, the 3143 Å CH feature weakens. Otherwise, early R stars have IUE spectra strikingly similar to those of G and K giants dominated by iron-peak elements. Eaton *et al.* (1985) remark that cool-carbon stars have strong individual absorption features, whereas M stars have rather featureless UV continua, as expected if they are formed in an atmosphere with a small temperature gradient.

The only present attempt for an extensive absorption-line identification has been done for Arcturus (Carpenter *et al.*, 1985a); it may serve as reference for identifying the UV absorption spectra of the cold stars.

8. Temporal Variations in UV Lines of Cold Stars

Study of variability in cold stars has received scant attention through IUE. It is not surprising when we remind us the long pulsation period of these variables (around a few years for the bulk of stars). Too much repeated observations, all along the cycle, are needed for being easily feasible within the reach of IUE. The high resolution spectrograph (HRS) of the Hubble Space Telescope is best suited to such an objective.

Temporal variations in the Mg II lines are evident in the irregular M supergiant, α Ori (Figure 7c, and Table III), as well as a slight increase in the flux of the $\lambda 1300$ feature (S I and O I), corresponding to a true flux chromospheric enhancement (Dupree *et al.*, 1984).

In N stars, time variability in the strength of the strongest emission lines (C II, Fe II, Mg II) is conspicuous on a series of spectra of the SRa star, TW Hor (N0; C7,2; $p \sim 157$ days) (Querci and Querci, 1983, 1985a; Figure 8). The most important variations are shown by Mg II U1 doublet and the Fe II V1 + V II V7 line blend which varies by at least a factor of 10. Line variability is erratic and takes place on different time scales ranging from a month to hours. Line blends may vary with different strength for very similar phases in various periods (perhaps demonstrating the different behavior of the blend contributors). The shortest time scale (1 hr) is suggested only by the Al II U1 lines at $\lambda 2670$ (images LWR 12834 and 12835 on Figure 8) which are near the noise level ($\sim 3\sigma$); further observations are needed for confirmation. A recently high-resolution spectrum of TW Hor recorded through a continuous VILSPA/Goddard observational run is under reduction. Another example of the UV variability of emission

TABLE III

Observed Mg II fluxes from Alpha Ori[c]

Date	Image No.[a]		Disp.	k-line ($\lambda2795$)	h-line ($\lambda2803$)	Total
				(units of 10^{-11} erg cm^{-2} s^{-1})		
1978.63	LWR	2099	H	18.3	17.0	35.3
1978.63	LWR	2116	L			39.7
1978.96	LWR	3195	H	18.4	15.7	34.1
1979.22	LWR	4090	H	22.2	21.3	43.5
1981.63	LWR	11360	H	14.8	10.8	25.6
1982.15	LWR	12668	L			45.0
1984.02	LWP	2574a[b]	L			54.2
1984.02	LWP	2574b[b]	L			49.6
1984.02	LWP	2573	H	27.9	21.8	49.7
1984.07	LWP	2702	L			60.4
1984.07	LWP	2703	H	30.6	22.9	53.5
1984.13	LWP	2791	L			65.3
1984.13	LWP	2792	H	29.6	22.6	52.2
1984.16	LWP	2860	L			58.6
1984.16	LWP	2861	H	31.6	23.0	54.5

[a] All exposures in large aperture.
[b] Two exposures in large aperture.
[c] From Dupree et al. (1984).

lines is given by the irregular N star, TX Psc (N0; C6,2). The line flux appears to vary by a factor of at least 6—8 (Johnson et al., 1986). Interpretations of such a variability might be: either a variable transit time for wave pulses to reach the high chromospheric layers destroying the periodic character of the wave train (in relation to our above discussion on chromospheric heating by short-period acoustic shock-waves), either temporal and spatial fluctuations in the thermo-dynamical state at a given level in the chromosphere, or variable magnetic fields changing the effective area of the star covered with plages, those areas of enhanced emission (we deal now with the so-called acoustic slow magnetic shock-waves).

Changes in the Mg II lines in the postmaximum spectra of the Mira S star, χ Cyg (Cassatella et al., 1980b) have been noted above. Other striking changes are displayed in support of a shock-wave model. At visual phase 0.04, the Fe II (V1) emission lines are remarkably strong; the Fe II (V6) lines are also visible. At phase 0.18, the Fe II (V1) lines are yet conspicuous, whereas the Fe II (V6) lines become barely identifiable, and while emission lines from Fe II (UV1, UV32), not visible at phase 0.04, now occur.

Finally, we note an IUE archive search for variability in emission line fluxes in cool giants (including spectral type F—M and luminosity class I—IV) by Oznovich and Gibson (1984). Among the physical phenomena expected to produce the variations in the UV line fluxes and indicated by the authors, are the formation

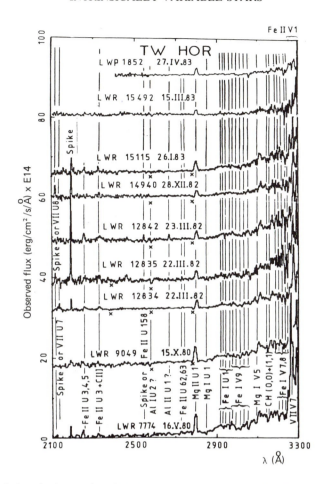

Fig. 8.   Line variations in time series LWR low-resolution spectra of the SRa carbon star, TW Hor (N0; C7,2). The LWR 12834 and LWR 12835 images are two successive exposures about 1 hr each. The crosses indicate reseau marks (from Querci and Querci, 1985a).

and decay of active regions, as we suggested for TW Hor. As also mentioned by Querci and Querci (1985a), rotational modulation is not appropriate to explain the observed line variations in the cold giant stars. The very low equatorial rotation velocity (< 12 km s^{-1}, giving a lower limit of about 7 yr to the rotational period) cannot influence the observed feature variability, while a 'stellar cycle' should do it.

9. Special Cold Stars

9.1. PARTICULAR CARBON-RICH OBJECTS

Besides the 'classical' R and N-type stars, there are other objects which are also rich in carbon, but that present an over- or under-abundance in one or more elements (e.g., see the review by Querci, 1986b). Among them, a few which are deficient in hydrogen — the so-called hydrogen-deficient carbon (HdC) stars — are the R Coronae Borealis (RCB) stars and the hydrogen-deficient R supergiants (the 'classical' R stars are giants).

IUE low dispersion spectra of R CrB, RY Sgr and XX Cam at maximum light (Holm and Wu, 1982) show strong absorption features of the Mg II doublet, Fe II, Mn II, Al I lines confirmed by high dispersion spectra of R CrB. Also C II λ1335 emission is observed. During the descent to minimum of R CrB in August— September 1983 (Holm *et al.*, 1984), the absorption features became filled in; Mg II acquired a strong emission core. The authors show how an obscuring cloud can produce the observed spectrum. The star is partially occulted by a dense, optically-thick cloud, in agreement with the observed initial absence of any UV extinction greater than that in the visual; the UV excess light arises from unocculted regions of the stellar atmosphere (in particular from a chromosphere giving emission lines and from an extended region giving resonance lines such as C II, Mg II). Recent observations of the two brightest RCB stars, R CrB and RY Sgr (Hecht *et al.*, 1984) confirm the hypothesis that the irregular and abrupt declines of these stars from maximum light is caused by dust obscuration (Loreta-O'Keefe's original hypothesis). These observations are consistent with particles of 5 to 60 nm, glassy or amorphous carbon rather than graphitic carbon.

The HdC supergiant R stars differ from the early (R0—R4) R classical giants by weakened lines of neutral metal (e.g., weaker Mg I (UV1), the absence of the large Fe I absorption feature at 3000 Å) and by strengthened lines of ionized metals (stronger Ti II (2) 3240 Å absorption), reflecting their higher luminosity (Eaton *et al.*, 1985). A high dispersion spectrum of such a star, HD 182040, has been analyzed by Johnson *et al.* (1984) confirming a strong enhancement of singly ionized iron-peak elements relative to normal G giants. Eaton *et al.* (1985) derive an equivalent oxygen-sequence spectral type of F8—G5 Ia for HD 182040.

Other particular carbon-rich objects are the subgiant CH stars, a group of high-velocity stars with strong CH bands and weak blue-violet atomic lines. Two CH stars have been observed at low dispersion, long-wavelength range by Johnson *et al.* (1985b): as in the optical range atomic lines are weakened, and the Mg II λ2800 line is a deep absorption feature with no hint of emission. The relative strengths of the absorption features give spectral types in the range F8—G5; however, the CH bandhead at 3144 and 3156 Å are stronger than in normal G giants.

At last, the Ba II stars, also belonging to the 'hot' group of particular giants with effective temperatures near those of the early R stars, are characterized by carbon molecular bands such as CH, CN, and C_2 bands, enhanced with respect to normal G and K giants, together with a Ba II λ4554 strong line. As summarized by Scalo (1981), there is evidence that these stars are single-line spectroscopic binaries with white dwarf companions; their abundance pecularities should be due to mass transfer from the companion when it was a cool giant before its pre-white dwarf evolution. The proposed binarity is supported by IUE short-wavelength spectra of the barium stars, Zeta Cap and 56 Peg, in which Böhm-Vitense (1980) and Schindler *et al.* (1982) show excess continuum flux shortward of 1600 Å indicative of a white dwarf companion.

9.2. COLD STARS WITH HIGHLY IONIZED EMISSION LINES

The unusual late R star, HD 59643 (R9, C6,2) appears to be in a stage a rapid evolution. In less than 10 yr, this normal giant (e.g., see Greene and Wing, 1971 and references therein) has evolved to a protoplanetary nebula-like object. An IUE spectrum (Carpenter and Wing, 1979) shows ions of a great range of ionization: Mg II, O I, C II, C IV, and N V. Further recent IUE spectra (Querci and Querci, 1985b; Querci *et al.*, 1986) demonstrate that the emission lines are variable. For example the 1550 Å C IV line appears narrow with a P-Cygni profile in 1979, broad in 1981, symmetric in 1982, and very asymmetric in 1983 indicating mass loss. Further high resolution IUE spectra (VILSPA/Goddard joint observations, i.e., Toulouse/Indiana collaboration) are planned for a near future to obtain more information on this fascinating carbon star. Besides the proto-planetary hypothesis, a variable accretion disk in a binary system should not be ruled out.

In the same vein, Jonhson and Ake (1984), observing the S giant star HD 35155 in the SWP and LWP ranges at low resolution, find a rich emission-line spectrum with ions as C IV, Si III—Si IV, N V, O I. Further IUE observations, 7 months later, showed that the continuum has decreased by about 0.5 mag and the emission lines by 15—50% with a most rapid decrease for the high excitation lines. The presence of a hot interacting companion, or of a compact companion, among other suggestions, is waiting further observations.

10. Conclusion

The study of cold stars in the far ultraviolet has only seen timid attempts with the IUE spacecraft, the sensitivity of which allows to survey only the few brightest stars. However, the obtained results are very stimulating and encourage further studies with more adapted instruments, like the HRS of the Hubble Space Telescope.

In this respect, to conclude, we should like to emphasize particular interesting points raised by the carbon stars and related carbon objects. The existence of these stars challenges theories of stellar structure and nucleosynthesis (Scalo, 1981). To understand the phenomena in their observable atmospheric layers (such as mass loss or nonthermal processes), may help to identify the physical processes at work in their interiors.

The first high-resolution LWR spectrum of an N star, TX Psc, shows clearly that the circumstellar absorption is high, suggesting very large mass loss rates, at least in certain stars. Let us note that no measurement of mass loss rate is available for carbon stars. Such a large rate of mass loss should be connected to the formation of protoplanetary nebulae, as we showed by an example. It could also influence the evolutionary status of these stars. Therefore, it is vitally important to understand the mass loss mechanism through UV high-resolution shell line profiles.

The SWP range of a cool carbon (N type) star cannot be reached with IUE. However, as with M stars, no high temperature lines are expected, and therefore no transition regions. Anyway, the far-UV regions are to be explored to study the time behavior of lines such as C I, C II, or S I.

Our quantitative approach of chromospheric heating in an N star, TW Hor, and surveys of some other cases, bring us to favor the so-called short-period acoustic heating theory (in which shock waves dissipate energy at the temperature minimum) which appears generally to work in M, S, and C semiregulars or irregulars, in contrast to Miras (in which shocks develop in the low photosphere). Is this dichotomy explaining, at least partly, that the Mira variables occupy a distinct zone with respect of the giant SR variables in the HR diagram? Larger samples of the various cold stellar types have to be prospected.

Chromospheric modeling in cool carbon stars is just at its beginning. The only attempts are from de la Reza (1986) and from Avrett and Johnson (1984) who tested their models on the Mg II and C II UV lines. The Fe II lines might also serve to constrain the models, since they take part in a substantial fraction of the total chromospheric radiative losses besides the Mg II lines, as shown by Querci and Querci (1985a) in TW Hor (as well as by Carpenter (1984a) in the M stars γ Cru and α Ori). At the same time, Mg II lines might be modeled through the co-moving frame spherical PRD code (Linsky, 1986).

Finally, the time variations of UV lines, found fully erratic in TW Hor, raise the question of the presence of active regions in these cool giant stars.

Acknowledgements

We are very grateful to W. Gieren, S. Parsons and E. Schmidt for helpful comments and for drawing our attention to additional references. We are also very grateful to the astronomers and the staff of the IUE observatories whose enthusiastic help, advice and support are invaluable for all IUE research. A NASA grant NSG 5398 to E. B.-V. is gratefully acknowledged.

References

Arellano Ferro, A. and Madore: 1985, preprint.
Avrett, E. H. and Johnson, H. R.: 1984, in S. L. Baliunas and L. Hartman (eds.), *Proc. Third Cambridge Workshop on Cool Stars, Sellar Systems, and the Sun*, Springer Verlag: New York, p. 330.
Ayres, T. R., Moos, H. W., and Linsky, J. L.: 1981, *Astrophys. J. Letters* **248**, L137.
Baliunas, S. L.: 1984, in J. M. Mead, R. D. Chapman, and Y. Kondo (eds.), *Future of Ultraviolet Astronomy Based on Six Years of IUE Research*, *NASA CP-2349*, p. 64.
Balona, L. A.: 1977, *Monthly Notices Roy. Astron. Soc.* **178**, 231.
Barker, T. *et al.*: 1971, *Astrophys. J.* **165**, 67.
Basri, G. S., Linsky, J. L., and Eriksson, K.: 1981, *Astrophys. J.* **251**, 162.
Becker, S. A. and Cox, A. N.: 1982, *Astrophys. J.* **260**, 707.
Becker, S. A., Iben, I., and Tuggle, R. W.: 1977, *Astrophys. J.* **218**, 633.
Bennett, P. D. and Johnson, H. R.: 1985, 'Cool Stars with Excesses of Heavy Elements', in M. Jaschek and P. C. Keenan (eds.), *Calcium Chloride in the IUE Spectra of Carbon Stars*, D. Reidel Publ. Co., Dordrecht, Holland, p. 249.
Bernat, A. P. and Lambert, D. L.: 1976, *Astrophys. J.* **204**, 830.
Böhm-Vitense, E.: 1974, *Astrophys. J.* **188**, 571.
Böhm-Vitense, E.: 1980, *Astrophys. J. Letters* **239**, L79.
Böhm-Vitense, E.: 1980, *Ann. Rev. Astr. Astrophys.* **19**, 295.
Böhm-Vitense, E.: 1985, *Astrophys. J.* **296**, 169.

Böhm-Vitense, E.: 1986, *Astrophys. J.* **303**, 262.

Böhm-Vitense, E. and Nelson, G.: 1976, *Astrophys. J.* **210**, 741.

Böhm-Vitense, E. and Dettmann, T.: 1980, *Astrophys. J.* **236**, 560.

Böhm-Vitense, E. and Proffit, Ch.: 1985, *Astrophys. J.* **296**, 175.

Böhm-Vitense, E., Proffitt, Ch., and Wallerstein, G.: 1984, in 'Future of Ultraviolet Astronomy based on Six Years of IUE Research', NASA Conf. Publ. No. 2349, pp. 348.

Bonnell, J. T. and Bell, R. A.: 1985, *Publ. Astron. Soc. Pacific* **97**, 236.

Borutzki, S., Böhm-Vitense, E., and Harris, H.: 1984, in A. Maeder and A. Renzini (eds.), *IAU Symp.* **105**, 449.

Brown, A. and Jordan, C.: 1980, *Monthly Notices Roy. Astron. Soc.* **191**, 37.

Brown, A. and Carpenter, K. G.: 1984, *Astrophys. J.* **287**, L43.

Brown, A., Jordan, C., and Wilson, R.: 1979, in A. J. Willis (ed.), *The First Year of IUE*, Univ. College London, p. 232.

Brown, A., Ferraz, M. C. de M., and Jordan, C.: 1981, in R. D. Chapman (ed.), *The Universe at Ultraviolet Wavelengths — The First Two Years of IUE*, NASA CP-2171, p. 297.

Carpenter, K. G.: 1984a, in J. M. Mead, R. D. Chapman, and Y. Kondo (eds.), *Future of Ultraviolet Astronomy Based on Six Years of IUE Research*, NASA CP-2349, p. 450.

Carpenter, K. G.: 1984b, *Astrophys. J.* **285**, 181.

Carpenter, K. G. and Wing, R. F.: 1979, *Bull. Amer. Astron. Soc.* **11**, 419.

Carpenter, K. G., Wing, R. F., and Stencel, R. E.: 1985a, *Astrophys. J. Supplement* **57**, 405.

Carpenter, K. G., Stencel, R. E., and Hagen, W.: 1985b, *Bull. Amer. Astron. soc.* **17**, 876.

Carpenter, K. G., Brown, A., and Stencel, R. E.: 1985c, *Astrophys. J.* **289**, 676.

Cassatella, A., Clavel, J., Gilra, D., Reimers, D., and Stickland, D. J.: 1980a, in *Proc. Second European IUE Conf.*, ESA SP-157, p. 233.

Cassatella, A., Heck, A., Querci, F., Querci, M., and Stickland, D. J.: 1980b, in *Proc. Second European IUE Conference*, ESA SP-157, p. 243.

Clube, S. V. H. and Dawe, J. A.: 1983, *Astron. Astrophys.* **122**, 255.

Cox, A. N.: 1980, *Ann. Rev. Astr. Astrophys.* **18**, 15.

Crawford, D. L.: 1978, *Astronomical J.* **83**, 48.

Crawford, D. L.: 1979, *Astronomical J.* **84**, 1858.

Christy, R. F.: 1966, *Astrophys. J.* **145**, 340.

Dean, C. A., Warren, W. H., and Cousins, A. W.: 1978, *Monthly Notices Roy. Astron. Soc.* **183**, 569.

de Jager, C., Kondo, Y., Hoekstra, R., van der Hucht, K. A., Kamperman, T.M., Lamers, H. J. G. L. M., Modisette, J. L., and Morgan, T. H.: 1979, *Astrophys. J.* **178**, 495.

de la Reza, R.: 1986, in H. R. Johnson and F. Querci (eds.), *CNRS/NASA Monograph Series on Nonthermal Phenomena in Stellar Atmosphere, The M, S, and C Stars*, Chap. 8 (in press).

Drake, S. A. and Linsky, J. L.: 1983, *Astrophys. J.* **273**, 299.

Drake, S. A. and Linsky, J. L.: 1984, *Bull. Amer. Astron. Soc.* **16**, 895.

Dupree, A. K.: 1979, *Highlights of Astronomy*, P. A. Wayman (ed.), **5**, 263.

Dupree, A. K., Sonneborg, G., Baliunas, S. L., Guinan, E. F., Hartman, L., and Hayes, D. P.: 1984, in J. M. Mead, R. D. Chapman, and Y. Kondo (eds.), *Future of UV Astronomy Based on Six Years of IUE Research*, NASA CP-2349, p. 21.

Eaton, J. A., Johnson, H. R., O'Brien, G., and Baumert, J. H.: 1985, *Astrophys. J.* **290**, 276.

Eichendorf, W., Heck A., Caccin, B., Russo, G. and Solazzo, C.: 1982, *Astron. Astrophys.* **109**, 274.

Eriksson, K., Gustafsson, B., Johnson, H. R., Querci, F., Querci, M., Baumert, J. H., Carlsson, M., and Olofsson, H.: 1986, *Astron. Astrophys.* (in press).

Evans, N. and Bolton, C. T.: 1986, preprint.

Evans, N., Böhm-Vitense, E., and Bolton, C. T.: 1984, *BANN* **17**, 559.

Fracassini, M. and Pasinetti, L. E.: 1982, *Astron. Astrophys.* **107**, 326.

Greene, A. E. and Wing, R. F.: 1971, *Astrophys. J.* **163**, 309.

Haisch, B. M., Linsky, J. L., Weinstein, A., and Shine, R. A.: 1977, *Astrophys. J.* **214**, 785.

Hartmann, L., Dupree, A. K., and Raymond, J. C.: 1982, *Astrophys. J.* **252**, 214.

Hecht, J. H., Holm, A. V., Donn, B., and Wu, C. C.: 1984, *Astrophys. J.* **280**, 228.

Henden, A., Cornett, R. H., and Schmidt, E.: 1984, *Publ. Astron. Soc. Pacific* **96**, 310.

Hinkle, K. H., Hall, D. N. B., and Ridgway, S. T.: 1982, *Astrophys. J.* **252**, 697.

Hodge, P. W. and Wallerstein, G.: 1967, *Astrophys. J.* **150**, 951.

Hodge, P. W. and Mateo, M.: 1986, *Astrophys. J. Supplement* **60**, 893.

Holm, A. V. and Wu, C. C.: 1982, in Y. Kondo, J. M. Mead, and R. D. Chapman (eds.), *Advances in Ultraviolet Astronomy: Four Years of IUE Research*, NASA CP-2238, p. 429.

Holm, A. V., Hecht, J., Wu, C. C., and Donn, B.: 1984, in J. M. Mead, R. D. Chapman, and Y. Kondo (eds.), *Future of UV Astronomy Based on Six Years of IUE Research*, NASA CP-2349, p. 330.

Johansson, S. and Jordan, C.: 1984, *Monthly Notices Roy. Astron. Soc.* **210**, 239.

Johnson, H. R. and O'Brien, G. T.: 1983, *Astrophys. J.* **265**, 952.

Johnson, H. R. and Ake, T. B.: 1984, in S. L. Baliunas and L. Hartman (eds.), *Proc. Third Cambridge Workshop on Cool Stars, Stellar Systems, and the Sun*, Springer Verlag: New York, p. 362.

Johnson, H. R. and Baumert, J. H.: 1986 (in prep.)

Johnson, H. R. and Luttermoser, D. J.: 1986, preprint (to appear in *Astrophys. J.*).

Johnson, H. R., Ameen, M. M., and Eaton, J. A.: 1984, *Astrophys. J.* **283**, 760.

Johnson, H. R., Ake, T. B., and Eaton, J. A.: 1985a, 'Cool Stars with Excesses of Heavy Elements', in M. Jaschek and P. C. Keenan (eds), *Ultraviolet Spectra of N, R, and S Stars*, D. Reidel Publ. Co., Dordrecht, Holland, p. 53.

Johnson, H. R., Baumert, J. H., and Eaton, J. A.: 1985b, *Bull. Amer. Astron. Soc.* **17**, 600.

Johnson, H. R., Baumert, J. H., Querci, F., and Querci, M.: 1986, *Astrophys. J.* **311**, 960.

Jordan, C. and Judge, P. G.: 1984, *Physica Scripta* **T8**, 43.

Judge, P. G.: 1984, in S. L. Baliunas and L. Hartman (eds.), *Proc. Third Cambridge Workshop on Cool Stars, Stellar Systems, and the Sun*, Springer Verlag: New York, p. 353.

Karovska, M., Nisenson, P., Noyes, R. W., and Stachnik, R. V.: 1986, in M. Zeilik and D. H. Gibson (eds.), *Proc. Fourth Cambridge Workshop on Cool Stars, Stellar Systems, and the Sun*, Springer Verlag: New York, p. 445.

Kurucz, R.: 1979, *Astrophys. J. Supplement* **40**, 1.

Linsky, J. L.: 1986, *The Irish Astron. J.* **17**, 343.

Linsky, J. L. and Haisch, B. M.: 1979, *Astrophys. J. Letters* **229**, L27.

Linsky, J. L., Ayres, T. R., Brown, A., Carpenter, K., Jordan, C., Judge, P., Gustafsson, B., Eriksson, K., Saxner, M., Engvold, O., Jensen, E., Moe, O. K., and Simon, T.: 1984, in J. M. Mead, R. D. Chapman, and Y. Kondo (eds.), *Future of Ultraviolet Astronomy Based on Six Years of IUE Research*, NASA CP-2349, p. 445.

Lloyd Evans, T.: 1982, *Monthly Notices Roy. Astron. Soc.* **199**, 9.

Lutz, T. E. and Kelker, D. H.: 1973, *Publ. Astron. Soc. Pacific* **85**, 573.

Madore, B. F.: 1977, *Monthly Notices Roy. Astron. Soc.* **178**, 505.

Mariska, J. T., Doschek, G. A., and Feldman: 1980, *Astrophys. J.* **238**, L87.

Matraka, B., Wassermann, C., and Weigert, A.: 1982, *Astron. Astrophys.* **107**, 283.

Miller, J. and Preston, G.: 1964, *Astrophys. J.* **139**, 1126.

Mullan, D. J.: 1984, *Astrophys. J.* **284**, 769.

O'Brien, G. T. and Johnson, H. R.: 1982, in Y. Kondo, J. M. Mead, and R. D. Chapman (eds.), *Advances in Ultraviolet Astronomy: Four Years of IUE Research*, NASA CP-2238, p. 255.

Oznovich, I. and Gibson, D. M.: 1984, *Bull. Amer. Astron. Soc.* **16**, 896.

Parthasarathy, M. and Parsons, S.: 1984, in *Future of Ultraviolet Astronomy Based on Six Years of IUE Research*, NASA CP-2349, p. 342.

Querci, M.: 1986a, in H. R. Johnson and F. Querci (eds.), *CNRS/NASA Monograph Series on Nonthermal Phenomena in Stellar Atmospheres, The M, S, and C Stars*, chp. 2 (in press).

Querci, F.: 1986b, in H. R. Johnson and F. Querci (eds.), *CNRS/NASA Monograph Series on Nonthermal Phenomena in Stellar Atmospheres, The M, S, and C Stars*, chp. 1 (in press).

Querci, F. and Querci, M.: 1983, in J. C. Pecker and Y. Uchida (eds.), *Proc. Japan-France Seminar on Active Phenomena in the Outer Atmosphere of the Sun and Stars*, p. 140.

Querci, M. and Querci, F.: 1986, in M. Zeilik and D. M. Gibson (eds.), *Proc. Fourth Cambridge Workshop on Cool Stars, Stellar Systems, and the Sun*, Springer Verlag: New York, p. 492.

Querci, M. and Querci, F.: 1985a, *Astron. Astrophys.* **147**, 121.

Querci, F. and Querci, M.: 1985b, 'Cool Stars with Excesses of Heavy Elements', in M. Jaschek and P. C. Keenan (eds.), *A New Photometric System for Monitoring Carbon Star Variability*, D. Reidel Publ. Co., Dordrecht, Holland, pp. 99.

Querci, F., Querci, M., Wing, R. F., Cassatella, A., and Heck, A.: 1982, *Astron. Astrophys.* **111**, 120.

Querci, M., Querci, F., Johnson, H. R. and Baumert, J. H.: 1986, Advances in Space Research (in press).

Sandage, A. R. and Tammann, G.: 1969, *Astrophys. J.* **157**, 683.

Savage, B. D. and Mathis, J. S.: 1979, *Ann. Rev. Astron. Astrophys.* **17**, 73.

Scalo, J. M. 1981, in I. Iben and A. Renzini (eds), *Proc. Second Workshop on Physical Processes in Red Giants*, D. Reidel Publ. Co., Dordrecht, Holland, p. 77.

Schindler, M., Stencel, R. E., Linsky, J. L., Basri, G. S., and Helfand, D. J.: 1982, *Astrophys. J.* **263**, 269.

Schmidt, E.: 1984, *Astrophys. J.* **285**, 501.

Schmidt, E. G. and Parsons, S. B.: 1982, *Astrophys. J. Supplement* **48**, 185.

Schmidt, E. R. and Parsons, S. B.: 1984a, *Astrophys. J.* **279**, 202.

Schmidt, E. R. and Parsons, S. B.: 1984b, *Astrophys. J.* **279**, 215.

Schommer, R. A., Olszewski, E. W., and Aaronson, M.: 1984, *Astrophys. J.* **285**, L53.

Schwarzschild, M. and Härm, R.: 1970, *Astrophys. J.* **160**, 341.

Steiman-Cameron, T. Y., Johnson, H. R., and Honeycutt, R. K.: 1985, *Astrophys. J.* **291**, L51.

Stencel, R. E.: 1986, *The Irish Astron. J.* **17**, 336.

Stencel, R. E. and Carpenter, K. G.: 1982, in Y. Kondo, J. M. Mead, and R. D. Chapman (eds.), *Advances in Ultraviolet Astronomy: Four Years of IUE Research*, NASA CP-2238, p. 243.

Stencel, R. E., Linsky, J. L., Brown, A., Jordan, C., Carpenter, K. G., Wing, R. F., and Czyzak, S.: 1981, *Monthly Notices Roy. Astron. Soc.* **196**, 47P.

Stencel, R. E., Linsky, J. L., Ayres, T. R., Jordan, C., Brown, A., and Engvold, O.: 1982, in Y. Kondo, J. M. Mead, and R. D. Chapman (eds.), *Advances in Ultraviolet Astronomy: Four Years of IUE Research*, NASA CP-2238, p. 259.

Stickland, D. J. and Sanner, F.: 1981, *Monthly Notices Roy. Astron. Soc.* **197**, 791.

Stickland, D. J., Cassatella, A., and Ponz, D.: 1982, *Monthly Notices Roy. Astron. Soc.* **199**, 1113.

Trimble, V.: 1978, *Observatory* **78**, 163.

Ulmschneider, P. and Stein, R. F.: 1982, *Astron. Astrophys.* **106**, 9.

VandenBerg, D. A.: 1984, *Astrophys. J. Supplement* **51**, 29.

Willson, L. A.: 1985, *Irish Astronomical Journal*.

Willson, L. A. and Bowen, G. H.: 1985, in R. Stalio and J. Zirker (eds.), *Proc. III Trieste Workshop on The Relationship between Chromospheric/Coronal Heating and Mass Loss*.

Wing, R. F.: 1978, in M. Hack (ed.), *Proc. 4th International Colloq. on Astrophysics, High Resolution Spectrometry*, Osservatorio di Trieste, p. 683.

Wing, R. F. and Carpenter, K. G.: 1978, *Bull. Amer. Astron. Soc.* **10**, 444.

Wing, R. F., Carpenter, K. G., and Wahlgren, G. M.: 1983, *Atlas of High Resolution IUE Spectra of Late-type Stars (2500—3230 Å)*, Special Publ. Perkins Obs. No. 1.

CHROMOSPHERES AND TRANSITION REGIONS

CAROLE JORDAN

Department of Theoretical Physics, Oxford University, Oxford, UK

and

JEFFREY L. LINSKY*

Joint for Laboratory Astrophysics, Boulder, CO, U.S.A.

1. Introduction

The existence of chromospheres on cool stars other than the Sun has been recognized for many years through the presence of emission components in the Ca II H and K lines and in He I 10 830 Å. While some trends in the properties of stellar chromospheres as a function of spectral type and luminosity (e.g. the Wilson—Bappu effect) were established on the basis of optical data, the field could not advance without observations at shorter wavelengths where many chromospheric lines and all easily observable lines produced by hotter material are located. The potential for the field was clear from developments in solar physics. The use of balloons (e.g. Kondo *et al.*, 1972) and rockets extended the spectral coverage to the Mg II h and k lines (2803 and 2796 Å, respectively), but satellites (e.g. *Copernicus* and TD—1) were required to observe down to H Ly α and to make broad surveys of stars (e.g. Weiler and Oegerle, 1979 and earlier references therein). The combination of fluxes and profiles in the Ca II, Mg II, and H Ly α lines allowed computation of simple, spherically-symmetric chromospheric models extending up to temperatures of 8000 K (see Part V). The first evidence for material at transition-region temperatures (20 000—200 000 K) came from observations of α CMi (F5 IV—V) with *Copernicus* (Evans *et al.*, 1975) and with the TD—1 satellite (Jamar, Macon-Hercot and Praderie 1976) and of α Aur (F9 III + G6 III) with *Copernicus* (Dupree, 1975) and a rocket-borne payload (Vitz *et al.*, 1976). Emission-line fluxes observed from space experiments are especially valuable for atmospheric modelling because they are well calibrated in absolute flux units.

The very different nature of the far-ultraviolet spectra of the chromospheres of cool giants was first indicated by rocket observations of α Boo (K2 III), in which McKinney *et al.* (1976) discovered a spectrum dominated by strong emission in the O I resonance lines, that Haisch *et al.* (1977) explained as fluorescent excitation of O I by H Ly β. Linsky (1980) has reviewed this early work. Nevertheless, observations of cool star chromospheres and transition regions prior to IUE could provide only a glimpse of the general trends that have been established through IUE observations of a wide variety of stars.

---

* Staff Member, Quantum Physics Division, National Bureau of Standards.

Y. Kondo (ed.), Chromospheres and Transition Regions, pp. 259—293.
© 1987 *by D. Reidel Publishing Company.*

The enormous increase in sensitivity and spectral resolution possible with IUE have allowed the study of a wide range of cool stars and detailed studies of the brighter objects. IUE Guest Observers have radically transformed our understanding of the structure and energy requirements of stellar chromospheres and transition regions. Only a few of the 'Accomplishments' of IUE in this area can be discussed below and a concise list, which excludes material in papers by Böhm-Vitense and Querci; Imhoff and Appenzeller; Dupree and Reimers; Vauclair and Liebert; Starrfield and Snijders; Córdova and Howarth; McCluskey and Sahade of this book, must include the following:

(1) The systematic exploration of the spectra of stars from late-A through mid-M has established the existence of chromospheres and transition regions as a function of fundamental stellar parameters such as effective temperature and luminosity. These parameters, in turn, are related to the physical conditions in the subphotospheric convective zone, the origin of the mechanical and magnetic energy which heats the outer atmosphere.

(2) Empirical correlations have been established between chromospheric and transition-region fluxes and stellar rotation rate and age. These stellar parameters play an important role in the generation of magnetic fields by dynamo action.

(3) Detailed observations of line fluxes and widths have been used to determine the atmospheric structure and energy balance and to put flux correlations and energy requirements on a quantitative scale involving such atmospheric parameters as the gas pressure. These data, together with empirical correlations of line flux and rotation, provide useful constraints on atmospheric heating processes. They suggest that the heating processes require magnetic fields, for stars later than about mid-F, while the least active and early-F stars may have acoustically heated atmospheres.

(4) Studies of stellar activity and rotational modulation have provided evidence of active regions on RS CVn and BY Dra stars, which are likely related to starspots.

(5) Observations of flare stars in quiescent and flaring states have established the energy balance and dynamics of the 10^4-10^5 K plasma.

(6) The presence of transition-region material and evidence for outflows from giant stars have shown that fundamental atmospheric changes begin near spectral type K1 for luminosity class III giants. The original proposal that a sharp division might occur between 'solar' type atmospheres (to the left of K1 III in the H—R diagram) and 'nonsolar' type atmospheres with massive cool winds (to the right of the dividing line) focussed attention on this region of the H—R diagram where, regardless of the 'sharpness' of the transition, the interesting changes in coronal structure occur. The more luminous stars are not consistent with the picture provided by the class III giants.

(7) 'Hybrid' stars have been discovered which show not only transition-region material at 10^5 K but also high velocity outflow in Mg II blue-shifted absorption components. (Although the Sun has a high velocity wind, such winds are not observable for other main sequence stars). The hybrid stars constitute an important new class of objects for studying the changes between 'solar' and 'nonsolar' types of atmospheres.

(8) Observations of late K and M giants and supergiants have shown spectra dominated by neutral and singly-ionized atoms and the presence of many lines controlled by the radiation fields in H Ly α and H Ly β. These observations have led to improved modelling of low-gravity atmospheres for which the physical properties are very different from the Sun.

Some of these topics will be described in more detail below. However, it is clear that after 8 yr of IUE operation, the field has moved from the stages of spectral exploration and systematic surveys of physical properties to the position where specific theories concerning the causes of stellar chromospheres and coronae can be tested.

2. Spectroscopy

The 1200—3200 Å region provides a rich set of spectral features from which to infer the atmospheric properties of cool stars.

2.1. MAIN SEQUENCE STARS

A substantial number of main sequence stars from early-F to mid-M have now been observed at low resolution with IUE. Subject to visibility above the photospheric and chromospheric continua, one can conclude that all of these stars show qualitatively similar emission-line spectra. The long wavelength region is dominated by chromospheric lines, e.g. the strong Mg II resonance lines, many multiplets of Fe II and lines of other species such as C II, Mg I, Al II, Si II, and other metals.

The short wavelength region includes lines formed from the low chromosphere (e.g. C I, O I) up to the chromosphere-corona transition region, (e.g. Si IV, C IV, and N V). Line identifications are straightforward for these stars because of previous experience in solar spectroscopy. Figure 1a—c shows a selection of short wavelength spectra of F, G, K, and M dwarf stars from Brown and Jordan (1981), Ayres, *et al.* (1981a), and Linsky *et al.* (1982). The immediate impression is that the relative line fluxes in the 1200—2000 Å region are very similar from star to star, with the decrease in the Si II (1808 Å, 1817 Å) and He II (1640 Å) fluxes in the active M dwarfs being the most obvious exceptions. The trends in ratios of chromospheric to transition region line fluxes are discussed in Section 3, but the decrease in the Si II flux can be understood in terms of a steeper temperature gradient and/or a higher pressure in the active M dwarfs.

Some of the most valuable lines for atmospheric analysis lie above 1800 Å and are best observed at high spectral resolution. The density regime of the main sequence stars ($N_e \geqslant 10^{10}$ cm^{-3} at $T_e \approx 50000$ K) is such that the C III] 1909 Å line is usually too weak to be useful, but the Si III] 1892 Å line flux can constrain N_e Lines sensitive to the chromospheric opacity, such as C I 1656 Å and 1994 Å are also observable in stars later than about G2 V (e.g. Ayres *et al.*, 1983). The application of diagnostic techniques to modelling the atmospheres is discussed in Section 5. Although many such methods have been developed in the context of solar physics, the IUE spectra have given a further stimulus to the calculation of

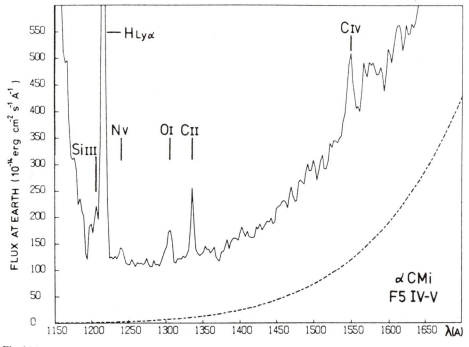

Fig. 1(a).

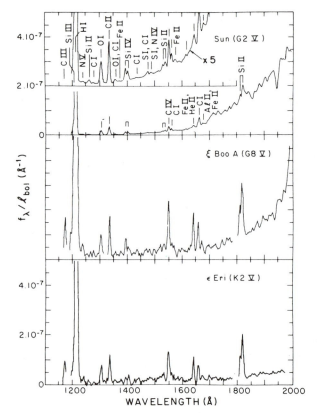

Fig. 1(b).

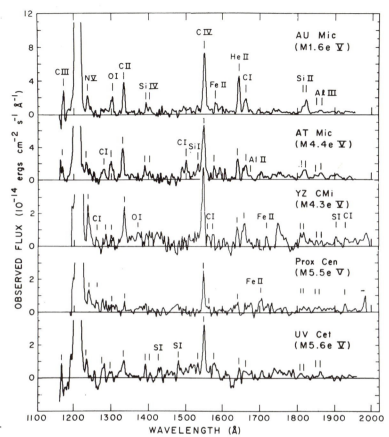

Fig. 1(c).

Fig. 1.   Low resolution IUE SWP spectra of cool dwarf stars. (a) α CMi (F5 IV—V), from Brown and Jordan (1981). (b) The Sun (G2 V), on the scale of IUE spectra, ξ Boo A (G8 V) and ε Eri (K2 V), from Ayres, Marstad and Linsky (1981a). (c) A selection of active dMe dwarfs between spectral types M1.6eV and M5.6eV, from Linsky *et al.* (1982).

collision cross-sections and transition probabilities which are essential for their interpretation.

2.2. GIANTS AND SUPERGIANTS

The early surveys of giants and supergiants (e.g. Linsky and Haisch, 1979; Dupree *et al.*, 1979; Brown *et al.*, 1979; Carpenter and Wing, 1979) showed the presence of C IV emission in F and G giants and early G supergiants but an absence of C IV emission in the giants later than about K1 III and supergiants later than mid-G.

The main characteristics of the spectra of giants earlier than about K0 III are illustrated at high and low resolution in Figure 2a—c from Brown *et al.* (1984a), Eriksson *et al.* (1983), and Hartmann *et al.* (1980). Emission lines associated with both chromospheric and transition-region material are present, with the latter on average decreasing with effective temperature and gravity. (The importance of

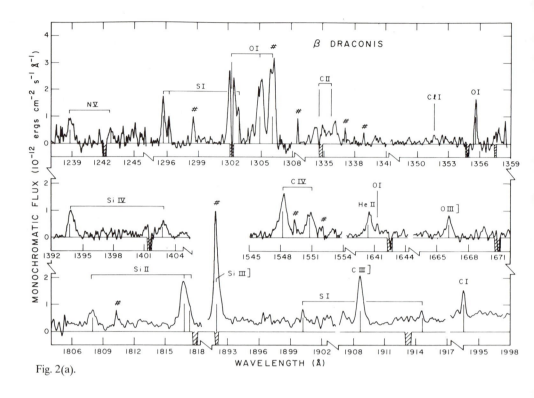

Fig. 2(a).

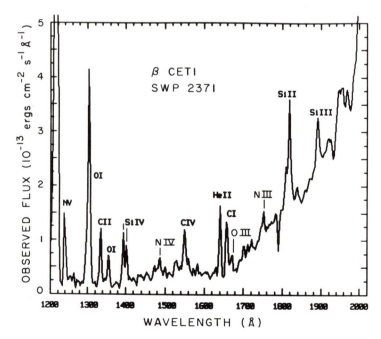

Fig. 2(b).

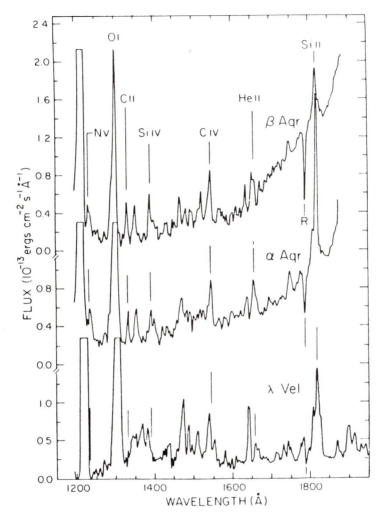

Fig. 2(c).

Fig. 2. IUE SWP spectra of cool giants and supergiants. (a) Sections of a high resolution spectrum of β Dra (G2 II), from Brown *et al.* (1984). (b) A low resolution spectrum of β Ceti (G9.5 III), from Eriksson *et al.* (1983). (c) Low resolution spectra of α Aqr (G2 Ib), β Aqr (G0 Ib) and λ Vel (K5 Ib), from Hartmann *et al.* (1980).

rotation is discussed in Section 3). The greater relative strength of Si III] 1892 Å indicates a lower electron density regime than in the main sequence stars. The major difference between G giants and dwarfs is the increase in the relative strength of the O I resonance lines. It has been recognized for some time that the H Ly β line both in the Sun and in giants such as α Boo (K2 III) can excite the O I $2p^3$ 3d ^{3}D level which subsequently decays through the upper level of the resonance lines (Haisch *et al.*, 1977). The relative increase in the giants can be understood in principle through a greater opacity in H Ly β in stars of lower gravity (see discussion of scaling laws by Ayres, 1979), since multiple scattering will enhance the probability of O I emission.

The early G supergiants [e.g. α Aqr (G2 Ib) and β Aqr (G0 Ib)] illustrated in Figure 2c and the K bright giant hybrid stars (Hartmann *et al.*, 1981; Reimers, 1982; Brown *et al.*, 1986) also show chromospheric and transition-region material. Except for trends expected from the lower gravity, the spectra of the early G-type supergiants are broadly similar to those of the giants.

The main area where observations with IUE have led to significant new developments in spectroscopy is in the understanding of the spectra of giants and supergiants later than about K2 III and G8 Ib, where fluorescent lines of O I, S I, Fe II, and CO dominate the spectra. More complete reviews of cool star spectroscopy have been given by Jordan and Judge (1984) and by Jordan *et al.* (1985).

Figure 2c shows a low resolution spectrum of λ Vel (K5 Ib). Figure 3a—c shows low resolution spectra of α Tau (K5 III), and β Gru (M5 III) from Brown and Jordan (1980) and α Ori (M2 Iab) from Basri *et al.* (1981) with line identifications brought up to date by Johansson and Jordan (1984). At low resolution it is difficult to determine which lines are present, a situation made more complex by low excitation species replacing transition-region lines such as those of He II, C II, and C IV. However, high resolution spectra have been used to establish the correct identifications. For example, the prominant feature at 1641.3 Å was shown to be not He II but O I (UV 146), which shares a common upper level with the O I resonance lines around 1304 Å (Brown *et al.*, 1981). Use of high resolution has also resolved blends whose dominant contribution changes with spectral type. For example, Carpenter and Wing (1979) found that in α Ori the main contribution to the feature near 1817 Å at low resolution is not Si II (UV 2) but SI (UV 2). Similarly, a blend in the blue wing of O I (UV 2) was found to be due to SI (UV 9) in high resolution spectra of α Tau and β Gru (Brown *et al.*, 1981).

As the spectra illustrated show, SI also contributes other strong lines in UV 1 (1900, 1914 Å) and UV 3 (1473, 1485 Å). The key to understanding these spectra seems to be the higher chromospheric opacity in the lower gravity stars. The overall strength of the SI spectrum appears to be a consequence of photo-excitation and photo-ionization by the strong broad H Ly α line (Brown *et al.*, 1981; Judge, 1984). The relative intensities of the SI multiplets then depend on a variety of processes including recombination and cascades, collisional excitation, and transfer of photons between multiplets (Judge, 1984). Accidental fluorescence is also important. Two of the broad optically thick O I resonance lines overlap two members of SI (UV 9) causing emission in three other members of the multiplet (Brown *et al.*, 1981).

Interlocking of optically thick and optically thin multiplets through a common upper level accounts for the strength of some intersystem lines, such as O I 1641.3 Å and C I 1994 Å. The relative intensities of multiplets from common upper levels can allow the chromospheric mass-column density to be determined, a parameter of use in modelling (Jordan, 1967; Judge, 1986a, b).

Other fluorescent processes occur. Ayres *et al.* (1981) proposed that features observed around 1380, 1510, and 1605 Å in low resolution spectra of α Boo are due to CO emission, pumped by the strong O I resonance lines. (Fluorescent excitation of CO is known to occur in the solar spectrum, Bartoe *et al.*, 1978.) This suggestion has been recently confirmed from high resolution spectra of α Boo (Ayres *et al.*, 1986b). The analysis of the CO emission indicates that chromospheric inhomogeneities may be important (Ayres, 1987). The presence of CO complicates the issue of how much C II and C IV emission is really present in early K giants and deep exposures and careful measurements of wavelengths are crucial.

In the M giants and supergiants fluorescent excitation of Fe II also becomes significant and its consequences appear in both long and short wavelength spectra. (For details see Johansson and Jordan 1984 and the review by Jordan, 1987.) Briefly, excitation by H Ly α from the a ^{4}D levels populates several 4p and 5p quartet levels around 11 eV. These decay to give emission around 1286—1299 Å, around 2505—2510 Å, and enhanced emission in multiplets between excited states such as UV 380, 391, and 399, which occur between 2800 and 2900 Å. Excitation by H Ly α also accounts for the feature observed at 1870 Å in low resolution spectra of M giants and supergiants, indicating that the Ly α line must be present even though the majority of the flux is not observed.

High resolution long wavelength spectra of a sample of K and M giants and supergiants have been presented in the form of an Atlas by Wing *et al.* (1983) and this is particularly useful for making systematic studies of the Fe II spectrum which dominates this region. The Fe II lines have a wide variety of line profiles which have considerable potential for studying mass motions in supergiant chromospheres (Carpenter, 1984). These spectra also show the lines of Fe I (mult 44) which are excited by Mg II (Gahm, 1974; Van der Hucht *et al.*, 1979). Although observations prior to IUE revealed the existence of fluorescent excitation, the dominance of such processes in producing the short wavelength spectra of cool giants and supergiants was not anticipated and is an important discovery with IUE.

The principle of using spectroscopic techniques for measuring the electron density is now well known. Methods developed for the Sun cannot be applied to low gravity stars since the electron density regime is lower. However, the inter-system lines of C II around 2325 Å provide a new example which can be used where the density is between about 10^7—10^9 cm^{-3} (Stencel *et al.*, 1981). Accurate collision strengths and transition probabilities are now available (Lennon *et al.*, 1985). This method has now been used to determine the density in a variety of cool stars (Brown and Carpenter, 1984; Judge, 1986a, b).

Several K bright giants have spectra which show C IV emission as well as indicators of high opacity chromospheres such as strong O I emission. They also have blue-shifted absorption components in Mg II which correspond to velocities of up to the escape velocity and have become known as 'hybrids' because they show simultaneous evidence of transition regions and winds. The early G supergiants α and β Aqr (G2 Ib and G0 Ib) are also hybrids in this sense (Hartmann *et al.*, 1980). The hybrid stars are particularly important in the context of the development of coronae and winds and are discussed further below.

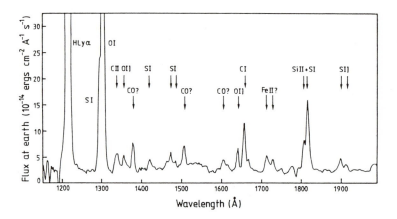

Fig. 3(a).

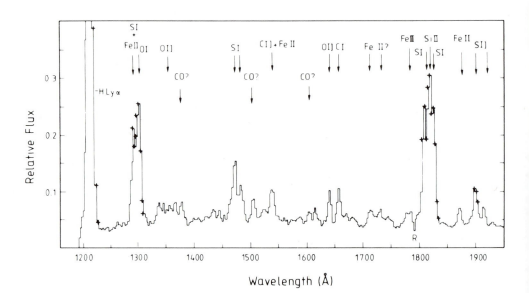

Fig. 3(b).

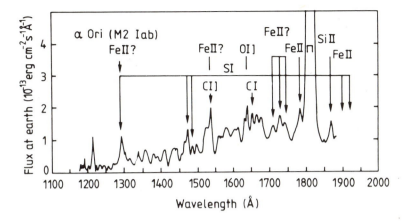

Fig. 3(c).

Fig. 3.   Low resolution IUE SWP spectra of cool giants and supergiants. (a) α Tau (K5 III) and (b) β Gru (M5 III) from Brown and Jordan (1980). (c) α Ori (M2 Iab), from Basri *et al.* (1981), all with line identifications brought up to date (Johansson and Jordan 1984).

3.   Global Properties of Chromospheres and Transition Regions

3.1.   MAIN SEQUENCE STARS

Across the surface of the Sun, the strength of emission lines formed in the chromosphere and transition region varies greatly with location but is tightly correlated with magnetic flux (cf. Skumanich *et al.*, 1975; Muller, 1985). This empirical fact has stimulated IUE observers to search for the dependence of emission line strengths on stellar parameters which can be related to the properties of the sub-photospheric convective zone where magnetic fields are generated. The depth of the convection zone depends on the effective temperature and luminosity and together with rotation determines the efficiency of the generation of magnetic fields (Belvedere, 1985). Linsky (1983, 1985) has reviewed the status of magnetic field measurements in cool stars as well as indirect indicators of magnetic fields. However, the existence of chromospheres with minimal (basal) surface flux levels in the Mg II and Ca II lines (Oranje and Zwaan, 1985; Schrijver, 1986) provides evidence for heating by a nonmagnetic mechanism (possibly acoustic waves). Stars with more active (i.e. brighter) chromospheres, for which the chromospheric emission above the basal level correlates tightly with the coronal X-ray flux, are thought to have an additional heating process which depends on the magnetic field.

Most dynamo theories predict that magnetic field generation should cease when the convection zone becomes very thin. It is difficult to predict accurately where in the H—R diagram this should occur, but it appears that the ratio of the convective turnover time (τ_c) to the rotation time (P_{rot}), which controls the equatorial acceleration and thus the dynamo efficiency, decreases precipitously somewhere in the early F stars (Gilman, 1980). However, the rapid increase in the background photospheric emission with increasing stellar effective temperature (decreasing

$B - V$ color) makes it difficult to detect emission lines in this important part of the H—R diagram. The photospheric background for F stars is weakest in the vicinity of the C II 1335 Å line, which is therefore the best line to choose when searching for high temperature gas against the background. In an early survey Böhm-Vitense and Dettmann (1980) observed C II and C IV emission from the F1 V ($B - V = 0.34$) star α Cae and subsequently Walter et al. (1984) detected C II and Si IV emission from the young F2 IV ($B - V = 0.32$) star α Crv. Walter and Linsky (1986) have extended this survey with heavily overexposed SWP spectra of 69 F and 2 A-type stars and detected C II emission in stars as blue as ($B - V) = 0.25$. Wolff et al. (1986) find C IV flux, as well as He I D3 absorption, in stars as early as ($B - V) = 0.29$. While the C II and C IV surface fluxes increase systematically towards the hotter stars, suggesting that plasma at 10^5 K exists even in the late A stars, repeated attempts to observe these emission lines in the early A stars have not been successful (e.g. Freire Ferrero, 1986). Stars redder than $B - V = 0.45$ (spectral type F6 V) share the correlation of emission line surface flux with rotational velocity seen in the cooler stars, indicating the onset of a solar-like dynamo near $B - V = 0.45$ (cf. Walter, 1983). The hotter stars do not show this strong correlation, suggesting that other processes such as the dissipation of acoustic waves may be a more important heating mechanism.

Observations at even shorter wavelengths may in principle detect emission from hotter stars. For example, emission in the wings of Ly α (outside of the interstellar absorption core) was detected from the A7V ($B - V = 0.22$) star Altair (Blanco et al., 1980; Landsman et al., 1984), which also has weak coronal X-ray emission (Schmitt et al., 1985). However, detections of X-ray emission from A-type stars suffer from the ambiguity that the emission may be instrumental ultraviolet light contamination or originate from previously unknown M dwarf companions which are powerful X-ray sources.

Theoretical models of stellar structure (Copeland et al., 1970) predict that stars later than about M3 V should be fully convective. Several models of the dynamo place the region of magnetic flux generation at the base of the convection zone, and Galloway and Weiss (1981) speculate that a different type of dynamo will probably operate in fully convective stars. IUE has been used by Linksy et al. (1982) and by Ambruster et al. (1986) to search for evidence of a change in the character of the dynamo in the mid-M dwarfs. Ambruster et al. (1986) concentrated on M5—6 stars (with and without H α emission) using coordinated optical observations to record flare events. Both surveys found much larger fractional luminosities (L_{line}/L_{bol}) for young (dMe) stars than for old (dM) stars. The latter survey also found that the fractional luminosities of young, active stars later than dM5e appear to be lower than those of active M dwarfs of earlier spectral type by nearly an order of magnitude, perhaps reflecting the difference in the type of dynamo between the partly and fully convective M dwarfs.

IUE observations have been used to explore the correlation between emission-line surface fluxes and stellar age and rotation. The aim is to relate the observed energy losses, and thus the nonthermal heating, to the generation of magnetic flux from dynamo action. Similarly, the effects of the braking of stellar rotation by the torque of winds with frozen-in magnetic fields can be investigated. X-ray data and ground-based measurements of Ca II fluxes and stellar rotation rates have been combined with IUE data to elucidate these complex phenomena as discussed below.

There is an obvious decrease in emission-line surface fluxes and fractional luminosities with stellar age among main sequence and pre-main sequence single stars (or members of wide binaries) of spectral types F—M. This general trend of decreasing nonradiative heating rate with age (the activity-age correlation) is seen in the soft X-ray flux, transition-region lines, chromospheric lines, and even the 1700 Å continuum flux (Simon *et al.*, 1985) formed at the top of the photosphere. For example, the surface fluxes of transition-region lines in the Hyades cluster (age $= 6 \times 10^8$ yr) for F5 V—G1 V stars are roughly 30 times larger than for the quiet Sun (Zolcinski *et al.*, 1982), while the surface fluxes for F2 IV—K3 V stars in the Ursa Major cluster (age $= 3 \times 10^8$ yr) are on average even higher (Walter *et al.*, 1984). Using a sample of 31 F7—G2 dwarfs of known age and rotational velocity, Simon *et al.* (1985) showed that the decline in emission-line surface flux with age is exponential with an e-folding time that is shorter for the transition-region lines than for the low chromospheric lines. They interpreted this decrease in flux with age as a decrease in the contribution from bright active regions (plages) through both the plage area and surface brightness decreasing with age. The differential age effect may explain in part the steep slope in the correlation diagrams of transition region versus chromospheric lines (i.e. Ayres *et al.*, 1981a; Schrijver, 1987).

Support for the activity-rotation correlations comes from the observation that both single stars and synchronously rotating subgiants (i.e. the RS CVn systems) obey the same activity-rotation relation, regardless of the age of the latter. Noyes *et al.* (1984) found that the mean fractional luminosity in the Ca II H and K lines (R_{HK}) tightly follows a unique curve when plotted against the nondimensional parameter P_{rot}/τ_c. A very similar functional relation holds for the Mg II lines (Hartmann *et al.*, 1984, see Figure 4a), the transition region lines (Simon *et al.*, 1985; Simon, 1986), and X-rays (Mangeney and Praderie, 1984). The significance of this result is that the Rossby number, $R_o = P_{rot}/\tau_c$, measures the importance of Coriolis forces that introduce helicity into convective motions and thus play an important role in the dynamo generation of magnetic fields (Noyes *et al.*, 1984). However, detailed conclusions concerning the dynamo properties should not yet be drawn from the activity-rotation correlations, because the relation between P_{rot} and differential rotation (which presumably drives the dynamo) is not yet clear, dynamo theory is still quite primitive, the correlations with R_o may not be unique (Rutten and Schrijver, 1987) and the correlations appear to change abruptly below a critical value (Simon, 1986). Also, Basri (1986) argued that plotting normalized fluxes against rotational period produces no more scatter than do plots against R_o.

Provided that other energy losses are not important, the relative amount of nonradiative heating in different atmospheric layers may be inferred by comparing line surface fluxes (F_{line}) or fractional luminosities (L_{line}/L_{bol}) in emission lines formed at different temperatures. Plots of F_{line} (or L_{line}/L_{bol}) for one line against another for many stars are commonly called flux—flux diagrams. In the first such study based on SWP low dispersion spectra of 28 cool dwarfs and giants, Ayres

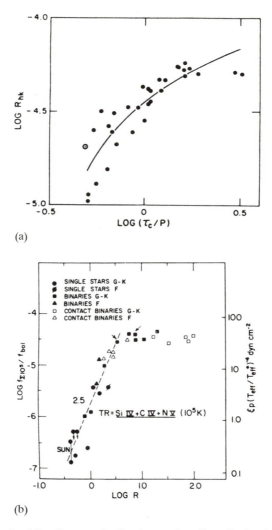

(a)

(b)

Fig. 4.   Plots of normalized line fluxes vs the Rossby number, $R_o = P_{rot}/\tau_c$. (a) Normalized fluxes in the chromospheric Mg II h and k lines ($R_{hk} = f_{Mg II}/f_{bol} = L_{Mg II}/L_{bol}$) for a sample of late-type dwarf stars (from Hartmann *et al.*, 1984). (b) Dependence of the normalized transition-region line flux (sum of Si IV, C IV, and N V lines) for a sample of F—K single stars, detached binaries and contact binaries (from Vilhu, 1984). Saturation occurs for log $R_o > 0.5$.

et al. (1981) found power law slopes of unity when comparing two different transition-region lines (e.g. C IV versus C II), about 1.5 when comparing transition-region lines to chromospheric lines (e.g. C IV versus Mg II), and about 3 when comparing the coronal X-ray emission to the chromospheric Mg II flux. They interpreted the steep power-law slopes as indicating that the heating rates in the chromosphere and the outer layers have a different functional dependence on the stellar parameters. This approach has been extended to include additional stars (e.g. Oranje *et al.*, 1982), while Bennett *et al.* (1984) showed that the flux—flux plots for the Sun observed as a star by the SME satellite have similar power law slopes to those that fit the IUE data for the solar-type (F8—G5) dwarfs. Hammer *et al.* (1982) proposed that the different pressure dependence of terms in the energy balance equation at coronal and chromospheric temperatures might account for the observed gradient of the power-law fits to the flux—flux diagrams.

Schrijver (1987) found that the power-law fits become much tighter when lower limit (basal) fluxes are subtracted from the chromospheric Ca II, Mg II, and Si II fluxes, and the transition-region or coronal fluxes are plotted against the excess chromospheric flux. Except for the M dwarfs, for which the excess Ca II emission is low compared to the higher temperature emissions (Schrijver and Rutten, 1987), the power-law fits are tight over a large range of activity (as measured by the surface fluxes). The log—log fits exhibit some curvature, however, for the most active stars when $F_x > 2 \times 10^6$ ergs cm^{-2} s^{-1} (Schrijver, 1987). This could result from different lines reaching their maximum (saturated field) values at different values of the magnetic field. Vilhu (1984), Vilhu *et al.* (1986) and Simon (1986) have presented evidence (see Figure 4b) for an empirical upper limit to the emission-line fluxes for the most rapidly rotating stars, which are generally in short period binary systems.

3.2. POST-MAIN SEQUENCE STARS

As stars evolve from the main sequence towards the cool giant and supergiant region of the H—R diagram, their atmospheres must respond to changes in the convective zone, the surface gravity and rotation rate. The gravity decreases as the stellar radius increases and the rotation rate decreases as angular momentum is either redistributed internally or lost in a stellar wind. IUE has provided the first important clues as to the evolution of stellar chromospheres and transition regions during this phase of stellar evolution (cf. reviews of this topic by Brown, 1984; Jordan, 1986a).

As discussed above the spectra of the mid-F to late-G giants are similar to those of the main sequence stars, apart from indicators of lower electron densities and higher chromospheric opacities. A similar dependence of fluxes on dynamo action might therefore be expected. In his IUE survey of F5 III—G8 III giants, Simon (1984) noted two clear trends in the normalized C IV flux ($F_{C\,IV}/F_{bol}$): the flux first increases from F5 III to G0 III and then drops dramatically by about a factor of 100 as stars approach spectral type K0 III. He attributed the first trend to increasing dynamo generation of magnetic fields with increasing convective zone depth, and the second trend to decreasing dynamo activity with decreasing rotation velocity. Gray (1982) detected a rapid drop in rotational velocity near spectral type G5 III, and Gray and Nagar (1985) measured a similar decrease in rotation among the subgiants at spectral type G0 IV. Even though Simon's stars are likely to be giants making their first crossing of the H—R diagram, the range in C IV flux at each spectral type suggests a dependence on more individual stellar properties. For example, Baliunas *et al.* (1983) interpreted the large differences in the C IV and X-ray flux of four presumably coeval K0 III Hyades giants with similar rotational velocities as due to observing the stars at different phases of their activity cycles.

Although the detailed geometric structure of the atmospheres of these giants is not known, it might be expected that magnetic fields produce the same range of topological features as in the solar atmosphere. In this sense one can think of them as 'active' stars. No evidence of winds is observed but hot, solar-type mass loss is probably occurring. The supergiant β Dra (G2 Ib—II) studied by Brown *et al.* (1984a) is an example of such a star.

One of the important discoveries of IUE is that fundamental change in atmospheric structure occurs between the yellow giants (spectral types G and up to K1 III, $V - R$ color index < 0.80) and the red giants (spectral types later than K2 III, $V - R$ color index > 0.80) and supergiants. In an early survey Linsky and Haisch (1979) noted a trend in which the yellow giants show emission lines formed at all temperatures up to 10^5 K, whereas the red giants show only chromospheric emission lines. On this basis they proposed a nearly vertical dividing line in the H—R diagram near $V - R = 0.80$ (see Figure 5) separating the yellow giants that

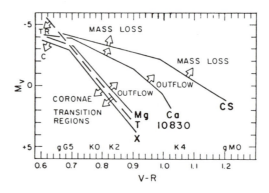

Fig. 5. An H—R diagram (from Mullan and Stencel, 1982) showing the rough location of boundaries where different phenomena typically occur in stars of different absolute visual magnitude (M_v), $V - R$ color and spectral type (indicated for giants). X-ray emission from coronae and C IV emission from transition regions are observed from stars to the left of the lines marked X and T, respectively. Mass loss, as indicated by asymmetries in the Mg II and Ca II emission lines and circumstellar absorption features, is detected in stars to the right of the lines marked Mg, Ca, and CS, respectively.

typically have transition regions from the red giants that typically do not. Subsequently, Ayres *et al.* (1981b) and Haisch and Simon (1982) showed that the *Einstein* soft X-ray observations are consistent with the presence of hot coronae in stars to the left of a similar boundary and the absence of coronae in single stars to the right (also shown in Figure 5). Also Reimers (1977), Stencel (1978), and Stencel and Mullan (1980a, b) presented evidence for the onset of massive cool winds (observed by asymmetries in the Ca II and Mg II lines and through the existence of circumstellar absorption lines in optical spectra) in stars lying to the right of a similar boundary.

The boundary as first proposed by Linsky and Haisch was based on a relatively small sample of stars and short exposure times. A large sample of 39 single stars was used by Simon *et al.* (1982) who found that the yellow giants show a range of C IV normalized flux ($F_{C_{IV}}/F_{bol}$). Some did not show C IV emission with low upper limits. These were attributed to the mixed evolutionary state of the sample. The absence of C IV emission in some yellow giants was also noted by Hartmann *et al.* (1982). Simon *et al.* (1982) concluded that the boundary proposed by Linsky and Haisch is genuine in that single stars to the right of K1 III have significantly less material at 10^5 K than do stars to the left. The sample of 15 red giants discussed by Stickland and Sanner (1981) did not show C IV emission, but most of the exposures times used in obtaining the spectra were very short.

Although there does seem to be a distinct different between the late G and early-M giants, the situation regarding the K giants is now less clear. There are a number of 'hybrid' K bright giants which show both C IV emission and evidence of cool winds (Hartmann, Dupree and Raymond 1980, 1981; Reimers 1982). Brown (1986a, b) has studied a wider sample of early KII stars and finds that a substantial proportion do show C IV emission. At present it is not known whether the KII hybrids are the result of the different evolution of these more massive stars or whether those which do show C IV emission are faster than average rotators for their class. While the brightest stars between K1 III and K5 III (i.e. α Boo and α Tau) do not show C IV emission, a hybrid giant star, δ And (K3 III), has recently been discovered which shows C IV and other transition-region lines as well as Mg II absorption in a wind (Judge *et al.*, 1987). Judge points out (private communication) that only seven stars between K1 III and K5 III have been observed with exposure times of equivalent depth to that used for δ And. Also Drake (1986) has recently discovered a number of K giants showing evidence of higher than average velocity cool winds, but deeper exposures are needed to investigate whether or not C IV emission is present. Thus K stars with a range of properties seem to exist. It is not yet certain whether binarity plays a significant part in the hybrid phenomenon, perhaps through causing higher than average rotation rates; δ And is a member of at least a triple system, although the components are too far apart for direct interactions. Reimers (1982) noted that some of the hybrid bright giants are binaries. Whatever the outcome there is no doubt that the proposal by Linsky and Haisch has led to attention being concentrated on this important part of the H—R diagram where significant changes in coronal structure are taking place.

The K and M giants and supergiants which do not show C IV emission even in deep exposures (Ayres *et al.*, 1982) may be regarded as 'quiet stars. α Boo (K2 III) is an example of such a giant. The overall trend for the line fluxes in these stars is for the chromospheric energy flux to decrease with decreasing effective temperature, with only a weak dependence on surface gravity (e.g. Linsky and Ayres, 1978; Stencel *et al.*, 1980). Correlations with rotation rate have not been established for these slow rotators, but the trends with T_{eff} and g_* are consistent with the dissipation of energy from acoustic slow mode waves (e.g. Stein, 1981; Ulmschneider and Stein, 1982). At present neither the magnetic field strength nor topology are known. However, Schwarzschild (1975) has argued, on the basis of models of the convective zone in giants and supergiants, that both the granulation and supergranulation will exist on larger spatial scales, with only a few super-granulation cells being present on a given hemisphere at any one time. Thus the proportion of the surface where significant magnetic fields may be present would be smaller and a natural explanation for variations with rotation is provided. The low gravity stars show evidence of outflows and winds, a topic discussed in the paper by Dupree and Reimers (p. 321).

We conclude by mentioning that the carbon rich R and N-type stars have been observed by IUE (Eaton *et al.*, 1985; Johnson *et al.*, 1986). The R stars show weaker Mg II emission than the K giants of similar effective temperature and the N-type star TX Pyx shows strong variations in the Mg II flux with phase.

4. Dynamic Phenomena and Structures in Chromospheres and Transition Regions

Although stellar surfaces cannot yet be directly resolved, IUE spectra have been used to deduce the presence of atmospheric inhomogeneities and to study what seem to be flows within the atmosphere.

4.1. SYSTEMATIC FLOWS

Following a suggestion by Stencel *et al.* (1982) that transition-region lines in β Dra (G2 Ib—II) are redshifted by up to 40 km s^{-1} relative to the low excitation lines formed at the base of the chromosphere, Ayres *et al.* (1983) reanalyzed high resolution spectra, separating optically thin lines from those that could be optically thick. A shift of about 26 km s^{-1} remained between the transition-region lines and the optically thin cool chromospheric lines, but when only intersystem (and thus optically thin) transition-region lines are used the difference reduces to about 10 km s^{-1}. The intersystem lines for which radiative transfer effects cannot be important therefore provide marginal evidence for a relative downflow of transition-region material, as occurs in some regions of the solar atmosphere (e.g. Brueckner, 1981; Dere, 1982). Ayres *et al.* (1983, 1986a, 1987) found a similar effect in seven other stars (β Cet, Capella, λ And, and four dwarfs). Figure 6 illustrates the most recent redshift measurements of Ayres *et al.* (1987b) which show significant redshifts in the high-excitation lines of β Dra and α Aur Ab, except for the C IV 1548 Å line which may be marginally optically thick in these stars. In the solar chromosphere downflows are associated with material within or between magnetic flux tubes (see Athay, 1985). It is therefore tempting to interpret stellar downflows in a similar model. This line of argument has been supported by the work of Ayres (1984) who obtained spectra of Capella with precise wavelength scales to measure the redshifts of optically thin and thick lines primarily from the F9 III star. The downflowing material could dominate in the stellar spectrum if it is denser and thus for the same temperature brighter than the upflowing component. In seeking evidence of wind expansion Hartmann *et al.* (1985) and Mendoza-Ortega (1985) noted that there are no detectable redshifts in the intersystem Si III and C III lines in the hybrid star α TrA compared with lines of SI formed in the same order. Brown *et al.* (1986) suggested that these lines are not formed in regions of strong magnetic fields and thus that magnetic flux tubes cover only a small portion of the surfaces of hybrid stars. This would be consistent with Schwarzschilds' argument for low gravity stars if magnetic fields are confined

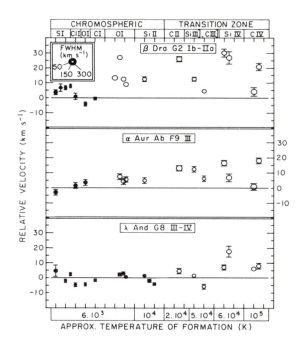

Fig. 6. Doppler shifts of emission lines relative to the low excitation lines (filled circles) formed low in the atmosphere of these giant stars. Error bars represent one standard deviation. See Ayres *et al.* (1987b) for more details.

to supergranulation cell boundaries. However, the production of transition-region lines in nonmagnetic regions would still require explanation.

4.2. ATMOSPHERIC INHOMOGENEITIES

Although stars are observed as point sources, three powerful techniques have provided some insight into the inhomogeneous structure of cool star atmospheres, mainly making use of active stars in binary systems. The first, *rotational modulation*, identifies relatively bright features (active regions) from changes in the integrated line fluxes as these structures rotate onto and off the stellar disk. This method was first used by Baliunas and Dupree (1982) in their study of λ And. Observations of the C IV line, for which the contrast between active and quiescent regions is significant, have been used by Boesgaard and Simon (1984) to identify an active region on the young star χ^1 Ori (G0 V) with a lifetime in excess of 1 yr. Also Butler *et al.* (1987) have presented evidence for an active region on BY Dra, and Rodono *et al.* (1987) detected active regions on the RS CVn systems II Peg (K2 IV + ?), AR Lac (G2 IV—V + K0 IV) and perhaps on V711 Tau (K1 IV + G5 V). An interesting aspect of these studies is that the C IV flux is usually greatest when the visual flux is least, indicating that the active regions and starspots are spatially correlated (see Figure 7) as one would expect if they are both magnetic structures. Observations of the RS CVn system λ And (Baliunas and Dupree, 1982) and the BY Dra system EV Lac (Andersen *et al.*, 1986) are

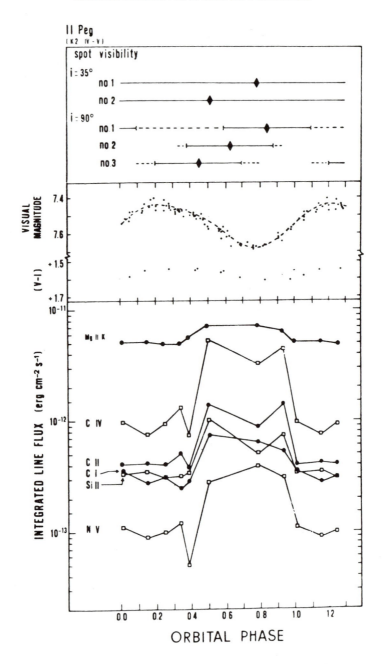

Fig. 7. Comparison of visual photometry and $V - I$ color (middle panel) with IUE integrated emission line fluxes (bottom panel) of the RS CVn system II Peg (K2 IV + ?) from Rodono *et al.* (1987). The top panel shows the location of the starspots (diamonds) and their visibility (solid lines) for two assumed values of the inclination of the rotational pole to the line of sight. It appears likely that the active region with the bright emission lines lies above spot No. 2 assuming an inclination of 90°.

consistent with this picture. Dorren *et al.* (1986) have followed the complex long term trend of the C IV flux with changing spot coverage of V711 Tau.

The emission lines from these stars can be modelled in the ways described below, but since neither the active/quiet flux ratio nor the relative areas of quiet and active regions are accurately known one of these quantities must be assumed. For example, Linsky *et al.* (1984) computed two component models for the active and quiescent transition regions of II Peg as a function of the assumed area of the active region. The mechanical flux input to the active transition region is then 40—150 times that of the quiet region for assumed active region areas of 6 to 1% of the visible surface.

The second technique, *Doppler imaging*, allows one to construct a crude image of a star from the wavelength and shape of discrete components in a rotationally-broadened line profile as regions of relative darkness (spots) or brightness (active regions) move across the stellar disk by rotation and thereby move across an emission line profile from the blue to the red wing (cf. Vogt and Penrod, 1983). Sequences of high-resolution spectra with high signal-to-noise, well distributed in rotational phase, permit accurate measurement of the longitude of individual active regions and cruder information on their latitude and area. The first systematic application of this technique to IUE spectra was by Walter *et al.* (1986) who observed the Mg II lines in the eclipsing RS CVn system AR Lac observed in October 1983. They fitted the observed line profiles with Gaussians (see Figure 8 for an example) representing the G star, K star, two active regions on the K star, and a flare also on the K star. They found that the G star flux was constant, that the two active regions together cover about 2% of the area of the K star, and that they both lie close to the equator.

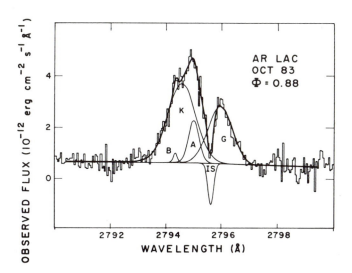

Fig. 8.   The Mg II k line profile of AR Lac fitted by components for the K0 IV star (component K), the G2 IV—V star (component G), two plages on the K0 IV star (components A and B), and the interstellar absorption feature (component IS). Walter *et al.* (1986) have obtained a Doppler image of AR Lac by modelling a set of Mg II line profiles obtained at many orbital phases.

The third technique uses the *occultation* of one star by another during an eclipse to obtain spatial resolution. The study of supergiants by this method is discussed by Hock and Stickland (p. 445). For example, Walter *et al.* (1986) used the eclipses in the AR Lac system to infer the short-wavelength spectrum of each star and of the active regions separately. Baliunas *et al.* (1986a) observed Si IV and C IV in absorption against the hot continuum of an sdOB star as this star passed behind the G8 III primary in the FF Aqr system. They concluded that 10^5 K plasma extends to at least 1.5 stellar radii above the surface of the G8 III star. Using a similar observing technique, Guinan *et al.* (1986) studied V471 Tau (K2 V + DA) to obtain the first directly spatially-resolved information on a cool dwarf star other than the Sun (see Figure 9). They observed C IV absorption out to nearly one stellar radius above the K2 dwarf only when a starspot (identified from visual photometry) was near the stellar limb. From this they concluded that loops containing 10^5 K plasma, extend outwards from the starspots for about a stellar radius. Such a structure is far more extensive than any solar counterpart.

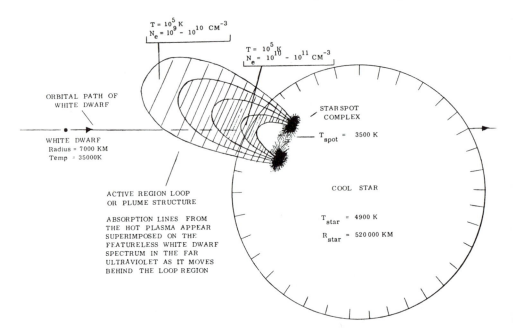

Fig. 9. A schematic picture of an active region loop system extending one stellar radius above the limb of the star V471 Tau (K2 V + DA), deduced by Guinan *et al.* (1986) from observations of C IV absorption as the white dwarf passed behind the K2 V star on the indicated trajectory.

4.3. INTRINSIC VARIABILITY

In addition to the changes in emission line fluxes and profiles produced by the rotation of active regions onto and off the stellar disk, the study of intrinsic variability on many time scales is needed to assess the dynamic properties of stellar atmospheres. IUE observations of flux variations on the multiyear time scale of stellar magnetic cycles is becoming feasible as the duration of the IUE data set expands, and Dorren *et al.* (1986) have already called attention to the variations in the transition-region line fluxes of V711 Tau over the 1978—84 period as the fractional coverage by spots has changed. In a similar way IUE has monitored the long term behavior of chromospheric emission from α Ori (Dupree *et al.*, 1984). However, so far variability studies have concentrated on the time scales of flares, which are less than an hour for dMe stars and 5—10 hr for the RS CVn systems.

IUE has now observed rapid brightenings of emission lines in a large number of dMe stars including BY Dra, YY Gem, AU Mic, EV Lac, AD Leo, AT Mic, EQ Peg B, Prox Cen, and UV Cet. Some of this work has been summarized by Bromage *et al.* (1983) and Butler (1986). The magnitude of the flares, as measured by the fractional enhancement of the C IV flux, extends from sporadic 50% enhancements seen in AU Mic and BY Dra (Butler *et al.*, 1987) and YY Gem (Baliunas *et al.*, 1986b), to a factor of 3 enhancement seen in Prox Cen (Haisch *et al.*, 1983), to an unknown enhancement (due to saturation), but for that reason very large, seen during a $\Delta U = 4.5$ flare on AD Leo (Pettersen *et al.*, 1986). For all of these flares the transition-region line flux enhancements could be much larger than measured because the typically 30—60 min exposures smear out the peak increase. During these flares the high temperature line fluxes increase by larger amounts and decay more quickly than for the chromospheric lines, as in solar flares. A bright continuum is detected in the SWP and LWP spectral ranges for the strongest flares such as that on AD Leo. For the flares recorded by IUE the radiative luminosity of the transition-region plasma can be as much as 5000 times larger than observed during solar flares.

Rapid increases in the ultraviolet emission line fluxes are also observed in RS CVn systems (see review by Catalano, 1986), but the observed fluxes are much larger than for the dMe stars and IUE has been able to obtain high resolution spectra in several cases to study the dynamics of the flaring plasma. Flares have now been observed on AR Lac, UX Ari, λ And, and V711 Tau with enhanced flux in the Mg II line present for at least 15 hr in AR Lac (Walter *et al.*, 1986). The bulk of the radiative output appears in the Mg II h and k and Ly α lines (Baliunas *et al.*, 1984), but the transition-region lines are enhanced by larger factors than the chromospheric lines (Simon *et al.*, 1980b) as is seen for the dMe flares. A high-dispersion long-wavelength spectrum obtained during the January 1, 1979 flare on UX Ari showed enhanced emission in the red wings of the Mg II lines extending to $+475$ km s^{-1}, which Simon *et al.* (1980b) interpreted as gas flowing from the K0 IV to the G5 V star perhaps along flux tubes connecting the two stars. High-dispersion long and short wavelength spectra obtained during the October 3, 1981 flare on V711 Tau indicated that the flare occurred on the K1 IV star but there was no evidence for flows (Linsky *et al.*, 1986). The total emission in all ultraviolet emission lines during this flare was a remarkable 4×10^{35} ergs.

5. Modelling of Structure and Energy Balance

While flux correlations between selected lines can establish general trends, the complementary approach of modelling individual stars allows these trends to be interpreted in terms of atmospheric parameters. The use of emission-line fluxes and resulting emission measures to make models of the chromosphere, transition region and corona has been developed in the context of solar physics with Pottasch (1964) pioneering the development of emission measure distributions. Methods of atmospheric modelling starting from the emission-measure distribution are described by Jordan and Wilson (1971) and the book by Athay (1976) discusses many relevant aspects. The review by Jordan and Brown (1981) gives the methods of analysis in the general context of high gravity stars. Withbroe and Noyes (1977) have reviewed questions concerning energy and mass transfer which also are important for main sequence stars. Space does not permit a review of atomic physics, but the interpretation of stellar or solar spectra would not be possible without the extensive developments that have taken place over the past ten years or so.

The structure of chromospheres was investigated prior to observations from IUE by combining information from the profiles and fluxes of Ca II H and K, Mg II h and k, and for some stars, the H Ly α line. This type of modelling effort has continued by including the observations from IUE.

5.1. MAIN SEQUENCE STARS

The methods of modelling the fluxes and profiles of lines formed in the low chromosphere have been described by Ayres *et al.* (1976) and Kelch *et al.* (1978), who applied them to quiescent main sequence stars, such as α Cen A (G2 V) and α Cen B (K0 V), and a number of active dwarfs including ξ Boo A (Kelch *et al.*, 1979). In later applications of these techniques, including some modelling also of transition-region lines, models were constructed based on IUE fluxes. The stars considered include α Cen A and B (Ayres and Linsky, 1980), and ε Eri (K2 V) (Simon *et al.*, 1980). There is a tendency for these models to give lower electron densities than models based on transition-region lines and X-ray fluxes. The methods described below for analyzing transition-region line fluxes, however, cannot be applied to chromospheric lines; thus the two types of modelling are complementary and both are essential. In future work it should be possible to improve the combination of the chromospheric and transition-region models. The region where the fraction N_e/N_H is changing rapidly is of particular importance since the gradient of the correlation between chromospheric and transition-region fluxes seems to be determined by changes from star to star in this part of the atmosphere.

The starting point for modelling is the emission line surface flux F_{12}. For a transition between the ground state, 1, and an excited state, 2, expressed in terms of the atomic constants, temperature-dependent functions and the spatial extent of the emitting region,

$$F_{12} = \text{const} \int g(T) N_e N_H \, \mathrm{d}h, \tag{1}$$

where

$$g(T) = \frac{N_{\text{ion}}}{N_e} T_e^{-1/2} \exp(-W_{12}/kT_e). \tag{2}$$

Above about 20 000 K where the emission measure method is most applicable, $N_H \approx 0.8 N_e$. Two approaches can be used. When a line is formed over a limited temperature range, an average $g(T)$ may be used and the emission measure

$$E_m(0.3) = \int_{\Delta R} N_e^2 \, dh \tag{3}$$

is calculated with ΔR corresponding to a region $\Delta \log T = 0.3$. When the contribution function $g(T)$ is broader, a locus of possible solutions, giving upper limits to $E_m(T_e)$ is useful, i.e.

$$E_m(T_e) = \int N_e^2 \, dh. \tag{4}$$

Both techniques can be applied to establish the optimum emission measure distribution to fit all the observed lines. In practice multi-level calculations are performed. Figure 10 shows the emission-measure distribution derived for ξ Boo A (G8 V) by Jordan *et al.* (1987).

Assumptions must now be made as there are insufficient spectroscopic diagnostic techniques to separate N_e and Δh in the emission measure at more than one temperature. Although for high gravity stars the electron pressure is nearly

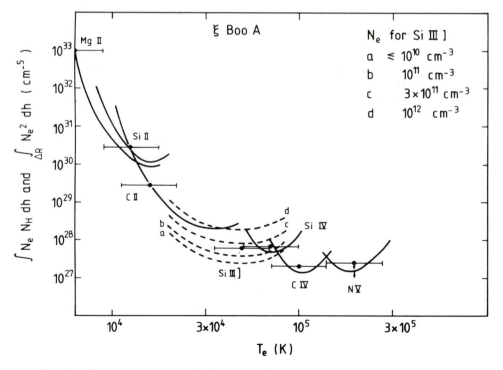

Fig. 10.   The emission measure distribution for ξ Boo A (G8 V), from Jordan *et al.* (1987).

constant between 20 000 and 200 000 K, there is no difficulty in using the better approximation of hydrostatic equilibrium, i.e.

$$\frac{\mathrm{d}P_e}{\mathrm{d}h} = -7 \times 10^{-9} P_e g_* / T_e. \tag{5}$$

The temperature gradient can now be expressed in terms of the emission measure and pressure such that

$$\frac{\mathrm{d}T_e}{\mathrm{d}h} = \frac{P_e^2}{\sqrt{2}} E_m(0.3) T_e, \tag{6}$$

where P_e is in units of cm^{-3} K. Equations (5) and (6) can be combined to give

$$P_e^2 = P_{\mathrm{ref}}^2 \pm 2 \times 10^{-8} g_* \int_{T_e}^{T_{\mathrm{ref}}} E_m(0.3) \, \mathrm{d}T_e \tag{7}$$

and thus provided the pressure is known at some reference point the pressure can be found as a function of T_e. For main sequence stars the most widely applicable method of measuring N_e is to compare the emission measures derived from the fluxes in the C III] and Si III] lines, as a function of N_e and T_e, with the mean distribution established from other lines.

Models have now been computed using this method for a number of main sequence stars including α CMi (F5 IV−V) by Brown and Jordan (1981) and Jordan et al. (1986); ι Vir (F7 V), χ Her (F9 V), β Com (G0 V) and κ Cet (G5 V) by Fernandez-Figueroa et al. (1981) and de Castro, Fernandez-Figueroa and Rego (1982); and χ^1 Ori (G0 V), α Cen A (G2 V), ξ Boo A (G8 V), α Cen B (K0 V), and ε Eri (K2 V) by Jordan et al. (1987). Some general conclusions can be drawn. First, the emission-measure distributions of early-G to early-K dwarfs between 20 000 and 200 000 K are remarkably similar in shape as implied also by the gradient of one in the C IV−C II flux−flux correlations (e.g. Oranje, 1986). Secondly the range of electron pressures is about 0.03 to 2 dynes cm^{-2}, similar to the spread of values in the solar atmosphere. Thirdly, the stars all have steep transition regions, indicating the presence of an overlying hot corona even when X-ray observations are not available. Moreover, the variation of dT_e/dh with T_e is similar in all of the stars (see Equation (6)), strongly suggesting that the same physical processes control the structure. The transition-region fluxes, for the same gravity, are roughly proportional to the pressure, but in general there is also a dependence on gravity to about a power of $-3/2$. As indicated also by the gradient of the flux−flux correlation between chromospheric and low temperature transition-region lines such as C II, the stars do not have the same shape of emission measure below 20 000 K. Full chromospheric models are required, including a treatment of radiative transfer in the stronger lines and calculations of the electron pressure to find the physical cause of this difference in structure.

Because little is known at present concerning supergranulation structure on these stars, the models represent only spatially-averaged conditions; as on the Sun, regions of higher and lower temperature gradients will occur. In the average models, conduction is not important below about 10^5 K and nonthermal energy input balances radiation losses. A simple balance between radiation and conduction will not reproduce the form of the emission measure distributions. The energy input requirements for the transition region are not large and only a small fraction of energy passing through to heat the corona is needed; the form of the emission measure distribution then mainly reflects the inverse of the radiative power loss curve (Jordan *et al.*, 1987), i.e. the quantity

$$\int_{\Delta R} \frac{\mathrm{d}F_R}{\mathrm{d}h} \, \mathrm{d}h \approx E_m(0.3) P_{rad}(T_e), \tag{8}$$

which is approximately constant. But the heating required $(\mathrm{d}F_m/\mathrm{d}h)$ is not constant since it depends on the temperature gradient. The steep rise in emission measure below $10\,000$ K or the large chromospheric radiative loss requires a separate region of energy deposition.

The available emission line widths of main sequence stars have been used to examine how much flux could be carried in simple acoustic and Alfven wave modes, by expressing the flux as

$$F_m = \rho V_{rms}^2 V_{prop}, \tag{9}$$

where V_{rms}^2 is related to the observed nonthermal line width and V_{prop} is the appropriate propagation velocity. As in the Sun, acoustic waves fail to carry sufficient flux above $20\,000$ K, but only modest magnetic field strengths are required for sufficient Alfven wave flux. Since wave periods are not known, the problems of damping cannot be addressed, but empirically the variation in V_{rms}^2 and flux predicted matches the energy losses required within present uncertainties. There is therefore no difficulty in accounting for the rise in the emission measures below 10^5 K as often claimed (e.g. Athay, 1985).

Coronal modelling and associated theory are beyond the scope of the present review, but in practice ultraviolet and X-ray data can be combined in simple spherically-symmetric models to relate transition-region properties (pressure and temperature gradient) and stellar parameters (gravity and rotation rate) to coronal properties (pressure and temperature). (See Jordan and Brown, 1981; Jordan *et al.*, 1987). The results suggest that the average coronal temperature is proportional to the rotation rate, while the pressure is proportional to the square root of the rotation rate.

It should be noted that although the simple models do not take varying area factors into account neither do the flux—flux correlations that fit a wide range of types of stars. The reason for this is not yet understood.

5.2. POST MAIN SEQUENCE STARS

The techniques developed for main sequence stars can be applied to giants and supergiants showing C IV emission, provided the effects of the lower gravity and differences in metallicity are taken into account and provided stellar winds are not significant in the momentum equation. The plane-parallel geometry is no longer assumed, and g is allowed to decrease with height. Then models can still be made by iteration, including also wave support implied by the larger nonthermal widths of lines in the lower gravity stars (Brosius and Mullan, 1986; Brown, Ferraz and Jordan, 1984b; Jordan, 1986b). So far this technique has been applied to only a few giants [e.g. β Cet (G9.5 III) by Eriksson *et al.* (1983), and β Gem (K0 III) by Jordan and Brown (1981)], to a few bright giants [e.g. β Dra (G2 Ib—II) by Brown *et al.* (1984a); and α TrA (K4 II), ι Aur (K3 II), θ Her (K3 II), and γ Aql (K3 II) by Hartmann *et al.* (1985)], and to the pre-main sequence star T Tau by Brown, Ferraz and Jordan (1984b).

The coolest active star to be modelled so far is β Cet (G9.5 III), which lies immediately to the left of the transition-region and coronal boundaries. In their study of this star Eriksson *et al.* (1983) noted a trend of rapidly decreasing pressure at the top of the chromosphere between stars immediately to the left (e.g. β Cet and β Gem) and those to the right (e.g. α Boo and α Tau) of the boundaries.

The apparent trends in flux ratios follow those for the main sequence stars as discussed above. However, for the same emission-line surface flux the electron density is lower, as expected from the lower gravity, and the temperature gradients are correspondingly less steep. Nonthermal line widths are observed to be larger in the early-G giants than in main sequence stars of similar effective temperature. This is expected since even when the nonthermal energy flux is similar, the density is lower and the magnetic field is not expected to be larger. Work is still in progress relating the flux-rotation correlations for giants to the chromospheric and transition-region structure and energy balance.

In β Dra, which we use as an example of a star for which a detailed model has been made, the chromospheric and transition-region fluxes are about a factor 60 greater than in the Sun, but the X-ray flux is weaker relative to the low temperature material. Because the electron pressure and temperature gradients are smaller, thermal conduction is not significant compared with the radiation losses and this is reflected in the shape of the emission-measure distribution between 10^5 and 10^7 K. As for the Sun, pure acoustic waves cannot provide enough energy to account for the radiation losses.

In stars later than about K1 III (excluding hybrids) the C II line at 1335 Å or Si III line at 1892 Å may be the highest excitation lines observed but even these are not always detected. The methods of analyzing ultraviolet emission lines through emission measures cannot be simply applied to lines formed predominantly in a chromosphere for several reasons: for example, the ratio N_H/N_e varies rapidly with temperature, the temperature region over which a line is formed is not obvious, some lines are effectively optically thick, and for atoms and first ions the ionization fraction can depend on the ambient radiation field in both continua and strong lines such as H Ly α. As discussed in Section 2, the far ultraviolet spectra of the low gravity giants and supergiants are quite unlike those of main sequence stars of similar effective temperature, and a major effort has gone into identifying lines and understanding excitation mechanisms. Atmospheric modelling directly from these lines is therefore at an early stage. As an initial step a variety of techniques using line ratios and absolute line fluxes have been used to constrain the conditions in the chromospheres of these stars. The methods have been described by Judge (1986a) and applied so far to α Boo (K2 III) by Judge (1986a) and to α Tau (K5 III) and β Gru (M5 III) by Judge (1986b).

The methods are as follows. The electron density where the C II] 2325 Å multiplet is formed can be determined from line ratios within the multiplet as described in Section 2, Figure 11, from Judge (1986b), shows these lines in α Tau,

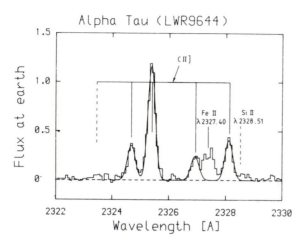

Fig. 11. The C II lines in the LWR high resolution spectrum of α Tau (K5 III), whose relative intensities lead to an electron density of 10^9 cm^{-3}, using the atomic data of Lennon *et al.* (1985). From Judge (1986b).

where using the atomic data of Lennon *et al.* (1985), they imply a density of 10^9 cm^{-3}. The mass column density can be found from the flux ratio of two lines from a common upper level, F_1 and F_2, where one is optically thick and the other optically thin,

$$F_1/F_2 = b_1 q_1 \lambda_2 / b_2 q_2 \lambda_1, \tag{10}$$

where q is the probability of escape and b is the branching ratio for emission. If q_1 is 1.0 (an optically thin line), then q_2 can be found and hence τ_2, the optical depth at line center of the optically thick line, can be found from

$$q = 1 - \mathrm{erf}(\ln \tau)^{1/2} \tag{11}$$

and

$$\tau = 6 \times 10^{-15} \lambda(\text{Å}) f_{12} M_i^{1/2} \int N_1 T_i^{-1/2} \, dh, \tag{12}$$

where f_{12} is the oscillator strength, M_i the atomic weight, N_1 the population density of the lower level and T_i the temperature that describes the line width. The pairs of lines which have been used so far are O I(1304 Å)/O I(1641.3 Å) and C I(1656 Å)/C I(1994 Å). Although the ion balance C I/C II is uncertain because of the influence of the H Ly α radiation field (Judge, 1986a), the two ratios suggest that the method gives values of the mass column density to within factors of two to three. The emission measure $E_m(T_e) = \int N_e N_H \, dh$ can be calculated as a function of T_e, using ion balance calculations, but this method gives only an upper limit to

$E_m(T_e)$ as a function of T_e. Thus the structure of the atmosphere, effectively Δh (range of formation) for a given line, cannot be found at a particular temperature. This procedure will certainly overestimate the thickness of the region.

Models of the chromosphere for some cool giant stars have been made by obtaining an optimum fit to both the line profiles and fluxes in Ca II H and K, Mg II h and k, and in some cases H Ly α. In particular, models have been made for α Boo by Ayres and Linsky (1975) and for α Tau by Kelch *et al.* (1978). These models can be used to compute the fluxes in the wider range of lines now observed, and Judge finds that they give a satisfactory flux to lines formed predominantly below 7000 K. However, because these models concentrated on the lower part of the chromosphere, they underestimated the flux in the higher temperature lines such as Si II (1808 Å, 1817 Å), and C II (1335 Å). An ab initio approach solving the radiative transfer and ion balance equations simultaneously is the next step to improve the details of the models. In these stars which do not have strong winds or evidence of large nonthermal widths the extent of the atmosphere is close to that expected in hydrostatic equilibrium, i.e. compared with a main sequence star it is larger by about the ratio of stellar gravities. In the stars studied so far the electron densities, surface fluxes and maximum temperature of the material decrease between K2 III and M5 III and the electron density and mass-column density follow roughly the scaling laws proposed by Ayres (1979).

Complementary information on the more distant parts of the giant atmospheres where wind velocities are significant may be obtained by comparing profiles of the Mg II and other lines with solutions of the partial redistribution radiative transfer equation in the comoving frame (Drake and Linsky, 1983; Linsky, 1986).

Some aspects of the spectra, e.g. the density measured from C II] and the CO fluorescence in α Boo, suggest that the effects of inhomogeneities might also be detectable (Judge 1986a; Ayres 1987). In a star such as α Ori (M2 Iab), where the wind is an important factor (see p. 321), departures from hydrostatic equilibrium and a greater extent are expected (e.g. Hartmann and Avrett, 1984).

The observations with IUE have considerably improved our understanding of the chromospheres of cool giants and supergiants and, in particular, have shown the importance of the H Ly α radiation field in determining ion fractions (e.g. S I/S II, C I/C II, Si I/Si II) as well as the emergent spectrum through fluorescence processes. The next stage is a closer examination of potential heating mechanisms.

6. Conclusion

Prior to IUE, ground-based observations of the visual spectra and the few glimpses of the ultraviolet spectra of cool stars provided by spectrographs on rockets, balloons and *Copernicus* served to whet our appetites concerning the properties and essential physical processes occurring in the outer atmospheres of these stars. More than 9 yr of IUE observations have provided answers to a great many questions but have, at the same time, posed a new group of more detailed questions:

(1) While we now have a good idea of which types of stars have chromospheres and transition regions and how their radiative output depends on effective

temperature, gravity and rotation, we do not yet understand the causes of the differences between the solar atmosphere and the analogous atmospheric regions in luminous stars, hybrid stars, the early-F dwarfs and the dMe stars.

(2) While we have acquired evidence that atmospheres of G—M dwarfs are heated by magnetic processes and the early-F stars perhaps by acoustic waves, we do not yet understand these heating mechanisms in detail.

(3) While we have constructed first order atmospheric models and described mechanisms for line formation and fluorescent excitation processes, we have not yet computed models in which the atmospheric structure, detailed heating processes, and line formation are solved self-consistently based on first principles.

(4) While we now know that the atmospheric structures are often inhomogeneous, we do not yet know the scales of these inhomogeneities or the thermal structure and energy balance in the atmospheric structures.

(5) While we now know that cool, luminous stars have both winds and chromospheres, we do not yet know the relationship between winds and chromospheres or why some types of stars have hot material, and very little mass loss, others have massive winds and no apparent hot material and still others appear to have both winds and hot material.

(6) While we are beginning to study how the outer atmospheres of cool stars change with time, we have very far to go in identifying the fundamental time scales or the physical processes responsible for the variability.

The enormous increase in our understanding of stellar chromospheres and transition regions made possible by the IUE has occurred because of the large increase in throughput for high resolution spectroscopy compared to previous space experiments, in particular *Copernicus*. The users of IUE, however, are keenly aware of IUE's limitations and await the enhanced sensitivity, spectral resolution, short wavelength and long slit capabilities, and high time resolution expected in future missions. Both the Hubble Space Telescope (HST) and the proposed LYMAN Far Ultraviolet Spectroscopic Explorer will have capabilities far beyond those of IUE and will thereby provide the data with which many of these new questions will be answered and the next generation of questions posed.

These enhanced capabilities will eventually become available, but for a considerable period of time IUE will remain essential for the study of cool stars because of its wide simultaneous spectral coverage, ability to monitor variable phenomena and compare with an existing 9-yr data set, its known and stable calibration, and because it remains our only means for studying the universe at ultraviolet wavelengths.

Acknowledgements

We wish to thank our colleagues Drs. T. Ayres, C. Ambruster, A. Brown, A. Dupree, P. Judge, and C. Schrijver for their comments on the manuscript. JLL wishes to acknowledge support by NASA through grant NAG5-82 to the University of Colorado.

References

Ambruster, C. W., Kunkel, W. E., Moreno, H., and Basri, G. S.: 1986, in 'Advances in Space Research', (in press).

Anderson, B. N., Kjeldseth-Moe, O., and Pettersen, B.: 1986, in 'New Insights in Astrophysics: Eights Years of UV Astronomy with IUE', ESA SP-263, p. 87.

Athay, R. G.: 1976, *The Solar Chromosphere and Corona: Quiet Sun*, D. Reidel Publ. Co., Dordrecht, Holland.

Athay, R. G.: 1985, *Solar Phys.* **100**, 257.

Ayres, T. R.: 1979, *Astrophys. J.* **228**, 509.

Ayres, T. R.: 1984, *Astrophys. J.* **284**, 784.

Ayres, T. R.: 1987, *Astrophys. J.* (in press).

Ayres, T. R., Engvold, E., Jensen, E., and Linsky, J. L.: 1986a, in 'Cool Stars, Stellar Systems, and the Sun', M. Zeilik and D. M. Gibson (eds.), Berlin, Springer-Verlag, p. 94.

Ayres, T. R., Jensen, E., and Engvold, O.: 1987, *Astrophys. J. Suppl.* (submitted).

Ayres, T. R., Judge, P., Jordan, C., Brown, A., and Linsky, J. L.: 1986b, *Astrophys. J.* **311**, 947.

Ayres, T. R. and Linsky, J. L.: 1975, *Astrophys. J.* **200**, 660.

Ayres, T. R. and Linsky, J. L.: 1980, *Astrophys. J.* **235**, 76.

Ayres, T. R., Linsky, J. L. Rodgers, A. W., and Kurucz, R. L.: 1976, *Astrophys. J.* **210**, 199.

Ayres, T. R., Linsky, J. L., Vaiana, G. S., Golub, L., and Rosner, R.: 1981b, *Astrophys. J.* **250**, 293.

Ayres, T. R., Marstad, N. C., and Linsky, J. L.: 1981a, *Astrophys. J.* **247**, 545.

Ayres, T. R., Moos, H. W. and Linsky, J. L.: 1981c, *Astrophys. J. Letters* **248**, L137.

Ayres, T. R., Simon, T., and Linsky, J. L.: 1982, *Astrophys. J.* **263**, 791.

Ayres, T. R., Stencel, R. E., Linsky, J. L., Simon, T., Jordan C., Brown, A., and Engvold, O.: 1983, *Astrophys. J.* **274**, 801.

Baliunas, S. L. and Dupree, A. K.: 1982, *Astrophys. J.* **252**, 668.

Baliunas, S. L., Guinan, E. F., and Dupree, A. K.: 1984, *Astrophys. J.* **282**, 733.

Baliunas, S. L., Hartmann, L., and Dupree, A. K.: 1983, *Astrophys. J.* **271**, 672.

Baliunas, S. L., Loeser, J. G., Raymond, J. C., Guinan, E. F., and Dorren, J. D.: 1986a, in 'New Insights in Astrophysics: Eights Years of UV Astronomy with IUE', ESA SP-263, p. 185.

Baliunas, L. L., Raymond, J. C., and Loeser, J. G.: 1986b, in 'New Insights in Astrophysics. Eight Years of UV Astronomy with IUE', ESA SP-263, p. 181.

Bartoe, J. D. F., Brueckner, G. E., Sandlin, G. D., Van Hoosier, M. E., and Jordan, C.: 1978, *Astrophys. J. Letters* **223**, L51.

Basri, G. S.: 1986, in M. Zeilik and D. M. Gibson (eds.), 'Cool Stars, Stellar Systems, and the Sun', Berlin: Springer-Verlag, p. 184.

Basri, G. S., Linsky, J. L., and Eriksson, K.: 1981, *Astrophys. J.* **251**, 162.

Belvedere, G.: 1985, *Solar Phys.* **100**, 363.

Bennett, J. O., Ayres, T. R., and Rottman, G. J.: 1984, in 'Future of Ultraviolet Astronomy based on Six Years of IUE Research, NASA, CP 2349, p. 437.

Blanco., C., Catalano, S., and Marilli, E.: 1980, in Proc. Second European IUE Conf., Tubingen, ESA SP-157, p. 63.

Boesgaard, A. M. and Simon, T.: 1984, *Astrophys. J.* **277**, 241.

Böhm-Vitense, E. and Dettmann, T.: 1980, *Astrophys. J.* **236**, 560.

Bromage, G. E., Patchett, B. E., Phillips, K. J. H., Dufton, P. L., and Kingston, A. E.: 1983, in P. B. Byrne and M. R. Rodono (eds.), 'Activity in Red Dwarf Stars', D. Reidel Publ. Co., Dordrecht, Holland, p. 245.

Brosius, J. W. and Mullan, D. J.: 1986, *Astrophys. J.* **301**, 650.

Brown, A.: 1984, in 'Cool Stars, Stellar System, and the Sun', S. L. Baliunas and L. Hartmann (eds.), Berlin, Springer-Verlag, p. 282.

Brown, A.: 1986a, in M. Zeilik and D. M. Gibson (eds.), 'Cool Stars, Stellar Systems, and the Sun', Berlin, Springer-Verlag, p. 454.

Brown, A.: 1986b, in 'Advances in Space Research', (in press).

Brown, A. and Carpenter, K. G.: 1984, *Astrophys. J. Letter* **287**, L43.

Brown, A., Ferraz, M. C. de M., and Jordan, C.: 1981, in 'The Universe of Ultraviolet Wavelengths: The First Two Years of IUE', NASA CP-2171, p. 297.

Brown, A., Ferraz, M. C. de M., and Jordan, C.: 1984b, *Monthly Notices Roy. Astron. Soc.* **207**, 831.

Brown, A. and Jordan, C.: 1980, *Monthly Notices Roy. Astron. Soc.* **191**, 37 P.

Brown, A. and Jordan, C.: 1981, *Monthly Notices Roy. Astron. Soc.* **196**, 757.

Brown, A., Jordan, C., and Wilson, R.: 1979, in 'The First Year of IUE', A. Willis (ed.), UCL, p. 232.

Brown, A., Jordan, C., Stencel, R. E., Linsky, J. L., and Ayres, T. R.: 1984a, *Astrophys. J.* **283**, 731.

Brown, A., Reimers, D., and Linsky, J. L.: 1986, in 'New Insights in Astrophysics: Eight Years of UV Astronomy with IUE', ESA SP-263, p. 169.

Brueckner, G. E.: 1981, in 'Solar Active Regions', F. Q. Orrall (ed.), Boulder, CO: Assoc. Univ. Press, p. 113.

Butler, C. J.: 1986, in *Proc. of RAL Workshop on Astronomy and Astrophysics*, Flares: Solar and Stellar', P. M. Gondhalekar (ed.), RAL-86-085, p. 81.

Butler, C. J., Doyle, J. G., Andrews, A. D., Byrne, P. B., Linsky, J. L., Bornmann, P. L., Rodono, M., Pazani, V., and Simon, T.: 1986, *Astron. Astrophys.* **174**, 139.

Carpenter, K. G.: 1984, *Astrophys. J.* **285**, 181.

Carpenter, K. G. and Wing, R. F.: 1979, *Bull. AAS* **11**, 419.

Catalano, S.: 1986, in *Proc. RAL Workshop on Astronomy and Astrophysics*, Flares: Solar and Stellar', P. M. Gondhalekar (ed.), RAL-86-085, p. 105.

Copeland, H., Jensen, J. O., and Jorgensen, H. E.: 1970, *Astron. Astrophys.* **5**, 12.

de Castro, E., Fernandez-Figueroa, M. J., and Rego, M.: 1982, *Astron. Astrophys.* **113**, 94.

Dere, K. P.: 1982, *Solar Phys.* **77**, 77.

Dorren, F. D., Guinan, E. F., and Wacker, S. W.: 1986, in 'New Insights in Astrophysics: Eight Years of UV Astronomy with IUE', ESA SP-263, p. 201.

Drake, S. A.: 1986, in 'New Insights in Astrophysics: Eight Years of UV Astronomy with IUE', ESA SP-263, p. 193.

Drake, S. A. and Linsky, J. L.: 1983, *Astrophys. J.* **273**, 299.

Dupree, A. K.: 1975, *Astrophys. J. Letters* **200**, L27.

Dupree, A. K., Black, J. H., Davis, R. J., Hartmann, L., and Raymond, J. C.: 1979, in A. Willis (ed.), 'The First Year of IUE', UCL, p. 217.

Dupree, A. K., Sonneborn, G., Baliunas, S. L., Guinan, E. F., Hartmann, L., and Hayes, D. P.: 1984, in 'The Future of UV Astronomy based on Six Years of IUE Research', NASA CP-2349, p. 462.

Eaton, J. A., Johnson, H. R., O'Brien, G. T., and Baumert, J. H.: 1985, *Astrophys. J.* **290**, 276.

Eriksson, K., Linsky, J. L., and Simon T.: 1983, *Astrophys. J.* **272**, 665.

Evans, R. G., Jordan, C., and Wilson R.: 1975, *Monthly Notices Roy. Astron. Soc.* **172**, 585.

Fernanez-Figueroa, M. J., de Castro, E., and Rego, M.: 1981, *Astron. Astrophys.* **99**, 141.

Freire Ferrero, R.: 1986, *Astron. Astrophys.* **159**, 209.

Gahm, G. F.: 1974, *Astron. Astrophys. Suppl.* **18**, 259.

Galloway, D. J. and Weiss, N. O.: 1981, *Astrophys. J.* **243**, 945.

Gilman, P. A.: 1980, in D. F. Gray and J. L. Linsky (eds.), 'Stellar Turbulence', Berlin, Springer-Verlag, p. 19.

Guinan, E. F., Wacker, S. W., Baliunas, S. L., Loeser, J. G., and Raymond, J. C.: 1986, in 'New Insights in Astrophysics: Eight Years of UV Astronomy with IUE', ESA SP-263, p. 197.

Gray, D. F.: 1982, *Astrophys. J.* **262**, 682.

Gray, D. F. and Nagar, P.: 1985, *Astrophys. J.* **298**, 756.

Haisch, B. M., Linsky, J. L., Bornmann, P. L., Stencel, R. E., Antiochos, S. K., Golub, L., and Vaiana, G. S.: 1983, *Astrophys. J.* **267**, 280.

Haisch, B. M., Linsky, J. L., Weinstein, A., and Shine, R. A.: 1977, *Astrophys. J.* **214**, 785.

Haisch, B. M. and Simon, T.: 1982, *Astrophys. J.* **263**, 252.

Hammer, R., Linsky, J. L., and Endler, F.: 1982, in 'Advances in Ultraviolet Astonomy: Four Years of IUE Research', NASA CP 2238, p. 268.

Hartmann, L. and Avrett, E. A.: 1984, *Astrophys. J.* **284**, 238.

Hartmann, L., Baliunas, L. S., Duncan, D. K., and Noyes, R. W.: 1984, *Astrophys. J.* **279**, 778.

Hartmann, L., Dupree, A. K., and Raymond, J. C.: 1980, *Astrophys. J. Letters* **236**, L143.

Hartmann, L., Dupree, A. K., and Raymond, J. C.: 1981, *Astrophys. J.* **246**, 193.

Hartmann, L., Dupree, A. K., and Raymond, J. C.: 1982, *Astrophys. J.* **252**, 214.

Hartmann, L., Jordan, C., Brown, A., and Dupree, A. K.: 1985, *Astrophys. J.* **296**, 576.

Jamar, C., Macon-Hercot, P., and Praderie, F.: 1976, *Astron. Astrophys.* **52**, 373.

Johansson, S. and Jordan, C.: 1984, *Monthly Notices Roy. Astron. Soc.* **210**, 239.

Johnson, H. R., Luttermoser, D. G., Baumert, J. H., Querci, F., and Querci, M.: 1986, in 'New Insights in Astrophysics: Eight Years of UV Astronomy with IUE', ESA SP-263, p. 149.

Jordan, C.: 1967, *Sol. Phys.* **2**, 441.

Jordan, C.: 1986a, in 'New Insights in Astrophysics: Eight Years of UV Astronomy with IUE', ESA SP-263, p. 17.

Jordan, C.: 1986b, *Irish Astron. J.* **17**, 227.

Jordan, C.: 1987, in M. Friedjung, A. Vittone, and R. Viotti (eds.), *Proc. IAU Colloq.* **94**, Capri, 1986, (in press).

Jordan, C., Brown, A., Walker, F. M., and Linsky, J. L.: 1986, *Monthly Notices Roy. Astron. Soc.* **218**, 465.

Jordan, C., Ayres, T. R., Brown, A., Linsky, J. L., and Simon T.: 1987, *Monthly Notices Roy. Astron. Soc.* (in press).

Jordan, C. and Brown A.: 1981, in R. M. Bonnet and A. K. Dupree (eds.), 'Solar Phenomena in Stars and Stellar Systems', NATO ASIC **68**, D. Reidel Publ. Co., Dordrecht, Holland, p. 199.

Jordan, C. and Judge, P. G.: 1984, *Physica Scripta* **T8**, 43.

Jordan, C., Judge, P. G., and Johansson, S.: 1985, in *IAU Colloq.* **86**, G. A. Doschek (ed.), Washington D.C., p. 51.

Jordan, C. and Wilson, R.: 1971, in C. J. Macris (ed.), 'Physics of the Solar Corona', D. Reidel Publ. Co., Dordrecht, Holland, p. 211.

Judge, P. G.: 1986a, *Monthly Notices Roy. Astron. Soc.* **221**, 119.

Judge, P. G.: 1986b, *Monthly Notices Roy. Astron. Soc.* **223**, 239.

Judge, P. G.: 1984, in S. L. Baliunas and L. Hartmann (eds.), 'Cool Stars, Stellar Systems and the Sun', Berlin: Springer-Verlag, p. 353.

Judge, P. G., Jordan, C., and Rowan-Robinson, M.: 1987, *Monthly Notices Roy. Astron. Soc.* **224**, 93.

Kelch, W. L., Linsky, J. L., Basri, G. S., Chiu, H. Y., Chang, S. H., Maran, S. P., and Furenlid, I.: 1978, *Astrophys. J.* **220**, 962.

Kelch, W. L., Linsky, J. L., and Worden, S. P.: 1979, *Astrophys. J.* **229**, 700.

Kondo, Y., Giuli, R. T., Modisette, J. L., and Rydgren, A. E.: 1972, *Astrophys. J.* **176**, 153.

Landsman, W. B., Henry, R. C., Moos, H. W., and Linsky, J. L.: 1984, in 'Local Interstellar Medium', NASA Conf. Publ. 2345, p. 60.

Lennon, D. J., Dufton, P. L., Hibbert, A., and Kingston, A. E.: 1985, *Astrophys. J.* **294**, 200.

Linsky, J. L.: 1980, *Ann. Rev. Astron. Astrophys.* **18**, 439.

Linsky, J. L.: 1983, in J. O. Stenflo (ed), 'Solar and Stellar Magnetic Fields: Origins and Coronal Effects', *IAU Symp.* **102**, 313.

Linsky, J. L.: 1985, *Solar Phys.* **100**, 333.

Linsky, J. L.: 1986, *Irish Astron. J.* **17**, 343.

Linsky, J. L. and Ayres, T. R.: 1978, *Astrophys. J.* **220**, 619.

Linsky, J. L., Bornmann, P. L., Carpenter, K. G., Wing, R. F., Giampapa, M. S., Worden, S. P., and Hege, E. K.: 1982, *Astrophys. J.* **260**, 670.

Linsky, J. L., Brown, A., Marstad, N. C., Rodono, M., Andrews, A. D. Butler, C. J., and Byrne, P. B.: 1984, in *Proc. of the Fourth European IUE Conference*, ESA SP-218, p. 351.

Linsky, J. L. and Haisch, B. M.: 1979, *Astrophys. J. Letters* **229**, L27.

Linsky, J. L., Neff, J. E., Gross, B. D., Simon, T., Andrews, A. D., and Rodono, M. R.: 1986, in 'New Insights in Astrophysics: Eight Years of UV Astronomy with IUE', ESA SP-263, p. 161.

Mangeney, A. and Praderie, F.: 1984, *Astron. Astrophys.* **130**, 143.

McKinney, W. R., Moos, H. W., and Giles, J. W.: 1976, *Astrophys. J.* **205**, 848.

Mendoza-Ortega, B. M.: 1985, DPhil. Thesis, Univ. of Oxford.

Mullan, D. J. and Stencel, R. E.: 1982, 'Advances in Ultraviolet Astronomy: Four Years of IUE Research', NASA CP-2238, p. 235.

Muller, R.: 1985, *Solar Phys.* **100**, 237.

Noyes, R. W., Hartmann, L. W., Baliunas, S. L., Duncan, D. K., and Vaughan, A. H.: 1984, *Astrophys. J.* **279**, 763.

Oranje, B. J.: 1986, *Astron. Astrophys.* **154**, 185.

Oranje, B. J. and Zwaan, C.: 1985, *Astron. Astrophys.* **147**, 265.

Oranje, B. J., Zwaan, C., and Middlekoop, F.: 1982, *Astron. Astrophys.* **110**, 30.

Pettersen, B. R., Hawley, S. L., and Andersen, B. M.: 1986, in 'New Insights in Astrophysics: Eight Years of UV Astronomy with IUE', ESA SP-263, p. 157.

Pottasch, S. R.: 1964, *Space Sci. Rev.* **3**, 816.

Reimers, D.: 1977, *Astron. Astrophys.* **57**, 395.

Reimers, D.: 1982, *Astron. Astrophys.* **107**, 292.

Rodono, M., Byrne, P. B., Neff, J. E., Linsky, J. L., Simon, T., Butler, C. J., Catalano, S., Cutispoto, G., Doyle, J. G., Andrews, A. D., and Gibson, D. M.: 1987, *Astron. Astrophys.* (in press).

Rutten, R. G. M. and Schrijver, C. J.: 1986, in M. Zeilik and D. M. Gibson (eds.), 'Cool Stars, Stellar Systems, and the Sun', Berlin: Springer-Verlag, p. 454.

Schmitt, J. H. M. M., Golub, L., Harnden, Jr., F. R. Maxson, D. W., Rosner, R., and Vaiana, G. S.: 1985, *Astrophys. J.* **290**, 307.

Schrijver, C. J.: 1987, *Astron. Astrophys.* **172**, 111.

Schrijver, C. J. and Rutten, R. G. M.: 1987, *Astron. Astrophys.* (in press).

Schwarzschild, M.: 1975, *Astrophys. J.* **195**, 137.

Simon, T.: 1984, *Astrophys. J.* **279**, 738.

Simon, T.: 1986, in 'New Insights in Astrophysics: Eight Years of UV Astronomy with IUE', ESA SP-263, p. 53.

Simon, T., Herbig, G., and Boesgaard, A. M.: 1985, *Astrophys. J.* **293**, 551.

Simon, T., Kelch, W. L., and Linsky, J. L.: 1980, *Astrophys. J.* **237**, 73.

Simon, T., Linsky, J. L., and Schiffer, F. H. III: 1980, *Astrophys. J.* **239**, 911.

Simon, T., Linsky, J. L., and Stencel, R. E.: 1982, *Astrophys. J.* **257**, 225.

Skumanich, A., Smythe, C., and Frazier, E.: 1975, *Astrophys. J.* **200**, 747.

Stein, R. F.: 1981, *Astrophys. J.* **246**, 966.

Stencel, R. E.: 1978, *Astrophys. J. Letters* **233**, L37.

Stencel, R. E., Linsky, J. L., Ayres, T. R., Jordan, C., and Brown, A.: 1982, 'Advances in UV Astronomy: Four Years of IUE Research', NASA CP-2238, p. 259.

Stencel, R. E., Linsky, J. L., Brown, A., Jordan, C., Carpenter, K. G., Wing, R. F., and Czyzak, S.: 1981, *Monthly Notices Roy. Astron. Soc.* **196**, 47 P.

Stencel, R. E., Mullan, D. J., Linsky, J. L., Basri, G. S., and Worden, S. P.: 1980, *Astrophys. J. Suppl.* **44**, 383.

Stencel, R. E. and Mullan, D. J.: 1980a, *Astrophys. J.* **238**, 221.

Stencel, R. E. and Mullan, D. J.: 1980b, *Astrophys. J.* **240**, 718.

Srickland, D. J. and Sanner, F.: 1981, *Monthly Notices Roy. Astron. Soc.* **197**, 791.

Ulmschneider, P. and Stein, R. F.: 1982, *Astron. Astrophys.* **106**, 9.

Van der Hucht, K. A., Stencel, R. E., Haisch, B. M., and Kondo, Y.: 1979, *Astron. Astrophys. Suppl.* **36**, 377.

Vilhu, O.: 1984, *Astron. Astrophys.* **133**, 117.

Vilhu, O., Neff, J. E., and Walter, F. H.: 1986, in 'New Insights in Astrophysics: Eight Years of UV Astronomy with IUE', ESP SP-263, p. 113.

Vitz, R. C., Weiser, H., Moos, H. W., Weinstein, A., and Worden, E. S.: 1976, *Astrophys. J. Letters* **205**, L35.

Vogt, S. S. and Penrod, G. D.: 1983, *Publ. Astron. Soc. Pacific* **95**, 565.

Walter, F. M.: 1983, *Astrophys. J.* **274**, 794.

Walter, F. M. and Linsky, J. L.: 1986, in 'New Insights in Astrophysics: Eight Years of Astronomy with IUE', ESA SP-263, p. 103.

Walter, F. M., Linsky, J. L., Simon, T., Golub, L., and Vaiana, G. S.: 1984, *Astrophys. J.* **281**, 815.

Walter, F. M., Neff, J. E., Gibson, D. M., Linsky, J. L., Rodono, M., Gary, D. E., and Butler, C. F.: 1986, *Astron. Astrophys.* (submitted).

Weiler, E. J. and Oegerle, W. W.: 1979, *Astrophys. J. Suppl.* **39**, 357.

Wing, R. F., Carpenter, K. G., and Wahlgren, G. M.: 1983, Perkins Obs. Spec. Publ. No. 1, Ohio State University and Ohio Wesleyan University.

Withbroe, G. L. and Noyes, R. W.: 1977, *Ann. Rev. Astron. Astrophys.* **15**, 363.

Wolff, S. C., Boesgaard, A. M., and Simon, T.: 1986, *Astrophys. J.* **310**, 360.

Zolcinski, M. C., Antiochos, S. K., Stern, R. A., and Walker, A. B. C.: 1982, *Astrophys. J.* **258**, 177.

PRE-MAIN SEQUENCE STARS

CATHERINE L. IMHOFF[1,2]

Computer Sciences Corporation, Greenbelt, MD, U.S.A.

and

I. APPENZELLER[1]

Landessternwarte, Königstuhl, Heidelberg, F.R.G.

1. Introduction

The IUE satellite has inaugurated a new era in the study of pre-main sequence (PMS) objects. IUE observations have revealed not just a new wavelength range but also a new temperature regime embracing physical phenomena that could not be studied before. Prior to 1978, virtually all data on PMS objects were obtained in the visual and infrared; thus investigations neccesarily concentrated on the moderate- and low-temperature phenomena associated with dust, the stellar photosphere, lower chromosphere, cool winds, and low-excitation shocks. IUE researchers now study the chromosphere and transition region, hotter temperature winds, and high-temperature shocked gas, while X-ray satellites have added data on coronae and flares. Thus IUE has provided the first important data on the high-temperature processes which are responsible for many of the unique characteristics of the pre-main sequence objects. Without IUE, progress toward understanding this critical phase of stellar evolution would have been severely limited.

Prior to IUE there were a few attempts to detect the faint PMS stars. The Chameleon T-assocation was included in the target list for the Skylab UV objective prism experiment, but no observations were obtained due to lack of time (Henize, private communication). Later de Boer (1977) reported observations of several PMS stars with ANS, but these were at best marginal detections. OAO—2 may have barely detected one of the brighter T Tauri stars, RW Aur (Herbig, private communication).

The PMS stars are of enormous interest because of what they may tell us about stellar evolution, including the roles played by angular momentum, magnetic fields, accretion, and mass loss. Of special interest is the insight they may provide concerning the processes involved in the formation of the Sun and solar system. In addition, the study of stellar physics benefits by the intercomparison of phenomena observed in a variety of stars under a range of conditions.

Before discussing the IUE observations of these objects, the various members of the pre-main sequence menagerie may be briefly noted.

— The T Tauri stars are late-type subgiants with strong H I and Ca II emission

---

[1] Guest Observer with the *International Ultraviolet Explorer* satellite.
[2] Staff member of the *International Ultraviolet Explorer* Observatory, at the Laboratory for Astronomy and Solar Physics, NASA Goddard Space Flight Center.

lines, infrared excesses, and variability. They are believed to have masses in the range of 0.5 to 2 $M_\odot$ and ages of 10^5 to 10^7 yr, thus probably resembling the Sun at an early age. Previous reviews on IUE results for these stars include those by Gahm (1981), Lago *et al.* (1984), Giampapa (1984), Imhoff (1984), and Giampapa and Imhoff (1985). Bertout (1984), Kuhi (1983), and Appenzeller (1986) have recently given more general reviews of these stars. In the discussion below, we refer to 'strong-emission' T Tauri stars and 'weak-emission' T Tauri stars. This refers to the strength of the visual emission lines, primarily the Balmer hydrogen series and Ca II lines.

— The Herbig Ae stars appear to be higher mass, hotter analogues of the T Tauri stars. A recent catalog and study of the stars is given by Finkenzeller and Mundt (1984).

— The FU Orionis stars are a small group of PMS stars that have experienced major nova-like outbursts persisting over years. They are discussed by Herbig (1977) and by Hartmann and Kenyon (1985).

— The 'weak-emission' PMS stars lack the strong emission characteristics of the T Tauri stars. We use this term to encompass both the 'post-T Tauri' stars, which are PMS stars somewhat older than the T Tauri stars, and the 'naked' T Tauri stars, PMS stars with no appreciable circumstellar material (Mundt *et al.*, 1983; Walter, 1984, 1986).

— The Herbig—Haro (HH) objects are a class of shocked nebulae, apparently ejected by embedded infrared PMS stars. They are discussed in general by Schwartz (1983 a, b). The IUE results have been reviewed by Böhm (1983).

In Section 2 we present a catalog of the PMS objects observed by IUE. The T Tauri stars are discussed in Section 3, and the Herbig Ae stars are examined in Section 4. The Herbig—Haro objects are considered in Section 5. Section 6 discusses the astrophysical conclusions and new insights derived from the IUE data. Finally in Section 7 we present the outlook and scientific potential of future UV instruments and programs.

2. A Catalog of PMS Objects Observed by IUE

The first IUE observations of PMS stars were approached rather cautiously, since there was considerable skepticism that these cool, relatively faint stars could be observed in the ultraviolet. This caution proved to be unfounded; in the first few years many spectra were obtained of the brighter T Tauri stars, Herbig Ae stars, and Herbig—Haro objects. During the long lifetime of the IUE satellite, efforts to observe a wider variety of characteristics and fainter objects have extended the list of observations to almost 150 objects! This list is especially impressive when one considers that these objects are generally faint in the ultraviolet and that many of them require 1 to 7 hr exposures.

In Table I we present a catalog, listed by right ascension, of all the PMS objects observed by IUE through early 1986. The catalog information includes coordinates, visual magnitude, spectral type, and so forth, taken almost exclusively from the IUE Merged Log. These entries were provided by the observers and have not

TABLE I

A Catalog of Pre-Main Sequence Objects Observed by IUE

Object Name	RA (1950)	Dec. (1950)	V	Spectral Type	Class	LW		SW		Comments
						LO	HI	LO	HI	
BD + 61°154	0ʰ40ᵐ23.0	+61°38′00	10.6	B8	HAE	5	0	2	0	
H–H 12F	3 25 53.8	+31 09 28			HH	1	0	1	0	
SSS–107	3 25 52.5	+31 07 45				1	1	0	0	
HD 283447	4 11 07.2	+28 04 41	10.7	K2 IV		1	0	0	0	
V410 Tau	4 15 24.8	+28 20 02	10.9	K3 V	TT	5	0	1	0	
BP Tau	4 16 08.5	+28 59 16	12.1	K7 IV	TT	15	2	2	0	
RY Tau	4 18 50.8	+28 19 34	10.8	G5 IV	TT	26	4	8	0	HD 283571
DE Tau	4 18 51.2	+27 48 16	12.9	M1 IV	TT	3	0	2	0	
HD 283572	4 18 52.5	+28 11 07	9.1	G0 IV	WE	5	0	0	0	
T Tau	4 19 04.2	+19 25 05	10.4	K1 V	TT	28	4	6	3	
DF Tau	4 23 59.6	+25 35 43	11.7	M0 IV	TT	8	0	0	0	
DG Tau	4 24 00.0	+25 59 36	12.7	G5 IV	TT	18	3	3	0	No detection
DI Tau	4 26 38.0	+26 26 20	12.6	M0 IV	TT	1	0	0	0	
UX Tau	4 27 09.8	+18 07 22	11.3	K2 IV	TT	6	0	0	0	
H–H 29	4 28 33.2	+18 00 00			HH	0	0	1	0	No detection
L1551IR/S5	4 28 40.0	+18 01 41			EMB	0	0	1	0	
H–H 30	4 28 43.6	+18 06 03			HH	1	0	4	0	No detection
HL Tau	4 28 44.3	+18 07 35	13.5		TT	2	0	1	0	
X0429 + 179	4 29 21.0	+17 55 24	11.6	K7 IV	WE	1	0	0	0	
X0429 + 182	4 29 22.9	+18 13 53	11.8	K7 IV	WE	2	0	0	0	
GG Tau	4 29 37.0	+17 25 22	12.4	K8 IV	TT	2	0	0	0	
X0430 + 245	4 30 11.0	+24 27 58	12.0	K7 IV	WE	2	0	0	0	
DL Tau	4 30 35.7	+25 14 51	12.2	G5 IV	TT	2	0	0	0	
HN Tau	4 30 50.4	+26 07 14	13.4	K5 IV	TT	1	0	0	0	
AA Tau	4 31 54.1	+24 23 16	12.6	K7 IV	TT	3	0	1	0	
DN Tau	4 32 25.5	+24 08 52	12.5	M0 IV	TT	6	0	0	0	No detection

Number of Spectra

Table I (continued)

Object Name	RA (1950)	Dec. (1950)	V	Spectral Type	Class	LW		SW		Comments
						LO	HI	LO	HI	
DR Tau	4 44 14.0	+16 53 00	11.5	K5 IV	TT	19	0	5	0	
DS Tau	4 44 38.9	+29 19 56	12.3	K3 IV	TT	2	0	0	0	
GM Aur	4 52 00.1	+30 17 11	12.0	K7 IV	TT	4	1	3	0	
AB Aur	4 52 34.1	+30 28 22	7.2	B9	HAE	12	61	54	10	HD 31293
SU Aur	4 52 48.1	+30 29 20	9.2	G2 III	TT	21	4	7	0	HD 282624
RW Aur	5 04 37.7	+30 20 14	10.8	K1 IV	TT	18	3	6	0	
CO Ori	5 24 20.8	+11 23 12	10.6	G5 IV	TT	3	1	0	0	
GW Ori	5 26 20.8	+11 49 53	9.7	K3 IV	TT	11	2	3	0	
V649 Ori	5 26 35.5	+11 49 45	13.2	K4 IV	TT	1	0	0	0	
HK Ori	5 28 39.9	+12 06 54	11.9		HAE	1	0	1	0	
AN Ori	5 30 47.4	−5 32 07	11.4		TT?	1	0	1	0	
P 1404	5 31 49.3	−5 38 44	11.7		TT?	1	0	1	0	
IU Ori	5 32 08.0	−5 44 00	9.2		TT?	1	0	0	0	
BD + 9°880	5 32 24.0	+10 00 28	10.0		HAE	0	0	1	0	
H−HI	5 32 44.0	−5 24 00			HH?	2	0	3	0	M42
V360 Ori	5 33 03.9	−5 11 20	12.5	K6 IV	TT	1	0	0	0	
T Ori	5 33 22.0	−5 30 21	10.4	A3 V	HAE	4	0	1	0	
BN Ori	5 33 47.8	+6 48 12	9.9		HAE?	2	0	2	0	
H−H 1	5 33 54.4	−6 47 01			HH	4	0	6	0	Includes H−H 1F
C−S star	5 33 55.8	−6 47 28	17.0		EMB	0	0	4	0	Cohen-Schwartz
H−H 2	5 33 59.6	−6 49 01			HH	4	0	5	0	
V380 Ori	5 34 00.0	−6 44 26	10.3	A1	HAE	10	0	5	0	NGC 1999
Pl 2441	5 34 23.0	−4 27 00	10.8		TT	1	0	1	0	
BF Ori	5 34 46.9	−6 36 46	9.7	A1	HAE	2	1	2	0	
H−H 43	5 35 45.4	−7 11 04			HH	1	0	5	0	
RR Tau	5 36 23.0	+26 23 32	11.3	K1 IV	HAE	0	0	1	0	
DL Ori	5 39 01.1	−8 07 05	12.9		TT	1	0	0	0	
FU Ori	5 42 38.2	+9 03 02	9.5	F5 II	FU	8	3	3	0	

Table I (continued)

Object Name	RA (1950)	Dec. (1950)	V	Spectral Type	Class	LW LO	LW HI	SW LO	SW HI	Comments
H–H 24A	5 43 35.5	−0 11 32			HH	1	0	3	0	
HD 250550	5 59 06.3	+16 30 58	9.7	B8 V	HAE	4	4	6	1	
LKH α 215	6 29 56.0	+10 11 54	10.8		HAE	1	0	1	0	
HD 259431	6 30 19.3	+10 21 37	8.7	B5	HAE	3	1	3	0	
R Mon	6 36 26.0	+8 47 00	11.7		EMB	1	0	1	0	No detection
Walker 20	6 36 43.4	+9 44 49	10.3	F2 V		1	0	0	0	
VSB20	6 37 20.7	+9 38 37	11.2	F5 V		1	0	1	0	
Walker 43	6 37 36.4	+9 44 40	10.5	A7 V		1	0	0	0	
Walker 46	6 37 39.7	+9 48 58	9.2	A3 V		2	0	1	0	
Walker 79	6 37 56.3	+9 36 49	15.6	K0 V	TT	2	0	1	0	
Walker 84	6 37 57.3	+9 36 29	12.0	G0 V	TT	1	0	0	0	
Walker 90	6 37 59.5	+9 50 53	12.8	B4	HAE	5	0	4	0	LkH α 25
Walker 92	6 38 00.1	+9 52 39	11.7	K		0	0	1	0	
Walker 100	6 38 03.7	+9 54 35	10.0	A0		3	0	2	0	
Walker 108	6 38 06.1	+9 47 37	11.9	F7	TT	1	0	0	0	
Walker 159	6 38 18.9	+9 40 02	10.9	A0 IV		1	0	1	0	
Walker 158	6 38 19.3	+9 57 37	10.4	A7 V		1	0	0	0	
LX Mon	6 38 20.5	+9 51 10	14		TT	1	0	0	0	
Walker 178	6 38 26.7	+9 30 45	7.1	B2 IV		0	1	0	1	
Walker 189	6 38 28.3	+9 30 26	11.2	F2 V		1	0	1	0	
Walker 215	6 38 45.0	+9 52 42	9.3	A0 V		2	0	2	0	
M0 Mon	6 38 46.3	+9 29 53	13.5	K3 IV	TT	1	0	1	0	
Walker 220	6 38 48.5	+9 21 53	9.7	F2 V		1	0	0	0	
HD 52721	6 59 28.6	−11 13 40	6.7		HAE	2	4	2	4	
Z CMa	7 01 23.0	−11 29 00	9.8	B8	HAE	4	1	3	0	
HD 53367	7 02 03.6	−10 22 44	7.0	B0 IV	HAE	7	1	5	1	HD 53179
CoD − 44°3318	7 17 56.0	−44 30 00	10.0		HAE	2	0	2	0	
H–H 46	8 24 16.9	−50 50 34	10.0		HH	0	0	3	0	

Table I (continued)

Object Name	RA (1950)	Dec. (1950)	V	Spectral Type	Class	Number of Spectra				Comments
						LW		SW		
						LO	HI	LO	HI	
H–H 47	8 24 22.8	−50 50 00			HH	1	0	2	0	
HD 76534	8 53 20.6	−43 16 29	8.1		HAE	0	1	0	0	
SY Cha	10 55 18.4	−76 55 34	12.8	M0	TT	1	0	0	0	No SWP detection
LH α 332−20	10 57 50.7	−76 45 33	11.1	K2	TT	4	0	3	0	CoD − 34°7151
TW Hya	10 59 30.3	−34 26 10	10.4	K7	TT	9	3	2	3	
HM 7	11 01 07.7	−77 17 25	11.5	M0	TT	2	0	1	0	
HM 9	11 02 41.8	−76 11 00	12.6	K7	TT	1	0	0	0	
Sz 19	11 05 58.0	−77 22 02	10.9		TT	3	0	1	0	
VW Cha	11 06 38.1	−77 26 12	12.8	K2	TT	2	0	1	0	CPD − 76°486
HD 97048	11 06 40.0	−72 23 00	8.5	A0	HAE	6	1	5	2	
VZ Cha	11 07 51.9	−76 07 02	13.4	K6	TT	1	0	1	0	
Sz 41	11 10 50.1	−76 20 45	11.2	K0	TT?	1	0	0	0	LH α 332−21
Sz 42	11 10 53.7	−76 28 01	10.9	K0	TT	5	2	3	0	
X-ray 22	11 11 03.5	−77 06 13	10.5	K2	TT?	2	0	0	0	
HD 142560	15 23 24.0	−37 40 58	11.1		TT	2	0	1	0	
Sz 68	15 41 59.0	−33 58 12	10.5	G	TT	1	0	0	0	CoD − 33°10685
Sz 71	15 43 32.6	−34 21 17	13.9		TT	1	0	0	0	LH α 450-6
Sz 75	15 45 15.0	−35 30 00	12.5		TT	1	0	1	0	CoD − 35°10525
Sz 77	15 48 32.3	−35 47 43	12.5	M0	TT	1	0	0	0	
Th 12	15 52 51.0	−37 47 20	12.2	M0 IV	TT	2	0	0	0	
RU Lup	15 53 24.0	−37 41 00	11.0	G	TT	18	6	15	2	
Sz 126	15 53 57.2	−42 31 24	14.1	M IV	TT?	1	0	0	0	
He 1125	15 55 50.5	−41 48 40	13.6	K7 IV	TT	1	0	0	0	Sz 129
RY Lup	15 56 05.0	−40 14 00	10.7		TT	2	0	2	0	
EX Lup	15 59 42.0	−40 10 00	13.5		TT	1	0	1	0	
Th 18	16 03 39.4	−38 54 19	13.6		TT	1	0	1	0	Sz 88
Sz 96	16 04 51.0	−39 00 32	13.7	M1 IV	TT	1	0	0	0	
Sz 98	16 05 00.7	−38 56 48	12.4	K7 IV	TT	1	0	0	0	

Table I (continued)

Object Name	RA (1950)	Dec. (1950)	V	Spectral Type	Class	LW		SW		Comments
						LO	HI	LO	HI	
HR 5999	16 05 12.8	−38 58 23	7.3	A7 III	HAE?	15	7	19	3	HD 144668
Sz 108	16 05 21.0	−38 58 20	13.2	M0 IV	TT?	1	0	0	0	
Th 33	16 05 31.6	−39 29 50	14.5	M0 IV	TT	2	0	0	0	Sz 124
Th 43	16 08 31.1	−38 54 32	13.1	M0 IV	TT?	1	0	0	0	
AS 205	16 08 37.0	−18 31 00	12.0		TT	1	0	1	0	
Haro 1−1	16 18 30.7	−26 05 24	13.3	K5 IV	TT?	1	0	0	0	
ROX−3	16 22 47.0	−24 44 11	13.2	M IV	TT?	1	0	0	0	
S−R 4	16 22 54.3	−24 13 58	12.8	K7 IV	TT	1	0	0	0	No detection
Do−Ar 21	16 23 02.0	−24 16 43	14.1			1	0	0	0	
S−R 12	16 24 53.7	−24 35 04	13.3	M2 IV	TT	1	0	0	0	
S−R 9	16 24 38.3	−24 15 21	11.3	K7 IV	TT	1	0	0	0	
S−R 10	16 24 53.7	−24 19 40	14.1	M2 IV	TT	1	0	0	0	
S−R 13	16 25 43.2	−24 21 43	13.1	M2 IV	TT	1	0	0	0	
Do−Ar 44	16 28 31.3	−24 21 18	12.6	K3 IV	TT?	1	0	0	0	
H−H 57	16 28 56.7	−44 49 07	16.0		EMB	1	0	1	0	No detection
Do−Ar 51	16 29 09.5	−24 33 59	13.6	M1 IV	TT?	1	0	0	0	
AS 209	16 46 25.6	−14 18 21	11.5	K2 IV	TT	2	0	0	0	V1121 Oph
AK Sco	16 51 23.0	−36 48 00	9.0		WE	2	0	1	0	
HD 163296	17 53 20.6	−21 56 57	6.8	A2	HAE	6	2	7	1	
FK Ser	18 17 37.6	−10 12 48	10.6	K5 IV	WE	2	0	1	0	BD − 10°4662
S CrA	18 57 48.0	−37 01 00	11.5		TT	1	1	1	0	
TY CrA	18 58 18.0	−36 57 00	9.5		HAE	2	0	2	0	
R CrA	18 58 32.5	−37 02 08	10	A	HAE	2	0	2	0	−37°13024
H−H 32	19 18 07.9	+10 56 21			HH	1	0	2	0	Includes H−H 32A
AS 353A	19 18 09.3	+10 56 15	12.5		TT	0	0	1	0	
HD 190073	20 00 34.4	+5 35 50	7.8	A0	HAE?	2	0	1	0	
BD + 40°4124	20 18 42.0	+41 12 00	10.6	B2	HAE	2	0	1	0	

Table I (continued)

Object Name	RA (1950)	Dec. (1950)	V	Spectral Type	Class	Number of Spectra LW LO	LW HI	SW LO	SW HI	Comments
V1057 Cyg	20 57 06.3	+44 03 49	12.0	G2 Ib	FU	1	0	0	0	
V1331 Cyg	20 59 31.0	+50 09 00	12.0		TT?	3	0	0	0	LkH α 120
HD 200775	21 00 59.0	+67 58 00	7.4		HAE	4	2	5	3	
BD + 65°1637	21 41 42.1	+65 52 51	10	B2e	HAE	2	0	1	0	
BD + 46°3471	21 50 39.0	+46 59 00	10.1	B9	HAE	2	2	3	0	
DI Cep	22 54 06.0	+58 24 00	11.3		TT	2	0	4	0	

Description of Table I

Column 1: Object name, from IUE Merged Log.

Column 2: Right ascension (epoch 1950), from IUE Merged Log.

Column 3: Declination (epoch 1950), from IUE Merged Log.

Column 4: Typical visual magnitude, from IUE Merged Log. Note that most PMS objects are variable.

Column 5: Spectral type, from IUE Merged Log.

Column 6: Class of PMS object, as described in Section I. TT = T Tauri star (from Herbig and Rao 1972, Appenzeller, Jankovics, and Krautter 1983); WE = weak emission PMS star; HAE = Herbig Ae star (from Herbig 1960, Finkenzeller and Mundt 1984); FU = FU Orionis star (Herbig 1977); HH = Herbig—Haro object (from Herbig 1974, Schwartz 1977); and EMB = embedded infrared object.

Columns 7, 8, 9, and 10: Number of IUE spectra of different types obtained of the object, as given by the IUE Merged Log as of mid 1986. LW = long-wavelength (2000—3200 Å), SW = short wavelength (1200—2000 Å), LO = low resolution (7 Å), and HI = high resolution (0.2 Å).

Column 11: Comments, such as alternative names, lack of IUE detection.

been checked for accuracy or consistency. Where possible the objects have been placed into one of the previously mentioned categories of PMS objects.

In the first eight years of the IUE mission, a great diversity of PMS objects has been observed, including objects located in a number of nearby star-formation regions, such as the Taurus—Auriga, Orion, Monoceros, Chameleon, and Lupus dark cloud regions. Many of the brighter members of the various PMS classes have been extensively observed, for instance RU Lup, RW Aur, and T Tau among the T Tauri stars, HR 5999 and AB Aur among the Herbig Ae stars, and H—H 1 and H—H 2 among the Herbig—Haro objects. Given the diverse characteristics of the objects in each class, observations of the fainter members of the classes are also very important. Another eight years of IUE observations would be welcomed by PMS researchers!

3. The T Tauri Stars

3.1. GENERAL APPEARANCE OF THE SPECTRA

The UV spectra of several T Tauri stars obtained during the first few years of IUE were described by Appenzeller and Wolf (1979), Appenzeller *et al.* (1980), Gahm *et al.* (1979), Gondhalekar *et al.* (1979), and Imhoff and Giampapa (1980). The UV spectra are dominated by emission lines, but a continuum is also seen at longer wavelengths. The strongest emission lines are the Mg II h and k doublet near 2800 A. These lines dominate the long-wavelength IUE spectra, but the lines of Mg I, Fe II, and Cr II may also be seen in emission or absorption. In the 1200—2000 A wavelength range important lines are O I, S I, C II, C III, C IV, Si II, Si III, Si IV, and occasionally He II and N V. The identification of these lines in the T Tauri stars, by comparison with chromospheric spectra of the Sun and other late-type stars, has given support to the idea that the 'T Tauri phenomenon' is primarily due to stellar surface activity. Hydrogen Ly-α emission can sometimes be observed, although it tends to be masked by geocoronal emission and interstellar absorption. Molecular hydrogen was first seen in T Tau and may appear in a few other stars as well; it is probably fluorescent emission excited by Lyman alpha (Brown *et al.*, 1981). IUE spectra for three stars are depicted in Figure 1, showing the basic characteristics of the UV spectra for T Tauri stars of various observational properties.

The continuum emission is much stronger than one might expect for typical G, K, or M stars. The 'UV excess', long recognized in the near-ultraviolet (Herbig, 1962), continues into the far-ultraviolet range. In a few of the T Tauri stars with weak emission characteristics, such as SU Aur, some indication of the photospheric spectrum can also be seen. Figure 2 shows a sequence of long-wavelength IUE spectra illustrating the range of observational characteristics, a normal late G dwarf, a weak-emission PMS star, a weak-emission T Tauri star, and a strong-emission T Tauri star. There is a clear progression from a normal ultraviolet photospheric spectrum to spectra with increasing contributions from the continuous UV excess and emission lines.

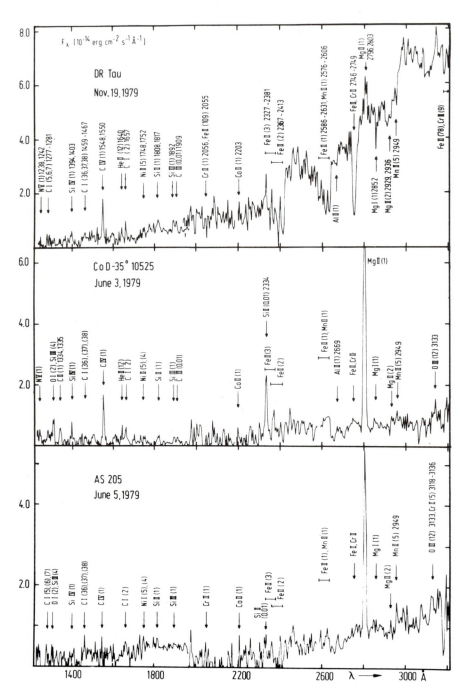

Fig. 1.   IUE spectra for three T Tauri stars. The absorption and emission lines in the ultraviolet spectra are indicated (from Appenzeller *et al.*, 1980). Note the envelope absorption features in the spectrum of DR Tau.

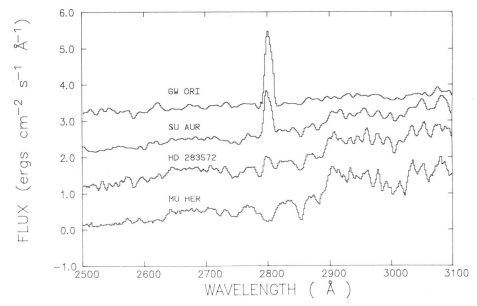

Fig. 2. A sequence of low-dispersion, long-wavelength IUE spectra in the region of the Mg II lines. Observed fluxes are given, in units of 10^{-13} erg cm^{-2} s^{-1} A^{-1} with offsets of 10^{-13} between the spectra. A range of spectral characteristics is depicted for a normal G dwarf (Mu Her), a weak-emission PMS star (HD 283572), a weak-emission T Tauri star (SU Aur), and a moderate-emission T Tauri star (GW Ori). A clear progression from a normal photospheric spectrum to spectra with increasing contributions from the continuous UV excess and emission lines is seen.

3.2. THE Mg II h AND k LINES

The Mg II lines are quite strong in nearly all of the observed T Tauri stars. These lines are the best studied spectral feature of the T Tauri stars in the IUE spectral range. Mg II surface fluxes (i.e. fluxes normalized to unit stellar surface area) have been computed for a number of T Tauri stars (Giampapa *et al.*, 1981; Calvet *et al.*, 1985). These surface fluxes are typically 10^7 to 10^8 erg cm^{-2} s^{-1}, or roughly 50 times larger than for the Sun, and contain about 0.1% of the stellar luminosity. The Mg II surface fluxes are among the highest seen in late-type stars with active chromospheres, including the active chromosphere dwarfs and the RS CVn binaries.

The Mg II and Ca II surface fluxes have been compared for a number of stars (Giampapa *et al.*, 1981; Calvet *et al.*, 1985). Among the lower-mass T Tauri stars, the Mg II and Ca II fluxes scale proportionally; that is, the stars with high Mg II surface fluxes also have high Ca II fluxes and vice versa. The relationship follows the extrapolation of the relation for the active chromopheric late-type dwarfs. Thus for these stars, the Mg II and Ca II lines arise in the chromosphere (albeit a very active one). In the higher mass T Tauri stars (in which a high level of mass loss occurs), a different relationship occurs between the lines. The Mg II fluxes are uniformly large and nearly unrelated to the Ca II fluxes. In addition, Ca II shows more variability for a given star than Mg II. Thus in the higher mass stars, the

Mg II flux is probably dominated by the contribution from the stellar wind and envelope, while Ca II is still primarily chromospheric. These same stars generally have strong emission characteristics and other spectral signatures of strong mass loss.

Many of the T Tauri stars with the strongest Mg II emission have been observed in the high resolution mode using multi-hour exposures. Good line profiles are available for several stars (Appenzeller *et al.*, 1981; Brown *et al.*, 1984a; Giampapa and Imhoff, 1985; Imhoff and Giampapa, 1981; Jordan *et al.*, 1982; Penston and Lago, 1983). The Mg II lines (see Figures 3 and 4) are generally quite broad, with

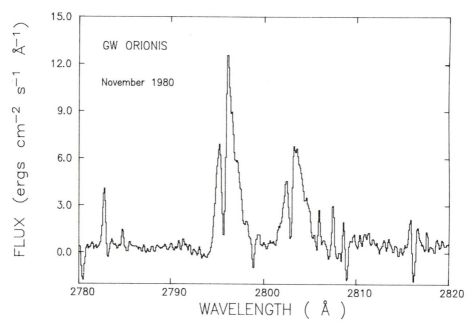

Fig. 3.   Mg II h and k line profiles for the T Tauri star GW Ori. Observed fluxes are given, in units of 10^{-13} erg cm^{-2} s^{-1} A^{-1}. The lines are broad, and each emission component is cut by a narrow absorption component. The other small, narrow spikes are noise.

widths of several hundred km sec^{-1}. The profiles are usually asymmetric, with the blueward absorption characteristic of mass loss. A relatively sharp interstellar absorption component is usually seen. In most stars, the h and k lines are of equal or nearly equal intensity, indicating that the lines are optically thick. A few stars have been observed more than once. Some degree of variability is usually seen (Figure 4), primarily in the blueward absorption component. The underlying emission-line profile does not seem to vary, even when the overall strength of the line varies.

3.3. The Far-Ultraviolet Emission Lines

The far-UV (1200—2000 Å) emission lines can be studied in the T Tauri stars because of their extreme intrinsic strength. Surface fluxes of the stronger lines have

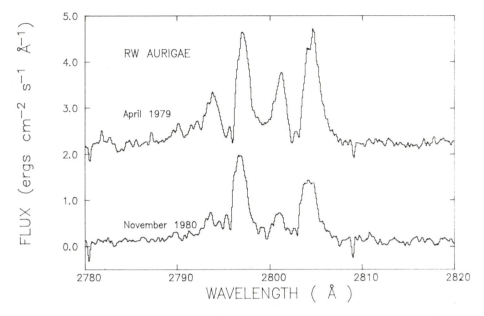

Fig. 4. Mg II h and k line profiles for the T Tauri star RW Aur at two epochs. Observed fluxes are given, in units of 10^{-12} erg cm^{-2} s^{-1} A^{-1}. The spectra are offset by 2×10^{-12}. The later spectrum shows reduced emission line strength, as seen from the redward portion of the line profile, plus increased absorption on the blueward side.

been computed for several stars (Imhoff and Giampapa, 1982). They are typically 10^6 to 10^7 erg cm^{-2} s^{-1}, which correspond to several thousand times the surface fluxes for the Sun. These surface fluxes equal or exceed those for the most active late-type stars, and together they contribute about 0.1 to 0.2% of the stellar luminosity.

It was quickly recognized that the relative strengths of the lines do not follow precisely the behavior seen in other active chromospheric stars (Imhoff and Giampapa, 1980, 1981). In many PMS stars, the lines corresponding to the highest temperatures are weak or even absent. For example N V, which corresponds to the highest temperature regime of about 2×10^5 K, is present in only a handful of PMS stars (Imhoff and Giampapa, 1982; Lago *et al.*, 1984). He II appears in only a few stars, although its origin is a complex issue (Brown *et al.*, 1984b). Since the high-temperature emission lines are normally not strong lines, one might consider whether the noise in the spectra might account for the apparent absence of these lines. This argument, however, cannot explain the relative weakness of C IV which is normally the strongest emission line in the short-wavelength spectral range for late-type stars. Using the ratio of the fluxes for C IV to Si IV as an indicator, the T Tauri stars show a range of values from those typical for active chromospheres (2 to 3) to values less than unity (Imhoff and Giampapa, 1982). In normal main-sequence stars, this ratio is rarely less than 1.5 and never less than 1 (cf. Ayres *et al.*, 1981). The stars with weakened high-temperature lines also tend to be those with strong emission characteristics and evidence for strong winds. Thus it is thought that mass loss plays a role in carrying off the energy that would have normally gone into heating the gas. This effect may be seen in Figures 5 and 6. GW Ori is a moderate-emission T Tauri star with strong C IV emission, while RW Aur is a strong-emission T Tauri star with weak C IV emission.

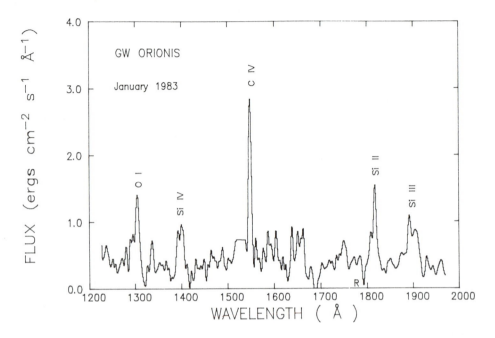

Fig. 5. The far-UV spectrum of GW Ori, a moderate-emission T Tauri star. Observed fluxes are given, in units of 10^{-14} erg cm^{-2} s^{-1} A^{-1}. The strongest emission lines are indicated. *R* denotes a camera reseau mark.

The far-UV emission-line fluxes have been converted to emission measures to permit analysis and modelling of the stellar atmospheres (Brown and Jordan, 1983; Brown *et al.*, 1984a; Cram *et al.*, 1980, Jordan *et al.*, 1982; Lago *et al.*, 1985). These studies confirm the relative weakness of the higher temperature lines in many T Tauri stars. This result is further strengthened when the X-ray fluxes obtained with the *Einstein* satellite are included (e.g. Feigelson and DeCampli, 1981; Gahm, 1980; Walter and Kuhi, 1981). Some of the strong-emission T Tauri stars apparently show no evidence for high-temperature coronal gas. The consistency of these results argues against the hypothesis that the missing X-ray flux has been absorbed in the gaseous envelopes of these stars (Walter and Kuhi, 1981).

Due to the long exposures required, there are only limited data on variability. In RW Aur, the higher temperature lines are the most variable, becoming much stronger when the star brightens (Imhoff and Giampapa, 1981, 1983), at least over yearly time-scales. This behavior resembles the changes seen during flares of other late-type stars, thus it might indicate different levels of flaring activity. Some other

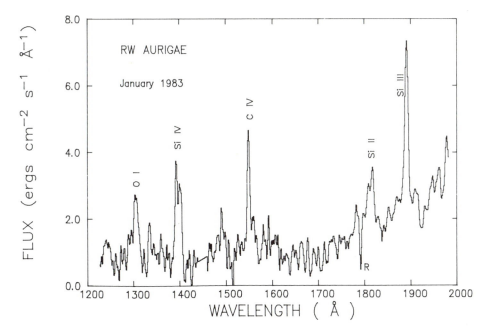

Fig. 6. The far-UV spectrum of RW Aur, a strong-emission T Tauri star. Observed fluxes are given, in units of 10^{-14} erg cm^{-2} s^{-1} Å^{-1}. The same emission lines as in Figure 5 are indicated. Note the relative weakness of C IV.

studies of variability have not yet been fully analyzed and completed (e.g. Walter and Brown, 1983).

Correlations of the far-UV emission-line behavior with quantities such as stellar rotation and X-ray luminosity are of interest in helping to explain the extreme nature of the T Tauri chromospheres. Thus far uncertainties in evolutionary status of individual stars, lack of rotational period determinations, and lack of simultaneity with X-ray observations have limited progress. The situation is much better when observations of main-sequence stars of various ages and of the T Tauri stars are combined to study the long time-scale effects (Simon *et al.*, 1985).

For most stars, the far-UV low-dispersion spectra described above require in excess of 3 hr integrations. Nevertheless, some high dispersion observations have been attempted and have yielded useful data. The line profiles and widths of the strongest far-UV emission lines have been described for RU Lup (Penston and Lago 1982, Brown *et al.*, 1984b) and TW Hya (Lago *et al.*, 1984). The transition region lines C IV and Si IV are narrower than the chromospheric lines, and some blueward extension of the emission is seen. These line profiles appear to yield kinematic information about the transition region and wind, but the analysis has not yet been completed.

3.4. THE ULTRAVIOLET CONTINUUM

The UV continuum can be understood as partly photospheric emission and partly an extension of the UV excess (originally seen near 3500 Å) into the far-ultraviolet. The UV excess is thought to be due to hydrogen continuum emission originating in the dense chromosphere (Kuhi, 1966, 1974; Calvet *et al.*, 1984). Some simple models of black body plus hydrogen free-free and free-bound emission have been fitted to continuum points in several T Tauri stars (Lago *et al.*, 1984). The fits reproduce the continuum reasonably well, yielding electron temperatures of 10 000 to 50 000 K, with typical values near 30 000 K. Fits using spectra from standard stars and hydrogen emission have also been used to reproduce the UV continuum, with similar success (Herbig and Goodrich, 1986).

4. The Herbig Ae Stars

Two Herbig Ae stars, AB Aurigae and HR 5999, have been studied in some detail with IUE; thus most of the discussion below is based on the results for these two stars. AB Aur is a Herbig Ae star of moderate characteristics. Since it is bright and relatively unreddened, it often serves as a prototype for the class. HR 5999 is a unique star. It has the observational characteristics of a Herbig Ae star and is located in the Lupus dark cloud; therefore it is thought to be a pre-main sequence object. However it also exhibits episodic veiling by dust as occurs in the helium-weak R CrB variables (Hecht *et al.*, 1984). The role of HR 5999 in pre-main sequence evolution is thus uncertain.

High resolution Mg II line profiles have been obtained for AB Aur, HD 250550, and BD + 46°3471 (Talavera *et al.*, 1982; Praderie *et al.*, 1982; Catala *et al.*, 1986a). The stellar spectra exhibit remarkable line profiles, characterized by broad, deep blueward absorption and some redward emission. The absorption reaches near zero residual intensity, and the blueward edge extends to 300 to 600 km sec^{-1} blueward of the rest velocity of the line. The emission is redshifted and asymmetric, with a wing extending redward. Narrow absorption components are probably interstellar in origin. These characteristics may be seen in Figure 7, which shows the Mg II h and k line profiles for AB Aur in July 1978.

Semiempirical modelling of the Mg II lines has demonstrated that both chromospheric and wind contributions are required to explain the line profiles (Catala *et al.*, 1984; Catala, 1984). The chromosphere must be relatively dense and extends out to 1 to 2.5 stellar radii. The wind must also extend to perhaps 50 stellar radii. The estimated mass loss rate for AB Aur is 4×10^{-11} to 7×10^{-9} $M_\odot$ yr^{-1}.

The Mg II lines show significant variability in AB Aur (Praderie *et al.*, 1985, 1986). Two sets of observations obtained over periods of several days each revealed variations of the Mg II wind velocity of 30 to 50%. The variations were found to occur with a period of about 45 hr, apparently related to the stellar rotational period. However similar variability of the Ca II H and K lines was found to follow a period of about 32 hr over the same period of observation (Catala *et al.*, 1986b)! The Fe II absorption lines showed long-term changes but did not vary with similar periodicity. A new series of observations was obtained in high resolution for the C IV lines and the Mg II lines for AB Aur (Catala *et al.*, 1986). The C IV lines showed broad, wind-dominated line profiles. The blueward edge of the absorption was seen to vary over several days in both the C IV lines and the Mg II lines. However the observations were too few to allow the determination of a period.

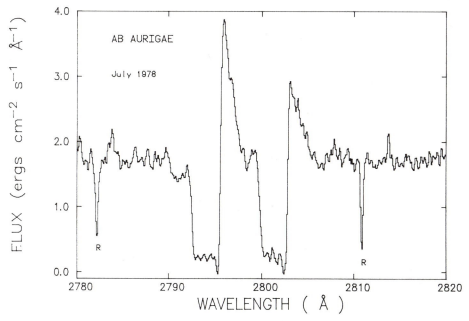

Fig. 7. The Mg II h and k line profiles for the Herbig Ae star AB Aurigae. Observed fluxes are given, in units of 10^{-12} erg cm^{-2} s^{-1} A^{-1}. R denotes a reseau mark.

The observers suggested an explanation for the unusual periodic behavior of the Mg II, Ca II, and C IV lines (Catala *et al.*, 1986b; Praderie *et al.*, 1986). The Ca II lines are expected to originate in the chromosphere very near the stellar surface, so their variations would follow closely the stellar rotational period. The Mg II and C IV lines may be formed in an outer atmosphere which is organized into fast and slow streams analogous to those seen in the Sun, thus originating in a region which may be rotating around the star at a speed less than that of the stellar surface. The Fe II lines are most likely formed at some distance from the stellar surface, where the streams have merged and there is no appreciable rotational velocity.

The occurence of the chromospheric and wind phenomena in the Herbig Ae stars clearly divides them from main sequence A stars, which normally do not possess appreciable chromospheres. The surface activity seen in these stars is apparently due to their pre-main sequence nature, but exactly how this arises has not yet been addressed.

The ultraviolet spectrum of HR 5999 is significantly different from that of AB Aur. The Mg II lines each show a broad emission line with deep central absorption (Tjin A Djie *et al.*, 1982). There is some indication of additional absorption blueward of the emission component. The lines lack the broad, nearly saturated absorption component seen in other Herbig Ae stars; instead they are more reminiscent of the line profiles seen in the T Tauri stars. The far-UV spectrum is notable in revealing strong emission lines similar to those seen in the T Tauri stars (Tjin A Djie and The, 1982; Tjin A Djie *et al.*, 1982). Emission lines of O I, C II, C IV, and Si IV are prominent, even in low resolution. The surface fluxes for these lines are quite large, and comparable to the T Tauri stars.

It is tempting to explain the high degree of chromospheric emission in HR 5999 as due to some unique (though as yet unknown) quality of the star. It should be noted, however, that HR 5999 is cooler (spectral type A7) and more rapidly rotating ($v \sin i = 180$ km sec^{-1}) than the other Herbig Ae stars studied so far. At present the reason for the differences between HR 5999 and the other Herbig Ae stars is unclear.

5. Herbig—Haro Objects

Although the Herbig—Haro (H—H) objects are probably shocked interstellar matter, it is appropriate to comment on these highly interesting objects since it is well established that H—H objects result from the interaction of PMS stellar winds with ambient dark cloud material. A detailed description of our present knowledge about the H—H objects can be found in several excellent recent reviews (Schwartz, 1983a, b, Böhm, 1983; Jones, 1983; Canto, 1983; Lada, 1985). As described in these reviews, H—H objects are faint, low-excitation emission nebulae with only very little visual continuum emission. These objects were not expected to be suitable IUE targets. It was a considerable surprise when Ortolani and D'Odorico (1980) in their first IUE observations of H—H 1 found not only high-excitation UV emission lines but also a relatively strong 1200—2000 Å continuum. This basic character of the far-UV spectrum of most H—H objects was confirmed by subsequent observations of other H—H objects by Böhm *et al.* (1981), Böhm-Vitense *et al.*, 1982; Brugel *et al.* (1982), Schwartz (1983c) and Böhm and Böhm-Vitense (1984). However, as noted by Schwartz (1983c), the far-UV line spectra of H—H objects differ among each other depending on the excitation character of the visual spectrum. In the case of 'high-excitation' visual spectra, the far-UV line spectrum is dominated by high-excitation atomic or ionic lines. Low-excitation H—H objects like H—H 43 and H—H 47 have far-UV spectra that contain only few atomic lines and are dominated by emission lines of the Lyman band of H_2. Figure 8 depicts the UV spectrum of H—H 2, a high-excitation Herbig—Haro object.

Another unexpected property of the far-UV spectra of the H—H objects is the occurrence of large amplitude and rapid time variations of the far-UV emission lines (Brugel *et al.*, 1985). These variations, which are not observed in the visual spectral range, provide important constraints on the dimensions of the emitting volumes and on other parameters of the theoretical models of the H—H objects.

The unexpectedly strong far-UV continuum emission and its spectral-energy distribution are usually interpreted as collisionally enhanced two-photon emission of hydrogen (cf. Dopita and Schwartz, 1981; Dopita *et al.*, 1982; Brugel *et al.*, 1982). Interstellar reddening and interstellar scattered light make a precise measurement of the far-UV energy distribution of the H—H continua difficult. For this reason the true nature of the emission mechanism remains somewhat uncertain.

6. The Impact of IUE on our Physical Understanding of PMS Stars

During the past decades many basic facts about PMS stars and PMS stellar

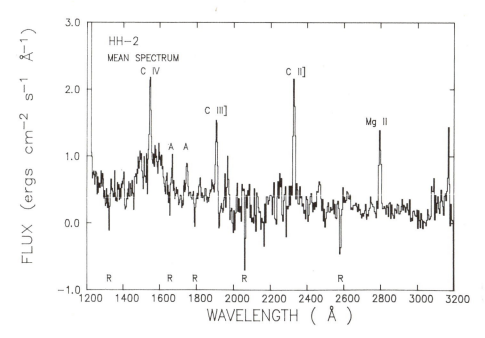

Fig. 8. The UV spectrum of H—H 2, a high-excitation Herbig—Haro object. Seven spectra were coadded to produce a spectrum with improved signal-to-noise (IUE images SWP 10218, 10246, 16671, and 19685 and LWR 8888, 8909, and 10450). Several reseau marks (*R*) and semi-permanent camera artifacts (A) are noted. The strongest lines (C IV, C III], C II], and Mg II) are indicated.

evolution have been clarified by means of new observations in many different parts of the electromagnetic spectrum and by continuously improving theoretical models. However, several fundamental questions and many quantitative details of the PMS stellar evolution still remain unknown or poorly understood. One of the most serious unsolved problems is the determination of accurate ages and evolutionary status of individual PMS stars and young associations. For main-sequence and post-main sequence stars the surface properties (luminosity, color, spectrum, or location in the HR diagram) are normally almost unique functions of the stellar mass and evolutionary age. In the case of PMS stars the spectral properties are in addition influenced by the stellar surface activity and the local environment. In fact, model computations indicate that at least in the very young PMS stars the activity may have a stronger effect on the spectral properties than differences in age or mass. Moreover, the surface activity (by changing the luminosity) may also strongly effect the evolutionary time scale of PMS stars, which makes any comparison of the observations with theoretical evolutionary tracks rather uncertain. For these reasons, a better understanding of the surface activity of PMS stars is a prerequisite for a better understanding of PMS stellar evolution. Fortunately, it is this field where the most important new results can be derived from the IUE observations. The most conspicuous evidence for surface activity consists of the chromospheric line and continuum emission, irregular and quasiperiodic (rotational) light variations, flare activity, and dense and massive stellar winds. As pointed out in the reviews cited above, there is indirect but reliable evidence that at least part of the T Tauri activity is caused by extended surface magnetic fields which are presumably enhanced by the deep convection zones and differential rotation expected for a PMS star. Heating of the outer atmospheric layers by hydromagnetic waves originating in the convective zones and the large extent of the magnetic surface regions provide a plausible explanation for the great strength of the chromospheric emission in the PMS stars. Also, in those PMS stars where spectroscopic evidence for supersonic mass infall is observed (the "YY Orionis" stars), large scale or small scale shock heating will probably also contribute to or even dominate the energy balance of the chromospheres. In both cases we expect to observe dense and moderately hot (10^4-10^5 K) plasma, although the details of the temperature and density stratification will depend on the heating mechanism.

In his poineering paper on the T Tauri stars, Joy (1945) noted the great similarity of the T Tauri emission-line spectra to the visual spectrum of the solar chromosphere. Later Herbig (1970) postulated that T Tauri stars should have chromospheres with a basic structure like the solar chromosphere but with densities and optical depths higher by typical factors of 10 to 100. Consequently, even before the IUE data became available, various authors had begun to analyze visual spectroscopic observations with the methods developed to model the solar chromosphere in order to derive hydrostatic models of T Tauri chromospheres (e.g. Dumont *et al.*, 1973; Cram, 1979; Heidman and Thomas, 1980; Calvet, 1980). These models were able to reproduce many but not all features of the emission spectrum of the T Tauri stars reasonably well. The visual emission spectra show, in addition to the Balmer lines, only lines of relatively uniform low excitation potential. In the strong emission-line T Tauri stars the Balmer lines are produced not only in the chromosphere but in an extended envelope, complicated by velocity fields and geometric structure. Therefore, these models yielded little information on the detailed structure and stratification of the T Tauri chromospheres.

Only after the launch of the IUE satellite did it become possible to observe chromospheric emission lines with a great range of ionization and excitation potential. These data made possible a new generation of considerably more reliable and detailed chromospheric models for T Tauri stars (Bisnovatyi-Kogan and Lamzin, 1980; Cram *et al.*, 1980; Calvet *et al.*, 1984; Brown *et al.*, 1984a; Lago *et al.*, 1985; Herbig and Goodrich, 1986) and Herbig Ae stars (Praderie *et al.*, 1982, 1985). For several T Tauri stars the analysis of IUE data has led to a rough description of the temperature stratification of the chromosphere and lower corona. These studies showed that the lower layers of the T Tauri chromospheres do not differ qualitatively from the solar chromosphere, except for much higher densities. Like the solar chromosphere, T Tauri chromospheres appear to be highly nonhomogenous (Herbig and Soderblom, 1980; Brown *et al.*, 1984; Vrba *et al.*, 1985).

The outer chromospheres of the T Tauri stars, especially those with strong visual emission lines and thus strong winds, seem to differ from those of main sequence active chromospheric stars. The new models noted above, applied to observations of these stars, show that the outer chromospheric layers of strong emission-line T

Tauri stars are cooler than the corresponding layers in the solar chromosphere. Apparently, the maximum temperature and the outward temperature gradient in the outer chromospheric layers are smaller than in the Sun. Together with the weakness of the X-ray radiation in these strong-emission T Tauri stars (cf. Feigelson, 1984), the low temperature of the outer chromospheric layers seems to confirm De Campli's (1981) theoretical prediction, based on energy arguments, that the strong emission-line T Tauri stars cannot have solar-type hot coronae and solar-type thermally driven stellar winds (as it has been generally assumed for many years). The likely absence of hot coronae in such T Tauri stars is supported by the absence or weakness of detectable coronal emission lines (see e.g. Gahm and Krautter, 1982; Lago et al., 1985), although as pointed out by Lamzin (1985) in his careful analysis, the observed upper limits may not yet be conclusive. We note that X-ray *flares* have been observed in a few of the strong emission-line T Tauri stars.

The T Tauri chromospheric structure indicated by the IUE observations and the models based on these data have indicated that the wind acceleration mechanism in these stars must be different from the solar mechanism. However, one must keep in mind that in these models the presence of a wind and a velocity field have been neglected. Therefore for this and other reasons these models should still be regarded as rough approximations only and they cannot be used to obtain detailed information on the acceleration mechanism itself and the resulting velocity law. As pointed out (e.g. by Hartmann, 1984) a solution of this problem requires dynamic chromospheric models where the wind velocity field is included and where a detailed comparison of observed and predicted line profiles can be carried out. High resolution profiles of the Mg II resonance lines obtained with IUE (described in Section 3) show clear evidence for velocity fields in the layers where these lines are formed. In the absence of detailed information on the wind acceleration mechanism, it is only possible to construct models with an *assumed* velocity field or acceleration law. It is now usually assumed that the T Tauri winds are driven mechanically by hydrodynamic waves (Lago, 1979; De Campli, 1981; Hartmann et al., 1982). This assumption is attractive because it provides the direct in-situ heating required to produce the very strong transition-region lines and also because extended magnetic fields of the PMS stars seem to be likely from other observations (see above). On the other hand, the detailed properies of such magnetically driven winds are unknown and depend on uncertain quantities such as the field strength and geometry. Nevertheless, a very interesting first attempt to combine the chromospheric emission-measure distribution derived from IUE observations and a velocity law derived for a magnetically driven wind was made by Brown et al. (1984a). Unfortunately, their simple assumptions about the magnetically driven wind did not lead to a self-consistent solution. They find that a two-component atmosphere is required.

A quite different (and self-consistent) non-static model was proposed by Bertout (1982) who showed that the relative far-UV line strengths of the T Tauri emission spectrum can also be explained by assuming a low density thermally driven wind and the presence of *two* transition zones, one (as in the case of the Sun) near the stellar surface, and one due to the interaction of the hot wind with ambient interstellar or circumstellar gas. Because of the low coronal density, Bertout's model is not incompatible with the X-ray results. As cool circumstellar matter is frequently observed near T Tauri stars, it would certainly be worthwhile to follow up Bertout's suggestion in more detail.

Because of the large concentration of resonance lines and lines originating from metastable energy levels, the 1200—2000 Å spectral range is particularly well suited to study stellar winds by means of line profile analysis. As noted in Section 3, the T Tauri stars are too faint to allow IUE high spectral-resolution observations with sufficient signal-to-noise to pursue these detailed studies. However, even the low-dispersion IUE observations of those T Tauri stars with particularly dense envelopes clearly show envelope absorption effects which are variable with time. Best examples for this effect are the deep absorption features caused by merging Fe II envelope lines in the two extreme T Tauri stars RW Aur and DR Tau (Imhoff and Giampapa, 1980, 1981; Appenzeller *et al.*, 1980). It is important to observe these features with high spectral resolution when more sensitive FUV instruments like the HST High Resolution Spectrograph become available.

7. Conclusion and Outlook

The IUE satellite was initially designed to obtain high-resolution spectra of hot stars and low-resolution spectra of extragalactic objects. PMS stars were not listed among the targets which motivated the construction of this instrument, and so the important IUE results concerning PMS stellar evolution were unexpected.

The IUE observations confirmed the presence of extensive chromospheric layers in PMS stars and allowed the derivation of the temperature stratification of these chromospheres. For typical T Tauri stars, the maximum chromospheric and coronal temperatures turned out to be considerably lower than in main-sequence late-type stars. This result showed that the T Tauri mass loss cannot be driven by a solar-wind-like thermal acceleration mechanism. Although at present there exists no fully self-consistent theory of the T Tauri winds, magnetic fields and hydromagnetic waves appear to be the most likely wind acceleration mechanism compatible with the UV observations. The IUE observations have provided a unique data base for detailed investigations of the relation between chromospheric activity and the basic stellar parameters (age, rotation, mass) of the PMS objects. Observations of the H—H objects resulted in important new information on the interaction of the PMS winds with the surrounding interstellar medium.

These studies have demonstrated the usefulness of far-UV observations but also the limitations of low-resolution spectra for studying strong stellar winds and atmospheric velocity fields. Further progress requires high spectral-resolution observations of these faint PMS stars by far-UV instruments with much greater sensitivity than IUE. Such observations will be possible with the Hubble Space Telescope High Resolution Spectrograph (HRS). The instrument will produce a new 'quantum jump' of our information on T Tauri winds, but the small number of spectral elements which can be observed simultaneously with the HRS will make such observations very time-consuming. Thus progress in the field of high-resolution spectroscopy of PMS stars with Space Telescope will be very slow. Another satellite which, like IUE, has a high resolution far-UV spectrometer but with enormously enhanced sensitivity would be of great value for further studies of PMS stars. The proposed Lyman satellite, which is presently being studied by NASA and by ESA, would be such an instrument. Unlike HST, Lyman could observe the particularly line-rich and informative 900—1200 Å spectral range totally unexplored in PMS objects, which will allow us to study a broader range of ionization stages. For this reason it seems safe to predict that Lyman or a similar dedicated and sensitive spectroscopic far-UV satellite will have a particularly great potential for providing progress in the field of PMS stars and their evolution.

Acknowledgements

We would like to express appreciation for the contributions of the many IUE researchers who have studied the pre-main sequence objects; this body of work can only be briefly summarized here. In addition, IA would like to acknowledge IUE research support from the Deutsche Forschungsgemeinschaft (SFB 132). CI would like to acknowledge support from NASA IUE research contracts NAS 5-25774 and NAS 5-28749 with CSC.

References

Appenzeller, I.: 1986, *Physica Scripta T11*, 76.
Appenzeller, I., Bertout, C., Mundt, R., and Krautter, J.: 1981, *Mitt. Astr. Gesellschaft* **52**, 15.
Appenzeller, I., Chavarria, C., Krautter, J., Mundt, R., and Wolf, B.: 1980, *Astron. Astrophys.* **90**, 184.
Appenzeller, I., Jankovics, I., and Krautter, J.: 1983, *Astron. Astrophys. Suppl.* **53**, 291.
Appenzeller, I., and Wolf, B.: 1979, *Astron. Astrophys.* **75**, 164.
Ayres, T. R., Marstad, N. C., and Linsky, J. L.: 1981, *Astrophys. J.* **247**, 545.
Bertout, C.: 1982, in *The Third European IUE Conference*, ESA SP-176, p. 89.
Bertout, C.: 1984, *Rep. Progr. Phys.* **47**, 111.
Bisnovatiy-Kogan, G. S. and Lamzin, S. A.: 1980, *Sov. Astron. Letters* **6**, 17.
Bohm, K. H.: 1983, *Rev. Mex. Astron. Astrof.* **7**, 55.
Bohm, K. H. and Bohm-Vitense, E.: 1984, *Astrophys. J.* **277**, 216.
Bohm, K. H., Bohm-Vitense, E., and Brugel, E.: 1981, *Astrophys. J. Letters* **245**, L113.
Bohm-Vitense, E., Bohm, K. H., Cardelli, J. A., and Nemac, J. M.: 1982, *Astrophys. J.* **262**, 224.
Brown, A., Ferraz, M. C. de M., and Jordan, C.: 1984a, *Monthly Notices Roy. Astron. Soc.* **207**, 831.
Brown, A., Jordan, C., Millar, T. J., Gondhalekar, P., and Wilson, R.: 1981, *Nature* **290**, 34.
Brown, A., Penston, M. V., Johnstone, R., Jordan, C., Kuin, N. P. M., Lago, M. T. V. T., Gross, B., and Linsky, J. L.: 1984b, in *Future of UV Astronomy*, NASA CP-2349, pp. 388.
Brugel, E. W., Bohm, K. H., Shull, J. M., and Bohm—Vitense, E.: 1985, *Astrophys. J. Letters* **292**, L75.
Brugel, E. W., Shull, J. H., and Seab, C. G.: 1982, *Astrophys. J. Letters* **262**, L35.
Calvet, N.: 1980, Thesis, University of California, Berkeley.
Calvet, N., Basri, G., Imhoff, C. L., and Giampapa, M. S.: 1985, *Astrophys. J.* **293**, 575.
Calvet, N., Basri, G., and Kuhi, L. V.: 1984, *Astrophys. J.* **277**, 725.
Canto, J.: 1983, *Rev. Mex. Astron. Astrof.* **7**, 109.
Catala, C.: 1984,:, in *The Fourth European IUE Conference*, ESA SP-218, pp. 227.
Catala, C., Czarny, J., Felenbok, P., and Praderie, F.: 1986a, *Astron. Astrophys.* **154**, 103.
Catala, C., Felenbok, P., Czarny, J., Talavera, A., and Boesgaard, A. M.: 1986b, *Astrophys. J.* (in press).

Catala, C., Kunasz, P. B., and Praderie, F.: 1984, *Astron. Astrophys.* **134**, 402.
Catala, C., Praderie, F., and Felenbok, P.: 1986, in *New Insights into Astrophysics*, ESA SP-263, pp. 125.
Cram, L. E.: 1979, *Astrophys. J.* **234**, 949.
Cram, L. E., Giampapa, M. S., and Imhoff, C. L.: 1980, *Astrophys. J.* **238**, 905.
de Boer, K. S.: 1977, *Astron. Astrophys.* **61**, 605.
De Campli, W. M.: 1981, *Astrophys. J.* **244**, 124.
Dopita, M. A. and Schwartz, R. D.: 1981, *Publ. Astron. Soc. Pacific* **93**, 546.
Dopita, M. A., Binette, L., and Schwartz, R. D.: 1982, *Astrophys. J.* **261**, 183.
Dumont, S., Heidmann, N., Kuhi, L. V., and Thomas, R. N.: 1973, *Astron. Astrophys.* **29**, 199.
Feigelson, E. D.: 1984, in S. L. Baliunas and L. Hartmann (eds.), *Cool Stars, Stellar Systems and the Sun*, Springer Verlag: Berlin, pp. 27.
Feigelson, E. D. and De Campli, W. M.: 1981, *Astrophys. J. Letters* **243**, L89.
Finkenzeller, U. and Mundt, R.: 1984, *Astron. Astrophys. Suppl.* **55**, 109.
Gahm, G. F.: 1980, *Astrophys. J. Letters* **242**, L163.
Gahm, G.: 1981, in *The Universe at Ultraviolet Wavelengrths*, NASA CP-2171, pp. 105.
Gahm, G. F., Fredga, K., Liseau, R., and Dravins, D.: 1979, *Astron. Astrophys.* **73**, L4.
Gahm, G. and Krautter, J.: 1982, *Astron. Astrophys.* **106**, 25.
Giampapa, M. S.: 1984, in S. L. Baliunas and L. Hartmann (eds.), *Cool Stars, Stellar Systems and the Sun*, Springer Verlag: Berlin, pp. 14.
Giampapa, M. S., Calvet, N., Imhoff, C. L., and Kuhi, L. V.: 1981, *Astrophys. J.* **251**, 113.
Giampapa, M. S. and Imhoff, C. L.: 1985, in D. C. Black and M. S. Matthews (eds.), *Protostars and Planets II*, Univ. Arizona: Tucson, pp. 386.
Gondhalekar, P. M., Penston, M. V., and Wilson, R.: 1979, in A. J. Willis (ed.), *The First Year of IUE* pp. 109.
Hartmann, L. 1984, in S. L. Baliunas and L. Hartmann (eds.), *Cool Stars, Stellar Systems and the Sun*, Springer Verlag: Berlin, pp. 60.
Hartmann, L., Edwards, S., and Avrett, A.: 1982, *Astrophys. J.* **261**, 279.
Hartmann L. and Kenyon, S. J.: 1985, *Astrophys. J.* **299**, 462.
Hecht, J. H., Holm, A. V., Ake, T. B., III, Imhoff, C. L., Oliversen, N. A. and Sonneborn, G.: 1984, in *Future of Ultraviolet Astronomy*, NASA CP-2349, pp. 318.
Heidmann, N. and Thomas, R. N.: 1980, *Astron Astrophys.* **87**, 36.
Herbig, G. H.: 1960, *Astrophys. J. Suppl.* **4**, 337.
Herbig, G. H.: 1962, *Adv. Astron. Astrophys.* **1**, 47.
Herbig, G. H.: 1970, *Mem. Roy. Soc. Sci. Liège, Ser. 5*, **9**, 13.
Herbig, G. H.: 1974, *Lick Obs. Bull.*, No. 658.
Herbig, G. H.: 1977, *Astrophys. J.* **214**, 747.
Herbig, G. H. and Goodrich, R. W.: 1986, *Astrophys. J.* **309**, 294.
Herbig, G. H. and Soderblom, D. R.: 1980, *Astrophys. J.* **242**, 628.
Herbig, G. H. and Rao, N. K.: 1972, *Astrophys. J.* **174**, 401.
Imhoff, C. L.: 1984, in *Future of Ultraviolet Astronomy*, NASA CP-2349, pp. 81.
Imhoff, C. L. and Giampapa, M. S.: 1980, *Astrophys. J. Letters* **239**, L115.
Imhoff, C. L. and Giampapa, M. S.: 1981, in *The Universe at Ultraviolet Wavelengths*, NASA CP-2171, pp. 185.
Inhoff, C. L. and Giampapa, M. S.: 1982, in *Advances in Ultraviolet Astronomy*, NASA CP-2238, pp. 456.
Imhoff, C. L. and Giampapa, M. S.: 1983, *Bull. Amer. Astr. Soc.* **15**, 928.
Jones, B. F.: 1983, *Rev. Mex. Astr. Astrof.* **7**, 71.
Jordan, C., Ferraz, M. C. de M., and Brown, A.: 1982, in *The Third European IUE Conference*, ESA SP-176, pp. 83.
Joy, A. H.: 1945, *Astrophys. J.* **102**, 168.
Kuhi, L. V.: 1966, *Publ. Astron. Soc. Pacific* **78**, 430.
Kuhi, L. V.: 1974, *Astron. Astrophys. Suppl.* **15**, 47.
Kuhi, L. V.: 1983, *Rev. Mex. Astron. Astrof.* **7**, 127.
Lada, C. J.: 1985, *Ann. Rev. Astron. Astrophys.* **23**, 267.

Lago, M. T. V. T.: 1979, Thesis, University of Sussex.

Lago, M. T. V. T., Penston, M. V., and Johnstone, R.: 1984, in *The Fourth European IUE Conference*, ESA SP-218, pp. 233.

Lago, M. T. V. T., Penston, M. V., and Johnstone, R. M.: 1985, *Monthly Notices Roy. Astron. Soc.* **212**, 151.

Lamzin, S. A.: 1980, *Sov. Astron.* **29**, 176.

Mundt, R., Walter, F. M., Feigelson, E. D., Finkenzeller, U., Herbig, G. H., and Odell, A. P.: 1983 *Astrophys. J.* **269**, 229.

Ortolani, S. and D'Odorico, S.: 1980, *Astron. Astrophys.* **83**, L8.

Penston, M. V. and Lago, M. T. V. T.: 1982, in *The Third European IUE Conference*, ESA SP-176, pp. 95.

Penston, M. V. and Lago, M. T. V. T.: 1983, *Monthly Notices Roy. Astron. Soc.* **202**, 77.

Praderie, F., Talavera, A., Felenbok, P., Czarny, J., and Boesgaard, A. M.: 1982, *Astrophys. J.* **254**, 658.

Praderie, F., Simon, T., Boesgaard, A. M., Felenbok, P., Catala, C., Czarny, J., and Talavera, A.: 1985, in *Origin of Non-Radiative Heating/Momentum in Hot Stars*, NASA CP-2358, pp. 81.

Praderie, F., Simon, T., Catala, C., and Boesgaard, A. M.: 1986, *Astrophys. J.* **303**, 311.

Schwartz, R. D.: 1977, *Astrophys. J. Suppl.* **35**, 161.

Schwartz, R. D.: 1983a, *Ann. Rev. Astron. Astrophys.* **21**, 209.

Schwartz, R. D.: 1983b, *Rev. Mex. Astron. Astrof.* **7**, 27.

Schwartz, R. D.: 1983c, *Astrophys. J. Letters* **268**, L37.

Simon, T., Herbig, G. H., and Boesgaard, A. M.: 1985, *Astrophys J.* **293**, 551.

Talavera, A., Catala, C., Crivelli, L., Czarny, L., Felenbok, P., and Praderie, F.: 1982, in *The Third European IUE Conference*, ESA SP-176, pp. 99.

Tjin A Djie, H. R. E. and The, P. S.: 1982, in *The Third European IUE Conference*, ESA SP-176, pp. 113.

Tjin A Djie, H. R. E., The, P. S., Hack, M., and Selvelli, P. L.: 1982, *Astron. Astrophys.* **106**, 98.

Vrba, F. J., Rydgren, A. E., Zak, D. S., and Schmelz, J. T.: 1985, *Astrophys. J.* **90**, 326.

Walter, F. M.: 1984, in S. L. Baliunas and L. Hartmann (eds.), *Cool Stars, Stellar Systems, and the Sun* Springer Verlag: Berlin, pp. 75.

Walter, F. M.: 1986, *Astrophys. J.* **306**, 573.

Walter, F. M. and Brown, A.: 1983, *Bull. Amer. Astron. Soc.* **15**, 966.

Walter, F. M. and Kuhi, L. V.: 1981, *Astrophys. J.* **250**, 254.

MASS LOSS FROM COOL STARS

A. K. DUPREE

Harvard-Smithsonian Center for Astrophysics, Cambridge, MA, U.S.A.

and

D. REIMERS

Hamburger Sternwarte, Universität Hamburg, F.R.G.

1. Introduction

The veritable renaissance in the study of cool star winds has been well documented in many review papers. A study of winds from massive stars with particular emphasis on theoretical aspects has been published by Chiosi and Maeder (1986); a review of cool star winds with a focus on observational aspects is found in the paper by Dupree (1986). Proceedings of the *Fourth Cambridge Workshop on Cool Stars, Stellar Systems, and the Sun* contain a major section on Mass Loss and Pulsation, highlighted with reviews by Drake (1986b) on 'Mass Loss Estimates in Cool Giants and Supergiants', and Willson and Bowen (1986) on 'Stellar Pulsation, Atmospheric Structure and Mass Loss' and followed by over 25 contributed papers on this topic. In 1984 two conferences on mass loss were held: one at University of California at Los Angeles on *Mass Loss from Red Giants* (M. Morris and B. Zuckerman (eds.)); the other was held at Sacramento Peak Observatory, Sunspot, New Mexico on *Relations Between Chromospheric-Coronal Heating and Mass Loss in Stars* (R. Stalio and J. B. Zirker (eds.)). Both of these volumes contain a mix of invited reviews and contributions pertaining to various observational and theoretical aspects of the mass loss process including reviews specifically directed to results from ultraviolet and optical spectroscopy (Goldberg, 1985; Linsky, 1985).

This review discusses the advantages of the ultraviolet spectral region for studies of mass loss in cool stars (Section 2), and presents evidence, as gleaned from IUE spectra, for the presence and existence of mass outflow across the cool half of the H—R diagram (Section 3). The physical characteristics of cool stellar winds as determined principally from ultraviolet observations are discussed in Section 4. Derivation of quantitative rates of mass loss generally requires complex analysis; some of the most reliable values result from study of widely separated binary stars (See Section 5). A summary of the present state of our knowledge of mass loss rates for luminous cool stars is contained in Section 5; contributions from IUE are highlighted specifically in 6, and some outstanding problems are listed in Section 7.

Y. Kondo (ed.), Scientific Accomplishments of the IUE, pp. 321—353.
© 1987 *by D. Reidel Publishing Company.*

2. The Ultraviolet Spectrum

The ultraviolet spectrum contains both obvious and subtle signatures of ordered mass motions and mass loss from luminous cool stars. Most strong resonance lines of important, abundant species occur in the ultraviolet. Extended stellar chromospheres* where effects of mass motions and winds are likely to be detected radiate profusely in ultraviolet lines. Here the Mg II resonance doublet at $\lambda 2796$ and $\lambda 2803$ is the preeminent example. Additionally, a plethora of weaker lines such as Sr II, Fe I, and II contribute to the analysis of both velocity fields and atmospheric structure in cool star atmospheres. Absorption features arising from resonance transitions in neutral or singly ionized species in surrounding circumstellar material can act as a probe of mass lost to the interstellar medium.

High dispersion spectra are necessary to study the mass flow in cool stars since the velocities encountered are generally less than ~ 100 km s^{-1}. Prior to IUE, only a handful of ultraviolet spectra was available from the *Copernicus* satellite (Bernat and Lambert, 1976; Dupree, 1975, McClintock *et al.*, 1978) and the Balloon-Borne Ultraviolet Spectrometer and Spectrograph (Kondo *et al.*, 1976; 1979; de Jager, 1979). IUE has been outstandingly successful in obtaining high dispersion spectra of bright cool stars at longest wavelengths ($\lambda\lambda 2200-3200$), but high dispersion spectra in the short wavelength regime are restricted to the brightest objects, and even in those the signal-to-noise ratio is not high. Thus, while signatures of mass flow probably occur in ions such as N V, C IV, and Si IV, they are not well studied due to the lack of appropriate spectroscopic material. The hydrogen Lyman-alpha profile is expected to be a strong indicator of mass flow, much as the Mg II doublet, but in addition to strong interstellar absorption, large aperture observations introduce the geocoronal emission, making reduction difficult.

Study of line profiles or measurement of line positions can be used to infer the presence of motions in stellar atmospheres. Simple measures of Doppler shifts of lines can achieve the theoretical limit of one-tenth of the spectral resolution of IUE or ~ 2.5 km s^{-1} by fitting line profiles to data with high signal-to-noise ratios. Cross-correlation techniques can achieve similar high precision. We focus here on four spectral features: Mg II, H Ly-α, Fe II, and O I/S I.

2.1. THE Mg II TRANSITION

The resonance transition of the Mg II atom is a doublet (2795.5 Å — the k line, and 2802.7 Å — the h line) that appears as an extremely deep photospheric absorption feature with a generally centrally reversed (though not always measurable due to insufficient spectral resolution) emission core. The emission core holds particular fascination for the stellar physicist, because it signals the presence of a chromospheric temperature rise thought to reflect stellar magnetic activity. The

---

* The term *extended chromosphere* is used loosely here to mean a cool atmosphere ($T \sim 10^4$ K, by analogy with the solar chromosphere) extending several stellar radii above the photosphere (again in contrast to the Sun where the thickness of the 10^4 K region is much smaller than the solar radius; cf. Vernazza *et al.*, 1981). Judge (1986) has proposed that the phrase *extended chromosphere* should be replaced (at least for cool giant stars, based on his analysis of Alpha Boo) by a thin *flux creation* region where most of the line flux is produced, and a larger *flux scattering* region where optically thick lines are resonantly scattered.

profile of this emission core can indicate the mass motion of the material in which it is formed.

A constant motion of material towards or away from the observer produces a simple Doppler shift of the line; however, differential expansion of a chromosphere produces an asymmetric emission core. As Hummer and Rybicki (1968) first noted, differential expansion causes a shift of line opacity to shorter wavelengths, thus weakening the short wavelength side of a line and apparently strengthening the long wavelength side (see Figure 1). Calculations of chromospheric line

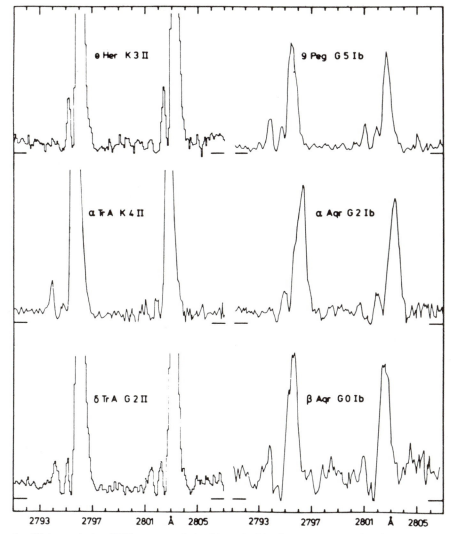

Fig. 1. High resolution IUE spectra of the Mg II doublet in a number of hybrid stars. Note the extreme asymmetry of the profiles as well as the narrow double absorption features. It is thought that the high velocity absorption feature arises in the stellar wind, and the low velocity feature may be composed of both atmospheric and interstellar components. The figure is reproduced from Reimers (1982) and incorporated spectra of α Aqr, β Aqr, α TrA from Hartmann *et al.* (1980, 1981) and of 9 Peg from Stencel *et al.* (1980).

profiles for stars of various gravities demonstrated that asymmetries easily occur in differentially expanding atmospheres (Drake and Linsky, 1983; Dupree, 1980, 1982; Hartmann and Avrett, 1984). If a star has an extended cool atmosphere or cool wind, the line profile can appear to be additionally attenuated by scattering from Mg II in the wind. Calculations suggest that Mg II is a dominant state of ionization in the far wind or envelope of low gravity stars due to its relatively high rate of recombination (see Dupree, 1982). Thus the Mg II profile, and in particular, the short wavelength portion, is especially sensitive to the structure of the outer atmosphere.

The narrow absorption feature at shortest wavelengths in the line profile is generally associated with the farthest reaches of a stellar wind, and the velocity indicated by this absorption is considered to be the asymptotic flow velocity of the wind. Whether such an absorption feature is detectable of course depends on three conditions: first, that sufficient Mg II is present at the 'limits' of the wind to produce absorption, second, that the terminal velocity lies within the velocity limits of the emission core in order to have a background continuum against which the absorption is visible. These velocity limits are typically less than ± 100 km s^{-1} for giants or ± 200 km s^{-1} for cool supergiants. Third: that the velocity does not coincide with interstellar absorption velocities in the direction of the star (which can range from 0 to $\sim \pm 25$ km s^{-1}).

Of serious concern to the interpretation of the Mg II profiles (and other strong stellar resonance lines) is the effect of *interstellar* Mg II and Mn I absorption. Velocities of material in the interstellar medium tend to values on the order of 20 km s^{-1} or less. These are similar to velocities found in many cool luminous stars and it is sometimes difficult to properly identify the source of absorption (Bernat and Lambert, 1976; Böhm-Vitense, 1981; Drake *et al.*, 1984). To understand the profile, researchers have relied on model profiles, a substantial velocity difference between the star and the interstellar medium, or detection of variability in the stellar line profile.

2.2. HYDROGEN LYMAN ALPHA

The resonance transition of hydrogen may be expected to show asymmetries similar to the Mg II line since it is an optically thick chromospheric transition, typically exhibiting a deep central reversal within its emission core. Again the line profile may be modified in appearance, both by interstellar hydrogen absorption and by emission from the Earth's geocorona. IUE observations taken through the small aperture cause the geocoronal contamination to appear as a narrow emission feature that can be removed. The effects of interstellar hydrogen absorption can be devastating of course, and effectively eradicate any sign of an emission core. However, most of the cool stars amenable to high resolution observations with IUE are nearby, with resultant low values of the hydrogen column density along the line of sight. With sufficient exposure time, the Ly-α emission has been detected. Since the early observations with *Copernicus* suggested that Lyman-α might be more sensitive than the Mg II profile in detecting low rates of mass flow (Dupree, 1976), these observations are particularly valuable.

2.3. Fe II LINES

A large number of Fe II emission lines occurs in the long wavelength spectral region of IUE, $\lambda\lambda 2300-3200$ (see identifications contained in Carpenter, 1984; Carpenter *et al.*, 1985). Since these lines represent a variety of excitation levels, and intrinsic line strengths, they can be used to map the extended atmosphere of a low gravity star. One advantage of the ultraviolet is the wealth of unblended Fe II lines thus giving statistical weight to the analysis.

2.4. O I/S I LINES

The O I multiplet ($\lambda 1302.17$, $\lambda 1304.86$, $\lambda 1306.03$) is strong in cool low gravity stars, and in a few cases high dispersion profiles have been observed to be asymmetric (Ayres *et al.*, 1982; Brown *et al.*, 1984; Hartmann *et al.*, 1981; Hartmann *et al.*, 1985). In this ion, there is no confusion with absorption in the interstellar medium since the $\lambda 1304$ and $\lambda 1306$ transitions arise from an excited state, and have not been generally observed in interstellar spectra. Both of these lines show absorption on the short wavelength side of the profile in the hybrid bright giant α TrA (K4 II) and in β Dra (G2 Ib—II). Such a profile is believed indicative of circumstellar material.

2.5. TRANSITION REGION LINES (C III] C IV, Si III] Si IV, N V...)

Lines formed in plasma of temperatures $\sim 10^5$ K, might be expected to show signatures of mass flow in cool stars. In the solar spectrum, Doppler shifts have been detected in such lines (Athay *et al.*, 1983; Dere *et al.*, 1984), but the large scale flow measured near active regions tends towards downflows, and not outflows. Studies of the solar transition region reveal outflow in the high transition region lines ($T \gtrsim 3 \times 10^5$ K) such as O V and Mg X (Orrall *et al.*, 1983). These transitions, of course are not accessible to IUE, but the magnitude of the Doppler shifts (-5 to -10 km s^{-1}) is so small as to be measurable only with IUE spectra of superb quality.

Brown *et al.* (1984) and Ayres *et al.* (1983b) claim to measure 'small, but statistically significant' redshifts of these lines relative to lines of lower stages of ionization in several chromospherically active giant stars. Although there is some ambiguity in the interpretation of the detected shifts — an outflow also can cause a red-shift in an *optically thick* line by the 'P Cygni' effect — a detailed study of optically thin lines of Si III] and C III] in the active binary Alpha Aurigae AB (*Capella*), comprised of a G6 IV and a F9 III star, suggests (Ayres, 1984) that the flow on the active secondary of the binary system is genuinely directed inwards. True downflows are perhaps similar to those detected in the Sun (Athay *et al.*, 1983; Dere *et al.*, 1984).

In the hybrid stars α TrA (K2 IIb—IIa) and γ Aql (K3 II), the intersystem emission lines of C III and Si III are essentially at their rest wavelengths and exceptionally broad (Brown *et al.*, 1986; Hartmann *et al.*, 1985). It appears that turbulent pressure is important at temperatures of $\sim 10^5$ K, in these atmospheres. While a fast wind is not detected at high temperatures, a slow expansion may be present and certainly cannot be ruled out by the existing observations.

3. Presence of Winds

The Mg II transition has been most easily and widely used as a signature of mass motions in the outer atmospheres of cool stars. The appearance of this line changes dramatically with luminosity (see Figure 2), becoming wider as the

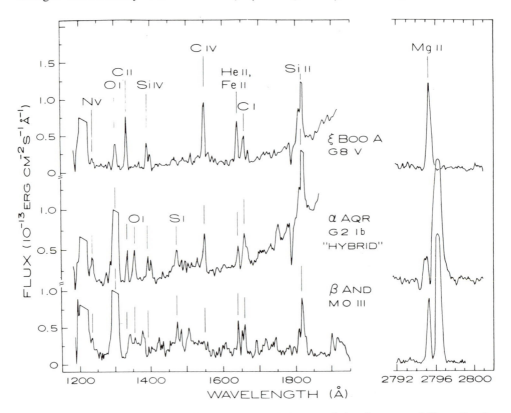

Fig. 2. Spectra from 3 cool stars exhibiting various different emission features and illustrating the detectability of winds from the Mg II line asymmetry. Xi Boo A is a magnetically active dwarf star with a spectrum similar to the solar spectrum and no indication of any asymmetry in the Mg II profile. Alpha Aqr (a hybrid star) exhibits C IV, low excitation species (O I, S I, and C I), and an asymmetric Mg II profile formed by the extended expanding atmosphere. Beta And lacks any detectable emission from high temperature species; in fact the emission near $\lambda 1350$ and $\lambda 1525$ may be CO, and the O I transition ($\lambda 1641.3$) is strong. Like Alpha Aqr, β And appears to possess a wind. The geocoronal Ly-α ($\lambda 1216$) and O I ($\lambda 1300$) peaks have been truncated where a flat top appears. The data are from Hartmann *et al.* (1982); the figure is taken from Dupree (1986).

intrinsic luminosity of the star increases (Dupree, 1976; Kondo *et al.*, 1976; Weiler and Oegerle, 1979), and exhibiting the typically short-wavelength asymmetry of mass outflow. Fundamental to the detection of outflow is the ability to resolve the line profile. In IUE spectra, dwarf stars exhibit narrow emission cores with generally unresolved profiles; more luminous stars show wider emission cores with central reversals, asymmetries, and in many cases circumstellar and interstellar absorption lines. Emission from gas at higher temperatures (detected in the accompanying short wavelength spectra) changes in character with stellar luminosity and temperature. Dwarf stars possess material at 10^5 K (inferred from the presence of N V and C IV), whereas more luminous stars can have either hot atmospheric material and massive winds (the *hybrid* stars, luminosity class I, II, and III) or cool atmospheres and winds (Linsky and Haisch, 1979; Hartmann *et al.*, 1980, 1981; Drake, 1986a). The character of these various atmospheres is discussed elsewhere in this volume.

The vast majority of cool stars observed with IUE belong to metal rich Population I, and so the dependence of atmosphere and wind characteristics upon metallicity has been virtually unexplored in the ultraviolet. Metal deficient giants are known to have chromospheric Mg II emission (Dupree *et al.*, 1984). Ground-based observations of the Ca II and H-α profiles of metal deficient giants (Mallia and Pagel, 1978; Peterson, 1981; Smith and Dupree, 1987; Spite *et al.*, 1981) give evidence for blue-shifts of line cores, so that a wind is likely to be present. Because stellar evolution calculations require knowledge of the rate of mass loss and its timing during a star's evolution (Iben and Renzini, 1983), it is important to define the mass loss process in various populations.

3.1. DWARF STARS

Although the IUE resolution is not sufficient to resolve fully the Mg II profile in cool dwarf stars (see Figure 2 and Ayres *et al.*, 1983a), observations of the Sun suggest that a consistent outward flow leading to mass loss is not detectable (and may not occur) until species of higher excitation ($T \gtrsim 3 \times 10^5$ K) are measured. Solar observations are usually made on a restricted portion of the solar disk enabling identification with atmospheric structures. Coronal holes or magnetically quiet regions revealed outflowing material (Brueckner *et al.*, 1977; Rottman *et al.*, 1982; Orrall *et al.*, 1983). These measurements suggest that a dwarf star may need to have a substantial fraction of its surface covered with ordered outward motion in order to be detectable. High resolution large aperture IUE observations of the bright stars Alpha Cen A (G2 V) and Epsilon Eri (K2 V) have failed to show any signs of outward movement in many chromospheric and transition region lines although there may be small inward motions relative to the low chromosphere (Ayres *et al.*, 1983b). These results suggest that the mass loss rate is not drastically different from that of the Sun — namely $2 \times 10^{-14} \, M_\odot \, \mathrm{yr}^{-1}$. An early *Copernicus* observation (McClintock *et al.*, 1976) of the H Ly-α line in Epsilon Eri hints at an asymmetry characterizing outflow, providing slight evidence for expansion in the atmosphere of a dwarf star.

Indirect evidence for mass loss from dwarf stars comes from the binary system Feige 24. In this system, composed of a white dwarf and a K-type dwarf, the spectrum indicates substantial column density of metal enriched material associated with the white dwarf. Since heavy elements are not expected to reside in the white dwarf atmosphere for a significant amount of time, this material must be constantly replenished. The K-dwarf component appears to be a likely source of this material (Dupree and Raymond, 1982). Although there may be a delicate balance between the accretion flow and radiation pressure from the hot white dwarf (Wesemael *et al.*, 1984), calculations suggest that a mass loss rate of

$\approx 10^{-13} \, M_\odot$ from the K-dwarf can accommodate the observations (Sion and Starrfield, 1984).

3.2. LUMINOUS STARS

The most extensive survey of the presence of winds in luminous cool stars was carried out by Stencel and Mullan (1980a, b) who searched for line asymmetries in the strong Mg II transition. These asymmetries, as discussed earlier, signify the presence of differential expansion (or contraction) in the region of line formation. Stencel and Mullan interpret the signatures as indicating outflow (or 'downdrafts' or downflow). When eliminating those spectra having possible confusion with interstellar features, they noted that detectable outflows consistently occur in the coolest, most luminous objects, and signatures of downflow were found in the less luminous stars. Based on the material in Figure 3, Stencel and Mullan (1980a)

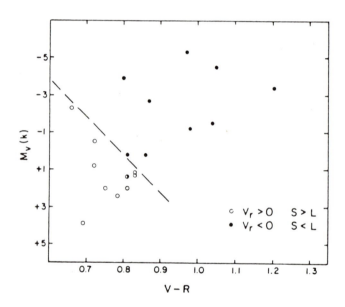

Fig. 3.   Observed asymmetry of the emission peaks of the Mg h and k lines. Filled circles denote stars with a negative radial velocity, V_r, in which the short-wavelength peak (S) is weaker than the long-wavelength peak (L). Open symbols designate the opposite: $S > L$ and $V_r > 0$, ($S/L \sim 1$ for the half-filled circle, α UMa). The broken line represents the locus along which the asymmetry reverses although it is ill-defined for stars brighter than $M_v = 0$ (figure from Stencel and Mullan, 1980b).

suggested that the presence of an outflow occurs abruptly (a 'dividing line') in a star's outer atmosphere. Note that the line is well defined only near $M_V = +1$, $V - R = 0.82$; the available data shown in Figure 3 do not rule out a vertical 'line' at $V - R \sim 0.8$ for luminous stars. Subsequent evidence from the four giant stars in the Hyades star cluster (Baliunas et al., 1983) suggests that the change is not as abrupt as first conjectured. These four stars show different asymmetries, although the interstellar absorption components should be identical. Clearly the chromospheres and the mass motions within it are different among these stars. The high temperature emission both in the ultraviolet (Baliunas *et al.*, 1983) and in X-rays (Stern *et al.*, 1981) also confirms the intrinsic differences between the stars. A further study of Mg II profiles in a large number of cool stars by Drake (1986a) confirms the lack of a sharp dividing *line* in atmospheric and wind characteristics. Of course, both time variability of a wind and an inhomogeneous distribution of outflow patterns across a stellar surface could contribute to the perceived blurring of transition among atmospheric characteristics. It appears difficult to conclude that the transition to a wind is abrupt.

The Stencel and Mullan survey (1980a, b) generally outlines the region of the HR diagram where mass outflow is sufficiently substantial to be detected in Mg II. Figure 4 delineates two spectroscopic signatures of mass loss, namely circumstellar Ca II absorption lines and the Mg II asymmetry and the position of their occurrence with respect to atmospheric characteristics. With decreasing surface gravity and temperature, the signatures of mass loss become more apparent, and the character of an atmosphere appears to change so that only signatures of cool plasma dominate the M-type giant and supergiant stars. None of these changes are abrupt across an HR diagram.

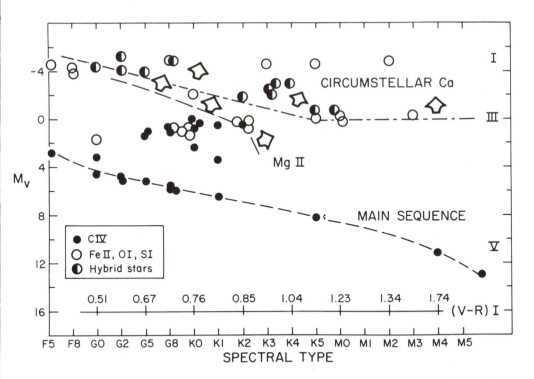

Fig. 4.   The presence of various spectral features in stars of different spectral types and luminosity classes. The broken line denotes the boundary above which variable circumstellar lines of Ca II appear (Reimers, 1977a, 1982). The position of an asymmetry change in the Mg II profiles as noted by Stencel and Mullan (1980b) is shown (see Figure 3, also).

We should be cautious however, because the ultraviolet transitions now available, and the spectral resolution attainable with IUE may only be sensitive to high mass loss rates. For instance, it is quite probable that the well-studied system, Alpha Aurigae (*Capella*), although exhibiting signs of 'downflow' in high temperature lines (Ayres, 1984), may indeed have a wind. The Lyman-alpha line — a sensitive diagnostic of the outermost regions of an atmosphere (Avrett, 1981; Dupree, 1982) — has been unwaveringly asymmetric (Dupree, 1975; Ayres, 1984) for over a decade. Interstellar absorption certainly affects the profile, yet the substantial asymmetry appears intrinsic to the star. This binary system has components classified as G6 III + F9 III and would lie to the left of the region in Figure 4 where mass outflow has been detected in Mg II. The Ly-α profile of α TrA (K4 II) also exhibits outward asymmetry (Hartmann *et al.*, 1985), and in this hybrid star, circumstellar absorption features in the Mg II profiles are quite apparent. We expect that measurement of many H Ly-α profiles will show the asymmetry to begin at higher temperature and lower gravities than denoted by inspection of the Mg II transition.

4. Physical Characteristics of Cool Stellar Winds

It is possible to describe the physical conditions of stellar winds as inferred from high resolution ultraviolet spectroscopy of bright stars with abundances similar to those of the Sun. Such a description can set constraints for theories of mass acceleration and mass loss in luminous cool stars. Five aspects of winds are addressed in this section: asymptotic flow speed, temperature structure, velocity structure, variability, and energy balance.

4.1. ASYMPTOTIC FLOW SPEED

The minimum intensity points of narrow absorption features on the short wavelength side of the Mg II resonance transitions are taken to indicate the terminal flow velocity of stellar winds. These features are thought to be produced in a circumstellar 'shell' that is formed from recombining ions at some distance (on the order of several stellar radii) from the star. The hybrid stars appear to show several components (Hartmann *et al.*, 1980; Reimers, 1982, and see Figure 1), of which the low velocity component may in part be interstellar in origin (Böhm-Vitense, 1981; Drake *et al.*, 1984; but see Reimers, 1982). Dwarf stars do not show circumstellar absorption lines in their spectra because there are insufficient numbers of Mg II atoms in the extended wind. In the Sun we know that the wind is hot ($T \approx 10^6$ K), thus preventing formation of much cool material; moreover, the mass loss rate is low ($\approx 10^{-14} M_\odot \text{ yr}^{-1}$). In more luminous stars the winds may also be warm, and the mass loss rates low as well, thus preventing detection of an absorption feature. The coolest giant and supergiant stars may have Mg II in their circumstellar material but this material is moving at low velocity. Thus interstellar absorption features can obscure the signature of a circumstellar shell unless the stellar radial velocity differs sufficiently from interstellar clouds along the line of sight. Table I contains values of the velocity as measured from the Mg II profile, and the high velocity components are displayed in Figure 5.

TABLE I

Wind velocity from Mg II lines

Star	Spectral Type	$V - R$	Velocity* $(- \text{km s}^{-1})$	Note	Reference
ε Sco	K1 III—IV	0.86	15	a	1
α Boo	K2 III	0.97	40.7		1
β Oph	K2 III	0.82	5.4	a	1
δ And	K3 III	0.92	300	c	11
α Tuc	K3 III	1.00	55.7		1
ζ Ara	K3 III	0.96	70, 11	a	12
β UMi	K4 III	1.11	30	b	1
β Cnc	K4 III	1.12	30	a	12
γ Dra	K5 III	1.14	65	c	2
γ Phe	K5 IIIa	1.26	100:	c	12
σ Pup	K5 III + G2 V	1.21	52	c	12
μ UMa	M0 III	1.28	55	c	2
γ Eri	M0 III	1.26	12		12
α Cet	M1.5 IIIa	1.35	13		12
δ Vir	M3 III	1.53	7		12
τ^4 Eri	M3 III	1.58	22		12
δ TrA	G2 II	0.58	90, 20	c	3
δ TrA	G2 II	0.58	81 ± 2; 10 ± 2	c	13
22 Vul	G3 II—Ib	0.74	160. ± 20	d	4
Θ Lyr	K0 II	0.87	25.7	b	1
γ Aql	K3 II	1.07	66, 15	c	3
γ Aql	K3 II	1.07	75, 29	c	5
γ Aql	K3 II	1.07	71 ± 3; 27 ± 2	c	13
γ Aql	K3 II	1.07	108	c	14
ι Aur	K3 II	1.06	72, 8	c	5
ι Aur	K3 II	1.06	77 ± 11; 10 ± 3	c	13
Θ Her	K3 II	0.90	100, 13	c	3
Θ Her	K3 II	0.90	91, 74, 5.6	c	5
Θ Her	K3 II	0.90	79 ± 8; 0.3 ± 2	c	13
α TrA	K4 II	1.05	180, 84, 15	c	5
α TrA	K4 II	1.05	93 ± 9; 20 ± 2	c	13
α TrA	K4 II	1.05	120	c	14
δ Sge	M2 II	1.44	28	d	6
α Her	M5 II	2.20	9.6	b	1
γ Cyg	F8 Ib	0.49	0.0	a	1
β Aqr	G0 Ib	0.61	80, 20	c	7
β Aqr	G0 Ib	0.61	135	c	1
β Aqr	G0 Ib	0.61	86 ± 4; 26 ± 2	c	13
α Aqr	G2 Ib	0.66	127, 20	c	7
α Aqr	G2 Ib	0.66	124 ± 5, 18 ± 2	c	13
9 Peg	G5 Ib	0.80	131, 7.5	c	1
9 Peg	G5 Ib	0.80	68 ± 3	c	13
56 Peg	K0 Ibp	0.97	21.4	b	1
12 Peg	K0 Ib	0.76	139 ± 6; 5 ± 4	g	13
ν' Sgr	K2 I	1.01	70 ± 4; 2 ± 4	g	13
ζ Aur	K4 Ib	1.13	40	d	8

Table I (continued)

Star	Spectral Type	$V - R$	Velocity* $(- km\ s^{-1})$	Note	Reference
32 Cyg	K4 Ib	1.20	60	d	8
31 Cyg	K5 Iab	1.97	80	d	8
ξ Cyg	K5 Ib	1.20	50, 37.5		1
α Sco	M1.5 Iab—Ib	1.55	17		9
Boss 1985	M2 ep Iab	1.34	10		10
α Ori	M2 Iab	1.64	20, 17, 11	e, f	15, 16

*   Velocity in stellar photospheric reference frame.
(a) Star with negative radial velocity and $V < R$ asymmetry.
(b) Inspection of Böhm-Vitense (1981) correlations of Mg II asymmetry with stellar radial velocity
     suggests circumstellar material is present.
(c) Hybrid star.
(d) Study of ultraviolet line profiles with binary phase and semi-empirical modeling leads to the
     terminal velocity.
(e) Based on semi-empirical modeling of atmosphere and wind.
(f) From circumstellar shell lines. ·
(g) Possible hybrid.

References:

(1) Stencel *et al.* (1980)
(2) Drake and Linsky (1984a).
(3) Reimers (1982).
(4) Reimers and Che-Bohnenstengel (1986).
(5) Hartmann *et al.* (1985).
(6) Reimers and Schröder (1983).
(7) Hartmann *et al.* (1980).
(8) Che, Hempe and Reimers (1983).

(9) Hagen, Hempe and Reimers (1986).
(10) Che and Reimers (1983).
(11) Judge, Jordan, and Rowan-Robinson (1986).
(12) Drake (1986a).
(13) Drake, Brown and Linsky (1984).
(14) Brown, Reimers and Linsky (1986).
(15) Hartmann and Avrett (1984).
(16) Reimers (1981).

The measured velocities are similar among the supergiants and the giant stars, with a general decrease in the value of the high velocity component with decreasing stellar effective temperature. With two exceptions, none of these values are near the escape velocity from the surface of the star which is typically 300 km s^{-1} for a K0 giant and 100 km s^{-1} for a M0 supergiant. This observation implies that a stellar wind is a gentle flow of material moving away from a star, and not a rapid discontinuous process.

The two exceptions show values approximating the escape velocity. One is the hybrid bright giant, α TrA in which a deep IUE exposure revealed a broad absorption trough extending out to 180 km s^{-1} (Hartmann *et al.*, 1985) as shown in Figure 6. The escape velocity at the surface of α TrA is 200 km s^{-1}, and decreases to 140 km s^{-1} at one stellar radius above the surface. The other star is δ And (K3 III) which shows a similar broad absorption trough extending to 300—350 km s^{-1} (Judge *et al.*, 1986). The escape velocity from the surface of δ And is ~ 175 km s^{-1}. These are remarkable observations for they directly demonstrate the presence of escaping material from a star and affirm the existence of relatively high-speed winds in luminous stars. There are indications that the flow speed may be variable since deep exposures of α TrA made two years later in 1984 show a

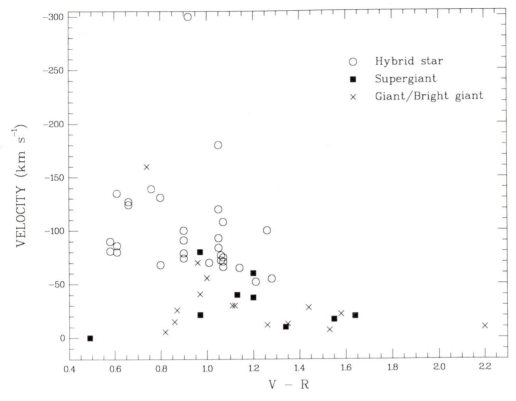

Fig. 5.   The velocity of the minimum intensity point, generally ascribed to a stellar wind in the Mg II profiles of luminous stars. Several components are plotted for hybrid stars.

terminal velocity of only 120 km s^{-1} (Brown *et al.*, 1986). The detection of asymptotic flow speeds of a few hundred km s^{-1} makes one mechanism to drive winds, namely Alfvén waves, more appealing, by assuaging the need to assume very specific and unique processes to damp out Alfvén waves in an extended atmosphere (Hartmann and MacGregor, 1980; Holzer and MacGregor, 1985). In the context of their wind characteristics, the hybrid stars may provide the connecting link between the fast winds of dwarf 'solar-type' stars and the slow outflows of the most luminous M-type supergiants (Hartmann *et al.*, 1980).

Some of these stars are known to be spectroscopic binaries. Delta And (K3 III), with its high wind velocity, is one such example although the separation of the components is sufficiently wide so that interaction is not expected (Judge *et al.*, 1986). Another star with high velocity is α TrA (K4 II) in which Ayres (1985) claimed detection of excess far ultraviolet emission. He suggested that α TrA possesses an F-dwarf companion and proposed that the high temperature emission features in *hybrid* stars originate not from the luminous supergiant or giant, but from the dwarf companion. Others have difficulty accepting this proposal because the atmospheric densities inferred from the spectrum indicate low values that are more typical of a giant star; moreover anomalously high surface fluxes result from

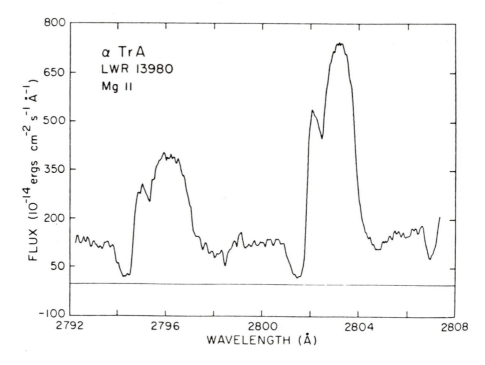

Fig. 6. A spectrum of the Mg II doublet in the hybrid star Alpha TrA. Note the absorption trough on the short wavelength side of each emission that occurs at high velocity relative to the Mg II lines and arises from the stellar wind. The peaks of both emission lines are saturated so that the fluxes are not correct. Figure from Hartmann *et al.* (1985).

assignment of the emission to the dwarf star (Hartmann *et al.*, 1985). Other stars, among them γ Phe, σ Pup, and stars of the VV Cep/ζ Aur type are well-documented binaries. To the extent that a companion can influence the properties of the wind, it should be remembered that the wind velocities listed in Table I are, in many cases, not observations of a single, isolated star.

4.2. TEMPERATURE

The temperature of stellar winds has been determined or inferred in several ways. For the Sun, the spatially resolved corona radiates thermally in ions characteristic of a hot plasma, $\approx 10^6$ K (Vaiana and Rosner, 1978). Moreover, the particle distribution function of components of the solar wind has been measured at the Earth's distance, and generally confirms electron and proton temperatures of $\approx 10^5$ K although the ionization temperatures may be much higher, on the order of 10^6 K (Neugebauer, 1983). We do not have such measurements for other dwarf stars, yet the similarity of their ultraviolet and X-ray emissions to the solar example (although in many instances, enhanced over the solar level) strongly suggests that the wind temperature is comparable that of the Sun.

The coolest and most luminous of the red giants [for instance, Alpha Boo (K2 III) and Alpha Ori (M2 Iab)] show no signs of high temperature emission features in the ultraviolet (Wing, 1978; Linsky and Haisch, 1979; Judge, 1986; Ayres *et al.*, 1987; Hartmann and Avrett, 1984; and see Figure 2). Additionally, stars of this type are not sources of X-rays to an upper limit of, in a few well-studied cases, three orders of magnitude in the stellar surface flux *below* that of a solar coronal hole (Rosner *et al.*, 1985). Many of these starts show circumstellar absorption in Ca II and Mg II as discussed in the previous section; these features are used to determine the asymptotic flow speed of cool stellar winds. In the well-studied supergiant Alpha Ori, the short-wavelength side of the Mg II doublet shows variability, signaling the changing opacity of the wind in an ion associated with 'low' temperatures (Dupree *et al.*, 1986). These observations suggest that the wind temperature is on the order of 5000 to 10 000 K. Detailed modeling of the atmosphere and wind of two prototype stars suggests similar low temperatures. Judge (1986) analyzed the IUE spectra of Alpha Boo (K2 III) using all available fluxes and line profiles that he combined with a variety of emission line diagnostics. The electron temperature in the atmosphere must be less than 5×10^4 K; even at these temperatures, the emission measure is down by a factor of ten with respect to the emission measure at ≈ 6000 K when compared to the structure of the outer solar atmosphere. The temperature in the 'wind' (or the scattering portion of the atmosphere) must be less than 5×10^4 K.

Alpha Ori is another well-studied star for which self-consistent semi-empirical models have been constructed by Hartmann and Avrett (1984). Adopting the theory of Alfvén-wave driven winds (Hartmann and MacGregor, 1980), they adjusted the temperature structure of the atmosphere until the calculated radiative losses agreed with the heating rate derived from Alfvén-wave dissipation. The model was also adjusted to match the extended distribution of emission as inferred from speckle imaging and radiofrequency observations. The temperature structure shows a local maximum of 8050 K at 4 stellar radii which decreases to 1200 K at distances greater than 30 R_*. Thus the spectroscopic results in conjunction with carefully constructed models suggest that cool winds, with temperatures less than $\approx 20\,000$ K are prevalent in the most luminous red giants and supergiants.

The situation is not so clearly understood for the *hybrid* stars, which were first discovered by Hartmann and colleagues (1980; 1981) to have both high temperature ultraviolet emission features (C IV, Si IV, N V) indicative of $\approx 2 \times 10^5$ K plasma in the presence of a massive wind (inferred from line asymmetries, and circumstellar Mg II and Ca II absorption). X-rays are generally not detected from this class of stars, although the closest example, Alpha TrA is an X-ray source as recently observed with EXOSAT (Brown, 1986) at a level consistent with a general decrease of material with increasing temperature in the atmosphere (see Figure 9 of Hartmann *et al.*, 1985). While it was noted (Hartmann *et al.*, 1980) that the hot gas and mass loss could arise from different part of the stellar atmosphere, the broad lines of C IV, Si IV, and C III observed in α TrA (K4 II), with widths comparable to the terminal velocity of the wind, suggested that these lines were formed over a large fraction of the stellar atmosphere and perhaps in the wind itself. A straightforward calculation invoking the dissipation of Alfvén waves showed that these waves could both heat (to temperatures in excess of 10^4 K) and drive the wind (Hartmann *et al.*, 1981). In this picture, the existence of hybrid atmospheres and winds act as the connecting link between the hot, fast winds of the dwarf stars, and the cool, low velocity winds of the red giants and supergiants.

Subsequently, much effort has been directed to define the character of the atmosphere and winds in the hybrid stars. At the furthest extent of the hybrid atmosphere, species such as Mg II and Ca II must exist, suggesting low terminal temperatures at the high asymptotic flow speed of these stars. In α TrA, simple atmospheric models suggest (Hartmann *et al.*, 1985) that the material at high temperatures of 2×10^5 K most probably has a scale height that is an appreciable fraction of a stellar radius. The forbidden lines of Si III] and C III] appear to be broadened by turbulent pressure rather than expansion. This observation of turbulent rather than expansion broadening for the forbidden lines was recently confirmed for α TrA and γ Aql, another hybrid star (Brown *et al.*, 1986). In fact, it has not been possible to ascertain the rate of expansion, if at all, of the material in the atmospheres of the hybrid stars. The centroids of the emission lines (both permitted and forbidden) appear to lie approximately at their rest wavelengths to within the measuring accuracy which is about 10 km s^{-1} in the weakly exposed spectra. A small expansion of the atmosphere in the line forming regions can be accommodated, which would be consistent with the turbulently broadened forbidden transitions, and a proposed model incorporating Alfvén-wave driven winds (Hartmann *et al.*, 1981).

One of the best ways to infer the temperature distribution of an extended atmosphere may result from study of binary systems in which a giant or supergiant star is one component, and a hot main sequence star acts as a background source of radiation observed through the atmosphere of the luminous companion. For two such systems, 32 Cyg and 22 Vul, the wind electron temperature T_e could be estimated from the *observed* population of excited Fe II levels. For 32 Cyg (K4 Ib), at distances of more than 5 K-star radii, Che-Bohnenstengel (1984) found $T_e = 4800$ K if $n_e/n_H = 0.01$, and $T_e \approx 10^4$ K for lower electron densities. The value of the temperature in LTE would be 4200 K. From his modeling of the 32 Cyg system, Schröder (1986) found a tendency for the electron temperature to increase with height in the atmosphere of the K supergiant, e.g. from 8500 K at 0.2 R_* to 11 000 K at 0.5 R_*. The hydrogen ionization appeared to increase over the same range for $N_e/N_H = 0.001$ to 0.01. These later results imply that extensive nonradiative heating occurs at heights where wind acceleration has started.

In the binary system 22 Vul (G3 II–Ib), the population of excited Fe II levels is much higher than in 32 Cyg, and a wind electron temperature of $3.0 \pm 1.0 \times 10^4$ K was estimated (Reimers and Che-Bohnenstengel, 1986). However, since radiative excitation via higher levels cannot be excluded, the estimate of T_e made with the assumption of pure electron collisions remains to be confirmed by improved methods. Such a high wind temperature is particularly remarkable in a star like 22 Vul which has characteristics in common with 'hybrid atmosphere' stars like α Aqr: a high wind velocity and position in the HR diagram. Twenty-two Vul is a young, intermediate-mass star, that has an extended chromosphere observed to 1 R_* above the photosphere during its 1985 eclipse (Schröder and Che-Bohnenstengel, 1985).

4.3. VELOCITY PROFILE

For the Zeta Aurigae binaries 32 Cyg and 31 Cyg Schröder (1985a) found from a combination of an empirical chromospheric density model with observed wind parameters, a velocity acceleration model (see Section 5):

$$V = \frac{\dot{M}}{4\pi r^2 \rho} (1 - R_*/r)^{2.5}.$$

The profile of the acceleration in cool star atmospheres has been studied in only a few stars. The Zeta Aurigae type eclipsing binary 22 Vul (G3 Ib—II + B9) has been studied by Reimers and Che-Bohnenstengel (1986) by constructing a three-dimensional non-spherical line transfer code to interpret the line profiles at various phases. A 'standard' velocity acceleration law, $v = v_\infty(1 - R_G/r)^\alpha$ (where $\alpha = 2.5$, $v_\infty = 160$ km s^{-1}, and R_G is the radius of the G supergiant star) gave satisfactory agreement with the lines of Multiplet 2 in Si II at $\lambda 1530$, but produced large discrepancies with Fe II multiplets. Much better agreement between the calculated profiles and observations were achieved by rather arbitrarily assuming two acceleration zones. A gradual acceleration to ≈ 100 km s^{-1} and then, at about $6 R_*$, the distance of the companion star, a rapid acceleration to the terminal velocity of ≈ 160 km s^{-1}. This terminal velocity exceeds the escape velocity from the supergiant which is ≈ 100 km s^{-1} at $6 R_*$. However, the somewhat artificial nature of the velocity law may be due to the neglect of the second ionization of Fe.

Alpha Ori (M2 Iab) has a rich spectrum of Fe II emission features from which Carpenter (1984) selected a number of unblended multiplets. Inspection of these features (see Figure 7) showed asymmetries that appear to correlate with the strength of the line. Within the same multiplet, lines of largest gf-value (the 'strongest' lines) are believed to form farthest out from the stellar photosphere; these profiles show a centrally reversed emission profile with the short wavelength side stronger than the long wavelength side, suggesting formation in a decelerating region. The profiles from intermediate strength transitions display the opposite asymmetry (short wavelength less than long wavelength component) indicating outward acceleration; profiles of the weakest lines (formed at the deepest levels of the atmosphere) are either symmetric or show no central reversal at all. This pattern suggested to Carpenter (1984) that the flow accelerates with distance from the photosphere, the velocity reaches a local maximum and then the wind decelerates.

Semi-empirical modeling of Mg II line profiles for single luminous stars by several authors (for instance, Dupree, 1982; Drake and Linsky, 1983, 1984b; Drake, 1985; Jasinski, 1986) or modeling of observed ultraviolet line fluxes (Hartmann et al., 1985; Brosius and Mullan, 1986) have considered relatively smooth velocity profiles which have given more or less reasonable agreement with the observed profiles, fluxes, or terminal velocities. It may be that line profiles other than Mg II are more sensitive to acceleration conditions in restricted atmospheric regions; it may also be that the two stars studied in detail (22 Vul and Alpha Ori) are somewhat anomalous, perhaps because both have companions that modify the flow characteristics.

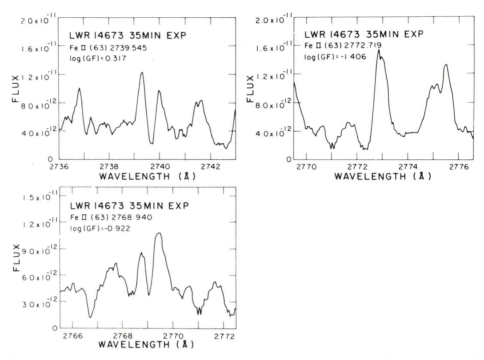

Fig. 7.   Several FeII profiles in Alpha Ori chosen to illustrate the change in line asymmetry with increasing line strength. Figure from Carpenter (1984).

4.4. WIND VARIABILITY

With repeated observations of cool stars with IUE, the variability of winds has been noticed. Ground-based observations first called attention to variable circumstellar lines in luminous cool star spectra, particularly in the CaII absorption features (Reimers, 1977a). The HeI λ10830 absorption feature appeared broad (several hundred kilometers per second) and variable in a few hybrid giant stars (O'Brien, 1980). The MgII profiles have been well-studied in a number of stars. The hybrid stars, in particular, exhibit irregular variability on a time scale of about one year (Dupree and Baliunas, 1979; Hartmann *et al.*, 1985; Brown *et al.*, 1986). Figure 8 displays an obvious case of variability in the wind of Alpha TrA, as the short wavelength portion of the MgII doublet varies. The deep exposure of this star shown previously (Figure 6) was obtained in August of 1982 when there was no high velocity emission. It appears that wind at highest velocities is the most variable. There were no concurrent changes in the fluxes of other lines or in several other hybrid stars, with the exception of Alpha Aqr (Dupree and Baliunas, 1979), or in the long wavelength portion of the MgII line profile, so that changes in the wind opacity appear to be the most likely interpretation of this variation.

The supergiant Alpha Ori (M2 Iab) also exhibits variability in the short wavelength side of MgII profiles and flux changes thta can amount to a factor of 2 in this line. Study of the profile indicates (Dupree *et al.*, 1984, 1986) both intrinsic chromospheric changes as well as changes in the wind opacity may be present.

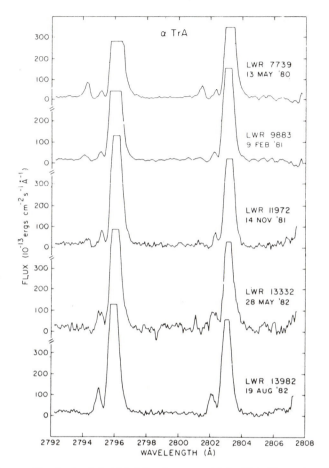

Fig. 8. Mg II line profiles in the hybrid bright giant star Alpha TrA illustrating the variability of the short wavelength emission on a time scale of ≈ 9 months or less. Figure from Hartmann *et al.* (1985).

The RS CVn system λ And (G7—G8, IV—III + ?) may have mass loss associated with surface inhomogeneities (Baliunas and Dupree, 1982). At times of low photospheric brightness, when spots are thought to be present on the visible hemispheres, both the Ca II and the Mg II lines exhibit asymmetries suggesting downflow in the atmosphere. When the system is bright, and spots are minimal or absent, the Mg II and Ca II lines indicate outward mass motions. This behavior is reminiscent of the phenomenology associated with solar coronal holes and active regions.

5. Mass Loss Rates

Actual values for the mass loss rates are of course required, and these are not easy to come by. Interpretation of line profiles must generally be made in the context of an atmospheric model that has been constructed to match as many of the observations as possible, with as physically realistic a model as one can construct. IUE has contributed quantitative measurements of line fluxes and line profiles which have been combined with ground-based measures of H-α, Ca II chromospheric transitions, or features from circumstellar envelopes, and when available, radio continuum observations or X-ray flux measurements, to evaluate the mass loss rate for a particular star. A review of mass loss determinations for cool stars using a variety of techniques (Figures 12 and 13 in Dupree, 1986) shows that giants and bright giants appear to have generally lower rates of mass loss (10^{-10} to $10^{-5} M_\odot$ yr^{-1}) than supergiant stars (10^{-9} to 10^{-4}). This simply may reflect the larger surface area of supergiants, with the same average energy per unit area devoted to extending the atmosphere and driving the wind. At the present stage of analysis, there is a large spread (about two orders of magnitude) in the inferred mass loss rate at a given position in the HR diagram.

5.1. DETAILED STUDIES OF SINGLE STARS

Two single stars, Alpha Bootes and Alpha Orionis, have observational material of sufficiently high quality to justify serious effort in modeling the atmosphere and wind. It is instructive to review briefly some of the recent results on these objects.

Alpha Bootes (*K*2 *IIp*): An early study of this star (Chiu *et al.*, 1977) resulted in a model based principally on the Ca II H and K lines (Ayres and Linsky, 1975), and obtained a mass loss rate ($\dot{M} = 8 \times 10^{-9} M_\odot$ yr^{-1}) with a plane-parallel atmosphere constructed to fit the Mg II and Ca II line profiles. Ayres *et al.* (1982) derived lower values ($\dot{M} = 3 \times 10^{-10}$ to $2 \times 10^{-9} M_\odot$ yr^{-1}) from the narrow circumstellar absorption features in the Mg II profiles. The value of $\dot{M}$ decreased still further ($\dot{M} = 10^{-10} M_\odot$ yr^{-1}) with a more sophisticated atmospheric code to interpret the Mg II lines that incorporated spherical geometry and partial frequency redistribution (Drake and Linsky, 1984b). Most recent studies using deep exposures of the ultraviolet spectrum of this star (Judge, 1986; Ayres *et al.*, 1987; Ayres, 1987) have led to a better quantitative understanding of the complex processes that occur in the line forming regions of Alpha Boo. Judge (1986) has recently constructed a detailed, static model of the atmosphere based on all available ultraviolet spectra; this model should provide a better reference with which to determine the mass loss rate for Alpha Boo.

Alpha Orionis (*Betelgeuse*; *M*1—2 *Ia—Iab*): To understand the energetics of the atmosphere, and to obtain a mass loss rate, Hartmann and Avrett (1984) constructed a self-consistent model of Alpha Ori. While adhering to the assumption that Alfvén waves are extending the atmosphere and driving the wind, these authors required the predictions of the model atmosphere to be in harmony with: a low terminal velocity; the continuum radio flux as a function of frequency; the cores of the H-α transition and the Ca II infrared triplet; fluxes and profiles of chromospheric lines; and speckle imaging of the circumstellar environment in H-α. Agreement with the various line profiles and fluxes are not as satisfactory as one would like, leading to the suggestion that the atmosphere has complex internal motions in the low chromosphere. The mass loss rate associated with this model is $1.4 \times 10^{-6} \, M_\odot \, \text{yr}^{-1}$, a value that is consistent with the most recent determination of Bowers and Knapp (1987), based on 21-cm emission observed with the VLA: $\dot{M} = 2.2 \times 10^{-6} M_\odot \, \text{yr}^{-1}$. Other determinations are listed in Table 1 of Dupree (1986).

Of particular interest to the mass loss process in this star is a recent result obtained from extensive monitoring with IUE. There appears to be an ≈ 1.1 yr periodic variation in the flux ratio of the Mg II h and k transitions (Sonneborn *et al.*, 1986; Dupree *et al.*, 1987). Because the appearance of the Mg II profiles can be modified by both interstellar and circumstellar absorption and scattering from the surrounding extended atmosphere, the observed periodic modulation in the Mg II ratio may result from a short-term variability in the extended atmosphere and wind, perhaps initiated by periodic photospheric variations. It has long been known that Alpha Ori is a semi-regular variable (see review by Guinan, 1984), and a much longer, 5.68 yr period in brightness and in radial velocity was identified approximately 50 yr ago. More recently, Karovska *et al.* (1986) have used speckle techniques to image two companions to Alpha Ori, the closest of which is only 0.06 arc seconds from Alpha Ori itself. The mean distance of 2.5 R_* for this companion star, places it deep within the atmosphere of the supergiant primary. The derived orbit indicates a ≈ 2.1 year period, and periastron shortly preceded a time when the wind of the star appeared disturbed in the IUE spectra of the Mg II lines (Dupree *et al.*, 1986). However, other two dimensional speckle images suggest that there is an elongated cloud of H-α emission, but no evidence for a close companion (Hebden *et al.*, 1986). If a close-in companion exists, interaction between the companion and the extended atmosphere of Alpha Ori would be expected.

5.2. BINARIES

High-resolution IUE spectra of ζ Aur eclipsing binary systems have provided some of the most accurate mass loss rates available for red supergiants (Table II). Zeta Aurigae binaries consist of a K supergiant and a B-type main sequence star with typical binary separation of 5 to 10 Au. The binary technique of determining mass loss rates was invented by Deutsch (1956) for visual binaries and applied to α Her (Reimers, 1977b, 1978) and α Sco (Kudritzki and Reimers, 1978). By using circumstellar absorption lines from a predominant ionization stage observed in the spectrum of a close visual companion, one can establish the scale of the system and avoid the difficulty always present for single stars, namely, to locate the circumstellar absorption in the line of sight.

With IUE, the apparent separation of red supergiants with hot companions is replaced by a separation of the complementary energy distributions of the components. At IUE wavelengths, in particular in the short-wavelength range, one observes a pure B star spectrum upon which is superposed numerous features from the red supergiant. These include circumstellar P Cygni type lines formed in the extended wind, and chromospheric absorption lines (near eclipse).

The B star serves as an astrophysical light source (a 'natural satellite') that moves around in the wind of the cool star (see Figures 9a and b). However,

TABLE II

Mass-loss rates determined from high-resolution IUE spectra of ζ Aur/VV Cep type binaries

Star	Spectral type (primary)	Mass $(M/M_\odot)$	Radius $(R/R_\odot)$	Luminosity $(\text{Log } L/L_\odot)$	Mass-loss rate, $\dot{M}$ $(M_\odot \text{ yr}^{-1})$	Wind velocity (km s^{-1})	Reference
ζ Aur	K4 Ib	8.3	140	3.41	6×10^{-9}	40	1
32 Cyg	K5 Iab	8	188	3.82	2.8×10^{-8}	60	1
31 Cyg	K4 Ib	6.2	202	3.91	4×10^{-8}	80	1
22 Vul	G3 II—Ib	4.3	40	2.99	6×10^{-9}	160	2
δ Sge	M2 II	8	140	3.43	2×10^{-8}	28	3
α Sco	M1.5 Iab—Ib	18	575	4.68	1×10^{-6}	17	4
Boss 1985	M2 Iab ep	—	—	—	$> 10^{-7}$	10	5

References

(1) Che *et al.* (1983).
(2) Reimers and Che-Bohnenstengel (1986).
(3) Reimers and Schröder (1983).
(4) Hagen *et al.* (1986).
(5) Che and Reimers (1983).

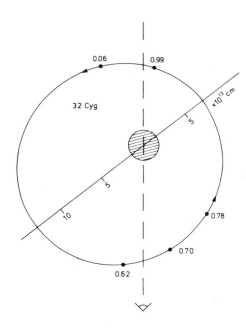

Fig. 9a.   Orbit and location of B star relative to K supergiant (roughly on scale) in the system 32 Cyg at phases of IUE observations.

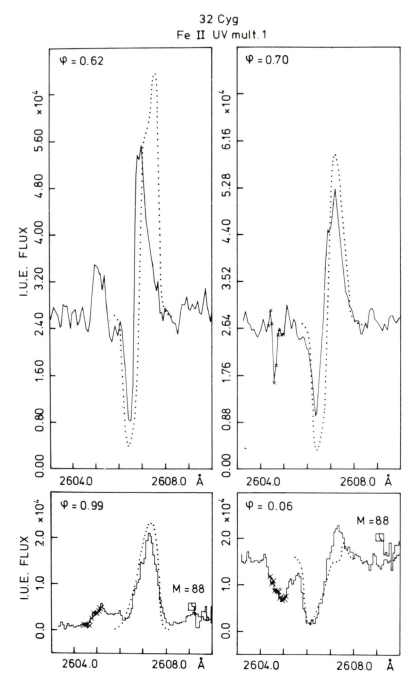

Fig. 9b. Profiles of wind lines superimposed upon B star spectrum in 32 Cyg at phases shown
in Figure 9a.

compared to widely separated visual binaries, a number of additional difficulties arise:

— a non-spherical, 3-dimensional line transfer problem has to be solved since the light source (B star) is offset from the center of symmetry of the wind. Computer codes to solve this problem have been developed in the 2-level approximation by Hempe (1982, 1984) and for the multilevel case by Baade (1986).
— the wind is disturbed in the immediate surrounding of the B star as it moves supersonically through the wind and forms an accretion shock front (Chapman, 1981). However, detailed study of the accretion shocks demonstrates that their geometrical size is very small compared to the circumstellar shell and can be neglected in line transfer calculations (Che-Bohnenstengel and Reimers, 1986).
— the hot B star ionizes the wind, i.e. an H II region is formed within the red supergiant wind. In 31 Cyg and α Sco, in particular, the size of the H II region is larger. This ionization structure must be taken into account quantitatively since ions like Si II and Fe II, which are used for the mass loss rate determination, may be ionized even further within the H II region.

On the other hand, ζ Aur binaries are one of the few stars in addition to the Sun where the winds and extended chromospheres can be studied with spatial (height) resolution.

Wind Lines

The wind is visible at all phases in P Cygni type profiles from ions such as Fe II, Si II, S II, Mg II, C II, Al II, and O I. (During total eclipse of the B star pure emission lines are found.) These lines are formed by scattering of B star photons in the wind of the red giant. A few wind lines such as Fe II Mult. 9 (~ 1275 Å) are seen in pure absorption due to the branching ratios of the upper levels which favor reemission as Fe II UV Mult. 191 photons (Hempe and Reimers, 1982; Baade, 1986).

Theoretical modelling of wind line profiles and of their phase dependency has yielded accurate mass loss rates and wind velocities for a number of systems (Table II). A good mass loss determination requires observations at phases with the B star in front of the red supergiant (which yields wind turbulence v_t) and phases with the B star behind the red supergiant (which yields the terminal wind velocity v_w). Typically, $v_w \approx 2v_t$. Che *et al.* (1983) showed that it was possible to match the circumstellar line profiles at all phases with *one* set of parameters v_w, v_t and — within a factor of 2 — one mass loss rate $\dot{M}$. This means that, at least in the orbital plane, the envelope asymmetries (in density) are within a factor of 2 on a length-scale of several K giant radii.

The example of α Sco shows that the influence of interstellar lines has to be considered carefully. As demonstrated in Figure 10 by Si II 1526 Å, *all* circumstellar 0.00 eV lines in the spectrum of α Sco B suffer from interstellar absorption on the longward side of the profile (the interstellar velocity is $\sim +5$ km s^{-1}, Kudritzki and Reimers, 1978), i.e. the reemission part of the circumstellar P Cygni profile appears absorbed at the resolution of IUE, while *all* other lines (like Si II 1533 Å) have the expected P Cyghi profiles. Both van der Hucht *et al.* (1980) and

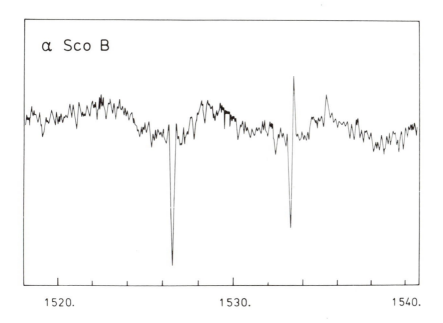

Fig. 10. High resolution IUE spectrum (5 single spectra summed) of α Sco B with Si II $\lambda 1526/33$ resonance doublet showing the interstellar contamination of the 0.00 eV ground state transition.

Bernat (1981) used only the 0.00 eV pure absorption lines (primarily interstellar, not circumstellar absorption) to determine mass loss rates which consequently overestimated the mass loss rate by one order of magnitude with a scatter of a factor of ~30 among rates from different ions. The correct treatment of the P Cygni profiles with Hempe's (1982) nonspherical line transfer code yields a mass loss rate of 10^{-6} $M_\odot$ yr^{-1} (Hagen, 1984; Hagen *et al.*, 1986), consistent with Hjellming and Newell's (1983) rate from radio emission from the H II region within the wind and with the rate from optical observations of Ti II lines in α Sco B (Kudritzki and Reimers, 1978).

Chromospheric Lines

The extended chromospheres — whre the wind starts to expand — were studied by means of atmospheric eclipses of the B stars by O. C. Wilson, H. G. Groth, Wright and others in the 1950's; and IUE data are a major advance in several respects: the B star provides a smooth continuum on which numerous absorption lines are observed up to heights (projected binary separations) of more than one supergiant radius above the photosphere. In addition to absorption lines, the wavelength and time dependence of totality in the UV can be used to construct a density model of the inner chromosphere (see Figure 11, Schröder, 1985a, b, 1986). Chromospheric densities could be represented by power laws of the form $\rho \sim r^{-2} \cdot h^{-a}$ with $a \approx 2.5$ where r is the distance from the center of the star and h is the height above the photosphere. The empirical density distribution indicates that after a steep decrease in the inner chromosphere, expansion starts

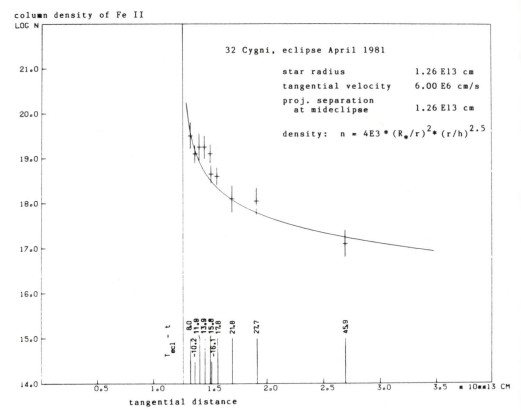

Fig. 11. Observed chromosphere column densities of Fe II versus tangential distance (crosses) in the system 32 Cyg. Solid line is calculated with density distribution as given by formula in the figure.

$(\rho \sim r^{-2})$ in the upper chromosphere at a height greater than 0.5 to 1 R_*. A typical density at a height of 2 R_* is 10^7 cm^{-3}.

Since one observes total particle densities in the expanding chromosphere up to $h \approx 1.5\ R_K$ and in addition the wind density and velocity outside of $\sim 5\ R_K$, one can try to look for consistency by assuming a steady wind, i.e. to apply the equation of continuity. Using $\dot{M} = 4\pi r^2 \cdot \rho(r) \cdot v(r)$ and

$$\rho(r) = \rho_0 \left[\frac{R_*}{r} \right]^2 \left[\frac{v}{r - R_*} \right]^a$$

for the chromosphere we find

$$v(r) = \frac{\dot{M}}{4\pi\rho_0 R^2} \left[1 - \frac{R_*}{r} \right]^a$$

and a wind terminal velocity

$$v_\infty = \dot{M}(4\pi\rho_0 \cdot R_*^2)^{-1}$$

which can be checked with observed values for ρ_0, $\dot{M}$ and v_∞ for consistency. For 32 Cyg and 31 Cyg, Schröder (1985a) found consistency. Thus, the empirical density distribution (when extrapolated by the equation of continuity to the outer wind) yields a mass loss rate consistent with the rate determined from circumstellar line profiles. In the case of ζ Aur, the chromospheric density distribution was far too steep — at least at that particular limit position during eclipse — to give the mean mass loss rate. In this star we might have the stellar analogue to a solar coronal hole.

Since these binaries represent a homogeneous sample, it is worthwhile to see whether a relation can be identified to relate the mass loss rate to fundamental physical parameters. Figure 12 exhibits the mass loss rates for the six systems listed in Table II and two relations (indicated by the broken lines) describing the mass loss rates: R_*^2 and L/gR_*. The top illustration contains the parameterization originally proposed by Reimers for red giants (1975); the lower illustration contains a simple surface area parameterization. Based on the data for these six systems, it may be that the dependence of the mass loss rate on L/gR_* is to be preferred, but the better agreement is certainly not overwhelming. Clearly, additional systems need to be carefully studied, and their results interpreted in conjunction with quantitative results for single stars.

6. Achievements of IUE

Access to ultraviolet spectroscopic observations of cool stars made possible by IUE has created a renaissance in cool star research and the study of stellar winds. Highlights of our new understanding include:

— the first comprehensive picture of mass loss and stellar winds across the HR diagram;
— discovery of a new class of stars with hot atmospheres and massive winds (the *hybrid* stars);
— incontrovertible evidence that stellar winds are variable;
— discovery of the connection between fast, hot, solar-type winds and slow, cold, massive winds of luminous cool stars;
— a powerful impetus to establish solar-stellar connections in magnetic activity and mass loss in stars;
— quantitative determination of relevant physical parameters of winds in cool stars;
— evidence for large scale mass motions in transition regions of low-gravity stars;
— mapping of wind flow patterns in luminous stars;
— detection of the start of wind accelerations in stellar chromospheres;
— elimination of a simple 2-dimensional categorization (temperature and luminosity) of stellar winds and their character;
— definition of a new field of study: the evolution of stellar atmospheres;
— unique investigations of stellar wind stratification from ultraviolet spectra of binaries;
— vital links between ground-based chromospheric studies and X-ray observations.

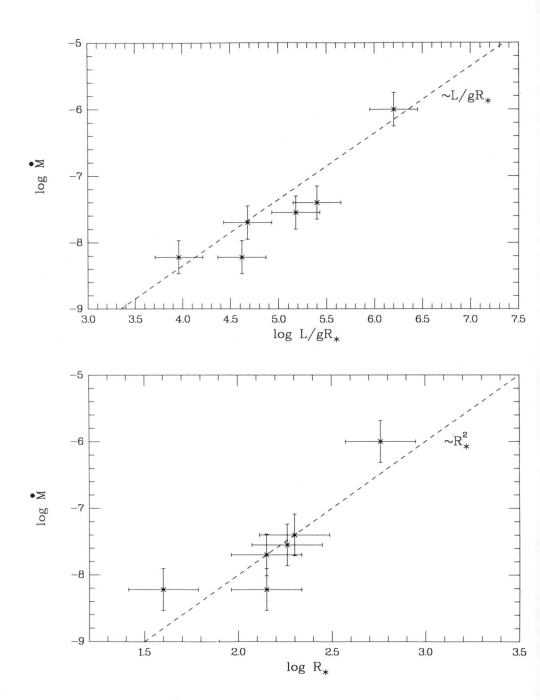

Fig. 12.   Mass loss rates ($\dot{M}$ in units of solar masses yr^{-1}) for several well-studied binary systems and 2 approximations (R_*^2 and L/gR_*, in solar units). The star and the source of $\dot{M}$ are given in Table II.

The results from IUE have been, and continue to be influential in the field of cool star winds. In the next Section, selected outstanding problems are discussed which will require not only extended capability in space instrumentation, but also vigorous ground-based measurements complemented by substantial theoretical activity.

7. Outstanding Problems

The great strides made by the IUE satellite have served to develop a comprehensive picture of mass loss across the HR diagram. Yet there are still problems to be addressed. We briefly discuss several below:

Lyman-α Observations: Mass loss most probably is ubiquitous in stars. The optically thick emission lines that we have discussed vary in their sensitivity to material motions in an atmosphere. Measurement of the Lyman-α profile will provide a more sensitive signature of motions than the Mg II line that is readily available. Only several of the brightest stars have been measured in Lyman-α; it will take renewed effort with IUE or undoubtedly HST to obtain these critical profiles. In most cool, luminous stars of Population I, one must address the serious problem of deconvolving the very strong interstellar Lyman-α component from the stellar profile.

Doppler Shift Measures: Direct measures of Doppler shifts of transition region lines are invaluable to determine the motions in an atmosphere. Spectral resolution on the order of 10 km s^{-1} will be necessary to detect the onset of a wind deep in a stellar atmosphere.

Metallicity Dependence: Observations of fainter stars will allow objects of different metallicity to be reached, and aid in identification of a driving mechanism for a wind. To date, almost all of the cool stars studied for mass loss belong to a metal rich population. It will be necessary to explore the effects of metallicity on mass loss for a true picture to extend our understanding to other galaxies.

Modeling: We have focussed principally on the detection of mass loss in cool stars, but the major physical result is a determination of the actual mass loss rates. For this one needs substantive modeling to complement the observations. Such semi-empirical modeling activities are time-consuming, but must be encouraged if we are to achieve an understanding of the energetics and influence of mass loss on stellar evolution.

Simultaneous Multi-frequency Observations: To understand deeply the physics of magnetic activity in stars, as evidenced by their atmospheres, winds, and mass loss, we can not focus on narrow ranges of emissions as measured in the optical, or the ultraviolet, or the X-ray, or the infrared. Every spectral region has its unique advantages, or probes complementary aspects of stellar physics. Our theoretical understanding will be woefully incomplete, or even wrong, if constructed on narrowly based observational material. The variety of dynamic processes that stars cleverly produce to challenge our theories needs to be measured at many frequencies, and in most instances measured contemporaneously, if not simultaneously.

Mass Loss Rates: Quantitative estimates of $\dot{M}$ have been derived for only a handful of stars, having relatively large mass loss rates and generally of high metallicity. Mass loss rates for a much broader sample of stars need to be obtained to understand the mass loss process and its effects on stellar evolution.

Acknowledgements

We have benefited from the enthusiastic comments proffered by T. Ayres, A. Brown, P. Judge, J. Linsky, and R. Stencel. Preparation of this review was supported in part by NASA Grant NAG5-87 to the Smithsonian Astrophysical Observatory.

References

Athay, R. G., Gurman, J. B., and Henze, W.: 1983, *Astrophys. J.* **269**, 706.
Avrett, E. H.: 1981, 'Energy Balance in Solar and Stellar Chromospheres', in R. M. Bonnet and A. K. Dupree (eds.), *Solar Phenomena in Stars and Stellar Systems*, D. Reidel Publ. Co., Dordrecht, Holland, pp. 173—198.
Ayres, T. R.: 1984, *Astrophys. J.* **284**, 784.
Ayres, T. R.: 1985, *Astrophys. J.* **291**, L7.
Ayres, T. R.: 1987, *Astrophys. J.* (in press).
Ayres, T. R. and Linsky, J. L.: 1975, *Astrophys. J.* **200**, 660.
Ayres, T. R., Simon, T., and Linsky, J. L.: 1982, *Astrophys. J.* **263**, 791.
Ayres, T. R., Linsky, J. L., Simon, T., Jordan, C., and Brown, A.: 1983a, *Astrophys. J.* **274**, 784.
Ayres, T. R., Stencel, R. E., Linsky, J. L., Simon, T., Jordan, C., Brown, A., and Engvold, O.: 1983b, *Astrophys. J.* **274**, 801.
Ayres, T. R., Judge, P., Jordan, C., Brown, A., and Linsky, J. L.: 1987, *Astrophys. J.* (in press).
Baade, R.: 1986, *Astron. Astrophys.* **154**, 145.
Baliunas, S. L. and Dupree, A. K.: 1982, *Astrophys. J.* **252**, 668.
Baliunas, S., Hartmann, L., and Dupree, A. K.: 1983, *Astrophys. J.* **271**, 672.
Bernat, A. P. and Lambert, D. L.: 1976, *Astrophys. J.* **204**, 830.
Bernat, A. P.: 1981, *Astrophys. J.* **252**, 644.
Böhm-Vitense, E.: 1981, *Astrophys. J.* **244**, 504.
Bowers, P. F. and Knapp, G.: 1987, *Astrophys. J.* (in press).
Brosius, J. W. and Mullan, D. W.: 1986, *Astrophys. J.* **301**, 650.
Brown, A.: 1986, in *Advances in Space Research*, Proc. COSPAR Meeting, No. 26, New York, Pergamon (in press).
Brown, A., Jordan, C., Stencel, R. E., Linsky, J. L., and Ayres, T. R.: 1984, *Astrophys. J.* **283**, 731.
Brown, A., Reimers, D., and Linsky, J. L.: 1986, in *New Insights in Astrophysics — 8 Years of UV Astronomy with IUE*, ESA SP-263, pp. 169—172, Noordwijk: ESA.
Brueckner, G. E., Bartoe, J. D. F., and Van Hoosier, M. E.: 1977, *Proc. OSO-8 Workshop*, Boulder, Univ. of Colorado, U.S.A., pp. 380—418.
Carpenter, K. G.: 1984, *Astrophys. J.* **285**, 181.
Carpenter, K. G., Wing, R. F., and Stencel, R. E.: 1985, *Astrophys. J. Suppl.* **57**, 405.
Chapman, R. D.: 1981, *Astrophys. J.* **248**, 1043.
Che-Bohnenstengel, A.: 1984, *Astron. Astrophys.* **138**, 333.
Che, A., Hempe, K., and Reimers, D.: 1983, *Astron. Astrophys.* **126**, 225.
Che, A. and Reimers, D.: 1983, *Astron. Astrophys.* **127**, 227.
Che-Bohnenstengel, A. and Reimers, D.: 1986, *Astron. Astrophys.* **156**, 172.
Chiosi, C. and Maeder, A.: 1986, *Ann. Rev. Astron. Astrophys.* **24**, 329.
Chiu, H. Y., Adams, P. S., Linsky, J. L., Basri, G. S., Maran, S. P., and Hobbs, R. W.: 1977, *Astrophys. J.* **211**, 453.
de Jager, C., Kondo, Y., Hoekstra, R., van der Hucht, K. A., Kamperman, T. M., Lamers, H. J. G. L. M., Modisette, J. L., and Morgan, T. H.: 1979, *Astrophys. J.* **230**, 534.
Dere, K. P., Bartoe, J. D. F., and Brueckner, G. E.: 1984, *Astrophys. J.* **281**, 870.

Deutsch, A. J.: 1956, *Astrophys. J.* **123**, 210.

Drake, S. A.: 1985, 'Modeling Lines Formed in the Expanding Chromospheres of Red Giants', in J. E. Beckman and L. Crivellari (eds.), *Progress in Stellar Spectral Line Formation Theory*, D. Reidel Publ. Co., Dordrecht, Holland, pp. 351—357.

Drake, S. A.: 1986a, in *New Insights in Astrophysics — 8 Years of UV Astronomy with IUE*, ESA SP-263, Noordwijk, ESA, pp. 193—196.

Drake, S. A.: 1986b, in D. Gibson and M. Zeilik (eds.), *Proc. Cambridge Workshop on Cool Stars, Stellar Systems, and the Sun, 4th*, Lecture Notes in Physics, No. 254, New York, Springer-Verlag, pp. 369—384.

Drake, S. A. and Linsky, J. L.: 1983, *Astrophys. J.* **273**, 299.

Drake, S. A. and Linsky, J. L.: 1984a, *Bull. Am. Astron. Soc.* **16**, 895.

Drake, S. A. and Linsky, J. L.: 1984b, in S. L. Baliunas, L. Hartmann (eds.), *Proc. Cambridge Workshop on Cool Stars, Stellar Systems, and the Sun, 3rd*, Lecture Notes in Physics, No. 193, New York, Springer-Verlag, pp. 350—352.

Drake, S. A., Brown, A., and Linsky, J.: 1984, *Astrophys. J.* **284**, 774.

Dupree, A. K.: 1975, *Astrophys. J. Lett.* **200**, L27.

Dupree, A. K.: 1976, in R. Cayrel and M. Steinberg (eds.), *Physique des Movements dans les Atmosphères Stellaires*, Colloques Internationaux du Centre National de la Recherche Scientifique, No. 250, p. 439.

Dupree, A. K.: 1980, in C. Chiosi and R. Stalio (eds.), 'Effects of Mass Loss on Stellar Evolution', *IAU Colloq.* **59**, 87.

Dupree, A. K.: 1982, in Y. Kondo, J. Mead, and R. Chapman (eds.), *Advances in Ultraviolet Astronomy: Four Years of IUE Research*, NASA CP-2238, pp. 3—16.

Dupree, A. K.: 1986, *Ann. Rev. Astron. Astrophys.* **24**, 377.

Dupree, A. K. and Baliunas, S. L.: 1979, *IAU Circ. No. 3435*.

Dupree, A. K. and Raymond, J. C.: 1982, *Astrophys. J. Lett.* **263**, L63.

Dupree, A. K., Hartmann, L., and Smith, G.: 1984, in S. L. Baliunas and L. Hartmann (eds.), *Proc. Cambridge Workshop on Cool Stars, Stellar Systems, and the Sun, 3rd*, Lecture Notes in Physics, No. 193, New York, Springer-Verlag, pp. 326—329.

Dupree, A. K., Sonneborn, G., Baliunas, S. L., Guinan, E. F., Hartmann, L.: 1984, in J. M. Mead, R. D. Chapman, and Y. Kondo (eds.), *Future of Ultraviolet Astronomy Based on Six Years of IUE Research*, NASA CP-2349, pp. 462—467.

Dupree, A. K., Baliunas, S. L., Guinan, E. F., Hartmann, L., and Sonneborn, G. S.: 1986, in D. Gibson and M. Zeilik (eds.), *Proc. Cambridge Workshop on Cool Stars, Stellar Systems, and the Sun, 4th*, Lecture Notes in Physics, No. 254, New York, Springer-Verlag, pp. 411—413.

Dupree, A. K., Baliunas, S. L., Guinan, E. F., Hartmann, L., Nassiopoulos, G., and Sonneborn, G. S.: 1987, *Astrophys. J.* (in press).

Goldberg, L.: 1985, 'Optical Spectroscopy of Red Giants', in M. Morris and B. Zuckerman (eds.), *Mass Loss from Red Giants*, Reidel Publ. Co., Dordrecht, Holland, pp. 21—27.

Guinan, E. F.: 1984, in S. L. Baliunas, L. Hartmann (eds.), *Proc. Cambridge Workshop on Cool Stars, Stellar Systems, and the Sun, 3rd*, Lecture Notes in Physics, No. 193, New York, Springer-Verlag, pp. 336—341.

Hagen, H. J.: 1984, Diplomarbeit Universität Hamburg.

Hagen, H. J., Hempe, K., and Reimers, D.: 1986, *Astron. Astrophys.* (submitted).

Hartmann, L. and Avrett, E. H.: 1984, *Astrophys. J.* **284**, 238.

Hartmann, L. and MacGregor, K. B.: 1980, *Astrophys. J.* **242**, 260.

Hartmann, L., Dupree, A. K., and Raymond, J. C.: 1980, *Astrophys. J. Lett.* **236**, L143.

Hartmann, L., Dupree, A. K., and Raymond, J. C.: 1981, *Astrophys. J. Lett.* **246**, 193.

Hartmann, L., Dupree, A. K., and Raymond, J. C.: 1982, *Astrophys. J.* **252**, 214.

Hartmann, L., Jordan, C., Brown, A., and Dupree, A. K.: 1985, *Astrophys. J.* **296**, 576.

Hebden, J. C., Christou, J. C., Cheng, A. Y. S., Hege, E. K., Strittmatter, P. A., Beckers, J. M., and Murphy, H. P.: 1986, *Astrophys. J.* **309**, 754.

Hempe, K.: 1982, *Astron. Astrophys.* **115**, 133.

Hempe, K.: 1984. *Astron. Astrophys. Suppl.* **56**, 115.

Hempe, K. and Reimers, D.: 1982, *Astron. Astrophys.* **107**, 36.

Hjellming, R. M. and Newell, R. T.: 1983, *Astrophys. J.* **275**, 704.

Holzer, T. E. and MacGregor, K. B.: 1985, 'Mass Loss Mechanisms for Cool, Low-Gravity Stars', in M. Morris, B. Zuckerman (eds.), *Mass Loss from Red Giants*, D. Reidel Publ. Co., Dordrecht, Holland, pp. 229—255.

Hummer, D. G. and Rybicki, G. B.: 1968, *Astrophys. J.* **153**, 107.

Iben, I., Jr. and Renzini, A.: 1983, *Ann. Rev. Astron. Astrophys.* **21**, 271.

Jasinski, M.: 1986, *Astron. Astrophys.* (in press).

Judge, P. G.: 1986, *Monthly Notices Roy. Astron. Soc.* **221**, 119.

Judge, P. G., Jordan, C., and Rowan-Robinson, M.: 1986, *Monthly Notices Roy. Astron. Soc.* (in press).

Karovska, M., Nisenson, P., and Noyes, R.: 1986, *Astrophys. J.* **308**, 260.

Kondo, Y., Morgan, T. H., and Modisette, J. L.: 1976, *Astrophys. J.* **207**, 167.

Kondo, Y., de Jager, C., Hoekstra, R., van der Hucht, K. A., Kamperman, T. M., Lamers, H. J. G. L. M., Modisette, J. L., and Morgan, T. H.: 1979, *Astrophys. J.* **230**, 526.

Kudritzki, R.-P. and Reimers, D.: 1978, *Astron. Astrophys.* **70**, 227.

Linsky, J. L.: 1985, 'Mass Loss from Red Giants: Results from Ultraviolet Spectroscopy', in M. Morris and B. Zuckerman (eds.), *Mass Loss from Red Giants*, D. Reidel Publ. Co., Dordrecht, Holland, pp. 31—54.

Linsky, J. L. and Haisch, B. M.: 1979, *Astrophys. J. Lett.* **229**, L27.

Mallia, E. A. and Pagel, B. E. J.: 1978, *Monthly Notices Roy. Astron. Soc.* **184**, 55.

McClintock, W., Henry, R. C., Moos, H. W., and Linsky, J. L.: 1976, *Astrophys. J. Lett.* **204**, L103.

McClintock, W., Moos, H. W., Henry, R. C., Linsky, J. L., and Barker, E. S.: 1978, *Astrophys. J.* **37**, 221.

Morris, M. and Zuckerman, B. (eds.): 1985, 'Thermal Radio Emission from Molecules in Circumstellar Outflows', in *Mass Loss from Red Giants*, D. Reidel Publ. Co., Dordrecht, Holland.

Neugebauer, M.: 1983, in M. Neugebauer (ed.), *Solar Wind Five*, NASA Conf. Pub. No. 2280, pp. 135.

O'Brien, G. T. Jr.: 1980, *A Study of the 10830 Å Line of Neutral Helium in the Spectra of Red Giants*, Ph.D thesis. Univ. Texas, Austin.

Orrall, F. Q., Rottman, G. J., and Klimchuk, J. A.: 1983, *Astrophys. J. Lett.* **266**, L65.

Peterson, R. C.: 1981, *Astrophys. J. Lett.* **248**, L31.

Reimers, D.: 1975, *Mem. Soc. R. Sci. Liège*, 6ᵉ Ser. 8, pp. 369—82.

Reimers, D.: 1977a, *Astron. Astrophys.* **57**, 395.

Reimers, D.: 1977b, *Astron. Astrophys.* **61**, 217.

Reimers, D.: 1978, *Astron. Astrophys.* **67**, 161.

Reimers, D.: 1981, in I. Iben and A. Renzini (eds.), *Physical Processes in Red Giants*, D. Reidel Publ. Co., Dordrecht, Holland, pp. 269.

Reimers, D.: 1982, *Astron. Astrophys.* **107**, 292.

Reimers, D. and Schröder, K. P.: 1983, *Astron. Astrophys.* **124**, 241.

Reimers, D. and Che-Bohnenstengel, A.: 1986, *Astron. Astrophys.* (in press).

Rosner, R., Golub, L., and Vaiana, G. S.: 1985, *Ann. Rev. Astron. Astrophys.* **23**, 413.

Rottman, G. J., Orrall, F. Q., and Klimchuk, J. A.: 1982, *Astrophys. J.* **260**, 326.

Schröder, K.-P.: 1985a, *Astron. Astrophys.* **147**, 103.

Schröder, K.-P.: 1985b, Ph.D. Thesis University of Hamburg.

Schröder, K.-P.: 1986, *Astron. Astrophys.* (submitted).

Schröder, K.-P. and Che-Bohnenstengel, A.: 1985, *Astron. Astrophys. Letters* **151**, L5.

Sion, E. M. and Starrfield, S. G.: 1984, *Astrophys. J.* **286**, 760.

Smith, G. H. and Dupree, A. K.: 1987, *Astron. J.* submitted.

Sonneborn, G. S., Baliunas, S., Dupree, A. K., Guinan, E. F., and Hartmann, L.: 1986, in *New Insights in Astrophysics — 8 Years of UV Astronomy with IUE*, ESA SP-263, Noordwijk: ESA, pp. 221—223.

Spite, M., Caloi, V., and Spite, F.: 1981, *Astron. Astrophys.* **103**, L11—13.

Stalio, R. and Zirker, J. B. (eds.): 1985, *Relations between Chromospheric-Coronal Heating and Mass Loss in Stars*, Trieste Workshop Series on Nonlinear Nonequilibrium Thermodynamics of Open Nonthermal Systems in Astronomy.

Stencel, R. E. and Mullan, D. J.: 1980a, *Astrophys. J.* **238**, 221.

Stencel, R. E. and Mullan, D. J.: 1980b, *Astrophys. J.* **240**, 718.

Stencel, R. E., Mullan, D. J., Linsky, J. L., Basri, G. S., and Worden, S. P.: 1980, *Astrophys. J. Suppl.* **44**, 383.

Stern, R. A., Zolcinski, M.-C., Antiochos, S. K., and Underwood, J. H.: 1981, *Astrophys. J.* **249**, 647.

Van der Hucht, K., Bernat, A. P., and Kondo, Y.: 1980, *Astron. Astrophys.* **82**, 14.

Vaiana, G. S. and Rosner, R.: 1978, *Ann. Rev. Astron. Astrophys.* **16**, 393.

Vernazza, J., Avrett, E. H., and Loeser, R.: 1981, *Astrophys. J. Suppl.* **45**, 635.

Weiler, E. J. and Oegerle, W. R.: 1979, *Astrophys. J. Suppl.* **39**, 537.

Wesemael, F., Henry, R. B. C., and Shipman, H. L.: 1984, *Astrophys. J.* **287**, 868.

Wing, R. F.: 1978, in M. Hack (ed.), *Proc. 4th Internat. Colloq. Astrophys.*, Trieste, pp. 683.

Willson, L. A. and Bowen, G. H.: 1986, in D. Gibson, M. Zeilik (eds.), *Proc. Cambridge Workshop on Cool Stars, Stellar Systems, and the Sun, 4th*, Lecture Notes in Physics, No. 254, New York, Springer-Verlag, pp. 385—396.

THE END STAGE OF STELLAR EVOLUTION: WHITE DWARFS, HOT SUBDWARFS AND NUCLEI OF PLANETARY NEBULAE

G. VAUCLAIR

Observatoire du Pic-du-Midi et de Toulouse, Toulouse, France

and

JAMES LIEBERT

Steward Observatory, University of Arizona, Tucson, AZ, U.S.A.

1. Introduction: Hot Evolved Stars and the IUE Observatory

The region of the H—R diagram to the left and below the main sequence (see Figure 1) is occupied by stars whose nuclear-burning phases are either ending or essentially ended. These objects for the most part include the degenerate stellar cores from the prior asymptotic giant branch (AGB) or double shell source burning phase, after the outer stellar envelope has been depleted by shell burning and extensive mass loss. Some remnants may evolve directly from the region of the horizontal branch (HB) after the core helium burning phase. After a brief phase dominated by gravitational contraction and residual shell burning, these dying stars enter the white dwarf cooling sequence, which was well defined more than twenty five years before the launch of the IUE satellite (Mestel, 1952). Recent results indicate that single stars with initial masses as large as about 8 $M_\odot$ manage to lose enough envelope mass primarily during their AGB and red giant phases that the stellar cores do not exceed the Chandrasekhar limit, permitting nonviolent evolution as a white dwarf (Reimers and Koester, 1982, Weidemann and Koester, 1983). Thus, an important task for IUE has been to search for and obtain spectra of pre-white dwarfs and white dwarfs from the young disk, old disk and halo populations.

The end of the AGB phase is accompanied by increasing mass loss, culminating with the likely ejection of a planetary nebula (PN). Thus, central stars or nuclei of planetary nebulae (PNN) are the most obvious examples of post-AGB stars. Also included are at least some of the hottest subdwarfs (Sd). While all but the hottest nebula-free Sd lie at lower temperatures than the PNN, a related helium-rich group of pulsating stars was discovered during the IUE era whose prototype is PG1159—035 (GW Vir); these lie at the juxtaposition of the hottest, high gravity Sd, the PNN and the white dwarf sequence. Hereafter, these shall be referred to as the 'PG1159' stars. In the subsequent discussion, the PNN, the Sd and the PG 1159 stars are covered in Sections 2, 3, and 4, respectively, and the white dwarfs in Section 5. Space limitations preclude discussion of the remarkable role of IUE in the discovery and direct analysis of hot evolved stars in all kinds of binaries from detached systems like Sirius B and V471 Tau, to the accreting systems (paper by Córdova and Howarth, pp. 395—426, see also Sion, 1986).

Y. Kondo (ed.), Scientific Accomplishments of the IUE, pp. 355—376.
© 1987 by D. Reidel Publishing Company.

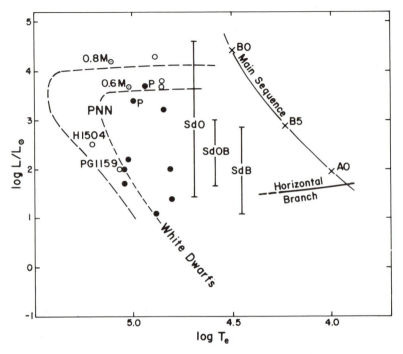

Fig. 1. An H—R Diagram including the classes of objects discussed in this review. Three classes of Sd stars are plotted schematically according to the T_e ranges of Table I, and with the luminosity ranges estimated from surface gravities, assuming $M = 0.6 \ M_\odot$. The wide temperature and luminosity ranges of the "hot Sd" class (Section 3) are illustrated by plotting individual hydrogen-rich objects estimated in a similar way from the parameters given in Schönberner and Drilling (1984) — filled circles. Two hydrogen-rich objects found to have residual (PN) nebulosities are noted with a 'P'. Three hot, helium-rich Sd stars are taken from the results of Husfeld (1986) near log $L/L_\odot \sim 4$ — open circles. Two helium-rich stars discussed in Section 4, denoted by 'PG1159' and 'H1504' are also plotted with open circles. The approximate locations of the main sequence, the horizontal branch and tracks for post-AGB stars of 0.6 and 0.8 $M_\odot$ are shown with labelled curves.

Since these groups include the hottest known stars, it is not surprising that IUE spectrophotometry has been important or fundamental in determining the atmospheric compositions, effective temperatures and surface gravities (i.e. radii, masses) for white dwarf stars and their various precursors. Most of their radiation is emitted in the ultraviolet (often too far in the UV even for IUE). Elements heavier than helium (the metals), if present in the atmosphere, are in high stages of ionization, for which most absorption lines occur in the UV.

In addition, the resonance lines of C IV, Si IV, and N V readily observed by IUE are often important in abundance determinations and may be crucial in understanding the ongoing mass loss often observed from evolved hot stars and in envelope processes (i.e. radiative accelerations, gravitational and thermal diffusion), which may be responsible for both surface abundances and the winds. Thus, IUE observations have been essential in developing our understanding of (1) the evolution of the stellar interiors at the end of the nuclear burning phases through improved determinations of the stellar parameters, and especially (2) the corresponding evolution of the stellar envelopes, surfaces and circumstellar environments.

2. Planetary Nebula Nuclei (PNN)

The paper by Köppen and Aller (pp. 589—602) includes some discussion of the central stars, with emphasis on their mass loss and the relationship of the stellar and nebular properties. It is appropriate here to consider the determination of the stellar parameters using IUE data and the relationship of the PNNs to the evolved, nebula-free stars.

2.1. PROBLEMS WITH EFFECTIVE TEMPERATURE DETERMINATIONS

Since the post-AGB evolutionary track positions and time scales are very sensitive to the remnant core mass, accurate determinations of the photospheric effective temperatures (T_e) and luminosities (surface gravities) are of crucial importance in understanding the remnant mass distributions. The fact that the PNN are surrounded by nebulae may be both a help and a hindrance. An optically thick (ionization bounded) nebula will reprocess the otherwise unobservable extreme UV photons from a hot PNN into observable emission lines, thus permitting T_e estimates from the Zanstra and Stoy techniques (c.f. Pottasch, 1984). Among the uncertainties involved in applying such methods, however, is the possibility that ongoing mass loss from the PNN may affect the ionization balance of the nebula, so that measurements of the latter may not reflect accurately the energy distribution of the photosphere. On the other hand, it is only for low surface brightness nebulae that the stellar photosphere may be observed accurately enough for standard model atmospheres techniques to give direct determinations of the temperatures and surface gravities.

Observations of the UV continua of exposed PNN (especially by IUE) have been widely used for estimating the stellar T_e values, usually by comparison with the energy distributions of models (c.f. Kaler and Feibelman, 1985). This procedure is similar to that used for the hot Sd stars (Section 3). However, since the PNN (and Sd) include some of the very hottest stars known, we note here some difficulties which rapidly become worse with increasing temperature:

(1) The IUE wavelength interval lies entirely in the Rayleigh—Jeans tail of the stellar energy distribution, and thus the measured spectral energy distribution is insensitive to the effective temperature. The uncertainties in the IUE flux calibrations therefore translate into large temperature uncertainties;

(2) The uncertainties in estimating the UV extinction may produce substantial errors in the temperature estimates. Note that the strength of the 2200 Å 'bump' may not correlate well with the general UV extinction (Greenberg and Chlewicki, 1983).

(3) The model stellar atmospheres may be too simplified to make sufficiently accurate predictions of the energy distributions: In particular, the line blocking (Section 3.5), line blanketing and NLTE effects from elements heavier than helium have not been widely incorporated into existing codes — a formidable task!

(4) For PNN undergoing substantial ongoing mass loss, such as those exhibiting Wolf—Rayet or O VI features, an extra source of continuum radiation, especially at higher frequencies and at bound-free continuum edges, may be a consequence of energy deposition in an extended envelope.

The fitting of existing NLTE atmospheric models to observed stellar hydrogen and helium absorption line profiles, where possible, may improve the T_e estimates, as well as providing surface gravities and atmospheric abundances. This modern approach has been pioneered by Mendez *et al.* (1983, and references therein) for PNN, but is limited by the accuracy of the atmospheric physics. For T_e above about 70 000 K, only one ion of hydrogen and helium is observed. The fitting of CNO or metallic features often observed with IUE should improve the temperature determinations in the future.

2.2. THE PARAMETERS AND EVOLUTION OF THE CENTRAL STARS

Despite the above difficulties, observations using IUE of dozens of PNN have added considerably to our knowledge of the positions of PNN in the H—R diagram. Of course the low surface brightness nebula stars amenable to photospheric analysis may themselves constitute a biased sample. First of all, we would expect this selection to emphasize the 'old' PN, in which the ejected nebula has had time to disperse and the PNN to evolve to high T_e and then downwards in the H—R diagram to low luminosity. Secondly, since the stellar core evolution time decreases sharply ($t \propto M^{-9.6}$) with increasing mass (Iben and Renzini, 1983), we expect that a low surface brightness PNN sample would be biased towards lower-than-average stellar masses. Conversely, the most massive PNN may evolve to very low luminosities while still buried in their nebulae. Nonetheless, for the selected samples of PNN and field Sd stars, the former have generally higher T_e, so that evolutionary model tracks indicate modestly higher masses. This suggests that even the low surface brightness PNN have larger mean masses than the normal Sd stars (Hunger *et al.*, 1981; Mendez *et al.*, 1983). Schönberner (1981, 1983) argues that the time scales for envelope dissipation and for the evolution of the stellar core agree for a mass distribution of observed PNN in the range 0.55—0.65 $M_\odot$. We should then expect a high mass 'tail' to this distribution.

The photospheric analyses also show that an upper limit of about 30% of the measured PNN of low surface brightness nebulae have atmospheres dominated by helium (Mendez *et al.*, 1986). This is not qualitatively different than the fractions for hot Sd or hot white dwarfs, but it should be noted that a hydrogen-burning post-AGB star should evolve about three times faster than a helium-burning star through the PNN region (c.f. Schönberner, 1986). There seem to be no clear differences in mean temperature or luminosity between the helium and hydrogen-rich PNN, or even in the abundances of their nebulae.

3. Hot Subdwarfs

IUE has been directly responsible for the development of a basic understanding of the hot Sd stars. The pioneering work of Greenstein and Sargent (1974) on a large sample of stars was an important first step. However, the use of optical spectra and

UBV colors with very limited grids of relevant stellar atmospheres (available at this time) did not permit the derivation of accurate physical parameters, especially for the hotter and more helium-rich stars. Access to the ultraviolet with its numerous metallic lines has led to more precise determinations of the atmospheric parameters — effective temperature, gravity and He/H abundance — and of the evolutionary status of these stars.

The discovery with IUE of winds and mass loss from the hottest subdwarfs is of great interest because of the importance of mass loss on stellar evolution and because these stars have very different physical parameters than the hot O and B main sequence stars and M supergiants, with which to learn how mass loss scales with stellar parameters. IUE has also revealed many details of the chemical composition of these stars which provide important clues for understanding the physical processes occuring in their outer layers. We will summarize these items in the final subsections.

The historic subdwarf spectral classification system generally consists of the SdB, SdOB and SdO types. SdB stars show strong H lines with weak He, while SdO stars show strong He lines, and SdOB stars are intermediate. However, since the hydrogen and helium lines used are sensitive to both temperature and to the He/H ratio, there has been confusion in the literature as to whether the sequence should connote abundance differences as well as (increasing) temperature. In the post-IUE picture — with the benefit of model atmospheres analysis — it now is possible to identify four subdwarf classes in order of increasing T_e, three of which are clearly restricted in He/H values. These may be identified with the historic definitions of SdB, SdOB, SdO and a new 'hot Sd' category, which overlaps directly with the PNN, the PG1159 stars (Section 4) and the hottest white dwarfs. The atmospheric parameter range found for each T_e class is summarized in Table I and discussed in Section 3.1. The consequences for the evolution of envelopes and interiors are discussed in Sections 3.2 and 3.3, respectively. Note in particular that if the term SdO is applied to stars hotter than about 60 000 K, then it must cover subdwarfs ranging from below solar He/H to nearly pure helium atmospheres and with temperatures from 40 000 K to over 100 000 K. The SdO term is in common usage for PNN, and this class includes a similar range in He/H and T_e (c.f. Pottasch, 1984). For clarity, we shall restrict the use of the term SdO to stars with $T_e < 60 000$ K.

3.1. FOUR SUBDWARF CLASSES: DERIVED RANGES IN ATMOSPHERIC PARAMETERS

In obtaining the results summarized in Table I low resolution IUE spectra are used, in conjunction with optical spectrophotometry or broad band photometry. Extreme ultraviolet spectrophotometry provided by the Voyager UV spectrometer is also used, when possible, for the hottest subdwarfs. The energy distribution covering the largest possible wavelength range is then fitted with model atmospheres to determine the effective temperature. The importance of access to the UV range is obvious. Visible spectra are useful in deriving surface gravity through Balmer line profiles fitting and He/H abundance ratios.

LTE atmospheric analysis has been shown to be adequate in the SdB tem-

TABLE I

Parameters for the four temperature classes of evolved subdwarfs

Subdwarf class	T Range	Log g Range	Helium mass fraction	References
SdB	25—35 000 K	5.0—6.0	< 0.005	1—9
SdOB	35—40 000 K	5.0—6.0	< 0.12	1—9
SdO	40—60 000 K	4.0—6.5	0.5—1	4—12
Hot Sd	60—100 000 K	4.0—7.0	0.1—1	13—16

References: 1. Heber *et al.* (1984c); 2. Wesemael *et al.* (1985a); 3. Heber (1986); 4—5. Baschek *et al.* (1982ab); 6. Kudritzki *et al.* (1982); 7—9. Heber *et al.* (1984abc); 10. Hunger *et al.* (1981); 11. Simon (1981); Gruschinske *et al.* (1983); 12. Simon (1982); 13. Schonberner and Drilling (1984); 14. Husfeld (1986); 15. Drilling *et al.* (1984); 16. Wesemael *et al.* (1985b).

perature range (Heber *et al.*, 1984c; Wesemael *et al.*, 1985a). However, for the higher temperature ranges of the SdO and hot Sd stars, it is necessary to include non-LTE effects in the model atmospheres and in the line formation computations (Kudritzki, 1976).

The extension of the subdwarfs to temperatures well above 60 000 K first became apparent from the analysis of IUE spectra of field stars discovered by Schönberner and Drilling (1984). In contrast with the 'traditional' field SdO stars, the majority of these are hydrogen-rich, although the indicated helium abundance varies greatly from star to star. The T_e estimates for the hottest stars are uncertain by at least 20%, for reasons outlined in Section 2. This group includes a few H-rich stars near 100 000 K, as confirmed by Voyager data (Drilling *et al.*, 1984). Two stars were also found to have residual planetary nebulae. Husfeld (1986) has analyzed the line spectra of three helium-rich stars from this sample using non-LTE atmosphere models. He finds that these have temperatures near 70—75 000 K, log g ~ 4.4—4.9, and hydrogen abundances of < 0.1 (by number). Thus, the hot Sd group overlaps with much of the T_e, log g and abundance ranges shown by PNN, and also with the stars discussed in Section 4.

3.2. THEORETICAL EXPLANATIONS OF THE SUBDWARF GROUPINGS IN He/H

The interpretation of this behavior of the He/H abundance ratio for the Sd groups has long been and remains a puzzle. The current idea is that this behavior results from surface physical processes (helium diffusion and convective mixing) and that the atmospheric He abundance allows only indirect conclusions about the internal structure and evolutionary state of these stars. Many authors have speculated that at the effective temperatures of SdB stars, helium diffuses downwards because (1) the radiative acceleration is not strong enough to counterbalance gravity and (2) there is no convection in these layers (Hunger and Kudritzki, 1981; Baschek *et al.*, 1982a. b; Kudritzki *et al.*, 1982; Heber *et al.*, 1984b). Consequently, helium settles downward while hydrogen floats at the stellar surface.

At about 32 000 K, the underlying helium becomes convectively unstable. When the total hydrogen fractional mass is small enough, helium mixes with the upper hydrogen layer and eventually reaches the photosphere in the He-rich SdO temperature range (Greenstein and Sargent, 1974; Winget and Cabot, 1980; Wesemael et al., 1982). At the hot end of the subdwarf temperature distribution, He becomes completely ionized and there is no more convective mixing. Helium could then sink by gravitational settling again, producing the hot hydrogen rich subdwarfs discovered by Schönberner and Drilling, (1984). The absence of hydrogen in other very hot post-AGB stars may require that the H envelope had been lost previously, for example, in convective mixing during final AGB thermal pulses, or in a late helium shell flash (Iben et al., 1983).

There is presently no detailed calculation of these processes relevant to subdwarf stars. However, computations of radiative acceleration on He (Michaud et al., 1979) and discussions of helium diffusion in hot horizontal branch star models similar to the coolest SdB stars (Michaud et al., 1983), suggest that these ideas are plausible. Such a scheme implicitly assumes an evolutionary link between He-poor SdB, SdOB and He-rich SdO stars.

This view has recently been challenged (Groth et al., 1985) on the basis of new evolutionary post-HB tracks, which do not cross from the SdB to the SdO stars. These authors rather suggest that (1) the SdB and SdOB stars form a unique sequence of evolved low mass stars lying on an extended horizontal branch (EHB), while (2) the SdO stars evolve quite differently from more massive HB (and AGB) progenitors. The former are He-poor as a consequence of gravitational settling, while the latter are probably He-rich as a consequence of convective mixing during the final thermal pulses. If this scenario is true, this leaves us without any explanation for the Schönberner and Drilling (1984) hot hydrogen rich subdwarfs. It may also be the case that He/H abundances observed in subdwarfs result from the interplay of more than only two physical processes. Mass loss, discussed in Section 3.4, may be an important ingredient in this respect.

3.3. EVOLUTIONARY STATUS: A SUMMARY

Greenstein and Sargent (1974) originally suggested that subdwarf stars are the extension at higher effective temperatures of the horizontal branch — the EHB. This impressive temperature spread would have to be caused by prior differential mass loss (perhaps at the core helium flash). The improved atmospheric parameter determinations now available as a result of IUE allow a better comparison with the evolutionary models.

Our present understanding of the evolutionary status of these stars can be summarized as follows. Following the core helium burning phase, post-HB stars ascend the AGB only if they have retained enough envelope mass ($M_e \gg 0.02\ M_\odot$) to support a double shell source phase. It is also likely that the C—O core mass for such AGB stars will exceed $0.5\ M_\odot$. It is not clear, especially for those stars with marginal envelope masses, that the AGB phase must end with a PN ejection. Schönberner (1981, 1983) showed that for stars with $0.55—0.65\ M_\odot$, the expansion time scale of the nebula and the evolutionary time for the stellar core are in reasonable agreement. Post-AGB stars with $M < 0.55\ M_\odot$ may simply be observed as Sd's, having already lost any ejected envelopes. With or without a PN, the post-AGB stars follow the familiar

Harman—Seaton path to the white dwarf sequence. Clearly, the hot Sd stars, the PG1159 stars discussed in Section 4, and probably the SdO stars are post-AGB objects.

When M_e is too small, nuclear burning in shells cannot develop and the stars evolve directly toward the white dwarf stage at high gravity — possibly even through the SdB, SdOB, SdO sequence. In the intermediate case where M_e is only slightly in excess of 0.02 $M_\odot$, the stars start evolving toward the AGB, but must turn around due to insufficient nuclear fuel, thus missing the AGB and PN phases. They subsequently evolve at constant luminosity to the SdO phase and then to the final white dwarf stage. In this picture, the SdB and SdOB stars evolve from low mass HB stars, while the SdO stars are a mixture of post-HB stars of lower gravity and of evolved SdB and SdOB stars. This would explain the large dispersion in surface gravity found in the SdO stars.

Analysis of a small sample led Drilling and Schönberner (1985) to the conclusion that ~ 80% of the white dwarf progenitors are PNN, the remaining 20% are Sd stars, among which only about 1% could come from the post-HB subdwarfs. However, these percentages probably remain very uncertain. The previous discussion of the He/H abundances shows that the evolutionary link between the He-poor and the He-rich subdwarfs is still a matter of controversy (Groth *et al.*, 1985).

3.4. WINDS AND MASS LOSS

An important early finding of IUE was the discovery of ongoing mass loss from hot subdwarfs. This was obtained from high resolution IUE spectra of a few hot, bright SdO stars (Heap, 1978; Darius *et al.*, 1979; Simon *et al.*, 1979) which exhibit N V P Cygni profiles. In HD 49798, a low gravity star (log g = 4.25), the N V profile indicates a maximum wind velocity of 1350—1500 km s^{-1}, and a mass loss rate in the range $-9.3 \leqslant \log(M/M_\odot \text{ yr}^{-1}) \leqslant -8.0$ (Hamann *et al.*, 1981; Bruhweiler *et al.*, 1981). Yet the CIV resonance doublet shows no indication of mass loss. HD 128220 B, another SdO star of similar temperature (45 000 K) and slightly higher gravity shows, in addition to the N V P Cygni profile, blue shifted C IV (Hamann *et al.*, 1981; Bruhweiler and Dean 1983). Other SdO stars exhibit P Cygni profiles in the N V and C IV resonance doublets, or extended blue wings arising in a wind (Rossi *et al.*, 1984). In addition, Bruhweiler and Dean (1983) note the existence of multiple, discrete, shortward-shifted components in the N V, C IV, and Si IV resonance doublets in some SdO stars, reminiscent of the similar discovery in hot O and B main sequence stars. Such observations may reveal the details of the mass loss process. They suggest that mass loss proceeds in some irregular way by blowing small clouds of gas. The new samples of hot subdwarfs studied by Schönberner and Drilling (1984) also contains three more mass-losing stars, as indicated by their N V and C IV line profiles; these authors subsequently discovered that the most luminous of their mass losing subdwarfs were indeed central stars of planetary nebulae. In contrast, none of the lower luminosity SdO (Hamann *et al.*, 1981; Schönberner and Drilling, 1984), SdOB (Baschek *et al.*, 1982a, b; Heber *et al.*, 1984b) or SdB stars (Lamontagne *et al.*, 1985) observed at high resolution has revealed any evidence for mass loss.

The discovery of mass loss from hot subluminous stars is important for many reasons. First, it provides a new group of mass-losing stars, different from the O and B main sequence stars and M supergiants, from which one determines the functional dependence of the mass loss rate with basic stellar parameters. Second, it reveals the existence of another important physical process which may compete with other processes (diffusion, convective mixing, etc.) in producing the observed variation in surface chemical composition. In attempting to understand He-weak and He-rich main sequence stars, Vauclair (1975) proposed a model in which mass loss and diffusion compete to produce the various helium abundances. It would be worthwhile to investigate the role of mass loss, in conjunction with gravitational settling and convective mixing.

There are still too few Sd stars with rough mass loss rates to settle the question of which physical mechanisms are responsible for mass loss. In comparing their derived mass loss rates with the theoretical predictions, Hamann *et al.* (1981) found that the radiation driven wind theory of Castor *et al.* (1975) does not predict correctly the mass flux differences seen between hot main sequence and SdO stars. From this point of view, IUE discoveries had the merit of pointing out the incompleteness of theory and the need for further basic research in this field.

3.5. CHEMICAL COMPOSITION

The brightest subdwarfs which have been observed with IUE at high resolution exhibit rich metallic spectra. In order to understand in detail the evolution of these stars, a knowledge of the surface chemical composition is crucial. The determination of whether the surface compositions are the result of physical processes at the stellar surface (diffusion, convection, mass loss etc.), or reflect the prior nuclear processing, can in principle be tested by accurate metal abundance measurements in the atmospheres.

Subdwarf stars are unfortunately faint stars for high dispersion spectroscopy, and in analyzing the IUE results we must deal with small number statistics. However, several general trends in the metal abundances are evident:

For the SdB stars, Lamontagne *et al.* (1985) derived abundances for C, N, and Si, but many additional species were identified in their spectra. Abundance determinations for these other species must await better signal/noise ratio spectra, unambiguous identifications, and especially more atomic physics parameters for highly ionized species. The C abundances are always below solar and tend to decrease slightly with increasing T_e, while N remains remarkably constant at nearly solar abundance. Si is strongly underabundant in all SdB stars. The Si underabundance varies from 2 to 5 orders of magnitude. At the hot end of the SdB sequence, Feige 110 and LB3459 differ greatly in their Si abundances, but these may be explainable: Feige 110 is in fact an SdOB star, while LB3459 is a short period binary system (Kilkenny *et al.*, 1981). The coolest SdB stars have solar C, N, Si abundances.

The large underabundances of Si, together with the He deficiency in the SdB stars, were considered by many authors to be produced by gravitational settling. However, Michaud *et al.* (1985) have shown that in the T_e range where Si deficiencies are observed, the radiative acceleration on Si is much too large to allow underabundances. They assume that small mass loss rates ($M < 10^{-14} M_\odot$ yr^{-1}), which are not detectable with IUE, compete with diffusion to produce the observed C, N, Si abundances. This promising idea suggests that the use of metal abundances could constrain mass loss rates in these stars. Another constraint which they did not consider could come from the necessity to preserve the helium deficiency in the SdB stars.

Only three SdOB stars have been observed at high resolution: HD 149382 (Baschek *et al.*, 1982a), Feige 66 (Baschek *et al.*, 1982b), and Feige 110 (Heber *et al.*, 1984b). These stars lie in a narrow T_e range and show quite similar spectra. They are characterized by large deficiencies in He, C (by 100 in HD 149382 and 3×10^5 in Feige 110), and Si (by at least 4000 in Feige 110), but almost normal N. Early interpretation of the C/N ratio suggested that the stellar surface shows the effects of prior CNO processing. However, this explanation is contradicted by the He depletion. If, on the other hand, the He depletion were due to gravitational settling, then one must consider how the gravitational settling would affect the CNO cycle products. The alternative explanation of diffusion in the presence of small mass loss seems again to be a promising idea for the SdOB stars (Michaud *et al.*, 1985).

Since it is difficult to derive simultaneously an abundance and a mass loss rate for the SdO stars from their N V or C IV P Cygni profiles, we do not know the abundances in the more luminous mass-losing SdO stars. In the other cases, high dispersion spectra taken with IUE and analyzed with NLTE model atmospheres show that C is slightly deficient, N is overabundant, and Si is normal (Heber *et al.*, 1984a). Whether this abundance pattern may be reconciled with the high He abundance in the framework of the diffusion theory and in presence of mass loss and/or convective mixing, will require detailed computations. In most (if not all) of the observed SdO stars, numerous absorption features are found, due to C III, C IV, N III, N IV, N V, O III, O IV, O V, Si II, Fe IV, Fe V (Rossi *et al.*, 1984; Schönberner and Drilling, 1984).

For several SdO stars, Bruhweiler *et al.* (1981), Schönberner and Drilling (1985) and Dean and Bruhweiler (1985) discovered extremely rich Fe spectra of Fe V, Fe VI, Fe VII lines, and the last authors note that lines from highly ionized Fe dominate the spectrum of hot SdOs stars in the far UV. Unfortunately, little is known about the relevant oscillator strengths for these ions. The discovery of this rich UV spectrum should encourage atomic physicists to provide oscillator strengths of highly ionized species. This would stimulate progress in both the determination of metal abundances in hot stars and in the accurate computation of radiative acceleration to check the validity of the diffusion theory. On the other hand, the discovery of all of these features adds to concerns about errors in the effective temperatures derived from model atmospheres which do not take into account the effect of the blanketing due to these numerous absorption lines, as previously discussed for PNN and hot subdwarfs.

4. Pulsating and Very Hot White Dwarf Precursors

McGraw *et al.* (1979) discovered a new type of very hot pulsating variable star,

PG1159—035 (GW Vir) in the Palomar Green (PG) Survey (Green *et al.*, 1986). The complicated, nonradial pulsation modes and their possible implications for the structure and evolutionary state of this object have been intensively investigated (c.f. Winget *et al.*, 1983a; Starrfield *et al.*, 1984; Kawaler *et al.*, 1985). In addition to the prototype, two other PG stars were later found to show pulsational instability (Bond *et al.*, 1984), and several other stars show similar spectra (as described below).

Optical and IUE spectrophotometry, supplemented by Voyager, *Einstein* and EXOSAT data at higher frequencies, permitted a T_e estimate for PG1159—035 of 120 000 K (+40 000/−20 000) by Holberg (1986, in preparation). The spectra of this class are uniquely defined by the appearance of absorption lines of He II, C IV, sometimes weak N V and usually O VI, often with reversed narrow emission cores (Bond *et al.*, 1984, Sion *et al.*, 1985a; Wesemael *et al.*, 1985b, WGL). The last paper applied LTE model atmospheres including hydrogen and helium to fit the optical and IUE data. The absence of hydrogen features at these temperatures allows only an upper limit of unity for the number ratio of H/He. The line profile fits indicate that the surface gravities are near $\log g = 7$. Pending more accurate analyses incorporating elements heavier than He and possible NLTE effects, these stars can be classified as either He-rich Sd stars of very high temperatures and gravities, or as white dwarfs according to the definition of Greenstein and Sargent (1974).

The central star of the planetary nebula K1—16 shows similar spectroscopic features and pulsation properties and is related to or a member of the PG1159 class (Grauer and Bond, 1984; Sion *et al.*, 1985a; Grauer *et al.*, 1986a). The last reference reports a similar, possibly pulsating PNN, belonging to the nebula VV47. The K1—16 star shows somewhat sharper absorption lines and somewhat longer periods than does PG1159—035; both may be a consequence of a higher luminosity for the central star. The appearance of narrow O VI 3811, 3830 Å emission (observed from the ground) strengthens the proposed link of the pulsating objects with the O VI class of PNN (Sion *et al.*, 1985a). On the other hand, once mass loss weakens and diffusive equilibrium prevails (Section 5.4), the surface abundances are fixed by the balance between gravity and radiative acceleration, and the photosphere 'forgets' its prior abundances. Hence, this similarity could be due to the similar T_e and gravities of the PG1159 and (some) O VI nuclei.

The IUE spectra of K1—16 reported by Kaler and Feibelman (1985) provide some evidence for both (1) UV light variations, which might be related to the optical pulsations, and (2) an extraordinarily fast wind (8500 km s^{-1}), as indicated by possible blueshifted C IV absorption. The latter would link K1—16 to several O VI — Wolf Rayet nuclei studied with IUE which show substantial mass outflows. The authors caution, however, that their K1—16 observations push the sensitivity limits of the IUE.

Starrfield *et al.* (1984) argue that a large envelope abundance of CNO species, especially O, provides the pulsational driving in ionization zones appropriately near to the surface. The atmospheric composition of these species has not yet been determined, due to the model limitations previously discussed, but the rather high space density of isolated PG1159 objects suggests that they are He shell burning objects (Fleming *et al.*, 1986), according to the model tracks of Iben and Tutukov (1984) and Koester and Schönberner (1986). Indeed, it now seems possible that pulsations may be driven by the He shell source (Kawaler *et al.*, 1986), although the predicted periodicities are too short.

Another isolated object possibly related to the pulsating stars is the stellar counterpart of the X-ray source H1504 + 65 (Nousek *et al.*, 1986). IUE, Voyager, and optical spectrophotometry show an energy distribution for this hydrogen-poor star which is even hotter than those for PG1159—035 and K1—16. The ratio of its soft X-ray (EXOSAT) flux to the optical flux is over 600 times that of PG1159—035. The optical and ultraviolet spectra show O VI and C IV features, but no He II lines and only interstellar Ly α. The absence of helium lines suggests that, if the atmosphere were helium-dominated, the temperature should exceed the 150 000 K limit of the Wesemael (1981) models. Nousek *et al.* (1986) also discuss the possibility that CNO species dominate the surface composition, an idea which is untestable by existing models. At the time of this writing, the star shows no evidence for soft X-ray (Nousek *et al.*, 1986) or optical pulsations (Grauer *et al.*, 1986b).

5. White Dwarfs

As with the hot subdwarfs, the IUE results have had a fundamental impact on our understanding of white dwarf stars, especially the physical processes and evolution of their envelopes (Liebert, 1984; Sion, 1986). In particular, these stars of highest surface gravity in the H—R diagram usually show only the dominant atmospheric constituent (H or He) at optical wavelengths, unless their temperatures are too low for the excited optical lines to appear. In the IUE bandpass, a wealth of new spectroscopic information has been revealed: (1) The hot stars usually show trace elements which are apparently supported through selective radiative accelerations and/or shortward-shifted lines indicative of modest mass loss (winds); (2) the cool He-rich white dwarfs generally show trace abundances of carbon dredged up from the core, or in some cases, accreted metals; (3) Finally, even the cool H-rich white dwarfs show spectra containing an intricate pattern of quasi-molecular hydrogen features. While the H-rich white dwarfs are generally of spectral type DA, the He-rich stars manifest a variety of spectral types — DO, DB, DC, DQ, DZ and hybrids (see Sion *et al.*, 1983). The DA stars are a large majority ($\sim$ 80%) of white dwarfs hotter than about 10 000 K.

5.1. EFFECTIVE TEMPERATURES AND STELLAR PARAMETERS FOR DA STARS

For the DA white dwarfs, which have nearly pure hydrogen atmospheres, the temperature scales and surface gravity, radius and mass distributions have been reasonably well determined from optical data and existing atmospheric models. The accuracy of the models is usually better than for hot PNN and Sd stars, since LTE is a much safer assumption at these high surface gravities, the atmospheric compositions are quite simple, and in particular, the atomic physics of hydrogen is well understood. The IUE energy distributions and Ly α profiles yield T_e estimates consistent with those derived from optical data (Greenstein and Oke, 1979), although there may be time-dependent changes in the IUE flux calibration (Finley *et al.*, 1984). Note, however, that the cases studied range up to 50—60 000 K, considerably cooler than the hottest PNN and Sd stars discussed earlier.

The fundamental studies of Koester *et al.* (1979) and Shipman (1979) indicate that the DA white dwarf mass distribution is rather narrow, with a peak near 0.6 $M_\odot$. This is in general agreement with the distribution of PNN and Sd masses discussed previously, to which IUE has contributed much more, and underscores presently accepted theory that the extensive massive loss is required in prior (especially AGB) stellar evolution phases.

IUE spectrophotometry has contributed substantially to refining the atmospheric parameters for the hottest and nearest of the DA stars. Holberg *et al.* (1986) analyzed IUE, Voyager and optical spectrophotometry, and Ly α line profiles to obtain improved T_e and surface gravity determinations, but such improvements over the parameters obtained only from optical data have thus far uncovered no DA star with a temperature above about 70 000 K. Likewise, Kahn *et al.* (1984) and Petre *et al.* (1986) have combined *Einstein* X-ray fluxes with IUE and other data in order to determine He/H atmospheric ratios. The latter authors have identified a possible trend of increasing He/H with increasing T_e which, if real, would be consistent with the predictions of diffusion theory. This trend may also be valid for the DAO white dwarfs, which may have T_e in the range 60—80 000 K, HeII lines detectable at optical wavelengths, and probably somewhat lower surface gravities (log $g \sim$ 7, WGL). This region of the H—R diagram (see Figure 1) is where the hydrogen atmosphere degenerate sequence approaches the lowest luminosity hydrogen-rich PNN (i.e. Abell 7 and the significantly hotter NGC 7293 (Mendez *et al.*, 1983).

5.2. TEMPERATURES AND STELLAR PARAMETERS FOR HELIUM-RICH DEGENERATES

Parameters for the He-rich white dwarfs are clearly not as well established as for the DA stars. This is primarily because the physics of the continuous opacities and line broadening of neutral helium are not well known. The lower continuum opacities of the He-rich stars relative to DA stars (except at very high T_e) means that the atmospheres usually extend to substantially higher densities and pressures.

Again, the basic temperature scales have been established primarily by fitting optical data — Oke *et al.* (1984, OWK) for DB stars, WGL for DO stars; Koester *et al.* (1982b, KWZ), and Wegner and Yackovich (1984, WY) for cooler DC—DQ stars, and numerous earlier and less comprehensive papers. There is no indication that the radii and masses differ substantially from the mean of the DA sample (Shipman, 1979; Wickramasinghe, 1983; OWK), but the dispersion is poorly established.

In the analysis of the hottest He-rich (DO) white dwarfs by WGL, IUE played a surprisingly peripheral role. For stars cool enough to show both optical HeI and HeII lines, the HeII/HeI line ratios are the most sensitive measure of temperature. For DO stars hotter than 70 000 K, the relative intensities and profile fitting of the HeII lines seemed to give safer, if not more accurate results, than do the UV energy distributions. On the other hand, the additional ions of CNO species observed in high dispersion IUE spectra of a few of these stars have not yet been fully used in a comprehensive analysis (c.f. Sion *et al.*, 1985b).

A recent analysis of DB stars with temperatures above 18 000 K by Liebert *et al.* (1986) using IUE spectrophotometry is only in rough agreement with the optically-based OWK analysis. Largely as a result, the temperature boundaries of the recently-discovered pulsational instability strip for DB stars (Winget *et al.*, 1983b) are uncertain by 3—4000 K. For the prototype pulsating DB star, GD358, the combined analysis of optical line profiles and the overall energy distribution suggests a 10% lower temperature (Koester *et al.*, 1985a) than does the analysis based only on IUE data. The uncertainty in the DB temperature scale is an important uncertainty in attempting to understand the apparent deficiency of DB stars in the 25—45 000 K temperature range (Liebert, 1986) and for evolution between DA and non-DA atmospheres.

The DBA stars are white dwarfs with He atmospheres and trace abundances of H, generally in the range 10^{-5}—10^{-4} (Shipman *et al.*, 1986). Nearly all of these stars have been analyzed using optical spectrophotometry of the HeI and Balmer lines. Yet the difficult Ly α observations, which are at the limit of the IUE for these faint and somewhat cool stars, offer up to an order of magnitude better sensitivity to the H abundance. In particular, Liebert *et al.* (1984) showed that the IUE energy distribution and the small aperture Ly α profile clearly indicate a discrepancy between 'photometric' T_e and that inferred from the line spectrum of the peculiar hybrid white dwarf GD323.

For cooler non-DA stars bright enough to be IUE targets, the UV energy distributions have proved valuable in improving the temperature determinations, at least down to the cool DC star Stein 2051B (WD0426 + 58). This important member of a nearby astrometric binary system has a derived Teff of 6800 ± 300 K (Wegner and Yackovich, 1983).

5.3. THE EVOLUTION OF WHITE DWARFS

A recent analysis of the large, complete sample of PG white dwarfs should provide an improved luminosity function and birthrate determination for white dwarfs hotter than about 10 000 K (Fleming *et al.*, 1986). Above about 40 000 K, the empirical results offer good agreement with the detailed evolutionary models of Iben and Tutukov (1984) and Koester and Schönberner (1986). IUE data have been obtained to help refine the luminosity function for the hottest DA stars. The derived birthrate for these color-selected DA and non-DA white dwarfs is $4.9 - 7.5 \times 10^{-12}$ stars pc^{-3} yr^{-1}. This is somewhat less than previous estimates, and lower than currently accepted values for the PN birthrate. However, the latter must be regarded as very uncertain, and the derived white dwarf number excludes those with companion stars bright enough to contribute at ultraviolet wavelengths.

A more curious result is the difference between the temperature distributions for the majority DA stars and the non-DA white dwarfs (Liebert, 1986). At the highest temperatures ($T_e \sim 80\,000$ K), the PG1159 group (Section 4) and hot DO stars have few, if any, DA/DAO counterparts of similar T_e. Below about 60 000 K, in contrast, the hot DA stars far outnumber their DO counterparts, and there are in fact no known DB or DO stars at all in the interval 30—45 000 K! At

$T_e < 25\,000$ K, the DB white dwarfs are numerous, accounting for 20—25% of all white dwarfs hotter than $12\,000$ K. Below $10\,000$ K, the non-DA white dwarfs are of order 50% of all white dwarfs (Sion, 1984; Greenstein, 1986).

The behavior described above, based on the T_e scales for DA and non-DA stars discussed previously, is spawning a variety of hypotheses to explain how the evolution of H-rich and He-rich stars might differ as they enter the white dwarf sequence, or how the atmospheres of these two types of stars may change as the stars cool. These ideas include (1) the assumption of thick hydrogen envelopes for the hot DA stars, (2) gravitational diffusion of any residual H in PG1159/DO stars, so that they can become DA stars, and (3) convective mixing of thin, outer H envelopes, so that they can become DB or cooler non-DA stars. It is also possible that accretion from the interstellar medium, or even quiescent nuclear burning of trace abundances of H may play important roles. A full discussion is beyond the scope of this summary. Instead, we focus here not on the dominant constituents (H, He), but rather upon the appearance of traces of heavier elements, to which the IUE has contributed enormously.

5.4. METALS AND MASS LOSS IN HOT WHITE DWARFS

The almost pure hydrogen or helium composition of the white dwarf outer layers is well understood as a result of gravitational settling in the high gravitational field present in these stars. It has been suggested, however, that at the hot end of the white dwarf sequence, the radiation field could balance gravity and play a role in determining whether metals sink (Fontaine and Michaud, 1979; Vauclair et al., 1979, VVG). Detailed calculations of the radiative acceleration on C, N, and O (VVG) showed indeed that they should be supported as trace elements in both hydrogen and helium envelopes. VVG predicted that larger abundances could be supported in He-rich envelopes where the higher pressure due to lower opacities makes the lines broader and less sensitive to saturation effects. Since then, IUE high dispersion observations have revealed the presence of highly ionized species in a number of hot white dwarf spectra (Bruhweiler and Kondo, 1981, 1983; Dupree and Raymond, 1982; Holberg et al., 1985; Sion et al., 1985a, b).

Whether these lines form in the stellar photospheres, in the circumstellar environment, or in the interstellar matter may be deduced, in principle, from the measurement of the radial velocities. If absorption lines from highly ionized species have radial velocities similar to the interstellar low ionization species, they probably form in a hot phase of the ISM along the line of sight. Otherwise, the redshift of these lines with respect to interstellar lines indicates a stellar or circumstellar origin. In those cases where metal lines are redshifted by the same amount as the photospheric H or He lines, one may safely conclude that they also form in the stellar photosphere. In a few intermediate cases, their smaller redshift may indicate that lines form in an expanding or stable shell. The absorbing material present at some distance from the white dwarf surface would then indicate that mass loss recently occurred, or is still occurring in these late stages of stellar evolution. This is reinforced by the discovery of shortward-shifted absorption components in some of these stars.

In hot DA stars, C, N and Si lines are present when $T_e \geqslant 33\,000$ K. In

G191B2B (T_e = 62 000 K) the redshifted lines show a residual velocity with respect to the Balmer lines. This, together with the observations of shortward-shifted components points to the existence of a wind (Bruhweiler and Kondo, 1981). Such a wind does not appear in Feige 24 (Dupree and Raymond, 1982) which has a similar T_e. The lines more probably originate in the stellar photosphere. An abundance analysis (Wesemael *et al.*, 1984) from a grid of model atmospheres and synthetic spectra (Henry *et al.*, 1985) led to C, N and Si abundances in rather good agreement with the predictions of the diffusion theory (VVG; Morvan *et al.*, 1986).

In DA stars cooler than ≈ 33 000 K, only Si is present in IUE spectra for GD394 (T_e = 33 000 K, Bruhweiler and Kondo, 1983), CD—38°10980 (T_e = 24 500 K, Holberg *et al.*, 1985), and Wolf 1346 (T_e = 21 500 K, Bruhweiler and Kondo, 1983). In addition, GD394 shows shortward-shifted components which is evidence for a wind. Radiative acceleration can no longer support CNO at these temperatures (VVG), whereas Si is still strongly supported (Morvan *et al.*, 1986) even more than is required to account for the small abundance derived for W1346 (Wesemael *et al.*, 1984). However, the diffusion theory predicts that the abundances of metals supported by the radiative acceleration should vary with atmospheric depth, while model atmospheres currently used to compute synthetic spectra and equivalent widths assume homogeneous composition. This could be the cause of the discrepancy.

Note that, while C, N and Si occur in the UV spectrum at temperatures where the theory qualitatively predicts that they should be supported, the presence of a detected wind is not so clearly related to the T_e: GD394 (33 000 K) has a wind, yet, for stars with T_e ≈ 60 000 K, G191—B2B shows evidence of a wind but Feige 24 does not. Likewise, the surface abundances of Feige 24 agree with diffusion theory; yet HZ43 with a similar T_e shows no trace of photospheric metals. HZ43 was found in the Holberg *et al.* (1986) analysis to have a higher than average log g ~ 8.3, while G191—B2B has low gravity. Both the delicate balance between radiative acceleration and gravity and the physical mechanism(s) leading to the wind may be sensitive to differences in surface gravity. Yet the observed sample is small enough that it may be possible that Feige 24 will prove to be anomalous: its high surface abundances might be maintained by accretion from the close, active dMe companion (Sion and Starrfield, 1984; Morvan *et al.*, 1986).

A few hot He-rich white dwarfs, have been observed at high dispersion with IUE. The PG1159 precursors (T_e ~ 10^5 K), are generally too faint to be observed at high dispersion, but their low dispersion IUE and optical spectra show strong C IV and O VI with only weak N V. Sometimes these lines show emission reversals suggesting the existence of heated atmospheric or circumstellar material (Sion *et al.*, 1985a; WGL). In the cooler (T_e ~ 80 000 K) DO stars like PG1034 + 001 (Sion *et al.*, 1985b) and KPD0005 + 5106 (Downes *et al.*, 1985) a spectrum featuring strong N V, and weaker O V and C IV is seen. The published CNO and Si abundances of the former are in good agreement with theory. Also as expected, HD 149499B with a cooler T_e (55 000 K) shows only C IV (Bruhweiler and Kondo, 1983; Sion *et al.*, 1982; Sion and Guinan, 1983). Remarkably, no evidence of wind has yet been found in any of the hot He-rich white dwarfs.

In summary, the theory of diffusive equilibrium between radiative acceleration and gravity has enjoyed limited success in accounting for the metals observed in several hot white dwarfs. More extensive observational and theoretical analysis of the metal abundances in hot white dwarfs are needed to understand their early evolution. Finally, the narrowness of the metallic lines could be used also to improve (1) gravitational redshift determinations (Sion and Guinan, 1983) and (2) limits on rotational velocities and magnetic fields.

5.5. COOL WHITE DWARFS

Understanding the evolution of cool helium atmosphere white dwarfs has long been a problem. The question was whether DC white dwarfs (which do not show any absorption features in the visual spectra), DQ white dwarfs (which exhibit molecular carbon bands), and DZ white dwarfs (showing metallic absorption lines), are descended from the same hotter helium rich white dwarf sequence (DB) and/or from DA stars. IUE observations have greatly changed our perspective on this problem since our last reviews (Vauclair, 1979; Liebert, 1980).

The major surprise from IUE was the discovery of C I resonance lines not only in all of the optical DQ stars (Weidemann et al., 1980; Wegner, 1981a; Weidemann et al., 1981) but also in essentially all observed optical DC stars (Wegner, 1981a, b; Vauclair et al., 1981; Vauclair et al., 1982; Koester et al., 1982a; Wegner, 1983a, b). This discovery led to the conclusion that all cool helium rich white dwarfs with T_e above about 6—7000 K probably have carbon as a trace element in their atmosphere. Furthermore, the dichotomy between metallic white dwarfs and carbon white dwarfs has disappeared: IUE revealed the coexistence of metals and carbon in some stars (Koester et al., 1982a; Zeidler-K.T. et al., 1986). Additional data from visual spectra of C_2 white dwarfs (WY) confirm the finding which emerges from the IUE observations of a correlation of the carbon abundance with T_e.

The discovery of carbon in a restricted T_e range and the existence of such a correlation lead to theoretical interpretations which, if correct, have far reaching consequences. First, the hypothesis that the carbon is accreted from the interstellar medium, as we believe is the case for the observed heavy elements in DZ stars, now appears difficult to reconcile with, among other things, the observed C/He $-T_e$ correlation. On the other hand, the idea of dredging-up a trace of the central carbon core is much more promising (KWZ; WY, Fontaine et al., 1984): In models with helium envelopes consistent with computations of pre-white dwarf evolution, the underlying carbon is located deep enough to be fully ionized, so that the helium convection zone never mixes with a carbon convection zone. Instead, the tail of the carbon abundance distribution in diffusive equilibrium at the core C/He interface is mixed from the bottom of the helium convection zone to the surface. Hence, the atmosphere reveals a smattering of the white dwarf's interior composition!

The expected C/He abundance after dredge-up shows good agreement with the observed ratio (Fontaine et al., 1984; Zeidler et al., 1986). The maximum observed C/He occurs near $T_e \sim 12\,000$ K, the effective temperature for which the convection zone reaches its maximum depth, with stars like G227—5 (Wegner and

Koester, 1985) and probably G35—26 (Liebert, 1983). However, observations so far have documented primarily the decline of C/He on the low temperature part of this correlation. One would like to see the confirmation of the other half part of the correlation where C/He decreases with increasing T_e, even though the hotter DBs observed with IUE so far show no trace of C (Wickramasinghe, 1983).

The existence of this correlation has at least three important consequences concerning stellar evolution: (1) it directly tests the chemical composition of the core, since the observed carbon comes from the upward tail of the He envelope/C core interface, and provides an *a posteriori* test of the previous stellar evolution stages which lead to a degenerate carbon/ oxygen core; (2) it provides another test of stellar evolution theory by allowing a direct measure of the He envelope mass; and (3), it places additional constraints on the efficiency of convection.

While carbon is surely dredged up from the core, other metals observed in cool He-rich (DZ) white dwarfs should have another origin. Diffusion time scales for metals in cool white dwarfs are much shorter than their cooling times (Fontaine and Michaud, 1979; VVG; Muchmore, 1984; Paquette *et al.*, 1986). Accordingly, the outer layers should be entirely deprived of metals if diffusion proceeds alone. The observed metals are, therefore, probably provided by accretion during encounters with interstellar clouds. They subsequently mix into and diffuse below the convection zone. Metal abundances and abundance ratios derived from IUE and optical observations of cool white dwarfs provide a crucial test of these ideas (Cottrell and Greenstein, 1980a, b; Wehrse and Liebert, 1980; Shipman and Greenstein, 1983; Zeidler-K.T. *et al.*, 1986). However, a direct comparison of the observations with the diffusion-accretion scenario is still premature. The theory remains incomplete in some respects. The accretion efficiency for gas and grains could be different. In particular, interstellar hydrogen accretion should be inhibited, perhaps by the propeller mechanism (Wesemael and Truran, 1982), by which the rotating magnetosphere of the star may repel ionized gas particles. Cool hydrogen atmospheres with metals are quite rare. Calcium has been detected definitely in only one DA star (Lacombe *et al.*, 1983) at optical wavelengths.

Strong absorptions at 1400 and 1600 Å have been discovered in all DA stars with $T_e < 13\,500$ K (for the $\lambda 1600$ Å absorption) and $T_e < 19\,000$ K (for the $\lambda 1400$ Å absorption). These features have been identified as quasi-molecular H_2 and H_{2+} absorption (Koester *et al.*, 1985b; Nelan and Wegner, 1985). The strengths of the molecular hydrogen absorption in the ultraviolet may eventually be used to identify white dwarfs with hydrogen atmospheres at temperatures (< 5000 K) too low for Balmer line absorption; such observations of fainter, cooler stars might be attempted with the Hubble Space Telescope.

5.6. MAGNETIC DEGENERATE STARS

About 1% of the known white dwarfs currently have detected magnetic fields, all in the range 10^6 to nearly 10^9 G (1—1000 MG). For stars in the range 1—50 MG, the basic Zeeman patterns due to hydrogen and neutral helium have been well established for many years, and have been used successfully to analyze optical data. For two recently discovered magnetic stars hotter than 30 000 K, IUE spectra enabled more accurate T_e estimates (Liebert *et al.*, 1983). However, it is for the magnetic stars with fields stronger than 50 MG that IUE has played a pivotal role, since a basic understanding of some of these stars having hydrogen-rich atmospheres has emerged in just the last several years.

The most interesting set of results concerns Grw + 70°8247, the first magnetic white dwarf, discovered from its optical polarization (Kemp *et al.*, 1970). Like several stars subsequently discovered, it shows unidentified optical absorption features. The literature was full of suggested interpretations when Angel (1979) first proposed that these are quasi-stationary hydrogen Zeeman features in a field of 180–350 MG. This interpretation received considerably more credibility when Greenstein (1984) identified a broad, Ly α $\sigma+$ component, because for this resonance transition the effect of the strong field was predictable. Subsequently, it has turned out that the complicated optical and near-infrared spectra fit the detailed new, theoretical predictions. A second such star, PG1031 + 234, has an even more clearly defined Ly α $\sigma+$ component, indicating a surface field in excess of 500 MG (Schmidt *et al.*, 1986).

The UV energy distributions of the white dwarfs with strong magnetic fields have led to temperature estimates discrepant from prior optical estimates. The Grw + 70°8247 continuum does not fit the predictions of any (nonmagnetic) hydrogen model (Greenstein and Oke, 1982). The best (least squares) fitting blackbody leads to a much higher T_e estimate (14 500 K) than previously indicated, which implies that the star has a small radius, and a large mass ($\gtrsim 1\ M_\odot$). Incorporation of the IUE energy distribution into the analysis drastically reduces the T_e estimates for PG1031 + 234 (Schmidt *et al.*, 1986) and for GD229 (Green and Liebert, 1981).

References

Angel, J. R. P.: 1979, in H. M. Van Horn and V. Weidemann (eds.), 'White Dwarfs and Variable Degenerate Stars', *IAU Coll.* **53**, p. 313.
Baschek, B., Hoflich, P., and Scholz, M.: 1982b, *Astron. Astrophys.* **112**, 76.
Baschek, G., Kudritzki, R. P., Scholz, M., and Simon, K. P.: 1982a, *Astron. Astrophys.* **108**, 387.
Bond, H. E., Grauer, A. D., Green, R. F., and Liebert, J.: 1984, *Astrophys. J.* **279**, 751.
Bruhweiler, F. C. and Kondo, Y.: 1981, *Astrophys. J. Letters* **248**, L123.
Bruhweiler, F. C. and Dean, C. A.: 1983, *Astrophys. J. Letters* **274**, L87.
Bruhweiler, F. C. and Kondo, Y.: 1983, *Astrophys. J.* **269**, 657.
Bruhweiler, F. C., Kondo, Y., and McCluskey, G. E.: 1981, *Astrophys. J. Suppl.* **46**, 255.
Castor, J. I., Abbott, D. C., and Klein, R. K.: 1975, *Astrophys. J.* **195**, 157.
Cottrell, P. L. and Greenstein, J. L.: 1980a, *Astrophys. J.* **238**, 941.
Cottrell, P. L. and Greenstein, J. L.: 1980b, *Astrophys. J.* **242**, 195.
Darius, J., Gidding, J. R., and Wilson, R.: 1979, in A. J. Willis (ed.), *The First Year of IUE*, University College London, p. 367.
Dean, C. A. and Bruhweiler, F. C.: 1985, *Astrophys. J. Suppl.* **57**, 133.
Downes, R. A., Liebert, J., and Margon, B.: 1985, *Astrophys. J.* **290**, 321.
Drilling, J. S. and Schönberner, D.: 1985, *Astron. Astrophys.* **146**, L23.
Drilling, J. S., Holberg, J. B., and Schönberner, D.: 1984, *Astrophys. J. Letters* **283**, L67.
Dupree, A. K. and Raymond, J. C.: 1982, *Astrophys. J. Letters* **263**, L63.
Finley, D., Basri, G., and Bowyer, S.: 1984, in *Future of Ultraviolet Astronomy Based on Six Years of IUE Research*, NASA CP-2349, p. 277.
Fleming, T., Liebert, J., and Green, R. F.: 1986, *Astrophys. J.* **308**, 176.

Fontaine, G. and Michaud, G.: 1979, *Astrophys. J.* **231**, 826.
Fontaine, G., Villeneuve, B., Wesemael, F., and Wegner, G.: 1984, *Astrophys. J. Letters* **277**, L61.
Grauer, A. D. and Bond, H. E.: 1984, *Astrophys. J.* **277**, 211.
Grauer, A. D., Bond, H. E., Liebert, J., Fleming, T., and Green, R. F.: 1986a, *Astrophys. J.* (in press).
Grauer, A. D., Bond, H. E., Green, R. F., Liebert, J., and Fleming, T.: 1986b, preprint.
Green, R. F. and Liebert, J.: 1981, *Publ. Astron. Soc. Pacific* **93**, 105.
Green, R. F., Schmidt, M., and Liebert, J.: 1986, *Astrophys. J. Suppl.* **61**, 305.
Greenberg, J. M. and Chlewicki, G.: 1983, *Astrophys. J.* **272**, 563.
Greenstein, J. L.: 1984, *Astrophys. J. Letters* **281**, L47.
Greenstein, J. L.: 1986, *Astrophys. J.* **304**, 334.
Greenstein, J. L. and Sargent, A.: 1974, *Astrophys. J. Suppl.* **28**, 157.
Greenstein, J. L. and Oke, J. B.: 1979, *Astrophys. J. Letters* **229**, L141.
Greenstein, J. L. and Oke, J. B.: 1982, *Astrophys. J.* **258**, 209.
Groth, H. G., Kudritzki, R. P., and Heber, U.: 1985, *Astron. Astrophys.* **152**, 107.
Gruschinske, J., Hamann, W. R., Kudritzki, R. P., Simon, K. P., and Kaufmann, J. P.: 1983, *Astron. Astrophys.* **121**, 85.
Hamann, W. R., Gruschinske, J., Kudritzki, R. P., and Simon, K. P.: 1981, *Astron, Astrophys.* **104**, 249.
Heap, S.: 1978, P. Conti and C. W. H. DeLoore (eds.), 'Mass Loss and Evolution of O-Type Stars', *IAU Symp.* **83**, 99.
Heber, U.: 1986, *Astron. Astrophys.* **155**, 33.
Heber, U., Hamann, W. R., Hunger, K., Kudritzki, R. P., Simon, K. P., and Mendez, R. H.: 1984b, *Astron. Astrophys.* **136**, 331.
Heber, U., Hunger, K., Jonas, G., and Kudritzki, R. P.: 1981c, *Astron. Astrophys.* **130**, 119.
Heber, U., Hunger, K., Kudritzki, R. P., and Simon, K. P.: 1984a, A. Maeder and A. Renzini (eds.), 'Observational Tests of Stellar Evolution Theory', *IAU Symp.* **105**, 215.
Henry, R. B. C., Shipman, H. L., and Wesemael, F.: 1985, *Astrophys. J. Suppl.* **57**, 145.
Holberg, J. B., Wesemael, F., Wegner, G., and Bruhweiler, F. C.: 1985, *Astrophys. J.* **293**, 294.
Holberg, J., Wesemael, F., and Basile, J. 1986, *Astrophys. J.* **306**, 629.
Hunger, K., Gruschinske, J., Kudritzki, R. P., and Simon, K. P.: 1981, *Astron. Astrophys.* **95**, 244.
Hunger, K. and Kudritzki, R. P.: 1981, *The Messenger* **24**, 7.
Husfeld, D.: 1986 Ph.D. Dissertation, the University of Munich.
Iben, I. and Renzini, A.: 1983, *Ann. Rev. Astron. Astrophys.* **21**, 271.
Iben I. and Tutukov, A.: 1984, *Astrophys. J.* **282**, 615.
Iben, I., Kaler, J. B., Truran, J. W., and Renzini, A.: 1983, *Astrophys. J.* **264**, 605.
Kahn, S. M., Wesemael, F., Liebert, J., Raymond, J. C., Steiner, J. E., and Shipman, H. L.: 1984, *Astrophys. J.* **278**, 255.
Kaler, J. B. and Feibelman, W. A.: 1985, *Astrophys. J.* **297**, 724.
Kawaler, S. D., Hansen, C. J., and Winget, D. E.: 1985, *Astrophys. J.* **295**, 547.
Kawaler, S. D., Winget, D. E., Hansen, C. J., and Iben, I.: 1986, *Astrophys. J. Letters* **306**, L41.
Kemp, J. C., Swedlund, J. B., Landstreet, J. D., and Angel, J. R. P.: 1970, *Astrophys. J. Letters* **161**, L77.
Kilkenny, D., Hill, P. W., and Penfold, J. E.: 1981, *Monthly Notices Roy. Astron Soc.* **194**, 429.
Koester, D. and Schönberner, D.: 1986, *Astron. Astrophys.* **154**, 134.
Koester, D., Schulz, H., and Weidemann, V.: 1979, *Astron. Astrophys.* **76**, 262.
Koester, D., Vauclair, G., Weidemann, V., and Zeidler-K. T., E. M.: 1982a, *Astron. Astrophys.* **113**, L13.
Koester, D., Weidemann, V., and Zeidler-K. T., E. M.: 1982b, *Astron. Astrophys.* **116**, 147. (KWZ)
Koester, D., Vauclair, G., Dolez, N., Oke, J. B., Greenstein, J. L., and Weidemann, V.: 1985a, *Astron. Astrophys.* **149**, 423.
Koester, D., Weidemann, V., Zeidler-K. T., E. M., and Vauclair, G.: 1985b, *Astron. Astrophys.* **142**, L5.
Kudritzki, R. P.: 1976, *Astron. Astrophys.* **52**, 11.

Kudritzki, R. P., Simon, K. P., Lynas-Gray, A. E., Kilkenny, D., and Hill, P. W.: 1982, *Astron. Astrophys.* **106**, 254.

Lacombe, P., Liebert, J., Wesemael, F., and Fontaine, G. 1983, *Astrophys. J.* **272**, 660.

Lamontagne, R., Wesemael, F., Fontaine, G., and Sion, E. M.: 1985, *Astrophys. J.* **299**, 496.

Liebert, J.: 1980, *Ann. Rev. Astron. Astrophys.* **18**, 363.

Liebert, J.: 1983, *Publ. Astron. Soc. Pacific* **95**, 878.

Liebert, J.: 1984, in *Future of Ultraviolet Astronomy Based on Six Years of IUE Research*, NASA CP-2349, p. 93.

Liebert, J.: 1986, K. Hunger, D. Schönberner and K. Rao, (eds.), 'Hydrogen Deficient Stars and Related Objects', *IAU Coll.* **87**, 367.

Liebert, J., Schmidt, G. D., Green, R. F., Stockman, H. S., and McGraw, J. T.: 1983, *Astrophys. J.* **264**, 262.

Liebert, J., Wesemael, F., Sion, E. M., and Wegner, G.: 1984, *Astrophys. J.* **277**, 692.

Liebert, J., Wesemael, F., Hansen, C. J., Fontaine, G., Shipman, H. L., Sion, E. M., Winget, D. E., and Green, R. F.: 1986, *Astrophys. J.* **309**, 241.

McGraw, J. T., Starrfield, S. G., Liebert, J., and Green, R. F.: 1979, H. M. Van Horn and V. Weidemann (eds.), 'White Dwarfs and Variable Degenerate Stars', *IAU Coll.* **53**, 377, Univ. of Rochester Press, Rochester NY.

Mendez, R. H., Miguel, C. H., Heber, U., and Kudritzki, R. P.: 1986, K. Rao and K. Hunger, D. Schönberner and K. Rao (eds.), 'Hydrogen Deficient Stars and Related Objects', *IAU Coll.* **87**, 323.

Mendez, R. H., Kudritzki, R. P., and Simon, K. P.: 1983, D. R. Flower (ed.),'Planetary Nebulae', *IAU Symp.* **103**, 343.

Mestel, L.: 1952, *Monthly Notices Roy. Astron. Soc.* **112**, 583.

Michaud, G., Bergeron, P., Wesemael, F., and Fontaine, G.: 1985, *Astrophys. J.* **299**, 741.

Michaud, G., Montmerle, T., Cox, A. N., Magee, N. H., Hodson, S. W., and Martel, A. C.: 1979, *Astrophys. J.* **234**, 206.

Michaud, G., Vauclair, G., and Vauclair, S.: 1983, *Astrophys. J.* **267**, 256.

Morvan, E., Vauclair, G., and Vauclair, S.: 1986, *Astron. Astrophys.* **163**, 145.

Muchmore, D. O.: 1984, *Astrophys. J.* **278**, 769.

Nelan, E. P. and Wegner, G.: 1985, *Astrophys. J. Letters* **289**, L31.

Nousek, J. A., Shipman, H. L., Holberg, J. B., Liebert, J., Pravdo, S. H., White, N. E., and Giommi, P.: 1986, *Astrophys. J.* **309**, 230.

Oke, J. B., Weidemann, V., and Koester, D.: 1984, *Astrophys. J.* **281**, 276. (OWK)

Paquette, C., Pelletier, C., Fontaine, G., and Michaud, G.: 1986, *Astrophys. J. Supp.* **61**, 197.

Petre, R., Shipman, H. L., and Canizares,C. R.: 1986, *Astrophys. J.* **304**, 356.

Pottasch, S. R.: 1984, in D. R. Flower (ed.), 'Planetary Nebulae', *IAU* **103**, 391.

Reimers, D. and Koester, D.: 1982, *Astron. Astrophys.* **116**, 341.

Rossi, L., Viotti, R., and Altamore, A.: 1984, *Astron. Astrophys. Suppl.* **55**, 361.

Schmidt, G., West, S., Liebert, J., Stockman, H. S., and Green, R. F.: 1986, *Astrophys. J.* **309**, 218.

Schönberner, D.: 1981, *Astron. Astrophys.* **103**, 119.

Schönberner, D.: 1983, *Astrophys. J.* **272**, 708.

Schönberner, D.: 1986, *Astron. Astrophys.* **16**, 189.

Schönberner, D. and Drilling, J. S.: 1984, *Astrophys. J.* **278**, 702.

Schönberner, D. and Drilling, J. S.: 1985, *Astrophys. J. Letters* **290**, L49.

Shipman, H. L.: 1979, *Astrophys. J.* **228**, 240.

Shipman, H. L. and Greenstein, J. L.: 1983, *Astrophys. J.* **266**, 761.

Shipman, H. L., Liebert, J., and Green, R. F.: 1986, *Astrophys. J.* (in press).

Simon, K. P.: 1981, *Astron. Astrophys.* **98**, 211.

Simon, K. P.: 1982, *Astron. Astrophys.* **107**, 313.

Simon, K. P., Gruschinske, J., Haman, W. R., Hunger, K., and Kudritzki, R. P.: 1979, in A. J. Willis (ed.), *The First Year of IUE*, University College London, p. 354.

Sion, E. M.: 1984, *Astrophys J.* **282**, 612.

Sion, E. M.: 1986, *Publ. Astron. Soc. Pacific* **98**, 821.

Sion, E. M. and Guinan, E. F.: 1983, *Astrophys. J. Letters* **265**, L87.

Sion, E. M. and Starrfield, S. G.: 1984, *Astrophys. J.* **286**, 760.

Sion, E. M., Guinan, E. F., and Wesemael, F.: 1982, *Astrophys. J.* **255**, 232.

Sion, E. M., Liebert, J., and Starrfield, S. G.: 1985a, *Astrophys. J.* **292**, 471.

Sion, E. M., Liebert, J., and Wesemael, F.: 1985b, *Astrophys. J.* **292**, 477.

Sion, E. M., Greenstein, J. L., Landstreet, J. D., Liebert, J., Shipman, H. L., and Wegner, G.: 1983, *Astrophys. J.* **269**, 253.

Starrfield, S., Cox, A., Kidman, R. B., and Pesnell, W. D.: 1984, *Astrophys. J.* **281**, 800.

Vauclair, G.: 1979, H. M. Van Horn and V. Weidemann (eds.), 'White Dwarfs and Variable Degenerate Stars', *IAU Coll.* **53**, 165, Univ. of Rochester Press, Rochester NY.

Vauclair, G., Vauclair, S., and Greenstein, J. L.: 1979, *Astron. Astrophys.* **80**, 79 (VVG).

Vauclair, G., Weidemann, V., and Koester, D.: 1981, *Astron. Astrophys.* **100**, 113.

Vauclair, G., Weidemann, V., and Koester, D.: 1982, *Astron. Astrophys.* **109**, 7.

Vauclair, S.: 1975, *Astron. Astrophys.* **45**, 233.

Wegner, G.: 1981a, *Astrophys. J. Letters* **245**, L27.

Wegner, G.: 1981b, *Astrophys. J. Letters* **248**, L129.

Wegner, G.: 1983a, *Astrophys. J.* **268**, 282.

Wegner, G.: 1983b, *Astron. Astrophys.* **128**, 258.

Wegner, G. and Yackovich, F. H.: 1983, *Astrophys. J.* **275**, 240.

Wegner, G. and Yackovich, F. H.: 1984, *Astrophys. J.* **284**, 257 (WY).

Wegner, G. and Koester, D.: 1985, *Astrophys. J.* **288**, 746.

Wehrse, R. and Liebert, J.: 1980, *Astron. Astrophys.* **86**, 139.

Weidemann, V. and Koester, D.: 1983, *Astron. Astrophys.* **121**, 77.

Weidemann, V., Koester, D., and Vauclair, G.: 1980, *Astron. Astrophys.* **83**. L13.

Weidemann, V., Koester, D., and Vauclair, G.: 1981, *Astron. Astrophys.* **95**, L9.

Wesemael, F.: 1981, *Astrophys. J. Suppl.* **45**, 177.

Wesemael, F. and Truran, J. W.: 1982, *Astrophys. J.* **260**, 807.

Wesemael, F., Holberg, J. B., Veilleux, S., Lamontagne, R., and Fontaine, G.: 1985a, *Astrophys. J.* **298**, 859.

Wesemael, F., Green, R. F., and Liebert, J.: 1985b, *Astrophys. J. Suppl.* **58**, 379. (WGL)

Wesemael, F., Henry, R. B. C., and Shipman, H. L.: 1984, *Astrophys. J.* **287**, 868.

Wesemael, F., Winget, D. E., Cabot, W., Van Horn, H. M., and Fontaine, G.: 1982, *Astrophys. J.* **254**, 221.

Wickramasinghe, D. T.: 1983, *Monthly Notices Roy. Astron. Soc.* **203**, 903.

Winget, D. E. and Cabot, W. 1980, *Astrophys. J.* **242**, 1160.

Winget, D. E., Hansen, C. J., and Van Horn, H. M.: 1983a, *Nature* **303**, 781.

Winget, D. E., Van Horn, H. M., Tassoul, M., Hansen, C. J., and Fontaine, G.: 1983b, *Astrophys J. Letters* **268**, L33.

Zeidler-K. T., E. M., Weidemann, V., and Koester, D.: 1986, *Astron. Astrophys.* **155**, 356.

OF GALACTIC NOVAE*

SUMNER STARRFIELD**

Joint Institute of Laboratory Astrophysics,
University of Colorado, Boulder, CO, U.S.A.
and
Theoretical Division, Los Alamos National Laboratory,
Los Alamos, NM, U.S.A.

M. A. J. SNIJDERS

Royal Greenwich Observatory,
Herstmonceux Castle, Hailsham, U.K.

1. Introduction

The classical nova outburst is the second most violent explosion that occurs in a galaxy and its violence is exceeded only by a supernova explosion. However, novae are relatively nearby and a nova outburst occurs much more frequently than a supernova outburst so that there are more than 200 outbursts tabulated and discussed by Payne-Gaposchkin (1957) and McLaughlin (1960). The first *ultraviolet* (hereafter: UV) study of a nova was that of the 1970 outburst of FH Ser (Gallagher and Code, 1974) and that was broad band photometry done by the OAO—A2 satellite. The first UV spectroscopic study was that of V1500 Cyg 1975 done with *Copernicus* (Jenkins *et al.*, 1977) but it was such a fast nova that the only line visible when the spectrum was taken was Mg II λ2800. Nova Cyg was also studied with the ANS satellite (Wu and Kester, 1975).

The whole picture changed with the launch of the International Ultraviolet Explorer satellite (hereafter: IUE) in 1978. Almost immediately after launch it began obtaining both low and high dispersion spectra of novae in outburst and old novae in quiescence. These data, taken over the last 8 years, have markedly increased our understanding of the nova outburst. Each outburst has proved to be unique and valuable and the IUE observations have shown that it is as important to obtain UV as optical, radio, and IR data. The data are complementary and all are necessary to understand the characteristics of the outburst.

The importance of the IUE data is that there are spectral lines in the 1200 Å to 3300 Å wavelength range that come from elements which do not have analyzable (or any) lines in the optical. In addition, the number of available diagnostic emission line ratios has been greatly expanded through the combination of optical and UV data. These lines can be used to determine elemental abundances, expansion velocities, and the amount of mass ejected. Many of these lines are the

* Supported in part by NSF grants AST83-14788 and AST85-16173 to Arizona State University, by NASA grant NAG5-481 to Arizona State University, and by the DOE.
** Permanent Address: Department of Physics and Astronomy, Arizona State University, Tempe, AZ.

Y. Kondo (ed.), Scientific Accomplishments of the IUE, pp. 377—393.
© 1987 *by D. Reidel Publishing Company.*

commonly observed and well understood medium ionization UV resonance and intercombination lines observed in most emission line objects. However, their time-dependant behavior in novae can be used to constrain the abundances that are determined for the ejected material. A table of such lines and the time variations of their fluxes for both a normal nova and also a 'neon' nova can be found in Stickland *et al.* (1981) and in Williams *et al.* (1985). Because of IUE data, we have recently been able to identify a new class of novae (Starrfield *et al.*, 1986). Finally, continuum flux distributions can be used to determine temperatures and compare the observations with predictions of accretion disc theory.

The classical reviews of the observed behavior of a nova in outburst are those of Payne-Gaposchkin (1957) and McLaughlin (1960). Reviews of novae in quiescence are Gallagher and Starrfield (1978; quiescence plus outburst), Cordova and Mason (1983), and Bode and Evans (1986). A compilation of the abundances of novae can be found in Truran and Livio (1986). The existence of these reviews allows us to skip the basic optical data and concentrate only on the observations from IUE.

2. The Cause of the Outburst

In this review we assume the commonly accepted model for a nova: a close binary system with one member a white dwarf and the other member a larger, cooler star that fills its Roche lobe. Because it fills its lobe, any tendency for it to grow in size because of evolutionary processes or for the lobe to shrink because of angular momentum losses (by some, as yet unknown, mechanism) will cause a flow of gas through the inner Lagrangian point into the lobe of the white dwarf. The size of the white dwarf is small compared to the size of its lobe and the high angular momentum of the transferred material causes it to spiral into an accretion disk surrounding the white dwarf. Some viscous process, as yet unknown, acts to transfer mass inward and angular momentum outward through the disk so that a fraction of the material lost by the secondary ultimately ends up on the white dwarf. Over a long period of time, the accreted layer will grow in thickness until the bottom reaches a temperature that is high enough to initiate thermonuclear burning of hydrogen. The further evolution of nuclear burning on the white dwarf now depends upon the mass and luminosity of the white dwarf, the rate of mass accretion, and the chemical composition of the reacting layer.

Given the proper conditions, a thermonuclear runaway (hereafter: TNR) will occur, and the temperature in the accreted envelope will grow to values exceeding 10^8 K. At this time the positron decay nuclei become abundant which strongly affects the further evolution of the outburst. Theoretical calculations demonstrate that this evolution will release enough energy to eject material with expansion velocities that agree with observed values and that the predicted light curves produced by the expanding material can agree quite closely with the observations (Sparks *et al.*, 1978; Starrfield *et al.*, 1978; Starrfield *et al.*, 1974a, b, 1985, 1986; Prialnik *et al.*, 1978, 1979; MacDonald, 1980; Prialnik *et al.*, 1982).

These theoretical studies have involved hydrodynamic calculations of TNR's in the accreted hydrogen rich envelopes of white dwarfs which are assumed to be the compact components of nova binary systems. This work has been extremely successful in reproducing the gross features of the nova outburst: ejected masses, kinetic energies, and light curves. More importantly, these calculations predicted: (1) that enhanced CNO nuclei would be found in the ejecta of fast novae, (2) that the isotopic ratios of the CNO nuclei would be far from solar, (3) that there should be a post maximum phase of constant luminosity lasting for months, or longer, and (4) that the properties of the outburst should be strong functions of the mass of the white dwarf. As discussed in Starrfield (1986a, b), observational confirmation of each of these points has now appeared in the literature. The theoretical studies of the nova outburst have also identified a number of parameters that strongly influence the characteristics of the outburst. These are: (1) the white dwarf mass, (2) the envelope mass, (3) the white dwarf luminosity, (4) the rate of mass accretion, and (5) the chemical composition of the envelope (see MacDonald, 1983; Paczynski 1983).

The nucleosynthesis that occurs during the outburst will produce ^{13}C, ^{14}N, ^{15}N, and ^7Li and these nuclei will be ejected by the nova explosion (Starrfield et al., 1978; Lazareff et al., 1979; Starrfield et al., 1978; Audouze et al., 1979; Wallace and Woosley, 1981; Hillebrandt and Thielemann, 1982; Wiescher et al., 1985). It has also been predicted that the ^{26}Al anomaly could be a result of nuclear burning during the nova outburst (Arnould et al., 1980) and the recent discovery of ONeMg novae (Starrfield et al., 1986) strengthens this prediction. These predictions, along with the observational confirmation of nonsolar CNO abundances in nova ejecta, demand that novae be included in studies of galactic nucleosynthesis.

The prediction that the ejecta of novae would be enhanced in CNO nuclei has been confirmed by a number of observational studies (Sneden and Lambert, 1975; Williams et al., 1978, 1981, 1985; Williams and Gallagher, 1979; Ferland and Shields, 1978; Tylenda, 1978; Gallagher et al., 1980; Stickland et al., 1981; Snijders et al., 1984). In addition, Sneden and Lambert (1975) determined that the $^{12}C/^{13}C$ isotopic ratio (in combination with the extreme overabundance of carbon) in DQ Her strongly supported a TNR as the cause of the outburst.

Studies of recent novae have reported very large enhancements of neon (V1500 Cyg: Ferland and Shields, 1978; V693 CrA: Williams et al., 1985; V1370 Aql: Snijders et al., 1984). Now, Gehrz, et al. (1985) have reported the discovery of [Ne II] emission at 12.8 μ in Nova Vul #2 1984 and Gehrz et al. (1986) report the condensation of SiO_2 grains in the same nova. These results, in combination with the IUE spectra that also show strong neon lines (Starrfield et al., 1987, in prep.; Snijders et al., 1986, in prep.; and this review), imply that at least four of the recent nova outbursts have ejecta rich in neon. The most likely explanation is that some novae are ejecting material which has been processed to neon and beyond during the prior evolution of the white dwarf (as suggested by Law and Ritter, 1983 and confirmed and expanded by Starrfield et al., 1986).

UV and IR studies confirm that most novae exhibit a phase of constant luminosity following the initial rise to maximum (Wu and Kester, 1977; Ney and Hatfield, 1978; Stickland et al., 1981; Sparks et al., 1982; Snijders et al., 1984; Williams et al., 1985; Snijders, 1986). This phase occurs because not all of the accreted material is ejected during the burst phase of the outburst and the radiated energy, the effective temperature, and the time scale of this phase of the outburst provide fundamental data about the white dwarf (Gallagher and Starrfield, 1976, 1978; Starrfield, 1979, 1980, 1986a; Truran, 1982; Sion and Starrfield, 1986; MacDonald *et al.*, 1985). These data have been used to show that the masses of white dwarfs in binary systems range from $\sim 0.6\ M_\odot$ (DQ Her 1934) to $\sim 1.2\ M_\odot$ or even higher (Nova V1500 Cyg 1975 and U Sco 1979). The inferred mass for DQ Her appears to be in substantial agreement with the values determined from radial velocity studies (Smak, 1980; Young and Schneider, 1980).

Recent calculations have demonstrated that a recurrent nova outburst can occur as a result of a TNR by examining the consequences of accretion of hydrogen rich material, with a solar abundance of the CNO nuclei, onto a massive white dwarf (Starrfield *et al.*, 1985). They found that a TNR resulted after ~ 33 yr of evolution and the evolution resembled that of U Sco. Sion and Starrfield (1986) have also modeled the observed behavior of Z And. Finally, a calculation of accretion onto ONeMg white dwarfs has produced extremely violent outbursts in which the entire accreted envelope was ejected at high velocities (Starrfield *et al.*, 1986). This study was an attempt to simulate the novae that are ejecting enhanced neon.

3. IUE Studies of Classical Novae in Outburst

In this section we review the published studies of novae in outburst as a function of the date of outburst. Table I gives a list of the novae studied by IUE.

TABLE I

Galactic novae in outburst

Name const.	Outburst year	R. A. hr	mn	sec	Dec. deg mn sec	l deg	b deg	V_{max}[a] mag	V_{min} mag
Muscae	(1983)	11h	49m	35s	−66° 55′ 43″	297.2	−05.0	6.5	14.5
U Sco	(1979)	16h	19m	37s	−17° 45′ 43″	357.7	+21.9	9.	14.3
RS Oph	(1985)	17h	47m	31s	−06° 41′ 47″	19.8	+10.4	6.5	11.7
Ser	(1983)	17h	53m	24s	−14° 00′ 56″	14.1	+05.6	11.9	
Sgr	(1982)	18h	31m	33s	−26° 28′ 25″	7.4	−08.3	8.	
V693 CrA	(1981)	18h	38m	30s	−37° 34′ 59″	357.8	−14.4	7.	14.
V1370 Aql	(1982)	19h	20m	50s	+02° 24′ 00″	38.8	−05.9	6.5	13.0
PW Vul	(1984)	19h	24m	3s	+27° 15′ 54″	61.1	+05.2	6.5	12.8
RR Tel	(1947)	20h	00m	18s	−55° 52′ 00″	342.2	−32.2	6.8	
WZ Sge	(1978)	20h	05m	18s	+17° 32′ 56″	57.5	−07.9	8.1	14.0
Ser	(1978)	20h	17m	59s	−14° 43′ 08″	29.4	−26.2	13.	
Vul #2	(1984)	20h	24m	41s	+27° 40′ 40″	68.5	−06.0	6.3	10.8
V1668 Cyg	(1978)	21h	40m	38s	+43° 48′ 00″	90.8	−06.8	6.5	14

[a] V_{max} and V_{min} refer to the maximum and minimum brightness of the nova at the times of IUE exposures.

3.1. V1668 CYGNI

This was the first classical nova to be studied with the IUE satellite. It was a moderately 'fast' nova which reached a V_{max} of ~6.2 on September 12, 1978. A brief review of the IUE data can be found in Sparks *et al.* (1980) who also present data on the outbursts of WZ Sge and U Sco. There were 194 spectra obtained for this nova during its outburst and these data consist of both low and high dispersion SWP and LWR images. The first spectrum was taken on day 1978/254 and the last on day 1980/363 (our dating system is year/day number). During this time the visual magnitude declined to ~14. Cassatella *et al.* (1979) also presented early spectra of this outburst. The data have been analyzed by Stickland *et al.* (1979, 1981). They restricted their analysis mainly to the nebular phase and found that the total CNO abundance in the ejecta was about 30 times solar and that nitrogen alone was about 200 times solar. They report an expansion velocity ~800 km s^{-1}, an ejected mass of ~6 × 10^{-5} $M_\odot$, and a kinetic energy of 6 × 10^{44} ergs. These results are in close agreement with calculations of Starrfield *et al.* (1978) for a CNO outburst on a 1.0 $M_\odot$ white dwarf.

3.2. WZ SGE

This object has had at least two prior recorded outbursts and, because its rise is from ~14 to ~8 mag, it is often referred to as a recurrent nova. In fact, both the optical and IUE data obtained during this outburst confirm that it is a dwarf nova with a very long interoutburst period (Sparks *et al.*, 1980; Fabian *et al.*, 1980). There were 76 spectra obtained during the outburst. The first was taken on day 1978/335 and the last on day 1981/326 (at least a year after its return to minimum). The analysis of the IUE data was done by Fabian *et al.* (1980) and Friedjung *et al.* (1980) who reported that the IUE data did not show strong emission lines as is normally seen in the outburst of a classical nova. They found that very little material was ejected, the UV spectra were characteristic of an accretion disk, and during the outburst it continued to show variations in the line profiles with phase. The line widths gave velocities expected for material orbiting in keplerian velocities at the outer edge of an accretion disk around the white dwarf.

3.3. U SCO

This was the first recurrent nova to be studied by the IUE and the analysis of its outburst has provided us with some very unusual findings. There were 18 spectra obtained for this nova from day 1979/179 to day 1979/237. Optical data for this outburst were analyzed by Barlow *et al.* (1981) who found that H/He in the ejecta was ~1/2. The analysis of the IUE spectra is in Williams *et al.* (1981) who report that nitrogen was overabundant while carbon and oxygen probably were not. In fact, the CNO abundances could be explained by solar CNO material being processed through a hot hydrogen burning region. The first spectra taken with the IUE showed P Cygni type profiles which upon analysis implied an ejected mass of ~10^{-7} $M_\odot$ to ~10^{-8} $M_\odot$. This low a value, plus the rapidity of the decline, suggests strongly that the outburst occurred on a very massive white dwarf

(Starrfield *et al.*, 1985). They evolved TNR's on 1.38 $M_\odot$ white dwarfs and found reasonably good agreement with the observations. One difficulty is that studies of the accretion disk, both before and after the outburst (Barlow *et al.*, 1981; Williams *et al.*, 1981; Hanes, 1985) show only lines due to helium. This nova appears to be transferring helium and ejecting both hydrogen and helium!

3.4. V693 CRA

This classical nova was discovered at ~7 mag. Optical spectra are published by Brosch (1982). There were 42 IUE spectra obtained during the outburst (day 1981/100 to day 1981/318). Williams *et al.* (1985) analyzed the IUE data and presented both a UV line list and line fluxes over the duration of the outburst. Brosch (1982) determined an ejection velocity of ~2200 km s⁻¹ from the FWHM of the hydrogen lines. This high a value is also derived from low dispersion IUE data. In addition, a recent analysis of a high dispersion SWP image (13712) taken on day 1981/122 shows C IV 1550 with the blue edge of the P Cygni profile extending to a velocity of almost 8000 km s⁻¹ (Sion *et al.*, 1986; and Figure 1).

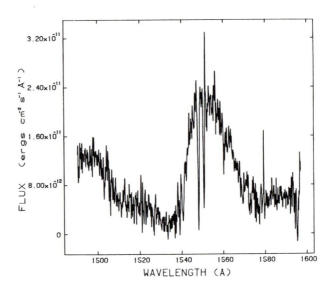

Fig. 1. The region around CIV 1550 taken from a high dispersion spectrum of V693 CrA. We point out the very narrow interstellar absorption lines which have recently been analyzed (Sion *et al.*, 1986). Note also the large blueward extent, −8000 km s⁻¹, of the P Cygni absorption profile.

The most important finding of the IUE analysis (Williams *et al.*, 1985) is that the abundances of all of the intermediate mass elements from nitrogen to aluminum are enhanced over a solar mixture by a factor of about 100 (by number). This implies that Z for the ejected material is ~0.4 (Truran and Livio 1986). V693 CrA is now identified as a member of a new class of novae which must be occurring on ONeMg white dwarfs (see Starrfield *et al.*, 1986, where the implications of this result are discussed).

3.5. V1370 Aql

This nova was discovered in outburst on January 28, 1982. The IUE spectra (22 from 1982/55 to 1982/181) have been analyzed by Snijders, *et al.* (1982, 1984). This nova was also observed in the optical by Andrillat (1983) and Rosino *et al.* (1983) and in the IR by Gehrz *et al.* (1984), Williams and Longmore (1984), and Bode *et al.* (1984). Snijders, *et al.* (1984) report that this nova ejected $\sim 5 \times 10^{-6}$ $M_\odot$ with expansion velocities of ~ 4000 km s^{-1}. A high velocity component with velocities up to $10\,000$ km s^{-1} was present for at least 35 days after the outburst (Snijders *et al.*, 1982, 1986a). The abundances determined for the ejecta were as unusual as reported for V693 CrA. Neon was the most abundant element in the ejecta and elements up to sulfur were enhanced. Truran and Livio (1986) calculate a Z for the ejecta of ~ 0.86. Starrfield *et al.* (1986) explain this as an outburst on an ONeMg white dwarf (but see Wiescher *et al.*, 1986). The IR data showed an excess at ~ 10 μ but no excess at ~ 20 μ (Gehrz *et al.*, 1984). Snijders *et al.* (1984) reported a grain mass of the same order of magnitude as the gas mass so that this material is an important part of the ejecta.

3.6. Nova Sgr (1982) and Nova Ser (1983)

These novae had 13 IUE spectra (1982/291 to 1982/304) and three spectra (day 1983/64) taken respectively. A discussion of these spectra can be found in Drechsel and Rahe (1982) and Drechsel *et al.* (1984).

3.7. Nova Muscae (1983)

There were 55 spectra obtained for this nova from day 1983/63 to day 1985/215. The analysis of the early IUE spectra can be found in Krautter *et al.* (1984) who also reported optical and IR data. They found a distance of ~ 5 kpc and noted that it was radiating at $\sim L_E$ for a 1.0 $M_\odot$ white dwarf in agreement with the TNR predictions. An abundance analysis showed that He/H was enhanced over solar, N/C was ~ 20, and N/O was ~ 2.4. These values are characteristic of hot hydrogen burning. Pacheco and Codina (1985) determined an expansion velocity of ~ 1000 km s^{-1} and confirmed, from optical spectra, the overabundance of He, and CNO. They also found an overabundance of Fe which is difficult to understand in terms of the TNR theory although iron enhancements have been suggested for other novae (V1500 Cygni: Ferland and Shields, 1978; V1370 Aql: Snijders *et al.*, 1984).

This nova was also detected by *Exosat* in the low energy detectors by Ogelman *et al.* (1984) who reported the discovery of a source with $T \sim 3 \times 10^5$ K at a late stage of the outburst. This is exactly what would be expected from a white dwarf which is finally burning out the remaining hydrogen envelope on its surface (Starrfield 1979; MacDonald *et al.*, 1985).

3.8. PW Vul (1984 No. 1)

This slow nova was discovered to be in outburst in July 1984 and the IUE began taking spectra almost immediately on day 1984/175. We have continued to obtain spectra up to the present time. Optical and IR data for this nova are presented in

Kenyon and Wade (1986) who find a distance of ~1.2 kpc and M_v ~5.5. They also report He/H of ~0.13 and that oxygen is enhanced in the ejecta but neon was not enhanced. The IUE data are currently being reduced and analyzed (Cassatella *et al.*, 1986, in prep.; Starrfield *et al.*, 1987, in prep.). We show, in Figure 2, spectra of PW Vul taken on June 24, 1985 and March 31, 1986.

3.9. NOVA VUL (1984 NO. 2)

This slow nova was discovered late in 1984 and the first IUE spectrum was obtained on day 1984/363. At the end of July 1986, it had declined only to ~10.8 mag and it is still being observed in the UV, optical, IR, and radio. More than 55 IUE spectra have been obtained, so far, and these data imply that the ejected material is very neon rich so that this must be a fourth member of the ONeMg class of outbursts (Starrfield *et al.*, 1986). The IUE data show that [Ne V] $\lambda 3346$ appeared during the fall of 1985 and that at the present time [Ne IV] $\lambda 1602$ is the strongest line in the SWP spectral region (Figure 3). In addition, [Ne IV] $\lambda 2422$ is also present which, in combination with the IR results (Gehrz *et al.*, 1985; Gehrz *et al.*, 1986) is strong evidence for enhanced neon in this nova.

3.10. RS OPH

This fact recurrent nova was discovered to be in outburst in January 1985. 73 spectra were obtained from day 1985/033 to day 1985/137. In addition, there are IUE images obtained, in quiescence, both before and after and outburst. Preliminary results from the IUE observations are now available (Cassatella *et al.*, 1985; Snijders, 1986; see also Bode, 1986). These data support TNR models for the outburst of this recurrent nova. A discussion of the relationship of this outburst to the outbursts of classical novae can be found in Sparks *et al.* (1986) and Starrfield *et al.* (1985; An alternative view is in Livio *et al.* [1986] but their hypothesis is in disagreement with the observations). A detailed analysis of the IUE observations is in progress (Snijders *et al.*, 1986b).

3.11. RR TEL

We briefly mention this slow nova which has been in outburst for nearly 40 yr and for which 106 spectra have been obtained by IUE. The early data have been analyzed by Penston *et al.* (1983) who complement the optical analysis of Thackery (1977). The line strengths imply a source of ionizing radiation for the nebula. Spectra since then have been taken under the ID: 'PHCAL' and are used for wavelength calibration of the IUE; they do not appear under the class '55' designation for classical novae in the IUE databank.

4. Novae at Quiescence

In this section we briefly discuss the IUE studies of old novae: those whose outburst began before the launch of the satellite and who have either returned to minimum or are on the way down. A list of these objects is given in Table II. A very wide range of nova speed class has been studied by IUE with a very broad range in time since outburst. The analysis of this data base should ultimately be

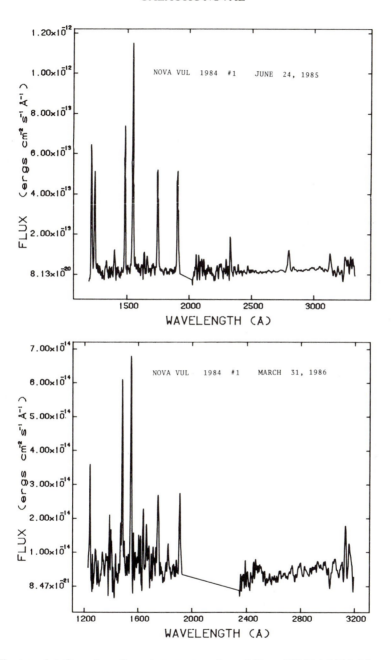

Fig. 2.   The top plot shows low dispersion spectra taken of Nova PW Vul 1984 #1 on June 24, 1985 and the bottom plot shows spectra obtained on March 31, 1986. We were not able to plot Ly on the same scale as the rest of the lines in the bottom plot so the first line is actually N v 1240. The other two strong lines are N iv] 1486 and C iv 1549. N iii] 1750 and C iii] 1909 are also present at both times. Note their peak fluxes have fallen by a factor of ~20 from June to March and that Mg ii 2800 is present in June but probably absent in March. The features longward of 3000 Å in March are probably noise.

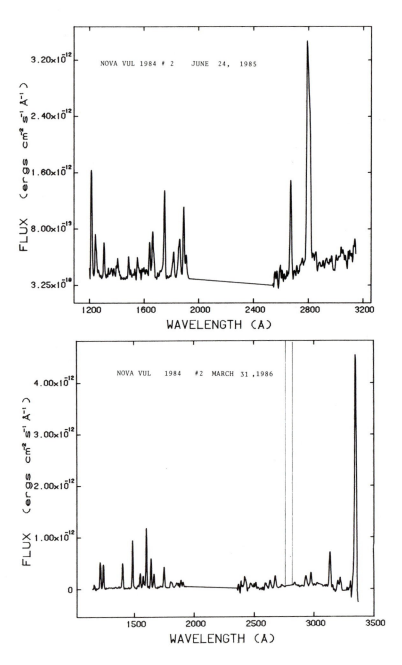

Fig. 3. These two low dispersion spectra show Nova Vul 1984 #2 on June 24, 1985 and March 31, 1986. We have only outlined Mg II 2800 on the bottom spectrum so as to be able to show the other lines on the same scale. The strong line at 3346 is [Ne V]. The strong line at 1602 is [Ne IV] and [Ne IV] 2426 is present in the March spectrum. These spectra resemble those taken of V693 CrA 1981 (Williams *et al.*, 1985) and a line list can be found in that paper.

TABLE II

Galactic novae in quiescence

Name const.	Outburst year	R. A. hr mn sec	Dec. deg mn sec	l deg	b deg	V_{max}[a] mag	V_{min} mag
GK Per	(1901)	3h 27m 48s	+43° 44′ 04″	151.0	−10.1	11.0	14.0
T Aur	(1891)	5h 28m 46s	+30° 24′ 35″	177.1	−01.7	15.2	
RR Pic	(1925)	6h 35m 10s	−62° 35′ 47″	272.4	−25.7	12.0	12.8
T Pyx	(1966)	9h 02m 37s	−32° 11′ 00″	257.2	+09.7	14.0	14.5
T CrB	(1946)	15h 57m 24s	+26° 04′ 00″	42.4	+48.2	9.9	10.8
RS Oph	(1967)	17h 47m 31s	−06° 41′ 47″	19.8	+10.4	10.7	11.7
DQ Her	(1934)	18h 06m 05s	+45° 51′ 01″	73.2	+26.4	14.0	14.8
V533 Her	(1960)	18h 12m 46s	+41° 50′ 21″	69.2	+24.3	14	
V603 Aql	(1918)	18h 46m 21s	+00° 31′ 00″	33.2	+00.8	11.0	12.0
CK Vul	(1670)	19h 45m 35s	+27° 11′ 18″	63.4	+01.0	16.0	
HR Del	(1967)	20h 40m 00s	+18° 59′ 00″	63.4	−14.0	12.0	12.5

[a] V_{max} and V_{min} refer to the maximum and minimum brightness of the nova at the time of IUE exposures.

very useful in determining the interoutburst characteristics of old novae. In this section, we order the discussion by right ascension and do not discuss those novae for which the IUE data have not been published. A description of the original outburst can usually be found in Payne-Gaposchkin (1957).

4.1. GK PER

This was a fast nova which showed pronounced oscillations during its decline. There were 34 spectra taken and some were obtained during a mini outburst which occurred in 1983. An analysis of the IUE data is in Bianchini and Sabbadin (1983) who found a UV continuum distribution characteristic of an accretion disk with a mass accretion rate of ~2 × 10^16 gm s^−1. However, the continuum slopes that they obtained do not agree with standard accretion disk predictions. Other UV studies have been reported by Selvelli and Hack (1983) and Rosino et al. (1982) who report a continuum fit that implies a temperature of ~12 000 K and $E(B − V)$ ~0.1.

4.2. RR PIC

There were 31 spectra taken from 1979 to 1982. Duerbeck et al. (1980) and Rosino et al. (1982) report that it shows very high ionization emission lines and that the continuum slope fits a temperature of ~3 × 10^4 K. Their spectrum shows that the strongest lines in the UV are N V λ1240, C IV λ1550, and He II λ1640. Other discussions of RR Pic in quiescence can be found in Selvelli (1982) and Selvelli and Cassatella (1982).

4.3. T CRB

There were 47 IUE spectra taken from 1979 until 1985. The UV spectra have been analyzed by Duerbeck *et al.* (1980) and Cassatella *et al.* (1982, 1986) who find large changes in the UV flux from one spectrum to another. The UV luminosity varies from about 5 $L_\odot$ to 40 $L_\odot$. Kenyon and Webbink (1984) attempted to fit the continuum by simulations of accretion onto main sequence stars at various rates. This binary system consists of an M giant and a massive (1.6 $M_\odot$: Kenyon and Garcia, 1986) compact component. The large mass of this star makes it unlikely that it is a white dwarf.

4.4. DQ HER

There were 15 IUE spectra obtained from 1979 to 1985. The early spectra have been analyzed by Lambert *et al.* (1981) and Ferland *et al.* (1984) who report $T_e < 500$ K, in agreement with the optical studies of the expanding nebula (Williams *et al.*, 1978). The extremely low temperature is probably caused by cooling by the enhanced CNO nuclei (Williams *et al.*, 1978).

4.5. V603 AQL

This is the brightest old nova in the sky and a prime target for both optical and UV studies. There were 32 IUE spectra taken in 1979 and 1980 and two were high dispersion. It is a strong X-ray source (Becker and Marshall, 1981). An interesting development is that Rahe *et al.* (1980) reported periodic light variations and, in addition, variations were found in the emission line strengths (Drechsel *et al.*, 1981). The light variations were not confirmed in an optical study (Slovak, 1981). The IUE spectra show the presence of a strong UV continuum ($T \sim 25\,000$ K) which is attributed to an accretion disk (Lambert *et al.*, 1981). Other IUE studies (Dultzin-Hacyan *et al.*, 1980) find that it is impossible to make a simple fit to the continuum flux distribution and suggest that there are at least two sources. Ferland *et al.* (1982) show that the abundances in the accretion disk are quite close to solar implying that the material being transferred from the secondary is normal and, therefore, that the enhanced abundances seen in novae ejecta must come from the core of the white dwarf.

4.6. HR DEL

This old nova is a bright object that could use regular monitoring both in the UV and optical. There were 32 spectra taken in 1978, 1979, and 1980. These spectra have been analyzed by Hutchings (1979, 1980) who found a variable P Cygni profile for C IV $\lambda1550$ and a continuum slope that rises steeply to the blue. He argues, by analogy with O star winds, that this old nova is losing mass at $\sim 10^{-8}$ $M_\odot$ yr^{-1}. He also obtained a UV luminosity of $\sim 25 L_\odot$. On the other hand, Andrillat *et al.* (1982) and Friedjung *et al.* (1982) interpret the steep UV continuum in terms of a disk accretion model and obtain a rate of mass accretion of $\sim 10^{-6}$ $M_\odot$ yr^{-1}. This value implies an accretion luminosity of better that 2500 $L_\odot$. In addition, another study of this nova by Dultzin-Hacyan *et al.* (1980) finds different temperatures from the previous authors. Finally, Wargau *et al.* (1982) do continuum fits for a number of CV's and find that none of them fit steady-state optically thick accretion disk predictions. This seems like an interesting and useful problem to solve.

5. Summary and Discussion

It is clear from this review of the IUE observations that the data from this satellite have been of paramount importance in extending our understanding of the characteristics and cause of the nova outburst and the structure of the nova binary. The ability of the satellite to make long, uninterrupted, exposures of a faint old nova or repeated exposures of a nova in outburst have allowed us to gather a wealth of material that is still being analyzed. Its ability to do Target-of-Opportunity observations on time scales of hours has made it possible to study some of these nova before maximum light when the expanding envelope is still optically thick. Finally, because all the data have been archived, IUE has provided material that will be useful for years.

The studies of old novae show that the continuum flux distributions fit power laws with some degree of accuracy. However, trying to estimate temperatures and mass accretion rates from these data is fraught with difficulties. The values of 10^{-8} to 10^{-9} $M_\odot$ yr^{-1} that have been found do not seem unreasonable and are roughly equivalent to the values obtained for dwarf novae in outburst. Studies of old novae such as V603 Aql, HR Del, and RR Pic show that some of their UV emission lines have P Cygni profiles which implies that mass loss is occurring. However dwarf novae also show wind mass loss during some phases of their outburst cycle so this result is not too surprising.

Old novae such as GK Per, HR Del, and T CrB, show variations in continuum flux, line profiles, and luminosity with time. Of these objects, only T CrB has been observed since 1981 or 1982 but all are prime candidates for continuous monitoring (Cassatella et al., 1986). It will be important to learn if these changes are caused by variations in mass loss, mass accretion, or the temperature of the compact component. Finally, a study of V603 Aql (Ferland et al., 1984) has shown that the enhancement of the CNO nuclei cannot come from the secondary.

Observations of novae in outburst have allowed us to determine abundances of the elements, determine masses of the ejecta, rates of mass loss, velocities of ejecta, and continuum flux distributions. Because many elements are only observable in the UV, we have been able to determine abundances for many more elements than were possible with only optical data. In addition, while obtaining abundances from optical data during outburst has had a shaky history, with many claims that have not held up under close scrutiny (Williams, 1977), the combination of optical and UV data has proved to be reasonably easy to interpret. *It is now clear that none of the novae studied with IUE have ejected material with a solar abundance.*

One of the most important recent results is the realization that three of the novae observed with the IUE (V693 CrA 1981, V1370 Aql 1982, and Nova Vul 1984 #2) are members of a class of novae in which the outburst is occurring on an ONeMg white dwarf rather than a CO white dwarf. The discriminating emission line is [Ne IV] $\lambda1602$. If it is present, at late times in the outburst, then the ejecta are rich in neon. In Nova Vul 1984 #2, it is currently the strongest line in the SWP spectrum (Figure 3).

Ultraviolet data taken with IUE also confirm that the flux peaks at shorter and shorter wavelengths as time progresses. This is a confirmation of the TNR theory of the outburst and demonstrates that nuclear burning is still occurring in the extended envelope of the white dwarf and that mass loss must still be going on inside the ejected shell. Finally, each nova that has been studied in the optical, UV, and IR has proved to be unique. We have observed both slow and fast novae, both classical and recurrent, and both CNO and ONeMg outbursts. There will always be new novae to study and this review is only a progress report to be continued by data on new novae in outburst.

Acknowledgements

We are grateful to a number of people who have contributed to this review in one form or another. S. Starrfield thanks the US Nova team: A. Cowley, G. Ferland, J. Gallagher, R. Gehrz, S. Kenyon, E. Ney, E. Sion, W. Sparks, J. Truran, R. Williams, and C.-C. Wu; and M. A. J. Snijders thanks the European Nova team: Y. Andrillat, G. Bath, A. Cassatella, H. Drechsel, A. Evans, M. Friedjung, B. Hassall, S. Pottasch, L. Rosino, M. J. Seaton, and D. Whittet for obtaining, analyzing, and interpreting the nova data that we have published over the last 6 years. Their encouragement and support have made this field enjoyable and worth pursuing. S. Starrfield is also grateful to Dr.'s George Bell, Arthur Cox, Stirling Colgate, Michael Henderson, and Jay Norman for the hospitality of the Los Alamos National Laboratory and support for an Association of Western Universities sabbatical leave fellowship. Finally, he thanks the Fellows of the Joint Institute of Laboratory Astrophysics for their support by awarding him a JILA Visiting Fellowship for 1986. Some of the material presented in this review was obtained through the facilities of the Boulder RDAF which is supported by NASA grant NAS5-26409 to the University of Colorado. Terry Armitage's help is gratefully acknowledged.

References

Andrillat, Y.: 1983, *Monthly Notices Roy. Astron. Soc.* **203**, 5p.
Andrillat, Y., Friedjung, M., and Puget, P.: 1982, in E. Rolfe, A. Heck, and B. Battrick (eds.), *Third European IUE Conference*, ESA SP-176, European Space Agency, Noordwijk, The Netherlands, p. 191.
Arnould, M., Norgaard, H., Thielemann, F.-K., and Hillebrandt, W.: 1980, *Astrophys. J.* **237**, 1016.
Audouze, J., Lazareff, B., Sparks, W. M. and Starrfield, S.: 1979, in M. Gabriel (ed.), *The Elements and Their Isotopes in the Universe*, Liege, University of Liege, p. 53.
Barlow, M. J., Brodie, J. P., Brunt, C. C., Hanes, D. A., Hill, P. W., Mayo, S. K., Pringle, J. E., Ward, M. J., Watson, M. G., Whelan, J. A. J., and Willis, A. J.: 1981, *Monthly Notices Roy. Astron. Soc.* **195**, 61.
Becker, R. H. and Marshall, F. E.: 1981, *Astrophys. J. Letters* **244**, L93.
Bianchini, A. and Sabbadin, F.: 1983, *Astron. Astrophys.* **125**, 112.
Bode, M. F.: 1987, *RS Ōph (1985) and the Recurrent Nova Phenomenon*, Utrecht, VNU Science Press.

Bode, M. F. and Evans, A. N.: 1987, *The Classical Nova*, New York, Wiley, (in press).

Bode, M. F., Evans, A. N., Whittet, D. C. B., Aitken, D. K., Roche, P. F., and Whitmore, B.: 1984, *Monthly Notices Roy. Astron. Soc.* **207**, 897.

Brosch, N.: 1982, *Astron. Astrophys.* **107**, 300.

Cassatella, A., Gilmozzi, R., and Selvelli, P. L.: 1986, in E. Rolfe (ed.), *New Insights in Astrophysics*, ESA SP-263, p. 277.

Cassatella, A., Harris, A., Hassall, B. J. M., and Snijders, M. A. J.: 1985, in *Recent Results in Cataclysmic Variables*, ESA SP-236, p. 281.

Cassatella, A., Benvenuti, P., Clavel, J., Heck, A., Penston, M., Selvelli, P. L., and Macchetto, F.: 1979, *Astron. Astrophys.* **74**, L18.

Cassatella, A., Patriarchi, P., Selvelli, P. L., Bianchi, L., Cacciari, C., Heck, A., Perryman, M., and Wamsteker, W.: 1982, in E. Rolfe, A. Heck, and B. Battrick (eds.), *Third European IUE Conference*, ESA SP-176, E. S. A., Noordwijk, The Netherlands, p. 229.

Cordova, F. A. and Mason, K. O.: 1983, in W. H. G. Lewin and E. P. J. van den Heuvel (eds.), *Accretion Driven Stellar X-ray Sources*, Cambridge, Cambridge University Press.

Drechsel, H. *et al.*: 1981, *Astron. Astrophys.* **99**, 166.

Drechsel, H. and Rahe, J.: 1984a, in *Proceedings of the Fourth European IUE Conference*, ESA SP-218, p. 371.

Drechsel, H., Wargau, W., and Rahe, J.: 1984b, *Astrophys. Space Sci.* **99**, 85.

Duerbeck, H. W., Klare, G., Krautter, J., Wolf, B., Seitter, W. C., and Wargau, W.: 1980, in *Proceedings of the Second European IUE Conference*, ESA SP-157, p. 91.

Dultzin-Hacyan, D., Andrillat, Y., Audouze, J., Friedjung, M., Gordon, C., Rocca-Volmerange, B., and Stasinska, G.: 1980, in *Proceedings of the Second European IUE Conference*, ESA SP-157, p. 87.

Fabian, A. C., Pringle, J. E., Stickland, D. J., and Whelan, J. A. J.: 1980, *Monthly Notices Roy. Astron. Soc.* **191**, 457.

Ferland, G. J., Lambert, D. L., McCall, M. L., Shields, G. A., and Slovak, M. H.: 1982, *Astrophys. J.* **260**, 794.

Ferland, G. J. and Shields, G. A.: 1978, *Astrophys. J.* **226**, 172.

Ferland, G. J., Williams, R. E., Lambert, D. L., Shields, G. H., Slovak, M., Gondhalaker, P. M., and Truran, J. W.: 1984, *Astrophys. J.* **281**, 194.

Friedjung, M., Andrillat, Y., and Puget, P.: 1982, *Astron. Astrophys.* **114**, 351.

Friedjung, M., Rocca-Volmerange, B., and Debeve, G.: 1980, in *Proceedings of the Second IUE Conference*, ESA SP-157, p. 85.

Gallagher, J. G., and Code, A. F.: 1974, *Astrophys. J.* **189**, 303.

Gallagher, J. S., Hege, E. K., Kopriva, D. A., Williams, R. E., and Butcher, H. A.: 1980, *Astrophys. J.* **237**, 55.

Gallagher, J. S. and Starrfield, S. G.: 1976, *Monthly Notices Roy. Astron. Soc.* **176**, 53.

Gallagher, J. S. and Starrfield, S. G.: 1978, *Ann. Rev. of Astron. and Astrophys.* **16**, 171.

Gehrz, R. D., Ney, E. P., Grasdalen, G. L., Hackwell, J. A., and Thronson, H. A.: 1984, *Astrophys. J.* **281**, 303.

Gehrz, R. D., Grasdalen, G. L., and Hackwell, J. A.: 1985, *Astrophys. J. Letters* **298**, L47.

Hanes, D. A.: 1985, *Monthly Notices Roy. Astron. Soc.* **213**, 443.

Hillebrandt, W. and Thielemann, F.-K.: 1982, *Astrophys. J.* **255**, 617.

Hutchings, J. B.: 1979, *Pub. Astron. Soc. Pac.* **91**, 661.

Hutchings, J. B.: 1980, *Pub. Astron. Soc. Pac.* **92**, 458.

Jenkins, E. B. *et al.*: 1977, *Astrophys. J.* **212**, 198.

Kenyon, S. J. and Garcia, M. R.: 1986, *Astron. J.* **91**, 125.

Kenyon, S. J. and Wade, R.: 1986, preprint.

Kenyon, S. J. and Webbink, R. F.: 1984, *Astrophys. J.* **279**, 252.

Krautter, J., Beuermann, K., Leitherer, C., Oliva, E. Moorwood, A. F. M., Deul, E., Wargau, W., Klare, G., Kohoutek, L., van Paradijs, J., and Wolf, B.: 1984, *Astron. Astrophys.* **137**, 307.

Lambert, D. L. *et al.*: 1981, in R. D. Chapman (ed.), *The Universe at Ultraviolet Wavelengths*, NASA Conference Publications CP 2171, p. 461.

Law, W. Y. and Ritter, H.: 1983, *Astron. Astrophys.* **123**, 33.

Lazareff, B., Audouze, J., Starrfield, S., and Truran, J. W.: 1979, *Astrophys. J.* **288**, 875.

Livio, M., Truran, J. W., and Webbink, R.: 1986, *Astrophys. J.* **308**, 736.

MacDonald, J.: 1980, *Monthly Notices Roy. Astron. Soc.* **191**, 933.

MacDonald, J.: 1983, *Astrophys. J.* **267**, 732.

MacDonald, J., Fujimoto, M. Y., and Truran, J. W.: 1985, *Astrophys. J.* **294**, 263.

McLaughlin, D. B.: 1960, in J. S. Greenstein (ed.), *Stellar Atmospheres: Stars and Stellar Systems VI*, Chicago University of Chicago Press, p. 585.

Ney, E. and Hatfield, B. F.: 1978, *Astrophys. J. Letters* **219**, L111.

Ogelman, H., Beuermann, K., and Krautter, J.: 1984, *Astrophys J. Letters* **287**, L31.

Pacheco, J. A. de Freitas and Codina, S. J.: 1985, *Monthly Notices Roy. Astron. Soc.* **214**, 481.

Paczynski, B.: 1983, *Astrophys. J.* **264**, 282.

Payne-Gaposchkin, C.: 1957, *The Galactic Novae*, New York, Dover.

Penston, M. V. *et al.*: 1983. *Monthly Notices Roy. Astron. Soc.* **202**, 833.

Prialnik, D., Livio, M., Shaviv, G. and Kovetz, A.: 1982, *Astrophys. J.* **257**, 312.

Prialnik, D., Shara, M., and Shaviv, G.: 1978, *Astron. Astrophys.* **62**, 339.

Prialnik, D., Shara, M., and Shaviv, G.: 1979, *Astron. Astrophys.* **72**, 192.

Rahe, J., Boggess, A., Drechsel, H., Holm, A., and Krautter, J.: 1980, *Astron. Astrophys.* **88**, L9.

Rosino, L., Iijima, T., and Ortolani, S.: 1983, *Monthly Notices Roy. Astron. Soc.* **205**, 1069.

Rosino, L., Bianchini, A., and Rafanelli, P.: 1982, *Astron. Astrophys.* **108**, 243.

Selvelli, P. L.: 1982, in E. Rolfe, A. Heck, and B. Battrick (eds.), *Third European IUE Conference*, ESA SP-176, European Space Agency, Noordwijk, The Netherlands, p. 197.

Selvelli, P. L. and Cassatella, A.: 1982, in E. Rolfe, A. Heck, and B. Battrick (eds.), *Third European IUE Conference*, ESA SP-176, European Space Agency, Noordwijk, The Netherlands, p. 201.

Selvelli, P. L. and Hack, M.: 1983, *Mem. Soc. Astron. Ital.* **54**, 467.

Sion, E. M. *et al.*: 1986, *Astron. Journ.* **92**, 1145.

Sion, E. M., and Starrfield, S.: 1986, *Astrophys. J.* **303**, 130.

Slovak, M.: 1981, *Astrophys. J.* **248**, 1059.

Smak, J.: 1980, *Acta. Astron.* **30**, 267.

Sneden, C. and Lambert, D. L.: 1975, *Monthly Notices Roy. Astron. Soc.* **170**, 533.

Snijders, M. A. J.: 1987, in M. Bode (ed.), *RS Oph (1985) and the Recurrent Nova Phenomenon*, Utrecht, VNU press, p. 51.

Snijders, M. A. J., Batt, T. J., Seaton, M. J., Blades, J. C., and Morton, D. C.: 1984, *Monthly Notices Roy. Astron. Soc.* **211**, 7.

Snijders, M. A. J., Batt, T. J., Seaton, M. J., Blades, J. C., and Morton, D. C.: 1986, preprint.

Snijders, M. A. J., Seaton, M. J. and Blades, J. C.: 1982, in Y. Kondo, J. M. Mead, R. D. Chapman (eds.), *Advances in Ultraviolet Astronomy: Four Years of IUE Research*, NASA Conference Publication 2238, p. 625.

Sparks, W. M., Starrfield, S., and Truran, J. W.: 1987, in M. F. Bode (ed.), *RS Oph (1985) and the Recurrent Nova Phenomenon*, Utrecht, VNU Press, p. 39.

Sparks, W. M., Starrfield, S., Williams, R. E., Truran, J. W., and Ney, E. P.: 1982, in Y. Kondo, J. M. Mead, R. D. Chapman (eds.), *Advances in Ultraviolet Astronomy: Four Years of IUE Research*, NASA Conference Publication 2238, p. 478.

Sparks, W. M., Starrfield, S., and Truran, J. W.: 1978, *Astrophys. J.* **220**, 1063.

Sparks, W. M., Wu. C.-C., Holm, A., and Schiffer, F. H.: 1980, *Highlights of Astronomy* **5**, 105.

Starrfield, S.: 1979, in H. M. Van Horn and V. Weidemann (eds.), *White Dwarfs and Variable Degenerate Stars*, Rochester, University of Rochester, p. 274.

Starrfield, S.: 1980, in A. N. Cox and D. S. King (eds.), *Stellar Hydrodynamics, Space Sci. Rev.* **27**, 635.

Starrfield, S.: 1986a, in N. Evans and M. Bode (eds.), *Classical Nova*, New York, Wiley, (in press).

Starrfield, S.: 1986b, in D. Mihalas, and K.-H. Winkler (eds.), *Radiation Hydrodynamics*, Springer-Verlag, New York, p. 225.

Starrfield, S., Sparks, W. M., and Truran, J. W.: 1974a, *Astrophys. J. Suppl.* **28**, 247.

Starrfield, S., Sparks, W. M., and Truran, J. W.: 1974b, *Astrophys. J.* **192**, 647.

Starrfield, S., Sparks, W. M., and Truran, J. W.: 1985, *Astrophys. J.* **291**, 136.

Starrfield, S., Sparks, W. M., and Truran, J. W.: 1986, *Astrophys. J. Letters* **303**, L5.

Starrfield, S. G., Truran, J. W., and Sparks, W. M.: 1978, *Astrophys. J.* **226**, 186.

Starrfield, S., Truran, J. W., Sparks, W. M., and Arnould, M.: 1978, *Astrophys. J.* **222**, 600.

Stickland, D. J., Penn, C. J., Seaton, M. J., Snijders, M. A. J., Storey, P. J., and Kitchin, C. R.: 1979, in A. J. Willis (ed.), *The First Year of IUE*, University College London, p. 63.

Stickland, D. J., Penn, C. J., Seaton, M. J., Snijders, M. A. J., and Storey, P. J.: 1981, *Monthly Notices Roy. Astron. Soc.* **197**, 107.

Thackery, A. D.: 1977, *Mem. Royal Astron. Soc.* **83**, 1.

Truran, J. W. and Livio, M.: 1986, *Astrophys. J.* **308**, 721.

Truran, J. W.: 1982, in C. A. Barnes, D. D. Claytom, and D. Schramm (eds.), *Essays in Nuclear Astrophysics*, Cambridge, Cambridge Univ. Press, p. 467.

Tylenda, R.: 1978, *Acta. Astron.* **28**, 333.

Wargau, W., Drechsel, H., and Rahe, J.: 1981, in E. Rolfe, A. Heck, and B. Battrick (eds.), *Third European IUE Conference*, ESA SP-176, European Space Agency, Noordwijk, The Netherlands, p. 215.

Wiescher, M., Gorres, J., and Thielemann, F.-K.: 1986, preprint.

Williams, P. M. and Longmore, A. J.: 1984, *Monthly Notices Roy. Astron. Soc.* **207**, 139.

Williams, R. E.: 1977, in R. Kippenhahn, J. Rahe, and W. Strohmeier (eds.), *The Interactions of Variable Stars with Their Environments*, Bamberg, Remeis-Sternwarte, p. 242.

Williams, R. E. and Gallagher, J. S.: 1979, *Astrophys. J.* **228**, 482.

Williams, R. E., Ney, E. P., Sparks, W. M., Starrfield, S., and Truran, J. W.: 1985, *Monthly Notices Roy. Astron. Soc.* **212**, 753.

Williams, R. E., Sparks, W. M., Gallagher, J. S., Ney, E. P., Starrfield, S., and Truran, J. W.: 1981, *Astrophys. J.* **251**, 221.

Williams, R. E., Woolf, N. J., Hege, E. K., Moore, R. L., and Kopriva, D. A.: 1978, *Astrophys. J.* **224**, 171.

Wu, C.-C. and Kester, D.: 1977, *Astron. Astrophys.* **58**, 331.

Young, P. and Schneider, D. P.: 1980, *Astrophys. J.* **238**, 955.

ACCRETION ONTO COMPACT STARS IN BINARY SYSTEMS

FRANCE A. CÓRDOVA

Los Alamos National Laboratory, Earth and Space Science Division,
Los Alamos, NM, U.S.A.

and

IAN D. HOWARTH

University College London, Dept. of Physics and Astronomy, London, U.K.

1. Introduction

In this chapter we review the IUE's contribution to our present understanding of accretion phenomena in compact binary systems.

White dwarfs, neutron stars, and black holes are possible endpoints of stellar evolution; they are called 'compact' stars because of their extreme mass to size ratio compared to normal stars. A compact binary consists of such a star in orbit around a 'companion' star, which is often on or near the main sequence. There are two principal ways in which matter is transferred from the companion star and accreted onto the compact star: by means of a stellar wind or by Roche-lobe overflow.

In a close binary containing a low-mass companion star, material is usually transferred via the latter mechanism. The dimensions of such a binary are on the order of a solar radius. When the accreting object is a white dwarf, such a binary is called a cataclysmic variable star (CV), and when the accreting object is a neutron star, the system is called a low-mass X-ray binary (LMXB). (Since CVs also emit X-rays, they could also be thought of as LMXB, but we continue to distinguish between these two kinds of system because the ratio of X-ray to optical light is so much higher for a neutron star LMXB than for a CV.) CVs include classical novae, dwarf novae, recurrent novae, novalike stars, and magnetic (AM Her) variables. Because of their relatively high space density and, therefore, close proximity to the Earth (i.e., distances less than a few hundred parsecs), many CVs have been studied well at all wavelengths, and much is known about their orbital elements. This is in contrast to the LMXB which have a lower space density and are often highly obscured and, hence, are more difficult to observe at optical and UV wavelengths.

Binaries in which the companion is a massive star of spectral class O and B are called high-mass X-ray binaries (HMXB). Because these massive stars are intrinsically blue and luminous, a substantial fraction of them have been studied at UV wavelengths, despite their generally large distances and often heavy reddening. The detection of rapid X-ray pulsations in many such systems is evidence for the presence of a rotating, magnetic neutron star. Among the nonpulsing HMXB (and LMXB) are a few black hole candidates. In the HMXBs with O or supergiant B companion stars mass is transferred from a stellar wind or by critical potential

Y. Kondo (ed.), Scientific Accomplishments of the IUE, pp. 395—426.
© 1987 by D. Reidel Publishing Company.

lobe overflow. In a such systems the size of the mass-donor is a large fraction of the orbital separation; thus, X-ray eclipses are common. The eclipses and the Doppler delays of X-ray pulse arrival times have allowed a determination of the orbits of some of the OB binaries (cf. review by Rappaport and Joss, 1983).

A separate class of HMXB are ones in which the mass-donor is a Be star (Maraschi *et al.*, 1976). A Be star is a B star of luminosity class III—V which shows, or has shown in the past, optical emission lines, usually of Hydrogen. The orbital periods of Be X-ray binaries are long (10—100 days) compared to the orbital periods of OB X-ray binaries (1—10 days), and the Be stars generally underfill their critical potential lobes by a large factor. The well-studied Be binaries span the range of X-ray pulse periods from 0.069 s (A 0538—66) to 835 s (X Per) and include steady as well as highly variable and transient systems. In Be-type binaries matter may be transferred to the compact star from an belt of dense material surrounding the (rapidly rotating) Be star in its equatorial plane.

Even before the IUE was launched it was expected that UV observations of compact binaries would be important in several areas: (1) the theoretical spectrum for an accretion disk surrounding a white dwarf predicted that most of the disk luminosity would be emitted in the UV (Bath *et al.*, 1974); (2) for an LMXB, X-ray heating of the disk and companion star was expected to have a substantial effect on the UV spectrum (Shakura and Sunyaev, 1973); (3) in CVs where the mass transfer rate was low and the disk, therefore, not UV bright, the spectrum of the white dwarf might be unmasked in the far UV; (4) variations in the stellar wind line profiles of the early-type stars in HMXBs might be used to probe the nature of the mass transfer and accretion process. It was also hoped that UV observations would shed some light on the mechanism for the dwarf nova outburst and perhaps the nature of the outbursts in some high-energy X-ray transients. All of these expectations have been borne out.

In addition to confirming the anticipated, the IUE has made some surprising discoveries. Among the foremost are detecting high velocity winds in CVs in which the mass transfer rate is high, and finding that the dwarf nova outburst is delayed in the UV with respect to the optical light. The former finding could have important consequences for the loss of angular momentum in CVs and, hence, the evolution of these binaries, while the latter discovery is important in the interpretation of the nature of the dwarf nova outburst.

Here we briefly touch on both the expected and unexpected results that have followed from studies of these objects with the IUE.

One of the most important things to keep in mind when reflecting on the accomplishments of the IUE is that it became a worldwide user facility at the same time that guest observer facilities were first being offered for X-ray, optical, infrared, and radio investigations. Because the spectra of compact binaries overlap these wavelength bands, all of the new facilities have been used more or less concurrently to investigate accretion phenomena. For example, the IUE frequently made simultaneous observations with such facilities as the European Space Agency's X-ray observing satellite, Exosat, with NASA's Infrared Telescope Facility, and with numerous optical telescopes. Therefore, it is nearly impossible to speak of IUE observations and the contribution of the IUE to the field of compact binaries without some mention of observations in other spectral regions.

2. The Spectrum of an Accretion Disk

The material exiting a low-mass companion star through the inner Lagrangian point has too much angular momentum to fall directly onto the compact star and so the matter rotates in circular orbits about that star. Shakura and Sunyaev (1973) showed that the transfer of angular momentum outwards, by means of turbulent or magnetic viscosity, for example, leads to the formation of a disk. Viscosity is also responsible for the dissipation of mechanical energy, and hence provides the source of the flux radiated away by the disk (Smak, 1984a). The structure and spectrum of the disk will depend on the nature of the accreting object and the rate of mass, $\dot{m}$, flowing into the disk.

There has been no single instrument as useful for examining the spectrum and structure of accretion disks as the IUE. An excellent place to study disks is in CVs, which have accretion disks that radiate predominantly in the UV, and the IUE has the sensitivity and spectral resolution necessary to compare observations with models in some detail. Here we review what we have learned from an examination of IUE data on CVs about the viscosity in disks, the spectra of disks, and the mass transport rate.

2.1. VISCOSITY OF DISKS

The nature and magnitude of the viscosity that drives the mass transport in a disk is not known. We observe that CVs are luminous sources of UV and X-radiation and therefore can assume that the matter orbiting the compact star eventually accretes onto it. Particles orbiting in rings about the star therefore must lose angular momentum due to momentum transfer between particles in adjacent rings, and spiral towards the compact star. The gravitational energy that is released in this process is converted in part to the kinetic energy of rotation and in part to thermal energy which is radiated from the surface of the disk. Shakura and Sunyaev (1973) characterize the efficiency of the momentum transport with a parameter, α; for turbulent viscosity, $\alpha = v/c_s H$, where v is the effective viscosity, c_s the sound speed in the disk, and H the disk half-thickness.

Although the spectrum of the radiation and the disk temperature are not dependent on α, the timescale for a mass transfer 'event', in which the mass transport changes significantly, does depend on α. Accretion events like the outbursts of dwarf novae give a timescale for variability that enables estimates of the viscosity to be made.

Bath and Pringle (1981) investigate the dependence of disk evolution time on α, following turn-on of mass transfer at a constant rate onto the white dwarf. In order for the reaction time of the disk to be 10^5-10^6 s as observed, large viscosities are required with $0.1 < \alpha < 1.0$.

More recent work at optical wavelengths shows that while $\alpha = 1$ is probably appropriate for the outburst itself, the value of α during the quiescent state may be smaller (Smak, 1984b; Cook, 1985). The observations imply that the viscous timescale in quiescence is on the order of the time between outbursts, so that any matter with low angular momentum which is added to the disk during quiescence must have the effect of making the disk smaller. Using the IUE, Hassall *et al.* (1985) show that the UV flux apparently decreases during the intervals between the outbursts of WX Hydri with an e-folding time of about 6 days, supporting the idea that the viscous timescale during quiescence (~ 1 week) is much longer than during outburst (~ 1 day).

The attempt to derive α depends on the choice of model considered for the cause of the outburst. The above modeling assumes that an episode of enhanced mass transfer from the companion causes the outburst. A competing, disk instability model for the outburst suggests that $\alpha = 0.1$ will reproduce the observed outburst decay rate (cf. Smak, 1984a, b; Cannizzo *et al.*, 1986). See Section V for more detailed discussion of models for the outburst.

2.2. THE SPECTRUM OF A STEADY-STATE DISK

A steady-state disk is one in which the viscosity adjusts itself to provide a steady mass flux, $\dot{m}$. To determine the emitted spectrum of such a disk, the spectrum of each element of the disk must be calculated, and then integrated over the entire disk (see the review of disks by Pringle, 1981). For a physically thin, optically thick disk, the problem simplifies and the effective disk temperature is

$$T(R) = [(3GM_*\dot{m})/(8\pi R^3\sigma)\,[1 - (R_*/R)^{1/2}]]^{1/4} \qquad (1)$$

where M_* is the mass of the compact star, R_* its radius, σ is the Stephan—Boltzmann constant, and R is the radius in the disk. For a disk in which each elemental area radiates with a blackbody spectrum, B_ν, the spectrum of the entire disk is

$$S_\nu \propto \int_{R_*}^{R_{out}} B_\nu(T(R)) \cdot 2\pi R \, dR \qquad (2)$$

where R_{out} is the outer disk radius. If $T_* = [(3GM_*\dot{m})/(8\pi R_*^3\sigma)]^{1/4}$, then for $\nu \gg kT_*/h$ (k is the Boltzmann constant, and h is Planck's constant), the spectrum is the exponential tail of the spectra of the hottest annuli in the disk. For $\nu \ll kT_*/h$, the radiation is from radii $R \gg R_*$ so that $T(R) \propto R^{-3/4}$. If T_{out} is the temperature of the outermost disk radius, for $kT_{out}/h \ll \nu \ll kT_*/h$, $S_\nu \propto \nu^{1/3}$ (i.e., the 'classic' disk spectrum). For $\nu \ll kT_{out}/h$, the Rayleigh—Jeans tail of the coolest disk elements dominate and $S_\nu \propto \nu^2$.

Approximately half of the accretion energy is radiated in the disk and half in the boundary layer between the inner disk and the compact star's surface.

Note that Equation (1) does not involve α; hence, fits of steady state disks to observations will not provide any information on the viscosity.

2.3. MODELING CV SPECTRA AND DERIVING THE MASS ACCRETION RATE

For low-luminosity CVs there is often a substantial dilution of the disk spectrum by other contributors in the system such as the component stars and the 'bright spot' where the mass stream from the companion impacts the outer disk. The spectrum of the companion star is rarely observable in the IUE bandwidth, its radiation being chiefly emitted in the infrared, but the white dwarf may have been detected by IUE in the far UV spectrum of several systems (Section 4). Combined IUE and optical spectra show that the bright spot is relatively cool, i.e., about 10^4 K, and therefore not a prominent feature in IUE light curves (Córdova and Mason, 1983; Panek and Holm, 1984; Szkody, 1985).

For high luminosity systems, modeling CV spectra is a test of our knowledge of accretion disk spectral distributions. As shown in Equation (1), fitting observed disk spectra can, given M_* and R_*, yield a measure of the mass accretion rate, $\dot{m}$, which is probably the single most important parameter in determining the short term (pulsation and fluctuation) and long term (evolutionary) behavior of CVs.

There are a number of steady-state accretion disk models in the literature, but few predict UV as well as optical spectra, and few are very general in terms of assumed parameters such as the disk size and the mass of the white dwarf, or give model spectra for a large range of mass accretion rates. Szkody (1985) has published extensive lists of CVs for which power law distributions have been fitted to the UV continua, and has compared these with the models of Williams and Ferguson (1982) to estimate $\dot{m}$. These models assume a constant, large disk size, $R = 6 \times 10^{10}$ cm, and a one solar mass white dwarf. Szkody finds values for $\dot{m}$ of 10^{-7} to 10^{-8} $M_\odot$ yr^{-1} for dwarf novae in outburst and the novalike stars, and $\dot{m}$ of 10^{-10} to 10^{-9} $M_\odot$ yr^{-1} for dwarf novae in quiescence.

The application of this method is hazardous for quiescent state spectra which may be contaminated by contributions from other sources in the system, and is severely limited by fitting UV data only (see examples in Figure 1a and 1b). In addition, there are a few dwarf novae that have the same UV slope at minimum as at maximum (Córdova et al., 1986; Szkody, 1985), and it is unreasonable to assume that $\dot{m}$ is the same in both cases. Córdova and Mason (1982) point out that while the UV spectrum of TW Virginis on the decline from an outburst mimics the $v^{1/3}$ distribution predicted for an optically thick blackbody disk, this distribution also fits very well a single temperature 18 000 K stellar (Kurucz 1979) atmosphere. Córdova and Mason point out that combined UV and optical data must be fitted to reach any conclusion about the true spectral distribution.

Wade (1984) computes spectra for steady-state models of accretion disks in CVs using two extreme assumptions: by summing Planck functions over a range of temperatures representing different annuli in the disk, and by similarly summing fluxes from model stellar atmospheres. Neither spectral distribution provides a completely satisfactory representation of observed spectra: the Kurucz atmosphere disks show Balmer jumps in absorption which are much higher than the ones observed, whereas the blackbody disks show no Balmer jump at all (e.g., Figure 1c and Figure 2). Both models can describe either the optical or UV spectrum alone, but neither can describe both wavelength regions simultaneously. In the UV, the observed power-law continua are reproduced better with model atmosphere disks.

Mass transport rates deduced from fitting the atmosphere models are smaller by a factor of 10 to 100 than rates deduced from the blackbody models. Wade shows, using stellar atmosphere models, that a large, bright disk can produce the same slope in the UV as a small, less luminous disk (Figure 2). The spectra differ from

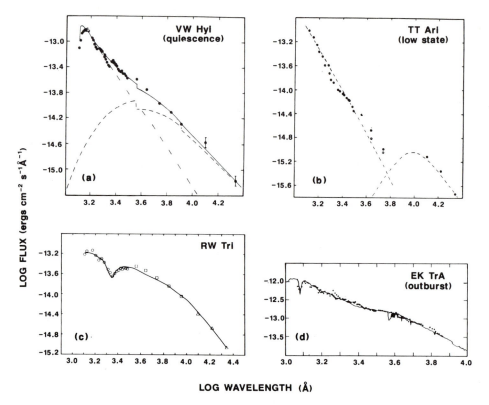

Fig. 1. Examples of model fits to the continuous ultraviolet-optical-infrared spectra of four very different CVs: (a) The spectrum of VW Hydri during quiescence shows a turnover at $\sim \lambda 1300$ Å which is fitted with a Wesemael *et al.* (1980) model white dwarf atmosphere with $T = 20\,000$ K and $\log g = 9$. The residual flux is fitted with a Williams and Ferguson (1982) steady-state disk model with $\dot{m} = 10^{-11}\ M_\odot$ yr^{-1}. From Mateo and Szkody (1984). (b) The spectrum of the novalike star TT Arietis, observed during a state of low mass accretion, is fitted with a 50\,000 K white dwarf atmosphere and a contribution due to a M2—M3 red dwarf. In this modeling, the disk may contribute about 20% of the optical emission. From Shafter *et al.* (1985). (c) The spectrum of the novalike 'disk' star RW Trianguli is fitted with a steady-state, blackbody disk model with $T_* = 9 \times 10^4$ K, and a ratio of outer to inner disk radii, $R_{out}/R_* = 50$, and $E(B - V) = 0.25$. The model underestimates the observed U band flux. From Córdova and Mason (1985). (d) The spectrum of the dwarf nova EK Trianguli Australis during an outburst is fitted with a Kurucz atmosphere disk with effective temperatures between 6000 K and 60\,000 K, plus a recombination spectrum at 10\,000 K. From Hassall (1985).

one another chiefly in the ratio of visual to UV flux. Stellar atmospheres fail to follow the $\nu^{1/3}$ law in detail over the UV-optical region because they have Balmer jumps and absorption features.

An example of the pitfalls in modeling disk spectra is the case of VW Hyi. Its outburst decline spectrum from the UV through the optical has a power-law distribution of $\nu^{1/3}$, and was originally fitted with a steady-state blackbody disk model, but Hassall *et al.* (1983) pointed out that this assumption leads to a disk size which extends beyond the white dwarf's Roche lobe!

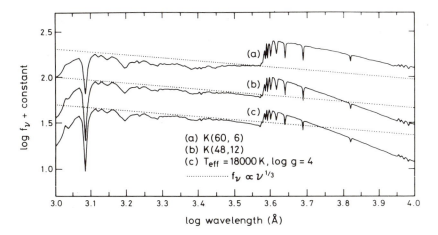

Fig. 2.   Three spectra that approximate $F_\nu \propto \nu^{1/3}$ in the ultraviolet region. The upper two spectra are of accretion disks comprised of Kurucz model stellar atmospheres with a range of temperatures, $K(T_*, T_{out})$; the bottom spectrum is a single Kurucz stellar atmosphere. The spectra differ from one another chiefly in the ratio of visual to ultraviolet flux. From Wade (1984).

A final example of the difficulties in modeling CV spectra, and, hence, the indeterminacy of $\dot{m}$, is illustrated by various attempts to model the spectrum of the eclipsing, novalike star, RW Trianguli. The model stellar atmosphere disks of Wade (1984) can be applied to RW Tri to derive $\dot{m} = 10^{-10.8}\ M_\odot\ \mathrm{yr}^{-1}$. Córdova and Mason (1985) use a blackbody disk to model the UV-optical-infrared flux distribution, incorporating the effects of reddening on the spectrum, and find $\dot{m} = 10^{-9.8}\ M_\odot\ \mathrm{yr}^{-1}$. Their fit to the UV depends on the combination $\dot{m}M_*/R_*^3$. Estimates also come from modeling the eclipse light curves. Frank and King (1981) fit a steady-state blackbody model to RW Tri's V and R light curves and derive $\dot{m} = 10^{-9.6\,\pm\,0.4}\ M_\odot\ \mathrm{yr}^{-1}$. Horne and Stiening (1985) determine $\dot{m}$ by modeling the surface brightness distribution of the disk. They map the eclipses in U, B, and R, and, using maximum entropy methods, determine that the disk is optically thick (i.e., $T \propto R^{-3/4}$) and $\dot{m} = 10^{-7.8\,\pm\,0.4}\ M_\odot\ \mathrm{yr}^{-1}$. (Note: The distance to RW Tri derived by Frank and King (1981) is half the distance assumed by Horne and Stiening, and this will affect a comparison of their results; however, the other comparisons are unaffected).

All of the attempts to fit models to observed spectra point to the need for improved physics to be incorporated into the models. One effort has been to combine the (Kurucz atmosphere) spectrum of an optically thick disk with a recombination spectrum representing an optically thin outer region of the disk (Hassall, 1985). This model fits the UV-optical outburst continuum spectrum of the dwarf nova EK TrA with moderate success (see Figure 1d), although it implies a luminosity for H_β much larger than is observed, under the assumption that the line is optically thin.

This discussion has concentrated on steady-state disks which have been used to model the peak and beginning of the decline in dwarf nova outbursts, and the quiescent period between outbursts. Clearly, during the rise to outburst the disk is not in a steady-state (i.e., the spectrum is often changing faster than the viscous timescale). We discuss the application of time-dependent models to this situation in Section 5.

3. The UV Continua of Neutron Star Binaries

3.1. DISK SPECTRA AND REPROCESSING

Low resolution IUE spectrophotometry of a number of LMXB has been obtained (Table I), in spite of their relative faintness compared to typical CVs (see examples in Figure 3 of the brightest LMXB). In some of these spectra at least part of the observed flux may be attributable to an accretion disk. The disk spectrum is most reliably seen in HZ Her (Howarth and Wilson, 1983a) and the 'Cen Xmas' transient V822 Cen (Blair *et al.*, 1984), while Mason and Córdova (1982b) have shown that a significant part of the UV flux from 2A 1822—371 originates in a disk. For Cyg X—2 (Chiapetti *et al.*, 1983), MXB 1735—444 (Hammerschlag-Hensberge *et al.*, 1982), and Sco X—1 (Willis *et al.*, 1980) the accretion disk may be observed, but Willis *et al.* (1980) argue that the long wavelength tail of bremsstrahlung emission from a hot plasma probably makes the principal contribution in the UV.

The dereddened UV spectra of all the LMXBs observed with the IUE are consistent with power law spectra, ($F_\nu \propto \nu^\alpha$), with α in the range 0—2. For HZ Her and Cen X—4 spectroscopic distances are available, and it is possible to compare the absolute disk luminosities with the predictions of simple models, using a similar approach to that discussed above for CVs. This comparison then gives an estimate of the mass transfer rate onto the neutron star, which can be compared with that obtained directly from the observed X-ray luminosity ($\dot{m} \sim L_x/0.1c^2$). In each case order of magnitude differences are found between the observed UV fluxes and those expected from simple viscous heating, in the sense that the disks are too bright. This has a straightforward interpretation in that the accretion disks are presumed to subtend substantial solid angles at the neutron star (demonstrated to be ~ 1 sr in the case of HZ Her), and in consequence intercept much of the emitted X-radiation. Although some of these X-rays will be scattered out of the disk, a fraction (of order 50%) will be absorbed, and act as the major heating agent.

The conversion of X-rays to less energetic radiation in the accretion disk has been shown to be consistent with IUE observations of HZ Her, Cen X—4, Cyg X—2, 2A 1822—371, and MXB 1735—444, as well as optical data for many low-mass systems (e.g., van Paradijs, 1983), and even some massive systems, where the giant star dominates the optical and UV flux (e.g., Howarth, 1981). The finite disk thickness implied by these observations confirms Milgrom's (1978) suggestion that X-ray eclipses are not generally seen in low mass systems because at inclinations from the disk normal that are high enough for eclipses to occur, the sightline to the neutron star is blocked by the geometrically thick disk.

TABLE I

IUE observations of systems which contain accreting neutron stars

X-ray I.D.	Optical I.D.	Approx. m_V	$P_{orbital}$	Selected References
(A) Low mass X-ray binaries				
Cen X—4	V822 Cen	13—18	8.2h?	(1)
Sco X—1	V818 Sco	13	18.7 h	(2)
Her X—1	HZ Her	14	1.7d	(3, 4)
1735—444	V926 Sco	18	4.3h?	(5)
1822—371	V691 CrA	15	5.6h	(6)
Cyg X—2	V1341 Cyg	15	9.8d	(7, 8)
(B) Massive X-ray binaries				
SMC X—1	Sk 160 (B0Ib)	13	3.9d	(9, 10)
LMC X—4	(07III—V)	14	1.4d	(9)
LMC X—1	(07—9III)	15		(11)
0501—704	Sk—70 36 (B1—2I)	15	6.9d	(27)
Vela X—1	HD 77581 (B0.5Ib)	7	8.9d	(12, 13)
1700—37	HD 153919 (06.5f)	7	3.4d	(14, 15)
Cyg X—1	HDE 226868 (09.7Iab)	9	5.6d	(16, 17)
(C) Be systems				
0053 + 604	Gamma Cas (B0.5III—Ve)	2		(18, 19)
0239 + 610	LSI + 61 303 (B4.5III)	11	26.5d?	(26)
0352 + 309	X Per (09.5III—Ve)	6	580d?	(18, 20)
0535 + 262	HDE 245770 (0.97IIIe)	9	111d?	(21)
A0538—66	(B2III)	15	16.6d	(23, 24)
1118—615	He 3—640 (09.5III—Ve)	12		(24)
1145—619	HD 102567 (B0—1Ve)	9	187d?	(18, 25)

Supplementary data mainly from van Paradijs (1983).

Columns are: (i) X-ray source; (ii) Optical counterpart; (iii) Appx. optical magnitude (all sources are optically variable); (iv) Orbital period; (v) refs to selected IUE studies, as follows:

1   Blair *et al.*, 1984
2   Willis *et al.*, 1980
3   Howarth and Wilson, 1983a, b
4   Gursky *et al.*, 1980
5   Hammerschlag—Hensberge *et al.*, 1982
6   Mason and Cordova, 1982a, b
7   Chiappetti *et al.*, 1983
8   McClintock *et al.*, 1984
9   van der Klis *et al.*, 1982
10   Hammerschlag-Hensberge *et al.*, 1984
11   Bianchi and Pakull, 1985
12   Dupree *et al.*, 1980
13   Sadakane *et al.*, 1985
14   Howarth *et al.*, 1986

15   Dupree *et al.*, 1978
16   Treves *et al.*, 1980
17   Davis and Hartmann, 1983
18   Hammerschlag-Hensberge *et al.*, 1980
19   Henrichs *et al.*, 1982
20   Bernacca and Bianchi, 1981
21   de Loore *et al.*, 1984
22   Raymond, 1982
23   Howarth *et al.*, 1984
24   Coe and Payne, 1985
25   Bianchi and Bernacca, 1980
26   Howarth, 1983
27   Bianchi and Pakull, 1984

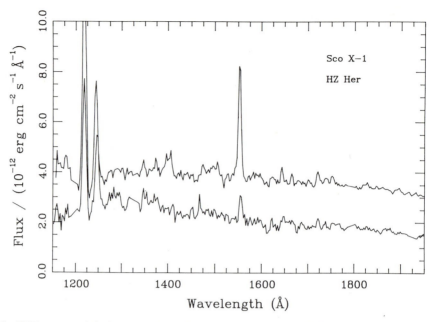

Fig. 3.   IUE spectra of the low-mass X-ray binaries Sco X—1 and HZ Her (SWPs 8744 and 22238), showing their blue continua and strong emission lines of N V $\lambda 1240$ and C IV $\lambda 1549$. The N V flux in Sco X—1 is attenuated by both interstellar extinction and absorption in the wings of L_a; it is *intrinsically* stronger than the Carbon line.

3.2. PHOTOSPHERIC HEATING

Just as reprocessing occurs in accretion disks around neutron stars, so X-ray 'heating' will occur in those parts of the mass donor's photosphere with an unobscured sightline to the regions of X-ray production. For many low mass systems such sightlines do not exist, simply because of the large disk thickness, which shields the nondegenerate star; only for HZ Her and 4U 2129 + 47 (too faint for observations with IUE) is photospheric heating obvious, while Mason and Córdova (1982b) and Blair *et al.* (1984) present evidence for its occurrence in UV observations of 2A 1822—37 and Cen X—4. X-ray heating has also been demonstrated in UV observations of some high-mass systems (van der Klis *et al.*, 1982); here the primary subtends a much larger angle than the disk, but the effects are nonetheless more subtle because the intrinsic luminosity of the massive companion star exceeds that of the X-ray source.

In the case of HZ Her ($L_x/L_{\mathrm{opt}} \sim 100$) X-ray heating dominates the phase 0.5 flux, and results in large orbital variations from the UV to the IR. For this system it is possible to construct a fairly detailed geometrical model which can be used to investigate the role of photospheric X-ray heating between about 1150 and 5500 Å (Howarth and Wilson, 1983a). The observations are quantitatively consistent with 'deep' heating (i.e. absorption of X-rays in the nondegenerate star's diffusion zone), with an albedo of ~ 0.5, in agreement with the predictions of model atmosphere calculations (by e.g., Chester, 1978; London *et al.*, 1981).

3.3. THE MASSIVE SYSTEMS

For all the HMXB observed by the IUE the nondegenerate star is (bolometrically) more luminous than the X-ray source, and completely dominates the UV flux. Thus although the continuum flux levels may show orbital variations at the 10% level (due to X-ray heating and to gravitational distortion of the shape of the nondegenerate star by the mass-accreting star), the shape and detailed structure of the photospheric spectrum are scarcely affected by the neutron star (or black hole). This makes the IUE useful in the study of the intrinsic, continuous energy distribution in HMXB in general, and the Be systems in particular.

By their nature, optical line spectra of Be stars contain peculiarities which can make accurate spectral typing difficult, hence rendering estimates of physical parameters (temperatures, luminosities) uncertain. Subject to an assumed form for the interstellar extinction, IUE spectrophotometry allows a direct determination of reddening, which can be used as a distance indicator in many circumstances (e.g., LSI + 61 303; Howarth, 1983), and, by comparison with model atmosphere fluxes, gives estimates of continuum temperatures. This in turn allows any infrared flux excess, which may be associated with an extended equatorial belt around the star rather than a stellar wind, to be accurately defined; such an excess is commonly found in Be systems. An example of this type of spectral deconvolution is the multispectral study of HDE 245770 reported by de Loore *et al.* (1984).

4. The Spectrum of the White Dwarf

The wavelength range covered by the IUE is well-suited to measuring the spectra of isolated white dwarfs (see pp. 355—376). But what of white dwarfs in interacting binaries such as CVs? How can the spectrum of the accretion disk and boundary layer be removed to reveal the white dwarf spectrum? The answer has been to examine CVs in which the accretion rate is extremely low, such as dwarf novae in quiescence or magnetic variables during low states. The flux distribution and the Hydrogen L_α absorption line (often seen as a turnover in the far UV) are fitted with model white dwarf atmosphere spectra, like those of Wesemael *et al.* (1980). This method has yielded approximate temperatures for the white dwarfs in several types of CV; (1) *dwarf novae* (U Gem, $T_{wd} = 30\,000$ K; Panek and Holm, 1984; VW Hyi, $T_{wd} = 18\,000$ K: Mateo and Szkody, 1984); (2) *novalike stars* (MV Lyr, $T_{wd} \geqslant 50\,000$ K: Szkody and Downes, 1982; TT Ari, $T_{wd} \geqslant 50\,000$ K: Shafter *et al.*, 1985); and (3) *AM Her stars* (AM Her, $T_{wd} \geqslant 50\,000$ K: Szkody *et al.*, 1982; CW 1103 + 254, $T_{wd} = 13\,400$ K: Szkody *et al.*, 1985). An example fit of a white dwarf atmosphere model to a UV turnover at ~ 1300 Å is shown in Figure 1a. We note, though, that a cool stellar atmosphere also exhibits such a turnover in its spectrum (e.g. a Kurucz 1979 stellar atmosphere with $T \sim 15\,000$ K and log $g = 4$, where g is the gravity). Thus we could be looking at a very cool disk, rather than a cool white dwarf. It is difficult to assess, using IUE data alone, the relative contributions of the accretion disk and white dwarf; one also needs X-ray data for information about the mass accretion rate, and EUV data for additional spectral line coverage useful in fitting white dwarf atmosphere models.

Using the UV and optically-derived temperatures of white dwarfs in CVs Smak (1984c) concludes that the temperature is correlated with the average accretion rate.

5. The Nature of the Outbursts in CVs and X-Ray Transients

5.1. THE DWARF NOVA OUTBURST

Every few weeks or months a dwarf nova brightens visually by a few magnitudes in a day or less, remains bright for days or weeks, and then returns to its quiescent flux level within a few days. What is the nature of these 'outbursts' and what triggers them? IUE light curves of the outbursts, when coupled with light curves at other spectral frequencies, are providing the data needed to answer these questions.

5.1.1. *The Outburst Rise*

Several dwarf novae have been observed in the UV and optical on their rise to maximum outburst brightness, and almost all of them show similar behavior: the UV rise lags the optical rise by up to a day (e.g., VW Hyi: Hassall *et al.*, 1983; CN Ori and RX And: Córdova *et al.*, 1986; WX Hyi: Hassall *et al.*, 1985). Figure 4

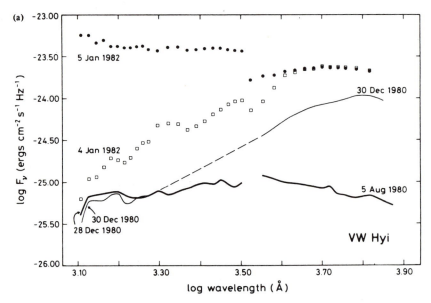

Fig. 4. Spectral continuua showing various stages in the outbursts of the dwarf nova VW Hydri. The quiescent spectrum (dark solid line) is a composite of data from 1980 December 28 and 1980 August 5. Two spectra taken on rises to outbursts on 1980 December 30 and 1982 January 4 show that the optical increases significantly while the UV is still low. The peak spectrum of 1982 January 5 is much flatter and resembles all the decline spectra of this star. (The shift between the optical and UV probably results because the optical data were taken a few hours after the UV data when the star's brightness was decreasing.) From Schwarzenberg-Czerny *et al.* (1985).

illustrates this for VW Hyi. The far UV region covered by the Voyager spacecraft (50—1200 Å) shows a similar lag with respect to the optical band (e.g., SS Cyg: Polidan and Holberg, 1984; VW Hyi: Polidan and Holberg, 1987). This puts the origin of the outburst in the cooler, outer disk rather than near the hot white dwarf, and therefore probably precludes any mini-nova model for the outburst. The two models for the outburst trigger that are at least superficially compatible with this observation are an instability in the companion star which results in more mass being suddenly transferred to the disk, and a thermal instability in the outer disk which results in material stored there being suddenly transported through the disk.

The mass transfer burst model assumes an as yet unmodeled ionization instability in the outer layers of the companion (Bath, 1975). As such, its application to the dwarf nova outburst light curve involves arbitrary assumptions about the magnitude and duration of the bursts of matter.

The disk instability model is based on the S-curve in the diagram of $\log T_e$ versus $\log \Sigma$: the curve denotes the position in the effective temperature — surface density diagram which a disk element at fixed radius must occupy to be in thermal equilibrium (Figure 5). The crucial feature in the curve is its 'bend' (negative slope) which occurs in the range of effective temperatures corresponding to hydrogen ionization. This branch is thermally unstable and unstable against perturbations in Σ. Stable curves are those with positive slope. S curves have been derived under a number of different assumptions about the viscosity and the importance of convection, and adopting different opacities and ways of treating optically thin conditions (Meyer and Meyer-Hofmeister, 1983; Faulkner *et al.*, 1983; Smak, 1984b; Mineshige and Osaki, 1985; Cannizzo and Wheeler, 1984). Good, introductory discussions of the curves are given in Smak (1984a) and in Hassall *et al.* (1985).

There have been several attempts to construct time-dependent accretion disks and compare them to observational data (Smak, 1984c; Meyer and Meyer-Hofmeister 1984; Cannizzo *et al.*, 1986, hereafter CWP; and Pringle *et al.*, 1986, hereafter PVW). These efforts variously seek to fit the length of the outburst, the interval time between outbursts, the optical outburst magnitude, the optical and UV rise and decay times, and the observed spectral behavior. (In fact, the different methods used to fit the data, such as calculating light curves or reproducing the spectral behavior, make a comparison between the various efforts difficult). The attempts to model the data make use of the disk instability behavior predicted by the multi-valued equilibrium $T_e - \Sigma$ curves, and also variations in the rate of mass transfer. It turns out that either model for the origin of the outburst can be made to fit roughly the spectra on the decline from outburst (CWP; PVW; but see Section 5.1.3). The crucial question is which, if either, of the models can reproduce the observed time dependence of the spectrum on the rise to outburst.

PVW find, using model atmosphere disks, that the disk instability models are unable to reproduce the large time lag of the UV with respect to the optical flux observed for the outbursts of VW Hyi. This is because the instability involves a sudden jump from a small region of the disk to a higher temperature ($\geqslant 10\,000$ K), while the observation is that the outburst begins in an extended,

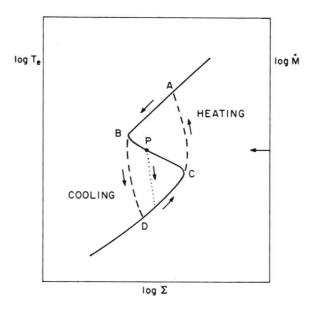

Fig. 5. Schematic of a thermal equilibrium curve in the $T_e - \Sigma$ plane. Models with cooling are located to the left, while those with heating are located to the right of the 'S curve'. The stable branches are AB and CD; the bend in the curve (BC) is unstable. To illustrate how a model might move along this curve: material is supplied to the outer disk at a constant rate by the companion star. Under this circumstance, the model could have a very low outer disk temperature (below point B), and be on the unstable BC curve at point P. On BC a random perturbation will force the model to go through the cooling area until it reaches the stable branch CD. On this branch the local accretion rate is lower than the rate at which material is supplied from the outside. Thus the local surface density Σ must increase and the model moves up along the CD branch until, at point C, the model is forced into the heating area and moves up onto the AB branch. The local accretion rate on AB is higher than the mass transfer rate, so the model moves down until it encounters the cooling domain and must drop down toward the CD branch. The cycle then repeats. From Smak (1984a), courtesy of the *Publications* of the Astronomical Society of the Pacific.

cool region ($\leqslant$ 7000 K), after which the disk temperature increases slowly. PVW can duplicate the observed rise light curves in the UV and optical with a single valued $T_e - \Sigma$ curve and an increase in mass transfer that happens on a timescale much faster than the viscous timescale in the disk. CWP are successful in fitting a similar lag in the far UV light curve of a different dwarf nova, SS Cygni, within the context of the disk instability model by using blackbody, rather than stellar atmosphere, disks.

In the rise light curve of CN Ori examined by PVW, the UV lag is not nearly so pronounced as in VW Hyi (CWP note there may be a inverse correlation between length of the rise time and the magnitude of the UV lag: i.e., the greatest UV lags are associated with the fastest optical rises). For the mass transfer burst model to fit this observation, the timescale for mass transfer must be much greater than the viscous timescale (PVW). The light curve is fitted in the disk instability model if $\dot{m}$ is low; then the instability starts in the inner part of the disk and quickly moves outward, rather than starting in the outer part of the disk and moving inward (CWP).

Smak (1984a) and PVW point out that the observance of different magnitudes for the lag in different stars, as well as very different ratios of outburst on-time to quiescent time interval, means that the prescription for α is probably not universal. If α is variable it loses its simple role in accretion disk theory, and it may be time to look for an alternate description of viscosity.

5.1.2. *The Outburst Peak*

While different dwarf novae exhibit quite different spectral distributions during their outburst peaks, for the same dwarf nova the continuum spectrum is nearly the same from outburst to outburst (Verbunt *et al.*, 1985). CWP are unable to reproduce the prolonged maximum of an outburst of SS Cygni using the disk instability model alone.

In the wavelength range covered by the IUE the flux of all dwarf novae at their peaks rises toward higher frequencies, suggesting that the bulk of the flux might be observed at even higher frequencies in the extreme ultraviolet. Voyager observations of dwarf novae in outburst, however, reveal that the spectrum is declining in the far UV (500 to 1200 Å), i.e., $F_\nu \propto \nu^{-2}$; furthermore, there is no flux observed shortwards of 912 Å (Polidan and Holberg, 1986). It is still unclear whether this lack of flux is due to interstellar absorption or absorption at the source, but it does seem at though the bulk of the detectable flux from CVs is emitted in the region observable by the IUE.

5.1.3. *The Outburst Decline*

After the peak of the outburst, the UV and optical apparently decline at the same rate; in addition, for the same dwarf nova, the time dependence of the spectral behavior on the decline is repeatable in different outbursts, so that observations taken during different outbursts can be scaled to the visual magnitude and interleaved (Verbunt *et al.*, 1984; Schwarzenberg-Czerny *et al.*, 1985; la Dous *et al.*, 1985).

This behavior also extends to the subclass of dwarf nova known as Z Cam stars, which undergo standstills in their brightness on the decline from outburst: the spectral distribution during the decline seems to depend more on the state of the decline (i.e., the V magnitude) than the rate of decline (Verbunt *et al.*, 1984; Szkody and Mateo, 1986). This means that a star undergoing an extended Z Cam type standstill will have the same overall spectral distribution, for the same V magnitude, as it would have during the decline from a normal outburst.

A standstill spectrum is that of a stationary disk. Given the similarity of such a spectrum with spectra of the same star during the decline from a normal outburst, and given the relatively slow rate of the decline, one can successfully model the observations during the decline with stationary disk models. The best fits to the observations are then obtained by assuming that the mass transfer rate subsides, and the disk decreases in size as the outburst declines (Hassall, 1985).

However, in PVW's attempt to model VW Hyi's outburst light curve, using both the mass transfer instability and the disk instability models, the former model predicts a decline for the outburst which is much longer than observed, a problem that might be circumvented by choosing a different $\log T_e - \Sigma$ curve than on the rise. The disk instability model, with its double-valued $T_e - \Sigma$ curve, actually gives a better description of the observed decline. In this case stellar atmosphere disks were used. When light curves were computed by CWP using blackbody models and the disk instability prescription, two features were generated that are not observed: the model light curves show a lag in the decline of the UV with respect to the optical light curve, which is similar in magnitude to the lag in the UV rise, and they show a changing spectral slope on the outburst decline.

5.1.4. *Between Outbursts*

Hassall *et al.* (1985) present some evidence (by piecing together IUE observations from different interoutburst epochs) that for the dwarf nova WX Hyi the UV flux decreases between outbursts. This is in agreement with the simple expectation of the mass transfer model that the flux decays toward the steady state. In the disk instability model the character of the outburst and the intervals between them is determined by the surface density distribution prior to the onset of instability which, in turn, is determined by the previous outburst history. Because of this hysteresis effect many different outburst patterns may be generated.

5.2. THE OUTBURSTS OF TRANSIENT X-RAY BINARIES: A 0538—66

Among the neutron star binaries, the periodic LMC transient A 0538—66 is unique in showing large flux increases in the UV/optical region in step with the X-ray outbursts. These outbursts, which are highly variable but which typically last for a few days, occur every 16.6 d when the system is in an active ('ON') state; however, 'OFF' states lasting many months can occur, when little or no activity is recorded.

Extensive observations of A 0538—66 have been made using the IUE, as well as optical and X-ray facilities, and from these studies a consensus model has emerged (Charles *et al.*, 1983; Brown and Boyle, 1984). This model consists of a neutron star in a highly eccentric orbit ($e \sim 0.5$) about an early B giant. Near periastron the companion's critical potential surface is comparable in size with the photospheric radius; thus lobe overflow and mass transfer onto the neutron star can take place. It is the resulting accretion energy which powers the X-ray outbursts.

During the OFF state (and the earliest cycles of an ON state) the giant companion appears as a relatively normal B or Be-type star. However, IUE observations have clearly demonstrated extensive changes in the temperature and surface area of the (UV/optical) continuum-emitting region, even between outbursts, as an ON state progresses; the temperature may drop by $\sim 30\%$, while the photospheric radius can apparently double. (The ON state data outside of outbursts are, however, all consistent with a constant bolometric luminosity of $\sim 10^{4.1}\ L_\odot$.) This suggests that not all the mass pulled away from the companion star at periastron is accreted onto the neutron star; sufficient material is left over from each orbital encounter to build up an extended, optically thick cloud, which eventually completely shrouds the B star (Howarth *et al.*, 1984; Densham *et al.*,

1983). Increases in the emitting area are then balanced by drops in temperature to match the bolometric flux generated by nucleosynthesis in the companion's interior.

The UV/optical outbursts can be quantitatively understood as the result of reprocessing of X-rays intercepted by the cloud cloaking the companion star. The amplitude of the outbursts can be more than two magnitudes at V, or $1\frac{1}{2}$ magnitudes in M_{BOL}, and the effective radius of the continuum-emitting cloud can be as great as $\sim 50\ R_{\odot}$, comparable to the orbital semimajor axis. In addition, strong, broad, emission lines develop, with line widths of up to ~ 5000 km s^{-1} (see Figure 6); material traveling with such velocities would traverse the orbit in only a few hours, and would be permanently lost from the system. This kinetic activity may be attributed to radiation pressure driven by the high X-ray luminosity of the neutron star, perhaps as great as 10^{39} erg s^{-1}.

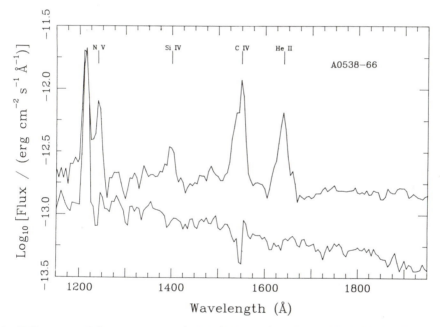

Fig. 6.   IUE spectra of the recurrent transient A 0538—66 in outburst (SWPs 13834—8 averaged) and quiescence (SWP 11042), showing large changes in the continuum slope and level, and in the lines of N v, C iv, Si iv, and He ii.

Although the eccentric binary model provides a satisfactory framework for interpreting many of the observations of A 0538—66, the system is by no means fully understood. An as yet unresolved puzzle is that the OFF state luminosity of the B star, $\sim 10^{4.4}\ L_{\odot}$, is about twice as great as the luminosity observed between outbursts during the ON state (Howarth, 1984). It is possible that the geometry of the cloud, affected by the nature of the orbit, is such that from our viewpoint its cross-section is smaller than its average cross-section; this will mean that the total ON state luminosity between outbursts is underestimated. It is also not clear what switches the system between OFF and ON states. One possibility is that the highly eccentric orbit drives oscillations in the companion star, so that it just under- or over-fills its critical lobe at periastron. Another possibility is that equatorial mass-loss associated with a Be phase initiates outbursts. In any event, the development of an extended cloud of material makes periastron accretion possible for several orbits, so that the ON state may be partially self-sustaining.

6. Interpretation of the Line Spectra of Compact Binaries

6.1. CV WINDS

The most prominent UV spectral lines in CVs are the alkali-like resonance lines N V $\lambda1240$, Si IV $\lambda1400$, and C IV $\lambda1549$, although weaker lines of C III $\lambda1176$, Si III/O I/Si II ~ $\lambda1300$, C II $\lambda1335$ (or violet-shifted O IV $\lambda1342$), He II $\lambda1640$, N IV $\lambda1718$, Al III $\lambda1855$, and Mg II $\lambda2800$ are also often detected. During low mass accretion states (e.g., for dwarf novae in quiescence), these lines usually appear as broad, fairly symmetric emission features.

The line profiles of the more luminous CVs, such as the novalike disk stars and dwarf novae during outbursts, often display asymmetric line profiles with broad absorption displaced to the violet and emission peaking near the rest wavelength. Similar line profiles viewed in the spectra of early-type stars such as P Cygni are interpreted as arising from an expanding, extended atmosphere: the outflowing wind between the star and the observer absorbs the stellar continuum at violet-shifted wavelengths and, for lines of sight not intercepting the stellar disk, scatters radiation towards the observer to produce an emission component at the rest wavelength. The term 'P Cygni profile' is commonly used to describe these lines. In fact, however, a CV wind-line profile can differ substantially from this profile, and from model OB star profiles (of e.g., Castor and Lamers, 1979): CV line profiles often consist only of absorption, or only of emission, or some admixture of absorption and emission components whose ratio appears to be a function of, among other things, orbital phase and the spectral energy distribution of the accretion disk.

In systems with disks viewed face-on, the profiles consist of broad, violet-shifted absorption, indicating that the wind is viewed in projection against the bright disk (Figure 7). A subset of these stars that have prominent UV emission lines during their low-luminosity states exhibit emission components in addition to the absorption components, and are thus the stars with profiles that most resemble those of P Cygni. In luminous CVs whose disks are viewed nearly edge-on, only emission lines are observed, suggesting that the wind is viewed projected against space, rather than against the bright disk (Figure 8). The emission line profiles appear to be asymmetric and redward-skewed (Córdova and Mason, 1985).

Although the wind lines are most discernable in the luminous (i.e., more highly mass-accreting) CVs, there is some evidence from the asymmetry in the line profiles that winds could be present in some CVs even during low-luminosity states (Raymond, 1984). In addition, winds are not necessarily peculiar to only CVs with extensive disks. A P Cygni-like profile may have been detected in the

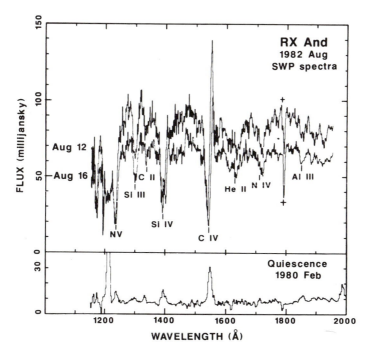

Fig. 7. IUE spectra of the dwarf nova RX Andromedae displaying broad, violet-shifted absorption lines (see, especially, N V and C IV) at the peak of an outburst (1982 August 12) and on the decline (1982 August 16). C IV λ1549 also has a prominent emission component. (Camera reseaux, which resemble narrow absorptions, are marked with crosses.) The spectrum during quiescence (1980 February) shows only emission lines superposed on a much weaker continuum. In this system the disk is viewed relatively face-on. From Verbunt *et al.* (1985).

AM Her star, E1405—451 (Nousek and Pravdo, 1983), and was clearly detected, together with violet-displaced N V and Si IV absorptions, in the DQ Her star TV Col, at the peak of a one hour UV-optical flare (Szkody and Mateo, 1984). (The AM Her and DQ Her subsets of magnetic CVs are discussed in Section 7).

An early review of the observational aspects of the wind is given in Córdova and Mason (1984); problems in interpreting the observations and attempting to model the wind are discussed in Drew and Verbunt (1985), Mauche and Raymond (1985), Raymond and Mauche (1985), Kallman and Jensen (1985), and Drew (1986). Here we summarize the characteristics of the wind phenomenon in CVs, and the efforts made to understand them.

(1) The terminal velocity of the wind is high. A lower limit to its value is provided by the observed 'edge' velocity, that is, the velocity where the blue edge of the absorption component intersects the continuum. Edge velocities ranging from 3000 km s^{-1} to $\geqslant$ 5000 km s^{-1} have been estimated. This is similar to the escape velocity from a white dwarf and has been used as an argument that the wind is emitted from the central regions of the disk (Córdova and Mason, 1982).

(2) The driving mechanism for the wind is uncertain, but the abundant supply of UV photons in the inner disk makes radiation pressure in the spectral lines a

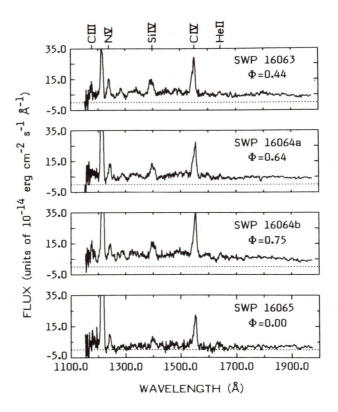

Fig. 8. IUE spectra of the 5.57 hr binary RW Trianguli taken at various orbital phases on 1982 January 18. This system is viewed nearly along the plane of the disk and thus deep eclipses of the UV and optical continuum from the disk are observed. Zero background is denoted with a short dashed line. The emission lines are prominent even during the eclipse of the continuum by the companion star ($\phi = 0.0$), indicating that the wind-emitting region is extensive. The Figure is from Córdova and Mason (1985), in which more expanded plots of the line profiles show that the emission lines are asymmetric and redward-skewed.

good candidate, by analogy to the early-type stars (Córdova and Mason, 1982). This assumes a rather modest mass loss rate, $\dot{m} = 10^{-11}$ to 10^{-10} $M_\odot$ yr^{-1}, so that the momentum rate of the wind is no larger than the momentum rate of the radiation or, equivalently, the mass loss rate in the wind is no larger than 1% of the mass accretion rate. If the mass loss rate is much higher (see item 3 in this list), then an alternate mechanism such as magnetic driving may apply (see Kallman and Jensen, 1985 for a discussion of this issue). We note that if the winds are radiatively driven, the presence of a wind in a DQ Her star such as TV Col (Szkody and Mateo, 1984) suggests the outflow may be hydromagnetically controlled in magnetic systems.

(3) Model line profiles are presently being constructed to estimate the size and shape of both the emitting region and the region responsible for the acceleration, the optical depths of the lines, the ion fractions, and, perhaps most importantly, the mass loss rate in the wind. The early attempts to derive $\dot{m}_{\text{loss}}$ in CVs used the

theoretical line profiles for early-type stars, which assumed spherically symmetry and a radiatively driven wind (i.e., the profiles calculated by Castor and Lamers, 1979). Using the observed ratios of the ion fractions in CV winds, and deriving an effective temperature for the radiation field from a comparison with similar data on O and B stars, Córdova and Mason (1982) deduce the ion fractions, and hence $\dot{m}_{loss}$ for a dwarf nova during outburst. All such similar exercises have yielded optical depths for the lines of less than unity, and mass loss rates on the order of a few times $10^{-11} M_\odot$ yr^{-1}. In these calculations the size of the wind-emitting region is assumed to be of order 10 white dwarf radii, which is the probable radius in the disk of the 1500 Å continuum.

More sophisticated attempts to model the profiles using an extended continuum source more appropriate to the disk in CVs have revealed that the emergent line profile is significantly weaker than it would be for a single star; thus, in a CV disk, whose radius is at least comparable to the length scale of the wind's acceleration, the lines may be optically thick even while evidencing shallow absorption (Drew, 1986). Thus the mass loss rate may be much higher than modeled under the assumption of acceleration by a small, spherical continuum source.

(4) The eclipsing systems provide information on the size of the wind and its optical depth. Phase-resolved spectroscopy of highly mass-accreting novalike stars shows that the UV resonance lines, in particular C IV and N V, suffer very little occultation of the line during the deep UV eclipses of the inner disk by the companion star (e.g., Figure 8). This indicates that the wind is extensive (Holm *et al.*, 1982; King *et al.*, 1983; Córdova and Mason, 1985). This observation, plus the relative and absolute strengths of the UV resonance lines, suggests that these lines are extremely optically thick to resonant scattering, thus requiring both high mass loss rates and the co-existence of sufficiently large fractions of N^{4+}, Si^{3+}, and C^{3+} (Drew and Verbunt, 1984).

(5) Modeling the wind usually starts, for simplicity, with the assumption that it is spherically symmetric and unclumped. Recent investigations, however, suggest a wind very different from this model.

(i) Córdova and Mason (1985) find a possible correlation between the blue edge velocity and the inclination, in the sense that v_{edge} is smaller when the inclination is higher. This effect would result if the wind were conical in shape with a cone angle $\leqslant 45°$: a lower velocity woud be measured in systems viewed perpendicular to the conical flow. Mauche and Raymond (1985) find, however, that to model the observed absorption profiles in low-inclination systems requires that the wind have an opening cone angle $\geqslant 60°$. They find that they can reproduce the observed profiles with a very slow velocity law, or with a velocity law, which can be either fast or slow, having a $\cos \theta$ dependence, where θ is the angle measured normal to the disk. Such a velocity law would come about if the planar disk, rather than the white dwarf, were the source of the acceleration. Mauche and Raymond suggest that the observed narrowing of the emission components at large inclination angles could result from a velocity law that varies as $\cos \theta$.

(ii) For $\dot{m} \leqslant 10^{-10} M_\odot$ yr^{-1}, the required ion fractions for N^{4+}, Si^{3+}, and C^{3+} are all too high ($\geqslant 0.01$) if the lines are required to be saturated (Drew and Verbunt, 1984). The presence of a highly ionizing UV/X-ray source in the boundary layer makes it unlikely that these ions are so abundant if the wind is spherically symmetric (Drew and Verbunt, 1984; Raymond, 1984; Kallman and Jensen, 1985). The wind must encompass a wide range of ionization conditions for these ions to coexist in large abundance. A non-spherical geometry alleviates the problem somewhat, but not enough (Raymond, 1984). Kallman and Jensen (1985) suggest that this problem resolves itself if $\dot{m}_{loss} \gtrsim \dot{m}_{accr}$. As noted previously, for disk-accreting stars up to one-half the accretion energy can be liberated in the boundary layer, which will radiate a thermal EUV/soft X-ray spectrum with $T \sim$ a few $\times 10^5$ K. In the model of Kallman and Jensen substantial O VI continuum absorption from the wind will reduce any boundary layer-produced soft X-ray flux. Drew and Verbunt (1985) note that a slow velocity law can increase the fraction of C IV somewhat and can account for the weakness of the C IV eclipse in edge-on systems.

An alternate and promising way out of the dilemma is suggested by recent attempts to understand O star winds as containing strong shocks; Raymond and Mauche (1985) find that, in compressing the gas in a CV wind, shocks can increase the fraction of C IV by 10^2. The shocked gas cools to an equilibrium temperature near 4×10^4 K and an ion fraction, $f_{C_{IV}}$, of about 2%.

(6) There are changes in the wind-line profiles on every timescale observed to be of importance in these systems.

(i) Outburst decay time. The wind line profiles develop at or near the peak of the outburst, when the spectrum steepens dramatically, heralding the appearance of a significant component at the shortest IUE wavelengths (see Figure 7). In some systems the absorption line troughs remains at nearly the same flux level throughout the decline, suggesting that in these systems the component providing much of the far UV enhancement is located behind the wind (Córdova et al., 1984). The equivalent widths of the absorption components, but not the emission components, scale with changes in the overall far UV continuum level as the outburst progresses. In addition, the ratios of the fluxes and equivalent widths of the lines of different elements are not the same for the emission and absorption components. Both of these facts suggest that the emission and absorption may be formed under different physical conditions, with much of the emission being produced other than in the wind. The flux distribution gradually approaches the quiescent state; the absorption component disappears and the emission profile (in systems with emission lines during quiescence and outburst) becomes more nearly symmetric.

In systems having outburst declines in which the flux remains constant for a prolonged time (i.e., the Z Cam standstills), there is weak evidence, based on comparing observations from different standstills, that the line profiles may change greatly through the course of the standstill even though there is little change in the UV flux level or flux distribution (Szkody and Mateo, 1986). This may indicate that the ionizing spectrum has changed during the long outburst, perhaps as a result of the continuous mass flow which has modified conditions of density and temperature in the mass loss region over time.

(ii) Orbital phase. The line profiles appear to change on the timescale of the binary orbital period (Hutchings, 1980; Holm et al., 1982; Szkody and Mateo,

1986), but what data exist on this phenomenon are patchy and poorly resolved in time, so that no conclusions can yet be made. This aspect of the wind phenomenon deserves much closer attention for what it can tell us about the structure of the disk and wind. It is plausible, for example, that the wind could have an unusual geometry, modified by rotation of the underlying white dwarf or the binary orbital motion. Alternatively, asymmetric structure on the outer disk, or material near the L_3 point, could periodically occult part of the wind-emitting region, causing phase-related changes in the line profiles.

(iii) Variations on timescales shorter than the orbital period. In high resolution IUE data, narrow (~ 60 km s^{-1}) features, whose velocities vary in less than 1 hr, have been observed in the broad, velocity-shifted wind line profiles of a novalike star (Córdova, 1986). Similar features are observed in the spectra of many Be stars and are also found superimposed on the broad P Cygni absorption wings of the luminous O and B stars. One interpretation is that they result from variable mass loss, or 'puffs', and subsequent expansion of a high-density layer (Henrichs, 1984). Another interpretation is that the wind is composed of high and low-density pockets, which are caused by instabilities, or shocks, in a radiation pressure driven flow (Lucy, 1983). Still another interpretation associates the transient distortions of the absorption lines with high-order non-radial pulsations in the Be stars; Vogt and Penrod (1983) suggest that mode-switching could release pulsational energy, perhaps triggering enhanced mass loss and the Be star phenomenon. This last conjecture is particularly interesting as CVs in high luminosity states exhibit rapid, soft X-ray pulsations which do not keep perfect coherence. It is not unreasonable to search for variability of the resonance line profiles on a pulsation timescale since both wind and pulsations are thought to originate from a similar location; a variation in, for example, the pulse phase of the X-ray oscillations could be communicated to the wind (see Córdova, 1985 for further discussion).

(7) A few dwarf novae, which have spectra that rise steeply towards shorter wavelengths during both quiescence and outburst, have less pronounced wind lines, with weak absorption and little or no emission, than dwarf novae with flatter spectra during quiescence and outburst (Klare et al., 1982; Córdova et al., 1984). A steeper spectrum could imply a smaller disk or a hotter white dwarf. The former possibility would mean less contribution to the optical and near UV luminosity, resulting in less absorbing background for the wind. The implied cone angle of the wind is narrow, however, or a prominent emission component would be observed in these systems. The latter possibility suggests a hotter ionizing spectrum which could reduce the ion fractions of the resonance lines detectable with the IUE.

6.2. THE EMISSION LINE SPECTRA OF LOW-MASS X-RAY BINARIES

Although attempts have been made to observe Sco X—1 and Her X—1 with the IUE in high resolution mode, in general the known LMXBs are too faint to secure kinematical information from UV spectroscopy. Thus information on UV emission lines in these systems is essentially confined to low resolution ($\leqslant 6$ Å) flux measurements. In almost all systems, the dominant features are N v $\lambda 1240$, C iv $\lambda 1549$ and Si iv $\lambda 1400$; with reasonable signal-to-noise numerous other lines can be detected (e.g., O iv $\lambda 1342$, O v $\lambda 1370$, He ii $\lambda 1640$, N iv $\lambda 1718$). For none of the LMXBs observed with the IUE has an entirely satisfactory quantitative analysis of the emission lines been carried out; indeed, for most systems even the location of the emission region is speculative. Observationally, the most common characteristic of UV line emission in LMXBs is the great strength of N V, which is usually the strongest line present (see Figure 3); this differs from the situation in quiescent dwarf novae and AM Her stars, where C IV is usually a few times stronger than N V. It is also inconsistent with the line ratios expected for simple, optically thin, X-ray heated plasmas (e.g. Hatchett *et al.*, 1976). One possible interpretation is that Nitrogen is relatively overabundant, as suggested for the recurrent nova U Sco (Barlow *et al.*, 1981; Williams *et al.*, 1981). For the LMXBs, however, the simplifying assumption of negligible line optical depth cannot be made without justification; if optical depth effects are to be taken into account, then the geometries of the line-forming regions must be approximately understood. In many cases, even this information is lacking (but see, e.g., Anderson, 1981; Howarth and Wilson, 1983b).

6.2. THE WINDS OF OB AND Be STAR BINARIES

All luminous OB stars, including those with collapsed companions, lose mass via stellar winds at rates of order $10^{-6} \, M_\odot$; this is testified by the ubiquity of P Cygni profiles on the UV resonance lines of abundant ions (see pp. 139—155). The gravitational capture of part of this wind may provide the material necessary to power the X-ray production in the HMXBs, although most OB stars fill or nearly fill their critical potential lobes (e.g. Bahcall, 1978), so that lobe overflow may also play a significant role. In practice, the two mechanisms become indistinguishable as a star approaches its critical surface (Friend and Castor, 1982), so that any distinction is arbitrary.

As noted earlier, the compact star has little effect on the photospheric spectrum of an HMXB. This is not true of the strong UV lines formed in the stellar wind, however, where the X-rays from the vicinity of the collapsed star can result in large variations in the P Cygni profiles.

This variation results from the substantial modification of the ambient ionization balance in the wind by the intense X-ray flux. Typically, the resonance lines of C IV and Si IV are observed; these are formed by ions one stage below the species which are dominant in OB winds. The addition of a dense X-radiation field selectively destroys these ions over a given volume, in a manner analogous to the familiar Stromgren sphere situation. As shown by Hatchett and McCray (1977), in a stellar wind, with its radial density dependence, surfaces of constant ionization can be approximately characterized by the parameter $q = R^2 v(R)/r^2 v(a)$, where R and r are the distances to the nondegenerate and degenerate stars, respectively, and $v(a)$ is the wind velocity at the orbit of the X-ray source. In a constant velocity flow, for $q > 1$ these surfaces resemble Stromgren-type spheres enclosing the collapsed star; for $q < 1$ they are open surfaces surrounding the mass donor. An accelerating flow gives qualitatively similar results, although the detailed shapes of the surfaces change (see Figure 9). To a first approximation, a given ion will be destroyed within some volume corresponding to a critical value of q. Then, as orbital motion presents different aspects of the system to the observer, the mass-

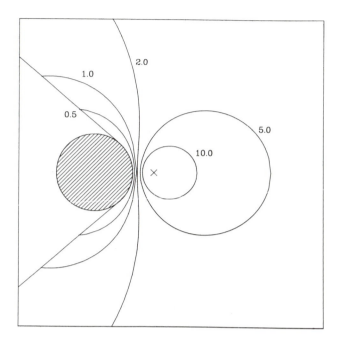

Fig. 9. Surfaces of constant q, drawn approximately to scale for the system HD 153919/4U 1700—37. The location of the neutron star is marked with a cross.

donor is viewed through different columns of that ion, and changes in the P Cygni profiles result. When observed at phase zero (superior conjunction of the compact object, usually X-ray eclipse) the absorption component of the profile is formed in an essentially undisturbed stellar wind; at phase 0.5 the absorption in C IV and Si IV should be greatly weakened.

These predicted effects have been most convincingly demonstrated in high resolution IUE observations of HD 77581, the optical counterpart to Vela X—1 (also called 4U 0900—40; Dupree *et al.*, 1980). They are confirmed by (more difficult) high resolution UV spectroscopy of HD 153919 (4U 1900—37: Figure 10), Sk 160 (SMC X—1; Hammerschlag-Hensberge *et al.*, 1984), and HDE 226868 (Cyg X—1; Davis and Hartmann, 1983), where they have been used to constrain the orbital inclination and hence the mass of the secondary; and by extensive low resolution results for several systems (e.g. Treves *et al.*, 1980; van der Klis *et al.*, 1982). However, even the most extensive modeling fails to match the observations in detail (McCray *et al.*, 1984). This is, in part, probably due to the difficulties encountered in reproducing P Cygni profiles even in isolated stars (e.g. Lucy, 1983), but it is clear that the presence of a compact object will influence the dynamics, as well as the ionization balance, of the wind (e.g. Fransson and Fabian, 1980; Friend and Castor, 1982; MacGregor and Vitello, 1982). Orbital line profile variability has yet to be exploited as a tool for probing these effects.

The IUE has also contributed to our understanding of the winds of Be X-ray

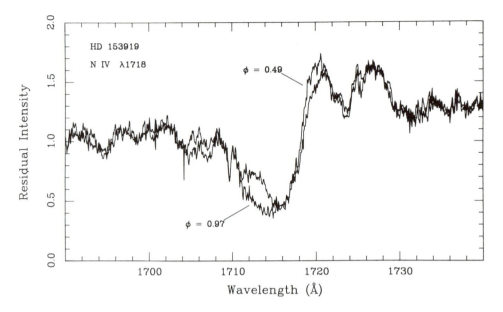

Fig. 10. The N IV $\lambda1718$ line in the spectrum of HD 153919 at orbital phases 0.0 and 0.5 (each spectrum is the average of 6 observations). The line is shallower at phase 0.5 due to the destruction of N^{3+} ions in the line of sight by X-rays.

binaries. Although low resolution data provide qualitative evidence for mass loss in many systems (e.g. Bianchi and Bernacca, 1980), it is high resolution spectroscopy that allows a more detailed study of this phenomenon. The sources studied to date are X Per, Gamma Cas, and HDE 245770 (Hammerschlag-Hensberge *et al.*, 1980; Bernacca and Bianchi, 1981; de Loore *et al.*, 1984). Although difficulties exist in estimating ion fractions (only minority ions are available in the IUE range for the hotter Be stars), mass-loss rates of order 10^{-8} to 10^{-9} $M_\odot$ yr^{-1} are representative; these values are typical of those thought appropriate for isolated Be stars of the earliest spectral types (e.g., Snow, 1981). In several cases it can be clearly demonstrated that the ambient stellar wind from the entire surface of the Be star cannot alone provide enough material to power X-rays by accretion (e.g., Coe and Payne, 1985); nor is lobe overflow plausible in these systems because of the small size of the Be star with respect to the size of the binary orbit. X-ray production in these binaries may therefore be powered by accretion onto the compact star from relatively dense, low-velocity material in the equatorial plane of the Be star, arising from enhanced surface mass flux and/or decreased flow velocities at low latitudes (probably a consequence of the relatively rapid rotation of the star). In some cases, X-ray emission could arise from the inter-action of a relativistic wind from a young pulsar with the Be star wind (Maraschi and Treves, 1981). The idea of nonspherically symmetric mass loss reconciles estimates of the mass loss rate based on UV P Cygni profiles (which sample the low density, high velocity wind) and the generally much higher values derived from IR excesses (which arise in relatively dense regions), and is consistent with the popular view of an isolated Be star as containing an equatorial belt of high density material (see papers in *IAU Symp.* **98**).

7. A New Class of Compact Binaries: the Magnetic CVs

The list of CVs in which accretion is thought to be strongly influenced by the presence of a magnetic white dwarf has grown, in the lifetime of the IUE, from a couple to almost two dozen. About half of these are systems in which the white dwarf has a field strength of $2 - 3 \times 10^7$ G and is rotating synchronously with the binary orbital period. The discovery of these 'AM Her' stars (named for the first such star found) is largely due to the detection of optical polarization from selected CVs with strong, high excitation optical and UV emission lines and having an appreciable soft X-ray flux. The remainder of the systems thought to be magnetic variables are CVs in which fairly rapid (i.e., about 10 s to 1 hr), strong X-ray or optical, coherent pulsations are detected. This is interpreted as emission from non-aligned magnetic poles on a white dwarf rotating more rapidly than the binary period. The field strengths of the white dwarfs in these, the 'DQ Her' systems, may be as much as ten times less than in the synchronously rotating AM Her stars (Lamb and Patterson, 1983).

In both the AM Her and DQ Her stars, or polars, magnetic fields channel the accretion flow onto one or more poles on the white dwarf. These stars therefore have complex optical and X-ray light curves which depend on the inclination of the binary system, the orientation of the accretion poles, and the amount of mass transferred from the companion star.

The relative faintness of these stars, coupled with the long readout time of the IUE images, have precluded both high spectral and high temporal resolution IUE observations. This is unfortunate because the emission line profiles in the AM Her stars are probably composites of emission spectra from several distinct regions in the binary (Liebert and Stockman, 1985), and, in the DQ Her stars, continuum and line variations would be expected to occur on the timescale of the white dwarf rotation period, which is on the order of minutes. Most IUE spectra are averages over at least a binary orbital period. In a few cases, however, short enough IUE spectra have been taken ($\leqslant 1$ hr) that the orbital period can be crudely resolved. These observations have produced especially interesting results, as we now illustrate.

7.2. AM HER AND THE SOFT X-RAY PUZZLE

In their high mass accretion states, AM Her and several other stars in its class produce intense soft X-ray fluxes. Time resolved IUE spectroscopy of AM Her showed a turnup ($F_\lambda \propto \lambda^{-4}$) in the far UV which, because it was present only outside of the X-ray eclipse, could be associated with the X-ray bright accretion pole (Raymond *et al.*, 1979). The simple interpretation of the far UV and soft X-ray spectra of AM Her was that they were the low and high energy tails, respectively, of a blackbody which peaked at 27 eV (Lamb, 1985; and references therein). The implied luminosity of the blackbody, L_{BB}, was far higher than the sum of the other emission components in the system, in contradiction to the theory for radial accretion onto a magnetic white dwarf (Lamb and Masters, 1979). This theory predicts that there are three components to the flux liberated close to the accreting star: (1) a blackbody-limited optical cyclotron component produced in the host, post-shock emission region of the accretion column, (2) a hard X-ray bremsstrahlung component also produced in this region, and (3) an EUV or soft X-ray blackbody component produced by cyclotron and bremsstrahlung radiation that is absorbed and remitted by the surface of the white dwarf, and whose luminosity is therefore equal to the sum of the luminosities of the cyclotron and bremsstrahlung components. However, for AM Her,

$$L_{BB} \gtrsim 50(L_{cyc} + L_{br}),$$

a conundrum which came to be know as the 'soft X-ray puzzle' (cf. Lamb, 1985, for a more complete discussion of this problem and its resolution).

Subsequent modeling of the Fe K_α line and the hard X-ray spectrum by Swank et al. (1984) gave a value of the optical depth to Thomson scattering across the diameter of the accretion column, a parameter from which the accretion luminosity could be derived if the cross sectional area of the emission region were known. The fractional area of the column, $f \leq 2 \times 10^{-4}$, has been derived from observations of the soft X-ray component, which is mesured to be a 40—46 eV blackbody by Tuohy et al. (1981) and Heise et al. (1984). Since $L_{acc} = L_{cyc} + L_{bb} + L_{br}$, and L_{cyc} and L_{bb} are observed quantities, L_{bb} can be derived. The resulting blackbody luminosity turns out to be consistent with the total soft X-ray flux measured and the simple magnetic accretion theory. Thus the soft X-ray and far UV spectra need not be parts of the same blackbody spectrum.

Arguments based on observations suggest that the optically polarized light which is a prominent characteristic of the AM Her stars is probably not the cyclotron component predicted by theory, but instead comes from a lower temperature, more extended region high up in the accretion column (Liebert and Stockman, 1985; Lamb, 1985). The implication from this deduction and the X-ray data is that the turnup in the far UV may be the predicted optically thick cyclotron component from the X-ray emission region.

Lamb (1985) shows that the earlier published flux distributions of AM Her and similar stars can be interpreted in the magnetic, radial accretion model, and that there is, therefore, no obvious soft X-ray puzzle. More recent X-ray light curves acquired with Exosat reveal, however, that the emission of AM Her can be exceedingly complicated: the hard and soft X-ray orbital modulations during one observation were completely out of phase, suggesting emission from two magnetic poles, rather than one (Heise et al., 1985). Whether or not this can be fitted in the magnetic, radial accretion picture requires better knowledge of the UV and soft X-ray spectrum than is presently available or published. If one pole had a stronger magnetic field than the other, the observation of enhanced soft X-rays and diminished hard X-rays when this pole was viewed could be understood in the context of the model if more UV (i.e., more cyclotron flux) were emitted at this time. At the pole with the weaker field, where more hard than soft X-rays were observed, the X-ray source is required to have been much softer than previously observed in order to satisfy the model (Lamb, 1986, private communication).

7.3. The Multiple Periodicities of DQ Her Stars

An illustration of the potential usefulness of time-resolved IUE observations of a DQ Her system is provided by the data on TV Col. IUE spectra of this suspected magnetic variable were influential in helping to solve a long-standing controversy over the origin of the three observed periods in this system. Optical data of TV Col revealed a photometric period of 5.2 hr, a spectroscopic period of 5.5 hr, and a 4 day photometric period, interpreted as the 'beat' period of the 2 shorter periods. The UV data of Bonnet-Bidaud *et al.* (1985) clearly resolved the 5.5 hr period, which they attributed to X-ray illumination of a broad disk bulge rotating at the binary orbital period. There was no indication in the UV data of the 5.2 hr period, previously hypothesized to be the rotation period of the white dwarf (Hutchings *et al.*, 1981). Nor did the light curve or spectrum of this modulation suggest white dwarf rotation. Bonnet-Bidaud *et al.* (1985) therefore suggested a model of a retrograde 'precessing' disk in the system, similar to the model proposed to account for the long-term X-ray and optical variations of Her X–1 and LMC X–4. In this model, the disk undergoes an apparent precession on a timescale of 4 days and the 5.2 hr optical modulation results from the sum of the precession and orbital periods, an effect detected in the X-ray binaries. Bonnet-Bidaud *et al.* predicted that a faster period, reflecting the rotation of the white dwarf, would still be found in the system. Their prediction was soon confirmed with the discovery of a 1943 s modulation in Exosat X-ray data on TV Col (Brinkmann and Schrijver 1984).

8. The Future

The discussion of the IUE's contribution to our understanding of accretion onto compact objects points out an important limitation: the X-ray binaries are characterized by rapid variability, which cannot be explored with the IUE. It remains for future space-based instrumentation to make observations that have both high temporal and spatial resolution. As amply demonstrated in this text, considerable benefit will be enjoyed from space platforms instrumented with detectors that cover the widest possible range of spectral wavelengths so that multifrequency observations can easily be made simultaneously.

Acknowledgements

The authors are grateful to Drs. Janet Drew, Richard Epstein, and Richard Wade for comments which improved the manuscript. F. A. C. received funding from the US Dept. of Energy.

References

Anderson, L.: 1981, *Astrophys. J.* **244**, 555.
Bahcall, J. N.: 1978, *Ann. Rev. Astron. Astrophys.* **16**, 241.
Barlow, M. J. *et al.*: 1981, *Monthly Notices Roy. Astron. Soc.* **195**, 61.
Bath, G. T.: 1975, *Monthly Notices Roy. Astron. Soc.* **171**, 311.

Bath, G. T. and Pringle, J. E.: 1981, *Monthly Notices Roy. Astron. Soc.* **194**, 967.
Bath, G. T., Evans, W. D., Papaloizou, J. C. B., and Pringle, J. E.: 1974, *Monthly Notices Roy. Astron. Soc.* **169**, 447.
Bernacca, P. L. and Bianchi, L.: 1981, *Astron. Astrophys.* **94**, 345.
Bianchi, L. and Bernacca, P. L.: 1980, *Astron. Astrophys.* **89**, 214.
Bianchi, L. and Pakull, M.: 1984, *Future of UV Astronomy Based on Six Years of IUE Research*, NASA CP-2349, p. 416.
Bianchi, L. and Pakull, M.: 1985, *Astron. Astrophys.* **146**, 242.
Blair, W. P., Raymond, J. C., Dupree, A. K., Wu, C.-C., Holm, A. V., and Swank, J. H.: 1984, *Astrophys. J.* **278**, 270.
Bonnet-Bidaud, J. M., Motch, Ch., and Mouchet, M.: 1985, *Astron. Astrophys.* **143**, 313.
Brinkmann, A. and Schrijver, T.: 1984, *IAU Circular*, No. 3980.
Brown, J. C. and Boyle, C. B.: 1984, *Astron. Astrophys.* **141**, 369.
Cannizzo, J. K. and Wheeler, J. C.: 1984, *Astrophys. J. Suppl.* **55**, 367.
Cannizzo, J. K., Wheeler, J. C., and Polidan, R. S.: 1986, *Astrophys. J.* **301**, 634.
Castor, J. I. and Lamers, H. J. G. L. M.: 1979, *Astrophys. J. Suppl.* **39**, 481.
Charles, P. A. *et al.*: 1983, *Monthly Notices Roy. Astron. Soc.* **202**, 657.
Chester, T. J.: 1978, *Astrophys. J.* **222**, 652.
Chiapetti, L., Maraschi, L., Tanzi, E. G., and Treves, A.: 1983, *Astrophys. J.* **265**, 354.
Coe, M. J. and Payne, B. J.: 1985, *Astrophys. Space Sci.* **109**, 175.
Cook, M. C.: 1985, *Monthly Notices Roy. Astron. Soc.* **216**, 219.
Córdova, F. A.: 1985, in Tanaka, Y. and Lewin, W. H. G. (eds.), *Galactic and Extragalactic Compact X-Ray Sources*, ISAS, Japan, p. 119.
Córdova, F. A.: 1986, in Mason, K. O., Watson, H. G., and White, W. E. (eds.), *Physics of Accretion onto Compact Objects Proceedings*, 1986, Tenerife, Spain, Springer Verlag, p. 339.
Córdova, F. A. and Mason, K. O.: 1982, *Astrophys. J.* **260**, 716.
Córdova, F. A. and Mason, K. O.: 1983, in Lewin, W. H. G., and van den Heuvel, E. P. J. (eds.), *Accretion-Driven Stellar X-Ray Sources*, Cambridge University Press, p. 147.
Córdova, F. A. and Mason, K. O.: 1984, in Mead, J. M., Chapman, R. D., and Kondo, Y. (eds.), *Future of Ultraviolet Astronomy Based on Six Years of IUE Research*, NASA CP-2349, p. 377.
Córdova, F. A. and Mason, K. O.: 1985, *Astrophys. J.* **290**, 671.
Córdova, F. A., Ladd, E. F., and Mason, K. O.: 1986, in Proc. of the Los Alamos Conference on *Magnetospheric Phenomena in Astrophysics*, 5—11 Aug. 1984, Taos, NM, Am. Inst. of Phys.: NY p. 250.
Davis, R. and Hartman, L.: 1983, *Astrophys. J.* **270**, 671.
de Loore, C., Giovanelli, F. *et al.*: 1984, *Astron. Astrophys.* **141**, 279.
Densham, R. H., Charles, P. A., Menzies, J. W., Van der Klis, M., and van Paradijs, J.: 1983, *Monthly Notices Roy. Astron. Soc.* **205**, 1117.
Drew, J.: 1986, *Monthly Notices Roy. Astron. Soc.* **218**, 41.
Drew, J. E. and Verbunt, F.: 1984, in *Proc. 4th European IUE Conf.*, ESA SP-218, p. 387.
Drew, J. E. and Verbunt, F.: 1985, *Monthly Notices Roy. Astron. Soc.* **213**, 191.
Dupree, A. K. *et al.*: 1978, *Nature* **275**, 400.
Dupree, A. K. *et al.*: 1980, *Astrophys. J.* **238**, 969.
Faulkner, J., Lin, D. N. C., and Papaloizou, J.: 1983, *Monthly Notices Roy. Astron. Soc.* **205**, 159.
Frank, J. and King, A. R.: 1981, *Monthly Notices Roy. Astron. Soc.* **195**, 227.
Fransson, C. and Fabian, C.: 1980, *Astron. Astrophys.* **87**, 102.
Friend, D. B. and Castor, J. I.: 1982, *Astrophys. J.* **261**, 293.
Hammerschlag-Hensberge, G. *et al.*: 1980, *Astron. Astrophys.* **85**, 119.
Hammerschlag-Hensberge, G., Kallman, T. K., and Howarth, I. D.: 1984, *Astrophys. J.* **283**, 249.
Hammerschlag-Hensberge, G., McClintock, J. E., and Van Paradijs, J.: 1982, *Astrophys. J.* **254**, L1.
Hassall, B. J. M.: 1985, *Monthly Notices Roy. Astron. Soc.* **216**, 335.
Hassall, B. J. M., Pringle, J. E., and Verbunt, F.: 1985, *Monthly Notices Roy. Astron. Soc.* **216**, 353.
Hassall, B. J. M., Pringle, J. E., Schwarzenberg-Czerny, A., Wade, R. A., Whelan, J. A. J., and Hill, P. W.: 1983, *Monthly Notices Roy. Astron. Soc.* **203**, 865.
Hatchett, S. and McCray, R.: 1977, *Astrophys. J.* **211**, 552.

Heise, J., Kruszewski, A., Chlebowski, T., Mewe, R., Kahn, S., and Seward, F. D.: 1984, *Physica Scripta* **T7**, 115.

Henrichs, H.: 1984, in *Proc. 4th European IUE Conference*, ESA SP-218, p. 43.

Henrichs, H. F., Hammerschlag-Hensberge, G., Howarth, I. D., and Barr, P.: 1983, *Astrophys. J.* **268**, 807.

Hill, P. W.: 1983, *Monthly Notices Roy. Astron. Soc.* **203**, 865.

Holm, A. V., Panek, R. J., and Schiffer, F. H.: 1982, *Astrophys. J. Letters* **252**, L35.

Horne, K. and Stiening, R. F.: 1985, *Monthly Notices Roy. Astron. Soc.* **216**, 933.

Howarth, I. D.: 1981, *Monthly Notices Roy. Astron. Soc.* **198**, 289.

Howarth, I. D.: 1983, *Monthly Notices Roy. Astron. Soc.* **203**, 801.

Howarth, I. D.: 1984, *Proc. 4th European IUE Conference*, ESA SP-218, p. 449.

Howarth, I. D. and Wilson, R.: 1983a, *Monthly Notices Roy. Astron. Soc.* **202**, 347.

Howarth, I. D. and Wilson, R.: 1983b, *Monthly Notices Roy. Astron. Soc.* **204**, 1091.

Howarth, I. D., Prinja, R. K., Roche, P. F., and Willis, A. J.: 1984, *Monthly Notices Roy. Astron. Soc.* **207**, 287.

Howarth, I. D., Hammerschlag-Hensberge, G., and Kallman, T.: 1986, *In New Insights in Astrophysics*, ESA SP-263, p. 475.

Hutchings, J. B.: 1980, *Publ. Astron. Soc. Pacific* **92**, 458.

Hutchings, J. B., Crampton, D., Cowley, A. P., Thorstensen, J. R., and Charles, P. A.: 1981, *Astrophys. J.* **249**, 680.

Kallman, T. R. and Jensen, K. A.: 1985, *Astrophys. J.* **299**, 277.

King, A. R., Frank, J., Jameson, R. F., and Sherrington, M. R.: 1983, *Monthly Notices Roy. Astron. Soc.* **203**, 677.

Klare, G., Krautter, J., Wolf, B., Stahl, D., Vogt, N., Wargau, W., and Rahe, J.: 1982, *Astron. Astrophys.* **113**, 76.

Kurucz, R. L.: 1979, *Astrophys. J. Supplement* **40**, 1.

la Dous, C., Verbunt, F., Schoembs, R., Argyle, R. W., Jones, D. H. P., Schwarzenberg-Czerny, A., Hassall, B. J. M., Pringle, J. E., and Wade, R. A.: 1985, *Monthly Notices Roy. Astron. Soc.* **212**, 231.

Lamb, D. Q.: 1985, 'Recent Developments in the Theory of AM Her and DQ Her Stars', in Lamb, D. Q. and Patterson, J. (eds.), *Cataclysmic Variables and Low-Mass X-Ray Binaries*, D. Reidel Publ. Co., Dordrecht, Holland, pp. 179—218.

Lamb, D. Q. and Masters, A. R.: 1979, *Astrophys. J. Letters* **234**, L117.

Lamb, D. Q. and Patterson, J.: 1983, 'Spin-up and Magnetic Fields in DQ Her Stars', in Livio, M. and Shaviv, G. (eds.), *Cataclysmic Variables and Related Objects*, D. Reidel Publ. Co., Dordrecht, Holland, pp. 229—237.

Liebert, J. and Stockman, H. S.: 1985, 'The AM Hercules Magnetic Variables', in Lamb, D. Q. and Patterson, J. (eds.), *Cataclysmic Variables and Low-Mass X-Ray Binaries*, D. Reidel Publ. Co., Dordrecht, Holland, pp. 151—177.

London, R., McCray, R., and Auer, L. H.: 1981, *Astrophys. J.* **243**, 970.

Lucy, L. B.: 1983, *Astrophys. J.* **274**, 372.

MacGregor, K. B. and Vitello, P. A. J.: 1982, *Astrophys. J.* **259**, 267.

Maraschi, L. and Treves, A.: 1981, *Monthly Notices Roy. Astron. Soc.* **194**, 1.

Maraschi, L., Treves, A., and Van den Heuvel, E. P. J.: 1978, *Nature* **259**, 292.

Mason, K. O. and Cordova, F. A.: 1982a, *Astrophys. J.* **255**, 603.

Mason, K. O. and Cordova, F. A.: 1982b, *Astrophys. J.* **262**, 253.

Mateo, M. and Szkody, P.: 1984, *Astron. J.* **89**, 863.

Mauche, C. and Raymond, J.: 1985, in Szkody, P. (ed.), *Ninth North American Workshop on Cataclysmic Variables*, University of Washington: Astronomy, p. 93.

McClintock, J. E., Petro, L. D., Hammerschlag-Hensberge, G., Proffit, C. R., and Remillard, R. A.: 1984, *Astrophys. J.* **283**, 794.

McCray, R., Kallman, T. R., Castor, J. I., and Olson, G. L.: 1984, *Astrophys. J.* **282**, 245.

Meyer, F. and Meyer-Hofmeister, E.: 1983, *Astron. Astrophys.* **128**, 420.

Meyer, F. and Meyer-Hofmeister, E.: 1984, *Astron. Astrophys.* **132**, 143.

Mineshige, S. and Osaki, Y.: 1985, *Publ. Astron. Soc. Japan* **37**, 1.

Nousek, J. A. and Pravdo, S. H.: 1983, *Astrophys. J. Letters* **266**, L39.
Panek, R. J. and Holm, A. V.: 1984, *Astrophys. J.* **277**, 700.
Poeckert, R.: 1982, in M. Jaschek and H-G. Groth (eds.), 'Be Stars', *IAU Symp.* **98**, 453.
Polidan, R. S. and Holberg, J. B.: 1984, *Nature* **309**, 528.
Polidan, R. S. and Holberg, J. B.: 1987, *Monthly Notices Roy. Astron. Soc.* **225**, 131.
Pringle, J.: 1981, *Ann. Rev. Astron. Astrophys.* **19**, 137.
Pringle, J. E., Verbunt, F., and Wade, R. A.: 1986, *Monthly Notices Roy, Astron. Soc.* **221**, 169.
Rappaport, S. A. and Joss, P. C.: 1983, in Lewin, W. H. G. and van den Heuvel, E. P. J. (eds.), *Accretion-driven Stellar X-ray Sources*, Cambridge Univ. Press, p. 1.
Raymond, J.: 1984, in Mead, J. M., Chapman, R. D., and Kondo, Y. (eds.), *Future of Ultraviolet Astronomy Based on Six Years of IUE Research*, NASA CP-2349, p. 301.
Raymond, J. C. and Mouche, C. W.: 1985, in Szkody, P. (ed.), *Ninth North American Workshop on Cataclysmic Variables*, U. of Washington, Astronomy, p. 128.
Raymond, J. C., Black, J. H., Davis, R. J., Dupree, A. K., Gursky, H., Hartmann, L., and Matilsky, T. A.: 1979, *Astrophys. J. Letters* **230**, L95.
Sadakane, K. *et al.*: 1985, *Astrophys. J.* **288**, 284.
Schwarzenberg-Czerny, A., Ward, M., Hanes, D. A., Jones, D. H. P., Pringle, J. E., Verbunt, F., and Wade, R. A.: 1985, *Monthly Notices Roy. Astron. Soc.* **212**, 645.
Shafter, A. W., Szkody, P., Liebert, J., Penning, W. R., Bond, H. E., and Grauer, A. D.: 1985, *Astrophys. J.* **290**, 707.
Shakura, N. I. and Sunyaev, R. A.: 1973, *Astron. Astrophys.* **24**, 337.
Smak, J.: 1984a, *Publ. Astron. Soc. Pacific* **96**, 5.
Smak, J.: 1984b, *Acta Astron.* **34**, 161.
Smak, J.: 1984c, *Acta Astron.* **34**, 317.
Snow, T. P.: 1981, *Astrophys. J.* **251**, 139.
Swank, J. H., Fabian, A. C., and Ross, R. R.: 1984, *Astrophys. J.* **280**, 734.
Szkody, P.: 1985, 'A Summary of the UV, Optical and IR Properties of Disks in CVs', in D. Q. Lamb and J. Patterson, *Cataclysmic Variables and Low-Mass X-Ray Binaries*, D. Reidel Publ. Co., Dordrecht, Holland, pp. 385—401.
Szkody, P. and Downes, R. A.: 1982, *Publ. Astron. Soc. Pacific* **94**, 328.
Szkody, P. and Mateo, M.: 1984, *Astrophys. J.* **280**, 729.
Szkody, P. and Mateo, M.: 1986, *Astrophys. J.* **301**, 286.
Szkody, P., Liebert, J., and Panek, R. J.: 1985, *Astrophys. J.* **293**, 321.
Szkody, P., Raymond, J. C., and Capps, R. W.: 1982, *Astrophys. J.* **257**, 686.
Treves, A. *et al.*: 1980, *Astrophys. J.* **242**, 1114.
Tuohy, I. R., Mason, K. O., Garmire, G.-P., and Lamb, F. K.: 1981, *Astrophys. J.* **245**, 183.
van der Klis, M. *et al.*: 1982, *Astron. Astrophys.* **106**, 339.
van Paradijs, J., 1983, in Lewin, W. H. G. and van den Heuvel, E. P. J. (eds.), *Accretion-driven Stellar X-ray Sources*, Cambridge Univ. Press, p. 189.
Verbunt, F., Pringle, J. E., Wade, R. A., Echevarria, J., Jones, D. H. P., Argyle, R. W., and Schwarzenberg-Czerny, A., la Dous, C., and Schoembs, R.: 1984, *Monthly Notices Roy. Astron. Soc.* **210**, 197.
Vogt, S. S. and Penrod, G. D.: 1983, *Astrophys. J.* **275**, 661.
Wade, R. A.: 1984, *Monthly Notices Roy. Astron. Soc.* **208**, 381.
Wesemael, F., Auer, L. H., Van Horn, H. M., and Savedoff, M. P.: 1980, *Astrophys. J. Supplement* **43**, 159.
Williams, R. E. and Ferguson, D. H.: 1982, *Astrophys. J.* **257**, 672.
Williams, R. E. *et al.*: 1981, *Astrophys. J.* **251**, 221.
Willis, A. J. and Wilson, R. *et al.*: 1980, *Astrophys. J.* **237**, 596.

INTERACTING BINARIES

GEORGE E. McCLUSKEY, JR.

Division of Astronomy, Department of Mathematics, Lehigh University, Bethlehem, PA, U.S.A.

and

JORGE SAHADE

Observatório Astronómico, Instituto Argentino de Radioastronomía, Villa Elisa,
Member of the Carrera del Investigador Científico, CONICET, Argentina

Introduction

The IUE observatory has provided a wealth of unique observational material to nearly every area of astronomy. The research effort and the extensive results engendered by this satellite are an extraordinary proof of the value of such orbiting observatories and of the IUE Guest Observer Program. In particular, this has certainly proven to be the case for the interacting binary systems.

The spectral signatures of interacting binaries range from barely detectable excess absorption to P Cygni-type absorptions, wind features, strongly asymmetric absorption, superionization and emission to systems dominated by circumstellar matter and strong emission lines. Since the very mechanisms which make a close binary interacting often give rise to accretional heating, and chromospheric and coronal activity, essentially all classes of interacting binaries can be profitably observed in the ultraviolet. A number of chapters in this book, including this one, are a testament to the importance of IUE to binary star research.

Although the present chapter is entitled 'Interacting Binaries', we shall discuss only those close binary systems which are not the subjects of other chapters of this book.

The scientific discoveries which the IUE satellite has contributed to the study of interacting binaries may be broken down into five main items, namely,

(a) the discovery of hot companions of late type giants and supergiants;
(b) the discovery of regions of high electron temperatures;
(c) the discovery of evidence for mass loss and velocity gradients;
(d) the discovery of intercombination lines in a few interacting binaries;
(e) the discovery of phenomena which so far remain unexplained.

1. Discovery of Hot Companions

1.1. One of the program objectives of the IUE satellite was to discover hot companions in peculiar systems like β Lyrae and ε Aurigae, where present ideas about their unobserved components suggest precisely that they may be hot objects. We shall come back to these two stars later as it is more appropriate to first stress the fact that the IUE has proven to be particularly successful in providing information about hot components in the cases of late type single-lined binaries

Y. Kondo (ed.), Scientific Accomplishments of the IUE, pp. 427–444.
© 1987 *by D. Reidel Publishing Company.*

and of some late type stars believed to be single but for which there were often indications of peculiar behavior. The majority of hot companions discovered by the IUE is due to the very productive efforts of Parsons (1981, 1983). Particularly interesting are HD 207739 (Parsons *et al.*, 1983) and 22 Vulpeculae (Ake *et al.*, 1985). The former has a highly complex and variable ultraviolet spectrum while the latter has characteristics similar to the ʒ Aurigae systems. In HD 207739 (Kondo *et al.*, 1985) variable optically thick extrastellar gas is clearly present and is seriously affecting the ultraviolet continuum. Table I lists the systems for which the IUE far-ultraviolet spectra indicate or prove the existence of a hot companion star with the addition of some relevant information. Newly determined orbital periods are indicated wherever known.

To the systems in Table I, we could add the symbiotic objects. The IUE continua suggest much hotter sources than those that would correspond to the late type giants observed in the photographic region (Sahade and Hernández, 1984).

TABLE I

Systems with hot companions discovered with the IUE

Star	HD	Spectral type	Orbital period (days)
AY Cet	7672	G5 III + wd	56.80
μ Per	26630	G0 Ib + B9—A0	283.299
KS Per	30353	A5 Iap + B2 Ib	360.0
—	37453	F5 II + B e	—
—	43246	F2 III + B8 V	23.2
—	57146	G2 Ib + B9	366.?
—	59067	G8 Ib + B:	—
—	101947	G0 Ia + B1 Ib	125.?
29 Dra	160538	K0 III + wd	30.?
ν Her	164136	F2 II + A0—A1	—
—	166612	F e + B1	—
β Sct	173764	G5 II + B8—B9	832.5
ν¹ Sgr	174974	K2 I + B	—
—	185110	K1 III + sd B	20.66
η Aql	187929	F6 Ib + A0 V	—
22 Vul	192713	G3 Ib + B8	249.083
35 Cyg	193370	F5 Ib + ?	2452.6
ξ Cyg	200905	K5 Ib + ?	113.464
—	203156	F3 II + B7	—
ʒ Cap	204075	G4 Ib + DA	—
—	207739	F8 IIe + B	—
ST Aqr	211965	G8 IV + A7	0.78
—	217476	G4 Ia + B1 V	—
56 Peg	218356	G8 Ib + wd	—

1.2. A few other systems have also been announced as having components hotter than those observed in the photographic region.

1.2.1. One of these discoveries refers to the system μ Sagittarii (Plavec, 1980a). The IUE continuum suggests a higher temperature than the B8 Ia supergiant observed from the ground. Unfortunately, the eclipse is not total and is very shallow. There is also no detectable secondary minimum. Therefore, there seems to be no way to isolate the continuum of the hotter component by using the IUE data alone.

1.2.2. The expectation of detecting the spectra of the disk-like or disk-surrounded fainter components in β Lyr and in ε Aur with IUE spectra has not been realized, mainly because the disks are thick and no evidence whatsoever exists that they become transparent in the ultraviolet.

Unfortunately, the existence of an opaque disk around the more massive component of β Lyr and the fact that the eclipses are partial do not enable us to isolate the ultraviolet spectrum of the fainter component, as Plavec (1980b), at one time, thought to be attainable by subtracting the IUE spectrum of β Lyr taken at secondary minimum from the IUE spectrum taken outside of eclipse.

The different IUE observations that have been reported (Boehm *et al.*, 1984) seem to suggest that in ε Aur, we are dealing with a variable continuum flux, rather than with evidence for the presence of a hot component.

2. Discovery of High Electron Temperature Regions

The IUE spectra of interacting binaries have disclosed the presence of high temperature resonance lines in every close pair that has been observed. These high temperature resonance lines are displayed either in emission or in absorption. A large portion of the material that has been secured with the IUE on interacting binaries has been taken in the low dispersion mode. Whenever possible, the observations were made at high dispersion and those systems displaying emissions have profiles of the P Cyg-type.

Table II lists the interacting systems — with the exclusion of those that are dealt with in other chapters of this book — for which, as far as the authors are aware, there are IUE observational data available in the literature. Our list has been completed with some of the unpublished results of Kondo and McCluskey (private communication).

We summarize the information that is contained in Table II and try to elaborate on it.

2.1. The Algol-type systems are systems containing a 'normal' late B-type or A-type main sequence star and a late type subgiant with no 'wind' outflow velocities. The presence of high temperature resonance lines of Si IV and C IV in absorption outside of eclipse in many Algol systems was first noted by Kondo *et al.* (1978, 1979, 1980, 1981) for U Cephei. Generally, the C IV doublet is much weaker than the Si IV doublet when detected. These lines are stronger on the following hemisphere of the primary than on the preceding hemisphere in all cases where such phase coverage has been obtained.

Some of these systems, namely, U Cep (Plavec, 1983a), RW Tauri (Plavec and

TABLE II
Interacting binaries

Star	HD/BD	Spectral type		Orbital period (days)	High temperature resonance lines present[a]		
					N V	Si IV	C IV
V Sge	—	WN5		0.51	e	e	e
—	134687	B3 IV		0.90		<u>A</u>	A
H Vel	76805	B5 V		0.91		<u>A</u>	A
V505 Sgr	187949	A2 V	+ G6—8 IV	1.18			A
AI Dra	153345	A0 V	+ (F7)	1.20		A	A?
Y Aql	178125	B5 V	+ B6	1.30		A	A
μ_1 Sco	166937	B1.5 V	+ B7	1.44		A	
V Pup	65818	B1 V	+ B2	1.45		A	A
γ_1 Vel	68243	B2 III		1.48		A	A
VV Ori	36695	B1 V		1.49		A	A
SV Cen	102552	B1 V	+ B6	1.66		A	A
δ Pic	42933	B0.5	+ B	1.67		A	A
TV Cas	1486	B9 V	+ (F8)	1.81		A	
V382 Cyg	228854	O6—7	+ O8 V	1.89		P	P
δ Lib	132742	A0 V	+ G—K	2.33		A	A
U Cep	5679	B7 V	+ G8 III	2.49	A/e	A/e	<u>A/e</u>
2 Lac	212120	B6 IV	+ B6 V	2.62		A	
RW Tau	25487	B9 V	+ K0 III	2.77	e?	e	e
β Per	19356	B8 V	+ (G8)	2.87	A	<u>A</u>	A
Y Cyg	198846	B0 IV	+ B0 IV	3.00	A	A	A
TX UMa	93033	B8 V	+ G0 III—IV	3.06	A	A	A
U Sge	181182	B8 V	+ G2 III—IV	3.38	A/e	<u>A/e</u>	A/e
U CrB	136175	B6 V	+ F8 III—IV	3.45	A	A	A
AO cas	1337	O9.5	+ O9.5	3.52	P	P	P
δ Cir	135240	O9 V	+ O	3.90	P	P	P
UW CMa	57060	O8	+ O8	4.39	P	<u>P</u>	<u>P</u>
R Ara	149730	B9 p		4.45	P	<u>P</u>	P
UX Mon	65607	A7 e	+ G2 IV	5.90	A/e?		<u>A/e</u>
CX Dra	174237	B2.5 Ve	+ F	6.70	<u>A</u>	A	A
RY Per	17034	B3 V	+ F0 III	6.86	<u>e</u>		e
TT Hya	97528	B9 V	+ G5	6.95	A/e	<u>A</u>	<u>A/e</u>
V861 Sco	152667	B0 Iae		7.85	P	<u>P</u>	<u>P</u>
V356 Sgr	173787	A2 II	+ B3 V	8.90	A/e	<u>A</u>	A/e
FF Aqr	− 03°5357	sd B	+ G8 III	9.21	A/e	A/e	<u>A/e</u>
AU Mon	50846	B6 p	+ F0	11.11	A	A	A
RY Sct	169515	B0 ep		11.12		e	e
V453 Sco	163181	B0.5 Ia	+ B1 e	12.01		P	P
β Lyr	174638	B8 II		12.93	<u>P</u>	P	P
RW Per	+ 41°0851	A5 III	+ G0 III	13.20	<u>e</u>		e
V373 Cas	224151	B0.5 III		13.42	P	<u>P</u>	<u>P</u>
W Ser	166126	F5 Ib		14.16	<u>e</u>	e	e
—	47129	O8 e		14.40	P	P	P
V367 Cyg	198287	A7 I	+ A0 III	18.60	e	e	e
AR Mon	57364	G8 III	+ K2—3	21.21	e?		e?

Table II (continued)

Star	HD/BD	Spectral type	Orbital period (days)	High temperature resonance lines present[a]		
				N V	Si IV	C IV
GG Car	94878	B e	31.00		A	A
RX Cas	+ 67°0244	A5 III + G3 III	32.32	e		e
SX Cas	232121	A6 III + G6 III	36.37	<u>E</u>	E	E
KX And	218393	B3 IVe + K1 III	38.90	P	P	
HR 2142	41335	B1 IV—Ve	80.9		A	A
φ Per	10516	B0 IV—Ve	126.70		A	P
v Sgr	181615	B8 p	138.00	A	A	A
—	207739	F8 IIe +B	140.80	P		
π And	3369	B5 V	143.60		A	A
μ Sgr	166937	B8 Iae	180.40		A	A
W Cru	105998	G1 Iab	198.50	e	e	e
22 Vul	192713	G3 Ib—II + B8—9	249.083		P	P
ε Aur	31964	F0 Ia	9883.00	A	A	A

[a] A(a): absorption detected at high (low) resolution. E(e): emission detected at high (low) resolution. P: P Cyg feature detected at high resolution. Underlined entry indicates ion of significantly greater strength.

Dobias, 1983) and TT Hydrae, U Sagittae and UX Monocerotis (Plavec *et al.*, 1984) were observed at principal eclipse, in the low dispersion mode and found to display a strong emission spectrum. At variance with the absorption spectra taken outside of eclipse, the strongest emission line during eclipse is the C IV resonance doublet. This change in the relative strengths of Si IV and C IV in absorption and emission will hereafter be referred to as the reversal of intensities.

Figure 1 illustrates the spectra of U Cep outside of (1a) and during eclipse (1b).

2.2. Systems which are not typical Algol systems should be dealt with separately from those in 2.1 because they do not seem to follow a homogeneous pattern. Hence, the interpretation may be different in each case. The information at our disposal refers to four systems.

2.2.1. FF Aquarii (9^m3, $A_1 = 0^m4$, $A_2 = -$; sd + G8 III—IV; Algol-type light curve)* appears to behave, according to Dorren *et al.* (1982), in a way similar to the systems discussed in 2.1, that is,

(a) at primary eclipse, the IUE spectrum is an emission spectrum where the resonance lines of C IV are very much stronger than Si IV and slightly stronger than N V;

(b) outside of eclipse, the IUE spectrum is an absorption spectrum where C IV

* In the parenthesis we give the apparent magnitude at maximum, the depths of the eclipses, the spectral types of the component, and the type of light curve.

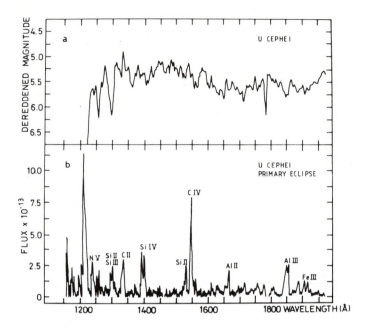

Fig. 1.   IUE spectra of U Cep: (a) outside of eclipse; (b) during eclipse, (from Plavec, 1983).

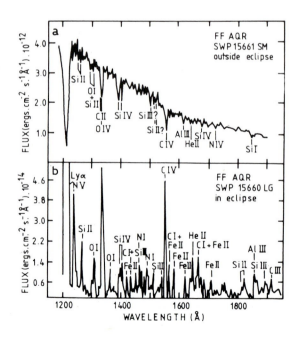

Fig. 2.   IUE spectra of FF Aqr: (a) outside of eclipse; (b) during eclipse, (from Dorren *et al.*, 1982).

is perhaps as strong as Si IV. In the Dorren *et al.* (1982) interpretation, the emission spectrum at eclipse is the spectrum of the G8 III—IV component.

2.2.2. R Arae (6^{m}0, A$_1$ = 0^{m}7, A$_2$ = 0^{m}2; B9 + ... ; Algol-type light curve) is characterized by variable emission. In 1980, the profiles of the high temperature resonance lines were of the P Cyg-type, while in 1982 the emission part of the profiles was not detectable (McCluskey, 1982; McCluskey and Kondo, 1983; Kondo *et al.*, 1985). McCluskey and Kondo (1983) detected a gas stream in Mg II and Si IV approaching the observer at −450 to −500 km s^{-1} near phase 0.4. This stream has not been seen at other phases. The ultraviolet continuum varied outside of eclipse by a factor of 2 in 10 days and by over 50% in one orbital cycle. These variations strongly indicate the presence of variable, optically thick material surrounding the primary star, and perhaps the entire system.

2.2.3. V356 Sagittarii (7^{m}0, A$_1$ = 0^{m}7, A$_2$ = −; A2 V + B3 V; Algol-type light curve), at eclipse, displays a spectrum in which the resonance lines of N V are, by far, the strongest high temperature features, Si IV is prominent and C IV appears as a much weaker feature. The observations of Plavec (1983b) and Plavec *et al.* (1984), suggest that in V356 Sgr, the relative intensities of these lines in the absorption spectrum, i.e., observed outside eclipse, are similar to those in the emission spectrum.

2.2.4. RY Persei (8^{m}6, A$_1$ = 1^{m}7, A$_2$ = 0^{m}1; B3 V + F0 III) appears to behave in a way similar to V356 Sgr (cf. Plavec, 1984). In both systems, the absorption spectrum secured at principal eclipse shows no reversal in the relative intensities of the high temperature resonance lines, and N V is, by far, the strongest high temperature resonance feature present.

2.3. For systems with one component of spectral type B0 or earlier, the high temperature resonance lines nearly always appear in emission throughout the orbital cycle. The same is true for the systems in Table II that have orbital periods over, say, 11—12 days, — certainly, over 30 days, — and also for systems with late supergiants, which are, of course, rather long period systems. The latter systems are discussed elsewhere in this volume.

Whenever the observations of these emission line objects have been made in the high dispersion mode, the line profiles are of the P Cyg-type. In those cases where emission is not clearly present, it appears that the absorptions are shortward-shifted, as though we are actually dealing with a sort of P Cyg profile where the emission does not reach above the level of the continuum. A good example of this assertion is provided by GG Carinae, a 31 day period peculiar binary where Brandi *et al.* (1986) have found that the emission of Mg II, O I (multiplet 2) and of several multiplets of Fe II have P Cyg profiles, while the resonance lines of Si IV and of C IV appear in absorption, shortward-shifted by 170 km s^{-1}.

2.4. The W Ursae Majoris systems display the high temperature resonance lines in emission, a conclusion that is derived from low resolution spectra. C IV is always the strongest of the high temperature resonance lines present in their spectra.

Two IUE high dispersion images taken with the long wave length camera definitely show, however, a P Cyg profile in Mg II, and this fact appears to be a sufficiently strong argument to think that the emissions, in the range of the SWP camera, may actually be P Cyg profiles. If this should be the case, the high resolution observations which could be made with the Space Telescope could provide information on outflow velocities in the W UMa systems.

2.5. The high temperature resonance lines require for their formation, temperatures that cannot be accounted for by the stellar radiation field and, therefore, their presence in the IUE spectra implies the existence of non-thermal sources. Almost fifteen years ago, Smak (1972) and Popper (1972) already suggested the existence of non-thermal energy sources in interacting binaries on the basis of the characteristics of the gaseous rings around the early type components in a number of systems. Moreover, radio observations of β Persei were explained by postulating the presence of non-thermal sources in the system (Woodsworth and Hughes, 1976) and similar explanations had to be proposed for the radio observations of UX Arietis (cf. Hjellming, 1976) and for the other RS Canum Venaticorum systems. The detection of X-rays from systems with non-compact components also points to the same conclusion.

Apparently, there may be more than one region where the high temperature resonance lines originate. This conclusion is suggested by the following facts:

(a) lines of ions like Si IV and C IV display large differences in profile (Figure 3);

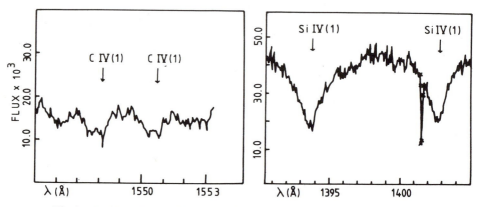

Fig. 3. Profiles of Si IV and C IV in γ_1 Vel (from Sahade and Hernández, 1984).

(b) in systems like V Puppis, UW Canis Majoris, V861 Scorpii and HD 47129, and γ_2 Velorum, the IUE spectra also display very sharp high temperature resonance lines that are very similar in profile to the lines arising in the interstellar medium and must originate in the very outer layers of the extended gaseous envelopes;

(c) some IUE spectra of interacting binaries display more than one component of the high temperature resonance lines, as is the case in AO Cassiopeiae (McCluskey and Kondo, 1981) and in V373 Cassiopeiae (McCluskey and Kondo private communication).

(d) the reversal in relative intensity between the high temperature resonance lines in spectra taken at principal eclipse and those taken outside of eclipse in the case of the Algol systems mentioned above.

Some investigators prefer to interpret feature (a) in terms of an underabundance of elements, but such an interpretation would seem to be ruled out by feature (d).

2.6. Now the question arises as to which is the origin and location of the regions where the high temperature resonance lines are formed. Sahade (1986a) has recently summarized the ideas that have been put forward by different authors and we are following him in what follows.

This problem was first discussed by Kondo et al. (1979) who, from their study of the U Cep system with the IUE, suggested that "the most probable place of production of Si IV and C IV is in the region of a hot spot or spots on the B star". They were followed by Plavec (1980b) who thought that the high temperature resonance lines were associated with accretion. Subsequently, Kondo et al. (1981) in a further investigation of U Cep were led to conclude that "Si IV and C IV originate in a pseudo-photosphere created by the material of the gaseous stream falling upon the surface of the B star".

Sahade and Ferrer (1982) published an investigation of the Algol-type system AU Monocerotis and concluded that in this system

(a) there are two regions of high T_e;

(b) N V and Si IV probably originate from the dissipation of shock waves produced when the material that is lost to the sytem through the region containing L_3, interacts with the envelope that surrounds the whole systems;

(c) C IV probably originates in the high temperature region that arises from the interaction between the gaseous stream from the late type subgiant component and the gaseous ring that surrounds the brighter component of the system. A similar conclusion was reached by Sahade and Hernández (1984) in the case of the non-eclipsing spectroscopic binary γ_1 Velorum.

The high temperature resonance lines of N V, Si IV, and C IV were first observed in emission in a binary too 'cool' to show these lines, namely, in β Lyr (Kondo and McCluskey, 1974; Hack et al., 1975). It was already clear from OAO—2 results (Kondo et al., 1971, 1972) that an optically thick plasma dominated the spectrum of β Lyr below $\lambda 2300$.

The existence of two high temperature regions in interacting binaries was first proposed in a study of β Lyr with the TD—1A satellite (Hack et al., 1976), and seems to be confirmed by an investigation of IUE spectra of β Per both from the points of view of the line intensities and of the radial velocities. In this case, Sahade and Hernández (1985) concluded that C IV arises in a region dynamically linked with the B8 V component while N V and Si IV form somewhere in the gaseous envelope that surrounds the whole system. To confirm such an important conclusion, additional IUE observations of β Per have recently been secured at the critical phases, but the results are still not available.

Further suggestions regarding the origin and locations of the high temperature regions were made by Polidan and Peters (1982), Plavec (1983a), Peters and Polidan (1984) and McCluskey and Kondo (1984).

Polidan and Peters (1982) note the presence of 'variable high ionization features' due to N v, Si IV, and C IV in the spectra of TX Ursae Majoris, U Coronae Borealis, CX Draconis and AU Mon.

Plavec (1983a) in a study of U Cep, suggested that "N v, Si IV and C IV arise in a 'hot' layer that probably surrounds the whole star around the equator" and added that "we do not know the mechanism that gives origin to the layer".

On the other hand, Peters and Polidan (1984), who have investigated several Algol systems, postulated the existence of 'high temperature accretion regions' located on the following side of the primary star in the system, where the lines we are discussing would form.

McCluskey and Kondo (1984a) have studied U Sge at high dispersion at numerous phases outside of totality. They found that U Sge is more active than was generally believed and relate the high temperature resonance lines to a 'pseudo-photosphere' surrounding the entire B-star and close to its surface. This material is more concentrated on the following hemisphere of the B-star. McCluskey and Kondo (private communication) found very similar results for TX UMa.

2.7. The question which arises next concerns explaining the fact that the high temperature resonance lines appear either in absorption or in emission and also explaining the additional fact of the reversal of intensities in relatively short-period Algols.

2.7.1. The behavior of the relatively short-period Algols, in regard to the appearance of spectrum outside of eclipse and at principal eclipse, as Sahade (1986a) has pointed out, is reminiscent of the behavior of the Hα emission in the photographic spectra of the same systems (cf. Sahade and Wood, 1978). This suggests that the explanation for the behavior described in 2.1 might be developed in terms of the dimensions and density of the gaseous formation around the brighter components of these systems.

As for the reversal of intensities, we may mention two possible explanations:

(a) different locations for the sources which give rise to the spectra at eclipse and outside of eclipse (the circumbinary envelope and the circumstellar envelope around the early type component, respectively);

(b) emission, strongest in C IV, reaching a level below the continuum and affecting the absorption lines outside of eclipse, i.e., partially filling in the absorption.

The right answer to the queries posed in 2.6 and 2.7 will remain pending until we are able to have much denser ultraviolet phase coverage than is currently available with high resolution spectra throughout the eclipse phases.

In interacting binaries of the Algol-type and similar configurations, the larger component probably cannot develop a normal atmospheric structure (Sahade, 1986b) and all of the phenomena that are observed are the result of interactions between the gaseous outflow and the early-type component and its surroundings.

It should be noted that a number of Algol binaries have been found to have X-ray luminosities of $10^{30-31.3}$ erg s^{-1} (White and Marshall, 1983; McCluskey and Kondo, 1984b) indicating the existence of coronal regions. It is not yet clear whether the X-rays originate in the vicinity of the cool star, the hot star or both.

2.7.2. The behavior of the systems discussed in 2.3 are most likely traced to the fact that radiation pressure [Schuerman (1972), Kondo and McCluskey (1976), Kondo *et al.* (1976), Vanbeveren (1977), Zorec and Niemela (1980a, b)] and/or strong matter outflow disrupt the "classical" Roche lobe configurations and give rise to sizeable expanding gaseous envelopes that surround the systems.

In very long-period binaries with no strong mass outflow, like ε Aur, which are neither W UMa nor semidetached systems and have no measurable winds, a component such as the FO Iab star in ε Aur, should have a normally structured atmosphere with transition and coronal regions, the former being responsible, at least in part, for the high temperature resonance lines observed in the spectrum.

2.8. Since the IUE material on the W UMa systems were essentially all secured in the low dispersion mode and since this group seems to have its own well-defined characteristics, we present a longer discussion on the present state of affairs.

2.8.1. The W UMa binaries have been the subject of intensive theoretical and observational effort in recent years. Recent reviews of our current understanding of these binaries are found in Shu (1980), Vilhu (1981), Rahumen and Vilhu (1982), Rucinski *et al.* (1982), Mochnacki (1981, 1985) and Rucinski (1985a).

No completely satisfactory theory of the internal, subsurface and atmosphere layers exists. All models are based on assumptions concerning the Roche equipotentials which are supposed to coincide in dimensions with the stars involved. It must be pointed out that if magnetic forces are significant, the Roche model could be invalid. At present, however, we know little about magnetic fields in these systems. Moreover, it is not certain that the components of W UMa systems fill their Roche lobes. The near equality of the temperatures of the two components regardless of the mass ratio has led to the idea of a common envelope enveloping both stars.

The IUE has the capability of examining the chromospheric and transition regions which are nearly ubiquitous among stars of spectral type F or later. In W UMa systems, the transition regions are characterized particularly by the presence of emission lines of C IV, N V, and Si IV. Figure 4 illustrates the ultraviolet spectra of several W UMa systems. Consequently, the study of IUE spectra of the W UMa systems can provide important diagnostics concerning their common envelope and any putative transition region. These studies are particularly valuable since there is no guarantee that the transition region of a W UMa binary is similar to that of a single star of the same spectral type or even that such a region exists.

2.8.2. The W UMa systems have been divided into two subclasses; the A-type and the W-type. The A-type systems have primaries with spectral types earlier than F9, mass ratios of 0.08–0.54 and relatively stable periods and light curves in many cases. The W-type systems have primary spectral classes later than F8, mass ratios

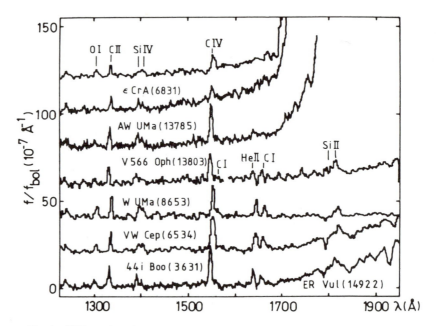

Fig. 4.   IUE spectra of several W UMa systems (from Rucinski and Vilhu, 1983).

of 0.33—0.88 and highly variable orbital periods and light curves. There are weak observational indications that the W-types have much more convective activity at or near their surfaces than do the A-types. This is expected from standard stellar structure theory. Star spots may play a role in the irregular activity of the W-type systems.

2.8.3. Eaton (1983) reported on low dispersion far ultraviolet IUE spectra of fourteen W UMa systems. All systems, except the A2 V binary GK Cephei and possibly the A8 V system AG Virginis, showed chromospheric and transition region absorption lines. The systems V535 Arae (A8—9) and S Antliae (A9) showed activity. Generally, the chromospheric lines of O I, C I, Si II, C II, and Mg II, and the transition region lines are seen in all active stars. It would seem that activity occurs for all stars of spectral type A9 or later, but not as early as A2. A greater sample of A-type (spectral) systems must be observed to verify and better define this 'cut-off' point. For single stars, the cut-off point for activity is F1—F2. This might imply that convective activity extends to slightly earlier spectral classes in W UMa binaries than in single stars. Usually, the tidally synchronous W UMa stars are rotating more rapidly than single stars of the same type and this makes comparisons difficult. It appears that the transition zone emission in W UMa stars is greater than in the RS CVn stars while the opposite is true of coronal activity. The ratio of chromospheric to transition region activity seems to be the same for both W-type and A-type systems.

Rucinski *et al.* (1984) observed the two extremes among W UMa binaries with IUE. These are AW Ursae Majoris, an extreme A-type with a mass ratio of

0.07—0.08, spectral type F0 + F2 and a very stable light curve, and SW Lacertae, an extreme W-type with mass ratio 0.88, spectral type G6 + G8 and a very unstable light curve. Here, extreme refers to mass ratio. These systems show similar N v, C IV, and Si IV emission indicating similar transition regions. Little or no phase dependence is seen in AW UMa. An insufficient number of spectra were obtained for SW Lac to determine the presence or absence of phase dependence. A large difference in the Mg II doublet was observed with strong emission in SW Lac and no emission in AW UMa. The authors interpret this to mean that the chromospheres of these two systems are quite different.

Rucinski (1985b) has discussed IUE spectra of Mg II for fifteen systems. He finds that the emission is phase independent and that it does not appear to depend on orbital period. It does increase with decreasing temperature. The late G-type and K-type systems follow a rough extrapolation of moderately rotating single G—K stars to higher rotation rates. Large uncertainties exist for the F-type and G-type stars and insufficient data is available to make any significant statements at the present time. High resolution spectra are badly needed.

Oranje (1986) measured ultraviolet fluxes from IUE spectra for one hundred thirty two stars including seven W UMa systems. The transition region lines behave similarly for all stars. Only Mg II and Si II showed fluxes different for W UMa systems than for other stars in that they were significantly weaker.

Rucinski (1985a) has reviewed the IUE and other observations and theory pertinent to W UMa systems. He finds that these binaries are the most active stars with respect to chromospheric emission. This emission does not depend significantly on orbital period or spectral type except for spectral types earlier than A8 or A9 where it rapidly weakens as effective temperature increases. He suggests that plasma can escape from the earlier type systems because no magnetic loops are present to confine it, as is the case for later types. The coronal activity of W UMa systems is weaker than the activity of single stars of the same spectral type. The F-types are much weaker than the G-types and the G-types are weaker than the K-types. The transition regions of W UMa stars are very similar for all except for those earlier than A8—A9.

If, as is likely, magnetic fields play a role in the evolution of W UMa systems, their activity might be related to magnetic braking (Rucinski, 1985a). Star spots have been invoked by many authors to explain various photometric and spectroscopic peculiarities, but the spots are required to be on the primary and it is not at all clear why this should be the case. The fact is that we do not understand convection, magnetic activity and energy transfer in W UMa systems.

One can draw several conclusions from the IUE observations of the W UMa binaries. A transition-like region exists and produces emission lines of C IV, N V, and Si IV. Generally, C IV is dominant. Little or no phase dependence of the relative line fluxes is seen, indicating that the emitting gas is relatively evenly spread around both stars. It would appear that a circumbinary transition-like region exists and that for all W UMa systems later than A8—A9, its properties are quite similar regardless of orbital period or mass ratio. Such a phenomenon may well be related to the unknown cause of the nearly equal temperatures of the two components of any W UMa system.

Relative to the bolometric flux of the stars, the emission line flux is stronger than in single stars. This may be related to the lower surface gravities, to the rapid rotation of the binary components, to complex magnetic interactions, etc.

Cruddace and Dupree (1984) surveyed seventeen W UMa systems observed with the Einstein Observatory. All nine of the W-type systems and five of the eight A-type systems were detected with X-ray luminosities of $8 \times 10^{28} - 2 \times 10^{30}$ ergs s^{-1}. These luminosities are typically an order of magnitude lower than those of the RS CVn stars. The X-ray spectra appear to be similar in these two types of binaries. Generally, the coronal activity of the W UMa systems is somewhat lower than expected by extrapolation from the longer period RS CVn systems and single stars. The A-type systems are particularly weak.

Radio detection of several Algol-type binaries and RS CVn binaries has been accomplished. Detection of W UMa binaries, particularly with the very high resolutions now or soon to be available could accomplish much in determining the structure of the transition and coronal regions. High spatial resolution and radio polarization studies could be very fruitful.

To improve our knowledge, we require high resolution spectroscopy with very good time resolution with respect to the orbital period.

3. Discovery of Evidence for Winds and Velocity Gradients in Close Binaries

3.1. The presence of winds in interacting binaries was not a straightforward outcome of the study of photographic spectra, except in the case of β Lyr, and a few others, where the diluted lines of He I (cf. Sahade et al., 1959) suggested an outward velocity of some 170 km s^{-1}. In other cases, the existence of large envelopes which surround the entire system, and have been known since the beginning of the "Struve revolution", merely implied the existence of winds but velocity information was not provided by the available observations in the photographic region. The presence of P Cyg profiles in the IUE spectra of a number of interacting binaries made up for such a lack of information. In this connection, it is worth mentioning that, in the case of β Lyr, the expansion velocities yielded by the diluted lines of He I in the photographic region coincide with the velocities suggested by the P Cyg profiles in the UV spectrum. Table II lists the relevant information; unfortunately, velocity measurements, have not been published in most cases.

3.2. Velocity gradients are indicated by the asymmetric absorption lines that are displayed by the IUE spectrum of binaries at certain phases in the orbital cycle. This statement is illustrated, for instance, by AU Mon, U Sge and TX UMa, among others, which display a shortward asymmetry in the resonance lines of Si IV. Figure 5 illustrates a shortward asymmetry in AU Mon.

4. Discovery of Intercombination Lines in Interacting Binaries

The IUE has also disclosed information regarding the presence of intercombination lines in the spectra of three objects in Table II, and has thus contributed to a

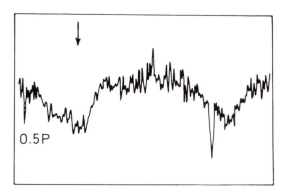

Fig. 5. Profile of Si IV in AU Mon at phase 0.5 P. The arrow indicated the rest wavelength of the Si IV 1393 Å line (from Sahade and Ferrer, 1982).

better knowledge of the structure of the gaseous envelope in close binaries. The densities implied should be of the order of some 10^{10} cm^{-3}.

UM CMa was found to display N IV] 1486 Å; FF Aqr, Si III] 1892 Å and C III] 1909 Å (Dorren *et al.*, 1982).

In the case of the famous system β Lyr, the IUE disclosed the presence of Si III] at 1892 Å (Plavec, 1980b), while observations made with the BUSS project, added the doublet of N II] at about 2140 Å (Hack *et al.*, 1983).

5. Discovery of New Phenomena

Here we report on the extra lines discovered in the IUE spectrum of γ_1 Vel by Sahade and Hernández (1984) at phase 0.376 P. The extra lines are very sharp, are displaced by some $+200$ km s^{-1} and arise from 0.00 eV, and perhaps from 0.01 eV, energy levels, as it is illustrated in Figure 6.

Unfortunately, only four IUE images were taken of the star and only one of the SWP images displays the extra lines. Thus, it is not possible to attempt an explanation at the present time. We most certainly need phase coverage throughout the orbital cycle to ascertain the phase interval or intervals at which this feature is present and to ascertain as to whether or not we are dealing with a property which is permanently or sporadically present.

McCluskey and Kondo (1980) found that in the IUE spectrum of δ Pictoris, sharp lines of Si II, Si III, S II, O I, C II, and Mg II are present with a mean velocity of -41 km s^{-1}. The interstellar lines have a velocity of $+9$ km s^{-1}. They attributed these negative velocities to an expanding circumbinary shell.

One of the most unusual findings with the IUE was the discovery by Koch *et al.* (1986) that the numerous emission lines in the ultraviolet spectrum of the nova-like binary V Sagittae all show an average radial velocity of $+700$ km s^{-1}. This velocity has persisted for at least 5 yr. No trace of absorption is detectable on IUE low resolution spectra. Even at low resolution, if sufficient P Cyg absorption were present to give the emission lines such a large positive velocity, then this absorption should be detectable. It is difficult to imagine that a decaying,

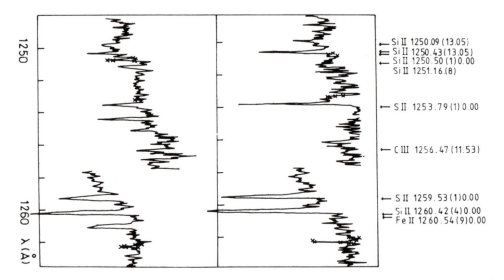

Fig. 6.   Extra sharp longward-displaced absorption lines in the spectrum of γ_1 Vel at phase 0.376 P
(lower panel) (from Sahade and Hernández, 1984, illustration unpublished).

collapsing disk could give rise to such a phenomenon for a period of years. Koch
et al. (1986) suggest that the emission lines might be forming sufficiently close
to a neutron star ($\approx$ 200 radii) to show a gravitational redshift equivalent to
700 km s^{-1}. Higher resolution observations of V Sge will be made with the Hubble
Space Telescope which should help to unravel this puzzling mystery.

6. Conclusions

We have discussed the scientific accomplishments of the IUE in the field of
'interacting binaries' as defined taking into account the assignments to the authors
of other papers of Part III of this book. The conclusions that stem from our
discussion appear to be
(a) every interacting binary displays the high temperature resonance lines either
in absorption and/or in emission;
(b) our objects can be grouped into three main broad categories, namely,
(i) *'windless' binaries or binaries with small or undetectable mass outflow
velocities*;
(ii) *binaries with important mass outflow*; and
(iii) *W UMa-binaries.*
In group (i), we have the Algol-type and similar objects, that is, the systems
discussed under 2.1, while group (ii) is represented by the systems dealt with in
2.3.
As a consequence of all this, and of the discussion that we have carried *ut
supra*, it would seem that it is inappropriate to try to set up a classification scheme
for interacting binaries based merely on whether the ultraviolet spectra are
characterized by emission or by absorption lines, as Sahade (1986b) has already
pointed out.

Another conclusion refers to the urgent need to carry out high dispersion observations with good phase coverage, extending the IUE domain in wavelength to a shorter wavelength range, and attaining good phase coverage. This is a must if we wish to make progress in our understanding of interacting binaries and answer the important questions that are pending and were brought to the limelight by the IUE.

Acknowledgements

We are indebted to Ms. Margarita Trotz and Mr. Guillermo Sierra for the illustrations and to Dr. Carolina Salas McCluskey for her assistance in the preparation of this manuscript.

References

Ake, T. B., Parsons, S. B., and Kondo, Y.: 1985, *Astrophys. J.* **298**, 772.
Boehm, C., Ferluga, S., and Hack, M.: 1984, *Astron. Astrophys.* **130**, 419.
Brandi, E., Gosset, E., and Swings, J. P.: 1986, private communication.
Cruddace, R. G. and Dupree, A. K.: 1984, *Astrophys. J.* **277**, 263.
Dorren, J. D., Guinan, E. F., and Sion, E. M.: 1982, in Y. Kondo, J. M. Mead and R. D. Chapman (eds.), *Advances in Ultraviolet Astronomy*, NASA CP-2238, p. 517.
Eaton, J. A.: 1983, *Astrophys. J.* **268**, 800.
Hack, M., Hutchings, J. B., Kondo, Y., McCluskey, G. E. Jr., Plavec, M., and Polidan, R. S.: 1975, *Astrophys. J.* **198**, 453.
Hack, M., Sahade, J., de Jager, C., and Kondo, Y.: 1983, *Astron. Astrophys.* **126**, 115.
Hack, M., van den Heuvel, E. P. J., Hoekstra, R., de Jager, C., and Sahade, J.: 1976, *Astron. Astrophys.* **50**, 335.
Hjellming, R. M.: 1976, in G. Setti (ed.), *The Physics of Non-thermal Radio Sources*, D. Reidel Publ. Co., Dordrecht, Holland, p. 203.
Koch, R. H., Corcoran, M. F., Holenstein, B. D., and McCluskey, G. E. Jr.: 1986, *Astrophys. J.* (in press).
Kondo, Y. and McCluskey, G. E. Jr.: 1974, *Astrophys. J.* **188**, L63.
Kondo, Y. and McCluskey, G. E. Jr.: 1976, in P. Eggleton, S. Mitton, and J. Whelan (eds.), 'Structure and Evolution of Close Binary Systems', *IAU Symp.* **73**, 277.
Kondo, Y., McCluskey, G. E. Jr. and Houck, T. E.: 1971, in L. Detre (ed.), *Proc. IAU Colloq. No. 15*, Budapest, Academic , p. 308.
Kondo, Y., McCluskey, G. E. Jr., and Houck, T. E.: 1972, in A. D. Code (ed.), *The Scientific Results from the Orbiting Astronomical Observatory (OAO—2)*, NASA SP-310, p. 485.
Kondo, Y., McCluskey, G. E. Jr., and Gulden, S. L.: 1976, *Proc. Symp. on X-ray Binaries*, NASA SP-389, p. 531.
Kondo, Y., McCluskey, G. E. Jr., and Stencel, R. E.: 1979, *Astrophys. J.* **233**, 906.
Kondo, Y., McCluskey, G. E. Jr., Rahe, J., Wolfschmidt, G. and Wu, C.-C.: 1978, in M. Hack (ed.), *Proc. 4th Internat. Colloq. Astrophys.* (Trieste), p. 486.
Kondo, Y., McCluskey, G. E. Jr., and Feibelman, W. A.: 1980, *Publ. Astron. Soc. Pacific* **92**, 688.
Kondo, Y., McCluskey, G. E. Jr., and Stencel, R. E.: 1980, in M. J. Plavec, D. M. Popper, and R. K. Ulrich (eds.), 'Close Binaries: Observations and Interpretation', *IAU Symp.* **88**, 237.
Kondo, Y., McCluskey, G. E. Jr., and Harvel, C. A.: 1981, *Astrophys. J.* **247**, 202.
Kondo, Y., McCluskey, G. E. Jr., and Parsons, S. R.: 1985, *Astrophys. J.* **295**, 580.
McCluskey, G. E. Jr.: 1982, in Y. Kondo, J. M. Mead, and R. D. Chapman (eds.), *Advances in Ultraviolet Astronomy*, NASA CP-2238, p. 102.
McCluskey, G. E. Jr. and Kondo, Y.: 1981, *Astrophys. J.* **246**, 464.
McCluskey, G. E. Jr. and Kondo, Y.: 1983, *Astrophys. J.* **266**, 755.

McCluskey, G. E. Jr. and Kondo, Y.: 1984a, in J. M. Mead, R. D. Chapman, and Y. Kondo (eds.), *Future of Ultraviolet Astronomy Based on Six Years of IUE Research*, NASA CP-2349, p. 382.

McCluskey, G. E. Jr. and Kondo, Y.: 1984b, *Publ. Astron. Soc. Pacific* **96**, 817.

Mochnacki, S. W.: 1981, *Astrophys. J.* **245**, 650.

Mochnacki, S. W.: 1985, 'Observational Evidence for the Evolution of Contact Binary Stars', in P. P. Eggleton and J. E. Pringle (eds.), *Interacting Binaries*, pp. 51—82.

Oranje, B. J.: 1986, *Astron. Astrophys.* **154**, 185.

Parsons, S. B.: 1981, *Astrophys. J.* **247**, 560.

Parsons, S. B.: 1983, *Astrophys. J. Suppl.* **53**, 553.

Parsons, S. B., Holm, A. V., and Kondo, Y.: 1983, *Astrophys. J.* **264**, L19.

Peters, G. J. and Polidan, R. S.: 1984, *Astrophys. J.* **283**, 745.

Plavec, M. J.: 1980a, in R. D. Chapman (ed.), *The Universe at Ultraviolet Wavelength*, NASA CP-2171, p. 397.

Plavec, M. J.: 1980b, in M. J. Plavec, D. M. Popper, and R. K. Ulrich (eds.), 'Close Binaries: Observations and Interpretations', *IAU Symp.* **88**, 251.

Plavec, M. J.: 1983a, *Astrophys. J.* **275**, 251.

Plavec, M. J.: 1983b, *J. Roy. Astron. Soc. Canada* **77**, 283.

Plavec, M. J. and Dobias, J. J.: 1983, *Astrophys. J.* **272**, 206.

Plavec, M. J., Dobias, J. J., Etzel, P. B., and Weiland, J. L.: 1984, in J. M. Mead, R. D. Chapman and Y. Kondo (eds.), *Future of UV Astronomy Based on Six Years of IUE Research*, NASA CP-2349, p. 420.

Polidan, R. S. and Peters, G. J.: 1982, in Y. Kondo, J. M. Mead, and R. D. Chapman (eds.), *Advances in Ultraviolet Astronomy*, NASA CP-2238, p. 534.

Popper, D. M.: 1972, in A. H. Batten (ed.), 'Extended Atmospheres and Circumstellar Matter in Spectroscopic Binaries', *IAU Symp.* **51**, p. 58.

Rahumen, T. and Vilhu, O.: 1982, 'Origin and Evolution of Contact Binaries of W UMa Type', in Z. Kopal and J. Rahe (eds.), *Binary and Multiple Stars as Tracers of Stellar Evolution*, D. Reidel Publ. Co., Dordrecht, Holland, p. 289.

Rucinski, S. M.: 1985a, in J. E. Pringle and R. A. Wade (eds.), *Interacting Binary Stars*, Cambridge: Cambridge University Press, p. 13.

Rucinski, S. M.: 1985b, *Monthly Notices Roy. Astron. Soc.* **215**, 615.

Rucinski, S. M., Brunt, C. C., and Pringle, J. E.: 1984, *Monthly Notices Roy. Astron. Soc.* **208**, 309.

Rucinski, S. M., Pringle, J. E., and Whelan, J. A. J.: 1982, in Z. Kopal and J. Rahe (eds.), *Binary and Multiple Stars as Tracers of Stellar Evolution*, D. Reidel Publ. Co., Dordrecht, Holland, p. 309.

Rucinski, S. M. and Vilhu, O.: 1983, *Monthly Notices Roy. Astron. Soc.* **202**, 1221.

Sahade, J.: 1986a, *Bol. Asoc. Argentina Astron.*, No. 31 (in press).

Sahade, J.: 1986b, in J. P. Swings (ed.), *Highlights of Astronomy*, D. Reidel Publ. Co., Dordrecht, Holland. (in press.)

Sahade, J. and Ferrer, O. E.: 1982, *Publ. Astron. Soc. Pacific* **94**, 113.

Sahade, J. and Hernández, C. A.: 1984, *Publ. Astron. Soc. Pacific* **96**, 88.

Sahade, J. and Hernández, C. A.: 1985, *Rev. Mexicana Astron. Astrophys.* **10**, 257.

Sahade, J., Brandi, E., and Fontenla, J. M.: 1984, *Astron. Astrophys. Suppl. Series* **56**, 17.

Sahade, J., Huang, S.-S., Struve, O., and Zebergs, V.: 1959, *Trans. Amer. Phil. Soc.* **49**, 1.

Sahade, J. and Wood, F. B.: 1978, *Interacting Binary Stars*, Pergamon Press, p. 45.

Schuerman, D. W.: 1972, *Astrophys. Space Sci.* **19**, 351.

Shu, F. M.: 1980, in M. J. Plavec, D. M. Popper, and R. K. Ulrich (eds.), 'Close Binaries: Observations and Interpretations', *IAU Symp.* **88**, 477.

Smak, J.: 1972, in A. H. Batten (ed.), 'Extended Atmospheres and Circumstellar Matter in Spectroscopic Binaries', *IAU Symp.* **51**, 57.

Vanbeveren, D.: 1977, *Astron. Astrophys.* **54**, 877.

Vilhu, O.: 1981, *Astrophys. Space Sci.* **78**, 401.

White, N. E. and Marshall, F. E.: 1983, *Astrophys. J.* **268**, L117.

Woodsworth, A. W. and Hughes, V. A.: 1976, *Monthly Notices Roy. Astron. Soc.* **175**, 177.

Zorec, J. and Niemela, V.: 1980a, *Comptes Rendus Acad. Sci. Paris, Ser. B.* **290**, 67.

Zorec, J. and Niemela, V.: 1980b, *Comptes Rendus Acad. Sci. Paris, Ser. B.* **290**, 95.

ATMOSPHERIC ECLIPSES BY SUPERGIANTS
IN BINARY SYSTEMS

MARGHERITA HACK

Dept. of Astronomy, University of Trieste, Italy

and

DAVID STICKLAND

Rutherford Appleton Laboratory, Chilton, U.K.

1. Introduction

The systems that we shall consider primarily in this paper consist of a late-type cool supergiant or bright giant and a hot main sequence companion which periodically passes behind the larger star and shines through its extended atmosphere. This gives us the opportunity of observing absorption lines along various sightlines through the cool star's atmosphere and hence permits us to study stratification effects and physical properties in these tenuous outer layers of the star.

Three classes of such stars have been studied extensively with IUE:

(a) The Zeta Aurigae group consisting of Zeta Aur itself, 31 Cyg and 32 Cyg, all composed of a K-type supergiant and a B-type main sequence companion.

(b) The VV Cephei group where the cool star is an M-type supergiant.

(c) Epsilon Aurigae, the F0 Ia supergiant which is unique among all binaries. It has a companion whose nature was largely unknown before the 1982—84 eclipse; what we observe is not the eclipse of the companion by the cool star as in (a) and (b), but the eclipse of the supergiant by its mysterious companion. In this case, therefore, we are using the F-type star as a probe of the structure of the companion.

During the lifetime of IUE thus far, we have had the good fortune to be able to observe not only the eclipses of all three members of group (a), but also the egress from eclipse of VV Cephei (20.3 yr period) and the full eclipse of Epsilon Aurigae (27.1 yr period). Moreover, two new members of the class — 22 Vul (G3Ib—II + B9) and Tau Persei (G5III + A2V) — have been discovered, and two other systems very similar to VV Cep have been observed in the far UV, namely, Delta Sge (M2II + B9V) and Boss 1985 (M2epIab + B).

Ultraviolet observations of these systems were originally planned with the objectives of obtaining a better determination of the spectral type of the companion and of studying the outer layers of the cool atmosphere and the circumstellar envelope. The latter was to be done by observing the ground level lines of the abundant ions which, with the exception of Ca I, Ca II, and Na I, all fall in the range of the spectrographs of IUE. The results obtained, however, have been much more rewarding than expected: the degree of interaction has been found much greater and the excitation much higher than previously suspected.

Y. Kondo (ed.), Scientific Accomplishments of the IUE, pp. 445—463
© 1987 *by D. Reidel Publishing Company.*

The results of optical region studies of atmospheric eclipses have been thoroughly reviewed by Wright (1970) and in the references given by him. However, to set the scene for a discussion of the new insights granted by use of IUE, it behoves us to describe briefly the ground-based spectroscopic results in order that we can be aware of the great benefits that IUE has bestowed in this area.

The common property of these eclipses (groups a and b) is that the resonance doublet of Ca II becomes enhanced several weeks before the geometrical eclipse starts and remains so several weeks after it has ended. A deep and sharp core appears, which is characteristic of the absorption in the outer atmosphere of the supergiant. This core often splits into two or more components with radial velocities clearly different from, and generally more negative than, the orbital one derived from the genuine photospheric lines (i.e., the excited lines of the neutral metals). Large differences are found between different eclipses and also between ingress and egress of the same eclipse.

VV Cep exhibits a more complicated scenario with changes in both emission and absorption features, suggesting the presence of a very extended envelope surrounding the whole system and perhaps an accretion disk around the hot companion (whose spectral class and luminosity are very poorly known).

Epsilon Aur presents us with a much stranger case. We have no evidence of a secondary eclipse, hence the contribution of the companion to the luminosity of the system in the optical region is negligible. The F0 Ia star is eclipsed every 27.1 yr with an average duration of 714 days including a span of 330 days at constant minimum brightness (Gyldenkerne, 1970) (notwithstanding some slight intrinsic fluctuations which are also observed out of eclipse and which are very common in such supergiants); it must be noted, however, that there are considerable differences in these timings from eclipse to eclipse, much more than can be attributed to the Cepheid-like variations.

The eclipse is most peculiar because:

(a) its amplitude is independent of wavelength in the spectral range accessible from the ground and is equal to 0.8 mag,

(b) during 'totality' we always see the spectrum of the F0 Ia star, and

(c) from the depth of the eclipse we might infer that the two stars have about the same magnitude, but only one spectrum is visible and no secondary eclipse is observed.

These peculiarities can be explained by assuming that the invisible companion is surrounded by an opaque disk or ring which is itself the occulting body and whose size must be very large to account for the duration of the eclipse. Furthermore, that the minimum amplitude is independent of wavelength can be explained by postulating that the disk is formed of dusty particles (whose dimensions must be much larger that those of normal interstellar dust in order to produce grey extinction) or by a rarefied ionized gas so that the opacity of the eclipsing body is due to electron scattering, or, perhaps, by some combination of the two.

The hypothesis that Thomson scattering might be responsible is suggested by the absorption line spectrum appearing before, and disappearing after, the geometrical eclipse and which, from its characteristics, we can call the 'shell' spectrum. It is rather similar to the photospheric spectrum of the F0 Ia star but the high excitation lines and those arising from non-metastable levels are much fainter or completely missing while the low level and metastable lines are strong. Additionally, the 'shell' lines are sharper than the photospheric ones and they are doppler-shifted with respect to the photospheric lines: at ingress by $20-30$ km s^{-1} to the red, and at egress by $30-50$ km s^{-1} to the blue. These traits tell us that the 'shell' spectrum is formed in a tenuous layer which eclipses the primary and rotates in the same sense as the orbital motion, so that at ingress it is the part of the disk rotating away from the observer which eclipses the supergiant while at egress it is the part rotating towards us which is responsible.

What could be the source of ionization of the 'shell'? Kraft (1954) suggested that it was the F0 Ia star itself, but certain characteristics of the 'shell' spectrum (the intensity of some of the non-metastable lines and also the degree of ionization) observed by Hack (1959) during the egress phase of the 1955—57 eclipse indicated that its radiation was insufficient. We had either to admit that the supergiant was very peculiar with a strong excess of (then unobservable) ultraviolet radiation, or, as Hack proposed in 1961, that a hot companion, embedded in the eclipsing body, was responsible for these 'shell' spectrum characteristics.

2. The IUE Observations

2.1. THE ZETA AURIGAE SYSTEMS

We shall discuss the typical Zeta Aur systems separately from those of the VV Cep group, since the members of the first group, at least in the optical region, have relatively simple spectra characterized by a G or K-type absorption spectrum and a B-type absorption spectrum. During the atmospheric phases of the eclipse, additional absorption lines, formed in the outer atmosphere of the cool star, appear. In the VV Cep systems, the situation is somewhat more confused: the M-type supergiant, and sometimes the hot companion, also exhibit their own emission spectra; then during eclipse, additional absorptions and emissions appear, making the interpretation very complicated.

The importance of studying the Zeta Aur binaries in the UV can be understood by considering that, for instance, in the case of the prototype itself, the ratio of fluxes F_K/F_B at $\lambda 4200$ Å is about 1, at $\lambda 3000$ Å is 0.1, at $\lambda 2000$ Å is 10^{-3} and at $\lambda 1000$ Å is 10^{-10}; similar values are found for 31 and 32 Cyg. Hence the ultraviolet is the appropriate region to investigate the absorption lines superimposed on the B-type spectrum by the atmosphere of the K-type star.

Most of the studies of Zeta Aurigae systems have been undertaken by two groups: one from the United States (Chapman, Kondo, McCluskey, Stencel, and collaborators) and one from Germany (Che, Hempe, Reimers, and Schröder). They have found previously unsuspected, highly ionized and extended regions surrounding the hot companion which is actually orbiting in the vast atmosphere of the cool supergiant.

The principal results obtained with IUE can be summarized as follows:

(a) UV observations imply a complex structure for the circumstellar envelope and suggest the presence of a shock front in the wind of the K-type star due to the proximity of the B-type star (Chapman, 1981), indicate clumpiness in the envelope (Reimers and Schröder, 1983; Faraggiana and Hack, 1980; Hack, 1981; Stencel *et al.*, 1979), and demonstrate the existence of transient high velocity clouds (Reimers and Schröder, 1983; Hempe, 1983),

(b) high excitation absorptions or P Cygni features, changing to pure emissions during totality, have been noted; their profiles change with orbital phase (Hagen *et al.*, 1980; Chapman, 1982; Stencel and Chapman, 1981),

(c) the presence of longward-shifted broad absorption components in the Mg II and C IV lines have been observed at some particular phases (Ahmad *et al.*, 1983),

(d) the mass-loss rate from the cool component has been derived with a method analogous to that employed by Deutsch for the visual binary Alpha Her (Che *et al.*, 1983),

(e) the expansion velocity of the late-type atmosphere has been measured; similar measurements using the Ca II H and K lines had been performed at ground-based observatories but they are disturbed by the strong photospheric absorptions, and finally,

(f) the opacity in the UV is higher than in the visual as indicated by a comparison of the eclipse depths (Hempe and Reimers, 1982; Schröder, 1983).

2.1.1. *Zeta Aurigae*

Two cycles of Zeta Aur have been witnessed with IUE, that of 1979—80 (Minimum I in December 1979 and Minimum II in August 1980; Chapman, 1981; Faraggiana and Hack, 1980; Ahmad *et al.*, 1983; Che *et al.*, 1983; Schröder, 1985) and that of 1982—83 (Minimum I in August 1982 and Minimum II in April 1983; Ahmad *et al.*, 1984). The principal phenomena observed with IUE are the following: multi-ionized resonance absorption lines, such as those of N V, C IV, and Si IV, are present in the spectrum; and when the B7 V companion is eclipsed, a large number of emission lines appears, matching almost perfectly the similarly large number of absorption lines produced by the atmosphere of the K4 II supergiant and seen in the B star's continuum at ingress and egress.

At first one might think that these emission features are the chromospheric lines observed in single late-type giants and which become visible when the overwhelming blue continuum of the companion is occulted. However, the fluxes of these lines are more than an order of magnitude stronger than those in single stars; moreover, late-type supergiants lack high excitation emission features (Linsky and Haisch, 1979; Ayres *et al.*, 1981; Simon *et al.*, 1982). Highly ionized, extended regions were a complete surprise before IUE observations of such systems. An open question is still whether these regions are Strömgren spheres induced in the supergiant chromosphere and wind, as suggested by Hjellming and Newell (1983), or the shock interface between the slower and denser wind from the cool star and the faster, thin wind from the hot one (Chapman, 1981; Ahmad *et al.*, 1983). This second hypothesis seems more capable of explaining certain details in the line profile variations with phase some months after the epoch of secondary eclipse which have been detected following both of the observed secondary minima in Zeta Aur and also in the corresponding events in 31 Cyg (Stencel *et al.*, 1984) and

32 Cyg (Ahmad *et al.*, 1984). They consist of the appearance of an absorption component, longward shifted by about 100 km s^{-1}, in the Mg II resonance doublet, which is not manifest before the hot star arrives directly in front of the cooler one (Figure 1). A similar pattern (broadening and deepening on the longward side) is also noted in the C IV line. These Mg II and C IV components can be attributed to

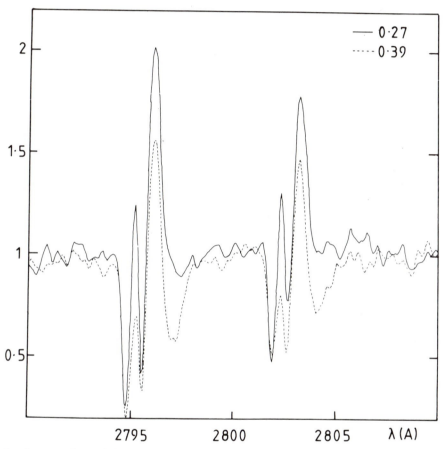

Fig. 1.   A comparison of the Mg II profiles in Zeta Aurigae close to secondary eclipse ($\phi = 0.27$) and at $\phi = 0.39$ when the absorption due to the wake can be seen.

absorption in an accretion column inside the shock wave, essentially a turbulent wake behind the B-type star; the line widths suggest a turbulence of up to 80 km s^{-1}. This absorption should be present as long as the line of sight to the B star passes inside the shock cone. However, since the absorption is stronger about 5 months after secondary minimum, it seems that the shock axis suffers an aberration effect because of the orbital motion of the B star and the motion of the wake (Figure 2). Specifically, the longward-shifted absorptions have been observed at phases 0.39 and 0.45 (secondary minimum occurs at phase 0.27), i.e., when the angle between the line joining the two components and the line of sight is 29° and

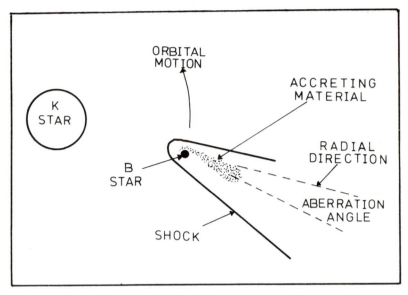

Fig. 2. A schematic diagram showing the shock zone around the B star and the wake
of accreting material.

45° (Ahmad *et al.*, 1983). Hence the shock cone has an aperture of at least 16°, and was observed during both eclipses with about the same aberration angle, around 35°. Similar values for the aberration angle and cone aperture have been found for 31 and 32 Cyg. Broad emission components of the C IV and Si IV lines have led Che-Bohnenstengel and Reimers (1986) to suggest that the B star in Zeta Aurigae (and also that in Delta Sge) may develop, from time to time, an accretion disk from the material being channelled towards it.

A comparison of the profiles of some of the highly ionized species' lines made by Hack (1981) at phases 0.91 and 0.27 shows that they are much stronger at secondary minimum than at phase 0.91. The effect is most marked for N V and is also conspicuous for C IV and Si IV. The C II resonance lines, on the other hand, do not exhibit any significant change. This points to the N V, C IV, and Si IV being formed close to the B companion while the C II is produced in a much greater volume. Adding to the confusion, and still to be evaluated, is the presence of multiple components in the resonance absorption lines with a range of radial velocities between -200 km s^{-1} and $+100$ km s^{-1}; this too is seen in 32 Cyg. Clearly the wind of the K-type star is far from being a smooth flow and work with IUE can give important insights relevant to all such supergiants.

2.1.2. *32 Cygni*

Stencel *et al.* (1979) observed 32 Cyg with IUE in September 1978 at phase 0.2, about 8 months after primary eclipse. P Cygni profiles with multiple absorption components with radial velocities between -50 and -400 km s^{-1} were seen; such high velocities have never been observed in the optical range so far from the epoch of the eclipse.

The general level of excitation of the UV absorption spectrum at that phase (0.2) is lower than that recorded in Zeta Aur, with no trace of the N V and C IV lines, the Si IV lines much fainter than in the spectrum of Zeta Aur, but the Fe II lines stronger. A comparison of the two systems observed by phase 0.91 (Zeta Aur) and 0.2 (32 Cyg) has been made by Faraggiana and Hack (1980): the emission wings were much stronger in 32 Cyg than in Zeta Aur, e.g., the ratio F_{em}/F_{cont} for the Mg II resonance lines was 1.8 in Zeta Aur and 3.7 in 32 Cyg. The separation of the two members of these systems at these phases was about 9 and 5.6 K-star radii for 32 Cyg and Zeta Aur respectively. Hence, at the epoch of these observations of the two systems, the excitation in the circumstellar environment of 32 Cyg was lower than that in Zeta Aur despite the possibly earlier spectral type of the companion (B4 V — Wright, 1970; although note that Che *et al.* (1983) find it similar to Zeta Aur's companion, about B8), while the intensity of the emission wings suggests that the cool star has a more extended outer atmosphere. 32 Cyg was observed in 1980—81 with IUE at various phases (Che *et al.*, 1983) and again in 1983 (Ahmad *et al.*, 1984) near the epoch of secondary eclipse in order to study the absorption effect due to the wake (produced by the shock wave generated by the interaction of the K and B star wind) and its extension behind the B star. The aberration angle is estimated by Ahmad *et al.* (1984) to be about 44°, but Che-Bohnenstengel and Reimers (1986) find little evidence for the wake, probably due to the inclination of the orbit, about 80°, which makes the eclipse a grazing one.

Che-Bohnenstengel (1984) used the Fe II absorption lines to derive the excitation temperature of the wind of the K5 Iab supergiant because they are strong and well-defined at all the observed phases (0.06, 0.62, and 0.78) compared with the same lines observed in Zeta Aur. An excitation temperature of 4200 K is revealed by comparing the relative populations of level ⁴F and ⁶D of Fe II. This estimate is valid for the inner wind zone of the supergiant atmosphere where these Fe II lines are formed. It is important to note that no appreciable change in the excitation temperature was found at the different phases, in spite of the fact that the separation at phase 0.62 was some 1.4 times larger than that at the other phases. However, Che remarks that we cannot exclude the possibility of this region being contaminated by material ejected from the chromosphere. Such an event, a stellar prominence from the K-type supergiant, seems to have been recorded as a transient absorption of the flux from the B star during the egress phase (Schröder, 1983) (Figure 3). The effect was almost negligible at $\lambda 2650$ Å but became very evident at $\lambda 1350$ Å and is ascribed to Rayleigh scattering by H I. The H I density in the prominence was estimated to be about 10^{12} cm^{-3}, 10 times higher than that of the surrounding chromosphere.

2.1.3. 31 *Cygni*

In the latter part of 1982, Stencel *et al.* (1984) secured simultaneous photometry and optical-UV spectroscopy of the primary eclipse of 31 Cyg, this being especially valuable since the photometry permits an accurate timing of the eclipse which is important for the interpretation of the spectral features. Altogether, about half of the 10 yr cycle has not been covered with IUE.

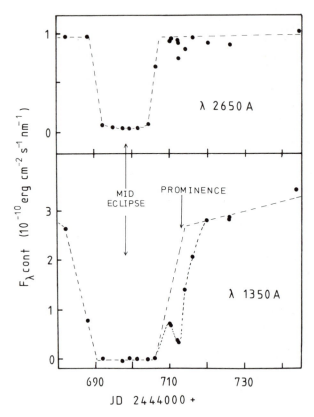

Fig. 3.   The eclipse light curve of 32 Cygni at two wavelengths in the UV. A transient absorption
event was seen during egress; it is attributed to a prominence.

The UV spectrum is similar to that of 32 Cyg: as in the optical range, the absorption lines of Fe II and other cool atomic species increase in strength during the ingress phase of the atmospheric eclipse because of the growing column density along the line of sight through the K star's atmosphere. The first absorption due to the K4 Ib atmosphere became detectable in June of 1982 at an apparent stellar separation of 2.3 K-star radii. This demonstrates the existence of an extended stellar chromosphere, comparable in size to those of single late-type supergiants (Stencel, 1982; Hartmann and Avrett, 1984). Striking variations are observed in the line spectrum from ingress to totality: the hot continuum disappears and emission lines, strictly matching the out-of-eclipse absorptions, become evident (Plate I). Stencel and Chapman (1981) note that the flux in the high excitation emission lines decreases more between total and partial eclipse than it does in the low excitation lines. This implies that the low excitation emissions are formed in a region surrounding the K-type star, while the high excitation lines are produced in a volume closer to the B4 V star. The strength of the Mg II emission lines, compared with similar single supergiants, requires an emitting volume at least 30 times larger than a one-stellar-radius-thick

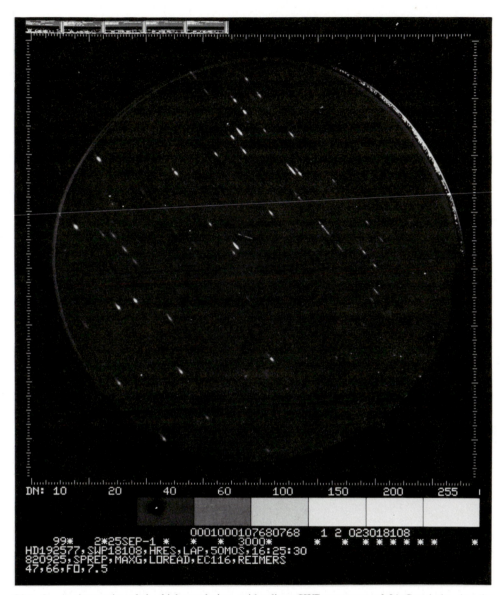

Plate I. A photowrite of the high-resolution, mid-eclipse SWP spectrum of 31 Cygni showing the
emission line spectrum from the K star's wind excited in the vicinity of the B star.

chromosphere; indeed, a zone with a radius of about 10^{14} cm is demanded, comparable with the separation of the two components.

2.1.4. *Mass Loss and the Chromospheric Density Distribution in Zeta Aurigae, 31 and 32 Cygni*

The main difficulty in obtaining the rate of mass loss, $\dot{M} = 4\pi r^2 \rho V$, is the impossibility of generally determining the location of the circumstellar shell, i.e., the value of r. In the case of visual binaries like Alpha Her or Alpha Sco B, where the circumstellar lines are seen in the spectrum of the hot companion (Deutsch, 1956), r is larger than the separation of the two stars and $\dot{M}$ can be derived. The same method can be used for the Zeta Aurigae stars if we can exploit the different spectral energy distributions of the two members instead of the spatial separation. The cool shell produced by the wind of the K-type star absorbs light from the blue continuum of the companion. To derive the mass-loss rate from the shell absorptions it is desirable that their equivalent widths are not affected by any contribution from the emission wings. Che *et al.* (1983), who have proposed this extension of the Deutsch method, observe that the Fe II lines of UV multiplet 9 have no emission components, the reason being that they have the same upper term, $x^6 P°$, as UV multiplet 191. Since the gf values of the lines of multiplet 191 are about 100 times higher than those of UV 9, recombination strongly favours the UV 191 triplet at $\lambda 1785$ Å, yielding pronounced emission. Meanwhile, UV 9 appears as pure absorption and completely disappears during totality when all the other shell absorptions change to emission lines (Figure 4). Using these lines, Che *et al.* (1983) compute mass-loss rates of 0.63×10^{-8}, 2.8×10^{-8} and $> 1.0 \times 10^{-8}$ $M_\odot$ yr^{-1} for Zeta Aur, 32 Cyg, and 31 Cyg with errors of about

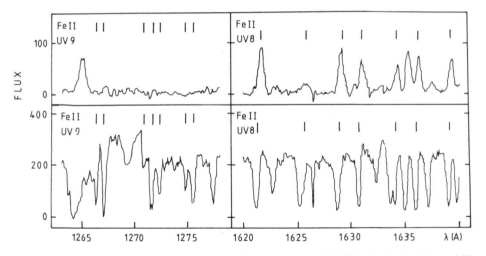

Fig. 4.   The upper left-hand panel shows the eclipse spectrum of 32 Cygni with the lines of Fe II UV 9 absent, while the lower one illustrates the situation shortly after eclipse when the same lines appear in absorption. The right-hand panels show the 'normal' behaviour of the Fe II UV 8 lines for comparison.

factors of 3, 3, and 5 respectively. Chapman (1981) found a value of 2×10^{-8} $M_\odot$ yr^{-1} for Zeta Aur, in reasonable agreement with Che *et al.* (1983), but for 32 Cyg, Stencel *et al.* (1979) earlier found a value of 4×10^{-7} $M_\odot$ yr^{-1}. This latter discrepancy may have various origins: different lines were used and crude assumptions were made concerning spherical symmetry, constancy of the expansion velocity, etc.

   The chromospheric density distributions for Zeta Aur, 31 and 32 Cyg have been derived by Schröder (1985) through the method of the curve of growth applied to chromospheric absorption lines observed at various phases and therefore various heights above their photospheres. The observed relation, log N vs height (h), can be fitted by the expression:

$$N(r) = N(R/r)^2 \cdot (r/h)^a$$

with $a = 2.5$ for 31 and 32 Cygni and 3.5 for Zeta Aurigae. The observed expansion velocities at different phases, and thereby different heights, permit a study of the wind accelerations in the three systems directly and subsequent comparison with appropriate models. The results by Schröder indicate that the observed acceleration takes place at greater heights and is more gradual than for Alfvén wave-driven winds.

2.1.5. 22 *Vulpeculae*

In April 1983 a new member of the Zeta Aur class was discovered by Parsons and Ake (1983): 22 Vul. It is composed of a G3 supergiant and a B9 main sequence companion and has a period of 249 days. Hence it has the earliest-type primary and the shortest orbital period among the Zeta Aur systems. It is thus important because firstly, it offers the chance of probing the atmosphere of a G-type supergiant, particularly so since it is one close to the hybrid chromosphere border. Secondly, as Ake *et al.* (1984, 1985) and Schröder and Che-Bohnenstengel (1985) observe, this system has an orbital period which is intermediate between those of the Zeta Aur systems and those of other interacting binaries such as the Serpentids (Plavec *et al.*, 1982). Because of the short period, 22 Vul is really always moving in the atmosphere of the G-type supergiant, even at the greatest separation, although the interaction is not so strong as for the Serpentids. That said, the continuum of the B star always exhibits some absorption lines formed in the extended atmosphere of the G star, and these become stronger and broader when the B star approaches primary eclipse.

Three eclipses have been monitored with IUE, in April 1983, December 1983 and April/May 1985; they were total, as shown by the complete disappearance of the spectrum of the B star and the revelation of the emission line spectrum, and lasted about 9.6 days. Comparison of this emission spectrum with that of Zeta Aur and that of the single star Beta Dra (G2 Ib—II) indicates that the emissions are generally stronger in 22 Vul; the peaks of the P Cygni profiles do not follow the orbital motion of the B star and remain nearly constant with phase at the same velocities found at totality. Hence they are not associated with the B star and are probably due to fluorescence of the B star photons in the cool G star wind. Indeed, according to computations by Hempe (1982), at totality, the emission arises from scattering in an uneclipsed region larger than the B star; at atmospheric eclipse phases, strong absorption obliterates much of the emission, producing a P Cygni profile.

IUE low resolution observations of the April 1983 eclipse show that partial eclipse effects around $\lambda 1650$ Å and $\lambda 2550$ Å persist at greater distances from the limb of the G star than at $\lambda 1850$ Å, $\lambda 2950$ Å or at optical wavelengths and are not symmetrical at ingress and egress. This reveals the greater opacity of the G star atmosphere at $\lambda 1650$ Å and $\lambda 2550$ Å due to the large number of Fe II lines crowding together at these wavelengths. The asymmetry can be explained by clumpiness of the G star atmosphere as well as by the presence of a jet from the cool star towards the hot component.

Being a hybrid supergiant, the G3 Ib star has a fast wind with a terminal velocity of about 200 km s^{-1}, although it is only 84 km s at the orbit of the B9 star (Ahmad and Parsons, 1985), where it is sufficient to generate a turbulent wake. The chromosphere is quite extensive, with column densities assessed by Schröder and Che-Bohnenstengel (1985) from the Fe II UV 9 lines to be similar to those of the larger K-type stars discussed earlier: log $N_H \simeq 21.6$. This again is significant in view of its hybrid type.

2.1.6. *Tau Persei*

Very recently, Ake *et al.* (1986) announced that the A2 V star in the Tau Persei system (4.15 yr period) had been observed with IUE passing behind the outer layers of its G5 III companion at a projected separation of 2 G-star radii. It suffered a decrease in light which reached 1.15 mag in the $\lambda 2500$ Å to $\lambda 1600$ Å region but was only 0.25 mag at $\lambda 1250$ Å. Numerous low excitation absorption lines were noted, rather similar to the corresponding event in 22 Vul.

2.2. THE VV CEPHEI SYSTEMS

Three systems of the VV Cep type are known and have been studied with IUE: VV Cep itself (M2 Ia + B) with a period of 20.4 yr and whose late egress from totality was observed in 1978 soon after the satellite was commissioned; Boss 1985 = KQ Pup (M2ep Iab + B) with a period of 26.7 yr; and Delta Sge (M2 Ib + B7) with a period of 10.2 yr.

2.2.1. *VV Cephei*

The most pronounced features of the spectrum of VV Cep in the optical region are:

    (a) the M2 spectrum of the primary,
    (b) the forbidden emission lines from the common envelope, and
    (c) the Balmer emissions formed in an H II region surrounding the hot companion.

Intermediate resolution optical spectra of the ingress and egress phases have been studied by Möllenhoff and Schaifers (1978, 1981): at ingress the violet emission wings of Hβ and Hγ started to decrease first, followed by the red wings. During totality these emissions were not seen and the spectrum was practically identical to that of Mu Cep. At egress, the opposite occurs: the violet emission wings started to increase first (in January 1978) while the red wings only reappeared in April 1978. These results suggest that the B-type star is surrounded by a rotating envelope or ring where this broad emission is formed, and its diameter can be deduced from the time difference between the contact dates. It turns out that the envelope has a radius of about 320 $R_\odot$, to be compared with the photometric radius of the B star, of about 13 $R_\odot$, and that of the primary, about 1600 $R_\odot$. From the radial velocity of these lines, and assuming that the ring is rotating with Keplerian velocities, a radius of 380 $R_\odot$ is derived, in reasonable agreement with the above value. Ingress lasts about 104 days while egress lasts roughly twice as long, 205 days. (Such asymmetries in the occulting atmosphere have been noted also in the Zeta Aur systems.) The last eclipse spanned about 700 days, with totality occupying some 400 days. Although the first contact happened early in November 1976, spectroscopic observations showed that the outer layers of the M star were already starting to occult the B star in 1975 (Wright, 1975; Faraggiana, 1976) as indicated by the strengthening of the lines of ionized metals and the chromospheric core of the Ca I $\lambda4226$ Å resonance line.

The system emerged from eclipse in March 1978. Several IUE spectra were secured between April 1978 and April 1979 (Hagen *et al.*, 1980; Faraggiana and Selvelli, 1979) and one of the most significant findings was that egress in the UV lagged behind the visible egress by two or three months because of the higher opacity of the M star atmosphere at the shorter wavelengths. This has permitted the observation of phases when the B star was contributing minimally to the far UV spectrum (April 1 to May 20, 1978) revealing an array of emission lines which, as for the Zeta Aur stars, were much stronger than would be seen in the spectrum of a single star of the corresponding type. These are produced by fluorescence in lines of abundant species which share the same upper levels as the resonance lines. For example, the surface brightness of the O I triplet at $\lambda1304$ Å is 3000 times brighter than that in the spectrum of the same type star, Alpha Ori (assuming for VV Cep a distance of 700 pc as derived by van der Kamp (1977)). Absorptions due to the Fe I and Fe II ions in the chromosphere of the M-type supergiant are apparent in the UV spectrum of the companion but at eclipse, the neutral lines disappear earlier than those of Fe II, pointing to their differentiation in the atmosphere.

A question that is still open pertains to the spectral type of the companion. Optical observations assign types ranging from O to A, the uncertainty being due to the overwhelming dominance of the M star continuum and its absorption lines and bands, and also to the forbidden and permitted emissions of the circumstellar envelope. The UV range ought to allow a much better determination of the spectral type, but the possible presence of a ring or envelope surrounding the hot companion, revealed by the work of Möllenhoff and Schaifers (1978, 1981), makes the task difficult. An analysis by Faraggiana and Selvelli (1979) yields a very flat energy distribution indicative of a temperature near 10 000 K. No help is forthcoming from the primary since its absolute magnitude is quite ill-determined, previous values wandering between -1.4 to -6.9! (Cowley, 1969). The base width of the Mg II k line is an indicator of luminosity and employing it, Faraggiana and Selvelli found $M_v = -4$, while Hagen *et al.* (1980) using various calibrations, obtained values between -3.2 and -4.4. We must sadly conclude that the absorption spectrum of the companion is very complex and it is impossible to assign a spectral type while it is still embedded in the extended M star atmosphere.

2.2.2. *Boss* 1985

The M1—2 Iab spectrum is characterized in the optical region by emission lines of H I, Fe II, and [Fe II]; the permitted lines rise to a maximum lasting several years around the time of periastron passage while those of [Fe II] are seemingly invariant. The UV spectrum, studied by Altamore *et al.* (1982), is represented by two systems of lines: broad absorption lines of Ly α, C IV, N V, Si IV, and Al III, perhaps originating in a disk around the hot companion, and sharp absorption and emission lines of Fe II, mainly together in P Cygni profiles, probably formed in the circumstellar envelope. The broad lines are shifted by about 85 km s^{-1} relative to the narrow lines. No He I or He II lines are observed and neither are the often strong intercombination lines of C III] and N III]. Che and Reimers (1983) have applied their method for determining the mass-loss rate, using the pure absorption lines of Fe II (UV 9), to Boss 1985; they find a value probably in excess of $1.0 \pm 0.5 \times 10^{-7} M_\odot$ yr^{-1}.

2.2.3. *Delta Sagittae*

Delta Sge is a spectroscopic binary well-known for its composite spectrum: M2II + late B. It must have an inclination greater than 70° since atmospheric eclipses have been observed, and with a period of 10.2 yr, the last one occurred around the beginning of March 1980 and lasted about 40 days; this event was observed with IUE by Reimers and his collaborators (Reimers and Kudritzki, 1980; Reimers and Schröder, 1983). The spectrum contains several double-peaked emission lines which suggest that the hot companion is girdled by an accretion disk or envelope in the manner of VV Cep. There is evidence for stratification in the envelope: in the cool outer zone where the lines of N I are formed the rotation velocity is about 65 km s^{-1}, where Fe II originates it is about 90 km s^{-1}, and where the Al III and C IV emission wings arise the velocity reaches 135 km s^{-1}. The hot gas producing these lines moves with the B star and remains observable even at mid-eclipse, implying that the envelope must have an extension larger than 1 AU.

While Fe II lines from UV multiplets 60—78 and Cr II lines from UV 5 apparently come from the rotating envelope, the ground state lines come from the wind of the M star expanding at 25—30 km s^{-1}. They yield a mass-loss rate of $2 \times 10^{-8} M_\odot$ yr^{-1}. Intermediate, i.e., the low excitation, non-resonance lines are generated in a region of the wind nearer the B star. As with the other Zeta Aur and VV Cep systems, multiple line structure is noted (Figure 5), telling us that the wind flow is not smooth and hence there are inhomogeneities in the circumstellar material.

2.3. Epsilon Aurigae

In order to test the hypothesis of the presence of a hot object in the system, Hack and Selvelli (1979) observed Epsilon Aur in April 1978 when IUE had just finished its commissioning in orbit. They used the low resolution mode and discovered that the flux at $\lambda < 1650$ Å was larger than expected from the spectral type of the primary (F0 Ia). Initially, however, there was some suggestion that this excess could be due to light scattered in the spectrograph from longer wavelengths. A first rough correction was applied at the time and more refined ones were made later when the properties of the instrument became better known (see the calibration by Stickland (1980) and the recent evaluation by Altner *et al.* (1984));

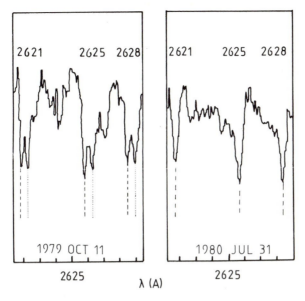

Fig. 5.   Circumstellar absorption lines of Fe II UV 1 in the spectrum of Delta Sagittae occasionally exhibit multiple structure indicative of transient absorbing clouds.

the excess flux seems to be real and it is encouraging that an excess, albeit from a rather larger area of sky, has been recorded by *Voyager I*.

Further exciting results came from the observations of the once-in-27 yr eclipse, not only in the far UV with IUE but in the infrared up to 60 μ with data from high altitude observatories and IRAS. These data, taken during the partial and total phases of the eclipse, have revealed that its depth from $\lambda 1600$ Å to 3.8 μ is almost constant (actually about 0.9 mag in the UV, 0.8 mag in the visual and 0.7 mag in the IR), while it decreases strongly in the far UV and in the far IR, becoming less than 0.2 mag at $\lambda < 1500$ Å and equal to 0.3 mag at 20 μ (Figure 6).

These findings suggest, at first sight, two apparently contradictory interpretations:

(a) that the depth of the eclipse becomes negligible in the far UV, hence a hot body is in front at the epoch of the eclipse, and

(b) that the depth of the eclipse is small in the far IR, hence a cool body is in front.

The UV flux at wavelengths shorter than $\lambda 2000$ Å has been found to be variable in time, with phases of 'activity' noted both inside and outside of eclipse and ranging up to 1—1.5 magnitudes (Boehm *et al.*, 1984; Ferluga and Hack, 1985). From all of these observations we can construct a plausible model for Epsilon Aur with the following components:

(1) The F0 Ia primary whose spectrum is always visible and dominating in the optical and near UV.

(2) A cool dusty body ($T_{eff} \simeq 750$ K (Stickland, 1985) or 500 K (Backman, 1985)) which is responsible for the photometric eclipse of the supergiant.

(3) A gaseous envelope, more extended than the cool body, which gives rise to

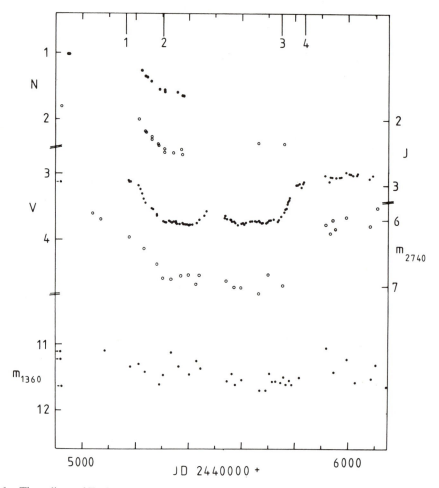

Fig. 6. The eclipse of Epsilon Aurigae as seen at a range of wavelengths from far UV to IR. The UV points have been re-extracted from archival data, account being taken of the scattered light at each phase. The optical and IR results came from the Epsilon Aurigae Campaign Newsletter No. 11. The times of contacts in the optical region have been adopted from Schmidtke *et al.* (1985).

the 'shell' spectrum characteristic of a gas with $N_e \simeq 10^{11}\,\mathrm{cm}^{-3}$ appearing before the start of the eclipse and disappearing after the end of it, and which is particularly evident during the partial phases.

(4) An extended zone surrounding the whole system where the emission lines of O I $\lambda 1304\,\text{Å}$ and Mg II $\lambda 2800\,\text{Å}$ are formed and whose intensities appear unaffected by the eclipse (although $\lambda 1304\,\text{Å}$ may be intrinsically variable).

(5) A possible chromospheric UV-emitting belt about the F star equator which may help to explain the progressive deepening of the UV eclipse during totality noted by Ake (1985) and is compatible with the polarization data of Kemp *et al.* (1985).

(6) A hot body which is not eclipsed and whose radiation dominates at $\lambda < 1500\,\text{Å}$. This hot object may be a single star, as suggested by Hack and

Selvelli (1979), or a close binary as proposed by Lissauer and Backman (1984), whose light, escaping from the poles, excites and ionizes the envelope producing both the 'shell' spectrum and at least part of the $H\alpha$ emission. Another possibility for the far UV source has been suggested by Parthasarathy and Lambert (1983): a hot spot on that part of the primary which is not occulted by the dusty disk.

Its high luminosity might lead us to assume that Epsilon Aur is a massive star, say $13 \ M_{\odot} < m_1 < 30 \ M_{\odot}$; but in this case the mass function, $f(m) = 3.34 \ M_{\odot}$, imposes the condition that $15 \ M_{\odot} > m_2 > 13 \ M_{\odot}$. It is difficult to reconcile this with a companion having so low a luminosity as to be undetectable in the visible and barely so in the UV, unless we postulate that its light is heavily dimmed by the disk. But in this case, the disk, heated by the companion, would become much hotter than is actually observed. For these reasons, Lissauer and Backman (1984) proposed that the companion is in reality a close binary whereby the total mass of the system reaches that demanded by the mass function but the total luminosity is compatible with the observations.

An alternative scenario is that Epsilon Aur may be a highly evolved star, as is suggested by its C_{12}/C_{13} ratio, and very undermassive for its luminosity (Lambert, 1986). In this situation, for a mass ratio $m_1/m_2 = 0.25$, we deduce $1 \ M_{\odot}$ for the supergiant and $4 \ M_{\odot}$ for the companion. A star leaving the AGB with a degenerate core, double shell sources and a hydrogen-rich envelope, will have a low mass and yet a moderately high luminosity (Paczinsky, 1971; Gingold, 1974; Schönberner, 1979; Takeuti, 1986), and this may be more compatible with van der Kamp's (1978) distance for the system.

The 'shell' spectrum has been observed through three eclipses, 1928—30, 1955—57 and 1982—84. It seems to us rather improbable that a hot spot, as the proposed source of UV radiation, has remained on the surface of the primary for such a long period. And what is the origin of this hot spot? The hypothesis seems more artificial than the others.

Detailed study of the radial velocity of the 'shell' spectrum and of its degree of excitation at different phases, produced by subtracting the visual out-of-eclipse spectrum from that taken in eclipse (Boehm and Ferluga, 1986), demonstrates that the internal regions of the 'shell' have lower density, higher rotational velocity and higher temperature than the outer regions. Hence the hot source causing the far UV excess should be at the centre of the 'shell'. The observed properties of the 'shell' spectrum can naturally be explained by the inclination of the gaseous and dusty disk with respect to the orbital plane, an interpretation also suggested by the polarization measurement during the eclipse.

In order to solve completely the mystery of Epsilon Aur and to understand the nature of the companion and the evolutionary stage of the system, we must learn the masses of the two components. Hence we need to obtain high resolution spectra at $\lambda < 1500$ Å where the light of the companion is dominant, and to measure its orbital radial velocity during an appreciable fraction of the 27.1 yr period. The object is too faint to be observed with IUE in the high resolution mode at $\lambda < 1700$ Å. Only the high resolution spectrograph on the Hubble Space Telescope will enable us to achieve our aim.

3. Conclusions

In many fields of astrophysical research, IUE has shed new light by allowing us to sample the wealth of atomic and ionic transitions occurring between $\lambda 1150$ Å and the Earth's atmospheric cut-off at about $\lambda 3200$ Å. A host of abundant chemical species has resonance lines in that range as well as subordinate and semi-forbidden lines linked to the ground state features in a physically valuable way. This rich 'tool box' has enabled astronomers to probe, as never before, the complex interactions taking place between the disparate components of binary stars of the Zeta Aurigae and VV Cephei types.

We have seen how the cool supergiant stars are losing mass at a rate of about $10^{-8} M_{\odot}$ yr^{-1} in probably none-too-smooth a manner. The hot main sequence star, in its leisurely orbit about the distended companion, is pouring not only UV radiation into the initially-cool but dense wind from the late-type star, but also its own fast, hot, but rather thin gusts of matter. The clash of these two conjures up a shock front curving around the B star and, in most cases, a wake of turbulent material streaming it; in favourable circumstances, this wake feeds gas into an accretion disk encircling the hot star. Each part of this drama gives rise to its own spectroscopic signature — and most of them could only be read by IUE.

Epsilon Aurigae is, in truth, still an enigma. IUE, together with a battery of other instruments from the astronomer's 'new technology' arsenal, has been used with enthusiasm and to great effect to cover the last eclipse. New constraints have been placed and new models, or refined versions of old models, proposed. The burning question still remains: what is at the centre of the disk? The current favourite is probably a close binary star but we can say no more at present. IUE has given us a tantalizing glimpse of an excess of radiation at the shortest wavelengths, but its little 45 cm telescope is just inadequate to deliver spectra good enough to further elucidate its nature. In many respects, IUE was a trial run for the Hubble Space Telescope; let us hope that the HST can finish the task of laying bare this most fascinating of binary systems.

References

Ahmad, I. A., Chapman, R. D., and Kondo, Y.: 1983, *Astron. Astrophys.* **126**, L5.
Ahmad, I. A., Chapman, R. D., Stencel, R. E., and Kondo, Y.: 1984, NASA CP. 2349, p. 357.
Ahmad, I. A. and Parsons, S. B.: 1985, *Astrophys. J.* **299**, L33.
Ake, T. B.: 1985, NASA CP. 2384, p. 37.
Ake, T. B., Parsons, S. B., and Kondo, Y.: 1985, *Astrophys. J.* **298**, 772.
Ake, T. B., Parsons, S. B., and Kondo, Y.: 1984, NASA CP. 2349, p. 369.
Ake, T. B., Fekel, F. C., Hall, D. S., Barksdale, W. S., Landis, H. J., Fried, R. E., Louth, H. J., and Hopkins, J. L.: 1986, *I.B.V.S.* No. 2847.
Altamore, A., Giangrande, A., and Viotti, R.: 1982, *Astron. Astrophys. Suppl.* **49**, 511.
Altner, B., Chapman, R. D., Kondo, Y., and Stencel, R. E.: 1985, NASA CP. 2384, p. 81.
Ayres, T. R., Marstad, N. C., and Linsky, J. L.: 1981, *Astrophys. J.* **247**, 545.
Backman, D. E.: 1985, NASA CP. 2384, p. 23.
Boehm, C., Ferluga, S., and Hack, M.: 1984, *Astron. Astrophys.* **130**, 419.
Boehm, C. and Ferluga, S.: 1986, in *Highlights of Astronomy* **7**, 201, New Delhi IAU, 1985.
Chapman, R. D.: 1981, *Astrophys. J.* **248**, 1043.
Chapman, R. D.: 1982, in *Binary and Multiple Stars as Tracers of Stellar Evolution*, p. 153.

Che-Bohnenstengel, A.: 1984, *Astron. Astrophys.* **138**, 333.
Che, A., Hempe, K., and Reimers, D.: 1983, *Astron. Astrophys.* **126**, 225.
Che, A. and Reimers, D.: 1983, *Astron. Astrophys.* **127**, 227.
Che-Bohnenstengel, A. and Reimers, D.: 1986, *Astron. Astrophys.* **156**, 172.
Cowley, A.: 1969, *Publ. Astron. Soc. Pacific* **81**, 297.
Deutsch, A. J.: 1956, *Astrophys. J.* **123**, 210.
Faraggiana, R.: 1976, *Astron. Astrophys.* **46**, 317.
Faraggiana, R. and Hack, M.: 1980, ESA SP. 157, p. 223.
Faraggiana, R. and Selvelli, P. L.: 1979, *Astron. Astrophys.* **76**, L18.
Ferluga, S. and Hack M.: 1985, *Astron. Astrophys.* **144**, 395.
Gingold, R. A.: 1974, *Astrophys. J.* **193**, 177.
Gyldenkerne, K.: 1970, *Vistas in Astron.* **12**, 199.
Hack, M.: 1959, *Astrophys. J.* **129**, 291.
Hack, M.: 1961, *Mem. Soc. Astron. Italiana* **32**, 351.
Hack, M.: 1981, *Astron. Astrophys.* **99**, 185.
Hack, M. and Selvelli, P. L.: 1979, *Astron. Astrophys.* **75**, 316.
Hagen, W., Black, J. H., Dupree, A. K., and Holm, A. V.: 1980, *Astrophys. J.* **238**, 303.
Hartmann, L. and Avrett, E.: 1984, *Astrophys. J.* **284**, 238.
Hempe, K. and Reimers, D.: 1982, *Astron. Astrophys.* **107**, 36.
Hempe, K.: 1982, *Astron. Astrophys.* **115**, 133.
Hempe, K.: 1983, *Astron. Astrophys. Supplement* **53**, 339.
Hjellming, R. M. and Newell, R. T.: 1983, *Astrophys. J.* **275**, 704.
Kemp, J. C., Henson, G. D., Kraus, D. J., and Beardsley, I. S.: 1985, NASA CP. 2384, p. 33.
Kraft, R. P.: 1954, *Astrophys. J.* **120**, 391.
Lambert, D. L.: 1986, in *Highlights of Astronomy* **7**, 131, New Delhi IAU, 1985.
Linsky, J. L. and Haisch, B. M.: 1979, *Astrophys. J. Letters* **229**, L27.
Lissauer, J. J. and Backman, D. E.: 1984, *Astrophys. J. Letters* **286**, L39.
Möllenhoff, C. and Schaifers, K.: 1978, *Astron. Astrophys.* **64**, 253.
Möllenhoff, C. and Schaifers, K.: 1981, *Astron. Astrophys.* **94**, 333.
Paczinsky, B.: 1971, *Acta Astron.* **21**, 417.
Parsons, S. B. and Ake, T. B.: 1983, *I.B.V.S.* No. 2334.
Parthasarathy, M. and Lambert, D. L.: 1983, *Publ. Astron. Soc. Pacific* **95**, 1012.
Plavec, M. J., Weiland, J. L., and Koch, R. H.: 1982, *Astrophys. J.* **256**, 206.
Reimers, D. and Kudritzki, R. P.: 1980, ESA SP. 157, p. 229.
Reimers, D. and Schröder, K.-P.: 1983, *Astron. Astrophys.* **124**, 241.
Schönberner, D.: 1979, *Astron. Astrophys.* **79**, 108.
Schröder, K.-P.: 1983, *Astron. Astrophys.* **124**, L16.
Schröder, K.-P.: 1985, *Astron. Astrophys.* **147**, 103.
Schröder, K.-P. and Che-Bohnenstengel, A.: 1985, *Astron. Astrophys.* **151**, L5.
Schmidtke, P. C., Hopkins, J. L., Ingvarsson, S. I., and Stencel, R. E.: 1985, *I.B.V.S.* No. 2748.
Simon, T., Linsky, J. L., and Stencel, R. E.: 1982, *Astrophys. J.* **257**, 225.
Stencel, R. E.: 1982, in *The 2nd Cambridge Workshop on Cool Stars, Stellar Systems and the Sun*, SAO Spec. Rep. 392, Vol. I, p. 137.
Stencel, R. E. and Chapman, R. D.: 1981, *Astrophys. J.* **251**, 597.
Stencel, R. E., Hopkins, J. L., Hagen, W., Fried, R., Schmidtke, P. C., Kondo, Y., and Chapman, R. D.: 1984, *Astrophys. J.* **281**, 751.
Stencel, R. E., Kondo, Y., Bernat, A., and McCluskey, G. E.: 1979, *Astrophys. J.* **251**, 621.
Stickland, D. J.: 1980, *ESA IUE Newsletter* **6**, p. 34.
Stickland, D. J.: 1985, *Observatory* **105**, 90.
Takeuti, M.: 1986, *Astrophys. Sp. Sci.* **121**, 127.
van der Kamp, P.: 1977, *Astron. J.* **82**, 750.
van der Kamp, P.: 1978, *Astron. J.* **83**, 975.
Wright, K. O.: 1970, *Vistas in Astron.* **12**, 147.
Wright, K. O.: 1975, *IAU Circ.* No. 2811.

PART IV

THE INTERSTELLAR MEDIUM AND NEBULAE
IN THE MILKY WAY AND MAGELLANIC CLOUDS

Edited by M. GREWING

THE INTERSTELLAR MEDIUM
NEAR THE SUN

FREDERICK C. BRUHWEILER

Department of Physics, Catholic University of America, Washington, D.C., U.S.A.

and

ALFRED VIDAL-MADJAR

Institut d'Astrophysique de Paris, Paris, France

1. Introduction

The local interstellar medium (LISM), especially the region within 100 pc, is extremely important to understanding the physical processes that arise in the general interstellar medium, especially the intercloud medium (ICM), that is, the interstellar medium away from cloud complexes and OB associations. Studies of the LISM are our only hope of directly probing what may be a typical volume element of the interstellar medium. The nature of the LISM is also crucial to the future of astronomy at Extreme-UV wavelengths. Observations with the International Ultraviolet Explorer (IUE) have been essential in determining the nature of the local interstellar medium, which prior to the IUE was, except for the nearest stars, largely unexplored.

In what follows, a brief picture of the local interstellar medium prior to the IUE will be presented. However, the thrust of this review will be to show how the results obtained with the IUE have shaped our current physical picture of the local interstellar medium. The IUE observations designed to probe absorption in the LISM have concentrated on three different types of objects: the hot white dwarfs, fast-rotating A and late-B stars, and late-type stars with chromospheric emission. We will discuss how observations of each of these types of objects have contributed to the present picture of the LISM. We will expand upon the previous reviews of the local interstellar medium by Bruhweiler (1982, 1984), Frisch and York (1983, 1986), and Paresce (1984). Of course, any review must build upon the views established in the recent *IAU Colloquium* No. 81 which was devoted to the subject of the local interstellar medium.

2. The Local Interstellar Medium Prior to the IUE

Until the advent of space astronomy, there were few means by which one could adequately probe the interstellar medium within 100 pc of the Sun (LISM). Ground-based investigations at visual wavelengths of interstellar absorption and at radio wavelengths of hydrogen 21-cm emission seemed ill-equipped to study this region. Only one star within 50 pc showed any hint of interstellar absorption features. These results inferred that the column densities within 100 pc were

Y. Kondo (ed.), Scientific Accomplishments of the IUE, pp. 467—484.
© 1987 *by D. Reidel Publishing Company.*

probably less than a few times 10^{19} cm^{-2}. Interpretations of the wide, low-intensity wings of 21-cm emission lines implied an intercloud medium (ICM) with temperatures near 7000 K. From these data combined with absorption line studies obtained primarily for lines-of-sight further than 100 pc, it was deduced that the ICM consisted of a uniform low density medium ($n \approx 0.3$ cm^{-3}) with regions of neutral and ionized hydrogen, or regions perhaps partially ionized with $n(\text{H\,I}) \approx 0.1$ cm^{-3} (see Shapiro and Field, 1976; and Field, 1975 and references therein). The credence of this interpretation was reinforced with observations made from space when Copernicus results for lines-of-sight toward stars within about 5 pc of the Sun (McClintock *et al.*, 1978, and references therein) and O and B stars beyond 100 pc away from OB associations (Bohlin *et al.*, 1978), as well as backscattering results (Bertaux and Blamont, 1971; Thomas and Krassa, 1971; Fahr, 1974, 1978), indicated that H\,I particle densities in the immediate vicinity of the Sun and in the ICM were near 0.1 cm^{-3}.

However, other evidence seemed to contradict this simple model for the ICM. For example, Copernicus observations over longer lines-of-sight had revealed the ubiquitous presence of O\,VI, apparently interstellar in origin, toward many O and B stars (Jenkins and Meloy, 1974). Of the stars yielding positive detections, four were on the order of 100 pc or less from the Sun. In addition, rocket observations indicated the presence of a pervasive soft X-ray background (Burstein *et al.*, 1976). Together, these results suggested the presence of a pervasive, hot ($T \sim 10^{5-6}$ K), low-density component to the interstellar medium. Furthermore, Field (1975) and Shapiro and Field (1976) showed that the simple model for the ICM would be unstable.

A new model for the interstellar medium, that of McKee and Ostriker (MO: 1977), synthesized all of these seemingly diverse data into a coherent picture. In their view, the ICM consists of a hot, low-density, coronal substrate ($T \sim 10^6$ K, $n \sim 10^{-2.5}$ cm^{-3}) in which interstellar clouds or 'cloudlets' are embedded. These cloudlets have an onion-like structure. The outer skins of these clouds contain the warm (8000 K) ionized and neutral components of the ISM. The cores of these clouds, depending on their size, may contain the cold (80 K), dense, neutral component. These small cloudlets, in the MO model, had characteristic dimensions of a few parsecs and a mean free path between these cloudlets of about 12 pc. Thus, a line-of-sight of 100 pc should intersect about eight such cloudlets. The average hydrogen particle density over this pathlength would be close to $n(\text{H\,I}) = 0.1$ cm^{-3}. Obviously, this theory did much to eleviate the discrepancies among the various data.

Further work modified this basic theoretical model. For example, evidence suggested that the O\,VI was predominately formed in the evaporative interfaces between the clouds and the pervasive hotter coronal susbtrate (Cowie and McKee, 1977; Cowie *et al.*, 1979). Despite modifications, the basic picture presented by McKee and Ostriker of clouds embedded in a hot coronal substrate remained the generally accepted model of the ISM.

A natural consequence of the McKee and Ostriker theoretical model was that the Sun was embedded inside a cloud. Further work by Vidal-Madjar *et al.* (1978), based upon H\,I column densities previously derived from Copernicus data toward nearby late-type stars, synthesized with other data, attempted to derive an observational model for the local ISM. They suggested that the Sun was near a dense cloud lying in the direction of the galactic center. However, McClintock *et al.* (1978) with new results, plus a reanalysis of previous Copernicus data, found that the Copernicus data could not support the conclusions of Vidal-Madjar *et al.* Even though the presence of a nearby cloud seemed to be unsupported by reanalysis of the observed data the work of Vidal-Madjar *et al.* has been shown to be unusually prophetic and has represented one of the first serious efforts to construct an observational picture for the local ISM.

Thus, we find that the observed morphology of the local ISM was far from established. The key to testing the theory of the ISM and determining the structure locally required observations over lines-of-sight in the range 5—100 pc, path-lengths, which before the IUE, were largely unprobed.

3. IUE Observations of the ISM Toward Nearbly White Dwarfs

Nearby, hot ($T > 20\,000$ K) white dwarfs with measured parallaxes provide an ideal means to study the intervening local interstellar medium. These objects possess almost featureless continua against which interstellar absorption features of such species as N I, O I, Si II, C II, Fe II, Mg II can be seen in high resolution spectra acquired with the IUE. Of these, the species of N I, Si II, and Mg II, with multiple interstellar features, can be used to define an interstellar curve of growth. Such studies have been carried out by Bruhweiler and Kondo (1981, 1982a) and by Dupree and Raymond (1982).

One must note that many hot white dwarfs are not completely featureless. Most, in fact, exhibit 'interstellar-like' features that either arise in the photosphere or in an expanding halo about these objects. However, these features can be differen-tiated from those of interstellar origin since they usually show fine-structure or higher excitation transitions, or velocities uncharacterisic of the LISM (Bruhweiler, 1984; Bruhweiler and Kondo, 1981, 1983, and in preparation).

One cannot directly measure neutral hydrogen column densities from the white dwarf IUE data, but good indirect determinations can be derived from the N I column densities. This can be accomplished since N I is neutral with an ionization potential (14.5 eV) not too different from that of hydrogen, and is essentially undepleted in the ISM. For example, a review of Copernicus data by Ferlet (1981) showed that the derived N I and H I column densities toward early-type stars were very closely correlated. These data further indicated that nitrogen exhibits very little depletion, only about 0.15 dex, in the ISM. Even though the N I resonance triplet lies at 1200 Å, a wavelength region of low IUE sensitivity, one finds sufficient flux at 1200 Å to estimate reliable N I column densities.

The deduced H I column densities for four white dwarfs from the IUE investiga-tions of Bruhweiler and Kondo (1981, 1982a), and for Feige 24 from that of Dupree and Raymond (1982) are presented in Table I. The data clearly show comparable column densities toward white dwarfs regardless of their distances and a dramatic decrease in average neutral hydrogen particle densities ($n_{\mathrm{H\,I}}$) for longer lines-of-sight. Specifically, the H I column densities imply $n_{\mathrm{H\,I}}$ of 0.006 and

TABLE I

Interstellar neutral hydrogen toward nearby white dwarfs

Star	d (pc)	$\log N(\text{H\,I})$ (cm^{-2})	$n_{\text{H\,I}}$ (cm^{-3})
Siruis B*	2.7	17.93	0.087
W 1346	13	18.04	0.024
HD 149499 B	34	18.09	0.012
G191—B2B	48	17.92	0.006
Feige 24	90	18.30	0.008

* Indicates NI data not available, value deduced from average of Mg II and Si II.

0.008 cm^{-3} for G191—B2B (48 pc) and Feige 24 (94 pc) respectively. These values contrast sharply with $n_{\text{H\,I}} = 0.087$ cm^{-3} for the much nearer Sirius B (2.7 pc).

Together with previous Copernicus and backscattering results, these results indicate a concentration or cloud of neutral hydrogen with $n_{\text{H\,I}} \approx 0.1$ cm^{-3} within 3 pc of the Sun with a sharp fall-off at greater distances (see Figure 2). Particularly, note in Figure 2 the remarkably vacant line-of-sight toward Beta CMa (200 pc) where a detailed Copernicus study by Gry, York, and Vidal-Madjar (1985) revealed one of the lowest average H I volume densities: $n_{\text{H\,I}} = 0.002$ cm^{-3} (see also Bohlin et al., 1978).

Unlike the other white dwarfs observed, only the most distant Feige 24 shows any indication of an H II region along the line-of-sight. It shows what appears to be interstellar C IV and Si IV in a Stromgren sphere ionized by Feige 24 (Dupree and Raymond, 1983). Unfortunately, due to the unusual characteristics of this short-period (4.2 days) white dwarf-red dwarf system, it is not clear if these features arise in circumbinary material produced by stellar mass loss or if they are truly interstellar in origin (see also Bruhweiler and Kondo, 1983; Sion and Starrfield, 1983; and Bruhweiler and Sion, 1986).

Extreme-UV observations obtained with the Voyager A and B spacecraft have provided direct measurements of H I column densities toward hot white dwarfs (Holberg et al., 1981a, b). Measurements of the hydrogen Lyman continuum shortward of 912 Å in the extremely hot white dwarfs HZ 43 (62 pc) and G191—B2B (48 pc), a white dwarf also studied by Bruhweiler and Kondo (1981, 1982a), show very little intervening hydrogen absorption toward these stars. They find that $n_{\text{H\,I}} = 0.002 \pm 0.001$ cm^{-3} for HZ 43 and an upper limit of $n_{\text{H\,I}} \lesssim 0.01$ cm^{-3} for G191—B2B. The HZ 43 result is consistent with unpublished IUE measurements of one of us (FCB), which indicate that $N(\text{H\,I}) = 6 \times 10^{17}$ cm^{-2} or $n_{\text{H\,I}} = 0.003$ cm^{-3}. Likewise, the upper limit for G191—B2B is consistent with the $n_{\text{H\,I}} = 0.006$ cm^{-3} deduced from the IUE data.

Observations of nearby white dwarfs present an unusual opportunity to study the true nature of the intercloud medium and adequately provide tests for current theoretical models. Although we must rely upon other results (i.e. Copernicus and diffuse soft X-ray observations) for information about the tenuous, coronal component of the ISM, we can, from the IUE data, point the way for future models describing the H I and H II components of the local ISM. The picture emerging from these data is that the Sun is embedded in a diffuse cloud, with $n_{H I} \approx 0.1$ cm^{-3} near the Sun. This cloud is then surrounded by the pervasive low density 10^6 K gas with no evidence of other clouds toward the white dwarfs studied.

From the McKee and Ostriker model, the predicted values for line-of-sight average total and H I particle densities (0.017 and 0.06 cm^{-3}) and the mean free path for cloudlets should not be taken too literally, since these values only represent 'educated guesses'. Nonetheless, the derived column densities for all ions pointed to a concentration near the Sun and did not suggest the presence of other H I or H II regions along any of the observed directions except for the most distant white dwarf studied, Feige 24. These data indicate that the predictions from the MO model overestimate by an order of magnitude the amount of H I and H II in the local ICM. At present, the IUE data show no evidence for additional cloudlets other than the cloud in which the Sun appears to be embedded.

If the vacant lines-of-sight sampled by the IUE are representative of the ICM and if cloudlets exist, then they must be larger and have larger mean free paths than those envisioned by McKee and Ostriker. A second possibility is that the local ISM is anomalous, displaying average particle densities much lower than those for longer lines-of-sight. There is some independent evidence of such an anomaly. Spitzer (1978) notes that Lyman-alpha and extinction results show that the overall mean $n_{H I}$ within 1000 pc of the Sun exceeds by a factor of 10 the value within 100 pc. A combination of both of the above possibilities is also plausible. If the local ISM were anomalous, it would imply that there has been recent or unusual supernova activity in the local ISM. In this case, many of the cloudlets would have been evaporated as depicted in the upper right of Figure 1.

The basic model for the interstellar medium may need to be modified to provide a more complete description. For instance, more recent theoretical studies have indicated that evaporative processes inside supernova remnants (Cowie, McKee and Ostriker 1981) and expanding shell structures, either driven by single or multiple supernova events or by the combined effects of stellar winds and supernovae (Cox, 1979; Weaver *et al.*, 1977; Bruhweiler *et al.*, 1980; Cox and Anderson, 1982), play important roles in shaping the morphology of the general ISM.

4. Interstellar Observations Toward Nearby A, B, and Late-Type Stars

Although the nearly featureless spectra of hot white dwarfs offer many advantages in probing the interstellar medium, there are only a few such objects which display enough ultraviolet flux to be useful probes at UV wavelengths. Thus, to probe the morphology in the 'local cloud' at shorter distances or to obtain a more complete picture of the distribution of gas at larger distances, one must use other objects. Several IUE studies have investigated the lines-of-sight toward nearby fast-rotating

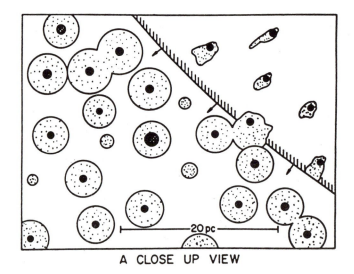

A CLOSE UP VIEW

Fig. 1.   A picture of the Interstellar Medium. The small clouds of McKee and Ostriker embedded in the hot coronal substrate are depicted. The dark central regions of the clouds represent the cold, dense component, while the surrounding material of each cloud represents the warm and ionized medium. A shock front is seen moving down from the upper right. Many of the clouds inside the hotter cavity, swept out by the expanding shock, have either been evaporated by the hotter coronal substrate of the cavity or disrupted and possibly swept away by the passing shock (From McKee and Ostriker, 1977; Courtesy of *The Astrophysical Journal*).

A and B stars. Due to the lack of flux at shorter wavelengths in A stars, any in-depth investigation using these stars must rely upon data acquired with the Long Wave Prime and Redundant (LWP and LWR) cameras of the IUE. Also, due to the rapid fall-off in sensitivity toward shorter wavelengths in these cameras, these studies have principally used the Mg II resonance doublet at 2800 Å as a probe of interstellar gas towards nearby A and B stars. These studies have included those of Bruhweiler and Kondo (1982b), Bruhweiler *et al.* (1984, 1986a, b), Freire Ferrero (1984), Freire Ferrero *et al.* (1984), Frisch (1981), and Skuppin *et al.* (1986).

There have also been several studies which have employed late-type stars, which greatly increased the number of potential targets. In these studies, velocity and equivalent width measurements were made of the interstellar Mg II overlying the chromospheric Mg II emission in these stars (Böhm-Vitense, 1981; Drake *et al.*, 1984; Franco *et al.*, 1984; Vladilo *et al.*, 1985). In a recent paper by Genova *et al.* (1986) results are given for the lines-of-sight to 91 cool stars. They found that 40% of the stars show no evidence of interstellar contamination. This suggests that the density of the gas in the immediate vicinity of the sun could in some directions be even lower than previously thought.

Studies using Mg II and Mg I, as well as those conducted at visual wavelengths, have also been used to determine the velocity vector of the bulk motion of the gas in the local cloud with respect to the Sun.

Finally, an interesting study by Molaro *et al.* (1984) revealed the presence of

interstellar-like features of C IV and Si IV signifying a possible semi-torrid
(log $T \approx 4.5-5.0$) region (Bruhweiler *et al.*, 1979, 1980) toward nearby late-type
fast-rotating B stars. If these lines are interstellar in origin as proposed also by
Ferlet and Freire-Ferrero (1986), these features might indicate an extremely
complex structure of the LISM. But such gas should cool rapidly (Shapiro and
Moore (1976) and be quite rare in the ISM. Also, it is not completely resolved
whether these features are interstellar or circumstellar in origin.

4.1. Mg II WITHIN 25 PC: PROBING THE LOCAL CLOUD

As mentioned above, the IUE results for white dwarfs pointed to a concentration
of gas about the Sun. If we wish to trace out the morphology of the cloud in which
the Sun is embedded, we must use species like Mg II, which is observable in a large
number of nearby stars. An examination of derived Mg II column densities toward
A and B stars within 25 pc revealed a definite asymmetry in the distribution of
Mg II about the Sun. Observations of stars in the hemisphere toward the galactic
anti-center indicated $N(\text{Mg II})$ of a few times 10^{12} cm^{-2}, while stars in the opposite
direction exhibited derived column densities an order of magnitude higher
(Bruhweiler, 1982).

The increase in Mg II column density seemed associated with a 'polarization
patch', or concentration of dust, centered near $1 = 5°$ and $b = -20°$ and extending
no further than 20 pc from the Sun (Bruhweiler, 1982). This patch was discovered
through extremely accurate polarization measurements of stars within 35 pc
(Tinbergen, 1982). If we adopt the gas-dust relation of Knude (1979), we find that
this patch corresponds to a $N(\text{H I}) = 1 - 2 \times 10^{19}$ cm^{-2}. It is noteworthy that the
location of this patch coincides closely with the body of the local cloud delineated
by enhanced Mg II column density (see Figure 2), and it also lies in the same
direction as Alpha PsA, where interstellar C I appears to have been detected
(Bruhweiler and Kondo, 1982b).

Further work by Bruhweiler *et al.* (1984; 1986a, b) has concentrated upon
combining very high signal-to-noise Copernicus data with highest quality IUE data
for A and B stars within 25 pc to probe the Mg II/Mg I ionization equilibrium,
spatial distribution of gas, and velocity structure within the local cloud.

High quality Mg II data were obtained using the IUE by first maximizing the
signal-to-noise about the interstellar Mg II, which was often located in the troughs
of corresponding broad photospheric features (see Figure 4). Then, multiple
images were coadded, incorporating observing techniques that minimized effects
of fixed-pattern noise. The derived Mg II equivalent widths were then combined
with existing high quality Copernicus data for Mg II and Mg I for the nearby stars.
The ratio, $N(\text{Mg II})/N(\text{Mg I})$, is quite sensitive to temperature since Mg I is largely
a result of dielectronic recombination near 10^4 K. The observed ionization ratios
suggested a temperature of $7500-10\,000$ K near the Sun. This result compares
favorably to the range deduced by Weller and Meier (1981) and Bertaux (1984)
based upon backscattering data for the region within 0.01 pc of the Sun (also see
Ripken and Fahr, 1983). The absence of Mg I toward Alpha PsA and the detection
of C I toward this star (Bruhweiler and Kondo, 1982b) strongly suggest much
lower temperatures in the cloud core.

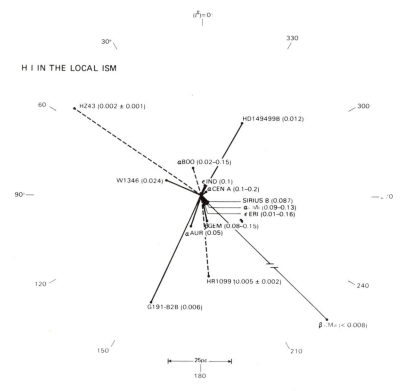

Fig. 2. Average Neutral Hydrogen Number Densities toward Nearby Stars. Vectors represent galactic longitudes for each object with the length proportional to its actual distance. Dashed lines indicate pathlengths with angles more than 45° out of the galactic plane, with distances greater than 5 pc. Notice the dropoff in n_{HI} at larger distances. The values for G191—B2B (48 pc), HD 149499 B (34 pc), W1346 (13 pc), and Sirius B (2.7 pc) are from the IUE results of Bruhweiler and Kondo (1982a). The values for late-type stars and Beta CMa are based upon Copernicus data, while that for HZ 43 is from Voyager data (Holberg, 1981b). See text for further discussion (From Bruhweiler and Kondo, 1982a; Courtesy of *The Astrophysical Journal*).

4.2. THE INTERSTELLAR WIND

The relative motion of the Sun with respect to the local cloud produces a measurable bulk motion of 'interstellar wind' (ISW) that can be detected either in backscattering data or in the radial velocity shifts of interstellar absorption measured against the continua of nearby stars. If this motion can be characterized by a uniform flow past the Sun, one can, once the velocity vector is determined, predict the radial velocity of absorption produced in the local ISM for any direction through this flowing gas. The deduced velocity vector from backscattering results for H I (Bertaux, 1984; Bertaux *et al.*, 1985) and for He I (Weller and Meier, 1981) differs from that derived from ground-based data for stars at distances typically 50—100 pc away (Crutcher, 1982). The differences in deduced vectors are too large to be explained by the uncertainties in the analyses.

A recent study (Bruhweiler *et al.*, 1986b) using Copernicus data supplemented

TABLE II

Observed versus predicted LISM velocities*

Star	d (pc)[1]	1(°)	b(°)	V^{obs} (km s^{-1})		V^{pred} (km s^{-1})	
				Mg II	Mg I	Crutcher[2]	Bertaux[2]
α Eri	27/38	291	−59	+8.7	+8.0	+5.2	+1.8
α Gru**	39/18	350	−52	−11.6	−8.0	−10.1	−7.1
ε Sgr**	23/43	359	−10	−22.5	−	−21.7	−17.9
α PsA**	7.4/6.7	20	−65	−8.0	−	−7.2	−2.8
α Oph**	15.7/14.9	36	+23	−	−24.8	−26.8	−17.2
α CrB**	18/22	42	+54	−20.6	−	−19.4	−13.4
α Aql**	4.7/5.0	48	−9	−25.7	−	−24.3	−12.8
α Lyr**	6.4/7.5	67	+19	−22.4	−20.0	−21.0	−10.0
α And	36/31	112	−33	+5.5	−	+1.4	+8.0
α Leo	22/22	226	+49	+6.6	−	+13.2	+5.2
α CMa	2.8/2.65	227	−9	+10.5	−	+26.0	+14.8
β Car	14/48	286	−14	−0.7	−	+5.4	+3.2

* The velocities are all heliocentric and are measured from Copernicus data supplemented with results from the IUE.
[1] The distances given are: (spectral classification and magnitude/trigometric parallax).
[2] The predicted velocities are based upon a vector of $(1 = 25°; b = 10°; v = −28$ km s^{-1}) and $(1 = 4°; b = 16°; v = −20$ km s^{-1}) for Crutcher and Bertaux respectively.
** Denotes good agreement with Crutcher vector. (Table from Bruhweiler *et al.*, 1986b)

with IUE data has derived the ISW velocity vector from Mg II and Mg I data for 12 A and B stars within 25 pc of the Sun. Lines-of-sight of seven of these stars seem to pass either through or near the main body of the local cloud. The measured radial velocities of the interstellar magnesium features toward these stars show excellent agreement with that expected from uniform flow and the deduced Crutcher velocity vector. However, stars lying in the opposite direction of the sky and toward regions of low column density show marked discrepancies from predictions based upon either the Bertaux or the Crutcher velocity vectors. The results can be fitted by a two component velocity flow: one velocity vector being that of Crutcher fits the data upward; the second vector fits the data downwind. The direction of this second vector seems to agree with that deduced from backscattering data (Bertaux *et al.*, 1985; Chassefiere *et al.*, 1986), but it has an amplitude of only −10 km s^{-1} instead of the −20 km s^{-1} implied from the backscattering results.

Prior to the Bruhweiler *et al.* (1986b) study, Lallement, Vidal-Madjar, and Ferlet (1986; and references therein), following Crutcher's approach, using ground-based data, studied interstellar Ca II toward 20 stars at distances within 100 pc of the Sun. These data were obtained at very high resolution (3 km s^{-1}) and high signal-to-noise (~ 1000) and they revealed multiple components to the interstellar wind. They also found that the Crutcher (1982) velocity vector represents an average of several distinct components to the interstellar wind. They

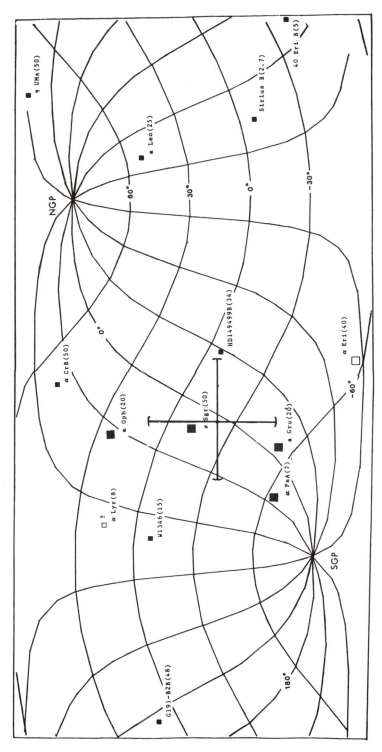

Fig. 3. H I Column Densities Toward Stars Within 50 pc. The stars with column densities derived from IUE or Copernicus data are displayed in galactic coordinates. Their distances in parsecs are given in parentheses. The stars represented with large filled squares likely have $N(\text{H I}) = 1 - 2 \times 10^{19} \text{ cm}^{-2}$ and delineate the main body of the 'local cloud'. This cloud is identified with the polarization patch of Tinbergen (1982) centered at $l = 5°$, $b = 20°$. The approximate extent of this patch is denoted. Stars with small filled squares have $N(\text{H I}) \approx 10^{18} \text{ cm}^{-2}$. The stars Eta UMa and 40 Eri B likely have $N(\text{H I}) \approx 10^{18} \text{ cm}^{-2}$. The stars with open squares appear to have contamination due to non-LTE photospheric line cores or circumstellar material. (From Bruhweiler, 1984).

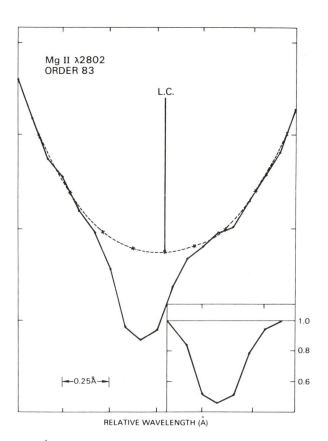

Fig. 4. The Mg II 2802 Å Feature in Alpha PsA. Data for order 83 of the LWR camera are presented from coadded spectra. These data were obtained by overexposing the stellar continuum by a factor of five in order to achieve good exposure levels in the trough of the photospheric feature. The photospheric line is denoted by the dashed line and its line center by L.C. The extracted residual intensities for the interstellar feature are at the lower right. The line centers for the interstellar and photospheric features are well-determined enabling excellent relative velocities to be obtained (From Bruhweiler and Kondo, 1982b; Courtesy of *The Astrophysical Journal*).

find excellent agreement (± 0.5 km s^{-1}) between measured radial velocities of interstellar features in the 20 stars studied and the four derived velocity vectors. Ferlet, Lallement, and Vidal-Madjar (1986) had previously found that two, possibly three, of these components were present in the short line-of-sight to Alpha Aql ($d = 5$ pc). All four of the deduced velocity vectors are shown in Figure 5. All but one of these (I) are nearly in the galactic plane. They are all different from that deduced by backscattering techniques as noted in Figure 5 by VLISM (the very local interstellar medium) (Bertaux, 1984; Bertaux *et al.*, 1985).

There are several interpretations to these results as pointed out in Bruhweiler *et al.* and Lallement *et al.* (1986) . One possibility is that we are seeing separate clouds within this short distance. However, as discussed in Bruhweiler *et al.*, if we maintain that clouds are embedded in the hot coronal substrate, which is

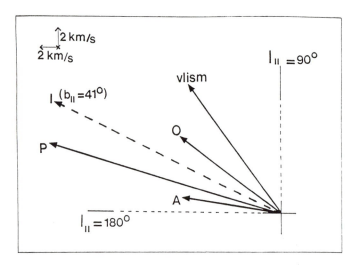

Fig. 5. The evaluated velocity vectors of the coherent motions observed to be present in the local ISM are shown in the LSR reference frame projected on the galactic plane. Only component I is clearly away from the plane and is indicated by a dashed arrow. (See text and Table III). The VLISM vector is the one observed inside the solar system by backscattering techniques (Bertaux *et al.*, 1985). The velocity vector found by Crutcher (1982) is not shown here because it seems to represent only a weighted average of the motions A, O, P, and I, discovered through much higher quality data (S/N > 1000 and resolution = 3 km s^{-1}) obtained from the ground by Lallement, Vidal-Madjar and Ferlet (1986) (From Lallement *et al.*, 1986; Courtesy of *Astronomy and Astrophysics*).

evidenced by the soft X-ray background, then these clouds would have extremely short lifetimes and would be very quickly evaporated within 10^6 yr by the surrounding hot plasma (Cowie and McKee, 1977).

An alternative view is that we are looking at the effects of a shock front moving through the local cloud. This shock could owe its origin to the putative supernova event occuring 1.4×10^5 yr ago suggested by Cox and Anderson (1982). They maintain that a supernova arose in an already evacuated cavity. Such an event would have produced a pressure increase in the coronal substrate. This increase would have then driven a pressure wave into the local cloud as well as other clouds embedded in this region (Jenkins, 1984; Bruhweiler *et al.*, 1986b). Even though the model of Cox and Anderson is still quite speculative and possibly not unique, it does show promise in explaining observations.

5. Direct IUE Measurements of H I and D/H Ratios

There have been several investigations to directly measure H I and D I column densities toward nearby late-type stars using the IUE (Landsman *et al.*, 1984, 1986; Murphy *et al.*, 1986). These studies attempted to disentangle the intervening interstellar absorption superimposed upon the chromospheric H I Ly-α emission of these stars. Since interstellar Ly-α lies on the saturated portion of the curve-of-growth, detailed profile fitting of high quality data is necessary in order to

TABLE III

Characteristics of the different motions observed in the local ISM ($d > 25$ pc)

Heliocentric Velocity*

Component	Modulus	α	δ	l^{II}	b^{II}	Reference
A	+25 km s^{-1}	85	−9	213	−19	1
O	22	83	5	199	−15	1
P	35	80	11	192	−14	1
I	29	111	27	192	19	1
V	20	74	17	184	−16	3
Crutcher	28	90	2	205	−10	2, 4
	10	74	17	184	−16	2

References:

(1) From the ground-based data of Lallement, Vidal-Madjar, and Ferlet (1986). Also see Figure 5.
(2) From the UV data of Bruhweiler *et al.* (1986b).
(3) From Backscattering data of Bertaux *et al.* (1985).
(4) From analysis of ground-based data by Crutcher (1982).

* Note that these vectors indicate the direction in which the wind is going and are opposite in sign to those given at the bottom of Table II.

determine the b-value of the H I curve-of-growth. The IUE observations complement the higher resolution, but lower signal-to-noise data obtained by the Copernicus satellite for all of the stars, except for Alpha Cen B, which was obtained only be IUE. The results for Alpha Cen A and B ($d = 1.3$ pc) show excellent agreement and indicate that $n_{H I} \lesssim 0.21$ cm^{-3} along their common line-of-sight (see Figure 6).

The study of Murthy *et al.* (1986) presents results for Procyon (3.5 pc), Epsilon Eri (3.5 pc), Altair (5.0 pc), Capella (13 pc), and HR 1099 (33 pc). The implied column densities all support the local hydrogen density being ~ 0.1 cm^{-3} near the Sun. The average H I density toward HR 1099 was found to be 0.009 cm^{-3} $\lesssim$ $n_{H I} \lesssim 0.016$ cm^{-3} and supported earlier Copernicus findings (Anderson and Weiler, 1978) that there was a sharp fall-off in $n_{H I}$ in the direction of HR 1099. It also supported the more general conclusion of Bruhweiler and Kondo (1982) that there was a fall-off in $n_{H II}$, in most directions, at larger distances from the Sun.

Of all the light elements produced in primordial nucleosynthesis of the early Universe, deuterium is most sensitive to current baryon density and the expansion rate of the Universe. We find deuterium, more specifically the present-day D/H ratio, to be one of our few probes of the physical conditions in the early Universe (see previous reviews by Vidal-Madjar, 1982; Laurent, 1983; Vidal-Madjar and Gry, 1984; Boesgard and Steigman, 1985). Since no current sites of deuterium production are known, and fragile deuterium is efficiently destroyed in stellar interiors, the observed D/H ratio is a lower limit to the primordial D/H ratio and an upper limit to the baryon density of the Universe.

Of the stars studied by Landsman *et al.* and Murthy *et al.* (1986) only Altair

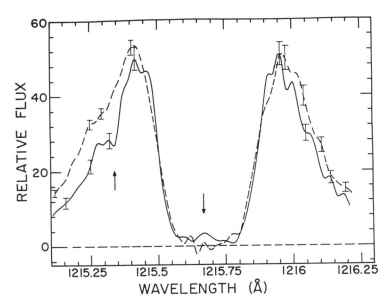

Fig. 6. The reduced, Co-added IUE Spectrum of Alpha Cen B (solid line) is compared with the IUE Ly-alpha profile of Alpha Cen A (dashed line). The data have been normalized at the red emission peak. Typical error bars are shown. The downward arrow denotes weak geocoronal/interplanetary emission, while the upward arrow denotes the predicted position of D I, as determined from the H I bulk motion. The Ly-α emission of the two stars shows some differences, but the interstellar H I absorption profiles observed toward the two stars are in excellent agreement (From Landsman *et al.*, 1986; Courtesy of *The Astrophysical Journal*).

shows no detectable interstellar deuterium. The non-detection of D I toward Altair was probably due to a combination of high column density and poor signal-to-noise ratio. The derived D/H ratios are consistent with previous determinations derived solely from Copernicus data. All lines-of-sight toward late-type stars point to $n_{D_I}/n_{H_I} \gtrsim 10^{-5}$. Furthermore, a particularly important result found in the direction of Capella ($n_{D_I}/n_{H_I} \gtrsim 1.8 \times 10^{-5}$) confirms previous Copernicus results (Dupree *et al.*, 1977; McClintock *et al.*, 1978) and indicates that apparently large n_{D_I}/n_{H_I} values are present in the nearby ISM. This contrasts with Copernicus observations of higher energy transitions of the Lyman series toward more distant early-type stars such as Lambda Sco (120 pc) in which $n_{D_I}/n_{H_I} < 10^{-5}$ are found (cf. York, 1983). This led Ferlet *et al.* (1984), Vidal-Madjar *et al.* (1986), and Murthy *et al.* (1986) to conclude that real variations of the D/H ratio might occur over relatively short distances ($d < 100$ pc) and possibly over distances even as small as a few parsecs.

The current determinations of the D/H ratio in the interstellar medium vary within a range of a factor of four ($5 \times 10^{-6} \lesssim D/H \lesssim 2 \times 10^{5}$). Nevertheless, they allow one to put a limit on the density of the Universe to less than 20% of the closure density (Boesgard and Steigman, 1985). If these determinations are compared with those for other light elements, strong constraints are also placed on the so-called standard Big-Bang model and galactic evolution models (Yang *et al.*, 1984; Delbourgo-Salvador *et al.*, 1985; Delbourgo-Salvador *et al.*, 1986).

6. Final Discussion

Observations made with the Copernicus and IUE satellites indicate that the Sun is embedded near the edge of a low column density cloud or cloudlet. Lines-of-sight toward nearby white dwarfs probed with the IUE, which only intersect the outer skin of the cloud, imply H I column densities $\lesssim 10^{18}$ cm^{-2} and a sharp dropoff in particle density after a few parsecs in these directions. Measurements toward the cloud core infer that the total gas column density is at least a magnitude higher, but still only a few times 10^{19} cm^{-2} (Bruhweiler, 1982). The H I particle density at the Sun is near 0.06 cm^{-3}, while the density seems enhanced toward the body of the cloud. The cloud probably has an extent less than 15 pc, and appears to be immersed in the hot coronal substrate, which gives rise to the soft X-ray background (Fried et al., 1980).

At distances within 50 pc of the Sun, the extremely low column densities toward white dwarfs, the absence of detectable interstellar reddening (Perry et al., 1982) and polarization (except for the small detectable polarization for the local patch), as well as the absence of cloud 'shadows' in the soft X-ray background indicate no other clouds near the Sun.

This evidence suggests that we are inside a cavity carved out by one or more supernovae, in which the cloud space density may be lower because of shock and thermal evaporation (Cowie et al., 1981; see also the review by Cowie, 1984). The local cloud appears to lie along the periphery of the 115 pc radius Loop I (Davelaar et al., 1980) and could be a shell fragment of the supernova remnant forming Loop I. Also, there is evidence that the local cloud has been shocked. Multiple velocity components observed in the interstellar gas within 5 to 7 parsecs of the Sun, as well as theoretical work, suggest that shocks are present, possibly within the local cloud (Cox and Anderson, 1982; Cowie, 1984).

At distances greater than 50 pc, the nature of the LISM is less certain (also see the review by Bruhweiler, 1984). We do see long lines-of-sight, toward Feige 24, GD 153, and Beta CMa, with very low column densities. However, analysis of Copernicus data for the line-of-sight toward Lambda Sco (105 pc) by York (1983), shows what is interpreted as five distinct H I and H II velocity components, with a total H I column density of 1.7×10^{19} cm^{-2}. Yet, this direction also intersects the local cloud and many of the components may originate within the immediate vicinity of the Sun. However, ground-based observations at millimeter and visual wavelengths do present strong evidence for a molecular cloud at a distance near 65 pc (Hobbs et al., 1986). Furthermore, ultraviolet and optical studies of interstellar lines also reveal significant gas toward stars such as Beta Cen (84 pc), Delta Cyg (50 pc), as well as toward 85 Vir, 69 Leo, and Alpha Vir, all at distances near 85 pc (again, see the review by Bruhweiler, 1984).

We cannot completely eliminate the possibility that the complex velocity structure seen in the very near LISM reflects the motion of several distinct, closely spaced clouds rather than one single cloud. Yet, the amplitude of the velocity variations and the short distances over which they arise imply significant interactions between the gaseous flow delineated by the derived bulk motions. Thus, the most likely interpretation favors the passage of shocks through possible one or two closely associated clouds (Lallement et al., 1986; Bruhweiler et al., 1986b).

Clearly, the IUE has played a fundamental role in the emergence of this much more elaborate physical description of the local interstellar medium. Although giant strides have been made, much work still needs to be done. The picture is not a simple one and many questions remain to be answered. Is the low concentration of clouds found in the LISM typical of the general interstellar medium? Is the complex velocity structure in the LISM due to a propagating shock front? If so, was it produced in a recent supernova event, possibly occurring as recently as 1.4×10^5 yr ago as suggested by Cox and Anderson (1982) and rediscussed by Arnaund and Rothenflug (1986)? Do the observed D I /H I ratios in the LISM signify real variations. Again, if so, what are the implications for galactic evolution?

References

Anderson, R. C. and Weiler, E. J.: 1978, *Astrophys. J.* **224**, 143.

Arnaud, M. and Rothenflug, R.: 1986, *Astron. Astrophys.* (in press).

Bertaux, J. L.: 1984, Y. Kondo, F. Bruhweiler, and B. Savage (eds.), 'The Local Interstellar Medium', *IUU Colloq.* **81**, 1.

Bertaux, J. L. and Blamont, J. E.: 1971, *Astron. Astrophys.* **11**, 200.

Bertaux, J. L., Lallement, R., Kurt, V. G., and Mironnova, E. N.: 1985, *Astron. Astrophys.* **150**, 1.

Boesgard, A. and Steigman, G.: 1985, *Ann. Rev. Astron. Astrophys.* **23**, 319.

Bohlin, R., Savage, B., and Drake, J.: 1978, *Astrophys. J.* **224**, 132.

Böhm-Vitense, E.: 1981, *Astrophys. J.* **244**, 504.

Bruhweiler, F.: 1982, in Y. Kondo, R. Chapman, and J. Mead (eds.), *Advances in Ultraviolet Astronomy: Four Years of IUE Research*, pp. 125.

Bruhweiler, F.: 1984, 'The Local Interstellar Medium', *IAU Colloq.* **81**, 39.

Bruhweiler, F., Gull, T., Kafatos, M., and Sofia, S.: 1980, *Astrophys. J. Letters* **238**, L27.

Bruhweiler, F. C. and Kondo, Y.: 1981, *Astrophys. J. Letters* **248**, L123.

Bruhweiler, F. C. and Kondo, Y.: 1982a, *Astrophys. J.* **259**, 232.

Bruhweiler, F. C. and Kondo, Y.: 1982b, *Astrophys. J. Letters* **260**, L91.

Bruhweiler, F. C. and Kondo, Y.: 1983, *Astrophys. J.* **269**, 657.

Bruhweiler, F. C., Kondo, Y., and McCluskey, G.: 1979, *Astrophys. J. Letters* **229**, L39.

Bruhweiler, F. C., Kondo, Y., and McCluskey, G.: 1980, *Astrophys. J.* **237**, 19.

Bruhweiler, F., Oegerle, W., Weiler, E., Stencel, R., and Kondo, Y.: 1984, in *Future of Ultraviolet Astronomy Based on Six Years of IUE Research*, NASA Publ., CP-2349, pp. 200.

Bruhweiler, F., Oegerle, W., Weiler, E., Stencel, R., and Kondo, Y.: 1986a, preprint.

Bruhweiler, F., Oegerle, W., Weiler, E., Stencel, R., and Kondo, Y.: 1986b, preprint.

Bruhweiler, F. C. and Sion, E. M.: 1986, *Astrophys. J. Letters* **304**, L21.

Bruston, P., Audouze, J., Vidal-Madjar, A., and Laurent, C.: 1981, *Astrophys. J.* **243**, 161.

Burstein, P., Borken, R. J., Kraushaar, W. L., and Sanders, W. T.: 1976, *Astrophys. J.* **213**, 405.

Chassefiere, E., Bertaux, J. L., Lallement, R., and Kurt, V. G.: 1986, *Astron. Astrophys.* (in press).

Cox, D. P.: 1979, *Astrophys. J.* **234**, 863.

Cox, D. P. and Anderson, P. R.: 1982, *Astrophys. J.* **253**, 268.

Cowie, L. L.: 1984, 'The Local Interstellar Medium', *IAU Colloq.* **81**, 287.

Cowie, L. L., Jenkins, E. B., Songalia, A., and York, D. G.: 1979, *Astrophys. J.* **232**, 467.

Cowie, L. L. and McKee, C. F.: 1977, *Astrophys. J.* **211**, 135.

Cowie, L. L., McKee, C. F., and Ostriker, J. P.: 1981, *Astrophys. J.* **247**, 908.

Crutcher, R. M.: 1982, *Astrophys. J.* **254**, 82.

Davelaar, J. A., Bleeker, J., and Deerenberg, A.: 1980, *Astron. Astrophys.* **92**, 231.

Delbourgo-Salvador, P., Gry, C., Malinie, G., and Audouze, J.: 1985, *Astron. Astrophys.* **150**, 53.

Delbourgo-Salvador, P., Audouze, J., and Vidal-Madjar, A.: 1986, *Astron. Astrophys.* (submitted).

Drake, S. A., Brown, A., and Linsky, J. L.: 1984, *Astrophys. J.* **284**, 774.

Dupree, A. K., Baliunas, S. L., and Shipman, H. L.: 1977, *Astrophys. J.* **218**, 361.

Dupree, A. K. and Raymond, J.: 1982, *Astrophys. J. Letters* **263**, L63.

Dupree, A. K. and Raymond, J.: 1983, *Astrophys. J. Letters* **275**, L71.

Fahr, H. J.: 1974, *Space Sci. Rev.* **15**, 483.

Fahr, H. J.: 1978, *Astron. Astrophys.* **66**, 103.

Ferlet, R.: 1981, *Astron. Astrophys.* **98**, L1.

Ferlet, R. and Friere Ferrero, R.: 1986, *Astron. Astrophys.* (submitted).

Ferlet, R., Gry, C., and Vidal-Madjar, A.: 1984, 'The Local Interstellar Medium', *IAU Colloq.* **81**, 84.

Ferlet, R., Lallement, R., and Vidal-Madjar, A.: 1986, *Astron. Astrophys.* (in press).

Field, G. B.: 1975, *Lectures at 1974 Les Houches Summer School in Theoretical Physics.*

Franco, M. L., Crivellari, L., Molaro, P., Vladilo, G., Ramella, M., Morossi, C., Allocchio, C., and Beckman, J. E.: 1984, *Astron. Astrophys. Suppl.* **58**, 693.

Fried, P. M., Nousek, J. A., Sanders, W. T., and Kraushaar, W. L.: 1980, *Astrophys. J.* **242**, 987.

Freire Ferrero, R.: 1984, Proc. 4th European IUE conference ESA SP-218, pp. 133.

Frisch, P.: 1981, *Nature* **293**, 377.

Frisch, P. and York, D. G.: 1983, *Astrophys. J. Letters* **271**, L59.

Frisch, P. and York, D. G.: 1986, in *The Galaxy and the Solar System*, The University of Arizona Press (in press).

Genova, R., Beckman, J. E., Molaro, P., and Vladilo, G.: 1986, in Symp. 7 of the XXVI COSPAR Meeting devoted to the Local Interstellar Medium, Toulouse, France, paper, 7.1.7.

Gry, C., York, D., and Vidal-Madjar, A.: 1985, *Astrophys. J.* **296**, 593.

Hobbs, L. M., Blitz, L., and Magnani, L.: 1986, *Astrophys. J. Letters* (in press).

Holberg, J. B., Sandel, B. R., Forrester, W. T., Broadfoot, A. L., Chipman, H. L., and Barry, D. C.: 1981a, *Astrophys. J.* **242**, L119.

Holberg, J. B.: 1981b, *Bull. Am. Astron. Soc.* **12**, 872.

Jenkins, E. B. and Meloy, D. A.: 1974, *Astrophys. J. Letters* **193**, L121.

Jenkins, E. B.: 1984, comments made at *IAU Colloq.* **81** in Madison, Wisconsin, U.S.A.

Knude, J.: 1979, *Astron. Astrophys.* **71**, 344.

Lallement, R., Vidal-Madjar, A., and Ferlet, R.: 1986, *Astron. Astrophys.* (in press).

Landsman, W. B., Henry, R. C., Moos, H. W., and Linsky, J. L.: 1984, *Astrophys. J.* **285**, 801.

Landsman, W. B., Murthy, J., Henry, R. C., Moos, H. W., Linsky, J. L., and Russell, J. L.: 1986, *Astrophys. J.* **303**, 371.

Laurent, C.: 1983, in Shaver *et al.* (eds.), *ESO Workshop on Primordial Helium*, pp. 335.

Laurent, C., Vidal-Madjar, A., and York, D.: 1979, *Astrophys. J.* **229**, 923.

McClintock, W., Henry, R. C., Linsky, J. L., and Moos, H. W.: 1978, *Astrophys. J.* **225**, 465.

McKee, C. F. and Ostriker, J. P.: 1977, *Astrophys. J.* **218**, 148.

Molaro, P., Beckman, J., Franco, M., Morossi, C., and Ramella, M.: 1984, 'The Local Interstellar Medium', *IAU Colloq.* **81**, 185.

Murthy, J., Henry, R. C., Moos, H. W., Landsman, W., Linsky, J. L., Vidal-Madjar, A., and Gry, C.: 1986, preprint.

Paresce, F.: 1984, *Astrophys. J.* **89**, 1022.

Perry, C. L., Johnston, L., and Crawford, D. L.: 1982, *Astron. J.* **87**, 1751.

Ripkin, H. W. and Fahr, H. J.: 1983, *Astron. Astrophys.* **122**, 181.

Shapiro, P. R. and Field, G. B.: 1976, *Astrophys. J.* **205**, 762.

Shapiro, P. R. and Moore, R. T.: 1976, *Astrophys. J.* **207**, 460.

Sion, E. M. and Starrfield, S. G.: 1984, *Astrophys. J.* **286**, 760.

Skuppin, R., Bianchi, L., de Boer, K. S., and Grewing, M.: 1986, *Astron. Astrophys.* (in press).

Spitzer, L.: 1978, *Physical Processes in the Interstellar Medium*, New York, Wiley, pp. 51.

Thomas, G. E. and Krassa, R. F.: 1971, *Astron. Astrophys.* **11**, 218.

Tinbergen, J.: 1982, *Astron. Astrophys.* **105**, 53.

Vidal-Madjar, A.: 1982, 'Interstellar Helium and Denterium', in Audouze *et al.* (eds.), *Diffuse Matter in Galaxies*, NATA. C 110, 57.

Vidal-Madjar, A., Ferlet, R., Gry, C., and Lallement, R.: 1986, *Astron. Astrophys.* **155**, 407.

Vidal-Madjar, A. and Gry, C.: 1984, *Astron. Astrophys.* **138**, 285.

Vidal-Madjar, A., Laurent, C., Bruston, P., and Audouze, J.: 1978, *Astrophys. J.* **223**, 589.

Vidal-Madjar, A., Laurent, C., Gry, C., Bruston, P., Ferlet, R., and York, D.: 1983, *Astron. Astrophys.* **120**, 58.
Vladilo, C., Beckman, J. E., Crivelli, L., Franco, M. L., and Molaro, P.: 1985, *Astron. Astrophys.* **144**, 81.
Weaver, R., McCray, R., and Castor, J.: 1977, *Astrophys. J.* **218**, 377.
Weller, C. S. and Meier, R. R.: 1981, *Astrophys. J.* **246**, 386.
Yang, J., Turner, M. S., Steigman, G., Schramm, and Olive, K. A.: 1984, *Astrophys. J.* **281**, 493.
York, D. G.: 1983, *Astrophys. J.* **264**, 172.

DIFFUSE AND DARK CLOUDS IN THE INTERSTELLAR MEDIUM

KLAAS S. DE BOER

Sternwarte, University of Bonn, F.R.G.

MICHAEL A. JURA

Dept. of Astronomy, Univ. of California, Los Angeles, CA, U.S.A.

and

J. MICHAEL SHULL

Joint Institute for Laboratory Astrophysics, Boulder, CO, U.S.A.

1. Introduction

This chapter deals with the IUE observations of interstellar matter in diffuse and dark clouds in the Milky Way and Magellanic Clouds. We will describe the data, its means of analysis, and the derived scientific knowledge about the structure, abundances, and physical state of the gas. We begin by discussing observations of diffuse and dark clouds in the Milky Way, in and near the disk. We then proceed to discuss depletion mechanisms and grain theories, and will review IUE observations of reflection nebulae, of gas in the Milky Way halo and of gas in the Magellanic Clouds.

2. Galactic Interstellar Clouds

2.1. CURVES OF GROWTH AND OSCILLATOR STRENGTHS

The usual technique for deriving interstellar elemental abundances is the single Maxwellian component curve of growth (Spitzer, 1978). Equivalent widths W_i (mÅ) of absorption lines are fitted to a family of curves (Figure 1) which depend on three parameters: the column density N (cm^{-2}), the Doppler velocity parameter b (km s^{-1}), and the set of absorption line oscillator strengths f_i. For extremely saturated lines, one requires the damping parameter γ, equal to the total radiative decay coefficient from the upper state. Generally, damping wings are only seen in the lower Lyman lines of H I, the R(0) and R(1) Lyman lines of H$_2$, and occasionally in C II $\lambda 1034$ and O I $\lambda 1302$. Once f_i and W_i are known, the pattern of equivalent widths can be used to derive b and N.

Curves of growth have three distinct portions, known as linear, Doppler (flat), and damped (square root). We define the optical depth at line center,

$$\tau_0 = (\pi e^2 / m_e c) N f / (\pi^{1/2} \Delta \nu_D) = N s \lambda / (\pi^{1/2} b) \tag{1}$$

where we have assumed a line cross section $\sigma_\nu = (\pi e^2 / m_e c) f \phi_\nu = s \phi_\nu$, where ϕ_ν is

Y. Kondo (ed.), Scientific Accomplishments of the IUE, pp. 485–515.
© 1987 *by D. Reidel Publishing Company.*

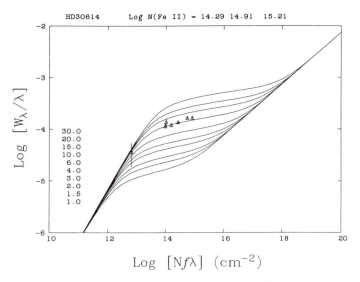

Fig. 1.   A family of curves of growth for Fe II towards HD 30614. Doppler velocities b (km s^{-1}) are listed at left. Note the importance of the weak line at 2260.08 Å. Horizontal scale is labelled with log $N = 14.91$ and log($f\lambda$) = 2.89, corresponding to the strongest line at 2382 Å.

the profile function,

$$\phi_\nu = [\pi^{1/2}(\Delta \nu_D)]^{-1} \exp[-(\nu - \nu_0)^2/(\Delta \nu_D)^2] \qquad (2)$$

with Doppler frequency width $(\Delta \nu_D) = b/\lambda$. The velocity width is $b = (2kT/m)^{1/2}$ if the broadening is thermal. However, it has been found empirically that micro-turbulence, shear flows, or multiple velocity components can produce non-Gaussian profiles and a b-value which substantially exceeds that expected from thermal broadening. The equivalent width (in wavelength units) for each portion of the curve of growth is given approximately by the formulae:

$$W_{\text{lin}} = (\pi e^2/m_e c^2)Nf\lambda^2 = (\lambda^2/c)\,(\pi^{1/2}\Delta \nu_D)\,\tau_0 \qquad (3)$$

$$W_{\text{Dopp}} = (\lambda^2/c)\,(2\Delta \nu_D)\,(\ln \tau_0)^{1/2} \qquad (4)$$

$$W_{\text{damp}} = (\lambda^2/c)\,(\Delta \nu_D \gamma/\pi^{1/2})^{1/2}\,\tau_0^{1/2}. \qquad (5)$$

The transition from linear to Doppler portions occurs at $\tau_0 \approx 1$, while the transition from Doppler to damped portions is given by the solution to the transcendental equation,

$$(\tau_0/\ln \tau_0) = (\Delta \nu_D/\gamma)\,(4\pi^{1/2}). \qquad (6)$$

For typical resonance lines with $b = 1 - 5$ km s^{-1}, this transition occurs at a central optical depth τ_0 between 10^3 and 10^4.

The single component curve of growth has come under attack, since realistic lines of sight contain many velocity components. Calcium II studies find 8 to 12 clouds/kpc (Marschall and Hobbs, 1972), while more sensitive *Copernicus* studies of UV resonance lines suggest 14 kpc^{-1} (Cowie and York, 1978). Because the large b values derived from UV data probably reflect the dispersion in component velocities, the standard curve of growth with moderately saturated lines may underestimate the total column density in cloud cores with small thermal b values (Nachman and Hobbs, 1973). We return to this point in Section 2.4.

The standard compilation of oscillator strengths used for analysis of *Copernicus* satellite data was Morton and Smith (1973), later updated in Morton (1978). For many lines new f-values were empirically determined from absorption lines in Copernicus and IUE interstellar studies. The most recent data are available for C I (Jenkins *et al.*, 1983), Si II (Shull *et al.*, 1981) and Fe II (Shull *et al.*, 1983), using an empirical method that attempts to minimize residuals of equivalent widths for 6 to 12 stars. Also, theoretical calculations brought improved f-values (Nussbaumer *et al.*, 1981; Dufton *et al.*, 1983; Hibbert *et al.*, 1985; Luo *et al.*, 1986). A few of these disagree with the empirical f-values but the discrepancies may be the result of multiple velocity components. As discussed in Section 2.4, systematic differences in column densities of Fe II determined from *Copernicus* and IUE data probably result from high velocity components which contaminate the wings of stronger lines. Table I lists the major lines used for analysis of IUE data on heavy elements, together with our current best estimates of oscillator strengths.

2.2. GENERAL STRUCTURE

The interstellar gas layer is generally confined to the galactic disk, although there is

TABLE I

Useful lines and oscillator strengths for interstellar work with IUE

Element	λ (Å)	f	Ref.	Element	λ (Å)	f	Ref.
C II	1334.532	0.118	1	Mn II	2576.107	0.288	1
	2324.69	4.5(−8)	2		2593.731	0.223	1
					2605.697	0.158	1
Si II	1808.012	0.0041	3				
	1526.708	0.116	3	Zn II	2025.512	0.412	1
	1304.372	0.10	3		2062.016	0.202	1
	1260.421	1.16	3				
	1193.289	0.580	3	Mg II	1239.925	2.67(−4)	5
	1190.416	0.294	3		1240.395	1.33(−4)	5
					2795.528	0.592	1
Fe II	1608.451	0.062	4		2802.704	0.295	1
	2343.496	0.108	4				
	2373.736	0.0395	4	Ni II	1317.38	0.12	6
	2382.038	0.328	4		1370.20	0.10	6
	2585.876	0.0573	4		1454.96	0.052	6
	2599.396	0.203	4		1741.56	0.068	6
	2260.08	0.0028	4		1751.92	0.040	6
	2249.2	0.0018	7		1773.96	0.0054	6

References: (1) Morton and Smith (1973); (2) Nussbaumer and Storey (1981); (3) Luo *et al.* (1986); (4) Shull *et al.* (1983); (5) Hibbert *et al.* (1983); (6) Kurucz and Peytremann (1975); (7) de Boer *et al.* (1986).

good evidence for high velocity clouds and small amounts of H I at high galactic latitudes. One popular model of the ISM (McKee and Ostriker, 1977) envisions a neutral medium almost entirely in clouds with cold, dense cores and tenuous outer envelopes, which mimic an intercloud medium. Most of the volume is filled with low density, hot ('coronal') gas produced by overlapping supernova remnants (SNRs). However, 21-cm and Ly α studies of the H I distribution (Heiles, 1980; Liszt, 1983; Lockman *et al.*, 1986) suggest that a portion of the H I is smoothly distributed.

Physical studies of the interstellar gas in the UV usually begin by determining the hydrogen content. The damping wings of the Ly α line can be fitted to yield $N(\text{H I})$, typically to 0.1 dex. IUE studies cannot directly measure H_2, whose resonance lines lie below 1120 Å. However, statistically, the molecular fraction makes a rapid transition from negligible values to sizeable fractions ($f > 0.1$) at $E(B-V) = 0.08$ (Savage *et al.*, 1977). One can use this correlation with $E(B-V)$ to estimate the molecular correction to the total hydrogen column density. In an IUE high resolution (0.1 Å) study, Shull and Van Steenberg (1985) determined $N(\text{H I})$, $E(B-V)$, and distances to 244 OB stars. The logarithmic column densities range from around 19.0 up to 21.81, and the data set includes stars up to $E(B-V) = 0.86$ and distance $r = 8.5$ kpc, with 68 stars in the galactic halo ($|b| > 20°$). Toward 75 stars also surveyed by *Copernicus*, the IUE values of $N(\text{H I})$ were in good agreement; toward the one exception, the star ρ Oph, the IUE column was 0.21 dex lower, a value confirmed by de Boer *et al.* (1986) in a reanalysis of the $N(\text{H I})$ from Copernicus.

For a statistical subset of 205 stars earlier than B2.5 (Figure 2), the survey yields a mean density $\langle N(\text{H I})/r \rangle = 0.46$ cm^{-3} and a ratio $\langle N(\text{H I})/E(B-V) \rangle = 5.2 \times 10^{21}$ cm^{-2} mag^{-1}, in excellent agreement with the *Copernicus* sample of 75 OB stars within 1 kpc. The exponential scale height of the H I layer is

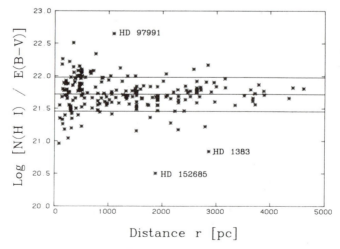

Fig. 2. The 'gas-to-dust' ratio, $N(\text{H I})/E(B-V)$, from a survey of 205 OB stars (Shull and Van Steenberg, 1985) shows no correlation with distance. Horizontal lines show the mean, 5.2 × 10^{21} cm^{-2} mag^{-1}, and 1 σ variations. Anomalous lines of sight are labelled.

144 ± 80 pc, and all 17 stars with vertical height $|z| > 1$ kpc have $N(\mathrm{HI}) > 10^{20}$ cm^{-2}, which suggests that lines of sight out of the Galaxy with EUV (100 Å) optical depth less than 1 will be rare. Both $\langle N(\mathrm{HI})/r \rangle$ and $\langle E(B-V)/r \rangle$ decrease for $r > 1$ kpc. This decrease may result largely from observational bias towards moderately reddened stars at distances beyond 2 kpc. However, within 2 kpc, some of the decrease may be due to interarm gas. A statistical analysis of the $E(B-V)$ distribution also suggests variations in the cloud density and the mean reddening per cloud.

Lockman *et al.* (1986) examined selected high latitude stars in the Ly α and 21 cm lines to determine the distribution of H I in the halo. The Ly α column, $N\alpha$, measures gas between us and the star, while the 21-cm column, N_{21}, includes gas beyond the star. The ratio, $N_{21}/N\alpha$, or the difference, $(N_{21} - N\alpha)$, provide information on H I at high latitudes. The 25 OB stars observed in detail lie between 60 and 3100 pc of the galactic plane. The total column density of H I at $|z| > 1$ kpc is, on the average, $(5 \pm 3) \times 10^{19}$ cm^{-2}, or 15% of the total H I. At low z, the data toward some stars suggest a low effective scale height and fairly high average foreground density, while toward others the effective scale height is large and the average density low. Also, the lack of large fluctuations in the ratio, $N_{21}/N\alpha$, is inconsistent with the McKee—Ostriker model in which most of the H I is in small cloud cores. These data can be understood if some H I is smoothly distributed. A satisfactory fit to the data can be found by placing half the H I mass in clouds of several parsecs radius and a Gaussian vertical distribution with $\sigma_z = 135$ pc, and the other half in a smooth exponential component with 500 pc scale height.

It is also instructive to note (Figure 3) that the IUE Ly α survey exhibits a range of mean densities, $\bar{n} = \langle N(\mathrm{HI})/r \rangle$. Cowie and Songaila (1986) pointed out that the minimum value of $\bar{n}$ should be greater than or equal to the average density of warm H I. The range of $\bar{n}$ in Figure 3, particularly for stars in the disk, suggests

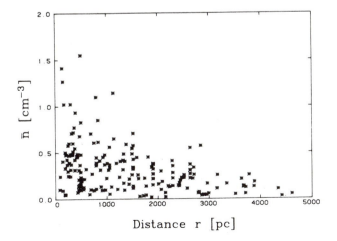

Fig. 3.   Mean line-of-sight H I density, $\bar{n} = N(\mathrm{HI})/r$, for 196 stars in IUE survey (Shull and Van Steenberg, 1985); 9 stars with $\bar{n} > 2$ cm^{-3} are excluded for clarity.

that $\bar{n}_{min} \approx 0.1$ cm^{-3}, somewhat lower than the values 0.25—0.35 cm^{-3} preferred on the basis of 21-cm data (Heiles, 1980; Payne et al., 1983). The difference between $\bar{n}_{min}$ from the two surveys is not understood, although there may be selection effects for lines of sight towards OB stars. If the intercloud substrate density is as low as 0.1 cm^{-3}, the greater size and lifetime of the radiative shells from SNRs may result in many radiative shocks per line of sight. The scarcity of high velocity components seen in UV resonance lines toward OB associations (Cowie et al., 1981) may be in conflict with this prediction.

2.3. HIGH VELOCITY GAS

The existence of high velocity clouds has been known since the Ca II studies by Adams (1949). The observation (Routly and Spitzer, 1952; Siluk and Silk, 1974) that interstellar clouds with high velocities show higher Ca II/Na I ratios has been widely interpreted as evidence of selective grain destruction, which returns the highly depleted calcium back to gas phase. The same correlation with cloud velocity has been seen in silicon and iron (Shull et al., 1977; Spitzer, 1976). While theoretical models of fractional grain destruction suffer from fundamental uncertainties about rates of sputtering, grain-grain destruction, and partial vaporization (Shull, 1977, 1978; Cowie, 1978; Draine and Salpeter, 1979; Seab and Shull, 1983, 1985), both models and data are consistent with the general conclusion that components with velocities greater than about 20 km s^{-1} have larger gas-phase abundances of refractory elements than low-velocity gas.

Ultraviolet extinction curves are also expected to show evidence of shock processing (Seab and Shull, 1983). Non-thermal sputtering and grain-grain collisions destroy relatively more large grains than small ones, and more silicates than graphite. Consequently, both the 2175 Å extinction feature and the far-UV rise in extinction are expected to increase in strength after the passage of a shock wave. IUE studies of nine stars near three supernova remnants (Monoceros Loop, Shajn 147, and Vela) exhibit these effects (Figure 4). Further work on these and other regions of activity are needed to see whether these effects are universal. If they are, then grain processing has significant implications for abundances, gas cooling, and molecular chemistry.

The probable sources of these high velocity components are SNRs, stellar wind bubbles, and the combined effects of many such events (superbubbles). Absorption line studies of stars behind SNRs can provide valuable information on velocities, masses, and pressures in the ISM. The SNRs that have been extensively observed include Vela (Jenkins et al., 1976a, b, 1984); Shajn 147 (Silk and Wallerstein, 1973; Gondhalekar and Phillips, 1980; Phillips et al., 1981; Phillips and Gondhalekar, 1983); and the Monoceros Loop (Wallerstein and Jacobsen, 1976). However, recent IUE studies of stars thought to lie behind the Mon Loop (Fesen and Shull, 1986) show no evidence for high velocity components. The one star (HD 47240) which shows evidence of high velocity gas lies at the edge of the Mon Loop, and could result from a separate dynamical interaction.

Absorption studies of OB associations would be expected to show numerous examples of high velocity gas and superbubbles. The Orion's Cloak structure is the best example (Cohn and York, 1977; Cowie et al., 1979) with many lines of sight

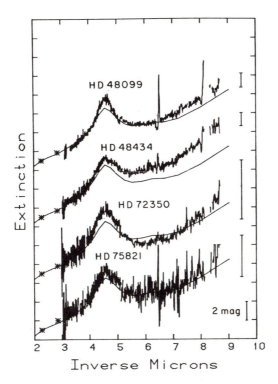

Fig. 4. Normalized selective extinction curves, $E(\lambda - V)/E(B - V)$, for four stars out of nine associated with SNR's (Seab and Shull, 1983). Solid curve shows Galactic average (Savage and Mathis, 1979).

showing negative velocity absorption components out to -100 km s^{-1}, but the Trapezium stars do not show such velocities (Franco and Savage, 1982). The spatial structure is patchy (Shore, 1982), but the 100 pc extent of the absorption across the Orion OB1 association, Barnard's Loop, and Lambda Ori region suggests that the region has been affected by a series of SNRs and stellar winds over the past million years. The low inferred hydrogen column density of the absorption component suggests that the mean density in the region, $< 3 \times 10^{-3}$ cm^{-3}, is typical of the coronal cavities in the hot phase of the ISM (McKee and Ostriker, 1977). However, except for Orion OB1, Carina OB1 and OB2, and I Per OB region, these structures appear to be rare. Cowie *et al.* (1981) found that only two out of thirteen OB associations studied with IUE in C IV and Si IV showed evidence for similar coherent structures. The reasons for this rarity are uncertain, but may involve the homogenizing effects of photoionization on the cloudy structure (Elmegreen 1976; McKee *et al.*, 1984).

2.4. HEAVY ELEMENT DEPLETION

Copernicus found that significant fractions of many heavy elements are depleted from the interstellar gas, presumably locked up in dust grains. Quantitatively, we define the depletion factor d_i of an element (i) by the relation,

$$\log d_i = \log(N_i/N_H) - \log(N_i/N_H)_\odot \tag{7}$$

where N_i and N_H are the column densities of (i) and of H, and where $(N_i/N_H)_\odot$ is the solar or cosmic abundance (Withbroe, 1971; Grevesse, 1984). Generally, the elements S, Ar, Cl, P, and Zn are not depleted; C, N, and O are depleted by factors of 2 or 3; and 'refractory' elements such as Si, Fe, Ca, Ti, Mn, Al, and Ni are depleted by factors ranging from 10 to 10^3 (Spitzer and Jenkins, 1975; Phillips *et al.*, 1982; de Boer *et al.*, 1986). Data for well studied elements are collected in Table II.

TABLE II

Abundance of elements in the diffuse interstellar medium from UV absorption lines

Element	Solar Abundance	Interstellar gas			
		lower column densities		higher column densities	
X	$\log(X/H)$	$\log(X/H)$	Ref.	$\log(X/H)$	Ref.
C	−3.4	−3.7	7	−4.2	7
N	−4.0	−4.2	14	−4.2	14
O	−3.16	−3.4	1, 14, 9	−3.4	1, 14, 9
Mg	−4.46	−4.8	10, 2, 8	−5.2	10, 2, 8
Al	−5.6	−6.6	11	−7.6	11
Si	−4.45	−5.0	11, 15	−5.8	11, 15
P	−6.5	−6.7	8	−7.1	8
S	−4.8	−4.8	3, 5	−5.4	3, 5
Cl	−6.5	−6.8	4, 8	−7.2	4, 8
Cr	−6.3	−7.8	2	−8.4	2
Mn	−6.58	−7.3	2, 8	−7.8	2, 8
Fe	−4.50	−5.9	12, 2, 8, 15	−6.7	12, 2, 8, 15
Ni	−5.72	−7.1	13	−7.7	8
Cu	−7.5	−.—		−9,1	8
Zn	−7.4	−7.5	2, 6	−7.9	2, 6

References: (1) de Boer, 1981; (2) de Boer *et al.*, 1986; (3) Gondhalekar, 1985; (4) Harris and Bromage, 1984; (5) Harris and Mas Hesse, 1986a; (6) Harris and Mas Hesse, 1986b; (7) Jenkins *et al.*, 1983; (8) Jenkins *et al.*, 1986; (9) Keenan *et al.*, 1985; (10) Murray *et al.*, 1984; (11) Phillips *et al.*, 1982; (12) Savage and Bohlin, 1979; (13) Savage and de Boer, 1981; (14) York *et al.*, 1983; (15) Van Steenberg and Shull, 1986a; solar abundances are from recent compilations.

The depletion factors of C, N, and O are difficult to measure, since they rely on weak lines whose oscillator strengths are uncertain. For an average of 53 lines sight, York *et al.* (1983) found that nitrogen and oxygen are depleted by about a factor of 2, and Hobbs *et al.* (1982) measured (at the 2σ level) a carbon depletion of a factor of 3 towards Delta Scorpii. However, each of these measurements is only as accurate as the assumed oscillator strengths and solar abundances. For instance, Hobbs *et al.* (1982) assumed an oscillator strength $f = 6.7 \times 10^{-8}$ for the 2324.69 Å spin-forbidden line of C II, based on calculations by Cowan, whereas Nussbaumer and Storey (1981) found $f = 4.5 \times 10^{-8}$ with an *ab initio* calculation. Hobbs *et al.* also assumed a solar carbon abundance of 3.7×10^{-4} (Withbroe, 1971), while a more recent compilation (Grevesse, 1984) gives 4.9×10^{-4}. The latter value may be too high if solar energetic particle abundances are the same as photospheric abundances (Breneman and Stone, 1985). Thus, the factors of 1.5 and 1.3 uncertainty in oscillator strength and solar abundance mean that the carbon depletion is uncertain by a factor of 2.

Although absolute depletions are uncertain, one can study the depletion pattern to understand the process by which heavy elements condense onto grains in cool stellar atmospheres or interstellar clouds. Field (1974) noticed a correlation between the relative depletions of most elements and the temperatures at which they would be expected to condense in cool stellar atmospheres. Snow (1975) noted a similar correlation with the first ionization potential of the elements, suggesting that some depletion may be related to gas-grain interactions in clouds. Elements depleted by more than a factor of 20 probably undergo some in-cloud depletion, since possibly as much as 5—10% of stellar mass loss to the ISM comes from O star winds, which apparently contain no grains (see next section).

Correlations have been found between depletion of refractory elements and mean density $\bar{n}$ along the line of sight (Savage and Bohlin, 1979; Shull and Van Steenberg, 1982; Phillips *et al.*, 1982; York and Jura, 1982; Harris *et al.*, 1983; Phillips *et al.*, 1984; Harris *et al.*, 1984; Murray *et al.*, 1984; Jenkins *et al*, 1986). It is tempting to interpret this correlation as evidence for some *in situ* depletion. However, Spitzer (1985) suggests that this correlation can be understood by an idealized model, based on random distributions of two types of cold clouds plus uniformly distributed warm H I (intercloud medium) at a density less than 0.2 cm^{-3}. Each component has a different depletion pattern, with heavy elements less depleted in intercloud matter. As $\bar{n}$ increases, the relative contributions of the three components change; the low-density intercloud gas dominates for $\bar{n} < 0.2 \text{ cm}^{-3}$ and the large clouds dominate at $\bar{n} > 3 \text{ cm}^{-3}$.

Alternatively, much of the depletion pattern may be determined by grain processing in shocks. We know that shocks of velocity 40 km s^{-1} and higher produce substantial grain destruction; 100 km s^{-1} shocks can destroy over half the grain material (Seab and Shull, 1983). Since the mean time between 100 km s^{-1} shocks in the diffuse ISM is about 10^8 yr (Seab and Shull, 1985), whereas the mean time for grain astration is about 2×10^9 yr (Dwek and Scalo, 1980), a typical grain population will be processed through 10—20 fast shocks if it lives long enough for astration. Although the calculations behind these numbers are approximate and depend on the model of the interstellar medium (Seab *et al.*, 1985), they suggest that the interstellar grain population and the heavy element depletion pattern is strongly influenced and perhaps determined by the physics of grain processing.

With its sensitivity to OB stars down to 11th magnitude, IUE has provided some information on these questions with a comprehensive survey of H I column densities (Shull and Van Steenberg 1985) and heavy element abundances (Van

Steenberg and Shull, 1986a, b). Figures 5 and 6 show the results for Fe and Si in 220 lines of sight towards stars out to about 5 kpc from the Sun. Several points are worth noting:

(1) Both Si and Fe show a correlation between depletion and $\bar{n}$.

(2) The Fe/Si abundance ratio is greater for halo stars; $\langle Fe/Si \rangle = 0.12$ and 0.22 for $|b| < 20°$ and $|b| > 20°$.

(3) Although depletions are generally large (10—1000), several lines of sight have low depletion factors (5 or less).

One could interpret point (1) as evidence for either in-cloud depletion or the Spitzer selection effect. A possible physical interpretation of the latter effect is that lines of sight with low mean density are more likely to have had extensive shock processing. But the scatter and the continuation of the correlation to large $\bar{n}$ may disagree with Spitzer's model in detail. Similarly, the enhanced Fe/Si abundances in the halo (point 2) may reflect the tendency for greater relative shock destruction of iron grains in the low density halo. Theoretical models (Shull, 1978) show some tendency for this effect, if the iron is contained in grains of greater than average mass-to-area ratio (greater density). In fact, the data on selective depletion may shed some light on the size-composition relation of grains. Finally, point (3) suggests that some lines of sight have undergone substantial shock processing. While this is expected for lines of sight with $\bar{n} < 1$ cm^{-3}, the existence of a few cases with large $\bar{n}$ is curious.

Before going too far with these interpretations, it is worth thinking about what could be wrong with the IUE data. On the observational side, many important weak lines are difficult to measure due to the finite dynamic range and fixed pattern noise of the IUE detector. The lines could be contaminated by stellar absorption, the continuum and background can be uncertain, and the 20—30 km s^{-1} IUE resolution could hide multiple velocity components. In the data analysis, uncertain f-values and the presence of multiple components makes accurate determinations of column density N and Doppler parameter b difficult. The latter effect deserves more discussion, since it may have led some authors to overinterpret correlations of depletion with $\bar{n}$.

The Fe II abundances were examined by Van Steenberg and Shull (1986a) for 45 lines of sight studied by both *Copernicus* (Bohlin *et al.*, 1983; Jenkins *et al.*, 1986) and IUE. The three far-UV lines used for the *Copernicus* survey (1122, 1134, and 1097 Å) have smaller f-values than the six lines generally used for the IUE survey (1608, 2343, 2374, 2382, 2586, 2599 Å). The chi-square fits to the IUE data yield systematically higher b-values and lower column densities log N(Fe II). For all 45 stars, the IUE columns are lower by 0.25 (dex), on average, with several stars lower by 0.5—0.8 (dex). We believe that this systematic difference results from the presence of high-velocity components which increase the equivalent width of strong lines faster than the $(\ln N)^{1/2}$ expected from a single component curve of growth. This effect is accentuated by the fact that Fe is more abundant in these components due to grain processing (the 'Routly-Spitzer effect' for Fe).

This bias can lead to erroneous correlations. Because the systematic under-estimate of Fe II column densities is greatest for strong lines, the IUE data yield

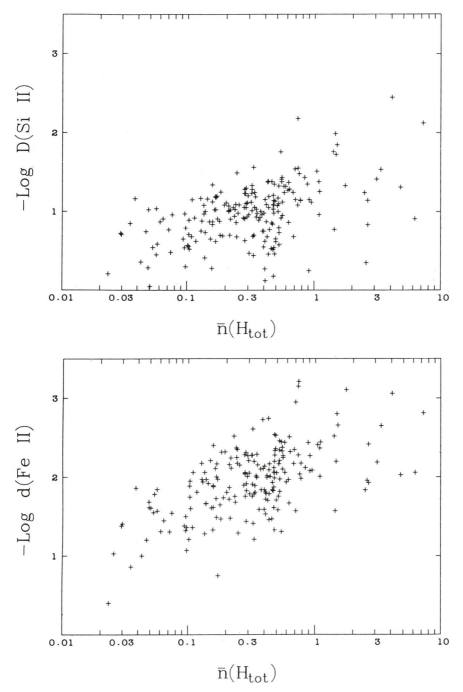

Fig. 5. Depletions of Fe and Si (Van Steenberg and Shull, 1986a) versus mean HI density, $\bar{n}$, for 220 OB stars earlier than B3. The correlation may result partly from a bias introduced by grain processing in high-velocity clouds.

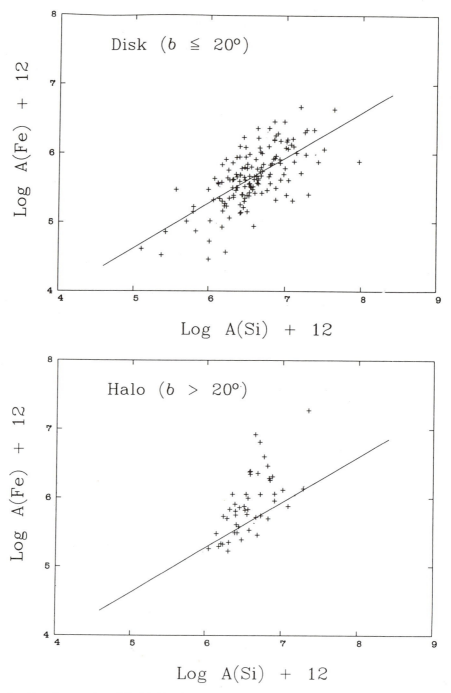

Fig. 6.   Abundance ratios of Fe/Si (Van Steenberg and Shull, 1986a) for stars in disk and halo. The diagonal, of slope 0.66, is the best fit for disk stars; halo stars have higher Fe/Si ratios by a factor of 2 on average.

greater depletions for lines of sight with larger column density and larger mean density $\bar{n}$. All correlations of depletion with mean density are therefore suspect. This bias should be significant for other heavy elements affected by grain processing in high velocity shocks: Si, Mn, Ca, Ni, etc. The IUE abundances for these elements probably say more about high velocity clouds and intercloud matter (with large effective b-values) than about cold cloud cores. Elements such as S, Cl, P, and Ar, which are thought to be depleted by small factors, should be less subject to the contamination of high velocity clouds.

With IUE, one can partially correct for this systematic effect (Figure 1) by measuring or obtaining upper limits for the weak Fe II 2260 Å line (Shull *et al.*, 1983) and the 2249 Å line (de Boer *et al.*, 1986). These lines have a smaller strength, $f\lambda$, than any other Fe II line in the *Copernicus* data, and effectively determine the curve of growth for the cold cloud cores. Evidently (Figure 5) some correlation of Fe depletion with $\bar{n}$ still exists. This is probably real, since some of the increase in $\bar{n}$ results from the contribution of clouds with greater physical density, in which many authors have demonstrated enhanced depletions (Stokes, 1978; Gondhalekar, 1985; Snow and Jenkins, 1980; Snow and Joseph, 1985).

2.5. DARK CLOUDS OBSERVED BY IUE

Some of the most significant interstellar IUE observations have been made towards heavily reddened stars in the outer portions of dark clouds. These observations push IUE to the limit, since the high selective extinction in the UV rapidly increases exposure times for $E(B-V)$ greater than 0.5. *Copernicus* showed (Morton, 1974, 1975; Meyers *et al.*, 1985) that depletions toward the moderately reddened star Zeta Oph, with $E(B-V) = 0.33$ and $\bar{n} = 1.3$ cm^{-3}, were larger than in typical diffuse lines of sight. In the IUE survey (Shull and Van Steenberg, 1985), there are only four stars with mean hydrogen densities greater than 3 cm^{-3}: ρ Oph A (6.1 cm^{-3}), 34 Cyg (4.8 cm^{-3}), σ Sco (4.5 cm^{-3}) and HD 37061 (3.2 cm^{-3}). These sightlines probably pass through one or more dense clouds with significant reddening. However, few ultraviolet studies have penetrated the outer regions of dark clouds, for obvious reasons.

The general result of abundance studies in dark clouds is that depletions are large, but uncertain owing to poor photometric quality and only approximate corrections for the amount of hydrogen in molecular form. Both *Copernicus* and IUE were able to observe the brightest stars in the ρ Oph cloud (Savage and Bohlin, 1979; Snow and Jenkins, 1980). The remarkable result of the latter study was a lack of variation in the relative depletions among lines of sight (Figure 7), over a range in log N(H) from 21.08 (β Sco) to 21.60 (ρ Oph A). This is surprising, since simple theoretical pictures of depletion predict that the most depleted elements should exhibit relatively more depletion in denser portions of the cloud. The range in column density, however, was not large enough to reach depths corresponding to the regime in which grains obtain ice mantles (Harris *et al.*, 1978) or where near-UV radiation ceases to play a role in the thermal balance and chemistry (Tielens and Hollenbach, 1985a, b).

Recent IUE work by Joseph *et al.* (1986) has shown that a unique pattern of relative depletion (especially Mn/Fe) occurs for interstellar clouds having large CN

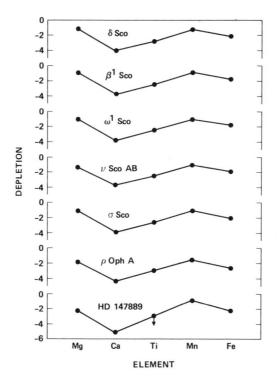

Fig. 7. Relative depletions in the ρ Oph dark cloud (Snow and Joseph, 1985); stars arranged in order of increasing $E(B-V)$, top to bottom. The general pattern appears constant with depth, except for Ca and Mn toward HD 147889.

abundances and anomalous extinction curves. This pattern is not seen in comparably reddened lines of sight with lower CN abundances. As we have seen, the correlation between depletions and mean density does not provide conclusive proof of gas-grain interactions. Since CN is a sensitive density indicator, this data may be the best observational evidence of grain accretion. Likewise, the anomalous extinction curves for these lines of sight are consistent with theoretical curves generated from a grain distribution which has experienced moderate accretion.

The embedded B2 V star HD 147889 provides the deepest IUE observation into the ρ Oph cloud. With $V = 7.92$ and $E(B-V) = 1.08$, this star is barely within the range of IUE (moderate S/N data requires in excess of 400 min exposures for both SWP and LWP cameras). The calcium depletion for this star appears to be in excess of a factor of 10^5 (Snow et al., 1983; Crutcher and Chu, 1984). Snow and Joseph (1985) have used IUE to infer that relative depletions are enhanced in the densest portions of the cloud toward HD 147889 (Figure 7).

Further work on the outer regions of dark clouds will have to await the *Hubble Space Telescope* and *FUSE/Lyman* satellites, whose sensitivity will allow studies of stars with $E(B-V)$ up to 1.5 or 2.0. Key problems are awaiting solution, such as the $^{13}CO/H_2$ ratio, the relative depletion factors, important trace molecular species which serve as diagnostics of photochemical models, and the growth and physical properties of grains in dark clouds.

3. DEPLETION THEORY

3.1. HOMOGENEITY IN THE INTERSTELLAR MEDIUM

The interstellar medium in the solar neighborhood appears to have true abundances, as opposed to just gas-phase abundances, that are homogeneous to within at least a factor of two (Jura, 1982). This is shown by observations of the dust to gas ratios (Bohlin *et al.*, 1978) and of the abundances of elements that are not substantially depleted onto grains such as oxygen and nitrogen (York *et al.*, 1983). However, an even tighter constraint on true variations of interstellar abundances can be given by observations of the ratio of $^{12}C/^{13}C$ since we do not expect isotopes to be differentially depleted onto grains. The most direct determination of the isotope ratio is achieved by measuring optical absorption lines of $^{12}CH^+/^{13}CH^+$ since CH^+ is not subject to chemical fractionation or isotope-dependent photodissociation (Watson, Anicich and Huntress, 1976). Hawkins *et al.* (1985) reported that toward ζ Oph, the ratio of $^{12}C/^{13}C$ is 43 ± 6. Observations of 4 other stars are all within 20% of this value (Hawkins, 1986; Hawkins and Jura, 1987). At this time, however, the isotope ratio determined from optical measurements of CH^+ differs by about a factor of 2 from the value of ~ 75 derived from radio observations of various molecules (see, for example, Wilson *et al.*, 1981). Once this discrepancy is understood, it may be possible to argue with confidence that the interstellar medium is homogeneous to within a factor of 1.2. At the moment, however, given all the uncertainties, it is more conservative to think that it is homogeneous only to within a factor of 2. We suspect that all well-established measurements performed on stars within 1 kpc of the Sun that display more than a factor of 2 variations from solar abundances reflect depletion onto grains.

3.2. SOURCES OF INTERSTELLAR MATTER AND DUST

The discovery of depletion of interstellar matter raises the question of whether the depletion occurs at the sources of the interstellar matter — mainly stars — or whether it occurs within the interstellar medium itself. As mentioned before, Field (1974) noted that there is a strong correlation between the condensation temperature of an element and the amount of its depletion. Therefore, Field proposed that the observed interstellar depletions occur mainly within the envelopes of mass losing red giants. This idea is attractive because red giants are the main source for the replenishment of the interstellar medium and these stars are known to contain large amounts of dust (Zuckerman, 1980; Olofsson, 1985). A traditional view of interstellar grains is that of refractory cores surrounded by mantles composed of relatively volatile material. In Field's picture, the cores are synthesized in the outflows from red giants, while the grains accrete mantles within interstellar clouds. However, quantitative evaluation of this picture shows that it is not likely to be correct. While condensation of elements certainly does occur in the outflows from red giants, this does not seem to explain the full pattern of depletions observed in the interstellar medium.

There are several arguments against the idea that grains are ejected into the interstellar medium but are not subsequently substantially processed. Although there are some uncertainties in the analysis, it seems that the red giants generally have dust to gas ratios by mass of about 1% (Knapp, 1985; Jura, 1986), a value similar to that which is found in the general interstellar medium (Spitzer, 1978). Therefore, if grains only came from red giants, there could be a close correlation between circumstellar and interstellar grains. However, mass loss from early type stars accounts for more than 10% of all the matter ejected into the interstellar medium, and there is no reason to think that grains form in the outflows from these stars, with the exception of the WC9 stars (Dyck *et al.*, 1984). Therefore, any element which displays a depletion which is smaller than 10%, such as iron, calcium or titanium, or any of the particularly refractory species that Field thought were mainly depleted in the atmospheres of red giants must actually stick to grains in the interstellar medium itself. Specifically, Knapp and Morris (1985) and Abbott (1983) give the rates for mass loss from red giants and blue giants as $2 \times 10^{-4}\ M_\odot\ \mathrm{yr}^{-1}\ \mathrm{kpc}^{-2}$ and $9 \times 10^{-5}\ M_\odot\ \mathrm{yr}^{-1}\ \mathrm{kpc}^{-2}$, respectively. There is such a relatively large mass loss rate from hot stars, that much of the interstellar matter must be processed through these objects.

Additional evidence that grain processing occurs in the interstellar medium is given by:

(i) Gas phase abundances of normally depleted elements like calcium, silicon and iron are much larger in high velocity clouds. This presumably occurs because of the removal of these elements from grains by sputtering and grain-grain collisions (see, for example, Seab and Shull, 1983).

(ii) The depletion of markedly refractory elements such as iron is highly correlated with cloud density (Savage and Bohlin, 1979 and others, see Section 2).

(iii) Phosphorus and iron have nearly the same condensation temperature (Grossman and Olsen, 1974; Wai and Wasson, 1977) yet phosphorus displays a much lower depletion than does iron (Jura and York, 1978). It seems that thermodynamic equilibrium does not govern the depletion of these two elements.

3.3. DEPLETION THEORY

The simple theory of depletion onto grains is very straightforward. The gas phase abundance of an element, $n(X)$, can be written (see, for example, Spitzer, 1978) as:

$$\mathrm{d}n(X)/\mathrm{d}t = -n(X)\langle n_{\mathrm{gr}}\sigma_v\rangle S(X) + J. \tag{8}$$

In Equation (8), the first term on the right hand side describes the rate of material adhering onto grains while J is the rate of removal. In this equation, n_{gr} is the density of grains, σ is the cross section for grain-atom collisions, v is the relative velocity of the grains and S is the probability of sticking onto a grain surface.

The amount of material that sticks onto a grain depends upon the rates in Equation (8) and, also, the initial conditions. The initial conditions depend upon the mass lost from stars and are sketched above in Section 3.2. Here, we note that there appears to be ample time in the interstellar medium for substantial depletion to occur. The best evidence for this point is the existence of large amounts of H_2. Molecular hydrogen is synthesized on the surfaces of grains (Shull and Beckwith, 1982), and the widespread presence of this molecule indicates that most clouds survive sufficiently long that there are appreciable numbers of gas-grain collisions. To be quantitative, Jura (1975a) found from analysis of *Copernicus* observations of H_2 that the rate of formation of H_2 on grains, R, is about 3×10^{-17} cm^3 s^{-1} in agreement with theoretical expectations. Here $R = \langle n_{gr}\sigma_v \rangle S$ for hydrogen, the equivalent to the first term on the right hand side in Equation (8). The time scale for the conversion of atomic to molecular hydrogen is $(Rn)^{-1}$ which, for $n = 100$ cm^{-3}, equals 10^7 yr. This time is apparently shorter than the lifetime of a diffuse interstellar cloud against disruption.

By analogy with the formation of H_2, we might expect that there could be considerable condensation onto grains of atoms in the interstellar medium. There are, however, significant differences between hydrogen and heavier atoms. First, the thermal speed of heavier atoms is, of course, smaller than that of hydrogen so the rate of atom-grain collisions is correspondingly reduced. To counterbalance this, however, the sticking probability of a heavy atom is likely to be close to 1 (Watson and Salpeter, 1972; Leger, 1983), while the sticking probability of hydrogen atoms onto grains is perhaps only about 1/3 (Hollenbach and Salpeter, 1971). One major difference might occur, however, if grains are positively charged. In this case, the rate of collisions with positively charged ions would be much smaller than given by Equation (8). Grains are charged by the photoeffect, and the yield from this for interstellar grains of different sizes and compositions is quite uncertain (Spitzer, 1978). Because some species which are normally positively charged, like iron, are observed to be depleted in the interstellar medium, it seems likely that a significant number of the grains are not positively charged. Some grains might even have a small negative charge (Spitzer, 1978) which would enhance the depletion rate of positively charged ions. It is observed that species with a higher ionization potential are less depleted than those with a lower ionization potential (Snow, 1975).

3.4. WHY ARE ONLY CERTAIN ELEMENTS DEPLETED?

As discussed in the section on observations, it seems that only certain elements are substantially depleted. While it seems that nearly every element should collide with and adhere to grains, in order to explain the observed pattern of depletion, we need to consider selective removal of atoms from grains. Although we do not yet fully understand the processes on grain surfaces, some progress has been made.

3.4.1. *Non-Selective Removal from Grains*

There is evidence that in high velocity clouds, even the most refractory species such as calcium and titanium are removed from grains. It is likely that this results from sputtering, and to a lesser extent, from grain-grain collisions (Seab and Shull, 1983). While these processes can produce some segregation by mass, the effect is not dramatic (Shull, 1978). Sputtering is probably not the main pathway off grains for volatile species such as oxygen.

3.4.2. *Selective Removal*

The process which most obviously leads to selective removal and which is straight-forward to calculate is classical evaporation as discussed by Watson and Salpeter (1972), Duley (1973, 1976), Allen and Robinson (1976), Purcell (1976), Aannestad and Kenyon (1979), and Leger (1983). The rate of classical evaporation of a species from a grain, R_{evap}, depends mostly upon the binding energy of the species onto the grain. Very roughly, we can write:

$$R_{evap} = 10^{13} \exp(-\Delta E/kT) \text{ s}^{-1}. \tag{9}$$

In Equation (9), ΔE is the binding energy of atom or molecule onto the grain and T_{gr} is the temperature of the grain. In diffuse clouds where T_{gr} is close to 20 K, the evaporation time is very sensitive to the binding energy. An important molecule is H_2 whose binding energy is about 500 K (Hollenbach and Salpeter, 1971). Therefore, the evaporation time is on the order of 0.1 s, and we do not expect these molecules to remain long on the surfaces on grains in diffuse clouds.

Unfortunately, at the moment, the binding energies of single atoms onto grain surfaces is not well known. We do know the binding of stable molecules onto themselves, and we can use these values to estimate the amount of these species that might remain in the gas phase or condense onto grains in molecular clouds (Leger, 1983; Leger *et al.*, 1985). Some species are not at all tightly bound such as H_2; others, such as H_2O are strongly bound to the grains and are not easily evaporated. These models can also be applied to mass-losing stars, and the theory actually is reasonably successful in explaining which oxygen-rich mass-losing stars have detectable ice band features in their outflows (Jura and Morris, 1985).

One variation of the simple evaporation model is that the grains undergo temperature fluctuations in a variety of ways, and when the temperature is high, evaporation proceeds very rapidly. For small grains, the heat capacity is sufficiently low that when a high-Z cosmic ray such as iron passes through the grain, there can be substantial temperature excursions (Leger *et al.*, 1985). For tiny grains, photon-induced temperature fluctuations are important (Sellgren, 1984). As can be seen from Equation (9), the rate of evaporation increases dramatically at higher temperatures. Even though a grain does not spend much time at elevated temperatures, most evaporation may occur at these periods. Another variation of the classical evaporation occurs for large grains. In these objects, the heat capacity of the whole grain is so large that a cosmic ray does not produce a significant temperature rise in the entire material, but there is a tube of impulsively heated solid that follows the trail of the cosmic ray through the gas (Watson and Salpeter, 1972; Leger *et al.*, 1985). At the surface of the grain, this 'spot heating' can lead to significant evaporation of adsorbed material.

Finally, chemical explosions may be another important removal mechanism (d'Hendecourt *et al.*, 1982). The basic idea is that molecules adhere to the surface of the grain. When the material is subject to an ultraviolet radiation field, radicals form. If the temperature of the mantle material is sufficiently high, the radicals can react with each other, release latent heat, and this results by a chemical explosion in the removal of most of the mantle material. This sort of processing could be selective in the sense that species which are unusually tightly bound may remain on the surface of the grains after the explosion.

While it is clear that there are processes on the surfaces of grains which can lead to the selective removal of different species and therefore lead to the observed pattern of differential depletions, a detailed explanation for this phenomena is still not clearly established.

4. Reflection Nebulae

New insights into the nature of dust scattered light were sought in a few well directed studies using data obtained with the IUE. The observational conditions are, however, not very favourable. Although the smallness of the IUE entrance aperture allows one to carefully select fields free of stars (in contrast to observations with the earlier UV photometric satellites OAO−2 and TD1) and very close to the star, the generally low surface brightness of reflection nebulae imposes to limit the observations with the IUE to the brightest nebulae. After measurements in the UV on a large scale of the Orion field (OAO−2: Witt and Lillie, 1978; TD1: Morgan *et al.*, 1982; rocket: Onanka *et al.*, 1984), of the Pleiades (ANS: Andriesse *et al.*, 1977), and of the ζ Ori field (ANS: de Boer, 1983a), the following reflection nebulae have been investigated with the IUE.

NGC 7023, a well known bright reflection nebula of about 6' diameter was investigated by Witt *et al.* (1982). The nebula is associated with a dark molecular cloud and the star HD 200775, of type B3 V, may be illuminating its immediate surrounding dense (200 cm^{-3}) gas. Intensities were recorded at two positions at about 20" from the star. Having corrected for possible severe contamination of light scattered in the IUE telescope, Witt *et al.* (1982) find that the surface brightness follows a radial power law $r^{-2.0}$ out to 30" in the reflection nebula NGC 7023 (see Figure 8). The grain albedo then changes from 0.4 to 0.6 at 2200 to 1400 Å and the dust scattering is more isotropic towards smaller wavelengths.

Scattered light near ζ Ori was further probed using the IUE, extending the photometry with the ANS (de Boer, 1983a) to very small distances from the star. A measurement with the IUE at the same location of an ANS measurement showed perfect agreement in the scattered light intensity. In this field the intensity drops as $I(r)/I(0) = r^{-1.7}$ (see Figure 8), but if light scattered in the IUE telescope was present, the slope could be more gentle at −1.4, out to 30' distance (de Boer, 1986). The diffuse light here must be due to scattering by foreground dust, suggesting enhanced forward scattering towards shorter wavelengths.

The nebulosities in the Pleiades were further probed by Witt *et al.* (1986a). Including information from extensive interstellar line work it is found that the Pleiades nebulosities are very close (10^{-2} pc) to the stars, and the measured diffuse light then suggests that scattering is more isotropic towards the smaller UV wavelengths. Small particles (10^{-6} cm) play an important role in the scattering of light in the Pleiades.

Pronounced spectral structure in scattered light was found by Witt *et al.* (1986b) from measurements of Ced 201 (22h 12, +69°) and IC 435 (5h 40, −2°). The normalised nebular intensities show distinct peaks in the 2200 to 2400 Å

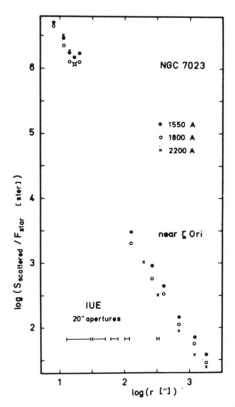

Fig. 8. The normalised nebular intensity of the scattered light near NGC 7023 (Witt *et al.*, 1982) and near ζ Ori (de Boer, 1986) are collected into one figure to demonstrate the extent of the scattered light measurements. The NGC 7023 data were shifted by 0.31 in log(S/F) and by −0.15 in log(r) to normalise to the same distance (500 pc) as the reflecting dust near ζ Ori. The scattered light intensities drop near both stars following a power law with $-1.4 < p < -2.0$, in spite of differing star-dust geometries. Light scattered in the IUE telescope drops with a power of $p = -3$ for r in arcsec (Witt *et al.*, 1982).

range, indicative of enhanced albedo at these wavelengths. This feature does not follow from models for dust extinction (e.g. Mathis *et al.*, 1979) so that the physical properties of dust in the interstellar medium may have to be reconsidered.

5. Gas in the Milky Way Halo

5.1. Observations

The IUE can achieve high dispersion spectra with fair signal to noise ratio of OB stars brighter than $V = 13$ mag. This capability allowed one to probe the outer reaches of the Milky Way, in particular away from the disk, by looking at UV-bright objects such as Magellanic Cloud stars and blue stars in galactic globular clusters, but also at extragalactic supernovae and at quasars. Such spectra yield information on gas along the entire line of sight. The spectra of extragalactic sources were the first to show convincingly the true interstellar nature (in contrast to circumstellar) of a major fraction of the C IV and Si IV absorption seen towards hot stars (Jenkins, 1981).

The pilot low dispersion spectrum of R136 in the LMC (de Boer *et al.*, 1978; Savage and de Boer, 1979), with the full 400 km s^{-1} wide absorption in the C II line, implied absorption at all velocities intermediate to the LSR and that of the LMC and therefore probably also absorption due to gas along the entire 55 kpc line of sight. Subsequently, many more stars in the LMC and the SMC were observed (e.g. Gondhalekar *et al.*, 1980) and a large body of data with comprehensive analysis was presented by Savage and de Boer (1981). Further experience with IUE observations of faint objects allowed to observe quasars with redshift such that the intrinsic Ly α emission appears near the velocity of galactic Si IV or C IV absorption (York *et al.*, 1984), extragalactic supernovae bright enough to see the galactic absorption line spectra (e.g. Pettini *et al.*, 1982; Jenkins *et al.*, 1984), UV-bright stars in galactic globular clusters (de Boer and Savage, 1983, 1984) and B stars outside the Milky Way disk (Pettini and West, 1982; Berger *et al.*, 1985). Visual observational work (Ca II and Na I interstellar absorption lines) received an impetus (studies by York, Songaila, Blades, Meaburn, Pettini, and collaborators) and new 21-cm HI observations helped to identify the absorption components seen in the UV (McGee, Newton and Morton, 1983). Reviews of the developments in the study of the halo, since the prediction by Spitzer (1956), are available from de Boer and Savage (1982), York (1982) and de Boer (1985a).

5.2. HIGH VELOCITY CLOUDS

The absorption seen on halo lines of sight is manifest in discrete components of the low ionization stages (O I, C II, Mg II, Fe II, etc.) and in more diffuse absorption features of highly ionized interstellar gas (Si IV, C IV, N V). The low ionization stages represent clouds of neutral hydrogen, which have a metal content close to that of the interstellar gas in the solar vicinity (Savage and de Boer, 1981; McGee *et al.*, 1983). Evidently, the gas outside the disk of the Milky Way is rather processed material, and not primordial. The gas seen, therefore, originates in the Milky Way disk. It is likely the cooled, recombined and descending part of a galactic fountain type flow as postulated by Shapiro and Field (1976), but other mechanisms to bring disk gas into the halo may operate as well.

The gas clouds seen in the halo in UV absorption may be similar to the high-velocity clouds (HVCs) discovered in 21-cm emission in the mid 1960's. After their discovery (see review by van Woerden *et al.*, 1985) efforts were made to determine their distances by searching spectra of stars in the same direction for Ca II and Na I interstellar lines, but without success (Habing, 1969). Also UV observations of halo stars within 2 kpc (Pettini and West, 1982) did not show neutral high-velocity gas. These failures imply a lower limit for the distance of the HVCs from the disk of up to 2 kpc. After the UV absorptions (at +60 and +120 km s^{-1}) were found towards the LMC stars, suggesting gas z-distances well over 2 kpc, new radio measurements confirmed these clouds in H I 21 cm (McGee *et al.*, 1983). The radial velocities found in absorption against the globular cluster M13 (de Boer and Savage, 1983) suggested rather large distances for the absorbing gas (but with upper limit of $z = 4$ kpc of M13), unless the gas has extremely large peculiar motion (de Boer, 1983b).

A consistency calculation was performed by Kaelble *et al.* (1985). They showed that the collection of HVCs in the survey of Giovanelli (1980) has velocities which are consistent with a gentle return of neutral gas in the fountain-type flow (Figure 9). The HVCs typically are at $z = 2$ to 5 kpc, rotate along with the Milky Way disk at a velocity of only 100 km s^{-1} (instead of the 220 km s^{-1} of the LSR), have a galactocentric velocity component of up to 100 km s^{-1} and a velocity towards the disk of up to 100 km s^{-1}.

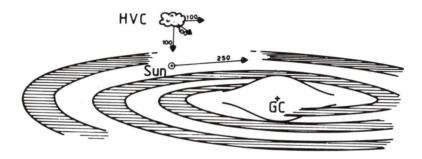

Fig. 9. Schematic representation of the typical motion of high velocity clouds in the halo of the Milky Way as derived by Kaelble *et al.* (1985) from an analysis of H I 21-cm data and IUE UV absorption-line data. At a mean z-distance of 3 kpc they have a space velocity with components of up to 100 km s^{-1} towards the disk and towards the galactic center, and of about 100 km s^{-1} in the direction of galactic rotation.

Dynamical models for the flow of halo gas had been presented by Bregman (1980), who also reached consistency with HVCs, and by Habe and Ikeuchi (1980, 1982).

5.3. IONIZATION STRUCTURE

The absorption lines of Si IV and C IV are very strong on lines of sight penetrating the halo. The amount of pertinent information is, however, still rather limited.

The Si IV and C IV lines are rather wide, extend to velocities well away from 0 km s^{-1} LSR and generally do not show discrete absorption components at higher radial velocity. This implies (a) that the gas is likely hot with a large velocity dispersion, and (b) that it does not spatially coincide with the gas that produces the O I, C II, Mg II, etc. absorption lines. Column densities from all available lines of sight indicate that the C IV gas, if with an exponential height distribution, has a scale height in the range of 2 to 6 kpc (de Boer, 1985a), possibly in the lower range or with small disk densities outside the solar circle, and in the higher range or with larger disk gas densities inside the solar circle (Figure 10).

A lot of effort has gone into finding the mechanism for the production of C IV and related ions in the halo. It may be that the high ions represent the cooling phase of the descending branch of the halo flow, the rapidly cooling C IV being constantly replenished from the hot, 10^6 K coronal phase. Ionization may also be

TABLE III

Observations of C IV and Si IV in the galactic halo

Object	l°	b°	d (kpc)	Reference
16 stars			⩽ 3	Pettini and West, 1982
M3	42	+79	10	de Boer and Savage, 1984
M13	59	+41	6	de Boer and Savage, 1983
HD 100340	259	+61	5	Berger et al., 1984
2 stars	280	−34	1	Savage and de Boer, 1981
LMC	280	−34	55	Savage and de Boer, 1981
				Gondhalekar et al., 1980
				de Boer and Nash, 1982
3C273	290	+64	—	Ulrich et al., 1980
				York et al., 1984
SMC	302	−45	70	Savage and de Boer, 1981
				Fitzpatrick and Savage, 1983
				Fitzpatrick, 1984

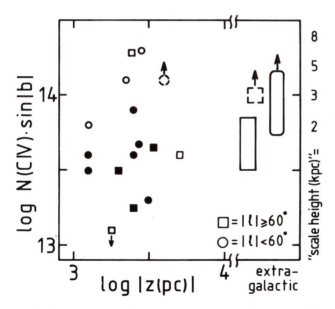

Fig. 10. The normalised C IV column density in relation to path through the Milky Way halo indicates that the C IV gas has a scale height of 2 to 6 kpc in the solar neighborhood. There is a slight tendency to larger values in directions towards the inner Milky Way. The filled symbols are 'accurate' data; the tall box represents the range of column densities found looking toward the LMC, the rounded box is for the SMC direction; the dashed square is for 3C273 and the dashed circle the lower limit on the path to M13. Figure from de Boer (1985a).

produced by energy transfer of Alfvèn waves (Hartquist and Tallant, 1981), by hard UV photons from QSOs (York, 1982), by cosmic rays (Hartquist et al., 1984; Fransson and Chevalier, 1985), by white dwarfs in the disk (Panagia and

Terzian, 1984), by globular cluster post-AGB stars (de Boer, 1985b), and by isolated disk OB stars (Bregman and Harrington, 1986). A thorough discussion of all those possibilities and their relative contribution to the ionization in the halo shows that all radiation sources play a role, each contributing essential photons at specific wavelength ranges (Bregman and Harrington, 1986).

6. Gas in the Magellanic Clouds

6.1. DIFFUSE GAS

The diffuse gas of the Magellanic Clouds (MCs) has been probed with the IUE in spite of the large distance (50—70 kpc) and the concurrent apparent faintness of the background lightsources employed. At low dispersion, exposure times for early type stars run from 10 to 60 m and extinction studies are easily possible (see pp. 517—529). Exposure times for high-dispersion spectra of the brightest and bluest Magellanic Cloud stars are of the order of one shift or more, which makes interstellar work in the MCs rather costly in observing time.

For a total of several dozen of MC stars high-dispersion spectra have been collected, but only data for a few lines of sight have been analysed and data been published. The statistics of the data show an interesting selection effect. In Table IV, which is updated from de Boer (1984a), the MC stars are listed for which interstellar lines have been reported in the visual and in the UV. A recent publication by Songaila *et al.* (1986) on Ca II lines, containing a list of 48 stars, could not be included in the Table; the publication contains new Ca II profiles only for 5 stars in the 30 Dor region and for 8 SMC stars. The visual and UV sample have only a small overlap, showing that in each wavelength range the stars with the largest apparent brightness in that range have been selected for observation. Because of the moderate signal to noise ratio in the IUE high-dispersion spectra of MC stars, only few lines of sight have been investigated in detail. In addition, part of the data was obtained to pursue the nature of the stars themselves or the nature of the gas in the galactic halo, and only profiles have been published (see Savage and de Boer, 1981, Gondhalekar *et al.*, 1980). This leaves only 4 lines of sight for which an in depth investigation has been carried out.

6.2. SPECIAL DIRECTIONS

The brightest object, R136, proved to be the easiest target for getting good spectra. The UV absorption lines revealed a complex pattern of velocity components for the gas in the surrounding 30 Doradus H II region. There are absorption lines from neutral gas, probably in the foreground of the H II region, there is absorption by lines from excited levels of O I and Si II likely to originate in a transition zone of dense and warm gas, and there are lines from well ionized gas such as the Al III, Si IV and C IV lines (de Boer *et al.*, 1980). The velocities of the absorption fit into the pattern of velocities seen in radio and visual absorption and emission lines. This picture was confirmed in a later study by Feitzinger *et al.* (1984). After several spectra of R136 became available, the sum of them allowed to search for weak absorption lines: C I, O I, Mg I, Cl I, Cr II, Mn II, Ni II, and Zn II

TABLE IV

Magellanic Cloud stars toward which interstellar absorption lines have been recorded

Star	name		Emission Nebula	Reference to data in visual	UV (IUE)
SMC:					
Sk	31			SY, C	
Sk	78	HD 5980	e66	C	BS, SB, FS
Sk	80		e66	C	SB
Sk	108	R 31	e76		BS, SB
Sk	159		e85	SY	P9, F
Sk	191			SY	
28	others			C	
LMC:					
−65	40			SCY	
−65	54	R 94	e43	S	
−66	28	R 64	e11B	S	
−66	169		e65	SY	
−67	5	HD 268605	e3	SY	SB
−67	18		e9		BS, SB
−67	104	HD 36402	e51D		BS, BN
−67	174		e57A	MB	
−67	250		e70	BEM	
−68	52	HD 269050	e100	FTW	
−68	82	HD 269546	e144	FTW, SCY	SLG
−69	91	eS 95		SY	
−69	102	HD 269357	e119		BS
−68	114	HD 269700	30 Dor	FDM	
−69	220	HD 269858F	30 Dor	FDM	
−69	221	HD 269859	30 Dor	FDM	
		R 134	30 Dor	FTW	
−69	243	HD 38268	30 Dor	FTW, BM, B, W,	BKS, BS, SB, FHS, BFS
		R 139	30 Dor	FTW, B	
		R 140	30 Dor	FTW	
−69	246	HD 38282	30 Dor	SY, SCY	BS, SB, GWMN
−69	248	R 145	30 Dor	B	
−71	3	eS 155		SY, SCY	
−71	42	R 112	e206	SCY	
−71	45	R 113	e206A		GWMN

Stars are identified with their Sanduleak number: SMC Sandulaek (1968), LMC Sanduleak (1970); star names with R are from FTW, with eS from Henize (1956); nebulae seen in projection with the stars are identified with their number from Henize (1956) preceeded by e(mission).

B = Blades 1980, BEM = Blades *et al.*, 1980, BFS = de Boer *et al.*, 1985, BKS = de Boer *et al.*, 1980, BM = Blades and Meaburn, 1980, BN = de Boer and Nash, 1982, BS = de Boer and Savage, 1980, C = Cohen, 1984, F = Fitzpatrick, 1984, FDM = Ferlet *et al.*, 1985, FHS = Feitzinger *et al.*, 1984, FS = Fitzpatrick and Savage, 1983, FTW = Feast *et al.*, 1960, GWMN = Gondhalekar *et al.*, 1980, MB = Meaburn and Blades, 1980, P9 = Prévot *et al.*, 1980, S = Songaila, 1981, SB = Savage and de Boer, 1981, SCY = Songaila *et al.*, 1981, SLG = Schulz-Lüpertz and Grewing, 1982, SY = Songaila and York, 1980, W = Walborn, 1980.

all were detected at the velocity of the neutral, dense component. The metal abundances appeared to follow a pattern like that of Milky Way interstellar gas (de Boer *et al.*, 1985). The level of the abundances is below that of the Milky Way, demonstrating the intrinsically lower metal content of the LMC, but the depletion pattern is similar (see Figure 11).

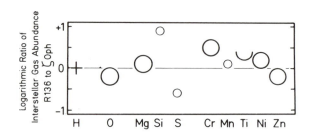

Fig. 11. Comparison of the abundance of free metals in gas in front of R136 in the LMC with gas in front of ζ Oph in the Milky Way. Small circles represent less certain values. Note that both lines of sight have the same reddening in the respective galaxies, $E(B-V) = 0.32$, while to R136 log N(H) = 21.8 and to ζ Oph log N(H) = 21.15. Since the average ratio of the abundance of free metals is unity, it follows that the depletion pattern in the LMC is similar to that in the Milky Way, but that the LMC has a larger gas-to-dust ratio. Data are from absorption lines only, for R136 from de Boer *et al.* (1985) and for ζ Oph from Morton (1975) and de Boer *et al.* (1986).

N51D in the LMC and its foreground gas was studied using IUE spectra of HD 36402 (de Boer and Nash, 1982). This rather isolated and almost spherical H II region contains a small cluster of stars. The velocity structure of the N51D bubble and its dimensions fitted well with the total energy radiated by the stars, as available from OAO—2 photometry (Koornneef, 1984). Abundances agreed with those from visual emission line studies. N51D was found to lie in a region of the LMC with rather dilute neutral gas.

In the SMC HD 5980 and its H II region were studied in detail by Fitzpatrick and Savage (1983). A difficulty, compared with lines of sight to the LMC, is that the SMC intrinsic velocity at only about 150 km s^{-1} causes blending of absorption by gas of the Milky Way with that of the SMC. Nevertheless, velocity patterns were determined and the abundances of metals derived were yet lower than in the LMC.

A puzzling aspect of the SMC HD 5980 absorption spectra is the presence of gas with velocity of 300 km s^{-1}, thus moving away from us towards HD 5980 at about 150 km s^{-1} (de Boer and Savage, 1980; Fitzpatrick and Savage, 1983). The gas has roughly normal metal ratios, has a low overall column density, but no emission by H I at 21 cm is known. This gas may be a normal HVC of the SMC approaching the SMC, or a very high-velocity cloud receding from the Milky Way.

Sk159 in the SMC had been observed in the early days with the IUE by Prévot *et al.* (1980). Fitzpatrick (1984) took several more spectra and combined them to improve the signal to noise in the data. He was able to identify, in spite of local severe blending with stellar lines, the main neutral gas absorption of the SMC and the absorption due to photoionized gas near the star. Abundances for SMC gas were found to be about 1/10 of that of Milky Way gas. Strong C IV absorption without Si IV at velocities receeding from the SMC provided support for the existence of a SMC coronal gas halo.

6.3. MAGELLANIC CLOUD HALOS

The presence of a halo of coronal type gas surrounding the MCs has been postulated on the basis of IUE data by de Boer and Savage (1980). Their work was subsequently criticised by Feitzinger and Schmidt-Kaler (1982). The difficulty in the analysis is the same as when one looks toward Milky Way early type stars: these stars may have generated their own H II regions containing C IV and Si IV gas, which then may blend with or even mimic absorption due to halo Si IV and C IV. The chance of the presence of such a superposition is particularly large since the most luminous of the MC stars are used to probe the gas. The various arguments were reviewed by de Boer (1984b), who concludes that the IUE data are consistent with the MCs possessing halos of coronal gas. In particular, on those LMC lines of sight where the C IV absorption is separable from that of the stellar environment, it occurs rather at the velocity of the LMC itself and does not follow the rotation curve of the LMC neutral gas. The existence of coronal gas around the SMC has been convincingly demonstrated by Fitzpatrick (1984), who found C IV at 100 km s^{-1} (heliocentric) with C IV/Si IV like that in the Milky Way halo. Here the 300 km s^{-1} cloud seen towards HD 5980 must be recalled, which may be an example of 'fountain' like conditions in the SMC halo, as they exist in the halo of the Milky Way.

Acknowledgements

The studies by J. M. Shull were supported by NASA/IUE grant NAG5-193 at Colorado. K. S. de Boer thanks Adolf Witt for valueable suggestions on the section on reflection nebulae, and Max Pettini for discussions on Magellanic Cloud abundances. J. M. Shull thanks Michael Van Steenberg for his help in accumulating data, results, and figures based on his Ph.D. thesis.

References

Aannestad, P. A. and Kenyon, S. J.: 1979, *Astrophys. J.* **230**, 771.
Abbott, D. C.: 1982, *Astrophys. J.* **263**, 723.
Adams, W. S.: 1949, *Astrophys. J.* **109**, 354.
Allen, M. and Robinson, G. W.: 1975, *Astrophys. J.* **195**, 81.
Andriesse, C. D., Piersma, T. R., and Witt, A. N.: 1977, *Astron. Astrophys.* **54**, 841.
Berger, J., Freire Ferrero, R., Fringant, A. M., Gerbaldi, M., and Morguleff, N.: 1985, in *Fourth European IUE Conference*, ESA SP-218, p. 161.
Blades, J. C.: 1980, *Monthly Notices Roy. Astron. Soc.* **190**, 33.
Blades, J. C., Elliot, K. H., and Meaburn, J.: 1980, *Monthly Notices Roy. Astron. Soc.* **192**, 101.
Blades, J. C. and Meaburn, J.: 1980, *Monthly Notices Roy. Astron. Soc.* **190**, 59.
Bohlin, R. *et al.*: 1983, *Astrophys. J. Suppl.* **51**, 277.

Bohlin, R. C., Savage, B. D., and Drake, J. F.: 1978, *Astrophys. J.* **224**, 132.

Bregman, J. N.: 1980, *Astrophys. J.* **236**, 577.

Bregman, J. N. and Harrington, P. J.: 1986, *Astrophys. J.* **309**, 833.

Breneman, H. H. and Stone, E. C.: 1985, *Astrophys. J. Lett.* **299**, L57.

Cohen, J. G.: 1984, *Astron. J.* **89**, 1779.

Cohen, H. and York, D. G.: 1977, *Astrophys. J.* **216**, 408.

Cowie, L. L.: 1978, *Astrophys. J.* **225**, 887.

Cowie, L. L., Hu, E. M., Taylor, W., and York, D. G.: 1981, *Astrophys. J. Lett.* **250**, L25.

Cowie, L. L. and Songaila, A.: 1986, *Ann. Rev. Astron. Astrophys.* **24**, 499.

Cowie, L. L., Songaila, A., and York, D. G.: 1979, *Astrophys. J.* **230**, 467.

Cowie, L. L. and York, D. G.: 1978, *Astrophys. J.* **223**, 876.

Crutcher, R. M. and Chu, Y.-H.: 1984, *Astrophys. J.* **290,**, 671.

de Boer, K. S.: 1981, *Astrophys. J.* **244**, 848.

de Boer, K. S.: 1983a, *Astron. Astrophys.* **125**, 258.

de Boer, K. S.: 1983b, in R. M. West (ed.), 'Highlights of Astronomy', 6, p. 657, Reidel.

de Boer, K. S.: 1984a, in *4th European IUE Conference*, ESA SP-218, p. 179.

de Boer, K. S.: 1984b, in S. van den Bergh and K. S. de Boer (eds.), 'Structure and Evolution of the Magellanic Clouds', *IAU Symp.* **108**, 375.

de Boer, K. S.: 1985a, *Mitt. Astron. Ges.* **63**, 21.

de Boer, K. S.: 1985b, *Astron. Astrophys.* **142**, 321.

de Boer, K. S.: 1986, in *New Insights in Astrophysics*, 8 years of IUE, ESA SP-263, 551.

de Boer, K. S. and Nash, A. G.: 1982, *Astrophys. J.* **255**, 447 (erratum **261**, 747).

de Boer, K. S. and Savage, B. D.: 1980, *Astrophys. J.* **238**, 86.

de Boer, K. S. and Savage, B. D.: 1982, *Scientific American* **247**, No. 2.

de Boer, K. S. and Savage, B. D.: 1983, *Astrophys. J.* **265**, 210.

de Boer, K. S. and Savage, B. D.: 1984, *Astron. Astrophys.* **136**, L7.

de Boer, K. S., Fitzpatrick, E. L., Savage, B. D.: 1985, *Monthly Notices Roy. Astron. Soc.* **207**, 115.

de Boer, K. S., Koornneef, J., Savage, B. D.: 1978, *Bull. Am. Astron. Soc.* **10**, 726.

de Boer, K. S., Koornneef, J., and Savage, B. D.: 1980, *Astrophys. J.* **236**, 769.

de Boer, K. S., Lenhart, H., van der Hucht, K. A., Kamperman, T. M., Kondo, Y., and Bruhweiler, F. C.: 1986, *Astron. Astrophys.* **157**, 119.

d'Hendecourt, L. B., Allamandola, L. J., Baas, F., and Greenberg, J. M.: 1982, *Astron. Astrophys.* **109**, L12.

Draine, B. D. and Salpeter, E. E.: 1979, *Astrophys. J.* **231**, 77.

Dufton, P. L., Hibbert, A., Kingston, A. E., and Tully, J. A.: 1983, *Monthly Notices Roy. Astron. Soc.* **202**, 145.

Duley, W. W.: 1973, *Astrophys. Space Sci.* **23**, 43.

Duley, W. W.: 1976, *Astrophys. Space Sci.* **46**, 261.

Duley, W. W.: 1980, *Monthly Notices Roy. Astron. Soc.* **190**, 683.

Dwek, E. and Scalo, J. M.: 1980, *Astrophys. J.* **239**, 193.

Dyck, H. M., Simon, T., and Wolstencroft, R. D.: 1984, *Astrophys. J.* **277**, 675.

Elmegreen, B.: 1976, *Astrophys. J.* **205**, 405.

Feast, M. W., Thackeray, A. D. and Wesselink, A. J.: 1960, *Monthly Notices Roy. Astron. Soc.* **121**, 337.

Feitzinger, J. V. and Schmidt-Kaler, T.: 1982, *Astrophys. J.* **257**, 587.

Feitzinger, J. V., Hanuschik, R. W., and Schmidt-Kaler, T.: 1984, *Monthly Notices Roy. Astron. Soc.* **211**, 867.

Ferlet, R., Dennefeld, M., and Maurice, E.: 1985, *Astron. Astrophys.* **152**, 151.

Fesen, R. and Shull, J. M.: 1986, unpublished IUE data.

Field, G. B.: 1974, *Astrophys. J.* **187**, 453.

Fitzpatrick, E. L.: 1984, *Astrophys. J.* **282**, 436.

Fitzpatrick, E. L. and Savage, B. D.: 1983, *Astrophys. J.* **267**, 93.

Franco, J. and Savage, B. D.: 1982, *Astrophys. J.* **255**, 541.

Fransson, C. and Chevalier, R. A.: 1985, *Astrophys. J.* **296**, 35.

Giovanelli, R.: 1980, *Astron. J.* **85**, 1155.

Gondhalekar, P. M.: 1985, *Astrophys. J.* **293**, 230.

Gondhalekar, P. M. and Phillips, A. P.: 1980, *Monthly Notices Roy. Astron. Soc.* **191**, 15.

Gondhalekar, P. M., Willis, A. J., Morgan, D. H., and Nandy, K.: 1980, *Monthly Notices Roy. Astron. Soc.* **193**, 875.

Grevesse, N.: 1984, *Physica Scripta*, T8, 49.

Grossman, L. and Olsen, E.: 1974, *Geochim. Cosmochim Acta.* **38**, 173.

Habe, A. and Ikeuchi, S.: 1980, *Progress Theor. Phys.* **64**, 1995.

Habe, A. and Ikeuchi, S.: 1982, *Progress Theor. Phys.* **68**, 1131.

Habing, H. J.: 1969, *Bull. Astron. Inst. Neth.* **20**, 177.

Harris, A. W. and Bromage, G. E.: 1984, *Monthly Notices Roy. Astron. Soc.* **208**, 941.

Harris, A. W. and Mas Hesse, J. M.: 1986a, *Astrophys. J.* **308**, 240.

Harris, A. W. and Mas Hesse, J. M.: 1986b, *Monthly Notices Roy. Astron. Soc.* **220**, 271.

Harris, A. W., Bromage, G. E., and Blades, J. C.: 1983, *Monthly Notices Roy. Astron. Soc.* **203**, 1225.

Harris, A. W., Gry, C., and Bromage, G. E.: 1984, *Astrophys. J.* **284**, 157.

Harris, D. H., Woolf, N. J., and Rieke, G. H.: 1978, *Astrophys. J.* **226**, 829.

Hartquist, T. W., Pettini, M., Tallant, A.: 1984, *Astrophys. J.* **276**, 519.

Hartquist, T. W., Tallant, A.: 1981, *Monthly Notices Roy. Astron. Soc.* **196**, 527.

Hawkins, I.: 1986, Ph.D. thesis, UCLA (in prep.).

Hawkins, I., and Jura, M.: 1987, *Astrophys. J.* (in press).

Hawkins, I., Jura, M., and Meyer, D. M.: 1985, *Astrophys. J. Letters* **294**, L131.

Heiles, C.: 1980, *Astrophys. J.* **235**, 833.

Henize, K. H.: 1956, *Astrophys. J. Suppl.* **2**, 315.

Hibbert, A., Dufton, P. L., and Keenan, F. P.: 1985, *Monthly Notices Roy. Astron. Soc.* **214**, 721.

Hibbert, A., Dufton, P. L., Murray, M. J., and York, D. G.: 1983, *Monthly Notices Roy. Astron. Soc.* **205**, 535.

Hobbs, L. M., York, D. G., and Oegerle, W.: 1982, *Astrophys. J. Lett.* **252**, L21.

Hollenbach, D. and Salpeter, E. E.: 1971, *Astrophys. J.* **163**, 155.

Jenkins, E. B.: 1981, in R. D. Chapman (ed.), *The Universe at Ultraviolet Wavelengths*, 2 Years of IUE, NASA CP 2171, p. 541.

Jenkins, E. B., Jura, M., and Loewenstein, M.: 1983, *Astrophys. J.* **270**, 88.

Jenkins, E. B., Rodgers, A. W., Harding, P., Morton, D. C., and York, D. G.: 1984, *Astrophys. J.* **281**, 585.

Jenkins, E. B., Savage, B. D., and Spitzer, L.: 1986, *Astrophys. J.* **301**, 355.

Jenkins, E. B., Silk, J., and Wallerstein, G.: 1976a, *Astrophys. J. Suppl.* **32**, 681.

Jenkins, E. B., Silk, J., and Wallerstein, G.: 1976b, *Astrophys. J. Lett.* **209**, L87.

Jenkins, E. B., Wallerstein, G., and Silk, J.: 1984, *Astrophys. J.* **278**, 649.

Joseph, C. L., Snow, T. P., Seab, C. G., and Crutcher, R. M.: 1986, *Astrophys. J.* **309**, 771.

Jura, M.: 1975, *Astrophys. J.* **197**, 581.

Jura, M.: 1982, in Y. Kondo, J. M. Mead, and R. Chapman (eds.), *Advances in Ultraviolet Astronomy: Four Years of IUE Research* (NASA), p. 54.

Jura, M.: 1986, private communication (also discussed on pp. 499—500).

Jura, M.: 1986, *Astrophys. J.* **303**, 327.

Jura, M. and Morris, M.: 1985, *Astrophys. J.* **292**, 487.

Jura, M. and York, D. G.: 1978, *Astrophys. J.* **219**, 861.

Kaelble, A., de Boer, K. S., and Grewing, M.: 1985, *Astron. Astrophys.* **143**, 408.

Keenan, F. P., Hibbert, A., and Dufton, P. L.: 1985, *Astron. Astrophys.* **147**, 89.

Knapp, G. R.: 1985, *Astrophys. J.* **293**, 273.

Knapp, G. R. and Morris, M.: 1985, *Astrophys. J.* **292**, 640.

Koornneef, J.: 1984, in S. van den Bergh and K. S. de Boer (eds.), 'Structure and Evolution of the Magellanic Clouds', *IAU Symp.* **108**, 105.

Kurucz, R. L. and Peytremann, E.: 1975, Smithsonian Astr. Obs. Spec. Report No. 362.

Léger, A.: 1983, *Astron. Astrophys.* **123**, 271.

Léger, A., Jura, M., and Omont, A.: 1985, *Astron. Astrophys.* **144**, 147.

Liszt, H. S.: 1983, *Astrophys. J.* **275**, 163.

Lockman, F. J., Hobbs, L. M., and Shull, J. M.: 1986, *Astrophys. J.* **301**, 380.

Luo, D., Pradhan, A. K., and Shull, J. M.: 1986, unpublished calculations.

Marschall, L. A. and Hobbs, L. M.: 1972, *Astrophys. J.* **173**, 43.

Mathis, J. S., Rumpl, W., and Nordsieck, K. H.: 1979, *Astrophys. J.* **217**, 425.

McGee, R. X., Newton, L. M., and Morton, D. C.: 1983, *Monthly Notices Roy. Astron Soc.* **205**, 1191.

McKee, C. F. and Ostriker, J. P.: 1977, *Astrophys. J.* **218**, 148.

McKee, C. F., Van Buren, D., and Lazareff, B.: 1984, *Astrophys. J. Lett.* **278**, L115.

Meaburn, J. and Blades, J. C.: 1980, *Monthly Notices Roy. Astron. Soc.* **190**, 403.

Meyers, K. A., Snow, T. P., Federman, S. R., and Berger, M.: 1985, *Astrophys. J.* **288**, 148.

Morgan, D. H., Nandy, K., Thompson, G. I.: 1982, *Monthly Notices Roy. Astron. Soc.* **199**, 399.

Morton, D. C.: 1974, *Astrophys. J. Lett.* **193**, L35.

Morton, D. C.: 1975, *Astrophys. J.* **197**, 85.

Morton, D. C.: 1978, *Astrophys. J.* **222**, 863.

Morton, D. C. and Smith, W. H.: 1973, *Astrophys. J. Suppl.* **26**, 333.

Murray, M. J., Dufton, P. L., Hibbert, A., and York, D. G.: 1984, *Astrophys. J.* **282**, 481.

Nachman, P. and Hobbs, L. M.: 1973, *Astrophys. J.* **182**, 481.

Nussbaumer, H., Pettini, M., and Storey, P. J.: 1981, *Astron. Astrophys.* **102**, 351.

Nussbaumer, H. and Storey, P. J.: 1981, *Astron. Astrophys.* **96**, 91.

Olofsson, H.: 1985, in P. A. Shaver (ed.), *Workshop on Submillimeter Astronomy*, European Southern Observatory.

Onaka, T., Sawamura, M., Tanaka, W., Watanabe, T., and Kodaira, K.: 1984, *Astrophys. J.* **287**, 359.

Panagia, N. and Terzian, Y.: 1984, *Astrophys. J.* **287**, 315.

Payne, H. E., Salpeter, E. E., and Terzian, Y.: 1983, *Astrophys. J.* **272**, 540.

Pettini, M. and West, K. A.: 1982, *Astrophys. J.* **260**, 561.

Pettini, M. *et al.*: 1982, *Monthly Notices Roy. Astron. Soc.* **199**, 409.

Phillips, A. P. and Gondhalekar, P. M.: 1983, *Monthly Notices Roy. Astron. Soc.* **202**, 483.

Phillips, A. P., Gondhalekar, P. M., and Blades, J. C.: 1981, *Monthly Notices Roy. Astron. Soc.* **195**, 485.

Phillips, A. P., Gondhalekar, P. M., and Pettini, M.: 1982, *Monthly Notices Roy. Astron. Soc.* **200**, 687.

Phillips, A. P., Pettini, M., and Gondhalekar, P. M.: 1984, *Monthly Notices Roy. Astron. Soc.* **206**, 337.

Prévot, L. *et al.*: 1980, *Astron. Astrophys.* **90**, L13.

Purcell, E. M.: 1976, *Astrophys. J.* **206**, 685.

Routly, P. M. and Spitzer, L.: 1952, *Astrophys. J.* **115**, 227.

Sanduleak, N.: 1968, *Astron. J.* **73**, 246.

Sanduleak, N.: 1970, C.T.I.O. Contrib. No. 89.

Savage, B. D. and Bohlin, R. C.: 1979, *Astrophys. J.* **229**, 136.

Savage, B. D., Bohlin, R. C., Drake, J. F., and Budich, W.: 1977, *Astrophys. J.* **216**, 291.

Savage, B. D. and de Boer, K. S.: 1979, *Astrophys. J.* **230**, L77.

Savage, B. D. and de Boer, K. S.: 1981, *Astrophys. J.* **243**, 460.

Savage, B. D. and Mathis, J. S.: 1979, *Ann. Rev. Astron. Astrophys.* **17**, 73.

Schulz-Lüpertz, E. and Grewing, M.: 1982, *Mitt. Astron. Ges.* **55**, 123.

Seab, C. G. and Shull, J. M.: 1983, *Astrophys. J.* **275**, 652.

Seab, C. G. and Shull, J. M.: 1985, in *Interrelationships among Circumstellar, Interstellar, and Interplanetary Dust*, NASA Conf. Publ. 2403, p. 37.

Seab, C. G., Hollenbach, D. J., McKee, C. F., and Tielens, A. G. G. M.: 1985, in *Interrelationships among Circumstellar, Interstellar, and Interplanetary Grains*, NASA Conf. Publ. 2403, p. A-28.

Sellgren, K.: 1984, *Astrophys. J.* **277**, 623.

Shapiro, P. R. and Field, G. B.: 1976, *Astrophys. J.* **205**, 762.

Shore, S.: 1982, in Y. Kondo, F. Bruhweiler and B. Savage (eds.), 'The Local Interstellar Medium', NASA CP-2345, *IAU Colloq.* **81**, 370.

Shull, J. M.: 1977, *Astrophys. J.* **215**, 805.

Shull, J. M.: 1978, *Astrophys. J.* **226**, 858.

Shull, J. M. and Beckwith, S.: 1982, *Ann. Rev. Astron. Astrophys.* **20**, 163.

Shull, J. M. and Van Steenberg, M. E.: 1982, *Bull. Ann. Astron. Soc.* **13**, 855.
Shull, J. M. and Van Steenberg, M. E.: 1985, *Astrophys. J.* **294**, 599.
Shull, J. M., Snow, T. P., and York, D. G.: 1981, *Astrophys. J.* **246**, 549.
Shull, J. M., Van Steenberg, M., and Seab, C. G.: 1983, *Astrophys. J.* **271**, 408.
Shull, J. M., York, D. G., and Hobbs, L. M.: 1977, *Astrophys. J. Lett.* **211**, L139.
Silk, J. and Wallerstein, G.: 1973, *Astrophys. J.* **181**, 799.
Siluk, R. S. and Silk, J.: 1974, *Astrophys. J.* **192**, 51.
Snow, T. P.: 1975, *Astrophys. J. Lett.* **202**, L87.
Snow, T. P. and Jenkins, E. B.: 1980, *Astrophys. J.* **241**, 161.
Snow, T. P. and Joseph, C. L.: 1985, *Astrophys. J.* **288**, 277.
Snow, T. P., Timothy, J. G., and Seab, C. G.: 1983, *Astrophys. J. Lett.* **265**, L67.
Songaila, A.: 1981, *Astrophys. J.* **248**, 945.
Songaila, A., Blades, J. C., Hu, E. M., and Cowie, L. L.: 1986, *Astrophys. J.* **303**, 198.
Songaila, A., Cowie, L. L., and York, D. G.: 1981, *Astrophys. J.* **248**, 956.
Songaila, A. and York, D. G.: 1980, *Astrophys. J.* **242**, 976.
Spitzer, L.: 1956, *Astrophys. J.* **124**, 20.
Spitzer, L.: 1976, *Comments on Astrophys.* **6**, 157.
Spitzer, L.: 1978, *Physical Processes in the Interstellar Medium*, New York, Wiley-Interscience.
Spitzer, L.: 1985, *Astrophys. J. Lett.* **290**, L21.
Spitzer, L. and Jenkins, E. B.: 1975, *Ann. Rev. Astron. Astrophys.* **13**, 133.
Stokes, G. M.: 1978, *Astrophys. J. Suppl.* **36**, 115.
Tielens, A. G. G. M. and Hollenbach, D.: 1985a, *Astrophys. J.* **291**, 722.
Tielens, A. G. G. M. and Hollenbach, D.: 1985b, *Astrophys. J.* **291**, 747.
Ulrich, M. H. *et al.*: 1980, *Monthly Notices Roy. Astron. Soc.* **192**, 561.
Van Steenberg, M. E.: 1986a, Ph.D. Thesis, University of Colorado.
Van Steenberg, M. E.: 1986b, *Astrophys. J.* (to be submitted).
Van Steenberg, M. E. and Shull, J. M.: 1986a, *Astrophys. J.* (to be submitted).
van Woerden, H., Schwarz, U. J., and Hulsbosch, A. N. M.: 1985, in H. van Woerden, R. J. Allen, and W. B. Burton (eds.), 'Milky Way Galaxy', *IAU Symp* **106**, 387.
Wai, C. M. and Wasson, J. T.: 1977, *Earth Planet Sci. Letters* **36**, 115.
Walborn, N. R.: 1980, *Astrophys. J.* **235**, L101.
Wallerstein, G. and Jacobsen, T.: 1976, *Astrophys. J.* **207**, 53.
Watson, W. D., Anicich, V. G., and Huntress, W. J.: 1976, *Astrophys. J. Letters* **205**, L165.
Watson, W. D. and Salpeter, E. E.: 1972, *Astrophys. J.* **174**, 321.
Wilson, R. W., Langer, W. D., and Goldsmith, P. F.: 1981, *Astrophys. J. Letters* **243**, L47.
Withbroe, G.: 1971, in K. B. Gabbie (ed.), *The Menzel Symposium* (NBS SP-353), Washington: Government Printing Office.
Witt, A. N., Bohlin, R. C., and Stecher, T. P.: 1986a, *Astrophys. J.* **302**, 421.
Witt, A. N., Bohlin, R. C., and Stecher, T. P.: 1986b, *Astrophys. J.* **305**, L23.
Witt, A. N., and Lillie, C. F.: 1978, *Astrophys. J.* **222**, 909.
Witt, A. N., Walker, G. A. H., Bohlin, R. C., and Stecher, T. P.: 1982, *Astrophys. J.* **261**, 492.
York, D. G.: 1982, *Ann. Rev. Astron. Astrophys.* **20**, 221.
York, D. G. and Jura, M.: 1982, *Astrophys. J.* **254**, 88.
York, D. G., Spitzer, L., Bohlin, R. C., Hill, J., Jenkins, E. B., Savage, B. D., and Snow, T. P.: 1983, *Astrophys. J. Lett.* **266**, L55.
York, D. G., Ratcliff, S., Blades, J. C., Cowie, L. L., Morton, D. C., and Wu, C.-C.: 1984, *Astrophys. J.* **276**, 92.
Zuckerman, B.: 1980, *Ann. Rev. Astron. Astrophys.* **18**, 263.

INTERSTELLAR DUST AND EXTINCTION

JOHN S. MATHIS

Washburn Observatory, The University of Wisconsin, Madison, WI, U.S.A.

1. Introduction

The ultraviolet (UV) region of the spectrum has been crucial in providing informa-tion on the nature of the material and size distribution of the particles of inter-stellar dust. Before there were any measurements of the UV properties of inter-stellar extinction, interstellar particles were believed to be composed primarily of dirty ices. The maximum of the interstellar extinction was believed to be at 0.3 μm, the shortest wavelength then observable. Both of these predictions were quite wrong (as are probably many of our present ideas regarding dust). The first rocket measurement (Stecher, 1965) showed that there is a very strong extinction feature at 0.22 μm. Now we also know the extinction increases dramatically towards the shortest wavelengths which can be reliably measured to date.

Studies of UV extinction using the Orbiting Astronomical Observatory-2 (Bless and Savage, 1972; Code *et al.*, 1976) and the TD-1 satellite (Nandy *et al.*, 1975, 1976) established that there is a well-defined average UV extinction law. However, the sensitivity and spatial resolution of these instruments did not allow an unam-biguous determination of variations of UV extinction along various lines of sight through the diffuse interstellar medium.

It is now clear that interstellar dust has properties which differ from one region of the sky to another. The wide variety of UV extinction was clearly recognized by Meyer and Savage (1981), who considered the extinctions of 1367 early-type stars observed by the ANS satellite. They found many stars with UV extinctions widely different from the average. There are now more precise studies down to shorter wavelengths with the IUE.

The wavelength dependence of interstellar extinction can be considered to be one or more smooth, continuous components plus some spectral features which are signatures of particular atomic or molecular transitions. There are surprisingly few of the spectral features, especially in dust outside of molecular clouds: a few infrared bands in the 3—18 μm interval and the very strong 0.2174 μm 'bump'. Historically, most substances which have been suggested as components of dust have been rejected because they predict spectral features which are *not* seen in addition to those which are. This is especially true in the UV spectral region, where many substances have strong absorption bands. In spite of careful searches (e.g., Massa *et al.*, 1983; hereafter MSF), no UV spectral features other than the bump have been found.

The average 'normal' (i.e., diffuse dust) extinction law is given by Savage and Mathis (1979) for 3.4 μm ⩾ λ ⩾ 0.1 μm and by Seaton (1979) for λ ⩽ 0.37 μm. Figure 1 shows it (heavy solid line) as well as the extinction in two separate portions of the the Large Magellanic Cloud (see below: heavy dashed and light

Y. Kondo (ed.), Scientific Accomplishments of the IUE, pp. 517—529.
© 1987 *by D. Reidel Publishing Company.*

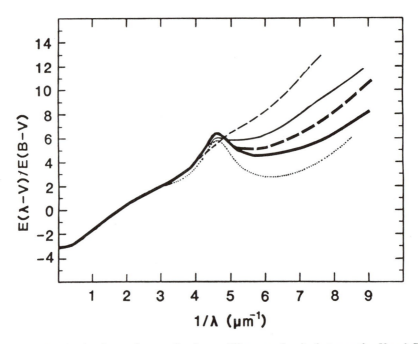

Fig. 1.   Several extinction laws, all normalized to a difference of unity between the V and B filters, plotted against wave number. Heavy solid line: the average curve in the diffuse regions of the Milky Way galaxy. Heavy dashed: the average in the Large Magellanic Cloud outside of the 30 Doradus region (Fitzpatrick 1985). Light solid: Average for stars in the 30 Doradus region of the Large Magellanic Cloud. Light dashed: the extinction in the Small Magellanic Cloud (Prevôt *et al.*, 1984). Dotted: HD 147889, a B star within the ρ Ophiuchus dark cloud, emphasizing that dust in dark clouds is not the same as that in diffuse regions.

solid lines). Also shown in Figure 1 (dotted line) is the extinction for HD 147889, which is rather deep within the ρ Oph dark cloud. Note that its extinction differs markedly from that of diffuse galactic or LMC dust. The Seaton law is a simple analytical expression. The Savage and Mathis table has a slight feature at $\lambda \approx 0.16$—0.14 μm which is spurious. It was introduced by using published extinctions using a filter including the very strong C IV $\lambda 1550$ doublet. The extinction was determined by comparing reddened and unreddened stars of different C IV strengths. Both average curves are primarily based on distant, very luminous stars with rather large extinctions. The dust towards these objects arises in both the diffuse ISM and in denser clouds. Subsequent IUE studies, discussed below, show that the extinction law is somewhat different in these two regions.

The differences among extinctions in various lines of sight are now being pursued vigorously with IUE. There are at least four quite different environments within which the dust properties are statistically distinct: (a) the diffuse interstellar medium; (b) dense but quiescent environments, such as in and near dense clouds; (c) dense but rather disturbed regions, such as in or near H II regions or regions of active star formation; and (d) deep within molecular clouds, in which there are extensive mantles of ices of CO, formaldehyde, ammonia, and water coated onto the surfaces of grains. Very little is known about the UV extinction properties of this last regime.

2. Properties of Infrared and Optical Extinction

The near-infrared (NIR) region is important because it contains some spectral features which provide direct evidence regarding the composition of dust. An excellent review of the infrared properties of interstellar dust has been given by Willner (1984).

There are 9.7 and 18 μm bands seen in all four dusty environments mentioned above, sometimes in emission and sometimes in absorption. They occasionally appear in circumstellar dust. They are usually attributed to amorphous silicates because of their width and the lack of structure in the bands as compared to crystalline silicates.

Another NIR feature is a 3.39 μm band. It is associated with the C—H stretching mode, so it probably indicates the presence of hydrocarbons. It may be found in all types of dust. Unfortunately, it is so weak that it is seen only in sources with very large amounts of extinction, such as IRS 7 near the galactic center (Jones *et al.*, 1983; Allen and Wickramasinghe, 1981). The line of sight to IRS 7 apparently does not include any molecular clouds, since its spectrum contains none of the absorption features of molecular ices found in clouds. The 3.39 μm band is not very specific as to which molecules produce it (Duley and Williams, 1979), and the fraction of the cosmic abundance of carbon in its carrier is poorly constrained.

I will not consider the spectral features found deep in molecular clouds, such as the 3.07 μm 'ice' band, because of the difficulty of getting UV extinction measurements within the same molecular clouds.

Aitken (1981) has reviewed the properties of several unidentified infrared (UIR) features in emission between 3.3 and 11.9 μm. They are found in a wide variety of sources in which dust in dense regions is being exposed to rather intense radiation from early-type stars. They are seen in bright reflection nebulae excited by B stars (Sellgren *et al.*, 1985, and references therein). In some cases the energy in the UIR emission is a substantial fraction (10—20%) of the total energy emitted throughout the entire infrared region, so these bands must be produced by a strong absorption in the UV region of the spectrum, where the exciting stars energy is concentrated. The strength of the UIR bands in planetary nebulae correlates with the carbon abundance (Cohen *et al.*, 1986).

3. The Galactic UV Extinction

3.1. PROBLEMS OF ITS DETERMINATION

Interstellar extinction is most precisely studied by the 'pair method', in which the flux of a reddened star is compared to the flux of an unreddened or lightly reddened star of the same intrinsic energy distribution. The ratio of the two fluxes gives the extinction directly.

The problems associated with determining UV extinction have been discussed by MSF. For work of the highest precision at wavelengths shorter than about 0.25 μm, stars of spectral type earlier than about B5 should be used. One advantage of hot stars is that the UV flux cannot be dominated by binary companions which are unseen in the visual region of the spectrum. Furthermore, the intrinsic UV flux of cooler stars is sensitive to their temperatures and luminosities. The reddened and the comparison stars have to be very close to the same spectral type and luminosity to avoid spectral differences which are erroneously attributed to the extinction (so-called 'mismatch errors'). One spectral subclass as derived from optical data allows a spread of UV intrinsic colors which is several times the photometric accuracy of IUE. Fortunately, the resolution of IUE even in the low-resolution mode is sufficient to provide information regarding the equivalent widths of strong stellar absorption lines. Some of these lines are sensitive to effective temperature and luminosity, so that a close match of reddened and comparison star is possible. An atlas of the IUE spectra of O stars in the range 0.12—0.19 μm is given by Walborn *et al.* (1985) and one of a large number of objects of various spectral types, both reddened and unreddened, is given by Heck *et al.* (1984). The dependence of spectral lines on luminosity and spectral type is discussed in Panek and Savage (1976).

There are three major problems with using stars of high luminosity (luminosity classes III—I) to study extinction. (1) Giants and and supergiants have appreciable differences in intrinsic fluxes even within the same spectral type. (2) In luminous stars some stellar features are very sensitive to luminosity. (3) There are no suitably unreddened comparison stars. This problem applies to all luminosity classes for stars earlier than about O6. Thus, for the highest precision work one must use reddened and unreddened pairs of types O6 IV—V to B5 IV—V. This rather severe restriction has made the IUE satellite uniquely valuable for the precise study of extinction. It is still impossible to fulfill the restriction to stars of classes IV or V when one wishes to discuss extinction in other galaxies.

3.2. NORMALIZATION USED IN DISPLAYING THE RESULTS

For historical reasons, $E(B - V)$, the difference of extinction in the V and B bandpasses, has assumed a great importance in the presentation of the results of interstellar reddening even in the UV portion of the spectrum. It has become traditional to present UV extinction curves in terms of $E(\lambda - V)/E(B - V)$, so that there is a unit difference between B and V. In some ways this tradition is not a useful one, because the grains producing the B and V extinction are different in size and possibly composition from those producing the extinction in other spectral regions. It is well known that there are definitely variations of $R \equiv A(V)/E(B - V)$ from one direction to another (e.g., Johnson, 1968), showing that the slope of the extinction between V and B can differ substantially from place to place. Extinctions in the UV, if presented as $E(\lambda - V)/E(B - V)$, may be seriously affected by this 'standard' normalization. For example, Figure 2a (adapted from MSF) shows the extinctions of four stars normalized to have the same *excess* extinction at 0.2175 μm (i.e., the 'bump') over the underlying smooth extinction, assumed linear. Note that the stars form a regular progression in *both* short

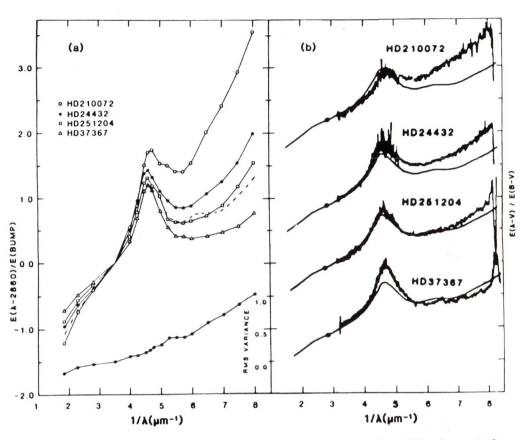

Fig. 2 The effects of normalizing in the usual way, to unity between the V and B bandpasses (as in Figure 1), can sometimes obscure physical correlations. Figure 2a shows the extinctions of four stars normalized to zero at $\lambda^{-1} = 0.33\,\mu m^{-1}$ and with an excess extinction of the bump, over the "background" extinction underlying the bump, of unity. Note that the extinctions form a regular progression over both the visual and UV wavelengths. The RMS variance is shown below the extinctions. Fig. 2b shows the same four stars, in the same order as in Figure 2a, but with the conventional normalization: unity between the V and B bandpasses. The average galactic curve from Savage and Mathis (1979) is shown as a light solid line. Note that it is not clear that there is any particular sequence among the stars with this normalization: for instance, the lowest looks quite normal between the visual and $\lambda^{-1} = 7\,\mu m^{-1}$, but it appears to have a very strong bump. Its extinction is really not so much peculiar as extreme, in the sense that it fits easily into the regular progression shown in Figure 2a.

wavelengths ($\lambda^{-1} > 5\,\mu m^{-1}$) and in the visual ($\lambda^{-1} < 3\,\mu m^{-1}$). This regular progression strongly suggests that there is a one-parameter sequence of grain properties which affects both visual and UV extinction. The curve at the bottom of Figure 2a, marked 'RMS variance', will be discussed below. This important regular progression in extinction properties is masked by the conventional normalization (Figure 2b). Note that HD 37367, the lowest curve in Figure 2b, appears to have an excess bump strength and a normal extinction at short wavelengths (i.e., a normal 'far-UV rise'). When normalized to the excess bump extinction (Figure 2a), it seems to be simply an extreme example of the sequence which varies continuously among the four stars.

It is premature to prescribe what is the 'best' normalization for UV extinction. We need to understand better the physical causes of extinction in various regions of the spectrum, or at least have an empirical knowledge of correlations between various wavelength regions. However, one should be well aware that the normalization of extinction results, especially the standard one, $E(\lambda - V)/E(B - V)$, is rather arbitrary and can influence their appearance.

3.3. VARIATIONS OF UV EXTINCTION

There are certainly variations of the UV extinction along various lines of sight (Koornneef, 1978; Kester, 1981; Meyer and Savage, 1981; Greenberg and Chlewicki, 1983, hereafter GC; MSF; Witt et al., 1984a, hereafter WBS), even within the diffuse ISM. The use of the average extinction law is, therefore, subject to considerable errors, especially if applied by considering the *visual* colors of the object to be dereddened. Dereddening will be discussed further in Section 3.4, below, after the properties of the 2175 Å bump have been considered.

In a given direction, the extinction law is usually rather well defined. The law for dust in the Perseus spiral arm near h and $\varkappa$ Persei has been especially well studied (Morgan et al., 1982; Franco et al., 1985). The results are that the extinction in the Perseus arm is similar within the cluster itself and in the nearby spiral arm.

Variations in extinction provide very important clues to the nature of grains. For instance, one can attribute the variance of the extinctions of the four stars in Figure 2 to the relative excess or deficiency of a material with a wavelength dependence of extinction shown in the curve marked 'RMS Variance' at the bottom of Figure 2a. This interpretation does not directly provide the wavelength dependence of grain components uniquely, but it puts powerful constraints on the properties of the component(s) which are actually producing the variations among the stars shown. Naturally, one should consider a larger sample of various extinction curves, all determined as accurately as possible. One can then see how few components are needed to fit each star's extinction to within the photometric accuracy.

3.4. THE λ2175 "BUMP"

By far the strongest and best studied spectral feature of extinction is the famous 'λ2175 bump'. WBS have determined its strength (excess extinction at $\lambda = 2175$ Å over a linear underlying extinction interpolated between 0.30 and 0.17 μm). Its strength, central wavelength, and width have recently been studied in five clusters (Massa and Fitzpatrick, 1986) and in 45 field stars (Fitzpatrick and Massa, 1986; hereafter FM). In the wavelength region 0.30—0.17 μm (3.3 μm$^{-1} \leqslant \lambda^{-1} \leqslant$ 5.9 μm^{-1}), FM assumed an underlying linear continuum, not associated with the bump, plus a bump with either a Lorentzian or a Drude profile (Bohren and Huffman, 1983). The parameters for the analytic fits were determined by least-squares procedures.

The observational features of the bump (FM, unless otherwise noted) are:

(a) The mean area of the bump per unit $E(B - V)$ is 5.17 ± 0.71 μm^{-1}. The mean FWHM is 0.992 ± 0.058 μm^{-1}, which corresponds to 470 ± 27 Å. The dispersions mainly represent real variations rather than photometric errors.

(b) The central wavenumber, λ_0^{-1}, of the fit to the bump is amazingly constant: $\lambda_0^{-1} = 4.599 \pm 0.019$ μm^{-1}, corresponding to $\lambda_0 = 2174 \pm 9$ Å. However, the variations of λ_0^{-1} among the stars in all five of the clusters which Massa and Fitzpatrick (1986) studied are significantly less than the variations of λ_0^{-1} among the field stars. The variations within a cluster establish an upper limit to the observational errors. Thus, there are *real* variations of λ_0 from one direction to another.

(c) There are large variations in the width of the bump (expressed as the FWHM) from one line of sight to another. The most extreme cases (HD 93028 and ζ Oph) have FWHM's of 359 and 768 Å, respectively. Figure 3 shows their profiles. It is interesting that the FWHM is smaller in dense disturbed regions (like the Orion Nebula and M8) than in dense quiescent regions.

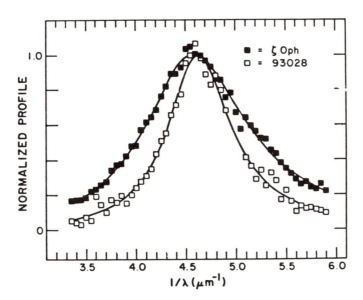

Fig. 3 Extreme examples of the widths of the $\lambda 2175$ feature as determined by Fitzpatrick and Massa (1986); narrow (ζ Oph) and wide (HD 93028). The features are normalized to have the same height a maximum.

(d) There is also very considerable variation in the strength of the bump (FM; WBS). How much variation there is depends on which of three reasonable ways one chooses to describe the strength of the bump: (1) by the coefficient of the profile function (Drude or Lorentzian), required; (2) by the central strength of the bump profile; or (3) by the equivalent width of the bump relative to the underlying continuum. The last description is independent of the normalization of the extinction law. These descriptions differ by successive powers of the FWHM, so that there must be a correlation of at least one of them with the FWHM. The area of the bump varied from 2.2 to 7.1 μm^{-1}, a factor of more than three, among the stars in FM's sample.

It is important to subtract an underlying extinction from the bump in order to study the strength of the bump. The excess extinction in the bump is comparable to the underlying extinction, so that variations in the bump strength and background are confused if one considers only the total extinction at $\lambda 2175$ instead of the excess of the bump above the background.

(e) The profile of the bump is fitted somewhat better by a Drude rather than a Lorentzian function. The Drude fit is excellent in all cases. This is interesting because the Drude profile is not symmetrical about λ_0^{-1}.

(f) FM found a correlation between the FWHM and the average density of the dust along the line of sight, as measured by $E(B - V)/r$, where r is the distance. The sense is that the bump is wider towards regions of higher dust density. This observation is contrary to what is expected if the bump arises from absorption through an electron resonance. There are two reasons for believing that the particles in a dense environment are larger than those in diffuse dust: (a) $\lambda(\max)$, the wavelength of maximum linear polarization for a given direction, is larger in dense regions, and (b) $A(V)/E(B - V)$ is, in general, larger in dense regions. Both of these are readily explained by having larger particles in the denser region. Large particles should have a narrower bump absorption than small ones because the electron responsible for the resonance encounters the edge of the grain less often (see Draine and Lee, 1984).

(g) There is no correlation between the variations of the FWHM and the small variations of λ_0^{-1}(FM), nor between the bump strength with the $\lambda 4430$ diffuse interstellar band (WBS). The characteristics of the bump are poorly correlated (GC) with the extinction law below 0.16 μm (the 'FUV rise'), which is discussed in the next section.

(h) Cox and Leene (1986) have investigated the relation of the bump strength with the unidentified infrared (UIR) bands at 3.3, 6.2, 7.7, 8.7, and 11.3 μm. They concluded that there is no correlation. They selected early-type stars with various bump strengths which are distant enough to (probably) lie behind most of the infrared emission observed by the IRAS satellite. The 12 μm IRAS band contains a considerable contribution from the 11.3 μm UIR band. The 100 μm IRAS channel contains mainly thermal emission from warm dust. There was no correlation between the bump strengths in the stellar spectra and the 12/100 μm ratios, strongly suggesting that absorption in the bump does not power the UIR emission bands.

The properties of the bump listed above do not depend on the assumption of linearity of the underlying continuum absorption. FM found that higher order polynomials fitted to the extinction outside of the bump wavelengths also give the same results. A very peculiar form of the underlying continuum would be needed if even the small variations in λ_0^{-1} were caused by the assumption of linearity of the underlying extinction. MSF considered the residuals arising from fitting their analytic expression for the bump, plus a linear underlying extinction, to the observed extinctions of several stars. These residuals show a very smooth increase with λ_0^{-1} right across the bump to well past the bump, showing that the linear assumption for the underlying extinction must be very close to correct.

In view of the variations of the UV extinction about the average extinction law, what is the best way to deredden an arbitrary object? One commonly used procedure is to assume enough of the average extinction law (say, Seaton, 1979 or Savage and Mathis, 1979) to make the dereddened flux smooth across the bump. MSF show that this method can lead to an error of a factor of ten for the reddening correction at 0.13 μm on a moderately reddened star! This example assumed an extreme deviation of the extinction bump from the average, but substantial errors (three or more) are very possible. Fortunately, the variations of extinction within a given region of the sky are considerably smaller than the overall variations. The best procedure is to use as much information as possible on the extinction law as measured for stars in the same general region of the sky, and if possible at similar distances, as the star whose dereddening is required.

The interpretation of the above properties of the bump is still not clear. What is clear is that the previously most common identification of the material causing the bump, graphite, is incomplete or incorrect. The reasons have been long understood (Gilra, 1972; Savage, 1975); the graphite explanation of the bump depends on a plasma resonance whose wavelength depends on both the size and the shape of the grain. The maximum extinction shifts to longer wavelengths as the size increases. Until the recent IUE studies, it was possible to explain the bump as being caused by a collection of graphite grains of various particle sizes. The power-law particle size distribution of Mathis *et al.*, (1977) fits the average extinction from 0.1 to 20 μm. However, large graphite grains have a broader absorption, peaked at larger wavelengths than small grains. Therefore, the graphite explanation predicts a strong positive correlation of the bump width with λ_0. In fact, λ_0 is almost constant while the FWHM varies by about a factor of two for extreme cases.

Alternative explanations for the bump still usually involve carbon in some form (Duley and Najdowsky, 1983 is an exception). A feature similar to the bump has been produced in the laboratory by Sakata *et al.* (1983) by depositing the products of a discharge through hydrocarbons. The residue is referred to as QCC, or 'quenched carbonaceous composite'. This material may be responsible for the UIR emission features (Sakata *et al.*, 1984). This idea seemed attractive because the bump absorbs enough energy to power them, if the conversion efficiency is high, but the recent work of Cox and Leene (1986) militates against it. It is not at all clear how the FWHM of the bump can be as variable as it is, even with a mixture of varying composition of QCC's, while the central wavelength is so constant. Amorphous hydrocarbons are another material which has been suggested (Duley and Williams, 1983; Hecht, 1986). In Hecht's model the central wavelength is almost constant because it is produced by very small carbon particles which do not contain hydrogen. There is also a population of larger amorphous carbonaceous material which contains hydrogen as well, possibly like QCC. The width is somewhat variable because it is produced by a broad absorption from these larger grains, but it is not clear whether the width can be as variable as is actually observed. The idea of having very small particles (not necessarily carbon) present has been invoked in another context: explaining the excess NIR continuum emission seen in some reflection nebulae (Sellgren 1984 and references therein; Witt *et al.*, 1984b). The excess radiation is most easily explained as the thermal emission following the heating of a tiny ($\lesssim 50$ Å) grain by a single UV photon (Draine and Anderson, 1985).

Another possibility for the bump is polycyclic aromatic hydrocarbons ('PAH's'), suggested independently by Léger and Puget (1984) and Allamandola *et al.* (1984). These are carbon-ring molecules, possibly ionized in the interstellar medium, of perhaps some 60 carbon atoms. In some ways they behave like small grains; in other ways, like molecules. However, one of the simplest of them, coronene, has very strong absorption features in the UV (Donn 1968) which are not seen in the ISM. Perhaps the varying width of the bump is caused by a distribution of PAH's of varying molecular weights, but one would expect the central wavelength of the absorption to vary with the width.

To date, the bump has always been observed to be very near 0.2175 μm. However, it is seen at larger wavelengths (0.24 μm) in two objects. One is R Cor Bor (Hecht *et al.*, 1984), which is a carbon star with a dusty circumstellar shell. The other is the peculiar planetary nebula Abell 30 (Greenstein, 1981). Both objects are carbon-rich and hydrogen-poor. The 0.24 μm bump is readily explained in terms of amorphous carbon grains (see Hecht, 1986). The infrared emissions from circumstellar grains in heavily reddened carbon stars (Jura, 1986) strongly suggest exactly the same thing: that amorphous carbon is being injected into the interstellar medium. It is estimated that roughly half of the mass of dust is contributed by the ejection from envelopes of carbon stars (Knapp and Morris, 1985).

It is safe to say that at the present time, the observation that the λ_0 is almost invariant but the FWHM varies appreciably is very difficult to interpret.

3.5. THE "FAR-UV RISE"; $\lambda < 0.16$ μm

There is a very steep increase of extinction from about 0.16 to 0.11 μm, the shortest wavelength which IUE can observe, which varies considerably from one object to another. GC found that in the diffuse ISM the 0.13—0.17 μm color index is poorly correlated with either $E(B - V)$ or with the strength of the bump as measured by the 0.22—0.25 μm index. Their result applies to dust in the Perseus arm (Morgan *et al.*, 1982; Franco *et al.*, 1985) and to dense quiescent and dense disturbed regions as well (MSF). GC obtained another interesting result: that the residuals of their FUV extinctions from the average law in the FUV range increases quadratically with λ^{-1}, beginning at about 1.6 μm^{-1}, so that the curvature is constant. Massa and Savage (1984) obtained a similar result for the extinctions in the cluster Cep OB3, in which there is a variation of extinctions within the cluster. This result suggests the wavelength dependence of the particles responsible for the extinction is roughly parabolic, and that varying amounts of this material are responsible for the residuals from the average.

The interpretation of the FUV rise is not clear. One possible candidate for the FUV-rise material, small silicate particles, does not seem likely. The stars in the reflection nebulae which show the strong unidentified NIR emission bands

(Sellgren *et al.*, 1985) show a reasonably strong FUV rise. The NIR emission shows a drop at 9.7 μm, which is certainly not what one would expect in the emission from hot silicate grains.

4. Extragalactic UV Extinction

With the IUE, the only galaxies near enough to provide reliable UV extinction are the Large Magellanic Cloud (LMC) and the Small Magellanic Cloud (SMC). Both have been extensively studied.

The LMC: Fitzpatrick (1985) has found that the UV extinction in the LMC is qualitatively different for stars seen projected within about 300 pc (20′) of the giant 30 Doradus Nebula (30 Dor) than for stars spread throughout the rest of the LMC. He was careful to use pairs of stars of about the same luminosity and temperature. The precaution of using pairs of about the same luminosity is important because of the dependence of the intrinsic energy distribution of B stars on the luminosity. The usually quoted 'LMC extinction law' as derived by Nandy *et al.* (1981) and Koornneef and Code (1981) pertains to the 30 Dor region. This region has a weaker bump than the average galactic extinction and a much stronger FUV rise. It is shown in Figure 1 as the light solid line. The differences among the individual stellar extinctions are explicable in terms of the expected random photometric uncertainties, so the results are consistent with a single UV law holding for all stars in the vicinity of 30 Dor. Outside of the 30 Dor region, Fitzpatrick (1985) found a law which is intermediate between the average galactic law and the 30 Dor law. It has a distinct bump, which definitely varies in strength from star to star but has the same FWHM as the average galactic law (see Figure 1, heavy dashed line). Unlike in the Milky Way galaxy, the FUV rise does not vary from star to star more than expected from random photometric errors. The gas-to-dust ratio in the LMC is usually expressed as $N(H)$, the H column density (from Lyman-α in the IUE spectra), relative to $E(B - V)$. It is four times larger than the average galactic value. This is consistent with the ratio of heavy elements relative to hydrogen.

The SMC: The extinction in the SMC is difficult to measure because there are no heavily reddened stars. The law was determined by Rocca-Volmerange *et al.* (1981), Nandy *et al.* (1982), and Bromage and Nandy (1983), and has been recently discussed by Prevót *et al.* (1984). The law is almost linear in λ^{-1}, with hardly a trace of the bump (or any other features). It is shown in Figure 1 as a light dashed line. The $N(H)/E(B - V)$ is about eight times the galactic average (Bouchet *et al.*, 1985).

References

Allamandola, L. J., Tielens, A. G. G. N., and Barker, J. R.: 1984, *Astrophys. J. Letters* **290**, L25.
Allen, D. A. and Wickramasinghe, D. T.: 1981, *Nature* **294**, 239.
Aitken, D. K.: 1981, *IAU Symp.* **96**, 207.
Bless, R. C. and Savage, B. D.: 1972, *Astrophys. J.* **171**, 293.
Bohren, R. C. and Huffman, D. R.: 1983, *Absorption and Scattering of Light by Small Particles*, New York, Wiley-Interscience, Chapter 12.

Bouchet, P., Lequeux, J., Maurice, E., Prevót, L., and Prevót-Burnichon, M. L.: 1985, *Astron. Astrophys.* **149**, 330.

Bromage, G. E. and Nandy, K.: 1983, *Monthly Notices Roy. Astron. Soc.* **204**, 29.

Code, A. D., Davis, J., Bless, R. C., and Hanbury Brown, R.: 1976, *Astrophys. J.* **203**, 417.

Cohen, M. *et al.*: 1986, *Astrophys. J.* **302**, 737.

Cox, P. and Leene, A.: 1986, *Astron. Astrophys.* (submitted).

Donn, B.: 1986, *Astrophys. J. Letters* **152**, L129.

Draine, B. T. and Anderson, N.: 1985, *Astrophys. J.* **292**, 494.

Draine, B. T. and Lee, H. M.: 1984, *Astrophys. J.* **285**, 89.

Duley, W. W. and Najdowsky, I.: 1983, *Astrophys. J.* **285**, 89.

Duley, W. W. and Williams, D. A.: 1979, *Nature* **277**, 40.

Duley, W. W. and Williams, D. A.: 1983, *Monthly Notices Roy. Astron Soc.* **205**, 67.

Fitzpatrick, E. L.: 1985 *Astrophys. J.* **299**, 219.

Fitzpatrick, E. L. and Massa, D.: 1986, *Astrophys. J.* **307**, 286.

Franco, M. L., Magazzu, A., and Stalio, R.: 1985, *Astron. Astrophys.* **147**, 191.

Gilra, D. P.: 1972, in A. D. Code (ed.), *Scientific Results from OAO—2* NASA SP-310, p. 295.

Greenberg, J. M. and Chlewicki, G.: 1983, *Astrophys. J.* **272**, 563.

Greenstein, J. L.: 1981, *Astrophys. J.* **245**, 124.

Heck, A., Egret, D., Jaschek, M., and Jaschek, C.: 1984, 'IUE Low-Dispersion Spectral Reference Atlas', ESA SP-1052.

Hecht, J. H.: 1986, *Astrophys. J.* **305**, 817.

Hecht, J. H., Holm, A. V., Donn, B., and Wu, C.-C.: 1984, *Astrophys. J.* **280**, 228.

Johnson, H. L.: 1968, in B. M. Middlehurst and L. H. Aller (eds.), *Nebulae and Interstellar Matter*, Chicago, U. of Chicago Press, p. 167.

Jones, T. J., Hyland, A. R., and Allen, D. A.: 1983, *Monthly Notices Roy. Astron. Soc.* **205**, 187.

Jura. M.: 1986, *Astrophys. J.* **303**, 327.

Kester, D.: 1981, *Astron. Astrophys.* **99**, 375.

Knapp, G. R. and Morris, M.: 1985, *Astrophys. J.* **292**, 640.

Koornneef, J.: 1978, *Astron. Astrophys.* **68**, 139.

Koornneef, J. and Code, A. D.: 1981, *Astrophys. J.* **247**, 860.

Léger, A. and Puget, J. L.: 1984, *Astron. Astrophys.* **137**, L5.

Massa, D. and Fitzpatrick, E. L.: 1986, *Astrophys. J. Suppl.* **60**, 305.

Massa, D. and Savage, B. D.: 1984, *Astrophys. J.* **279**, 310

Massa, D., Savage, B. D., and Fitzpatrick, E. L.: 1983, *Astrophys. J.* **266**, 662.

Mathis, J. S., Rumpl. W., and Nordsieck, K. H.: 1977, *Astrophys. J.* **217**, 425.

Meyer, D. M. and Savage, B. D.: 1981, *Astrophys. J.* **248**, 545.

Morgan, D. H., McLachlan, A., and Nandy, K.: 1982, *Monthly Notices Roy. Astron. Soc.* **198**, 779.

Nandy, K., Thompson, G. I., Jamar, C., Monfils, A., and Wilson, R.: 1975, *Astron. Astrophys.* **44**, 195.

Nandy, K., Thompson, G. I., Jamar, C., Monfils, A., and Wilson, R.: 1976, *Astron. Astrophys.* **51**, 63.

Nandy, K., Morgan, D. H., Willis, A. J., Wilson, R., and Gondhalekar, P. M.: 1981, *Monthly Notices Roy. Astron. Soc.* **196**, 955.

Nandy, K., McLachlan, A., Thompson, G. I., Morgan, D. H., Willis, A. J., Wilson, R., Gondhalekar, P. M., and Houziaux, L.: 1982, *Monthly Notices Roy. Astron. Soc.* **201**, 1.

Panek, R. J. and Savage, B. D.: 1976, *Astrophys. J.* **206**, 167.

Prevót, M. L., Lequeux, J., Maurice, E., Prevót, L., and Rocca-Volmerange, B.: 1984, *Astron. Astrophys.* **132**, 389.

Purcell, E. M.: 1969, *Astrophys. J.* **158**, 433.

Rocca-Volmerange, B., Prevót, L., Ferlet, R., Lequeux, J., and Prevót-Burnichon, M. L.: 1981, *Astron. Astrophys.* **99**, L5.

Sakata, A., Wada, S., Okutsu, Y., Shintani, H., and Nakada, Y.: 1983, *Nature* **301**, 493.

Sakata, A., Wada, S., Tanabe, T., and Onaka, T.: 1984, *Astrophys. J. Letters* **287**, L51.

Savage, B. D. 1975, *Astrophys. J.* **199**, 92.

Savage, B. D. and Mathis, J. S.: 1979, *Ann. Rev. Astron. Astrophys.* **17**, 73.

Seaton, M. J.: 1979, *Monthly Notices Roy. Astron. Soc.,* **187**, 73.

Sellgren, K.: 1984, *Astrophys. J.* **277**, 623.

Sellgren, K., Allamandola, L. J., Bregman, J. D., Werner, M. W., and Wooden, D. H.: 1985, *Astrophys. J.* **299**, 416.

Stecher, T. P.: 1965, *Astrophys. J.* **142**, 1683.

Walborn, N. R., Nichols-Bohlin, J., and Panek, R. K.: 1985, NASA Ref. Publ. 1155.

Willner, S. P.: 1984: in M. F. Kesslu and J. P. Phillips (eds.), *Observed Spectral Features of Dust*, D. Reidel Publ. Co., Dordrecht, Holland, p. 37.

Witt, A. N., Bohlin, R. C., and Stecher, T. P.: 1984a, *Astrophys. J.* **279**, 698.

Witt, A. N., Schild, R. E., and Kraiman, J. B.: 1984b, *Astrophys. J.* **281**, 708.

OBSERVATIONS OF ABSORPTION LINES FROM HIGHLY IONIZED ATOMS*

EDWARD B. JENKINS

Princeton University Observatory, Princeton, NJ, U.S.A.

1. Beginnings

Well before space-borne instruments were launched to observe ultraviolet absorption lines from the interstellar medium, it was suggested that features from highly ionized atoms might provide a valuable insight on the magnitude and character of ionizing and heating processes which were responsible for the observed physical states of gases in space. Early proposals centered on the notion that primarily neutral gas (H I regions) occupied most of the space, and that low energy cosmic ray particles or X-rays were responsible for the observed temperature and small traces of ionization (Pikel'ner, 1967; Spitzer, 1968; Spitzer and Tomasko, 1968; Field *et al.*, 1969; Spitzer and Scott, 1969, Goldsmith *et al.*, 1969; Silk and Werner, 1969; Dalgarno and McCray, 1972). Since low energy cosmic rays are shielded from us by interplanetary magnetic fields and low energy X-rays were difficult to measure with existing experiments, we could not measure the strength of these proposed sources of heating and partial ionization. Instead, it was suggested that the fluxes could be ascertained indirectly by observing the interstellar abundances of some highly ionized forms of particular heavy elements, such as C III, C IV, N III, N V, Si III, Si IV, S III, and S IV, all which had strong resonance lines in the ultraviolet (Silk, 1970; Silk and Brown, 1971; Weisheit, 1973).

A few years after a search for such absorption features in stellar spectra had been initiated by the *Copernicus* satellite, Steigman (1975) cautioned that the highly charged ions could be severely depleted in H I regions if, as seemed likely, the cross section for charge exchange with the neutral atoms were reasonably high. This effect could be much more important than recombinations with free electrons considered earlier. In the years which followed, Steigman's warning was shown to be valid: Blint *et al.* (1976) calculated $\langle \sigma v \rangle \approx 10^{-9}$ cm^3 s^{-1} for charge exchanges between C IV and H for temperatures between 1000 and 20 000 K, and similar conclusions for N IV, the primary reservoir for the observable N V, were derived by Christensen *et al.* (1977). These calculations using the scattering approximation were valid only for impact energies ~ 1 eV, but Watson and Christensen (1979) carried out quantal calculations which showed that the large cross sections persisted even to energies which were characteristic of temperatures in ordinary, dense interstellar clouds ($T \sim 100$ K).

* This article is an updated version of one which appeared in *Local Interstellar Medium* (IAU Colloquium 81), Y. Kondo, F. C. Bruhweiler, and B. D. Savage (eds.), NASA CP-2345, p. 155 (1984). Much new material has been added in the text which follows Section 3.

While these theoretical conclusions muted one particular incentive for observing highly ionized atoms in the interstellar medium,* the *Copernicus* researchers were also interested in the prospect of detecting low density, *thermally ionized* gases which could conceivably inhabit some voids between the clouds of ordinary, cool H I. Their efforts were rewarded by an early indication of absorption near the wavelength of the 1032 Å feature of O VI in the spectra of α Eri and λ Sco, reported by Rogerson *et al.* (1973). There soon followed definitive studies which showed that both members of the O VI doublet could be seen in many stars, that the absorptions were very likely interstellar, and that the high state of ionization was created by collisions in a hot plasma, instead of cosmic ray or X-ray bombardment in an ordinary cool gas (York, 1974; Jenkins and Meloy, 1974). From the ubiquity of the O VI lines and their large apparent widths, coupled with a lack of conspicuous features from other ions having a slightly lower ionization potential, these authors concluded that low density gases having a temperature somewhere in the interval 2×10^5 to 2×10^6 K were pervasive.

The O VI findings were published concurrently with an article by Williamson *et al.* (1974), who suggested that the diffuse, soft X-ray background radiation observed from their sounding-rocket experiments originated from the same very hot medium, although their measurements were mostly sensitive to emissions from gases at the upper end of the temperature range, 1 to several $\times 10^6$ K. These two fundamentally different experimental results, the detections of UV absorption features and the emissions at X-ray energies, were important milestones which precipitated a new era of thought on the state of the interstellar medium. In many respects, the widespread existence of hot material was reminiscent of a proposal by Spitzer (1956) that the corona of our galaxy is comprised of a very low density medium in a temperature regime similar to that of the solar corona.[†]

2. Investigation of the O VI-Bearing Gas

2.1. Origins: stellar, circumstellar or interstellar?

The first, and perhaps most critical challenge for those who advocate an interstellar interpretation for certain absorption lines is to convince the skeptics that the stars themselves, or their influences on the environment, are not responsible for the observed features. This problem is not a new one; controversy in this area extends back to the turn of the century when the first interstellar lines, those of Ca II, were identified in the spectra of binary stars (Hartmann, 1904; Slipher,

---

* Other ways were found to measure indirectly the cosmic ray and X-ray ionization rates in the interstellar medium, using the abundances of HD and OH which are influenced by the proton densities (O'Donnell and Watson, 1974; Jura, 1974; Black and Dalgarno, 1973).

[†] Unfortunately, the *Copernicus* satellite lacked the sensitivity to measure the spectra of sources more than 1 or 2 kpc from the galactic plane, so the structure of a corona of our galaxy could not be investigated. Spectra from the IUE satellite show evidence for an excess of Si IV and C IV in high latitude sources (and occasionally some N V), indicating the presence of material at 'transition layer' type temperatures. However the wavelength coverage of IUE does not extend down to the O VI lines, so the structure of material at 'coronal' temperatures has yet to be determined, see Section 5.

1909). For the O VI features, which must arise from material which is subjected to strong heating, a resolution of circumstellar versus interstellar origins is particularly relevant, inasmuch as the O and B type stars which produce enough radiation to show features at 1032 and 1038 Å are the very stars which must deliver a large amount of energy to the environment. Plausible theoretical interpretations of the observed O VI column densities were advanced for both viewpoints: The O VI could reside in zones associated with the target stars or whole groups of stars in the association (the 'interstellar bubbles' of Castor *et al.*, 1975; also see Weaver *et al.*, 1977) or, alternatively, there could be an interconnected network of regions in space heated, on successive occasions, by supernova blast waves (the 'interstellar tunnels' of Cox and Smith, 1974).

While the atmospheres of hot stars are responsible for broad O VI absorption features, often with P-Cygni type characteristics, the radial velocities of the narrower features identified as 'interstellar' show no correlation with stellar velocities (Jenkins and Meloy, 1974; Jenkins, 1978b), nor are the O VI lines influenced in any way by the stars' projected rotational velocities, $v \sin i$. In addition, York (1977) studied the O VI feature in the spectrum of the spectroscopic binary λ Sco and found no variation with orbital phase.

The possibility that O VI is circumstellar is a more difficult issue to address. A detailed discussion of the observational clues was presented by Jenkins (1978c), based on a comprehensive survey of O VI lines (Jenkins, 1978a) which followed the earlier discoveries. Briefly, Jenkins concluded that *most* of the O VI came from truly interspersed material, but that *some* the absorptions might be attributable to the evaporation zones within the very hot cavities created by the high velocity mass loss from the stars being observed, as described by Castor *et al.* (1975) and Weaver *et al.* (1977).

An interesting outgrowth of Jenkins's (1978c) analysis which is relevant to the local interstellar medium is that the general rate of increase of the O VI column density $N(\text{O VI})$ with distance r, while indicating an average density $n(\text{O VI}) = 1.7 \times 10^{-8}$ cm^{-3} but with considerable scatter, favors the existence of a positive intercept with $N(\text{O VI})$ of around 5×10^{12} cm^{-2} at $r = 0$.* This offset could be interpreted either as an excess resulting from the circumstellar bubbles around many of the stars, or as a contribution from a local concentration of hot gas in our neighborhood which has a disproportionate influence because all of the lines of sight, by necessity, emanate from a single point instead of being randomly scattered about.

2.2. General Properties

A striking feature of the O VI profiles seen in the survey of 72 stars reported by Jenkins (1978a) is that the dispersion of radial velocities is much smaller than that expected from the shock speeds of $v \sim 150$ km s^{-1} needed to heat the gas to the temperatures $T \sim 3 \times 10^5$ K for collisionally ionizing oxygen to +5. A composite

---

* The few lines of sight with anomalously high column densities (containing Jenkins's (1978b) 'second population' regions) were excluded in this study. If one were to include these regions and also insist the intercept is really zero, the average $n(\text{O VI})$ increases to 2.8×10^{-8} cm^{-3}.

of all the profiles extending over a total distance of about 50 kpc to all of the stars had an *rms* spread in radial velocity of 28 km s^{-1}, after applying small corrections to compensate for differential galactic rotation. Moreover, Jenkins (1978b) examined the statistical behavior of the velocity centroids and widths of profiles seen in the individual lines of sight, and he concluded that the distributions were consistent with the notion that individual packets of O VI, each with a column density of about 10^{13} cm^{-2} and internal broadening of 10 km s^{-1}, moved about at random with an *rms* dispersion of radial velocities of only 26 km s^{-1}. Large fluctuations in average O VI densities to different stars independently supported the idea of discrete domains of O VI-bearing gas; such variations were in accord with the expected Poisson statistical fluctuations.

While appreciable concentrations of O VI can exist in collisional equilibrium for temperatures in the range 1.7×10^5—2.0×10^6 K (Shull and Van Steenberg, 1982), the relative lack of accompanying N V absorptions (York, 1977) indicates that most of the material is at $T > 2.5 \times 10^5$ K, since N V has its peak fractional abundance at $T \sim 1.8 \times 10^5$ K. The narrowness of many of the individual profiles ($\langle v^2 \rangle^{1/2} \lesssim 14$ km s^{-1}), however, puts a restriction on the amount of thermal Doppler broadening present, which means $T < 16 m_p \langle v^2 \rangle / k = 3.8 \times 10^5$ K. Actually, the proposition that the hot gas is nearly isothermal is probably overly simplistic and has no compelling theoretical justification.* Jenkins (1978b) showed that the effects from the above mentioned observational constraints, when integrated over the peak of the curve for the O^{+5} ionization fraction with temperature, still permitted a power-law distribution in temperature for the electron density n_e, with $dn_e/d \ln T = T^{0.5 \pm 0.5}$ for $4.7 \lesssim \log T \lesssim 6.3$.

The most difficult attribute to measure for the O VI-bearing gas is the average volume filling factor $f \equiv \langle n_e \rangle^2 / \langle n_e^2 \rangle$. Contrary to what is seen for the low energy X-ray background (McCammon *et al.*, 1983), there seems to be no anticorrelation between the excesses or deficiencies of N(O VI) per unit distance and variations of normal (cool) interstellar material, as evidenced by the reddenings to the respective stars. From an upper limit to this anticorrelation, Jenkins (1978b) concluded that $f < 0.2$, provided most of interstellar space is filled with the gas responsible for the reddening and not some other phase at low density. If f were considerably less than this upper limit, we would conclude from the overall average for n(O VI) that the pressure of the coronal gas would exceed a reasonable upper bound for the pressure of most of the interstellar medium, $p/k = 10^4$ cm^{-3} K.

---

* Paresce *et al.* (1983) interpreted the diffuse ultraviolet emission observed by Feldman *et al.* (1981) to come from gas within a very narrow range of temperatures: 1.6—2.0×10^5 K. Strictly speaking, their conclusions would not permit the existence of any power law distribution over a broad temperature range. However it is quite possible that some of the spectral features have been misidentified; the resolution of the apparatus which observed the emission was only 60 Å, and the feature identified as N IV] $\lambda 1488$ might be something else, such as H$_2$ fluorescence (Duley and Williams, 1980); see, e.g. the emission spectrum shown by Brown *et al.* (1981) which comes from the Lyman bands of vibrationally excited interstellar H$_2$ exposed to L-α emission. If the N IV identification is indeed incorrect, a power law in temperature would be permitted (and the inferred emission measure could be lower).

2.3. COMPARISON WITH DIFFUSE BACKGROUND EMISSIONS

In simplest terms, a measure of $N(O VI)$ can be translated into an integral of the electron density n_e over a well defined path for gases at temperatures near the peak of the curve for the O VI ion fraction. The intensities of the soft X-ray [130—850 eV] and EUV [50—100 eV] emission backgrounds, which arise chiefly from collisional excitations and radiative decay of the electronic levels of highly ionized atoms, allow us to determine the integral of n_e^2 for the high-temperature phases over paths of somewhat indefinite length, in large part defined by absorption phenomena from intervening or interspersed denser material ('normal' interstellar matter). Thus, the observations of absorption and emission by these atoms are complementary, since they sample the hot material in different ways. We must acknowledge, of course, that the conclusions will have some errors attributable to not only the observations themselves, but also to uncertainties in temperatures of the gas, the relative abundances of key elements, and the calculations of ion fractions (and, for the X-ray and EUV results, the excitation cross sections). Indeed, the ion fractions are probably not governed by an *equilibrium*, since the time scale for heating or cooling of the gas could be substantially shorter than the equilibrium time for collisional ionization and recombinations.

It is generally believed that the immediate neighborhood of the Sun contains a gaseous medium which is partially ionized with $n(H) \sim 0.1$ cm^{-3} and $T \sim 1$—2×10^4 K (Weller and Meier, 1981; Dalaudier *et al.*, 1984). In turn, this local complex of warm gas is surrounded by a volume roughly 100 pc in diameter which may be almost completely filled with hot gas (Sanders *et al.*, 1977; Tanaka and Bleeker, 1977; Hayakawa *et al.*, 1978; Kraushaar, 1979; Fried *et al.*, 1980; Arnaud *et al.*, 1984). This conclusion follows from the observation that a good portion of the X-ray background at lower energies is fairly uniform over the whole sky, and this emission seems not to have undergone any (energy dependent) absorption short of reaching known dense clouds which are approximately 100 pc away. A general picture where neutral hydrogen and hot plasmas are intermingling over small scales seems to be disfavored by the steepness of the spectrum of the EUV/soft X-ray emission between 50 and 190 eV (Paresce and Stern, 1981). The X-ray results indicate that the emission measure of the hot bubble is generally of order 1—3×10^{-3} cm^{-6} pc if there is virtually no absorption [N(H) $\ll 10^{20}$ cm^{-2}] and $T \sim 10^6$ K (a temperature giving the most emission in the B and C X-ray bands).* On the basis of the X-ray data alone, one must conclude that the pressure of the gas $p/k \gtrsim 10^4$ cm^{-3} K. While this local pressure seems to exceed by a substantial margin the general value of about 2—3×10^3 cm^{-3} K, it is not out of the question that the Sun is inside a supernova remnant 100 pc in diameter having an internal temperature $T \sim 10^6$ K, age $t \sim 1.4 \times 10^5$ yr and a thermal energy $\sim 3 \times 10^{50}$ erg (Cox and Anderson, 1982).

When we try to reconcile the O VI data with the X-ray and EUV emission measures, we find that there is too little O VI for the amount of emission seen, if the gas were distributed fairly uniformly. One way to resolve the problem is to say

---

* In certain directions, not necessarily associated with identifiable supernova remnants, the emission measure reaches about 10^{-2} cm^{-6} pc. (McCammon *et al.*, 1983; Rocchia *et al.*, 1984).

that the hot gas is very patchy (i.e., f is low) and rather dense within the patches, but then the inferred pressure of the medium rises. A discussion of this idea was originally presented by Shapiro and Field (1976) who derived remarkably high pressures, but more recent, refined interpretations seem to indicate that the apparent deficiency of O VI is not too severe.

Burstein *et al.* (1977) found that the relative strengths of the X-rays in three different energy bands could only be explained by having the emission come from gases at two very different temperatures (Also see de Korte *et al.*, 1976. A dissenting view has been presented by Hayakawa [1979] however.) They estimated that the lower temperature regime ($T \lesssim 10^6$ K) was responsible for about half of the emission sensed by their lowest energy [B band: 130—190 eV] detector; these temperatures are not too far removed from the range where the O VI is produced. For the power-law temperature distribution defined earlier (Section 2.2), one can make half of the Wisconsin group's average B-band count rate consistent with $\langle n_e \rangle$ from the O VI data if $p/k = 4 \pm 1.5 \times 10^4$ cm^{-3} K, if one assumes for the X-rays that there is very little foreground absorption [N(H) $< 2 \times 10^{19}$ cm^{-2}] and $d = 100$ pc.*

A comparison of the O VI data with EUV emission intensities is less sensitive to uncertainties in temperatures, since the most EUV emission comes from a plasma at T best suited for O VI production. Paresce and Stern (1981) found n(O VI) too low by a factor 5—10 if p/k is set to 10^4 cm^{-3} K, again assuming the emission is virtually unabsorbed by intervening neutral gases and the existence of a power-law distribution for hot gas temperatures. However they noted that the *local* n(O VI) could be somewhat higher than the overall average (see last para. of Section 2.1); by stretching their lower limit for the EUV emission downward by 50% and assuming the local n(O VI) $= 1.7 \times 10^{-7}$ cm^{-3}, they could arrive at $f = 0.4$ and $p/k = 1.5 \times 10^4$ cm^{-3} K.

2.4. INTERPRETATION

A principal theme of the theoretical description of the interstellar medium by McKee and Ostriker (1977) is that a very inhomogeneous mixture of gas phases is subjected to shocks from supernova explosions. These shocks propagate rather freely through the intercloud medium and heat it to temperatures of around 5×10^5 K. Since the isothermal sound speed in this gas is about 80 km s^{-1} and there is probably considerable turbulence, one might expect that there may be much more O VI in the medium than observed, but that this additional O VI has not been seen because its large velocity dispersion makes the profiles difficult to distinguish from the undulating continua in the spectra of the target stars. Thus the ultraviolet absorption measurements may be badly biased toward only those O VI

---

* This result is obtained by evaluating

$$p/k = \frac{(1.91)\, 25 \text{ counts s}^{-1}}{(100 \text{ pc})C_e \ln 10 \int_{5.15}^{6.35} T^{\eta-1} R_B d \log T},$$

where $\eta = 0.5 \pm 0.5$ and C_e are as defined by Jenkins (1978b) from the O VI survey, and R_B is the B-band emissivity *vs* temperature shown in Fig. 11a of McCammon *et al.* (1983).

components which have a low velocity dispersion, and this may explain why there is an apparent deficiency relative to the soft X-ray and EUV backgrounds.

There is some indication that the components which are observed in O VI come only from interfaces undergoing conduction and evaporation between the cool clouds and the hot intercloud medium, which could explain why the profiles appear to be narrow. Cowie *et al.* (1979) found correlations between the end-point velocities of O VI profiles and those of some N II and Si III components which seem to support this interpretation. It is also interesting to note that the statistical description of Jenkins (1978b) for packets of O VI gas, each with N(O VI) $\sim 10^{13}$ cm^{-2}, implies that roughly 6 of these regions will be intercepted by a random line of sight 1 kpc long. Although possibly coincidental (or attribut-able to observational effects), this figure is remarkably close to the 7 to 8 clouds kpc^{-1} deduced from surveys of visual interstellar lines and extinction by dust (Spitzer, 1968). If the interpretation that we are seeing conductive interfaces is correct, one must employ the theory of such interfaces in concert with models for the clouds and intercloud medium to ascertain the conditions in the medium. It is important to emphasize, however, that we badly need more definitive observations to confirm more directly that the O VI indeed arises from evaporation zones around clouds.

While there are formidable observational and theoretical problems associated with derivations of average densities of collisionally ionized, interstellar Si IV, C IV, and N V from IUE observations of their absorption doublets (discussed in some detail in Section 4), it is interesting to note that the most recent, and probably best controlled, measurements by Savage and Massa (1987) for the average densities of these ions give number-density ratios, relative to O VI, of 0.07, 0.25, and 0.11, respectively. Except for a marked overabundance of Si IV*, these numbers are in satisfactory agreement with the theoretically computed ratios for evaporative interfaces of 0.010, 0.16, and 0.063 (Weaver *et al.*, 1977). However, again excluding Si IV, the measurements are an even better match to the values 0.018, 0.28, and 0.078 computed by Jenkins (1981) for an ensemble of regions which are undergoing isochoric, radiative cooling from initially much higher temperatures. Thus, we could be viewing specific regions within the cavities of hot gas created by supernova explosions, as described by Cox and Smith (1974) and Cox (1986). An even moderate skeptic could claim, however, that the improvement in the fit is not a really justifiable point, in view of the uncertainties in both the observations and theoretical calculations.

Finally, one might ask whether we should be embarrassed by the apparent mismatch in pressure between the very local, partially ionized medium with $T \approx 1-2 \times 10^4$ K and $n \approx 0.1$ cm^{-3}, discussed earlier, and the much hotter, X-ray emitting gas which surrounds it. The answer is, 'probably not', or at least, 'not yet'. When our local cloud was suddenly exposed to a big increase in external pressure as it was overtaken by a supernova blast wave, an isothermal shock started to progress toward the cloud's interior. If the ratio of external to internal pressures was, say, a factor of 5, this shock moved at a velocity $5^{1/2}$ times the local

---

* The extra Si IV is probably created by photoionization from the light of hot stars; see Section 4.

sound speed, $v_s = 10 - 20$ km s^{-1}. Over the 1.4×10^5 yr lifetime for the local supernova remnant calculated by Cox and Anderson (1982), this shock has probably eaten through only the outermost 3—6 pc of the perimeter of the local cloud. This length is only comparable to, or possibly much less than, the various estimates for the distance to the cloud's edge, which range from 3 or 4 pc (Bruhweiler and Kondo, 1982; Bruhweiler, 1982), out to about 50 pc in some directions (Paresce 1984).

3. Interstellar Shocks

Except for the Vela Supernova Remnant and one star (15 Mon) (Jenkins *et al.*, 1976; Jenkins and Meloy, 1974), components of O VI at high radial velocity ($v \gtrsim 50$ km s^{-1}) have not been detected. Those which have been seen may be confined within specific regions of enhanced density. The thickness of post-shock, O VI-bearing gas in the general intercloud medium could be so large that interactions with small clouds may create enough ricochet shocks and turbulence to wash out the profiles of O VI, as discussed above. Slower shocks which produce less energetic ionizations may be more coherent and thus more easily observed.

Cohn and York (1977) have observed high velocity features ($50 \lesssim v \lesssim 100$ km s^{-1}) of C II, C III, and Si III, mostly toward stars in active associations (e.g. Orion) or high-latitude stars. The absence of detectable Si II in these components indicates that photoionization is not the primary cause of ionization. Thus, they were not just viewing high speed clumps within ordinary H II regions. Instead, shocks which are strong enough to create temperatures of $3-8 \times 10^4$ K are the most likely explanation, according to the calculations of Shull (1977). More elaborate shock models by Shull and McKee (1979), which include the effects of ionizing radiation on the preshock gas, confirm that Si III should be abundant in the immediate post-shock region for a shock travelling at around 100 km s^{-1}, but that there should be virtually no Si III farther back where appreciable cooling and recombination have occurred, because of the rapid charge exchange with neutral hydrogen.

Cowie and York (1978a) performed additional *Copernicus* observations to look for more high velocity features; this survey extension also emphasized low ionization stages of abundant elements which had intrinsically strong lines in the ultraviolet. They synthesized velocity distribution functions based on the statistics of velocity extrema of the strong lines and concluded that shocks which had reached the isothermal phase were extremely rare (Cowie and York, 1978b). Combining their upper limits with estimates for supernova energies and birthrates, they concluded that densities for the intercloud regions with large f, the primary medium for shock propagation, should either be > 0.1 cm^{-3} or $< 7 \times 10^{-3}$ cm^{-3}. The high density alternative is based on the containment of high speed isothermal shocks to small enough radii that they are rarely seen, but this possibility seems pretty unlikely in view of the O VI and EUV/X-ray backgrounds and the theory of McKee and Ostriker (1977). The low density possibility is based on the notion that the shocks never have a chance to become isothermal before they escape from the galactic plane (Chevalier and Oegerle, 1979; Cox, 1981) or run into other supernova cavities (Cox and Smith, 1974; Smith, 1977). The upper limit given by Cowie and York (1978b) is consistent with McKee and Ostriker's (1977) estimate of 3.5×10^{-3} cm^{-3} for the typical hot medium.

4. Divergent Clues from C IV and Si IV

The spectrometer on the *Copernicus* satellite was not well suited for observing the lines of Si IV and C IV because at their wavelengths the instrument had low sensitivity and uncertain scattered light levels. Hence, in spite of its inferior wavelength resolution, the *International Ultraviolet Explorer* (IUE) has been more productive in exploring the properties of Si IV and C IV absorptions toward a broad selection of sources. Even though results for the narrow features of these ions toward many stars have appeared in the literature, and suitable spectra for hundreds of other stars exist in the archives, several factors have hampered our progress in achieving a clear interpretation on the lines' origins and significance. First, in many instances strong *stellar* (and possibly circumstellar) features have made it very difficult to achieve a clean differentiation of the narrow, supposedly interstellar components. (Sometimes a fairly drastic overexposure of a spectrum is needed to properly register an interstellar line in the trough of a stellar feature: see, for example, Figure 2 of Pettini and West (1982).) Second, and perhaps more important, is that unlike O VI which has a very high ionization potential, we can plausibly attribute the creation of Si IV, C IV, and to some extent even N V to either photoionization from very hot stars (and/or X-ray sources) or, alternatively, to gas which is either part of, similar to, or somehow connected with the hot, coronal phase which is responsible for the O VI, EUV and X-ray results discussed in Section 2. Finally, for gas at low radial velocities, we usually can not use the presence or absence of species at lower ionization to differentiate between these fundamentally different origins because of the inevitable contamination from H I regions or other H II regions of lower excitation.*

Initially, the results of many of the standard tests to discriminate between truly interstellar origins and effects from the target stars (or their immediate neighbors) had not been too satisfactory; the observations evidently were influenced by a complicated interplay of several phenomena requiring a careful interpretation to unravel them. As a first step, perhaps the simplest relationship to examine was that of column densities versus distance. Column densities toward stars more distant than 1 kpc should not suffer appreciably from the fluctuations due to Poisson statistics discussed at the beginning of Section 2.2, if the sizes and separations of discrete interstellar gas complexes are comparable to what had been seen in other studies of the local part of our galaxy. However early reports on assemblages of sight lines longer than 1 kpc showed that the average density of Si IV and C IV

---

* On many occasions, discrete features of Si IV and C IV at high velocity have been recorded by IUE for sources within or behind special, violently disturbed regions, such as those near O subdwarfs (Bruhweiler and Dean, 1983), some binary systems (Bruhweiler *et al.*, 1980), active OB associations (Phillips and Gondhalekar, 1981; Cowie *et al.*, 1981, Laurent *et al.*, 1982) or supernova remnants (Jenkins *et al.*, 1984). None have been identified with *generally distributed* material in the local neighborhood, however.

varied over almost *two orders of magnitude* (Jenkins, 1981; Cowie *et al.*, 1981). Likewise, a survey of Wolf—Rayet stars by Smith *et al.* (1980) showed virtually no correlation of the column densities of these ions with distance. However, a tantalizing and very convincing demonstration that *some* Si IV and C IV arises from the general interstellar medium was presented by Savage and de Boer (1979, 1981). In spectra of stars in the Magellanic Clouds, they found features which corresponded to velocities in our galaxy, in addition to absorptions from gases attributable to the halo and the Magellanic system. Galactic Si IV features were also seen in the spectrum of 3C273 by York *et al.* (1983). How much of these low velocity absorptions apply to highly ionized material within the galactic plane, rather than a transition layer below the halo, was unresolved at the time.

Another classic test to show interstellar origins, comparing the radial velocities of the features with the expected velocities produced by differential galactic rotation at the positions of the stars, was carried out by Cowie *et al.* (1981). These results, however, showed too much scatter to indicate whether or not the lines came from the positions of the stars or halfway between.

It is likely that the initial lack of success in demonstrating the genuine interstellar nature of Si IV and C IV was due to contamination by circumstellar zones containing material ionized to high stages by quasi-stationary shocks created by the stellar winds of the target stars (once again, the 'interstellar bubbles' discussed in Section 2.1) or by the stars' UV ionizing radiation (i.e., the inner parts of ordinary H II regions). To partially overcome this problem, Savage and Massa (1987) selected stars which (a) are more distant than about 1.5 kpc, (b) have effective temperatures below about 30 000 K and (c) are not projected onto obvious H II regions which could be seen on the Palomar sky survey prints. Their sample also emphasized stars which are less likely to have interfering, narrow stellar features. Savage and Massa's plots of log N(Si IV) and N(C IV) vs log distance still showed appreciable scatter, but considerably less than the earlier samples. Indeed, from the general distribution of the points in their diagram for C IV, it seems plausible that most of the scatter is caused by some lines of sight having a *deficiency* of these ions, rather than excesses near the target or foreground stars.

If we switch our perspective and ask what information we can extract from the data regarding the influence of the stars themselves on the strengths of narrow Si IV and C IV features, we again are confronted with complications which, while understandable, thwart our comparisons with theory. Many upper limits for hot stars with $d < 1$ kpc obtained from the *Copernicus* archives by Jenkins (1981) yielded column densities for Si IV and C IV which were lower than 10^{13} cm^{-2} and generally consistent with an extrapolation of the average N per unit distance from the stars at $d > 1$ kpc analyzed by Savage and Massa (1987). One might have expected such nearby stars to show a slight excess in the average density of these ions if contributions from the circumstellar zones were appreciable. Attempts to correlate the strengths of Si IV and C IV lines with properties of the target stars have not been very fruitful. Black *et al.* (1980) found that the lines did not seem to correlate with the velocities of stellar winds, and Jenkins (1981) found that N(C IV)/N(Si IV) showed no correspondence with the temperatures of the target stars. In support of the existence of collisionally ionized Si IV and C IV in gases at $T \sim 5 \times 10^4$ K, Bruhweiler *et al.* (1979, 1980) called attention to the fairly restricted range in the ratios of $N(\text{C IV})$ to $N(\text{Si IV})$ in their sample, extending between only 0.8 and 3.7 for most cases. This, they contended, was much less than the spread one would expect for photoionization from the widely divergent mix of stellar temperatures. Nearly all of their stars had $T_e < 35\,000$ K; the calculations of Cowie *et al.* (1981) indicate that $N(\text{C IV}) > N(\text{Si IV})$ in photoionized regions surrounding stars only with $T_e > 50\,000$ K. While it is possible that these results indicate that photoionization by stellar photons is not the dominant source of ionization, one could equally well propose that the density of ambient material near these stars is low enough that the lines of sight are almost always influenced by the starlight from nearly all of the stars in an association. This may explain why $N(\text{C IV})/N(\text{Si IV})$ does not seem to vary by large amounts and may also account for the hint that observations of different stars within single associations show some coherence (Cowie *et al.*, 1981).

A conclusive way to demonstrate that collisional ionization is not responsible for the production of Si IV and C IV is to find profiles with velocity dispersions lower than that expected for thermal Doppler broadening at temperatures needed to produce these ions (At equilibrium, $T \approx 10^5$ K; see Shull and Van Steenberg (1982). Slightly lower temperatures could be anticipated if the gas is cooling radiatively, because recombination times are slower than cooling times (Shapiro and Moore 1976)). Since the resolution of IUE is only 30 km s^{-1}, the dispersions must be inferred from the b-values derived from doublet ratios. Most of the observations reported in the literature give b-values comparable to or greater than the limiting values of 12 and 8 km s^{-1} for the Doppler motions of C IV and Si IV, respectively. A noteworthy exception, however, is Dupree and Raymond's (1983) measurement of C IV absorption in the spectrum of Feige 24. The preponderance of large b-values elsewhere does not mean, of course, that collisions at high temperatures *must* be the source of ionization, since the velocity dispersion could be produced by other means, such as turbulence, differential galactic rotation, etc.

Some, but not all, white dwarfs observed by IUE show prominent features of Si IV, C IV, and N V (Bruhweiler and Kondo, 1981; Dupree and Raymond, 1983; Malina *et al.*, 1981; Sion and Guinan, 1983). The positive measurements to these stars give column densities well above those generally seen within a few hundred pc (Jenkins, 1981). However, it is not completely clear whether these enhancements result from the presence of circumstellar material produced by the stars, or alternatively, from the action of these stars on nearby interstellar gas. Calculations by Dupree and Raymond (1983) indicate that hydrogen-rich, hot dwarfs ($T \approx 6 \times 10^4$ K) should be able to ionize enough Si IV and C IV to be seen with IUE if the ambient density is greater than about 0.1 cm^{-3}.* If a star has no appreciable helium cutoff at the high energy end of its spectrum, some N V and O VI will also be produced (O VI can not be seen with IUE however). More data

---

* No allowance was made for absorption of the ionizing photons by dust, however, so the actual yields may be somewhat lower than those estimated by Dupree and Raymond (1983); see, e.g. Sarazin (1977).

from local white dwarfs may give us a better insight on what proportion of the medium is filled with gas at moderate densities, as opposed to the very low density ($n \approx$ few $\times$ 10^{-3} cm^{-3}) hot material. As emphasized by Hills (1972, 1973, 1974), white dwarfs may be a very important source of ionizing radiation for the interstellar medium in the galactic plane. Dupree and Raymond's (1983) calculations combined with further results on the distribution of C IV and Si IV from IUE may give us more insight on this important topic.

5. Structure of the Galactic Halo

Of the ions we have discussed so far, O VI is probably the most suitable tracer of hot gas in the halo of our galaxy. This ion is most abundant at temperatures where the lifetime against radiative cooling is longer than that of gas at the somewhat lower temperatures which would favor Si IV or C IV. Unfortunately, the survey of O VI lines using the *Copernicus* satellite (Jenkins, 1978a) could not reach farther than about 1 kpc from the galactic disk because the instrument could observe only stars brighter than a V magnitude of 7. By comparing column densities toward a very few stars with $|z| \sim 1$ kpc against the general results in the plane, Jenkins (1978b) concluded that there was some drop off in O VI density away from the plane, roughly consistent with an exponential scale height of 300 pc (but with a large uncertainty in this number). It is important to emphasize that the scale height of O VI (or any other ion which may be observed) is not a measure of the extent of hot gas above or below the plane, since the primary influence may be a shift to a different stage of ionization.

Jenkins (1978b) found that there was no systematic flow of O VI-bearing gas in excess of 10 km s^{-1} away from the galactic plane. From this limit, he concluded that the rate of escape of hot material is less than $7 \times 10^{-4} M_\odot$ kpc^{-2} yr^{-1} in our region of the galaxy. Since this figure is an order of magnitude lower than Oort's (1970) estimate of $0.01 M_\odot$ kpc^{-2} yr^{-1} for the rate of infall of high-velocity, neutral gas at high latitudes, it appears that we can rule out the simple picture that Oort's clouds are *only* a direct result of cooling coronal gas which had spewed out of the plane at an earlier time (unless, of course, there is an appreciably higher flow from somewhere else, such as the galactic center).

IUE's ability to acquire high resolution spectra of stars as faint as $V \sim 13$ opened the way to studying the lines of sight toward high-latitude, field stars (Pettini and West, 1982), the brightest, hot stars in the Magellanic Clouds (Savage and de Boer, 1979, 1981) and globular clusters (de Boer and Savage, 1983) and, in some particularly favorable cases, even extragalactic targets: a supernova in NGC 6946 (Pettini *et al.*, 1982) and the brightest quasar, 3C273 (York *et al.*, 1983). This enabled us to probe the distribution of Si IV, C IV and occasionally even N V at large distances from the galactic plane. Savage and de Boer (1979, 1981) estimated the scale height for the densities of these ions to be in the range 2 to 4 kpc, based on a mapping of extended absorption wings to a velocity shear caused by differential galactic rotation. We now should probably question the validity of these determinations, in view of recent evidence which suggests that the rotation velocity of the halo decreases with $|z|$, relative to the disk material immediately below (or above) it (de Boer and Savage, 1983; Savage and Massa, 1987).

Another way to measure the distribution of ions with $|z|$ is to study a plot of $\log N(\mathrm{ion})|\sin b|$ vs $\log|z|$ for the observations at hand. Data assembled by Pettini and West (1982) indicated an abrupt increase in $n(\mathrm{C\,IV})$ for $|z| > 1$ kpc, instead of the expected exponential decrease, but much of this effect could be attributed to an abnormally low estimate for $n(\mathrm{C\,IV})$ at $z \sim 0$. Even so, the more complete and better controlled survey by Savage and Massa (1985, 1987) seems to substantiate Pettini and West's claim, and this effect, a jump by a factor of about 2 or 3, appears for Si IV and perhaps even N V. The plots of Savage and Massa (1986) show that over the much greater distances to the Magellanic Clouds the values of $N|\sin b|$ for the three highly ionized species level off, consistent with a scale height of 2—3 kpc. As with the O VI results, the envelope of points in Savage and Massa's (1987) plots at $|z| \sim 1$ kpc had a width of about 1 dex, indicating a very patchy distribution of highly ionized material.

The sudden increase in the amount of Si IV and C IV at about 1 kpc from the plane is in accord qualitatively with what one would expect for a gaseous halo at moderate temperatures (i.e., $T \approx 10^4$ K, well below that needed to create these ions collisionally) which is supported to a large scale height by other than thermal means and which is ionized by UV or soft X-ray radiation. At the higher energies, the primary source of this radiation is from extragalactic sources, such as quasars and active galactic nuclei. Chevalier and Fransson (1984) have obtained solutions for the structure of this type of galactic corona which is supported by the pressure gradient of outwardly diffusing cosmic rays. Fransson and Chevalier (1985) have explored the distribution and abundances of certain ions for a number of photo-ionization models. A key feature of the ionization structure with z is that from about 0.5 to 1.5 kpc the neutral fraction of atomic hydrogen undergoes a transition from being almost 100% down to 1%. Unlike the structure one would compute for the edge of a Strömgren sphere, this transition is not very sharp because the ionizing radiation is spread over a broad range of energies and corresponding absorption length scales. This same radiation, of course, can create the highly ionized species observed by IUE, but charge exchange with the neutral hydrogen (see Section 1) will quench these ions at the lower values of z.

Away from the neutral layer, the ion density is regulated by the balance of ionizations against radiative and dielectronic recombinations. In this regime, within certain very broad limits the observed column densities will be insensitive to total gas density since the increased number of recombinations will balance any enhancement of the total Si or C atoms present. This may explain why Savage and Massa (1985) do not see an appreciable increase in Si IV and C IV which coincides with any extension of the marked enhancement of H I away from the plane, at some 2.5 to 5 kpc from the galactic center, discussed by Lockman (1984).

Whether or not hot stellar populations in or near the disk are important sources of UV photons depends on whether the low-z gas has enough porosity to allow these photons to escape into the halo. If the photons from below are very influential sources of ionization, as advocated by Bregman and Harrington (1986), one might expect a pronounced enhancement of the highly ionized atoms at galactocentric distances $4 < R < 8$ kpc, because there are more H II regions present (Burton *et al.*, 1975; Lockman, 1976). This enhancement does not seem to be indicated by the survey of Savage and Massa (1985, 1987).

While an appreciable fraction of the determinations of $\log [N(\text{N V})/N(\text{C IV})]$ in the survey of Savage and Massa (1987) are upper limits, six actual measurements and one lower limit seem to cluster around the value -0.4. Unless the densities are extraordinarily low ($n_{\text{total}} \lesssim 3 \times 10^{-4}$ cm^{-3} — see Bregman and Harrington (1986)), this concentration of N V relative to that of C IV is 1 to 1.5 orders of magnitude higher than that predicted in the photoionization models. From this, one could conclude that the cosmic-ray supported gas at 10^4 K may coexist with a separate component of the galactic corona, namely, a thermally supported gas at $T \approx 10^6$ K in hydrostatic equilibrium in the galactic gravitational potential, with a scale height of about 8 kpc (Spitzer, 1956). Depending on the balance of heating and cooling in the corona, the gas may not be static: a hot wind may be escaping, or alternatively, radiative cooling may produce cool clouds which then fall towards the plane (Shapiro and Field, 1976; Chevalier and Oegerle, 1979; Bregman, 1980; Habe and Ikeuchi, 1980). The existence of such a hot halo is supported by Jakobsen and Kahn's (1986) analysis of the variations of intensity and spectral hardness of the soft X-ray background over different parts of the sky. They concluded that the X-ray emitting gas (at $T \approx 10^6$ K) has a scale height about 10 times that of an embedded population of absorbing H I clouds. Since a fair fraction of such clouds, especially the larger ones with low internal densities, have a scale height of 350—500 pc (Bohlin *et al.*, 1978; Lockman *et al.*, 1986), the large vertical extent for the hot gas seems reasonable, even though the O VI seems to drop off just below 1 kpc.

6. Future Prospects

In years ahead, our ability to investigate the ultraviolet absorption lines will improve dramatically. The High Resolution Spectrograph (HRS) (Brandt *et al.*, 1982) aboard the Hubble Space Telescope will have a wavelength resolving power of $\lambda/\Delta\lambda = 8 \times 10^4$, i.e., $\delta v = 4$ km s^{-1}, in its highest resolution mode (Bottema *et al.*, 1984). The tremendous increase in photometric accuracy and resolution over that obtainable with IUE should virtually eliminate the confusion in identifying different parcels of gas at low velocity and allow us to differentiate those which produce the features from highly ionized atoms from those which do not. Information on whether the Si IV and C IV is ionized by UV photons or by collisions in a hot plasma should come from good measurements of the velocity dispersions, provided, of course, that turbulence and/or velocity shears are not too large. Unfortunately, HRS will not be sensitive to wavelengths near the O VI features (unless one ventures to absorbing systems with a redshift $z \gtrsim 0.1$); to further study O VI we will need a facility such as Lyman (formerly called FUSE, for Far Ultraviolet Spectroscopic Explorer), now being considered by NASA and ESA for flight in the 1990's. The telescope and spectrographs on Lyman will be designed specifically to do spectroscopy at wavelengths below the efficiency cutoff of MgF$_2$ coated mirrors and conventional UV detector faceplates.

We desperately need absorption line data from ions which have their peak abundances in the range $10^6 \lesssim T \lesssim 10^7$ K where most of the X-ray emission occurs. One way to observe them is to use a good crystal or transmission grating spectrometer aboard some reasonably large, orbiting X-ray facility. York and Cowie (1983) have calculated that the continua of brighter X-ray sources (with intensities $\gtrsim$ a few keV cm^{-2} s^{-1} keV^{-1}) should give enough signal-to-noise in a reasonable integration time to permit detection of the strongest lines which might arise over a distance of 1 kpc. These authors have tabulated an assortment of strong transitions to levels $0.2 \lesssim E \lesssim 1.0$ keV above the ground states of appropriate ions.

Another approach for measuring highly ionized species is to record the extremely weak absorption features from some coronal forbidden lines at visible wavelengths, the most promising of which seem to be [Fe X] $\lambda6375$ and [Fe XIV] $\lambda5303$. Hobbs (1984a, b) and Hobbs and Albert (1984) have already made a bold attempt to detect these ions in the interstellar medium, but except for some initially tantalizing absorptions at 6367 Å interpreted as Fe X at a large negative velocity in front of two stars in Cephius, they were only able to report rather high upper limits because detector instabilities limited their signal-to-noise ratios to about 350. Further research by Hobbs (1985) showed that the earlier suggested Fe X absorption is more likely to be an exceptionally weak, unidentified molecular or diffuse interstellar line produced in relatively dense, cool clouds. A very important result from these first attempts, however, was an exploration of where one can expect to find interference from telluric absorption lines, diffuse interstellar bands, and stellar features. These unwanted features, while present, do not seem to offer serious problems. Pettini and D'Odorico (1986) were able to achieve signal-to-noise ratios roughly comparable to those of Hobbs (1984a, b) when they recorded spectra of some faint halo stars and stars in the LMC, using a directly illuminated CCD on the echelle spectrograph of the ESO 3.6M telescope. Here, the noise was very close to the limits imposed by photon counting statistics. This, together with the large aperture, gave a considerable increase in sensitivity for comparing the amount of Fe X to other highly ionized species observed by satellites, but still a factor of about 10 short of having noteworthy theoretical implications. In due course, when the practice of obtaining high dispersion spectra using CCDs becomes more refined and we can routinely acquire photon-limited accuracies with tens of millions of counts per velocity bin, there is a chance that the realm of absorption features from highly ionized atoms in the medium, now the private hunting ground of space astronomers, will soon open up to a much wider community of observers.

The writing of this paper was supported by NASA Grant NAGW-477.

References

Arnaud, M., Rothenflug, R., and Rocchia, R.: 1984, in Y. Kondo, F. C. Bruhweiler and B. D. Savage (eds.), 'Local Interstellar Medium', *IAU Colloq.* **81**, 301.
Black, J. H. and Dalgarno, A.: 1973, *Astrophys. J. Letters* **184**, L101.
Black, J. H., Dupree, A. K., Hartmann, L. W., and Raymond, J. C.: 1980, *Astrophys. J.* **239**, 502.

Blint, R. J., Watson, W. D., and Christensen, R. B.: 1976, *Astrophys. J.* **205**, 634.

Bohlin, R. C., Savage, B. D., and Drake, J. F.: 1978, *Astrophys. J.* **224**, 132.

Bottema, M., Cushman, G. W., Holmes, A. W., and Ebbets, D.: 1984, in A. Boksenberg and D. L. Crawford (eds.), *Instrumentation in Astronomy V*, Proc. SPIE 445, p. 436.

Brandt, J. C. and the HRS Investigation Definition and Experiment Development Teams: 1982, in D. N. B. Hall (ed.), *The Space Telescope Observatory*, NASA CP-2244, p. 76.

Bregman, J. N.: 1980, *Astrophys. J.* **236**, 577.

Bregman, J. N. and Harrington, P. J.: 1986, preprint.

Brown, A., Jordan, C., Millar, T. J., Gondhalekar, P., and Wilson, R.: 1981, *Nature* **290**, 34.

Bruhweiler, F. C.: 1982, in Y. Kondo, J. M. Mead, and R. D. Chapman (eds.), *Advances in Ultraviolet Astronomy: Four Years of IUE Research*, NASA Conf. Pub. 2238, p. 125.

Bruhweiler, F. C. and Dean, C. A.: 1983, *Astrophys. J. Letters* **274**, L87.

Bruhweiler, F. C. and Kondo, Y.: 1981, *Astrophys. J. Letters* **248**, L123.

Bruhweiler, F. C. and Kondo, Y.: 1982, *Astrophys. J.* **259**, 232.

Bruhweiler, F. C., Kondo, Y., and McCluskey, G. E.: 1979, *Astrophys. J. Letters* **229**, L39.

Bruhweiler, F. C., Kondo, Y., and McCluskey, G. E.: 1980, *Astrophys. J.* **237**, 19.

Burton, W. B., Gordon, M. A., Bania, T. M., and Lockman, F. J.: 1975, *Astrophys. J.* **202**, 30.

Burstein, P., Borken, R. J., Kraushaar, W. L., and Sanders, W. T.: 1977, *Astrophys. J.* **213**, 405.

Castor, J., McCray, R., and Weaver, R.: 1975, *Astrophys. J. Letters* **200**, L107.

Chevalier, R. A. and Fransson, C.: 1984, *Astrophys. J. Letters* **279**, L43.

Chevalier, R. A. and Oegerle, W. R.: 1979, *Astrophys. J.* **227**, 398.

Christensen, R. B., Watson, W. D., and Blint, R. J.: 1977, *Astrophys. J.* **213**, 712.

Cohn, H. and York, D. G.: 1977, *Astrophys. J.* **216**, 408.

Cowie, L. L., Jenkins, E. B., Songaila, A., and York, D. G.: 1979, *Astrophys. J.* **232**, 467.

Cowie, L. L., Taylor, W., and York, D. G.: 1981, *Astrophys. J.* **248**, 528.

Cowie, L. L. and York, D. G.: 1978a, *Astrophys. J.* **220**, 129.

Cowie, L. L. and York, D. G.: 1978b, *Astrophys. J.* **223**, 876.

Cox, D. P.: 1981, *Astrophys. J.* **245**, 534.

Cox, D. P.: 1986, in D. Pequinot (ed.), *Proceedings of the Meudon Workshop on Model Nebulae*, (in press).

Cox, D. P. and Anderson, P. R.: 1982, *Astrophys. J.* **253**, 268.

Cox, D. P. and Smith, B. W.: 1974, *Astrophys. J. Letters* **189**, L105.

Dalaudier, F., Bertaux, J. L., Kurt, V. G., and Mironova, E. N.: 1984, *Astron. Astrophys.* **134**, 171.

Dalgarno, A. and McCray, R. A.: 1972, *Ann. Rev. Astron. Astrophys.* **10**, 375.

de Boer, K. S. and Savage, B. D.: 1983, *Astrophys. J.* **265**, 210.

de Korte, P. A. J., Bleeker, J. A. M., Deerenberg, A. J. M., Hayakawa, S., Yamashita, K., and Tanaka, Y.: 1976, *Astron. Astrophys.* **48**, 235.

Duley, W. W. and Williams, D. A.: 1980, *Astrophys. J. Letters* **242**, L179.

Dupree, A. K. and Raymond, J. C.: 1983, *Astrophys. J. Letters* **275**, L71.

Feldman, P. D., Brune, W. H., and Henry, R. C.: 1981, *Astrophys. J. Letters* **249**, L51.

Field, G. B., Goldsmith, D. W., and Habing, H. J.: 1969, *Astrophys. J. Letters* **155**, L149.

Fransson, C. and Chevalier, R. A.: 1985, *Astrophys. J.* **296**, 35.

Fried, P. M., Nousek, J. A., Sanders, W. T., and Kraushaar, W. L.: 1980, *Astrophys. J.* **242**, 987.

Goldsmith, D. W., Habing, H. J., and Field, G. B.: 1969, *Astrophys. J.* **158**, 173.

Habe, A., and Ikeuchi, S.: 1980, *Prog. Theor. Phys.* **64**, 1995.

Hartmann, J.: 1904, *Astrophys. J.* **19**, 268.

Hayakawa, S.: 1979, in W. A. Baity and L. E. Peterson (eds.), *X-Ray Astronomy* (COSPAR Symposium), (Oxford: Pergammon), p. 323.

Hayakawa, S., Kato, T., Nagase, F., Yamashita, K., and Tanaka, Y.: 1978, *Astron. Astrophys.* **62**, 21.

Hills, J. G.: 1972, *Astron. Astrophys.* **17**, 155.

Hills, J. G.: 1973, *Astron. Astrophys.* **26**, 197.

Hills, J. G.: 1974, *Astrophys. J.* **190**, 109.

Hobbs, L. M.: 1984a, *Astrophys. J.* **280**, 132.

Hobbs, L. M.: 1984b, *Astrophys. J. Letters* **284**, L47.

Hobbs, L. M.: 1985, *Astrophys. J.* **298**, 357.

Hobbs, L. M. and Albert, C. E.: 1984, *Astrophys. J.* **281**, 639.

Jakobsen P. and Kahn, S. M.: 1986, preprint.

Jenkins, E. B.: 1978a, *Astrophys. J.* **219**, 845.

Jenkins, E. B.: 1978b, *Astrophys. J.* **220**, 107.

Jenkins, E. B.: 1978c, *Comments Astrophys.* **7**, 121.

Jenkins, E. B.: 1981, in R. D. Chapman (ed.), *The Universe at Ultraviolet Wavelengths*, NASA Conf. Pub. 2171, p. 541.

Jenkins, E. B. and Meloy D. A.: 1974, *Astrophys. J. Letters* **193**, L121.

Jenkins, E. B., Silk, J., and Wallerstein, G.: 1976, *Astrophys. J. Suppl.* **32**, 681.

Jenkins, E. B., Wallerstein, G., and Silk, J.: 1984, *Astrophys. J.* **278**, 649.

Jura, M.: 1974, *Astrophys. J.* **191**, 375.

Kraushaar, W. L.: 1979, in W. A. Baity and L. E. Peterson (eds.), *X-Ray Astronomy* (COSPAR Symposium), (Oxford: Pergammon), p. 293.

Laurent, C., Paul, J. A., and Pettini, M.: 1982, *Astrophys. J.* **260**, 163.

Lockman, F. J.: 1976, *Astrophys. J.* **209**, 429.

Lockman, F. J.: 1984, *Astrophys. J.* **283**, 90.

Lockman, F. J., Hobbs, L. M. and Shull, J. M.: 1986, *Astrophys. J.* **301**, 380.

Malina, R. F., Basri, G., and Bowyer, S.: 1981, *Bull. Am. Astron. Soc.* **13**, 873.

McCammon, D., Burrows, D. N., Sanders, W. T., and Kraushaar, W. L.: 1983, *Astrophys. J.* **269**, 107.

McKee, C. F. and Ostriker, J. P.: 1977, *Astrophys. J.* **218**, 148.

O'Donnell, E. J. and Watson, W. D.: 1974, *Astrophys. J.* **191**, 89.

Oort, J. H.: 1970, *Astron. Astrophys.* **7**, 381.

Paresce, F.: 1984, *Astron. J.* **89**, 1022.

Paresce, F., Monsignori Fossi, B. C., and Landini, M.: 1983, *Astrophys. J. Letters* **266**, L107.

Paresce, F. and Stern, R.: 1981, *Astrophys. J.* **247**, 89.

Pettini, M., Benvenuti, P., Blades, J. C., Boggess, A., Boksenberg, A., Grewing, M., Holm, A., King, D. L., Panagia, N., Penston, M. V., and Savage, B. D.: 1982, *Monthly Notices Roy. Astron. Soc.* **199**, 409.

Pettini, M. and D'Odorico, S.: 1986, *Astrophys. J.* **310**, 700.

Pettini, M. and West, K. A.: 1982, *Astrophys. J.* **260**, 561.

Phillips, A. P. and Gondhalekar, P. M.: 1981, *Monthly Notices Roy. Astron. Soc.* **196**, 533.

Pikel'ner, S. B.: 1967, *Astron. Zh.* **44**, 915 (English trans.: *Soviet Astron. A J* **11**, 737 (1968)).

Rocchia, R., Arnaud, M., Blondel, C., Cheron, C., Christy, J. C., Rothenflug, R., Schnopper, H. W., and Delvaille, J. P.: 1984, *Astron. Astrophys.* **130**, 53.

Rogerson, J. B., York, D. G., Drake, J. F., Jenkins, E. B., Morton, D. C., and Spitzer, L.: 1973, *Astrophys. J. Letters* **181**, L110.

Sanders, W. T., Kraushaar, W. L., Nousek, J. A., and Fried, P. M.: 1977, *Astrophys. J. Letters* **217**, L87.

Sarazin, C. L.: 1977, *Astrophys. J.* **211**, 772.

Savage, B. D. and de Boer, K. S.: 1979, *Astrophys. J. Letters* **230**, L77.

Savage, B. D. and de Boer, K. S.: 1981, *Astrophys. J.* **243**, 460.

Savage, B. D. and Massa, D.: 1985, *Astrophys. J. Letters* **295**, L9.

Savage, B. D. and Massa, D.: 1987, *Astrophys. J.* **314**, 380.

Shapiro, P. R. and Field, G. B.: 1976, *Astrophys. J.* **205**, 762.

Shapiro, P. R. and Moore, R. T.: 1976, *Astrophys. J.* **207**, 460.

Shull, J. M.: 1977, *Astrophys. J.* **216**, 414.

Shull, J. M. and McKee, C. F.: 1979, *Astrophys. J.* **227**, 131.

Shull, J. M. and Van Steenberg, M.: 1982, *Astrophys. J. Suppl.* **48**, 95.

Silk, J.: 1970, *Astrophys. Letters* **5**, 283.

Silk, J. and Brown, R. L.: 1971, *Astrophys. J.* **163**, 495.

Silk, J. and Werner M. W.: 1969, *Astrophys. J.* **158**, 185.

Sion, E. M. and Guinan, E. F.: 1983, *Astrophys. J. Letters* **265**, L87.

Slipher, V. M.: 1909, *Lowell Obs. Bull.*, No. 51.

Smith, B. W.: 1977, *Astrophys. J.* **211**, 404.

Smith, L. J., Willis, A. J., and Wilson, R.: 1980, *Monthly Notices Roy. Astron. Soc.* **191**, 339.

Spitzer, L.: 1956, *Astrophys. J.* **124**, 20.

Spitzer, L.: 1968, *Diffuse Matter in Space*, New York: Wiley Interscience.

Spitzer, L. and Scott, E. H., 1969, *Astrophys. J.* **158**, 161.

Spitzer, L. and Tomasko, M. G., 1968, *Astrophys. J.* **152**, 971.

Steigman, G.: 1975, *Astrophys. J.* **199**, 642.

Tanaka, Y. and Bleeker, J. A. M.: 1977, *Space Sci. Rev.* **20**, 815.

Watson, W. D., and Christensen, R. B.: 1979, *Astrophys. J.* **231**, 627.

Weaver, R., McCray, R., Castor, J., Shapiro, P., and Moore, R.: 1977, *Astrophys. J.* **218**, 377.

Weisheit, J. C.: 1973, *Astrophys. J.* **185**, 877.

Weller, C. S. and Meier, R. R.: 1981, *Astrophys. J.* **246**, 386.

Williamson, F. O., Sanders, W. T., Kraushaar, W. L., McCammon, D., Borken, R., and Bunner, A. N.: 1974, *Astrophys. J. Letters* **193**, L133.

York, D. G.: 1974, *Astrophys. J. Letters* **193**, L127.

York, D. G.: 1977, *Astrophys. J.* **213**, 43.

York, D. G. and Cowie, L. L.: 1983, *Astrophys. J.* **264**, 49.

York, D. G., Wu, C. C., Ratcliff, S., Blades, J. C., Cowie, L. L., and Morton, D. C.: 1983, *Astrophys. J.* **274**, 136.

SUPERNOVAE AND THEIR REMNANTS

WILLIAM P. BLAIR

The Johns Hopkins University, Baltimore, MD, U.S.A.

and

NINO PANAGIA*

Space Telescope Science Institute, Baltimore, MD, U.S.A.

1. Introduction

Supernovae (SNe) and supernova remnants (SNRs) represent an important area of research in astrophysics because they are central to our understanding of such diverse fields as the late stages of stellar evolution, mass loss from late-type stars, nucleosynthesis, and interstellar medium processes and abundances. In addition, they provide a laboratory for investigating the physics of explosion mechanisms, blast waves and dust grain formation and destruction. SNe and young SNRs provide information on the dynamics of the explosion, and the abundances in the ejecta provide direct evidence for nucleosynthesis. In some cases, this information can be used with stellar evolution models to infer the properties of the precursor star. Evolved remnants can be used as probes of normally invisible regions of the interstellar medium (ISM). The passage of a SN blast wave heats and compresses the ISM, and observation of the subsequent cooling flow provides information on abundances and physical conditions in both the pre-shock and post-shock gas. Extensive reviews on SNe and SNRs can be found in the proceedings of several recent conferences, including *IAU Symposium 101* (Danziger and Gorenstein, 1983), 'Supernovae as Distance Indicators' (Bartel, 1985), the NATO ASI on 'Supernovae: A Survey of Current Research' (Rees and Stoneham 1982) and the Cargese conference on 'High Energy Phenomena Around Collapsed Stars' (Pacini, 1986). Several recent reviews on SNRs have addressed both the observational (Raymond, 1983, 1984; Lozinskaya, 1984) and theoretical (McKee and Hollenbach, 1980) aspects of these objects.

The era of modern observations of supernovae (SNe) began in 1885 when the first spectroscopic measurements of S And, or SN 1885a, in the Andromeda galaxy were made. Since then several hundred SNe have been discovered in external galaxies and a number of them have been studied in detail both photometrically and spectroscopically. Until a few years ago observations were limited almost exclusively to the optical domain because the existing observing facilities at other wavelengths (either ground-based such as infrared observatories and radio telescopes, or airborne, i.e., satellites, rocket and balloon mounted telescopes) were not sensitive enough to detect the weak fluxes from distant SNe. On the basis of

---

* Affiliated with the Astrophysics Division, Space Science Department of ESA; on leave from the Instituto di Radioastronomia CNR, Bologna, Italy.

Y. Kondo (ed.), Scientific Accomplishments of the IUE, pp. 549–575.
© 1987 *by D. Reidel Publishing Company.*

optical information, SNe were subdivided into the two main categories of Type I and Type II SNe: Type II SNe display lines of hydrogen (Balmer lines: Hα, Hβ, etc.) and possibly neutral helium (e.g. He I 5875 Å), while the optical spectra of Type I SNe do not show the presence of any H nor He. Moreover, while Type I SNe appear to form a very homogeneous class of objects which are found in all types of galaxies, Type II SNe have been found only in spiral galaxies and have properties which vary widely from object to object.

The launch of the International Ultraviolet Explorer (IUE) satellite in early 1978 marked the beginning of a new era for SN studies because of its capability of measuring the ultraviolet emission of objects as faint as $m_B = 15$. In early 1979 ESA and SERC set up a Target-of-Opportunity program to observe bright SNe ($m < 12$) with IUE. This made it possible over the years to cover a number of SN events in detail. The new results acted as a catalyst to revive the interest in SNe and to stimulate new systematic studies at other frequencies. Luckily this occurred at a time when other powerful astronomical instruments had become available, such as the *Einstein* Observatory for X-ray measurements, the VLA for observations at radio wavelengths and a number of new telescopes either dedicated to infrared observations (e.g. UKIRT and IRTF at Mauna Kea) or equipped with new and highly efficient infrared instrumentation (e.g. AAT and ESO observatories). As a result, a wealth of new information has become available which, thanks to the coordinated efforts of astronomers observing at widely different wavelengths, has provided fresh insights into the properties and the nature of SNe of both types.

Likewise for SNRs, ultraviolet spectroscopic data provide much more leverage for the interpretation of physical conditions, abundances and the physical processes at work in the shocked gas than do optical spectra alone. The highest ionization lines in optical SNR spectra are usually those of O III, which are formed in a region of 20 000 to 40 000 K. The IUE spectral range contains lines of many higher ionization lines such as O IV, C IV, and N V, which sample temperatures up to 200 000 K or more. The presence or absence of these lines permit a much stricter interpretation of the peak post-shock temperature, and thus the shock velocity, responsible for the emission.

In addition to higher ionization stages of optically detected elements, ultraviolet spectra of SNRs contain emission lines of several elements not seen optically, the most important of which are carbon and silicon. These elements are usually depleted onto grains, so their absolute abundances are hard to estimate. Theoretical studies by Draine and Salpeter (1979) indicate that grains should be destroyed by thermal sputtering in fast SNR shocks. Hence, by observing a range of shock velocities, one should be able to not only determine the real abundances of these elements, but also investigate grain destruction properties as a function of shock velocity.

Prior to IUE, very little was known about the ultraviolet emission of SNRs. Carruthers and Page (1976) obtained low resolution far ultraviolet images of a portion of the Cygnus Loop with an electrographic Schmidt camera on Apollo 16 that showed emission corresponding roughly to the bright optical filaments. Jenkins *et al.* (1976) used the *Copernicus* satellite to observe stars behind the Vela SNR, finding high velocity components in many lines (including O VI λ1035) attributable to the remnant. Also, concurrent with the initial results from IUE, Shemansky *et al.* (1979) reported a *Voyager* spectrum of the Cygnus Loop in the range 600—1700 Å, showing a number of strong, but largely unresolved emission lines.

In what follows we will discuss the primary results from IUE as they pertain to Type I SNe, Type II SNe, young SNRs and evolved SNRs separately. For completeness, we will mention recent unpublished results as well as those in the literature whenever possible. (This is a testament to the fact that even in its old age, IUE continues to make significant contributions to SN and SNR research). We conclude with a section on the outlook for future ultraviolet observations of these objects.

2. Supernovae with IUE

So far eleven SNe have been observed with IUE, of which 5 were Type II and 6 Type I (see Table I). However, only four SNe, namely 1979c, 1980k, 1981b and 1983n, were bright enough to obtain high quality ultraviolet spectra and/or to follow their time evolution. Therefore, our discussion of individual objects will be concerned mostly with these four SNe while referring to the others just briefly.

TABLE I

IUE Observations of Supernovae

Supernova	Supernova				Number of exposures
	Galaxy	Type	Period of observation		
1978g	IC 5201	II	1978 Nov 30—Dec 11		2
1979c	NGC 4321/M100	II—L	1979 Apr 21—Aug 4		29
1980k	NGC 6946	II—L	1980 Oct 30—1981 Jan 5		33
1980n	NGC 1316	Ia	1980 Dec 11—1981 Jan 16		8
1981b	NGC 4536	Ia	1981 Mar 9—11		7
1982b	NGC 2268	Ia	1982 Feb 18		3
1983g	NGC 4753	Ia	1983 Apr 8—25		6
1983n	NGC 5236/M83	Ib	1983 Jul 4—Aug 18 26		
1984	NGC 1559	II	1984 Aug 13		1
1985f	NGC 4618	Ib	1985 May 17		1
1985l	NGC 5003	II	1985 Jun 28—Jul 17		3

2.1. TYPE I SUPERNOVAE

Recent observations of Type I SNe have led to the identification of two distinct subclasses, hereafter called Type Ia and Type Ib, respectively. The class of Type Ia SNe comprises all 'classical' SNI including all to those discovered within elliptical and S0 galaxies and a substantial fraction of those discovered in spiral galaxies. Recently, a separate class called Type Ib has been identified; the basic difference is that in the spectrum of Type Ib SNe the 6150 Å feature is missing while it is the strongest and deepest absorption observed in the spectrum of Type Ia SNe. Type Ib SNe are only found in spiral galaxies and are apparently associated with spiral arms. Moreover, Type Ib SNe appear to be underluminous by about 1.5 magnitudes relative to SNIa.

Here, we first present the observational results for 'classical' Type I, or Type Ia, SNe and the next section will be devoted to the 'subluminous' Type I, or Type Ib SNe.

2.1.1. 'Classical' Type I, or Type Ia, Supernovae

So far four Type Ia SNe have been observed with IUE (see Table I). However, none of them was followed for a long time mostly because of pointing constraints of the satellite. Therefore, in each case the observations are concentrated around the epoch of their maximum light but we know little about their time evolution.

The UV spectrum is clearly not a smooth continuum but rather displays a number of "bands" which are observed for all of the SNe (cf. Figure 1). The most prominent feature is the emission which peaks at $\sim$ 2950 Å with a half-power width of $\sim$ 100 Å, i.e. $\Delta v = 10^4$ km s^{-1}. Alternatively, this band may be the result of strong absorptions occurring on both sides of the apparent emission, i.e. centered at $\sim$ 2840 Å and $\sim$ 3060 Å and having half-power widths of the order of 100 Å. A similarly prominent emission band is seen at $\lambda \sim$ 1890 Å in the only spectrum obtained at short wavelengths for SN 1981b. Several other absorptions features can be recognized, which are present at all epochs of observation. Although some of these features might be identified with multiplets of Fe I, Fe II, and Mg II, no satisfactory identification has been found yet for the majority of the absorptions. Nevertheless, the very fact that the spectrum is virtually the same for all four SNe and at all epochs when observations were made is already an important result, since it confirms that the homogeneous properties of Type I SNe extend to the ultraviolet.

After the pioneering observations of SN 1972e (Kirshner et al., 1973) a large sample of Type I SNe has been studied in recent years at infrared wavelengths (Elias et al., 1981; Elias et al., 1985). Again a very close similarity is found in the spectral behavior and the light variation among the observed SNe. This extends their characteristics as 'standard candles' into the infrared. Two of the Type I SNe with infrared observations have been observed with IUE (i.e. 1980n and 1981b).

Combining different observations, one can construct the complete spectrum of a Type Ia SN from the ultraviolet to the infrared. As an illustration, the spectrum of SN 1981b at an epoch around the optical maximum is shown in Figure 2. We see that the ultraviolet spectrum declines very steeply at short wavelengths. Moreover, the ultraviolet emission is much lower than a blackbody extrapolation of the optical spectrum. This is just the opposite of what is found for Type II SNe (cf. Section 2.2). In particular, the ultraviolet flux is approximately 10 times lower than the blackbody curve at the color temperature appropriate to match the visual spectrum ($T = 15\,800$ K, upper curve). Also, in the infrared the observed spectrum is much weaker than a blackbody extrapolation from the optical. On the other hand, the infrared and ultraviolet spectra can smoothly be connected to each other by a lower temperature blackbody ($T \sim$ 9400 K, lower curve). These results

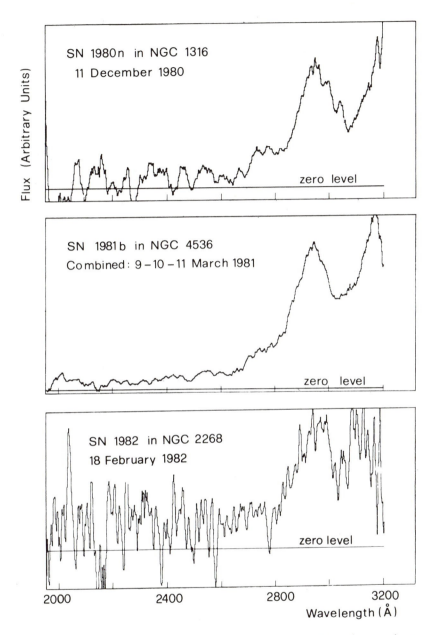

Fig. 1. The IUE long wavelength spectra of Type Ia Supernovae. Note the prominent emission feature centered at ~ 2950 Å.

indicate that the opacity in both ultraviolet and infrared must be much higher than at optical wavelengths, so that the radiation temperature is close to the effective temperature in the visual but reduces to the minimum temperature $T_{min} \sim 0.6\ T_{eff}$ both at shorter and at longer wavelengths.

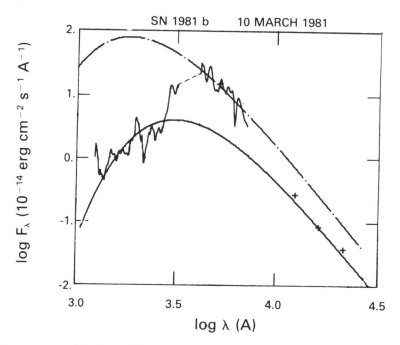

Fig. 2.   The spectrum of the Type Ia SN 1981b on March 10, 1981, dereddened with $E(B - V) =$ 0.24. The upper curve is a black body at 15 800 K which fits the optical spectrum. The lower curve is a black body at 9400 K which connects the IR and UV spectra.

Within this sample of Type Ia SNe only SN 1981b has been observed extensively in the radio. It has been monitored at 5 GHz with the VLA at regular time intervals during the period March 1981—February 1983 but it was never detected (Weiler *et al.*, 1986), the upper limit to the radio flux being as low as 50 μJy (1σ). From a comparison with the radio properties of Type II SNe, Panagia (1984a) argues that the lack of radio emission from SN 1981b implies that its progenitor was a star which in the late phase of its evolution underwent a low rate of mass loss ($\dot{M} < 10^{-7}\ M_\odot\ \mathrm{yr}^{-1}$) and, therefore, was originally a relatively low mass star ($< 3\ M_\odot$).

2.1.2. *Subluminous Type I, or Type Ib, Supernovae*

The existence of a separate subclass of Type I SNe became evident when SN 1983n was discovered in the southern spiral galaxy NGC 5236 (= M83) and its properties and evolution were studied in great detail. Discovered on 1983 July 3 the IUE observations were started on July 4th: the long wavelength spectrum was remarkably similar to that displayed by SN 1981b on 1981 March 9—11, i.e. at the epoch of its maximum in the B band (Barbon *et al.*, 1982).

It was soon apparent that this was not a 'normal' SNI. When the ultraviolet emission reached a maximum between the 8th and the 12th of July, the spectrum was markedly different from what it was on July 4 and, therefore, different from the spectrum of classical Type I SNe at maximum. In the optical spectrum, not only was the 6150 Å feature missing at all epochs but also other spectral features were quite dissimilar from those of Type Ia SNe as well. On the other hand, a reasonable resemblance with SNIa spectra was found if one compared the observations of SN 1983n near maximum light with those of SN 1981b (which is a prototype of SNIa) at much later epochs (Panagia *et al.*, 1986a). In other words, SN 1983n looked like it was 'born old' in both its ultraviolet and optical spectral characteristics. Another basic difference is that, while Type Ia SNe have infrared emission significantly lower than a blackbody extrapolation of the optical specturm (Panagia, 1984a), SN 1983n displayed emission at all epochs which could almost perfectly be fitted to a blackbody curve for all wavelengths longer than about 4000 Å (see Figure 3). Lastly, the FES light curve reached a maximum on ~ July 17th at m(FES) $\simeq$ 11.5 which, corrected for reddening (Panagia *et al.*, 1986a) and adopting the largest distance proposed for M83 (~ 8 Mpc; Sandage and Tammann, 1974), corresponds to M(FES) = -18.5. This makes it about 1.5 mag fainter than classical Type I SNe.

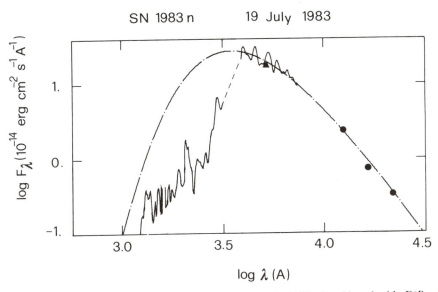

Fig. 3.   The spectrum of the Type Ib SN 1983n on July 19, 1983, dereddened with $E(B - V) =$ 0.16. Both the UV and the optical spectra have been smoothed with a 100 Å bandwith. The triangle represents the FES photometric point, the dots represent the J, H and K photometric data. The dashed-dotted curve is a black body curve at $T = 8300$ K.

The detailed light curves from ultraviolet to infrared are displayed in Figure 4 in the form of λF_λ as a function of time. Note, in particular, that the infrared light curves do not exhibit the secondary maximum which is found in all Type Ia SNe. As already apparent in Figure 3, most of the energy at all epochs is emitted at optical wavelengths. Only 13% of the total luminosity was emitted shortward of 3400 Å at the epoch when the ultraviolet emission attained a maximum. Moreover, there is no indication of any stronger emission in the ultraviolet at very early epochs: this implies that the initial radius of the SN (i.e. the radius of the stellar

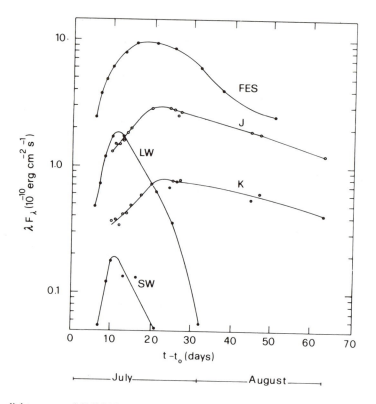

Fig. 4.   The light curves of SN1983n from the UV to the IR, in the form of log λF_λ as a function of the time elapsed since the explosion. The curves denoted as SW and LW represent the behaviour of the average flux measured with short and the long wavelength IUE cameras, respectively.

progenitor when the shock front reached its photosphere) was definitely much less than 10^{13} cm and probably lower than 10^{12} cm, which rules out a red supergiant as the progenitor.

The bolometric light curve presented in Figure 5 is computed by direct integration of the spectrum over all wavelengths: this is the first time that this has been done reliably for a Type I SN. In this context it is worth stressing the crucial role played by the FES measurements for the study of SNe. This 'side product' has been particularly important since it provided optical data simultaneously with the ultraviolet results. SN 1983n, for example, emitted most of its energy at optical wavelengths but almost no photometric data were obtained from ground based observatories. A comparison of the bolometric light curve with the results of model calculations involving the radioactive decay of ^{56}Ni into ^{56}Co as the main source for the luminosity (e.g. Arnett, 1982) leads to estimates of the ^{56}Ni mass in the range $0.1-0.5\ M_\odot$ for an adopted distance of 8 Mpc (and much less if the distance is smaller). This implies that models which involve the complete incineration of a white dwarf to transform it entirely into ^{56}Ni can be ruled out for this supernova.

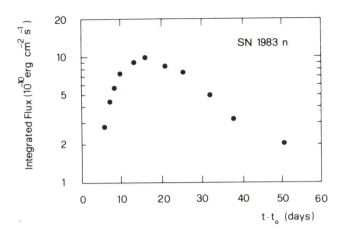

Fig. 5. The bolometric light curve of SN1983n obtained by direct integration of the spectra dereddened with $E(B - V) = 0.16$.

SN 1983n was also observed at radio wavelengths with the VLA and detected as early as July 6th (Sramek *et al.*, 1984). The maximum flux density was detected on July 27th (S = 28 mJy) followed by a steady decay at a rate of the type $S \propto t^{-1.6}$. Observations made at 20 cm suggest a power law spectrum with a strongly non-thermal spectral index $\alpha \sim 1$ (where $S_\nu \propto \nu^{-\alpha}$). The most plausible picture for this SN is that it was produced in a massive binary system: both stars had to originally have masses greater than 5 $M_\odot$ with the primary, which eventually became the SN, about $1-2$ $M_\odot$ more massive than the companion (Sramek *et al.*, 1984). The primary evolves first and ends up as a massive white dwarf, a little below the Chandrasekhar limit ($M_{wd} \sim 1.35$ $M_\odot$). It slowly accretes mass from the companion which, meanwhile, has reached the red supergiant phase. Thus, when the WD exceeds the critical mass and explodes as a SN, the explosion occurs within the envelope of the companion: the observed radio emission originates from the interaction of the ejecta with that envelope. Within this framework strong radio emission from Type I SNe is possible only when the involved binary system is relatively massive.

SNe with spectral characteristics similar to those of SN 1983n were noticed more than 20 yr ago by Bertola and collaborators (Bertola, 1964; Bertola *et al.*, 1965). However, only the wealth of information on SN 1983n has made the recognition of this new subclass a reality. Since its discovery, much attention has been devoted to an appropriate identification and classification of Type I SNe, with the lack of the 6150 Å feature from the optical spectrum being the main criterion to identify a Type Ib SN. The renewed interest has led to the recognition (for old events) and to the discovery (for new events) of a number of Type Ib SNe, Judging from the still quite limited statistics (7 objects; see Panagia, 1986) Type Ib SNe seem to constitute a class as homogeneous as Type Ia do. In particular, they are found only in spiral galaxies, they attain an optical magnitude at maximum which is ~ 1.5 mag fainter than that of SNIa and they display a distinctive infrared light curve and evolution. More detailed discussion is given in Panagia *et al.* (1986b) and Panagia (1986).

The fact that Type Ib SNe occur only in spirals lends support to the idea that SNIb represent the end products of the evolution of rather massive binary systems in which neither of the two components was massive enough to explode as a Type II SN. Panagia (1986) has estimated that ~ 20% of the Type I SNe discovered in spiral galaxies are Type Ib events. Since they are fainter than SNIa and since the SN searches are essentially magnitude limited, the true percentage of Type Ib events in spiral galaxies may be considerably higher. In particular, adopting $\Delta M = 1.5$ as the magnitude difference between the two subtypes, the observed fraction f(SNIb) ~ 20% transforms into a true percentage of SNIb events of ~ 50% of the total Type I SN explosions. Moreover, since in spiral galaxies one observes approximately as many Type II events as Type I (Tammann 1978), it is clear that Type Ib SNe not only are a substantial fraction of the Type I explosions but must account also for quite a fraction of all SN events in spiral galaxies, say, ~ 25%.

2.2. TYPE II SUPERNOVAE

Of the five Type II SNe which have been observed with IUE, the information is meager for three because they were rather faint. Thus, the present discussion is limited to the other two SNe (1979c and 1980k) for which the data are quite comprehensive (Benvenuti *et al.*, 1982).

Figure 6 displays the ultraviolet spectrum and the fluxes in the *U, B, V* bands of SN 1980k in NGC 6946 obtained on 1980 November 9, i.e. about 10 days after the optical maximum. The ultraviolet flux is higher than the extrapolation of the optical spectrum with a blackbody curve: a clear excess can be seen for $\lambda <$ 2000 Å. Such an excess is found at all epochs although it is less pronounced at early times. The characteristics of a high continuous ultraviolet emission and an excess at shortest wavelengths were also found for SN 1979c in NGC 4321 (Panagia *et al.*, 1980). Fransson (1984) has covincingly demonstrated that such an excess is the result of inverse Compton scattering of photospheric radiation by energetic, thermal electrons ($T \sim 10^9$ K) at the shock front where the ejecta interact with pre-existing circumstellar material. This model also provides a natural explanation for the high ionization implied by some emission lines (e.g. N IV] λ1486, C IV λ1550, etc.) observed in the spectrum of SN 1979c (Panagia *et al.*, 1980; see Figure 7).

From a comparison of the line profiles observed in the ultraviolet and in the visual with theoretical calculations (Fransson *et al.*, 1984) it is concluded that the ultraviolet emission lines of highly ionized species are produced in the upper atmosphere, as well as Hα or Mg II λ2800 although just in the outermost layers where the density is lower and the ionizing radiation flux is higher. The line profiles imply an expansion velocity of 8400 kms^{-1} (Fransson *et al.*, 1984), which is only marginally lower than that measured in the optical (i.e. 9200 kms^{-1}; Panagia *et al.* 1980). From an analysis of the N V λ1240, N IV] λ1486, C IV λ1550, N III] λ1750 and C III] λ1909 line intensities the abundance ratio of nitrogen to carbon has been estimated to be N/C ~ 8 (Panagia, 1980; Fransson *et al.*, 1984), i.e.

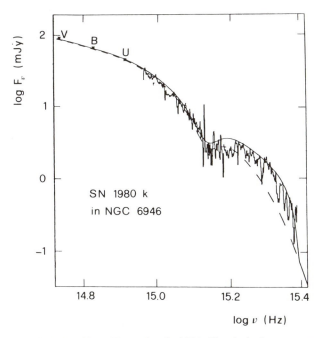

Fig. 6. The spectrum of SN 1980k on November 9, 1980. The dashed curve represents the best fit to the optical and near ultraviolet data in terms of a black body at 9900 K and a reddening of $E(B - V) = 0.32$. The solid line is the spectrum obtained by including also radiation produced by two photon emission (Benvenuti *et al.*, 1984).

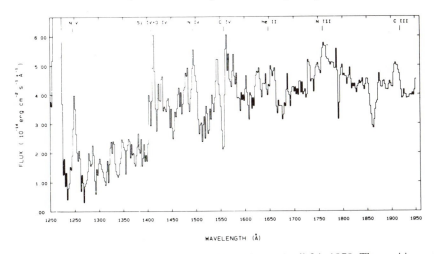

Fig. 7. The short wavelength IUE spectrum of SN 1979c on April 24, 1979. The positions of the identified emission lines are shown redshifted by $z(\text{NGC }4321) = 0.0054$.

~ 30 times higher than the cosmic value. This strong enhancement of nitrogen relative to carbon suggests that the pre-supernova star was a massive supergiant which had undergone a long period of mass loss, thereby exposing CNO processed material.

As for observations at other frequencies, both SNe have been detected in the radio and are currently being monitored with the VLA at wavelengths from 2 to 20 cm (Weiler *et al.*, 1981, 1986). The time behavior is qualitatively the same in the two SNe: the radio 'light' curve is characterized by a steep rise, with time scale of a few months which occurs at earlier times for higher frequencies, followed by a slow decay. The high brightness temperature and the rather steep spectrum at late times ($S_v \propto v^{-0.7}$) indicate a non-thermal origin for the radio emission. Also, the early and steep rise of the radio emission implies that it originates at the outer layers of the SN envelope because otherwise the high free-free opacity of the ejecta would have prevented any signal from becoming detectable for several years. These data are best explained in terms of a model in which the intrinsic emission is due to the synchrotron process and decreases steadily with time ($\propto t^{-0.7}$). The acceleration of the relativistic electrons may be produced by turbulence at the shock front of the SN, where the ejecta interact with pre-existing wind material shed by the red supergiant progenitor (Chevalier, 1981). The presence of wind material which is responsible for the free-free absorption accounts for both the steep rising branch of the radio light-curve and the delay of the rise at lower frequencies. From the amount of free-free absorption the mass loss rate of the stellar progenitor is estimated to be around 10^{-5} $M_\odot$ yr^{-1} for both SNe (Weiler *et al.*, 1986). The X-ray results for these two SNe, namely the missed detection of SN 1979c and the positive detection of SN 1980k followed by a quick decline (Panagia *et al.*, 1980; Palumbo *et al.*, 1981; Canizares *et al.*, 1982), confirm the general picture derived from the optical, ultraviolet and radio data of the interaction of the ejecta with pre-existing circumstellar material.

It is clear that, although qualitatively very similar to each other, these two SNe were considerably different in absolute luminosity. After allowance for extinction, SN 1979c at maximum appeared to be only about one magnitude fainter than SN 1980k even if the distance to NGC 4321 is almost three times greater than to NGC 6946. Therefore, SN 1979c became more than one magnitude intrinsically brighter than SN 1980k. This is not surprising since the spread of properties among different Type II SNe is large, not only in absolute luminosity but also in time evolution (i.e. light curve and colors).

3. Supernova Remnants with IUE

The ultraviolet spectral range is rich with emission lines of importance as plasma diagnostics, but the sample of SNRs that is available to IUE for direct observation is quite small. Inspection of the IUE merged log for class 75 [SNRs] shows that fully two-thirds of all IUE spectra on SNRs have been taken of two objects — the Cygnus Loop and the Crab Nebula! Most galactic remnants are located in or near the galactic plane at distances of a kiloparsec or more. Often they are faint to begin with, and interstellar reddening makes them difficult targets for IUE. Likewise, many extragalactic SNRs are too faint, too small in angular size, or too reddened for successful IUE observation (with several significant exceptions). IUE observers have used indirect observations as well (e.g. IUE observations of stars lying in the direction of galactic SNRs to search for absorption lines from the

SNR) to extend the sample of objects for which IUE has made significant contributions.

3.1. YOUNG SUPERNOVA REMNANTS: LEFTOVERS OF AN EXPLODED STAR

For our purposes, a young SNR will be defined simply as an object for which there is a historical record of the actual SN or for which kinematical information implies relative youth (e.g. $v_{exp} \geqslant 1000$ km s^{-1}; age $\leqslant 2000$ yr.). These objects are dominated by the rapidly expanding ejecta of the exploded star which have not yet encountered enough ISM to be decelerated or diluted significantly. Theoretically, Type I SNe are thought to come from exploded white dwarfs and should produce significant amounts of iron (cf. Woosley *et al.*, 1986, and references therein) while more massive stars produce Type II SNe with ejecta enriched in intermediate mass elements, depending on the mass of the precursor star (Weaver and Woosley 1980). Observationally, one would expect the remnants from Type I and Type II SNe to be quite different, and they are.

3.1.1. *The Crab Nebula*

The Crab Nebula is the most intensely studied young SNR both observationally and theoretically. Produced by the historical SN of 1054 A.D., the Crab Nebula was heralded for many years as the prototypical young remnant; however, this intense scrutiny has now identified the Crab as a rather exceptional event, perhaps involving a sub-luminous, low energy SN from an intermediate mass progenitor. We will concentrate on IUE's contribution to studies of the Crab Nebula; for a recent synposis of general research on the Crab Nebula and similar objects, see Kafatos and Henry (1985).

The Crab Nebula is located about 200 pc above the galactic plane at a distance of about 2 kpc, and subtends an angular size of 5 by 7 arcmin (roughly 3 by 4 pc), much larger than the IUE large aperture. Optically, the remnant consists of a diffuse continuum component and a complex filamentary component (perhaps distributed in a double-ring structure; see Clark *et al.*, 1983), expanding at a mean velocity of ~ 1500 km s^{-1}. Optical spectroscopy (Miller, 1978; Davidson, 1978, 1979; Fesen and Kirshner, 1982) has established that He is overabundant relative to H in the filaments, and that the He abundance is variable. Optical reddening estimates indicate $E(B - V) = 0.5$, which makes IUE observations difficult.

The motivation for observing the Crab Nebula in the ultraviolet is to estimate or place limits on the abundances of C, N and other elements in the ejecta and to determine the physical conditions in the filaments. While IUE exposures of the Crab have been obtained by many observers, most of the observational results are reported in Davidson *et al.* (1982). Most of the spectra have been taken at two general locations in the nebula, at or adjacent to the pulsar or about 40 arcsec to the southwest on a bright filament that Davidson *et al.* (1982) call Region I. Although faint, diffuse continuum and emission lines are visible on individual long exposures of Region I, Davidson *et al.* carefully added six SWP and six LWR spectra totaling 31 and 17 hr integration, respectively. Lines of C III] $\lambda 1909$, He II $\lambda 1640$ and C IV $\lambda 1550$ are clearly seen protruding from a diffuse, reddened continuum: no other lines in the SWP or LWR range are detected. Pairs of spectra at a position just northwest of the pulsar (Region II) show diffuse continuum only. Dereddening these observations with the ANS extinction curve derived by Wu (1981) demonstrates the non-thermal nature of the optical/ultraviolet continuum, although the slope of the power law is marginally different at the two positions. Benvenuti *et al.* (1981) report IUE spectra of the pulsar itself, but only claim detection of it in the LWR range. Using Seaton's (1979) extinction curve to de-redden the diffuse continuum left residuals around 2200 Å, providing evidence for peculiar extinction.

Even with the uncertainties in reddening correction, the ultraviolet line intensities provide constraints on the nucleosynthesis of the precursor star. Davidson *et al.* (1982) estimate a C/O ratio of roughly unity, and a nearly solar ratio of C/H. This can be used with stellar evolution calculations and the fact that the Crab SN produced a pulsar to estimate a precursor main sequence mass of about 8 $M_\odot$ (Nomoto *et al.*, 1982). Basically, less massive progenitors are expected to be completely disrupted by the SN explosion leaving no compact remnant, while stars much more massive are expected to produce large overabundances of C, O or other elements (see section below on O-rich SNRs).

The presence of strong C I lines in the near infrared (Henry *et al.*, 1984; Dennefeld and Pequignot, 1983) and C III and C IV in the ultraviolet attest to the wide range of physical conditions present in the filaments. Detailed photoionization models have recently appeared (Henry and MacAlpine, 1982; Pequignot and Dennefeld, 1983) that attempt to separate variations in chemical abundances and variations in physical conditions within the filaments. Models involving cylindrical filaments (i.e. ropes) with dense, neutral cores and extended, tenuous envelopes appear to be necessary to explain the range of ionization and the ratio of C IV/C III. The variation of He line intensities is consistent with both an enhanced and variable He abundance in the filaments, but the situation for other elements is model dependent. Unfortunately, variation of the C abundance in the Crab cannot be addressed with the current data at only one position.

3.1.2. *Oxygen-rich Remnants*

Model calculations indicate that explosions of stars with initial masses $\geqslant 10\ M_\odot$ should produce young SNRs with enhanced abundances of C, N, O, S, Ne, and possibly other elements, depending on the mass of the star. The element most readily observable in the optical is O, and the small class of objects belonging to this group have become known as 'O-rich remnants'. These objects are very strong X-ray sources because the densities are high and the X-ray emissivity is an increasing function of the heavy element abundance. Optically, they are marked by filaments with very enriched abundances relative to H and rapid expansion velocities of order several thousand km s^{-1}. The galactic remnant Cas A is apparently related to this class, showing some knots that are apparently pure O (cf. Chevalier and Kirshner, 1979), but many other knots rich in O-burning products (i.e. S, Ar, Ca, etc.). These knots are thought to be bits of material from the various interior layers of the precursor star. Ultraviolet spectra of Cas A would be of great interest for determining abundances of C, Si and Mg in the ejecta, but it is too heavily reddened ($A_v = 4.3$) for detection with IUE.

Until very recently, the only O-rich SNR detected at ultraviolet wavelengths was the powerful young SNR in the irregular galaxy NGC 4449. At a distance of 5 Mpc, the remnant is unresolved even with the VLA ($\theta = 0.07$", Bignell and Seaquist, 1983) and matters are further complicated by the fact that the SNR is embedded in an H II region (Kirshner and Blair, 1980). Optical spectra of the emission region allow the SNR lines to be distinguished kinematically from the H II region, however, because the remnant is expanding at 3500 km s^{-1}. The optical SNR is dominated by lines of O, although Ne and S lines are also detected (Blair *et al.*, 1983). Hence, the fast moving material shows similarities to the fast moving knots in Cas A. The large intrinsic luminosity and moderately low reddening ($E(B - V) = 0.17$) make it possible to detect the remnant with IUE.

Blair *et al.* (1984) obtained two long SWP exposures of the emission region containing the SNR, totaling 17.7 hr exposure. The individual line-by-line files were carefully inspected and the data were processed to remove hits and camera features. A broad feature was detected in both spectra at 1660 Å that was identified as O III] $\lambda 1663$ from the SNR. An extraction using only the lines of the images that contained the SNR feature is shown in Figure 8. The ultraviolet spectrum of the emission region is dominated by the hot stars that excite the H II component; a strong P Cygni profile at C IV $\lambda 1550$ is seen emerging from a noisy, but very blue continuum. Although the spectrum is a composite of many stars, the strength of this P Cygni profile argues for the presence of main sequence stars with mid-O spectral type ($M \sim 30\ M_\odot$) in the region directly adjacent to the SNR.

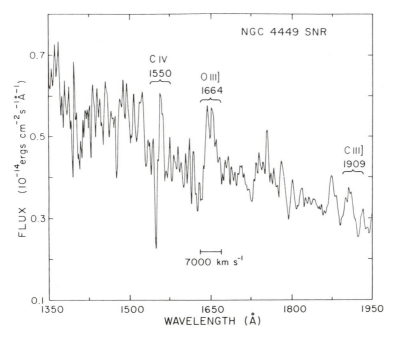

Fig. 8. Portion of SWP spectrum of emission region containing the powerful O-rich SNR in NGC 4449. Only the O III] feature is attributable to the SNR. (From Blair *et al.*, 1984.)

These are the kind of stars that are expected *a priori* to produce O-rich SNRs.

The integrated flux in the O III] feature is close to that predicted by Itoh's (1981) pure oxygen shock models. Blair *et al.* (1984) used upper limits on the strength of any C and Si lines with calculations of their own to place upper limits on the C and Si abundances in the ejecta. Comparison of the optical/ultraviolet data to model calculations of massive star evolution implies a precursor star of order 30 $M_\odot$ for this remnant.

Very recently, another O-rich SNR has been detected with IUE, and it shows a very different ultraviolet spectrum. The object, cataloged as 1E 0102—7219, is the brightest extended X-ray source in the Small Magellanic Cloud (SMC), and optical emission from the remnant was discovered by Dopita *et al.* (1981). The optical spectrum is dominated by lines of O and Ne (Dopita and Tuohy, 1984) and the ejecta are distributed in a distorted ring expanding at roughly 3200 km s^{-1} (Tuohy and Dopita 1983). The object was observed in 1986 March by W. Blair, J. Raymond, J. Danziger and F. Matteucci using collaborative ESA/US1 shifts to obtain long exposure times. Their 14.5 hr SWP spectrum is shown in Figure 9. In contrast to the NGC 4449 SNR, the SMC remnant does show lines of C IV $\lambda1550$ and C III] $\lambda1909$, while O III] $\lambda1663$ is quite weak. Other emission lines are seen at 1400 and 1355 Å, presumably belonging to O IV] (and/or Si IV and S IV]) and O I], respectively. No Si II or Si III lines are seen in the spectrum (especially Si III] $\lambda1892$) and S II is absent from the optical spectrum, so the 1400 Å feature is probably dominated by O IV]. If the O I $\lambda1357$ identification is correct, it is surprising, because the optical [O I] lines are very weak, and O-rich shock models currently available (Itoh, 1981; Dopita *et al.*, 1984) do not predict it to be strong. The 14 hr LWP spectrum shows two lines only, Mg II $\lambda2800$ and [Ne IV] $\lambda2423$. These spectra are so recent that analysis is still pending, but the combined optical

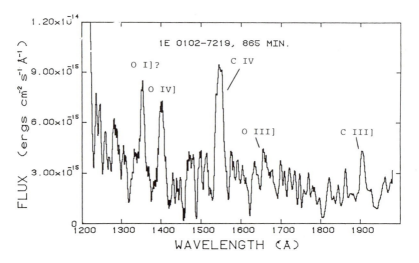

Fig. 9. SWP spectrum of the O-rich SNR 1E 0102—7219 in the Small Magellanic Cloud. Note the strength of C IV and C III] relative to C III], which is much different from that seen in the NGC 4449 SNR.

and ultraviolet data for 1E 0102—7219 are expected to provide the first serious test of O-rich shock model calculations.

3.1.3. *Other Young SNRs*

Direct IUE observations of other young SNRs have been mostly unsuccessful. The O-rich remnant N132D in the Large Magellanic Cloud (LMC) has been attempted by several observers without success. The LMC remnant 0540—69.3 shows characteristics of both O-rich and Crab Nebula-like remnants (see Part VI of Kafatos and Henry, 1985), but an SWP spectrum of it only showed faint continuum emission (T. Gull, priv. comm.). The galactic SNR Puppis A is known to show strong optical [N II] lines at several locations, and O-rich knots were recently identified (Kirshner and Winkler, 1985). An SWP spectrum of one of the strong [N II] positions showed very faint lines of N IV] $\lambda 1486$ and N III] $\lambda 1748$ (R. Fesen, priv. comm.), but because of patchy extinction in the region, observations at several other positions have been blank. Also, two SWP spectra of the LMC Balmer-dominated remnant 0505—67.9 (presumably a remnant of a Type I SN by comparison to Tycho's SNR and the remnant of SN 1006) produced null results (R. Fesen, priv. comm.).

One of the outstanding observational problems for remnants of Type I SNe is the apparent absence of Fe. Recent attempts to model light curves and spectra of Type I SNe using the radioactive decay of ^{56}Ni (Branch *et al.*, 1985, and references therein) have been very successful, but one of the ramifications is that 0.5—1 $M_\odot$ of Fe should be produced. Lines of Fe II and Si II are thought to be present in spectra of Type I SNe shortly after outburst (Branch *et al.*, 1983), but evidence for these elements in the resulting remnant has been hard to find.

In an interesting but somewhat controversial result, ultraviolet lines of Fe and Si may have been detected in absorption from the galactic remnant of SN 1006. Wu *et al.* (1983) obtained a 6 hr SWP and 7.3 hr LWR spectrum of the so-called Schweizer-Middleditch star (or SM star), which is a $V = 16.7$ sdOB star that lies in projection near the center of the SN 1006 remnant. Although faint, the high temperature ($\sim$ 38 000 K) and low reddening (i.e. no detectable 2200 Å absorption) permit the star to be detected. The spectroscopic distance of $\sim$ 1.1 kpc for this star puts it roughly at the suspected distance of the SNR, and Schweizer and Middleditch (1980) originally suggested it to be the stellar remnant of the SN. If Wu *et al.* (1983) are correct, the SM star is behind the SNR, placing an upper limit on the distance.

Wu *et al.* (1983) compare the ultraviolet spectrum of the SM star to the white dwarf HZ 29 and point out several anomalies. Although the spectra are noisy, broad absorption troughs at the positions of the strongest ultraviolet Fe II multiplets are apparent in the LWR spectrum of the SM star; nothing is apparent in the HZ 29 spectrum. Also, a number of narrow absorptions are present in the SWP spectrum of the SM star that are absent in the HZ 29 spectrum. While the identification of these features is uncertain, Wu *et al.* argue that they are consistent with highly redshifted lines ($v_r \sim$ 4000—6000 km s^{-1}) of Si II, III and IV. Since the Si lines are narrow but redshifted and the Fe II lines are broad, Wu *et al.* (1983) suggest a model with freely expanding Fe in the center of SNR 1006, and

Si absorption in a shell (of which we are seeing only the redshifted part along this line of sight).

The reality of these features in the ultraviolet spectrum of the SM star has been confirmed recently. As part of an eighth round program, R. A. Fesen, C.-C. Wu and M. Leventhal obtained additional long IUE exposures on the SM star as well as other comparison stars besides HZ 29 (R. A. Fesen, priv. comm.). The broad Fe lines, observed this time with the LWP camera, are present and the absorption lines in the SWP range are confirmed. Analysis of these data is in progress, but these new observations strengthen the argument in favor of Fe detection is SNR 1006.

Distance determinations for galactic SNRs are often difficult, and Tycho's SNR (SN 1572) is a case in point. Distance estimates from the radio $\Sigma - D$ relation place the remnant at about 5 kpc (Milne, 1979); also, H I 21 cm absorption at roughly -50 km s^{-1} in this direction, when interpreted assuming a standard galactic rotation model, also indicates a distance of 4—7 kpc (Schwarz et al., 1980). Proper motion studies of Tycho's SNR seem to indicate a much smaller distance of $\sim$ 2.3 kpc (cf. Strom et al., 1982). Black and Raymond (1984) have made high dispersion optical and ultraviolet observations of five stars in the direction of Tycho's SNR, all within 2.5 kpc. They find IS absorption components near -50 km s^{-1} in a number of these stars, indicating the presence of non-circular motions in this direction and strongly favoring the smaller distance estimates.

3.2. EVOLVED SNRs: PROBING THE ISM

As an operational definition of an evolved SNR, we simply mean those remnants that are old enough and large enough that they are no longer dominated by the ejecta and the initial kinematics of the SN explosion. By the time most SNRs have expanded to diameters of 10 pc or so, the ejecta have been either left behind by the expanding blast wave or diluted by swept-up IS matter to the point where they are no longer a factor.

3.2.1. *The Cygnus Loop*

The Cygnus Loop is the prototypical evolved SNR, with a diameter of 40 pc and an age estimate of 30 000 yr. At a distance of only 770 pc, the low reddening ($E(B - V) \leqslant 0.08$), and bright, well resolved filamentary structure (angular diameter $= 3°$) make it an excellent target for IUE. This remnant has been observed over a wide range of frequencies (radio through X-ray), so a broad base of information is available for understanding the structure of this remnant and its interaction with the ISM.

The picture that has developed for the Cygnus Loop over the last decade is based on the multiple-phase ISM theory of McKee and co-workers (McKee and Cowie, 1975; Mckee and Ostriker, 1977). The main blast wave expands through the low density ($N \sim 0.01$ cm^{-3}) ISM component at a velocity of 400 km s^{-1} (judging from the temperature in the X-ray gas). As it expands, the blast wave encounters denser 'clouds' in the ISM ($N = 1 - 10$ cm^{-3}); slower secondary shocks (of order 100 km s^{-1}) are driven into these clouds. Because of the higher density, the cooling times are short and these shocks become radiative, producing the bright optical filaments. The main blast wave is identified with a faint string of filaments seen in projection about 5′ to 10′ outside the bright filamentary structure. These outer filaments are dominated by Balmer-line emission and are called 'non-radiative' because radiative cooling is unimportant to the structure of the shock (Raymond *et al.*, 1980b, and references therein).

Quite naturally, IUE observers concentrated first on the bright optical filaments, in particular on three positions with good optical spectroscopic data (Miller, 1974). Benvenuti *et al.* (1979) and Benvenuti *et al.* (1980) reported SWP spectra of Miller's positions 1 and 2, while Raymond *et al.* (1980a) observed Miller's position 3. Qualitatively, the spectra at these three positions are quite similar, showing strong emission lines of C III, C IV, O III, Si IV/O IV, and He II, with weaker lines of N and Si, and a faint continuum component identified as two-photon emission of H (D'Odorico *et al.*, 1980; Raymond *et al.*, 1980a). The main differences in the spectra are a stronger N V line at position 1 and a weaker continuum component at position 3. Spectra with the LWR camera showed emission lines of C II, Ne IV, O II, and Mg II. A representative spectrum is shown in Figure 10.

Interpretation of these data has depended to a large extent on comparison to radiative shock model calculations such as those of Dopita (1977), Raymond (1979) and Shull and McKee (1979). These are all plane parallel, steady-flow models which predict the emission expected from the entire cooling and recombination zone behind shocks of various characteristics. While the models differ

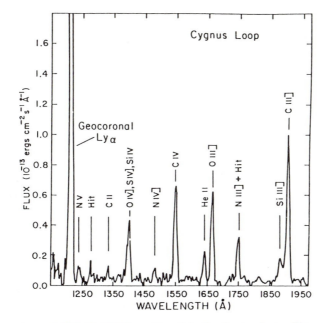

Fig. 10.   SWP spectrum of Miller's (1974) position 3 in the Cygnus Loop. The weak continuum is due to two-photon emission of H, and the line strengths are typical of many of the bright, radiative filaments. (From Raymond *et al.*, 1980.)

from each other in detail, the gross similarities between the models and observations serve to confirm the radiative shock picture for the Cygnus Loop, while pointing out some discrepancies between theory and observation.

One of the largest discrepancies was the strength of the $C\,\textsc{ii}\ \lambda1335$ line, which was expected to be strong from the models, but was observed to be weak or absent. Initially, the relative intensities of the C lines were interpreted as being due to progressive grain destruction in the shock (Benvenuti et al., 1979). However, Raymond et al. (1980a, 1981) point out that the relative intensities of the $C\,\textsc{ii}$ intercombination line at $\lambda2325$ and the permitted line at $\lambda1335$ imply non-negligible resonant scattering; this effect would be most prominent for a thin sheet seen edge-on, which is the orientation suggested at Miller's position 3. Thus, permitted lines in general may have a tendency to be weaker than expected from models because of this process.

Taking resonant scattering into account does not entirely explain the differences between the models and well observed filaments. One additional factor may be that the $10'' \times 20''$ aperture of IUE permits light from shocks of several velocities to enter the spectrograph simultaneously; this would be especially applicable to a region such as Miller's position 1, which is in the complex NE region of the Cygnus Loop. Benvenuti et al. (1980) fitted their spectra of this region using models with depleted abundances and a range of shock velocities from $80-130$ km s^{-1}.

Another limitation for the comparison of observations and models comes from the different slit sizes and orientations used for the ultraviolet and optical spectra. For instance, Miller (1974) used a $2.7'' \times 68''$ slit for his optical observations, compared to the $10'' \times 20''$ IUE large aperture. Interference filter photographs (cf. Fesen et al., 1982) show that many filaments change their appearance when imaged in different optical emission lines, and no one knows how closely the ultraviolet emission follows the optical. With observed variations over scales as small as several arcseconds, differing slit sizes cause substantial uncertainties in combining optical and ultraviolet data.

Optical interference filter photographs in $H\alpha$ and [O III] and optical spectra indicate that the assumption of steady flow may not apply to all filaments in the Cygnus Loop (Fesen et al., 1982). The steady flow shock models typically predict [O III] $\lambda\lambda5007, 4959/H\beta$ ratios of 9 or less, while many filaments (especially toward the outside edge of the bright filaments) show observed ratios in the range $10-40$ (i.e. very strong [O III] emission). These have been called non-steady flow filaments and are thought to arise from recently shocked IS clouds which have not yet established a complete cooling and recombination zone. This interpretation is confirmed by ultraviolet spectra (Raymond et al., 1981), which show the two-photon continuum and low temperature lines like Mg II, C II, C III, and Si III to be weaker than expected from a steady flow model at position 3 (which shows strong [O III] emission), while spectra of a 'normal' [O III]/$H\beta$ filament show line intensities in agreement with the steady flow predictions.

Taking the various difficulties into account, the bright optical filaments in the Cygnus Loop are typically due to shocks with $v_s = 130$ km s^{-1} in IS clouds with initial densities $n_0 = 1-5$ cm^{-3}. Despite the early claims of grain destruction and depleted abundances, models with cosmic or near-cosmic abundances can account for the line intensities when the effects discussed above are taken into account. Raymond *et al.* (1981) estimate a depletion of C relative to O of less than 25% for two filaments and normal N and Ne abundances. Even Si, which appeared somewhat over-abundant, seems to have been brought into line by updated collision strengths. However, with the advent of more sophisticated shock models and an improved understanding of the processes being observed, the problem of grain destruction in shocks is still open for investigation.

In addition to the bright radiative filaments, IUE has also been used to observe one of the optically faint non-radiative filaments in the NE section of the Cygnus Loop, about 5′ ahead of the bright filaments. This filament (position Q of Fesen *et al.*, 1982) is part of an extensive line of faint filaments (cf. Hester *et al.*, 1986) that are thought to demark the actual location of the SN blast wave. The emission from these filaments is governed by the non-radiative shock theory of Chevalier and Raymond (1978) and Chevalier *et al.* (1980).

The optical emission of this filament is dominated by the Balmer lines produced by neutral H being swept up by the shock, but faint lines of [Ne V], [O II], He II, [O III], [N II] and [S II] have also been detected (Raymond *et al.*, 1983; Fesen and Itoh, 1985). An echelle observation of Hα by Raymond *et al.* (1983) shows the characteristic two component line profile expected from a non-radiative shock; the narrow component reflects the thermal distribution in the pre-shock gas and the broad component represents the bulk and thermal velocity structure in the post-shock region. A shock velocity of 170 km s^{-1} is indicated for position Q.

The IUE spectrum of the filament is presented by Raymond *et al.* (1983) and it confirms the non-radiative shock interpretation. The spectrum (see Figure 11)

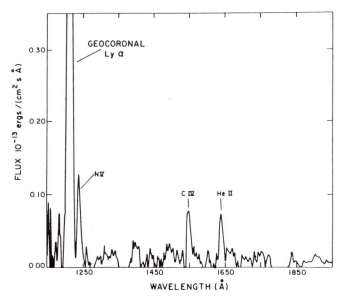

Fig. 11. SWP spectrum of a faint, non-radiative filament in the Cygnus Loop. The apparent negative values near 1370 and 1800 Å result from subtraction of particle events which were in the background. Note the absence of low ionization lines such as C III] λ1909. (From Raymond *et al.*, 1983.)

shows strong lines of N V, C IV and He II, and faint two-photon continuum emission; conspicuous by their absence are C III] $\lambda1909$, O III] $\lambda1663$, and other lower ionization lines which are prominent in the spectra of radiative shocks. The strength of the two-photon continuum and the relative strength of the Balmer and forbidden lines in the optical permit the pre-shock neutral fraction to be estimated at 30%. Comparing the ultraviolet and optical line intensities to models with instant electron/ion equilibration (by plasma turbulence) and with slower equilibration through Coulomb collisions, Raymond et al. (1983) conclude that the latter model produces a better match. Abundances in the pre-shock gas are apparently typical of diffuse IS clouds.

3.2.2. *Other Evolved Remnants*

The Vela SNR is similar to the Cygnus Loop in that it is large in angular size and only slightly reddened. However, much less supporting optical and X-ray data have been available in the past and Vela has been relegated to secondary status relative to the Cygnus Loop. Danziger et al. (1980) observed a bright optical knot near the center of Vela with the SWP camera and identified faint two-photon continuum and several emission lines. The relative intensities of C III], C IV, and C II make it clear that resonant scattering is occurring, as seen in the Cygnus Loop. The corrected line intensities are consistent with a 90 km s^{-1} shock at this position. Raymond et al. (1981) observed a Vela filament chosen from an [O III] photograph, and the combined SWP + LWR spectrum indicates $V_s = 130$ km s^{-1} and non-steady flow conditions, similar to some positions in the Cygnus Loop.

Several other evolved SNRs have been observed with IUE, but the data has been of lower quality and the analysis has been more limited. The LMC remnant N49 was observed by Benvenuti et al. (1980) and qualitatively compared to the Cygnus Loop, the main difference being a much weaker C III] $\lambda1909$ line, possibly due to a low C abundance. The galactic SNRs S147 and G65.3 + 5.7 were both marginally detected with full shift SWP exposures (R. Fesen, priv. comm.) and would make good potential targets for future, more sensitive ultraviolet telescopes. Also, several other bright filaments in Vela and the Cygnus Loop have been observed with IUE, but the analysis is still in progress at this writing.

The LMC remnant N63A was observed by Benvenuti et al. (1980) but very little was said about the spectrum, which shows lines of C IV $\lambda1550$, He II $\lambda1640$ and C III] $\lambda1909$ of nearly equal intensity emerging from a relatively strong continuum component. Benvenuti et al. (1980) claim that the continuum is dominated by stellar emission. Blair et al. (1985) recently obtained a new spectrum of N63A and reprocessed the old data; these data are shown in Figure 12. Inspection of the photowrite shows that the continuum emission in Figure 12 is not stellar, but *diffuse*. Also, it does not drop off as rapidly at short wavelengths as would two-photon emission and it is much stronger relative to the lines than two-photon continuum is expected to be. The relative intensity of the ultraviolet lines and of the line and continuum components are very similar to the Crab Nebula spectrum shown by Davidson et al. (1982); since continuum emission is also seen in (uncalibrated) optical spectra, it seems probable that the continuum is due to synchrotron emission and the object has Crab Nebula-like properties. If

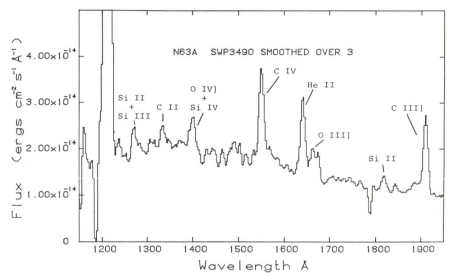

Fig. 12. Re-extracted version of SWP 3490, a spectrum of the LMC remnant N63A, first shown by Benvenuti *et al.* (1980). The spectrum was reprocessed from the line-by-line file and includes only emission from the SNR. A number of lines that were only marginally detected in the published spectrum are clearly present, and the diffuse continuum is much stronger than expected from two-photon emission.

verified, this would be very interesting, because the precursor star for N63A is thought to have been about 30 $M_\odot$ (van den Bergh and Dufour, 1980).

In addition to these direct observations of evolved SNRs, IUE has been used to observe stars behind some evolved remnants to search for absorption. Jenkins *et al.* (1981) find evidence for a compressed cloud from the Vela SNR along the line of sight to HD 72350. Using high dispersion data, they have identified many strong lines of C I arising from excited fine-structure levels and determine an unexpectedly high pressure in the cloud, which they identify with shock compression. In S147, Phillips and Gondhalekar (1983) have identified high velocity absorption due to the remnant along the line of sight to HD 37318. They find abundances somewhat less depleted than seen in low velocity gas, perhaps indicating grain destruction. Observations of other stars toward both of these remnants indicate that the ISM is inhomogeneous.

4. Future Prospects for Ultraviolet Observations

IUE has produced some very significant results in its studies of SNe and SNRs, but it has been pushed to its limit in many cases. The observations discussed above often involve careful processing of long exposures and still result in low signal-to-noise data. Bright extragalactic SNe typically reach $m_v = 11-12$, but their evolution can only be followed for a relatively short time. IUE continues to make significant strides in SN and SNR research but it is important to look ahead as well.

For SNRs, there is an obvious desire for more photons and photon counting detectors; also, wavelength coverage extended down to the Lyman limit and ultraviolet imagery are both needed. These will be supplied by the *Astro* Observatory, which is a shuttle-based suite of telescopes dedicated to ultraviolet astronomy. The *Astro* Observatory consists of the Hopkins Ultraviolet Telescope (HUT), the Ultraviolet Imaging Telescope (UIT), and the Wisconsin Ultraviolet Photo-Polarimeter Experiment (WUPPE). HUT is a 0.9 m telescope with a photon counting detector that is sensitive in the 800—1850 Å range with about 3 Å resolution. Thus, it will be able to seriously investigate the 900—1200 Å region for the first time, and look for O VI λ1035 emission from shocks of intermediate ($\sim$ 200 km s^{-1}) velocity in objects such as the Cygnus Loop and Vela. UIT is a 0.4 m telescope that will image a 40' circular field with 2" resolution using any of a number of ultraviolet filters. It will produce the first high quality ultraviolet imagery of SNRs. In addition to being a sensitive near ultraviolet (1500—3200 Å) spectrometer, WUPPE is also a polarimeter and may be able to measure polarization in ultraviolet lines. While the spectroscopic instruments of *Astro* are much more sensitive than IUE's, they will also be in low earth orbit and observing time will be limited to occasional one week missions. It will probably not be practical to extend the sample of galactic SNRs observed in the ultraviolet until these or similar instruments are placed on a free flying spacecraft or platform.

The Hubble Space Telescope (HST) will offer several advantages to SNR research. The exceptional spatial reolution should permit resolved imagery of bright filaments in nearby remnants and the ability to obtain calibrated optical and ultraviolet spectra of SNR filaments through the same size aperture will reduce the uncertainties in comparing these data to models. However, the small field of view of HST and small spectrograph apertures are not ideal for many galactic SNRs (which are faint and extended), so these observations will be difficult.

Actually, HST may be more generally applicable to studies of remnants in nearby galaxies. IUE has detected two SNRs in M33 (cf. Blair and Raymond, 1984), but the small angular size of these objects is a better match to HST. Imaging of nearby galaxies with 0.1" resolution and filters centered on emission lines will identify more complete samples of SNRs for follow-up ultraviolet and optical spectroscopy and studies of SNR evolution.

For SNe there also is the obvious need to improve the statistics, especially at ultraviolet wavelengths. More important, however, is the extension of observations to very late stages of the evolution. It is only at late stages (i.e. about 3 yr after the actual explosion) that one can recognize the importance of the nuclear decay of ^{56}Co and ^{56}Ni relative to the cooling of the ejected envelope and, therefore, identify the correct mechanisms that give rise to a SN explosion. These old SNe (or young SNRs) are simply too faint for IUE. In addition to the study of the SN phenomenon itself, SNe can be used as background sources against which one can measure the absorption lines produced by the intervening interstellar and intergalactic matter. This will permit a determination of their chemical and dynamical properties with great accuracy. These important aspects have been covered only marginally with IUE observations (e.g. see Pettini *et al.*, 1982) because ordinary extragalactic SNe are too faint for obtaining satisfactory high resolution ultraviolet spectra.

For both these aspects, we believe that HST will be able to make essential contributions because of its larger collecting area and the availability of more sophisticated instrumentation operating at both ultraviolet and optical wavelengths. The major limitation for studying SNe with HST will not be posed by their flux, which remains suitably high for quite some time, but rather by the difficulty of scheduling observations of newly discovered sources on short notice. Therefore, although HST will allow the study of individual events in much greater detail and with much higher accuracy than has been possible with IUE, we expect the statistics to grow as slowly (or as quickly!) as with IUE, i.e. at a rate of a few new events a year. This implies that it will take the whole lifetime of HST to obtain an order of magnitude improvement in both the quality and the statistical significance of the information on SNe.

Note added in proof: As this book goes to press, the occurrence of a bright SN in the LMC (Supernova Shelton 1987a) on 1987 February 24 has excited the astronomical community. Through target-of-opportunity programs both in the U.S. and in Europe, IUE has followed the rapid evolution of the ultraviolet spectral properties of this unique object in great detail. Initially, the SN was visible through-out the IUE spectral range, and both low and high dispersion spectra were obtained at this epoch. The low dispersion spectra showed a strong continuum at wavelengths above 1300 Å, punctuated by very broad P Cygni profiles of (so far) largely un-identified lines. Narrow interstellar absorption lines are apparent even at low dispersion, and the high dispersion spectra are providing a tremendous wealth of information about the halo of our Galaxy and the LMC along this line of sight: Some lines, such as Mg I $\lambda2852$, show at least six components. The far UV light faded rapidly to the point where the SWP spectra no longer showed the SN, but rather the underlying spectrum of the LMC B3 I star Sanduleak-69 202, which had previously been the leading contender for the SN precursor star. As the UV light continued to fade, many changes have been noted on fairly short timescales. Detailed modeling will be necessary to identify the species responsible for the UV lines and the variations that have been observed. At this writing (March 20), monitoring with the LWP camera continues. We once again mention the importance that the FES has played in providing concurrent photometric information through-out the IUE observing phase.

References

Arnett, W. D.: 1982, *Astrophys. J.* **253**, 785.

Barbon, R., Ciatti, F., and Rosino, L.: 1982, *Astron. Astrophys.* **116**, 35.

Bartel, N. (ed.): 1985, in *Lecture Notes in Physics, 224, Supernovae as Distance Indicators*, Springer-Verlag.

Benvenuti, P., Bianchi, L., Cassatella, A., Clavel, J., Darius, J., Heck, A., Penston, H. V., Macchetto, F., Selvelli, P. L., and Zamorano, J.: 1981, in *The Universe at UV Wavelengths: The First Two Years of IUE*, NASA CP-2171, p. 701.

Benvenuti, P., D'Odorico, S., and Dopita, M. A.: 1979, *Nature* **277**, 99.

Benvenuti, P., Dopita, M. A., and D'Odorico, S.: 1980, *Astrophys. J.* **238**, 601.

Benvenuti, P., Sanz Fernandez de Cordoba, L., Wamsteker, W., Macchetto, F., Palumbo, G. C., and Panagia, N.: 1982, ESA SP-1046.

Bertola, F.: 1964, *Ann. Astrophys.* **27**, 319.
Bertola, F., Mammano, A., and Perinotto, M.: 1965, Contr. Asiago Obs. No. 174.
Bignell, R. C. and Seaquist, E. R.: 1983, *Astrophys. J.* **270**, 140.
Black, J. H. and Raymond, J. C.: 1984, *Astron. J.* **89**, 411.
Blair, W. P., Kirshner, R. P., and Winkler, P. F.: 1983, *Astrophys. J.* **272**, 84.
Blair, W. P. and Raymond, J. C.: 1984, in Y. Kondo (ed.), *Future of Ultraviolet Astronomy Based on Six Years of IUE Research*, NASA CP-2349, p. 103.
Blair, W. P., Raymond, J. C., Kirshner, R. P., and Winkler, P. F.: 1985, *Bull. AAS* **17**, 546.
Blair, W. P., Raymond, J. C., Fesen, R. A., and Gull, T. R.: 1984, *Astrophys. J.* **279**, 708.
Branch, D., Doggett, J. B., Nomoto, K., and Thielemann, F.-K.: 1985, *Astrophys. J.* **294**, 619.
Branch, D., Lacy, C. H., McCall, M. L., Sutherland, P. G., Uomoto, A., Wheeler, J. C., and Wills, B. J.: 1983, *Astrophys. J.* **270**, 123.
Canizares, C. R., Kriss, G. A., and Feigelson, E. D.: 1982, *Astrophys. J. Letters* **253**, L17.
Carruthers, G. R. and Page, T.: 1976, *Astrophys. J.* **205**, 397.
Chevalier, R. A.: 1981, *Astrophys. J.* **251**, 259.
Chevalier, R. A. and Kirshner, R. P.: 1979, *Astrophys J.* **233**, 154.
Chevalier, R. A., Kirshner, R. P., and Raymond, J. C.: 1980, *Astrophys. J.* **235**, 186.
Chevalier, R. A. and Raymond, J. C.: 1978, *Astrophys. J. Letters* **225**, L27.
Clark, D. H., Murdin, P., Wood, R., Gilmozzi, R., Danziger, J., and Furr, A. W.: 1983, *Monthly Notices Roy. Astron. Soc.* **204**, 415.
D'Odorico, S., Benvenuti, P., Dennefeld, M., Dopita, M. A., and Greve, A.: 1980, *Astron. Astrophys.*, **92**, 22.
Danziger, J. and Gorenstein, P. (eds.): 1983, *Supernova Remnants and Their X-ray Emission, IAU Symp.* **101**.
Danziger, I. J., Wood, R., and Clark, D. H.: 1979, *Monthly Notices Roy. Astron. Soc.* **192**, 83.
Davidson, K.: 1978, *Astrophys. J.* **220**, 177.
Davidson, K.: 1979, *Astrophys. J.* **228**, 179.
Davidson, K. *et al.*: 1982, *Astrophys. J.* **253**, 696.
Dennefeld, M. and Pequignot, D.: 1983, *Astron. Astrophys.* **127**, 42.
Dopita, M. A.: 1977, *Astrophys. J. Suppl.* **33**, 437.
Dopita, M. A., Binette, L., and Tuohy, I. R.: 1984, *Astrophys. J.* **282**, 142.
Dopita, M. A. and Tuohy, I. R.: 1984, *Astrophys. J.* **282**, 135.
Dopita, M. A., Tuohy, I. R., and Mathewson, D. S.: 1981, *Astrophys. J. Letters* **248**, L105.
Draine, B. D. and Salpeter, E. E.: 1979, *Astrophys. J.* **231**, 438.
Elias, J. H., Frogel, J. A., Hackwell, J. A., and Persson, S. E.: 1981, *Astrophys. J. Letters* **251**, L13.
Elias, J. H., Matthews, K., Neugebauer, G., and Persson, S. E.: 1985, *Astrophys. J.* **296**, 379.
Fesen, R. A., Blair, W. P., and Kirshner, R. P.: 1982, *Astrophys. J.* **262**, 171.
Fesen, R. A. and Itoh, H.: 1985, *Astrophys. J.* **295**, 43.
Fesen, R. A. and Kirshner, R. P.: 1982, *Astrophys. J.* **258**, 1.
Fransson, C.: 1984, *Physica Scripta* **T7**, 50.
Fransson, C., Benvenuti, P., Gordon, C., Hempe, K., Palumbo, G. G. C., Panagia, N., Reimers, D., and Wamsteker, W.: 1984, *Astron. Astrophys.* **132**, 1.
Henry, R. B. C. and MacAlpine, G. M.: 1982, *Astrophys. J.* **258**, 11.
Henry, R. B. C., MacAlpine, G. M., and Kirshner, R. P.: 1984, *Astrophys. J.* **278**, 619.
Hester, J. J., Raymond, J. C., and Danielson, G. E.: 1986, *Astrophys. J. Letters* **303**, L17.
Itoh, H.: 1981, *Pub. Astron. Soc. Japan* **33**, 1.
Jenkins, E. B., Silk, J., and Wallerstein, G.: 1976, *Astrophys. J. Suppl.* **32**, 681.
Jenkins, E. B., Silk, J., Wallerstein, G., and Leep, E. M.: 1981, *Astrophys. J.* **248**, 977.
Kafatos, M. C. and Henry, R. B. C. (eds.): 1985, *The Crab Nebula and Related Supernova Remnants*, Cambridge: Cambridge University Press.
Kirshner, R. P. and Blair, W. P.: 1980, *Astrophys. J.* **236**, 135.
Kirshner, R. P., Willner, S. P., Becklin, E. E., Neugebauer, G., and Oke, J. B.: 1973, *Astrophys. J. Letters* **187**, L97.
Kirshner, R. P. and Winkler, P. F.: 1985, *Astrophys. J.* **299**, 981.
Lozinskaya, T. A.: 1984, *Sov. Sci. Rev. E Astrophys. Space Phys.* **3**, 35.

McKee, C. F. and Cowie, L. L.: 1975, *Astrophys. J.* **195**, 715.

McKee, C. F. and Hollenbach, D. J.: 1980, *Ann. Rev. Astron. Astrophys.* **18**, 219.

McKee, C. F. and Ostriker, J. P.: 1977, *Astrophys. J.* **218**, 148.

Miller, J. S.: 1974, *Astrophys. J.* **189**, 239.

Miller, J. S.: 1978, *Astrophys. J.* **220**, 490.

Milne, D. K.: 1979, *Australian J. Phys.* **32**, 83.

Nomoto, K., Sparks, W. M., Fesen, R. A., Gull, T. R., Miyaji, S., and Sugimoto, D.: 1982, *Nature* **299**, 803.

Pacini, N. (ed.): 1986, *High Energy Phenomena Around Collapsed Stars*, D. Reidel Publ. Co., Dordrecht, Holland (in press.)

Palumbo, G. G. C., Maccacaro, T., Panagia, N., Vettolani, G., and Zamorani, G.: 1981, *Astrophys. J.* **247**, 484.

Panagia, N.: 1984a, *Physica Scripta* **T7**, 15.

Panagia, N.: 1986, NATO ASI in F. Pacini (ed.), *High Energy Phenomena Around Collapsed Stars*, D. Reidel Publ. Co., Dordrecht, Holland (in press.)

Panagia, N. *et al.*: 1980, *Monthly Notices Roy. Astron. Soc.* **192**, 861.

Panagia, N. *et al.*: 1986a, (in prep.).

Panagia, N., Sramek, R. A., and Weiler, K. W.: 1986b, *Astrophys. J. Letters* **300**, L55.

Pequignot, D. and Dennefeld, M.: 1983, *Astron. Astrophys.* **120**, 249.

Pettini, M. *et al.*: 1982, *Monthly Notices Roy. Astron. Soc.* **199**, 409.

Phillips, A. P. and Gondhalekar, P. M.: 1983, *Monthly Notices Roy. Astron. Soc.* **202**, 483.

Raymond, J. C.: 1979, *Astrophys. J. Suppl.* **39**, 1.

Raymond, J. C.: 1983, *Adv. Space Res.* **2**, 145.

Raymond, J. C.: 1984, *Ann. Rev. Astron. Astrophys.* **22**, 75.

Raymond, J. C., Black, J. H., Dupree, A. K., Hartmann, L., and Wolf, R. S.: 1980a, *Astrophys. J.* **238**, 881.

Raymond, J. C., Black, J. H., Dupree, A. K., Hartmann, L., and Wolff, R. S.: 1981, *Astrophys. J.* **246**, 100.

Raymond, J. C., Blair, W. P., Fesen, R. A., and Gull, T. R.: 1983, *Astrophys. J.* **275**, 636.

Raymond, J. C., Davis, M., Gull, T. R., and Parker, R. A. R.: 1980b, *Astrophys. J. Letters* **238**, L21.

Rees, M. J. and Stoneham, R. J. (eds.): 1982, *A Survey of Current Research*, NATO ASI, D. Reidel Publ. Co., Dordrecht, Holland.

Sandage, A. and Tammann, G. A.: 1974, *Astrophys. J.* **194**, 559.

Schwarz, U. J., Arnal, E. M., and Goss, W. M.: 1980, *Monthly Notices Roy. Astron. Soc.* **192**. 67p.

Schweizer, F. and Middleditch, J.: 1980, *Astrophys. J.* **241**, 1039.

Seaton, M. J.: 1979, *Monthly Notices Roy. Astron. Soc.* **187**, 79p.

Shemansky, D. E., Sandel, B. R., and Broadfoot, A. L.: 1979, *Astrophys. J.* **231**, 35.

Shull, J. M. and McKee, C. F.: 1979, *Astrophys. J.* **227**, 131.

Sramek, R. A., Panagia, N., and Weiler, K. W.: 1984, *Astrophys. J. Letters* **285**, L59.

Strom, R. G., Goss, W. M., and Shaver, P. A.: 1982, *Monthly Notices Roy. Astron. Soc.* **200**, 473.

Tammann, G. A.: 1978, *Mem. S.A.It.* **49**, 315.

Tuohy, I. R. and Dopita, M. A.: 1983, *Astrophys. J. Letters* **268**, L11.

van den Bergh, S. and Dufour, R. J.: 1980, *Publ. Astron. Soc. Pacific* **92**, 32.

Weaver, T. A. and Woosley, S. E.: 1980, *Ann. N.Y. Acad. Sci.* **336**, 335.

Weiler, K. W., van der Hulst, J. M., Sramek, R. A., and Panagia, N.: 1981, *Astrophys. J. Letters* **243**, L151.

Weiler, K. W., Sramek, R. A., Panagia, N., van der Hulst, J. M., and Salvati, M.: 1986, *Astrophys. J.* **301**, 790.

Woosley, S. E., Taam, R. E., and Weaver, T. A.: 1986, *Astrophys. J.* **301**, 601.

Wu, C.-C.: 1981, *Astrophys. J.* **245**, 581.

Wu, C.-C., Leventhal, M., Sarazin, C. L., and Gull, T. R.: 1983, *Astrophys J. Letters* **269**, L5.

H II REGIONS

REGINALD J. DUFOUR

Department of Space Physics and Astronomy, Rice University, Houston, TX, U.S.A.

1. Introduction

By virtue of its spectrometers having simultaneous multichannel integration capability and relatively large input apertures, the *International Ultraviolet Explorer* satellite has been the first instrument capable of effective study of the ultraviolet spectra of the gas from extended H II regions. This review will concentrate on discussing the results of the studies made of the ultraviolet emission line and continuum spectra of the gas and dust of extended H II regions in the Galaxy and the Magellanic Clouds. However, it should be noted also that ultraviolet spectral studies with the IUE have contributed new information about the extinction and dust associated with H II regions, the nature of the highly ionized gas surrounding their exciting stars via ultraviolet absorption lines, and the integrated stellar + nebular spectra of more distant extragalactic H II regions. Since the latter three subjects are parts of other reviews in this book, herein we will limit our discussion to results related to the continuous and emission line ultraviolet spectra from the gas and dust within the H II regions themselves.

This is the first review concentrating on IUE results regarding nearby H II regions. Some of the early results of IUE studies of planetary nebulae, supernova remnants, and H II regions have been reviewed by Peimbert (1981) at the first Goddard IUE conference. In addition, a thorough discussion of the expected scientific benefits of studying the ultraviolet spectra of gaseous nebulae has been given by Osterbrock (1979). It should be noted that relative to planetary nebulae and supernova remnants, the ultraviolet spectra of H II regions show fewer emission lines useful as nebular diagnostics due to their usually lower degree of ionization. This, coupled with their usually low surface brightness and sometimes large reddening, make studies of the ultraviolet spectra of H II regions more difficult and less attractive to nebular spectroscopists than planetary nebulae and supernova remnants.

A search of the IUE archives in 1986 May indicated that some 355 SWP and 302 LWP/R spectra have been taken of H II regions (object class 72), of which approximately 90% are of the Orion Nebula and H II regions in the Large and Small Magellanic Clouds (LMC and SMC). In addition to these, many spectra of more distant extragalactic H II regions exist in other class categories (e.g., object classes 82 & 88 — irregular and emission line galaxies, respectively). While most of the spectra were taken with the IUE in the low dispersion mode, there exist a substantial number of high dispersion spectra of the brighter H II regions such as the Orion Nebula, 30 Doradus in the LMC, and SMC N81 & N66A.

Some example low dispersion SWP and LWR spectra of galactic and Magellanic Clouds' H II regions are shown in Figures 1 and 2. In comparison to

Y. Kondo (ed.), Scientific Accomplishments of the IUE, pp. 577–587.
© 1987 *by D. Reidel Publishing Company.*

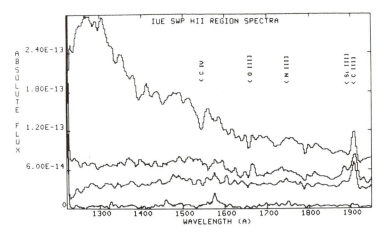

Fig. 1. Example IUE low dispersion SWP spectra of four H II regions. The top spectrum (SWP 5019, 10 min exp.) is a position near the center of the Orion Nebula (note that the absolute flux level is reduced by a factor of 20). Second from the top (SWP 5390, 180 min exp.) is of the bright knot N66A in the SMC H II region NGC 346. Third from the top (SWP 8929, 230 min exp.) is of a region near the center of the giant LMC H II complex, 30 Doradus, or NGC 2070. At the bottom (SWP 16973, 300 min exp.) is a spectrum of the brightest rim in the galactic H II region NGC 6611 (M16). The locations of several of the more prominent ultraviolet emission lines expected to be seen in H II region spectra are noted.

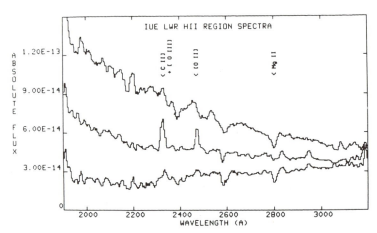

Fig. 2. Example IUE low dispersion LWR spectra of three H II regions. At the top (LWR 7699, 30 min exp.) is a spectrum of the small SMC H II region IC 1644 (N81, including the exciting star) for which the absolute flux has been divided by a factor of 2.5. The middle spectrum (LWR 4354, 17 min exp.) is a position near the center of the Orion Nebula. The bottom spectrum (LWR 7677, 150 min exp.) is of the center of LMC 30 Doradus. The locations of a few of the more prominent ultraviolet emission lines expected to be seen in H II region spectra are noted.

ground-based spectra, the UV spectra of H II regions are relatively devoid of prominent emission lines. In H II regions of moderate and high ionization (and/or high electron temperature), the inter-recombination lines of C III] $\lambda\lambda$1907,9 are usually the most prominent, with Si III] $\lambda\lambda 1883, 92$, O III] $\lambda\lambda 1661-6$, N III] $\lambda\lambda 1747-54$, and [O II] $\lambda 2470$ usually present, but weaker. Lower ionization objects, such as the Orion Nebula, also show weak lines of C II] $\lambda\lambda 2324-29$ and Mg II $\lambda\lambda 2795, 2800$. Some high ionization H II regions may even show the resonance lines of C IV $\lambda\lambda 1548, 50$ in emission (but this in part may be due to dust scattered light from the exciting stars). While the ultraviolet emission lines are relatively few and weak, they nonetheless provide important and even unique astrophysical information about the physical state and composition of the ionized gas in H II regions which cannot be obtained by any other means.

From Figures 1 and 2 it can be seen that the continuum is a dominant and quite variable feature of H II region spectra in the ultraviolet. The prominence and shape of the ultraviolet continuum for a given nebula depend on several factors: (a) the existence of any OB stars in the slit, (b) the amount of internal dust producing scattered starlight from the exciting stars, (c) the magnitude and wavelength dependence of the line-of-sight reddening, and (d) the nebular atomic continuum surface brightness. For distant or small H II regions, (a) and (c) dominate the continuum, while (b) and (c) dominate for the spectra of most galactic H II regions with slit positions off the exciting star(s). The true nebular continuum (d) is significant in only a few cases of the large extended (and dust-free) H II regions of the LMC and SMC with small foreground reddening. The UV absorption lines seen in some low dispersion spectra usually arise from the light from the direct or dust-scattered spectra of the exciting stars embedded in the H II regions themselves. However, some narrow and weak absorption lines seen in high dispersion spectra of giant H II regions or that of embedded stars may arise from absorption by the ionized gas.

2. Ultraviolet Continuum Studies

The large angular size and high surface brightness of the Orion Nebula make it the most accessible galactic H II region for ultraviolet continuum studies with the IUE without the direct contamination of stars in the spectrographic aperture. Nominally, exposures of only a few minutes are required at low dispersion to obtain adequate ultraviolet spectra of the brighter central regions of the nebula. Consequently, the first published study of the ultraviolet continuum of an H II region based on IUE observations was that of the Orion Nebula by Perinotto and Patriarchi (1980a).

Perinotto and Patriarchi observed two regions at low dispersion with the SWP and LWR cameras through the large aperture, one close to the Trapezium and another close to θ^2 Ori A. They found the ultraviolet continuum at both locations to be dominated by stellar light scattered by dust particles located within the nebula. This can be seen in Figure 3, which shows the observed UV continuum flux in the near Trapezium region compared to the estimated nebular atomic emission. It clearly demonstrates the advantages of ultraviolet data for studying the scattering properties of internal dust in H II regions, compared to visible light, as had been the case in all previous studies. Based on their corrected ultraviolet continuum data, they concluded that (1) the wavelength dependence of the gas-to-

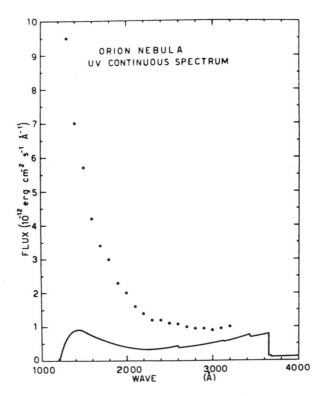

Fig. 3. A plot of the observed ultraviolet continuum fluxes (dots) compared to a model-derived reddened atomic continuum (solid line) for a region near the Trapezium in the Orion Nebula taken from the study of Perinotto and Patriarchi (1980a). The dominance of the dust-scattered continuum relative to the nebular atomic continuum at ultraviolet wavelengths permits more accurate studies of the properties of the internal dust in the nebula compared with visible-light studies.

dust ratio, $N(\text{gas})/\sigma(\lambda)N(\text{dust})$, in the central regions of the Orion Nebula is not large, and (2) the gas-to-dust ratios found for the two regions (between 17 and 47×10^{20} cm^{-2}; with a given value largely depending on the physical approximations made rather than the region observed or wavelength used) were not significantly different from that of the mean for the general interstellar medium ($\approx 20 \times 10^{20}$ cm^{-2}; cf. O'Dell *et al.*, 1966). The latter result was at odds with previously found larger gas-to-dust ratios for the Orion Nebula based on ground-based visible-light observations of the nebular continuum and hydrogen emission lines.

Mathis *et al.* (1981) studied the ultraviolet properties of dust in the Orion Nebula using IUE observations of the ultraviolet continuum at 16 positions varying from 30 arc sec to 5 arc min from the central star θ^1 Ori C. They modeled the ultraviolet atomic continuum based on the observed Hβ surface brightness and the dust scattered continuum based on several dust distribution models of Schiffer and Mathis (1974), which included the albedo and scattering phase function of the grains as well as their distribution. From the observed ultraviolet dust scattered continuum they concluded that (1) dust must be depleted near θ^1 Ori C, (2) the albedo of the dust is fairly constant (≈ 0.45) in the ultraviolet (including across the $\lambda 2200$ extinction feature), and (3) the grains are quite forward throwing at $\lambda 1300$ but more isotropic at $\lambda 2400$. They note that some of their results for the behavior of the albedo and scattering phase function of the dust in Orion are at odds with previous results for the general diffuse interstellar medium, but point out that while the dust in the Orion Nebula may be peculiar compared to the general ISM, previous studies of the properties of dust from reflection nebulae have lacked adequate treatment of possible geometrical effects and that this must be done to obtain realistic results. Finally, as a postscript, Mathis *et al.* note that scattering plays an exceedingly important role in affecting the emission line strengths in the Orion Nebula, both in the optical and ultraviolet. They suggest that most of the emission line strengths at distances greater than 2.5 arc min from the exciting star are due to scattering rather than simple gas emission along the line of sight.

IUE spectra of other galactic H II regions existing in the archives (M8, M16, the Eta Carinae nebula, etc.) are dominated by the strong UV continuum, presumably due to largely dust scattered starlight by internal dust. These spectra will undoubtedly be an important resource to future investigations of the properties of internal dust in galactic H II regions when adequate supporting optical, infrared, and radio data become available for model analyses. While the existing IUE spectra of H II regions in the Magellanic Clouds may provide useful information on the nature of internal dust (if any) from UV continuum studies, such analyses are relatively much more difficult due to the complications of accounting for faint blue stars in the aperture and geometrical complexities.

3. Emission Line Studies

3.1. THE ORION NEBULA

By virtue of its high surface brightness and relatively low reddening, the Orion Nebula has been the most extensively studied galactic H II region with the IUE. Its UV spectrum shows prominent emission lines of C III], C II], and [O II]. Surface brightness maps of the central region of the Orion Nebula in the light of the above three UV emission lines have been ingeniously constructed from a series of 69 large aperture low dispersion SWP and LWR spectra of 35 contiguous overlapping areas by Turnrose *et al.* (1980) and Perry *et al.* (1981). The maps indicate significant spatial variations in both the surface brightness and relative strengths of C III], C II], and [O II] lines. For example, the nebula's 'bar' just to the northwest of θ^2 Ori A is most prominent in C II]. The C III] emission seems to be concentrated in localized 'hot spots' in the bar and region towards θ^1 Ori C with little or none to the southeast. By contrast, there are several regions with localized C II] and [O II] emission in the region southeast of the bar, some of which seem to be concentrated around the Taylor—Munch cloudlets.

The prominent C III] and C II] ultraviolet emission lines in the IUE spectra of the Orion Nebula have provided astrophysicists with the first reliable information about the gaseous-phase carbon abundance in an H II region. Perinotto and

Patriarchi (1980b) and Torres-Peimbert *et al.* (1980) independently used IUE spectra of several regions in the Orion Nebula to derive a carbon abundance for the nebula.

Perinotto and Patriarchi (1980b) obtained low dispersion SWP and LWR spectra through the large aperture of two positions near the center of the nebula. From the observed absolute surface brightness of the C III] λ1909 and C II] λ2326 lines compared with ground-based observations with Hβ for the two regions, they derived a carbon abundance of $12 + \log \text{C/H} = 8.4 \pm 0.2$ for the nebula. They note that this is similar to the solar coronal value, but significantly lower than that found for the solar photosphere. They also noted the unexpected absence of the Mg II λ2800 lines in Orion compared to the planetary nebula IC 418, which is of similar ionization. This led them to conclude that the gaseous-phase magnesium abundance in the Orion nebula is at least ten times smaller than that in the Sun.

Torres-Peimbert *et al.* (1980) studied the UV spectra of three central regions of Orion and used the [O II] transauroral-auroral lines $\lambda\lambda$2470—7325 to tie together the IUE and ground-based spectra. Making extensive use of the results of previous ground-based optical and radio observations of the reddening, temperatures, densities, and ionization levels of the three IUE positions, they derived C$^+$ and C^{++} concentrations using $T(\text{C III}) = T(\text{O III})$ and $T(\text{C II}) = T(\text{N II})$. They also derived C^{++} concentrations from the optical C II λ4267 recombination line observed in six positions by Peimbert and Torres-Peimbert (1977). They found that the carbon abundances derived from the UV and optical lines are in good agreement if one adopts a root-mean-squared temperature fluctuation (Peimbert, 1967) of $t^2 = 0.016$, for which $12 + \log \text{C/H} = 8.5 \pm 0.1$. In addition, they found only a weak decrease in the C^{++}/H$^+$ ratio with increasing distance from θ^1 Ori C, which was more similar to the variation in the He$^+$/H$^+$ ratio than to that in O^{++}/H$^+$. It was also noted that while the gaseous-phase C/H ratio found for Orion is 0.15 dex lower than that of the solar photosphere by Lambert (1978), the C/O ratio in Orion is similar to that of the sun within the analytical errors — indicating that depletion of carbon into dust grains in the central regions of the nebula is probably negligible.

Carbon abundances based on the UV lines of C II] and C III] in the Orion Nebula observed by Torres-Peimbert *et al.* (1980) have been rederived using nebular photoionization models by Dufour *et al.* (1982) and by Mathis (1985). For comparison, the results for carbon, nitrogen, and oxygen from the four previously cited investigations are presented in Table I. Also given are mean abundances and the standard deviation computed by averaging the results from the three latter studies. For comparison, the solar photospheric abundances of Lambert (1978) are presented. The agreement for carbon abundances between the three studies is rather good, largely because most of the carbon in the Orion Nebula exists in the form of C^{++} and is observable through the relatively strong C III] λ1909 line. Moreover, since the ionization potentials of C$^+$ and O$^+$ are similar, C^{++} coexists with O^{++} inside the nebula, and using the electron temperature derived from the [O III] lines to derive the C^{++}/H$^+$ abundance is probably realistic for the situation in the Orion Nebula (but possibly not for higher temperature and ionization giant H II regions — as noted later).

TABLE I

CNO abundances in the Orion Nebula

$12 + \log N(X)/N(H)$					Reference
C	N	O	C/O	C/N	
8.5					Perinotto and Patriarchi (1980b)
8.52	7.65	8.62	−0.10	0.87	Torres-Peimbert et al. (1980)
8.46	7.48	8.60	−0.14	0.98	Dufour et al. (1982)
8.42	7.61	8.54	−0.12	0.81	Mathis (1985)
8.47	7.58	8.59	−0.12	0.89	Mean
±.05	±.09	±.04	±.02	±.09	1σ
8.67	7.99	8.92	−0.25	0.68	Sun: Lambert (1978)

It is important to note that while the C/H abundance in the Orion Nebula is about 0.2 dex lower than solar, O/H and N/H are found to be even lower relatively. Consequently, the C/O and C/N ratios in the Orion Nebula are slightly larger than solar. While the compositional disagreement (of metals relative to hydrogen) between the Orion Nebula (including those of other nearby H II regions) and the Sun is a current problem of concern among nebular and stellar astrophysicists, the IUE observations of the UV carbon lines suggest that C/O and C/N are at approximately solar ratios in the nebula and the (at least relative) depletion of gaseous-phase carbon via grains in the form of graphite or carbides is negligible. However, other refractory elements such as magnesium (and possibly also silicon since the UV Si III] lines are unexpectedly weak) do appear to be appreciably depleted by factors of at least ten relative to both oxygen and hydrogen.

High dispersion IUE observations of six regions of the Orion Nebula have been reported by Boeshaar et al. (1982). They presented emission line strengths for three regions at or near the bright bar near θ^2 Ori A and for three of the Taylor—Munch cloudlets nearby. In addition to the prominent lines of C III] $\lambda 1909$, C II] $\lambda 2326$, and [O II] $\lambda 2470$, they also find lines such as C II $\lambda 1335$, He I $\lambda\lambda 2764, 2829$, [Mg V] $\lambda 2784$, Mg II $\lambda 2800$, Mg I $\lambda 2852$, and O VI $\lambda 3067$ in both nebular and cloudlet regions. Relative to [O II] $\lambda 2470$, the strengths of the other UV lines generally do not differ markedly among the various regions. While some of these lines (most notably O VI) may originate from dust scattered starlight, these results demonstrate the advantages of high dispersion IUE spectra over low dispersion spectra in detecting weak nebular emission lines in H II regions where the continuum level is high. The IUE archives contain many such high dispersion spectra of Orion and other H II regions for which detailed analysis of should provide important astrophysical information about H II regions.

3.2. OTHER GALACTIC H II REGIONS

IUE low dispersion spectra of other galactic H II regions are notable for their lack of emission lines compared to the UV continuum level. Consequently, there have not been any published studies of abundances or physical conditions based on UV emission lines for any galactic H II regions besides the Orion Nebula. The relatively low ionization levels, apparent significant amounts of internal dust, and relatively low electron temperatures are undoubtedly responsible for the lack of detectable UV emission lines in the IUE spectra. Possibly future IUE high dispersion spectra of galactic H II regions would result in the detection and measurement of some UV emission lines for diagnostic purposes, but this would require long exposure times.

3.3. H II REGIONS IN THE MAGELLANIC CLOUDS

In contrast to galactic H II regions, the H II region complexes in the Large and Small Magellanic Clouds have been scientifically productive targets for the IUE. Compared with most galactic H II regions, the more prominent H II regions in the Magellanic Clouds generally show a number of ultraviolet emission lines useful for determination of abundances and physical conditions. A review of the abundance results from ground-based and pre-1983 IUE spectrophotometry of H II regions in both Clouds has been given by Dufour (1984a).

The first published results derived from IUE spectra of H II regions in the Magellanic Clouds was of a low carbon abundance in two SMC H II regions (NGC 346 and IC 1644) by Dufour *et al.* (1981). From low dispersion SWP observations of C III] in the two nebulae, they found C/H values lower by −0.8 dex compared to the Orion Nebula and −0.9 dex compared to the Sun. Since these carbon deficiencies were similar to those previously found for other elements such as oxygen, neon, sulphur, and argon, they concluded that the nucleosynthesis origins of carbon were similar to those of the other cited elements, namely from massive stars.

A subsequent more detailed study of gaseous-phase carbon abundances in three SMC and four LMC H II regions based on IUE observations was published by Dufour *et al.* (1982; hereafter called DST). They combined extensive ground-based spectrophotometry of each of the nebulae with IUE observations of the C III] $\lambda 1909$, C IV $\lambda 1549$, and C II] $\lambda 2326$ lines (where detected) and derived abundances using nebular photoionization models. The results for C/H among the studied H II regions of each Cloud were in good agreement: $12 + \log \text{C/H} = 7.16 \pm 0.04$ for the SMC and 7.90 ± 0.15 for the LMC. The C/O results for the nebulae in each Cloud showed even better consistency, possibly due to the lower sensitivity of the C/O result to the adopted electron temperatures of individual nebulae compared to C/H. They found $\log \text{C/O} = -0.89 \pm 0.02$ for the three SMC H II regions and -0.48 ± 0.04 for the four LMC H II regions.

DST noted that the C/H and C/O deficiency for the SMC was even lower than that of nitrogen — previously the most deficient element found in the two metal-poor galaxies. They suggested that this result indicated that the nucleosynthesis yield of carbon in massive stars is lower than previously thought, and that significant carbon enrichment may arise in intermediate mass stars, resulting in a later increase in C/O during the chemical evolution of a galaxy. They further note that C/N is lower in the SMC compared with the LMC and the Orion Nebula, possibly due to the fact that part of the nitrogen in the more metal-rich galaxies arise from secondary nucleosynthesis (of carbon) production in the ejecta from intermediate and low mass stars (e.g., planetary nebulae).

Mathis *et al.* (1985) used DST's IUE observations of the 30 Doradus nebula in the LMC with new extensive ground-based spectrophotometry and detailed nebular photoionization models to derive improved abundances and physical conditions in this giant H II region. Based on comparison of the spatial variations in the observed Balmer decrement reddening, the nebular emission, and the IR emission, they concluded that the dust which reddens this nebula is not well mixed with the ionized gas, and that this foreground extinction follows the standard $R_v = 3.1$ galactic law (Savage and Mathis, 1979; but cf., also Fitzpatrick and Savage, 1984). They also found that the C IV emission line strength reported by DST was too large to be reconciled with the somewhat lower effective temperatures that they derived for the exciting stars.

Mathis *et al.* (1985) further argue that temperature inhomogeneities in 30 Doradus and other H II regions require that the C^{+2} abundances derived largely from C III] using [O III] electron temperatures, as was done by DST, should be lowered by 10—20%. By contrast, they find relatively larger ionization correction factors for C^{+2} than DST. For 30 Doradus they derive $12 + \log C/H = 7.48$, which is −0.21 dex lower than the value of 7.69 by DST. However, they reanalyzed DST's data for the SMC H II region, N66NW (in NGC 346), and found $12 + \log C/H = 7.27$, a value +0.10 dex higher than the value of 7.17 by DST. The authors note in conclusion that their derived carbon abundances suggest (1) C/O = 0.16 for *both* the SMC and LMC (compared to C/O = 0.8 for the Orion Nebula) and (2) C/N is essentially the same for the SMC, LMC, Orion Nebula, and the Sun.

These carbon abundance results for the H II regions in the Magellanic Clouds compare favorably with those of other studies of extragalactic H II regions based on IUE observations (cf. the article by Hutchings, Lequeux, and Wolf, pp. 605—622). In particular, the overall results indicate that while C/O increase with O/H in H II regions, C/N is constant within observational errors. This is shown in Figure 4, which is taken from the recent review on CNO abundances by Pagel (1985) using the data of Torres-Peimbert *et al.* (1980), Peimbert *et al.* (1986), Dufour *et al.* (1984), and Mathis *et al.* (1985), which is partly based on IUE observations of the Orion Nebula, H II regions in the Magellanic Clouds, and other prominent extragalactic H II regions. The variable C/O but constant C/N variations with O/H are similar to the variations found in metal-poor dwarf stars by Tomkin and Lambert (1984). The simplest interpretation for these variations is that most of the carbon and nitrogen nucleosynthesis is primary, arising from stars of a lower mass range than those which produces the oxygen enrichment.

4. Concluding Remarks

The IUE observatory has enabled the first quantitative studies of the ultraviolet emission line and continuum spectra of the ionized gas and dust in H II regions to be made. These have resulted in providing us with our first reliable knowledge about the behavior of the astrophysically important element carbon in galaxies.

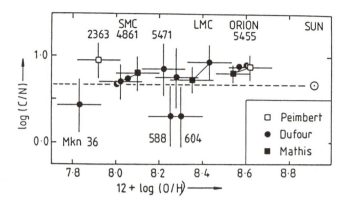

Fig. 4. Results of the variation between C/N and O/H for galactic and extragalactic H II regions resulting from several studies of carbon abundance determinations based on the ultraviolet carbon lines from IUE spectra. This diagram is taken from the review of CNO abundances in galaxies by Pagel (1985). Lines connect different investigations of the same object.

The large numbers of low and high dispersion spectra of H II regions obtained by the IUE available in the archives should be a valuable source for future investigations. For example, many spectra show lines of N III], Si III], and Mg II, which future model-based analyses could utilize to improve our knowledge of nitrogen abundances, as well as producing the first determinations of gaseous-phase abundances of the refractory elements silicon and magnesium in H II regions. The large number of existing high and low dispersion spectra of the Orion Nebula could be used to study spatial variations in the gas-to-dust ratio in the nebula, as well as spatial variations in the gaseous-phase abundances of carbon, silicon, and magnesium.

Future space-borne observatories such as the Hubble Space Telescope (HST) and the Shuttle *Astro* telescopes will undoubtably pursue more detailed investigations of the ultraviolet spectra of nearby H II regions motivated in part by IUE results. However, the small aperture sizes of the spectrographs on the HST will limit its capability for effective study of the ultraviolet spectra of nearby large extended H II regions. The spectrographs on the *Astro* HUT and WUPPE telescopes are better suited for extended nebular spectroscopy than HST, but these have more limited wavelength ranges than the IUE. Nonetheless, these instruments will have a very positive impact on ultraviolet spectroscopy of H II regions by their ability to effectively observe and measure much of the ultraviolet line and continuum spectra of fainter nebulae with improved sensitivity and dynamic range compared to the IUE.

While nebular spectroscopists can look enthusiastically to the future of ultraviolet nebular spectroscopy, they should also note the past fundamental contributions made by the IUE, and particularly note the current scientific possibilities afforded by the large IUE data archive base existing on H II regions.

References

Boeshaar, G. O., Harvel, C. A., Mallama, A. D., Perry, P. M., Thompson, R. W., and Turnrose, B. E.: 1982, in Y. Kondo, J. M. Mead, and R. D. Chapman (eds.), *Advances in Ultraviolet Astronomy: Four Years of IUE Research*, NASA CP 2238, p. 374.

Dufour, R. J.: 1984a, in S. van den Bergh and K. S. de Boer (eds.), 'Structure and Evolution of the Magellanic Clouds', *IAU Symp.* **108**, 353.

Dufour, R. J.: 1984b, in J. M. Mead, R. D. Chapman, and Y. Kondo (eds.), *Future of Ultraviolet Astronomy Based on Six Years of IUE Research*, NASA CP 2349, p. 107.

Dufour, R. J., Shields, G. A., and Talbot, R. J.: 1982, *Astrophys. J.* **252**, 461.

Dufour, R. J., Talbot, R. J., and Shields, G. A.: 1981, in R. D. Chapman (ed.), *The Universe at Ultraviolet Wavelengths: The First Two Years of International Ultraviolet Explorer*, NASA CP 2171, p. 671.

Fitzpatrick, E. L. and Savage, B. D.: 1984, *Astrophys. J.* **279**, 578.

Lambert, D. L.: 1978, *Monthly Notices Roy. Astron. Soc.* **182**, 249.

Mathis, J. S.: 1985, *Astrophys. J.* **291**, 247.

Mathis, J. S., Chu Y.-H., and Peterson, D. E.: 1985, *Astrophys. J.* **292**, 155.

Mathis, J. S., Perinotto, M., Patriarchi, P., and Schiffer, F. H.: 1981, *Astrophys. J.* **249**, 99.

Mathis, J. S., Perinotto, M., Patriarchi, P., and Peimbert, M.: 1966, *Astrophys. J.* **143**, 743.

O'Dell, C. R., Hubbard, W. B., and Peimbert, M.: 1966, *Astrophys. J.* **143**, 743.

Osterbrock, D. E.: 1979, in *Scientific Research With the Space Telescope*, NASA CP 2111, p. 99.

Pagel, B. E. J.: 1985, in I. J. Danziger, F. Matteucci, and K. Kjar (eds.), *Production and Distribution of the CNO Elements*, Garching: ESO, p. 155.

Peimbert, M.: 1967, *Astrophys. J.* **150**, 825.

Peimbert, M.: 1981, in R. D. Chapman (ed.), *The Universe at Ultraviolet Wavelengths: The First Two Years of the International Ultraviolet Explorer*, NASA CP 2171, p. 557.

Peimbert, M., Pena, M., and Torres-Peimbert, S.: 1986, *Astron. Astrophys.* **158**, 266.

Peimbert, M. and Torres-Peimbert, S.: 1977, *Monthly Notices Roy. Astron. Soc.* **179**, 217.

Perinotto, M. and Patriarchi, P.: 1980a, *Astrophys. J.* **238**, 614.

Perinotto, M. and Patriarchi, P.: 1980b, *Astrophys. J. Letters* **235**, L13.

Perry, P. M., Turnrose, B. E., Harvel, C. A., Thompson, R. W., and Mallama, A. D.: 1981, in R. D. Chapman (ed.), *The Universe at Ultraviolet Wavelengths*, NASA CP 2171, p. 601.

Savage, B. D. and Mathis, J. S.: 1979, *Ann. Rev. Astron. Astrophys.* **17**, 73.

Schiffer, F. H., III, and Mathis, J. S.: 1974, *Astrophys. J.* **194**, 597.

Tomkin, J. and Lambert, D. L.: 1984, *Astrophys. J.* **279**, 220.

Torres-Peimbert, S., Peimbert, M., and Daltabuit, E.: 1980, *Astrophys. J.* **238**, 133.

Turnrose, B. E., Perry, P. M., Harvel, C. A., Thompson, R. W., and Mallama, A. D.: 1980, *S. P. I. E.* **264**, 257.

PLANETARY NEBULAE

J. KÖPPEN

Institut für Theoretische Astrophysik, F.R.G.

and

L. H. ALLER

Department of Astronomy, University of California, Los Angeles, CA, U.S.A.

1. The IUE and Planetary Nebulae

Aside from giving us the fun of exploring a new spectral region, IUE observations of planetary nebulae (PN) have greatly enhanced our knowledge about these objects:

Let us consider first their nuclei (PNN): From most of these hot stars we can observe in the optical only the Rayleigh—Jeans tail of the energy distribution. The extension of the stellar continuum and line spectrum to shorter wavelengths substantially improves our understanding of the nature of hot stellar atmospheres. The spectra, which are so terribly dull in the visual, suddenly 'spring to life' in the UV: We know now many PNN to have P Cygni lines indicative of brisk stellar winds. Traditional optical region criteria for stellar temperatures can be improved with IUE data: The Zanstra method can be applied to the stonger stellar continuum and the brighter He II 1640 Å line. Using UV emission lines we can set tighter constraints for our model nebulae and improve our leverage on the unobservable stellar fluxes below 912 Å.

As for the PN themselves, IUE data allow

(a) improvements in determinations of interstellar extinction and give information on extinction by internal dust,

(b) a more thorough examination of physical processes, in particular dielectronic recombination, whose effects are much more pronounced in the UV,

(c) an opportunity to derive abundances of additional elements, particularly carbon which plays such a crucial role in stellar evolution. We get good data on Mg and Si, and improved information on nitrogen which in the optical is visible only as [N II].

(d) We can greatly improve our nebular models since with the additional data we can set more severe constraints on physical processes, excitation conditions, stellar fluxes, geometry etc.

2. Extinction

Interpretation of IUE PN data requires reliable estimates of the interstellar extinction (ISE). In addition to the Balmer decrement and radio continuum to $H\beta$ fluxes, we can use methods unique to the UV, viz.

(a) the strength of the 2200 Å interstellar feature,

Y. Kondo (ed.), Scientific Accomplishments of the IUE, pp. 589—602.
© 1987 *by D. Reidel Publishing Company.*

(b) a comparison of He II recombination lines 1640, 2733, and 3202 with 4686 Å (Seaton, 1978a), provided the fluxes are measured in the same areas of the nebula,

(c) a comparison of auroral and transauroral transitions in p^2, p^3, and p^4 configurations (see e.g. Draine and Bahcall, 1981). Among these are [O II] 2420/7330 Å in low excitation PN (Flower, 1980) and [O III] 2321/4363 Å in medium excitation PN.

For applications to interstellar extinction and consistency tests of theory and various methods see Seaton (1979b), Gathier *et al.* (1986), Harrington and Feibelman (1983) and Goharij and Adams (1984).

2.1. INTERNAL EXTINCTION

The infrared properties of PN suggested that in most instances, internal extinction is negligible compared to ISE (Pottasch *et al.*, 1977; Köppen, 1977; Gathier *et al.*, 1986). The C IV 1550 Å resonance line (observed in moderate and high excitation PN) provides a useful tool for measuring extinction by local dust grains. Forbidden line quanta escape as soon as they are produced ($\tau_0 \ll 1$), but resonance photons can be scattered many times before they reach the optically thin line wing or are extinguished by dust grains. Calculations of the radiative transfer problem in planar (Hummer and Kunasz, 1980) and spherical geometry (Köppen and Wehrse, 1983) show that for an attenuation of the C IV line by a factor 2 one needs a dust optical depth τ_D of 0.1 to 0.3. This amount of dust suffices to account for the IR emission.

To measure τ_D one compares the C IV line with an optically thin line. Köppen (1983) and Köppen and Wehrse (1983) looked at the C IV/C III] ratio, which turns out to be more affected by the nebular geometry than by dust extinction. Harrington *et al.* (1981, 1982) compare the C III 2297 dielectronic recombination line and the collisionally excited C IV 1550 line which both measure the concentration of C^{3+} ions.

3. The Spectrum of the Central Stars

Measurements of UV fluxes of PNN extend the wavelength baseline for colour temperature (T_c) determinations. Kaler and Feibelman (1985) found these T_c's to agree well with (or even exceed) He II Zanstra temperatures. The UV flux distributions for some PNN follow the Rayleigh—Jeans slope.

The great breadth of their *P Cygni lines* permit fast winds to be recognized even in the low spectral resolution mode. A large fraction of PNN show these mass loss manifestations (Perinotto, 1983). The resonance lines of N V, Si IV, C IV and the subordinate lines O IV 1342, O V 1371, N IV 1718 are identified (Perinotto *et al.*, 1982). The edge velocities are of the order 1000 to 3000 km s^{-1}. All low excitation PN (without nebular He II) have winds but some high excitation PN have no wind, viz. 1360, 2448, 3918, 4361, 7293, 7662. All nuclei with visual spectral type WR, Of+WR, Of, Ofp (Aller, 1976) — which have indications of extended atmospheres and mass loss in their optical spectra — have also P Cygni lines. Cerruti-Sola and Perinotto (1985) argue that if the surface gravity of the PNN is less than about log $g = 5.2$, the star always has strong mass loss.

Mass loss rates for these winds have been determined from low dispersion data (Pottasch *et al.*, 1981; Castor *et al.*, 1981; Cerruti-Sola and Perinotto, 1985) and from fitting calculated line profiles to high dispersion data (Perinotto and Benvenuti, 1981; Perinotto *et al.*, 1982; Harrington, 1982; Hamann *et al.*, 1984; Adam and Köppen, 1985). The rates are between 10^{-10} and several 10^{-7} $M_\odot$ yr^{-1}.

Little is known about the ionization mechanisms and stratification in the winds. Castor *et al.* (1981) show that a 'cold wind', i.e. one in radiative equilibrium with the stellar radiation field, is consistent with all the observational data for NGC 6543. In NGC 3242 Hamann *et al.* (1984) find that the level of N V ionization in the wind — due to stellar photons — is *lower* that could be expected for a stellar model atmosphere which gives a He II Zanstra temperature consistent with the nebular ionization. Adam and Köppen (1985) interpret the P Cygni lines of NGC 1535 by a 'warm' collisionally ionized wind, which by its thermal emission could provide enough He ionizing photons required by the nebula.

Narrow unshifted absorption lines with equivalent widths of about 1A have been found in a number of central stars: C IV 1550, N V 1238/42, O V 1371, He II 1640 (Köppen and Wehrse, 1980, 1981; Hamann *et al.*, 1984; Adam and Köppen, 1985; Kaler and Feibelman, 1985). The UV spectrum of the nucleus of NGC 40 shows numerous strong *emission lines*. The identifications and the strengths of the lines are consistent with Smith and Aller's (1969) visual classification as WC8 (Nussbaumer *et al.*, 1981). In the windy PNN of NGC 1535 and 6210 Adam and Köppen (1985) find a number of unidentified broad 'troughs' between 1270 and 1300 Å which are absent in windless NGC 4361.

4. The Nebular Spectrum

Up to now more than 130 PN have been observed in the low dispersion mode, and about half that number were observed with high dispersion. An atlas of line fluxes for 28 objects has been compiled by Boǵgess *et al.* (1981). In most instances the aperture was centered on the PNN, although a few offset observations have been obtained. Spatial resolution is used to separate nebular and stellar features in low dispersion observations. At high dispersion narrow nebular emission lines are easily distinguished from broad stellar wind features. Table I lists the spectral lines so far definitely observed. We give the approximate wavelength, identification, transition scheme, lower and upper excitation energies in eV, and excitation mechanism.

For several nebulae, observers have found the anticipated thermal *emission continuum* of the ionized gas, i.e. the hydrogen and helium bound-free continua and the H I two-photon emission. The latter depends on the optical depth of the nebula in the H I Lyman α line (Pottasch *et al.*, 1981, 1982). There is no contribution by starlight scattered by dust particles in the nebula (Benvenuti and Perinotto, 1981).

Direct *recombination lines* for the He I (2s ^{3}S—np series), the He II Paschen series, and He II Balmer α 1640 have been observed. Any thorough study of the relative line intensities will require a careful examination of ISE. Seaton (1979a) found good agreement of theory and observation for He II lines in NGC 7027.

Dielectronic recombination (inverse of autoionization) is important for the

TABLE I

Lines observed in planetary nebulae

Wavel.	Identification		Lo	Hi	Excitation
1175/76	C III	$2p\ ^3P^0{-}2p^2\ ^3P$	6	17	coll?
1239/43	N V	$2s\ ^2S{-}2p\ ^2P^0$	0	10	coll, wind
1309	Si II	$3p\ ^2P^0{-}3p^2\ ^2S$	0	9	coll
1335/36	C II	$2p\ ^2P^0{-}2p^2\ ^2D$	0	9	coll, diel.rec.
1371	O V	$2p\ ^1P^0{-}2p^2\ ^1D$	20	29	wind, rad.
1394/1403	Si IV	$3s\ ^2S{-}3p\ ^2P^0$	0	9	coll
1397—1407	[O IV]	$2p\ ^2P^0{-}2p^2\ ^4P$	0	9	coll
1483/87	[N IV]	$2s^2\ ^1S{-}2p\ ^3P^0$	0	9	coll
1548/50	C IV	$2s\ ^2S{-}2p\ ^2P^0$	0	8	coll
1575	[Ne V]	$2p^2\ ^3P{-}2p^2\ ^1S$	0	8	coll
1602	[Ne IV]	$2p^3\ ^4S^0{-}2p^3\ ^2P^0$	0	8	coll
1640	He II	Balmer alpha	41	49	rec
1658—66	[O III]	$2p^2\ ^3P{-}2p^3\ ^5P^0$	0	8	coll
1718	N IV	$2p\ ^1P^0{-}2p^2\ ^1D$	16	23	wind, diel.rec.
1711	Si II	$3p^2\ ^2D{-}5f\ ^2F^0$	7	14	coll?
1747—54	[N III]	$2p\ ^2P^0{-}2p^2\ ^4P$	0	7	coll
1760	C II	$2p^2\ ^2D{-}3p\ ^2P^0$	9	16	rec.
1815	[Ne III]	$2p^4\ ^3P{-}2p^4\ ^1S$	0	7	coll
1808/17	Si II	$3p\ ^2P^0{-}3p^2\ ^2D$	0	7	coll
1882/92	[Si III]	$3s^2\ ^1S{-}3p\ ^3P^0$	0	7	coll
1907/09	[C III]	$2s^2\ ^1S{-}2p\ ^3P^0$	0	7	coll
2253	He II	Paschen 6	48	54	rec
2297	C III	$2p\ ^1P^0{-}2p^2\ ^1D$	13	18	diel.rec.
2306	He II	Paschen epsilon	48	54	rec
2321/31	[O III]	$2p^2\ ^3P{-}2p^2\ ^1S$	0	6	coll
2325—29	[C II]	$2p\ ^2P^0{-}2p^2\ ^4P$	0	6	coll
2334—50	[Si II]	$3p\ ^2P^0{-}3p^2\ ^4P$	0	6	coll
2385	He II	Paschen delta	48	54	rec
2423/25	[Ne IV]	$2p^3\ ^4S{-}2p^3\ ^2D$	0	6	coll
2470	[O II]	$2p^3\ ^4S{-}2p^3\ ^2P$	0	6	coll
2511	He II	Paschen gamma	48	54	rec
2663	He I	$2s\ ^3S{-}11p\ ^3P^0$	20	24	rec
2696	He I	$2s\ ^3S{-}9p\ ^3P^0$	20	24	rec
2723	He I	$2s\ ^3S{-}8p\ ^3P^0$	20	24	rec
2733	He II	Paschen beta	48	54	rec
2763	He I	$2s\ ^3S{-}7p\ ^3P^0$	20	24	rec
2784/2929	[Mg V]	$2p^4\ ^3P{-}2p^4\ ^1D$	0	4	coll
2786	[Ar V]	$3p^2\ ^3P{-}3p^2\ ^1S$	0	4	coll
2796/2803	Mg II	$3s\ ^2S{-}3p\ ^2P^0$	0	4	coll
2791/2797	Mg II	$3p\ ^2P^0{-}3d\ ^2D$	4	9	coll?
2829	He I	$2s\ ^3S{-}6p\ ^3P^0$	20	24	rec
2837/38	C II	$2p^2\ ^2S{-}3p\ ^2P^0$	12	16	rec
2837	O III	$3p\ ^3D{-}3d\ ^3P^0$	36	41	Bowen fluo.
2852	Mg I	$3s^2\ ^1S{-}3p\ ^1P^0$	0	4	coll
2854/68	[Ar IV]	$3p^2\ ^4S{-}3p^2\ ^2P$	0	4	coll
2929/37	Mg II	$3p\ ^2P^0{-}4s\ ^2S$	4	9	coll?
2929	[Mg V]	$2p^4\ ^3P{-}2p^4\ ^1D$	0	4	coll
2945	He I	$2s\ ^3S{-}5p\ ^3P^0$	20	24	rec

Table I (continued)

Wavel.	Identification		Lo	Hi	Excitation
2973/79	N III	$3p\ ^2P—3d\ ^2P^0$	38	43	???
3023	O III	$3s\ ^3P^0—3p\ ^3P$	33	37	Bowen fluo.
3043/47	O III	$3s\ ^3P^0—3p\ ^3P$	33	37	Bowen fluo.
3063/71	[N II]	$2p^2\ ^3P—2p^2\ ^1S$	0	4	coll
3109/3005	[Ar III]	$3p^4\ ^3P—3p^4\ ^1S$	0	4	coll
3133	O III	$3p\ ^3S—3d\ ^3P^0$	37	41	Bowen fluo.
3188	He I	$2s\ ^3S—4p\ ^3P^0$	20	24	rec
3203	He II	Paschen alpha	48	54	rec

interpretation of IUE spectra of PN. Descriptions are given e.g. by Seaton and Storey (1976) and Aller (1984). In IC 418, Harrington et al. (1981) noted a discrepancy in the ionization balance C^+/C^{2+} which was explained by dielectronic recombination through low-lying autoionization levels. At nebular temperatures the calculation of the rate coefficients is rather involved (See e.g. Nussbaumer and Storey (1983, 1984) and references cited therein). This process also affects the populations of individual levels and hence line emissivities: e.g. C II 1335, C III 2297, O III] 1663, and probably the recombination lines C II 1760 and 2837/38 show this influence. (See Section 5, Storey (1981), Clegg et al. (1983), Nussbaumer and Storey (1984). The latter give effective recombination coefficients α_{eff}^R, α_{eff}^D for lines of CNO ions.)

Electron collisional excitation to levels a few eV above the ground followed by radiative decay is the mechanism responsible for most nebular emission lines. These include the resonance lines of C II, C IV, N V, Mg I, Mg II, Si II (1309, 1808/17), and Si IV, forbidden lines from the levels which also produce the well known and used optical lines: [N II], [O II], [O III], [Ne III], [Ne IV], [Ar III], [Ar IV], [Ar V] and also [Mg V]. In addition, there are intercombination lines of C II], C III], N III], N IV], O IV], Si II], Si III]. PN are optically thin in both forbidden and intercombination lines. Because of the low collisional deexcitation rate, they are usually effectively optically thin in most permitted resonance lines.

Consider resonance transitions of the $^2S—^2P^0$ type. Under nebular conditions the $^2P_{3/2}$ and $^2P_{1/2}$ levels will be populated in the ratio 4:2 according to their statistical weights and the intensity ratio of the two components of the doublet, e.g. C IV 1548/51 will be 2:1. This ratio is verified in normal PN (Aller et al., 1981b; Flower, 1982; Feibelman, 1983), whereas in 'protoplanetary nebulae', where the lines are asymmetrical and broadened, Feibelman (1983) finds the ratio to be smaller than expected.

Similarly the theoretical ratio of the C II transitions $(^2P^0_{3/2}—^2D_{3/2,5/2})/(^2P^0_{1/2}—^2D_{3/2}) = 1335.7/1334.5$ is 2:1. The nebular line profiles are usually strongly distorted by interstellar absorption lines and in some nebulae there may be an emission contribution to 1335.7 from a surrounding C^+ region of low electron density ($n_e < 10$ cm^{-3}; Köppen and Wehrse, 1980). Dielectronic recombination can be major contributor to the strength of C II 1335 as well as to that of O III] 1663 (Nussbaumer and Storey, 1984).

The O III 2837, 3023, 3045, and 3133 lines that fall in the long wavelength end of the IUE range are produced by the *Bowen fluorescent mechanism*. The calculated and observed numbers of photons seem to be in satisfactory agreement for NGC 7662 (Saraph and Seaton, 1980; Flower, 1982). The derived fluorescence efficiency 0.51 ± 0.10 agrees with a theoretical value by Harrington (1972) and falls in the range covered by observations in the optical region (Likkel and Aller, 1986 and papers therein cited).

5. Plasma Diagnostics

A determination of electron temperatures and densities in a nebula is a prerequisite for any study of its chemical composition. Aside from the inevitable observational errors the greatest practical source of uncertainty is the electron temperature. In the optical region, plasma diagnostics are normally obtained from the intensity ratios of forbidden lines. In conjunction with these, the UV forbidden and intercombination lines offer valuable additional opportunities to measure these vital quantities (see e.g. Aller (1984) and Czyzak *et al.* (1986) for a number of graphs showing the dependance of various line intensity ratios on (n_e, T_e)).

Electron temperatures can be measured by comparing the intensity of a collisionally excited line with that of an appropriate recombination line. As an example consider C IV 1550 which is collisionally excited. Its intensity depends on $N(C^{3+})$ and T_e. The intensity of the dielectronic recombination line C III 2297 likewise depends on $N(C^{3+})$, but a different function of T_e. Hence a measured value of the intensity ratio of these lines will fix a unique value for T_e. Another useful example is the ratio N V 1240/N IV 1718 (diel.rec.) (Nussbaumer and Schild, 1981). The C III] 1906/09 pair may be compared with C II 1335 (diel.rec. + coll.excit.) or C II 4267 (dir.rec.) (Clegg *et al.*, 1983).

The ratio of the intercombination lines C III] 1906/1909 Å is sensitive to electron densities above about 1000 cm^{-3} (Nussbaumer and Schild, 1979), so this ratio has been used to obtain n_e for a number of PN (see e.g. Feibelman *et al.*, 1980; Köppen and Wehrse, 1980; Harrington and Feibelman, 1983; Adam and Köppen, 1985). Other density indicators are the analogous pair N IV] 1483/87 in the same isoelectronic sequence (Nussbaumer and Schild, 1981), O III] 1666/[O III] 2321 and O IV 1404.8/1401.2 and 1404.7/1404.2 ratios (Nussbaumer and Storey, 1981b, 1982). The latter ratio, practical with high dispersion observations of bright objects is a good density indicator in two quite different domains: $100 < n_e < 10^5$ and also $10^8 < n_e < 10^{12}$ cm^{-3}. The [Ne IV] 2422/2424 ratio may also be employed to get densities (Lutz and Seaton, 1979). Hayes and Nussbaumer (1984) suggested to use the ratios of IUE lines to IR transitions to determine densities and temperatures, for example, O IV 1401 Å/25.4 μm and C II 2325 Å/158 μm.

A perennial problem we face in high excitation objects is the lack of available temperature diagnostics for the inner zones of the nebulae. One might compare [Ne V] 1575 or 2973 with the 3345 and 3428 lines. However the far UV lines are very faint. A more useful diagnostic is [Ne IV] $(2422 + 2424)/(4714/15 + 4724/26)$ (Aller and Keyes, 1980). Here, the limitation of accuracy lies in measuring the weak optical lines.

6. Chemical Compositions of Planetary Nebulae

In recent years studies of the element abundances in PN have greatly increased in scope and number, as the important role played by PN in stellar evolution and nucleosynthesis was realised more and more. PN represent the penultimate stage of evolution after a star has ascended the giant branch for the second time ('asymptotic giant branch') and ejects its envelope. In this manner the products of nuclear synthesis are added to the interstellar environment. Though a single, more spectacular supernova supplies more processed material to the interstellar medium than do many planetaries, the latter contribute important quantities of C, N, O, and s-product elements.

6.1. METHODS OF ANALYSIS

The chemical composition of the nebular gas is assessed from emission line data from all spectral regions accessible. Preferably these data should be from the entire nebula. Otherwise, one must exercise care when collating the fluxes obtained with different apertures, or must ensure that always the same parts of the nebula are observed. From observed line fluxes and continuum data we determine n_e and T_e and evaluate ionic abundance ratios by well known procedures (see e.g. Osterbrock, 1974; Aller, 1984). If present, temperature fluctuations (Peimbert, 1967; Rubin, 1969; Dinerstein *et al.*, 1985) can be very important. To get the element abundance ratio, one then must add up the contributions from all ionization stages — if available. But except for H and He and occasionally a few other elements, our observational data bank normally does not include all ionization stages of an element, and one must account for unseen ions: Simple empirical ionization correction formulae are often employed (e.g. Barker, 1983). One can calculate ionization models which represent the level of excitation and ionization in the nebula, reproducing 'critical' line intensities, such as [O III]/[O II], [Ne V]/[Ne III], and He II/He I (Shields *et al.*, 1981; Aller and Czyzak, 1983). For symmetrical, well observed objects one can build very detailed models which reproduce the observed line intensities (Harrington *et al.*, 1982; Harrington and Feibelman, 1983).

With bright PNN one can sometimes observe nebular resonance lines in absorption. Then a rather conventional curve of growth analysis yields the ionic column densities. In NGC 6543 Pwa *et al.* (1984) observed C II, C III, Si II, Si III, S II, S III, S IV, and Al III.

6.2. DIFFICULTIES AND DISCREPANCIES

The accuracy of abundance determinations from IUE data is limited by several factors. In compact and faint objects it is difficult to separate nebular and stellar components of some lines, e.g. C IV and N V, either by angular or spectral resolution. The C II and C IV resonance lines can be attenuated by dust grains in the nebula. Interstellar extinction is greater in the UV. Often the stellar continuum is strong, which makes measurements of weak superimposed nebular lines difficult. In some lines (e.g. O III] 1663, C II 1335) which we expect to be collisionally excited dielectronic recombination may play an important role. If numerical models are used to estimate ionization corrections, care must be taken to use a realistic type of model: A central cavity can have a large effect on predicted intensities (Köppen, 1983; Köppen and Wehrse, 1983) and on the correction factors (Aller *et al.*, 1985).

An annoying problem is that the C^{++} concentration as derived from the collisionally excited C III] 1908 lines is often systematically smaller than that inferred from the C II 4267 recombination line. Torres-Peimbert *et al.* (1980) find that the discrepancy can be removed by invoking temperature fluctuations. Aller *et al.* (1980) suggest that the population of the upper level of the 4267 transition is enhanced by radiative excitation by starlight. Torres-Peimbert and Peimbert (1977) also suspected this, as the C/H ratio derived from 4267 Å is correlated with the ionization level of the nebulae. In his study of extended nebulae, Barker (1985) found the C^{++} discrepancy to be largest near the central stars but to nearly vanish in more distant parts of the nebulae. Lacking information about the far UV PNN spectrum, a quantitative comparison is difficult. For 4267 $\alpha_{\text{eff}}^D \ll \alpha_{\text{eff}}^R$ (Storey, 1981). In low excitation nebulae such as NGC 40 or IC 418 this discrepancy may not exist (Clavel *et al.*, 1981; Clegg *et al.*, 1983). Perhaps the PNN are too cool.

6.3. SOME FEATURES OF NEBULAR ABUNDANCES

Some planetaries appear to have compositions very similar to that of the Sun although an apparent oxygen deficiency is found. If T_e fluctuations are important this discordance can be removed. We must emphasize however that although ratios such as C/H would be enhanced in the presence of temperature fluctuations, ratios like C/O would be less affected. A C/O ratio exceeding the solar value is found in a clear majority of 68 planetaries with reasonably well determined C/O ratios (Zuckerman and Aller, 1986). There seems to be no correlation with nebular morphology or central star multiplicity. Bipolarity is not a characteristic peculiar to carbon-rich evolved stars.

PNN with Wolf—Rayet type spectra are always of the carbon rich variety. For NGC 40, Aller (1943) found C/He = 0.1 with no trace of hydrogen; more recent results raise C/He to 0.2. Yet the nebula has nearly a 'normal' composition: Clegg *et al.* (1983) found log A(O) = 8.9 and C/O = 1.25, so that the nebula is only moderately carbon rich. Even more remarkable is SwSt1 where the infrared data indicated the presence of silicate grains in an O-rich nebula, wheres the central star is a C-rich object like the nuclei of NGC 40 and BD+30°3639. Flower *et al.* (1984) found C/O = 0.72 ± 0.1 i.e. a solar ratio.

Does the composition correlate with the temperature of the exciting star? Although highly excited or very luminous PN — with presumably more massive central stars — tend to show results of intensive nuclear processing, the very highest excitation object studied so far, M1—1, does not show markedly abnormal abundances (Aller *et al.*, 1986).

The nitrogen rich objects (Peimbert's class I) such as NGC 2818 (Dufour, 1984) constitute a particularly interesting group of nebulae. One of the most spectacular members is NGC 6302: In this intricately structured PN, He and N are enhanced, but Ne and heavier elements show an essentially solar abundance pattern (Aller *et al.*, 1981). C and O are depleted in NGC 6302 and also in NGC 6537 where Feibelman *et al.* (1985) suggest that the progenitor star had essentially solar composition, but that C and possibly O were converted into N. Presumably N-rich objects evolve from stars massive enough for even the CN cycle of the CNO bi-cycle to run.

For Al, Mg, and Si our only data come from the IUE. Harrington and Marionni (1981) studied Si III, Si IV, Mg II, and [Mg V] in NGC 2440, 7662, IC 418, 2003, 2165 and Hu1—1. They found silicon to be depleted compared to the Sun by about an order of magnitude, probably locked up in dust grains, as Shields (1978) suggested for iron. A more complicated situation is found for magnesium, as noted by Péquignot and Stasińska (1980) in NGC 7027, where the [Mg V] lines in the inner zones suggest a solar abundance, while Mg is depleted in the outer regions. For NGC 2440 and IC 2165 Harrington and Marionni get a similar result. In NGC 6543 Pwa *et al.* (1984) found for carbon and silicon lower than solar values by factors 2 and 4, but iron is depleted by at least a factor of 70, and aluminum even by a factor of 300!

Abundances in planetaries provide insights on nucleogenesis in stellar evolution. We can run assays on the actual material that is returned to the interstellar medium, not just the thin photospheric layers of an evolved star which may not always be a true sample of the material actually ejected. The presently available data show that PN are important contributors of carbon to the interstellar medium; oxygen must come from more massive stars.

In interpreting nebular abundances we must be ever mindful of the important role played by the dust. Silicate grains are found in some objects, but in most PN the grains must be amorphous carbon. This dust must serve as a trap for many heavy elements. The same elements Al, Si, Mg, Fe are depleted in PN as in the interstellar medium. Since the mass invested in grains is small, carbon is probably not strongly depleted. It and oxygen can however be tied up in molecules such as CO. Thus a proper assessment of the chemical composition of any complex PN requires the interplay of infrared, optical, and ultraviolet observations.

6.4. PLANETARY NEBULAE IN A GLOBULAR CLUSTER AND IN OTHER GALAXIES

In K648, the PN in the globular cluster M15, C/O = 11 (Adams *et al.*, 1984), C/H being enhanced by 10%, and O/H depleted with respect to the Sun. One dramatic accomplishment of IUE was the observations of PN in nearby galaxies. Previous optical studies yielded of course no information on the carbon abundance. To date, six PN have been observed in the LMC, five in the SMC (Maran *et al.*, 1982, 1986, Table II), and the one in the Fornax galaxy (Maran *et al.*, 1984). In the latter the C/O ratio was 3.7, as high as any value found in the Milky Way. Carbon is enriched over the ambient interstellar gas in P02 and P33 of the LMC by a factor 5, and in the SMC's N5 by a factor 50. P07 and P09 show the abundance pattern characteristic of Peimbert's type I nebulae with C and O

TABLE II

Abundances in planetaries of the Magellanic Clouds (log A(H) = 12.00) from Maran *et al.* (1986)

	LMC						SMC					
	P02	P07	P09	P25	P33	P40	N2	N5	N43	N44	N67	Sun
He	11.05	11.21	11.24	11.05	11.09	11.00	11.04	11.09	11.05	11.07	11.23	11.0:
C	8.67	7.48	7.52	7.26	8.60	8.36	8.20	8.88	8.54	8.53	6.3	8.66
N	7.91	8.36	8.21	7.57	7.95	7.20	7.61	7.50	7.49	7.55	7.6	7.98
O	8.24	8.00	7.91	8.01	8.41	8.20	8.33	8.34	8.05	8.24	7.2	8.91
Ne	7.40	7.40	7.28	7.38	7.71	7.53	7.36	7.67	7.20	7.57	7.7	8.05
S	6.59	6.66	6.70	6.48	6.73	6.43	6.30	6.85	6.35	6.60	6.5	7.23
Ar	6.02	6.08	6.18	6.0	6.18	6.16	6.00	6.00	5.45	5.68		6.6

depleted, while He and N are enriched. The oxygen depletion seems to be more severe than for galactic PN.

7. Nebular Models

Nebular models supply corrections for unseen ionization stages needed to find PN chemical compositions. However the advantage of theoretical models lies in the possibility of introducing physical concepts and processes deemed relevant to PN. Thus we can calculate an 'internally consistent' model.

We usually assume a static spherically symmetric gas cloud excited by a central star. Radiative equilibrium is postulated. Photoionization, direct and dielectronic recombination, free-free emission, collisional excitation and charge exchange with H and He are all included. We adopt atomic parameters, A-values, collision strengths, photoionization, recombination, and charge exchange coefficients. The adjustable parameters are the radius and energy flux distribution of the central star $F^*(\lambda)$, the distribution of mass in the nebula and its chemical composition. The objective is to modify these parameters to fit the observed data. This has been done either by trying to reproduce as many observational data as possible of a *single* PN by fine-tuning the model (Harrington *et al.*, 1982; Harrington and Feibelman, 1983; Adam and Köppen, 1985), or by interpreting observations of *several* objects by a restricted set of models which do not reproduce every detail (Aller, 1982; Köppen, 1983). One procedure is to reproduce the level of excitation by fitting excitation sensitive line ratios e.g. 4686/4861, 5007/3737, T_e, and then the intensities of individual lines as well as possible. With this method zero-order models for some fifty PN have been constructed. For most objects further refinement is not justified because the observational data are not good enough. We urgently need more accurate, detailed data on individual nebulae, isophote monochromatic images, greater wavelength range including UV and IR. As we attempt more sophisticated models tailored to individual nebulae the need for a greatly extended observational base is enhanced.

Consider the fundamental role played by IUE observations:

(a) In the realm of atomic physics: from attempts to explain the UV spectra, we have come to realize the importance of dielectronic recombination which affects emission line intensities, ionization balances etc.

(b) The UV spectrum provides us with many more diagnostic criteria which enable us in particular to explore the high excitation inner zones of PN.

(c) The UV lines supply precious criteria which enable us to tighten constraints on theoretical models. One that adequately represents the visual region may fail when the UV region is considered.

(d) Improved abundances particularly for carbon enable us to fix the contribution of this important cooling agent and set more constraints on the stellar energy distribution (Köppen, 1983). It also allows us to check whether there are other heating processes besides photoionization (Harrington and Feibelman, 1984).

(e) Most of the stellar flux that controls the ionization and excitation of a PN falls in the inherently inaccessible wavelengths shortward of 912 Å. The stellar flux in this region must be established by an interative procedure. We usually assume a type of stellar continuum, such as blackbody, LTE, or NLTE model atmospheres. Most often the stellar parameters are not known, so we select the one which results in the best match of the nebular spectrum. Often we are forced to 'bend' the stellar spectrum, especially for energies above 54 eV (Aller, 1983; Harrington and Feibelman, 1983; Köppen, 1983; Adam and Köppen, 1985). Heap (1977) already noticed that published LTE and NLTE atmospheres do not supply enough photons to ionize He^+. It seems that published NLTE models did not have the parameters appropriate to the central stars we observe (Köppen, 1986): Evolutionary tracks take the stars rather close to their Eddington limit, where Husfeld et al. (1984) find that NLTE effects will weaken the He II absorption edge. Unfortunately PNN are so hot that the observable flux distribution $F^*(\lambda)$ falls in the Rayleigh—Jeans regime. The required photometric accuracy in $F^*(\lambda)$ does not suffice to set meaningful constraints (Kaler and Feibelman, 1985). Stellar absorption lines interpreted by NLTE model atmospheres may help (Méndez et al., 1981; Adam and Köppen, 1985). A pool of information that has hardly been tapped is the soft X-ray range, where the high energy tail of the stellar continuum shines through the nebula which is transparent at these energies. Preliminary results from EXOSAT (Osborne, 1985) are not in violent disagreement with those flux distributions required to ionize the nebula.

(f) The role of internal dust which can be estimated from its effect on the C IV line was noted above. These results confirm optical and IR work that the optical depths are rather small. Very little is known about the optical properties and the spatial distribution of the grains, so UV data may give badly needed additional clues. One may envisage the use of the intensity variation across the nebula as an additional source of information.

As there is little evidence from the optical spectrum about a very exciting phenomenon, the interaction of stellar wind and nebula, we may hope that our improved understanding of ionization models could give us an indirect means to separate possible interaction effects from the thermal aspects of the PN.

Some information on spatial distribution of C^{2+} or C^{3+} ions may be obtained for extended PN from IUE data, thus to probe ionization zones different from ones observable in the visual region. Nevertheless, for high resolution pertaining to filamentary structure and clumpiness we must rely on optically region images and hopefully data obtained with the Hubble Space Telescope.

Acknowledgements

JK wishes to acknowledge financial support from the Deutsche Forschungsgemeinschaft (SFB 132). LHA is grateful for support from NASA grant NSG 5358 to UCLA.

References

Adam, J. and Köppen, J.: 1985, *Astron. Astrophys.* **142**, 461.

Adams, S., Seaton, M. J., Howarth, I. D., Aurière, M., Walsh, J. R.: 1984, *Monthly Notices Roy. Astron. Soc.* **207**, 471.

Aller, L. H.: 1943, *Astrophys. J.* **97**, 135.

Aller, L. H.: 1976, Mém. Soc. Roy. Sci. Liège, 6è série, tome IX, p. 271.

Aller, L. H.: 1982, *Astrophys. Space Sci.* **83**, 225.

Aller, L. H.: 1984, *Physics of Thermal Gaseous Nebulae*, D. Reidel Publ. Co., Dordrecht, Holland.

Aller, L. H. and Czyzak, S. J.: 1983, *Astrophys. J. Suppl.* **51**, 211.

Aller, L. H. and Keyes, C. D.: 1980, *Proc. Nat. Acad. Sic. USA* **77**, 1231.

Aller, L. H., Keyes, C. D., Ross, J. E., and O'Mara, B. J.: 1980, *Monthly Notices Roy. Astron. Soc.* **197**, 647.

Aller, L. H., Keyes, C. D., and Czyzak, S. J.: 1981a, *Astrophys. J.* **250**, 596.

Aller, L. H., Ross, J. E., O'Mara, B. J., and Keyes, C. D.: 1981b, *Monthly Notices Roy. Astron. Soc.* **197**, 95.

Aller, L. H., Keyes, C. D., and Czyzak, S. J.: 1985, *Astrophys. J.* **296**, 492.

Aller, L. H., Keyes, C. D., and Feibelman, W. A.: 1986, *Proc. Nat. Acad. Sci. USA* **83**, 2777.

Barker, T.: 1983, *Astrophys. J.* **267**, 630.

Barker, T.: 1985, *Astrophys. J.* **294**, 193.

Benvenuti, P. and Perinotto, M.: 1981, *Astron. Astrophys.* **95**, 127.

Boggess, A., Feibelman, W. A., and McCracken, C. W.: 1981, in R. D. Chapman (ed.), *The Universe at Ultraviolet Wavelengths*, NASA CP **2171**, 663.

Castor, J. I., Lutz, J. H., and Seaton, M. J.: 1981, *Monthly Notices Roy. Astron. Soc.* **194**, 547.

Cerruti-Sola, M. and Perinotto, M.: 1985, *Astrophys. J.* **291**, 237.

Clavel, J., Flower, D. R., and Seaton, M. J.: 1981, *Monthly Notices Roy. Astron. Soc.* **197**, 301.

Clegg, R. E. S., Seaton, M. J., Peimbert, M., and Torres-Peimbert, S.: 1983, *Monthly Notices Roy. Astron. Soc.* **205**, 417.

Czyzak, S. J., Keyes, C. D., and Aller, L. H.: 1986, *Astrophys. J. Suppl.* **61**, 159.

Dinerstein, H. L., Lester, D. F., and Werner, M. W.: 1985, *Astrophys. J.* **291**, 561.

Draine, B. T. and Bahcall, J. N.: 1981, *Astrophys. J.* **250**, 579.

Dufour, R. J.: 1984, *Astrophys. J.* **287**, 241.

Feibelman, W. A.: 1983, *Astron. Astrophys.* **122**, 335.

Feibelman, W. A., Boggess, A., Hobbs, R. W., and McCracken, C. W.: 1980, *Astrophys. J.* **241**, 725.

Feibelman, W. A., Aller, L. H., Keyes, C. D., and Czyzak, S. J.: 1985, *Proc. Nat. Acad. Sci. USA* **82**, 2202.

Flower, D. R.: 1980, *Monthly Notices Roy. Astron. Soc.* **193**, 511.

Flower, D. R.: 1982, *Monthly Notices Roy. Astron. Soc.* **199**, 15p.

Flower, D. R., Goharji, A., and Cohen, M.: 1984, *Monthly Notices Roy. Astron. Soc.* **206**, 293.

Gathier, R., Pottasch, S. R., and Pel, J. W.: 1986, *Astron. Astrophys.* **157**, 171.

Goharji, A. and Adams, S.: 1984, *Monthly Notices Roy. Astron. Soc.* **210**, 683.

Hamann, W.-R., Kudritzki, R.-P., Méndez, R. H., and Pottasch, S. R.: 1984, *Astron. Astrophys.* **139**, 459.

Harrington, J. P.: 1972, *Astrophys. J.* **176**, 127.

Harrington, J. P.: 1982, in Y. Kondo, J. M. Mead, and R. D. Chapman (eds.), *Advances in Ultraviolet Astronomy*, NASA CP-**2238**, p. 610.

Harrington, J. P. and Marionnni, P. A.: 1981, in R. D. Chapman (ed.), *The Universe at Ultraviolet Wavelengths*, Harrington, J. P. and Feibelman, W. A.: 1983, *Astrophys. J.* **265**, 258.

Harrington, J. P. and Feibelman, W. A.: 1984, *Astrophys. J.* **277**, 716.

Harrington, J. P., Lutz, J. H., and Seaton, M. J.: 1981, *Monthly Notices Roy. Astron. Soc.* **195**, 21p.

Harrington, J. P., Seaton, M. J., Adams, S., and Lutz, J. H.: 1982, *Monthly Notices Roy. Astron. Soc.* **199**, 517.

Hayes, M. A. and Nussbaumer, H.: 1984, *Astron. Astrophys.* **139**, 233.

Heap, S. R.: 1977, *Astrophys. J.* **215**, 864.

Hummer, D. G. and Kunasz, P. B.: 1980, *Astrophys. J.* **236**, 609.

Husfeld, D., Kudritzki, R.-P., Simon, K. P., and Clegg, R. E. S.: 1984, *Astron. Astrophys.* **134**, 139.

Kaler, J. B. and Feibelman, W. A.: 1985, *Astrophys. J.* **297**, 724.

Köppen, J.: 1977, *Astron. Astrophys.* **56**, 189.

Köppen, J.: 1983, *Astron. Astrophys.* **122**, 95.

Köppen, J.: 1986, in A. Acker (ed.), *Planetary Nebulae*, 8ème journée, Strasbourg.

Köppen, J. and Wehrse, R.: 1980, *Astron. Astrophys.* **85**, L15.

Köppen, J. and Wehrse, R.: 1981, *Etoiles Bleues en-dessous de la Séquence Principale*, 3ème journée, Strasbourg, p. 70.

Köppen, J. and Wehrse, R.: 1983, *Astron. Astrophys.* **123**, 67.

Likkel, L. and Aller, L. H.: 1986, *Astrophys. J.* **301**, 825.

Lutz, J. H. and Seaton, M. J.: 1979, *Monthly Notices Roy. Astron. Soc.* **187**, 1p.

Maran, S. P., Aller, L. H., Gull, T. R., and Stecher, T. P.: 1982, *Astrophys. J.* **253**, L43.

Maran, S. P., Gull, T. R., Stecher, T. P., Aller, L. H., and Keyes, C. D.: 1984, *Astrophys. J.* **280**, 615.

Maran, S. P., Gull, T. R., Stecher, T. P., Aller, L. H., and Keyes, C. D.: 1986 (in prep.).

Méndez, R. H., Kudritzki, R.-P., Gruschinske, J., and Simon, K. P.: 1981, *Astron. Astrophys.* **101**, 323.

Nussbaumer, H. and Schild, H.: 1979, *Astron. Astrophys.* **75**, L17.

Nussbaumer, H. and Schild, H.: 1981, *Astron. Astrophys.* **101**, 118.

Nussbaumer, H. and Storey, P. J.: 1981, *Astron. Astrophys.* **99**, 177.

Nussbaumer, H. and Storey, P. J.: 1982, *Astron. Astrophys.* **115**, 205.

Nussbaumer, H. and Storey, P. J.: 1983, *Astron. Astrophys.* **126**, 75.

Nussbaumer, H. and Storey, P. J.: 1984, *Astron. Astrophys. Suppl.* **56**, 293.

Nussbaumer, H., Schmutz, W., Smith, L. J., and Willis, A. J.: 1981, *Astron. Astrophys. Suppl.* **47**, 257.

Osborne, J.: 1985, in J. Köppen (ed.), *Workshop on Planetary Nebulae*, Frankfurt.

Osterbrock, D. E.: 1974, *Astrophysics of Gaseous Nebulae*, Freeman, San Francisco.

Peimbert, M.: 1967, *Astrophys. J.* **150**, 825.

Péquignot, D. and Stasińska, G.: 1980, *Astron. Astrophys.* **81**, 121.

Perinotto, M.: 1983, in D. R. Flower (ed.), 'Planetary Nebulae', *IAU Symp.* **103**, 323.

Perinotto, M. and Benvenuti, P.: 1981, *Astron. Astrophys.* **100**, 241.

Perinotto, M., Benvenuti, P., and Cerruti-Sola, M.: 1982, *Astron. Astrophys.* **108**, 314.

Pottasch, S. R., Wesselius, P. R., Wu, C. C., and van Duinen, R. J.: 1977, *Astron. Astrophys.* **54**, 435.

Pottasch, S. R., Gathier, R., Gilra, D. P., and Wesselius, P. R.: 1981, *Astron. Astrophys.* **102**, 237.

Pottasch, S. R., Gilra, D. P., and Wesselius, P. R.: 1982, *Astron. Astrophys.* **109**, 182.

Pwa, T. H., Ho, J., and Pottasch, S. R.: 1984, *Astron. Astrophys.* **139**, L1.

Rubin, R. H.: 1969, *Astrophys. J.* **155**, 841.

Saraph, H. E. and Seaton, M. J.: 1980, *Monthly Notices Roy. Astron. Soc.* **193**, 617.

Seaton, M. J.: 1978, *Monthly Notices Roy. Astron. Soc.* **185**, 5p.

Seaton, M. J.: 1979a, *Monthly Notices Roy. Astron. Soc.* **187**, 785.

Seaton, M. J.: 1979b, *Monthly Notices Roy. Astron. Soc.* **187**, 73p.

Seaton, M. J. and Storey, P. J.: 1976, in P. G. Burke and B. L. Moiseiwitsch (eds.), *Atomic Processes and Applications*, North Holland, Amsterdam, p. 133.
Shields, G. A.: 1978, *Astrophys. J.* **219**, 559.
Shields, G. A., Aller, L. H., Keyes, C. D., and Czyzak, S. J.: 1981, *Astrophys. J.* **248**, 569.
Smith, L. F. and Aller, L. H.: 1969, *Astrophys. J.* **157**, 1245.
Storey, P. J.: 1981, *Monthly Notices Roy. Astron. Soc.* **195**, 27p.
Torres-Peimbert, S. and Peimbert, M.: 1977, *Rev. Mex. Astron. Astrofis.* **2**, 181.
Torres-Peimbert, S., Peimbert, M., and Daltabuit, E.: 1980: *Astrophys. J.* **38**, 133.
Zuckerman, B. and Aller, L. H.: 1986, *Astrophys. J.* **301**, 772.

PART V

GALAXIES AND COSMOLOGY

Edited by W. WAMSTEKER

STARS AND H II REGIONS IN NEARBY GALAXIES

JOHN B. HUTCHINGS

Dominion Astrophysical Observatory, Victoria, BC, Canada

JAMES LEQUEUX

Observatoire de Marseille, Marseille, France

and

B. WOLF

Landessternwarte, Königstuhl, Heidelberg, F.R.G.

1. OB Stars in Local Group Galaxies

In this section, we review the work of IUE in investigating the hot luminous stars outside of our own Galaxy. Although it was not originally anticipated that such observations could even be made with IUE, they have been extensive and now represent a major data base in the study of OB stars. The investigations fall into three main categories, which are outlined separately below: OB star stellar winds in the Magellanic Clouds; X-ray and other massive binaries in the Magellanic Clouds; and the brightest hot stars in M31 and M33. In much of this work, observational discoveries have been unexpected, and to a significant extent they are still unexplained. Future work in both observation and theory will thus still determine the full significance of the pioneering work of IUE.

1.1. MAGELLANIC CLOUD OB AND W-R STARS

There have been several studies directed at OB stars in the Magellanic Clouds. An initial low dispersion study by Hutchings (1980) was later expanded to high dispersion observations and a larger sample of low dispersion data (Hutchings, 1982). Bruhweiler *et al.* (1982) studied SMC stars at low dispersion. Shore and Sanduleak (1984) combined IUE and ground-based low dispersion spectra to discuss another sample, and Garmany and Conti (1985) devoted another discussion to Magellanic Cloud star stellar winds, based largely on IUE data. In a number of these papers the same spectra have been used, as they have been available from the IUE archive. However, there is general agreement as to the principal results, although their interpretation is less unanimous.

The interstellar extinction, as discussed elsewhere, is generally unlike that of the Galaxy, so that the derivation of continuum flux distributions is not straight-forward. However, most of these authors have found that the fluxes and effective temperatures of the Magellanic Cloud stars are close to that expected from Galactic stars of this type. It is usually possible to obtain consistent estimates of M_{BOL} from ground-based and IUE fluxes, using appropriate bolometric corrections. The two approaches are complementary in that ground-based determina-

Y. Kondo (ed.), Scientific Accomplishments of the IUE, pp. 605—622.
© 1987 *by D. Reidel Publishing Company.*

tions are sensitive to the effective temperature, and the UV-based are sensitive to the extinction. The study of the continuum flux from hot stars has been an important aspect of IUE work, and is particularly so in the Magellanic Clouds, since extinction (although complex) is low, and distances are relatively well-known. In addition to the work already cited, Fitzpatrick and Savage (1984) have derived continuum fluxes for a number of OB stars near 30 Doradus. While there are complications in this region due to several layers of weak extinction, they find a *low* effective temperature for several stars, compared with expectations from models of Galactic stars. This tendency is not generally found, although some of the Shore and Sanduleak O star temperatures are also low.

Perhaps the most important results concern the UV line spectra in the Magellanic Cloud stars. Firstly, the line strengths are weak compared with Galactic, as is also found in the visible spectrum. In the luminous stars, the resonance lines of Si IV, C IV, and N V are all weaker both in emission and absorption, and *have lower outflow velocities* than comparable stars in the Galaxy. This difference is more pronounced in the SMC than the LMC. Figure 1 shows some illustrative line profiles. These results (as well as ground-based observations — see e.g. Hutchings, 1980a) indicate that the stellar winds in the Magellanic Cloud stars differ systematically from Galactic, and it is generally supposed that the principal cause is the radiative opacity due to metal abundance differences between the three galaxies. Thus, in a qualitative way, the idea that stellar winds are radiatively driven, is supported. However, there are no detailed models of how this works, and indeed not even certainty as to whether mass-loss is greater or less in the Clouds than in the Galaxy. While the wind-driving lines are weaker and slower moving and probably arise closer to the stars, the lower abundance and the unknown ionisation balance may well hide a higher overall mass-loss rate. A further clue in this direction lies in the very high X-ray luminosity and pulsar spin rates in the Magellanic Cloud X-ray binaries: both of which indicate a higher mass-accretion rate (lower opacity) than in similar Galactic binaries. Numerical estimates of mass-loss are given by Garmany and Conti (1985) and by Shore and Sanduleak (1984). The two groups of stars do not overlap in the H—R diagram, but the mass-loss rates quoted differ by more than an order of magnitude. Clearly, the subject needs more detailed modelling and higher S/N data before such numbers can be regarded as established.

There is less clearly a difference between Magellanic Cloud and Galactic stars in the weaker lines and lines of lower luminosity stars (Shore and Sanduleak, 1984; Nandy et al., 1983). This, too, suggests that radiative acceleration processes are most sensitive to the fundamental (metal abundance) difference with Galactic stars. Garmany and Conti point out in their paper the importance of matching spectral types in making such comparisons, and the difficulty in doing this when the Cloud stars have systematically weak lines.

Turning to the W—R stars, we note a paper by Garmany et al. (1984) who point out the difficulties of determining the continuum flux distribution of W—R stars with typical (strong) Galactic reddening. IUE spectra of Magellanic Cloud W—R stars, where extinction is lower, and distances are consistently known, are being studied to address this problem. Smith and Willis (1983) have combined IUE and

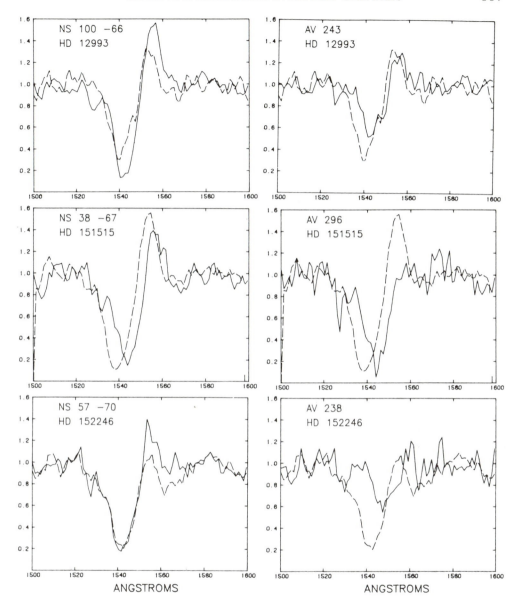

Fig. 1. Stellar wind line profiles of SMC OB stars and Galactic counterparts, from Garmany and Conti (1985). Dashed profiles are Galactic, and in general have larger velocities and emission intensities.

ground-based spectra in such a study of 9 LMC W—R stars, and they are able to place them on the H—R diagram. The late WN stars appear generally cooler and more luminous, and early WN and WC stars hotter than Galactic counterparts. Wind velocities are similar to Galactic, but emission line strengths are higher. They find no evidence for abundance differences from Galactic counterparts in these (evolved) objects (i.e. supporting the idea that stellar nucleosynthesis is the same, but that the Magellanic Clouds have undergone less of it). In an earlier IUE study, Nandy *et al.* (1980) suggest that the star Sk—65—22 is an example of the predicted class of stars intermediate between Of and late WN stars, from the UV line spectrum and flux levels, thereby supporting the evolutionary scenario between them.

The central object of 30 Doradus, R136, has received much attention in recent years, largely as a result of IUE observations (e.g. Cassinelli *et al.*, 1981, Feitzinger *et al.*, 1983). UV fluxes confirm that the object has very high bolometric luminosity ($M_{BOL} \leqslant -14$), and a very high temperature ($> 50\,000$ K). These results led to the much publicised suggestion that R136 is a supermassive star, with a mass-loss rate of $\sim 10^{-4}\,M_\odot\,\mathrm{yr}^{-1}$. Recent speckle interferometric images (Weigelt and Baier, 1985) have shown that there are in fact several stars within a few arcseconds here, so that these conclusions need some revision. Another bright LMC star, R122, has a similar spectrum, but lower luminosity (Hutchings, 1982; Fitzpatrick and Savage, 1984.) The spectra are still of considerable interest as they represent stars of very early type ($\sim$ O3), and are largely responsible for ionising large H II regions.

1.2. MAGELLANIC CLOUD X-RAY BINARIES

The study of binaries in the Magellanic Clouds has been restricted almost entirely to X-ray sources. Recently, some normal binaries have been studied in the visible (Sk 188, HD 5890, R31), and there are IUE data on them (e.g. Fitzpatrick and Savage, 1983: Figure 2), which would form a useful data base for examining interacting stellar winds or phase dependencies in the winds, and comparison with galactic counterparts. So far this does not appear to have been done. Among the

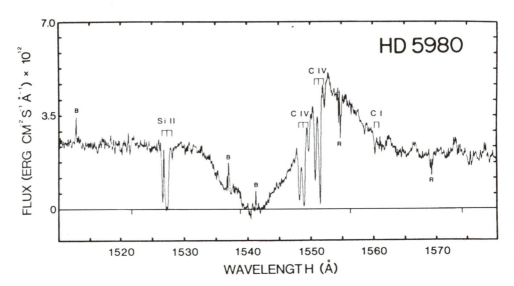

Fig. 2. High dispersion IUE line profile of SMC star HD 5890, showing high quality of data obtainable. (From Fitzpatrick and Savage, 1983).

X-ray sources, there are published IUE spectra of SMC X—1, SMC X—2, LMC X—1, LMC X—4, and 0535—66. Most of these are at low dispersion and have investigated phase dependence of resonance line strengths (X-ray ionisation bubbles), and the UV flux level (light-curves) (Figure 3). References are Tarenghi *et al.*, 1981; van der Klis *et al.*, 1982; Hammerschlag-Hensberge *et al.*, 1984; Bianchi and Pakull, 1985). While these studies show that the effects are present they are generally too noisy to enable very definitive modelling. Figure 4 shows SMC X—1 and LMC X—4 compared with two galactic sources, and it is not clear that there are marked systematic differences. However, we must remember that the Cloud sources have considerably higher X-ray fluxes than the Galactic ones.

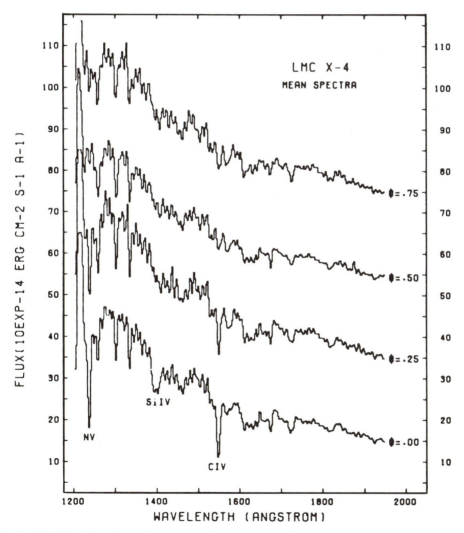

Fig. 3. LMC X—4 low dispersion spectra at four equally spaced binary phases showing the changes in wind line strengths caused by X-ray ionisation.

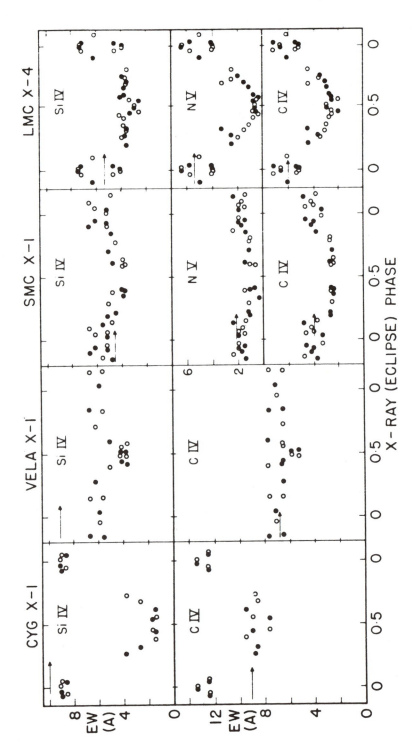

Fig. 4.   Phase-dependent wind line changes in OB star X-ray binaries in the Magellanic Clouds and the Galaxy. Values are reflected about phase 0.5, and horizontal arrows show galactic single star means.

There has been considerable work on the peculiar system 0535—66 (Charles *et al.*, 1983; Howarth *et al.*, 1984). This is a B star in an eccentric 16.5 day orbit with a rapid X-ray pulsar, which underwent an extended optical high state for more than 2 yr. IUE spectra showed that the UV fluxes do not undergo such changes (i.e. the effective temperature drops). However, the line spectra in the UV are unique in this state, with QSO-like profiles indicating large outflow velocities in a dense wind. Spectral changes are particularly dramatic through periastron passage, when the X-ray source probably passes through the envelope of the B star. Models of the orbit and interaction of the two stars were largely based on the IUE data, and have since been partially confirmed by a preliminary low-state orbital determination from ground-based data (Corbet *et al.*, 1985; Hutchings *et al.*, 1985).

There is another unique LMC X-ray source observed with IUE: CAL 83. This faint (17 mag) hot star is almost featureless in optical and UV spectra. IUE data are being used to help determine its luminosity and temperature, as well as look for changes and faint line features.

1.3. STELLAR WINDS IN M31 AND M33

In the past few years IUE has been able to investigate individual stars in the local group galaxies M31 and M33, mainly by devoting whole or even double shifts to an exposure. The results have been of considerable interest. We describe elsewhere the observations of Hubble—Sandage variables, and in this section we look at mass-losing hot stars: the central objects of giant H II regions, and the most luminous OB stars.

After the discovery that the giant H II regions of M33 contain stars with W—R spectra in the visible, the central object of IC132 was observed by D'Odorico and Benvenuti (1983). Their IUE spectra indicated a high temperature and bolometric luminosity. Massey and Hutchings (1983) observed the central objects of NGC 588, 595, and 604 in M33 and derived similar high temperatures and luminosities. These objects have sufficient far UV flux to ionise the giant H II regions and are comparable in luminosity (although lower) with R136. See also the discussion later in this chapter on H II regions. The spectra, too, are very similar to R136, although the lines are somewhat weaker (Figure 5). Large outflow velocities are seen (3000—4000 km s^{-1}), and a P Cygni profile in C IV. Such objects (presumably, like R136, tight clusters of hot stars) seem to be a characteristic of galaxies with active hot star formation, since they are not seen in M31, and appear to be rare in the Galaxy.

Massey *et al.* (1985, 1986) have observed a sample of hot stars in M31 and M33 which are not associated with H II regions. Ground-based and IUE spectra indicate that they are OB supergiants similar to those studied in the Magellanic Clouds. They find that the lines are weak and the wind velocities are low in all objects observed (admittedly a small number) in *both* galaxies (Figure 6). (In addition, as noted elsewhere, they and others find that an LMC-like extinction law is called for in both galaxies.) Thus, we now appear to have the situation that strong stellar winds (and Galactic extinction) are characteristic only of stars in our Galaxy, and the simple notion that the abundance of heavy elements is the

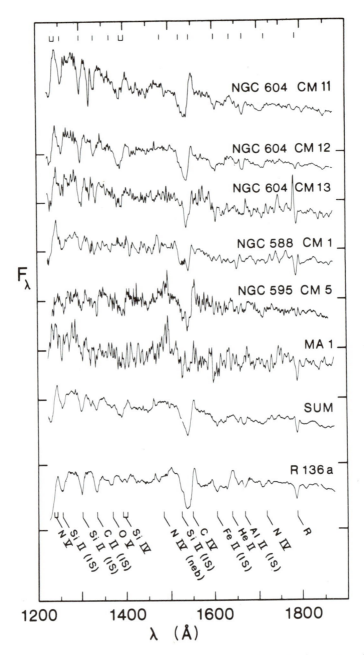

Fig. 5. IUE spectra of exciting stars of giant H II regions in M33, showing strong similarity with each other and with R136 in the LMC.

principal parameter, may not be entirely correct. The crucial data lie in the M31 stars, since that galaxy most resembles our own, and it may be that the stars observed so far (by necessity) are those with unusual extinction or other

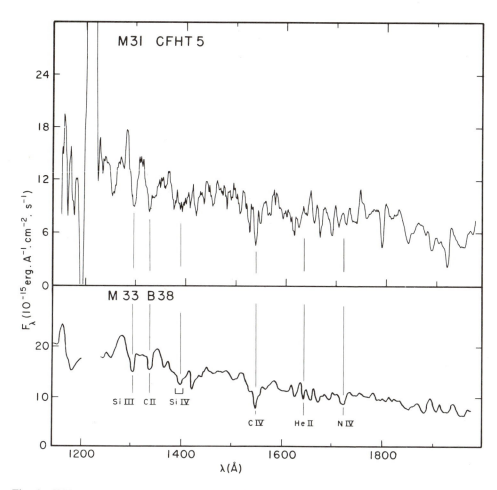

Fig. 6. IUE spectra of OB stars in M31 and M33, showing weakness of stellar wind lines, and absence of emission components. So far, this appears to be typical of these galaxies.

abnormalities connected with high luminosity. Nevertheless, the results so far are without exception. Clearly, further work is called for in this area, and will be pursued with IUE and HST. In the meantime, IUE results indicate that stellar wind phenomena are universally weak in M31 and M33, in OB stars of M_{BOL} −9 to −10. Their effective temperatures are not very well determined, but appear to be in the expected range for their spectral type.

2. H II Regions and Their Exciting Stars in Nearby Galaxies

The element abundances in galaxies are of interest to many aspects of research. Most fundamentally, the abundances are a record of the stellar history of a galaxy, and comparison between our own and other galaxies is of great significance. It has also become clear that these abundances affect many ongoing processes in a profound way — principally by affecting the opacity and scattering of radiation. Observed differences between galaxies in the behaviour of stellar envelopes, stellar mass-loss, interstellar extinction, the determination of absolute fluxes, and strong X-ray source fluxes, are all known examples of where this is important. The UV spectra of H II regions offer unique opportunities for abundance determinations, and in this section we review the fundamental contributions of IUE from observations of giant H II regions beyond the Milky Way. We have already discussed the central stellar aspects of some of them in the foregoing sections. Work on normal H II regions in the Magellanic Clouds and their important abundance results is also reported in another chapter.

UV low-resolution spectra have been taken with IUE of a number of H II regions in nearby galaxies. A catalogue and atlas of spectra processed uniformly has been conveniently published by Rosa et al. (1984). See also d'Odorico and Benvenuti (1983), and the bibliographies by Mead et al. (1984) and Feitzinger et al. (1984) for the Magellanic Clouds. The only emission line which is strongly detected in most of these spectra is C III] 1909 Å. However, a few weaker emission lines have also been seen in some of the LMC and SMC H II regions (Dufour et al., 1982), and in the compact blue galaxy I Zw 36 (Viallefond and Thuan, 1983). Carbon abundances have been derived from IUE observations of the Orion Nebula (Torres-Peimbert et al., 1980), and of H II regions in the Magellanic Clouds (Dufour et al., 1982). These data have been reanalysed by Mathis et al. (1985). Similar determinations in other extragalactic H II regions have been made by Gondhalekar (1983), Dufour et al. (1984), and Peimbert et al. (1985). The results are summarised by Pagel (1985), who notices that C/N is constant within the errors, while C/H and N/H do not vary like O/H, from galaxy to galaxy. This is a very important result in the context of the chemical evolution of galaxies, as discussed in the latter reference. The main problem with these abundance determinations is the correction for extinction of the UV line intensities. Fortunately, this correction does not appear to be very large in most of the cases studied.

The Magellanic Clouds provide a good illustration of the significance of these results. In the SMC, C is depleted (with respect to the galaxy) more than N or O, while in the LMC, N is depleted more than C or O. It is considered that N and O are initially generated by massive ($> 10\ M_\odot$) star evolution, and C from less massive (4 to 10 $M_\odot$) stars. Thus, C/N and C/O start low and rise with time as the less massive stars begin to evolve. Later still, when $< 4M_\odot$ stars have evolved, C/N will decrease again, since these stars produce N. Maran et al. (1982) used IUE to study planetary nebulae in the Clouds, and obtained interestingly different results. They found C abundances only slightly less than the Galaxy, while other elements are depleted by the 'normal' amounts. Their finding in other words is that in PN ejecta, C is 40 times the local ISM value in the SMC, and 6 times in the LMC. Thus, they find that at present planetary nebulae are the principal source of C enrichment in the Clouds, and that the C generation processes in Cloud stars are roughly the same as in the Galaxy — an important result for stellar evolution theory.

The IUE spectra of extragalactic H II regions also exhibit a few of the strongest interstellar absorption lines, in particular Ly α, C II 1335 Å, and O I 1302 Å. The lines visible at low resolution are very saturated and little can be extracted from them, except for Ly α, from which the column density of atomic hydrogen has been extracted in a few cases. More interesting information comes from the observation of the relatively bright stars embedded in the 30 Dor nebula of the LMC: high resolution IUE observations reveal a complex interstellar absorption spectrum which has been interpreted both in terms of column densities of ions up to C IV — hence of ionisation — and in terms of kinematics from their radial velocities. Some of the components certainly come from regions of the H II complex accelerated by stellar winds, but the actual situation is rather complicated (see e.g. Gondhalekar *et al.*, 1980, de Boer *et al.*, 1980, Feitzinger *et al.*, 1984). Similar observations exist for other H II regions in the LMC: e.g. N51D (de Boer and Nash, 1980). Note that high-velocity interstellar absorption or emission components are also seen in the visible in 30 Dor (Blades and Meaburn, 1980; Meaburn, 1984) as well as in other extragalactic H II regions (see e.g. Rosa and Solf, 1984).

Apart from these lines and a probably small contribution of continuum emission by the ionised gas (Lequeux *et al.*, 1981; Benvenuti, 1982), the UV spectrum of extragalactic H II regions is dominated by stellar light, giving a continuum over which stellar lines are often superposed. Clearly the hot component of the stellar cluster embedded in these H II regions is responsible for the UV spectrum. Of particular interest are the P Cygni profiles of C IV 1548, 51 Å, Si IV 1394, 1403 Å, and N V 1239, 43 Å (this latter blended with interstellar Ly α), which are visible in the H II regions NGC 604 and IC 132 of the nearbly spiral galaxy M33. This reveals the presence of hot stars with strong winds, which must be quite massive. The He II 1640 Å broad emission, characteristic of Wolf-Rayet stars, is also visible in these two objects. Figure 7 illustrates this. The presence of W—R stars is confirmed by optical observations of the same objects (d'Odorico *et al.*, 1983; d'Odorico and Benvenuti, 1983; Massey and Conti, 1983). P Cygni and W—R features are also seen in IUE spectra of starburst galaxies, which are discussed elsewhere in this volume. Unfortunately the angular resolution of IUE does not allow spatial discrimination of the stars in the ionising star clusters (Lequeux *et al.*, 1981), suggesting a similar stellar content responsible for these lines. There are exceptions in the Magellanic Clouds, particularly in 30 Dor. In 30 Dor, several individual objects have been observed to exhibit W-R characteristics, both optically and in the UV (see references in Mead *et al.*, 1983). As mentioned above, this is true for the central luminous object R136, now shown by speckle optical observations to be a composite of at least 8 stars (Weigelt and Baier, 1985). Similar objects with W—R features have been observed optically and with IUE in H II regions of M33 (Massey and Hutchings, 1983), but it is of course still impossible to decide if they are single supermassive stars or composite like R136.

Thus, most of the stars contributing to the UV fluxes are unresolved with IUE, and one must have recourse to population synthesis techniques to be able to draw conclusions about them. Fortunately, 30 Dor may serve as a guide for this kind of work. Data collected by Lequeux *et al.* (1981) show that 30 Dor is rather typical of giant extragalactic H II regions, although rather on the weak side. Melnick

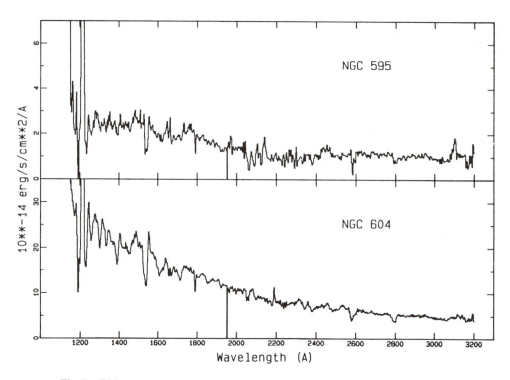

Fig. 7.   IUE merged spectra of NGC 595 and NGC 604, giant H II regions in M33.

(1985) has classified many stars in the core of the exciting cluster, including W—R stars, B supergiants and O stars (see also Moffatt *et al.*, 1985). Amongst the 49 objects classified as O stars, no less than 15 have spectral type earlier than O6, and 6 are O3 stars. The classified objects, plus the equivalent of 6 O3f stars which might represent R136, are enough to ionise the whole nebula. (Unfortunately, lower luminosity stars could not be studied and the integrated UV spectrum of 30 Dor is known only from ANS wide-band photometry.) Other extragalactic H II regions with IUE spectra have similar effective temperatures of their ionising star cluster (Lequeux *et al.*, 1981), suggesting a similar stellar content. Evolutionary continuum-spectrum synthesis which calculates the time evolution of the UV spectrum of stellar clusters with given initial mass functions and histories of star formation (Lequeux *et al.*, 1981, Viallefond and Thuan 1983), confirms that the exciting stars in the studied extragalactic H II regions must be very young and have a rather flat, although poorly determined, initial mass function (in other words, a large proportion of very massive stars). Other spectral synthesis models (Benvenuti, 1982; Sanz Fernandez de Cordoba and Benvenuti, 1984; Thuan, 1986) also indicate a large quantity of the hottest O stars, and the presence of B stars.

A major difficulty in this kind of work is caused by interstellar extinction and scattering. An evaluation of the scattered light is still very difficult: it may not be negligible (Lequeux *et al.*, 1981; Benvenuti, 1982), although all authors have neglected it for lack of quantitative evaluation. The problem of extinction in extragalactic H II regions is very involved, as discussed in detail by Lequeux *et al.*, 1981, Viallefond and Goss (1986), Caplan and Deharveng (1986), and others. It is impossible to present here a full discussion of this problem, but we give a few hints. In order to match the observed UV spectra with the synthesis results, a far-UV *reddening* must be introduced. Correcting for the galactic foreground UV reddening is usually not sufficient and some extra reddening in the galaxy containing the H II region must be introduced; the shape of the relevant UV reddening law can be derived roughly, and usually it lies between the Galactic and the SMC laws and seems to depend on the metallicity of the galaxy (Lequeux *et al.*, 1981; Thuan, 1986). However, the absolute amount of the corresponding extragalactic *extinction* often appears to be surprisingly small; much smaller than one would have expected from observations of the Balmer decrement or of the Hα/radio continuum intensity ratio. This is most clearly seen in NGC 604 (Benvenuti, 1982), and also in some starburst galaxies (Augarde and Lequeux, 1985; Thuan, 1986). In the case of NGC 604, comparison of high angular resolution Hα and radio photometry shows that a large amount of dust is mixed with the ionised gas (Viallefond, in prep.). A large amount of far-UV extinction, including a strong 2200 Å absorption, would be expected if the exciting stars are also mixed with the gas: this is not observed, and in particular the 2200 Å feature is not seen in IUE spectra of NGC 604. A possible explanation is that the exciting stars that contribute most to the UV emission are in front or outside of most of the dust related to the H II region. This might just be a selection effect: UV extinction is extremely strong and would greatly reduce the contribution to the UV spectra of stars which lie beyond a large amount of dust. Given this problem, it still seems somewhat premature to draw quantitative conclusions on the exciting clusters based upon the observed far-UV fluxes. Higher sensitivity and angular resolution observations with the Hubble Space Telescope are required for a better understanding of the situation.

Thus, IUE short wavelength spectra of extragalactic H II regions have provided unique results on the abundance of carbon in external galaxies, and interesting information on the young, massive stars embedded in these objects. However, much remains to be understood of the intricacies of the interplay between extinction, scattering, and stellar UV radiation in these complexes.

3. Peculiar Early-Type Emission Line Stars in Local Group Galaxies

The uppermost part of the H-R diagram is populated by early-type emission line supergiants, mainly of type B0 to B5. These objects are of major interest in connection with current theories of the evolution of very massive stars. They may represent important (yet scarcely investigated) short-lived phases of the evolution of the most massive stars, characterized by strongly enhanced mass-loss. Furthermore, these objects are — apart from supernovae — the brightest stars in the universe and are of considerable interest as distance indicators in extragalactic systems. Since these objects emit most of their radiation in the UV, IUE has been particularly useful in studying their properties. A considerable number of these blue emission-line stars are known to be members of the Magellanic Clouds, of M31 and M33, and are bright enough for IUE (cf. e.g. Shore and Sanduleak (1984), Humphreys *et al.* (1984), Wolf (1985), Zickgraf *et al.* (1986) and literature quoted in these papers).

3.1. S DOR VARIABLES IN THE LMC, M31 AND M33

The most fascinating subgroup of the blue emission-line supergiants form the S Dor variables or Hubble—Sandage variables, which are characterized by irregular photometric variations in the visual and photographic range of more than one mag on timescales of a year to decades. Their highly variable spectra show strong Balmer emission lines and pronounced forbidden lines ([Fe II], [N II] etc.) which dominate the maximum and minimum spectra, respectively (cf. e.g. Thackeray, 1974). The prototype S Dor in the LMC has been in a bright phase since the launch of IUE. Its ultraviolet spectrum has been studied by Wolf *et al.* (1980) in the low- and by Leitherer *et al.* (1985) in the high-dispersion mode. Combining these IUE observations with coordinated ground-based spectroscopic and photo-metric (particularly IR) observations has yielded much data concerning the strange wind characteristics of S Dor. It has been found that during bright phases, S Dor is surrounded by a slowly expanding cool ($T \sim 9000$ K) envelope, with a low terminal velocity ($v = 140$ km s^{-1}) being reached only at large distances ($> 10\ R_*$). The mass-loss rate is of the order $M_\odot \sim 5.10^{-5}\ M_\odot$ yr^{-1}. Similar wind characteristics were found for the LMC star R 127. This star was previously classified by Walborn (1977) as an extreme Of star with similarities to very late WN stars (WN 9—10). An S Dor-type outburst of this star by more than one magnitude in the visual was observed by Stahl *et al.* (1983). Since then, the visual brightness of R 127 has further increased and in the beginning of 1986 it was (visually) the second brightest star in the LMC. The IUE LWR high resolution spectrum taken in November 24, 1982 shows that R 127 has had a complex shell ejection event, evidenced by the multiple-component substructure of the Fe II lines (Figure 8). More recent IUE and ground-based spectroscopic observations during the very bright phase (Cassatella, 1986; Appenzeller *et al.*, 1986) have shown the spectrum to be practically indistinguishable from that of the prototype S Dor at maximum light. Interestingly, from long slit spectra, a well-resolved extended gaseous nebula similar to that around the galactic extreme supergiant AG Car was found around this distant LMC star. A kinematic age of this nebula of 20 000 yr is derived, which is of the order of the expected lifetime of the S Dor evolutionary phase.

The third known S Dor variable of the LMC, R 71, was observed with IUE during quiescence (Wolf *et al.*, 1981a). The derived continuum energy distribution is that of an early B supergiant. However P Cygni type Fe II lines in the LWR have shown that R71 is surrounded by very cool slowly expanding ($v_{max} \sim 127$ km s^{-1}) envelopes. R 71 has also been found to be an IRAS point source (Wolf and Zickgraf, 1986), indicating that R 71 is surrounded by a highly extended ($R = 8000\ R_*$) dust envelope of very low temperature ($T = 140$ K), with a total mass of $3.10^{-2}\ M_\odot$. A possible cause for the dust condensation around this hot star could be the comparatively low wind velocity which is characteristic for the S Dor variables.

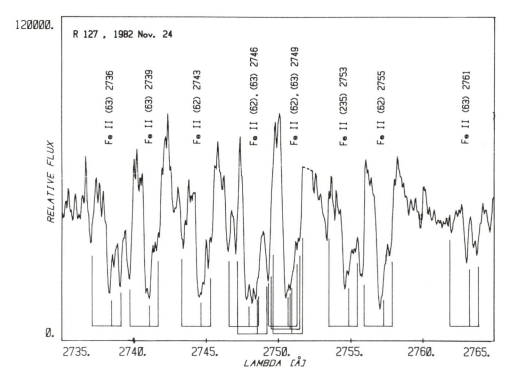

Fig. 8. Section of an IUE LWR spectrum of the LMC S Dor variable R 127. This wavelength region is dominated by absorption lines of the Fe II multiplets 62 and 63. For each line the mean heliocentric radial velocities +12, +158, and +232 km s⁻¹ are indicated. A complex shell phenomenon is indicated by the multiple substructure of these ultraviolet lines.

Related to the S Dor variables is the LMC blue emission line star R 81 which was observed in the IUE high-dispersion mode and which was shown by Wolf *et al.* (1981b) to be a close counterpart of the galactic star P Cygni. This star was recently found to be an eclipsing binary (Stahl *et al.*, 1986) and is therefore of particular interest since it offers the unique possibility of determining the mass of a P Cygni star, whose absolute luminosity is reliably known due to its membership of the LMC.

Combining the IUE and ground-based observations of the LMC S Dor variables leads us to some important conclusions concerning their nature. S Dor variables are hot OB supergiants which, during maximum, are surrounded by cool ($T_e \sim 8000-10\,000$ K), dense ($N \sim 10^{11}$ cm⁻³), slowly expanding envelopes. The mass-loss rate during maximum is of the order of 10^{-5} to 10^{-4} $M_\odot$ yr⁻¹; it is about a factor 10 to 100 lower during minimum phases. However the bolometric luminosity of the S Dor variables seems to remain almost constant. The observed brightness variations in the visual and photographic range are regarded as a consequence of the variable mass-loss rate and a correspondingly variable redistribution of the spectral flux.

IUE observations also provided reliable bolometric luminosities for a number of S Dor variables in the LMC, and in M31 and M33 (Humphreys *et al.*, 1984).

They are located around the upper envelope of known stellar luminosities (Figure 9). The S Dor variables are regarded as key objects for the understanding of the evolution of the very massive ($M \geqslant 50\ M_\odot$) stars and are supposed to represent a short-lived phase as immediate progenitors of the massive WR stars (cf. Maeder, 1983).

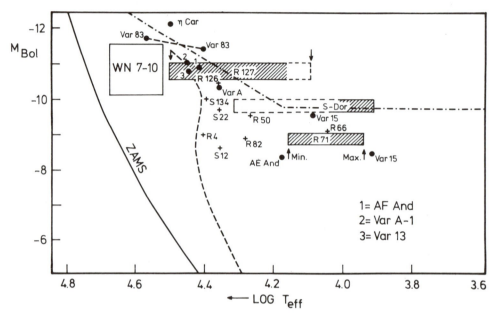

Fig. 9.   Location of the LMC S Dor variables R 71, S Dor, and R 127 in the HRD. Also included in the figure are the S Dor variables or Hubble-Sandage variables of M31 and M33 (Humphreys *et al.*, 1984), the B[e] supergiants (full dots) of the MCs and the upper envelope (dashed-dotted line) of known stellar luminosities (Humphreys and Davidson, 1979). The approximate position of the late WN-type stars is also given.

3.2. B[E] SUPERGIANTS WITH CIRCUMSTELLAR DUST IN THE MC'S

Another particularly interesting subgroup of the blue emission line stars are the dusty B[e] supergiants of the MC's. They are characterized by the following typical properties:

(a) strong Balmer emission lines, frequently with P Cygni profiles,

(b) permitted and forbidden emission lines of Fe II, [Fe II], [0 I] etc.,

(c) strong infrared excess due to thermal radiation of circumstellar dust. IUE spectra of these stars have been described by Shore and Sanduleak (1984) and by Zickgraf *et al.* (1985, 1986).

These stars belong to the most luminous and presumably most massive stars of the MC's and are located in the same region of the HR-diagram as the S Dor variables (Figure 9). These objects are particularly distinguished by the hybrid character of their line spectrum: broad UV resonance absorption lines and narrow emission lines. For the prototype R 126 (B0.5Ia$^+$) Zickgraf *et al.* (1985) suggested a two-component stellar wind model with a disk-like outer configuration. According to this model, the UV resonance lines of highly ionized species (N V, C IV, Si IV etc.) originate in a high velocity ($v_{max} \sim 1800$ km s^{-1}) line-driven 'normal' wind near the pole of the hypergiant star. The disk is supposed to be formed by a slow (~ 40 km s^{-1}), cool, dense wind. The observed narrow emission lines of Fe II, [Fe II] and other singly ionized elements in the IUE LWR spectrum and in the optical wavelength range, as well as the thermal dust emission, are produced in this disk. Stellar rotation at close to the break-up velocity is the driver of this two-component structure. In a recent paper, Zickgraf *et al.* (1986) extended this model to all eight known B[e] stars of the MC's.

As shown by Figure 9, the B[e] supergiants are located to the right of the main-sequence band and represent evolutionary stages of massive ZAMS O stars. Unlike the S Dor variables the B[e] supergiants of the LMC have not shown photometric variations; the mass loss from the B[e] stars appears to be much more stationary and stable. Possibly the rapid stellar rotation leading to an enhanced mass loss from the equatorial regions prevents them from becoming S Dor-type unstable.

References

Appenzeller, I., Wolf, B., and Stahl, O.: 1986, 'Instabilities in Luminous Early Type Stars', in Lamers and de Loore (eds.), *Workshop in Honour of C. de Jager* (in press).
Augarde, R. and Lequeux, J.: 1985, *Astron. Astrophys.* **147**, 273.
Benvenuti, P.: 1983, in R. M. West (ed.), 'Highlights of Astronomy', *IAU Symp.* **6**, 631.
Bianchi, L. and Pakull, M.: 1985, *Astron. Astrophys.* **146**, 242.
Blades, J. C. and Meaburn, J.: 1980, *Monthly Notices Roy. Astron. Soc.* **190**, 59.
Bruhweiler, F. C., Parsons, S. B., and Wray, J. D.: 1982, *Astrophys. J.* **256**, L49.
Caplan, J. and Deharveng, L.: 1986, *Astron. Astrophys.* (in press).
Cassatella, A.: 1986, private communication.
Cassinelli, J. P., Mathis, J. S., and Savage, B. D.: 1981, *Science* **212**, 1497.
Charles, P. A. *et al.*: 1983, *Monthly Notices Roy. Astron. Soc.* **202**, 657.
Corbet, R. H. D. *et al.*: 1985, *Monthly Notices Roy. Astron. Soc.* **212**, 565.
de Boer, K. S., Koornneef, J., and Savage, B. D.: 1980, *Astrophys. J.* **236**, 679.
de Boer, K. S. and Nash, A. G.: 1982, *Astrophys. J.* **255**, 447.
D'Odorico, S. D. and Benvenuti, P.: 1983, *Monthly Notices Roy. Astron. Soc.* **203**, 157.
d'Odorico, S., Rosa, M., and Wampler, J.: 1983, *Astron. Astrophys. Suppl.* **53**, 97.
Dufour, R. J., Schiffer, F. H., and Shields, G. A.: 1984, *The Future of Ultraviolet Astronomy Based on 6 Years of IUE Research*, NASA.
Dufour, R. J., Shields, G. A., and Talbot, R. J.: 1982, *Astrophys. J.* **252**, 461.
Feitzinger, J. V., Hanuschik, R. W., and Schmidt-Kaler, T.: 1983, *Astron. Astrophys.* **120**, 269.
Feitzinger, J. V., Hanuschik, R. W., and Schmidt-Kaler, Th.: 1984, *Monthly Notices Roy. Astron. Soc.* **211**, 867.
Fitzpatrick, E. L. and Savage, B. D.: 1983, *Astrophys. J.* **267**, 93.
Fitzpatrick, E. C. and Savage, B. D.: 1984, *Astrophys. J.* **279**, 578.
Garmany, C. D., Massey, P., and Conti, P. S.: 1984, *Astrophys. J.* **278**, 233.
Garmany, C. D. and Conti, P. S.: 1985, *Astrophys. J.* **293**, 407.
Gondhalekar, P. M.: 1983, *Adv. Space Res.* **2**, 163.
Gondhalekar, P. M., Willis, A. J., Morgan, D. H., and Nandy, K.: 1980, *Monthly Notices Roy. Astron. Soc.* **193**, 875.
Hammerschlag-Hensberge, G., Kallman, T. R., and Howarth, I. D.: 1984, *Astrophys. J.* **283**, 249.
Howarth, I. D., Prinja, R. K., Roche, P. F., and Willis, A. J.: 1984, *M.N.* **207**, 287.

Humphreys, R. M., Blaha, C., D'Odorico, S., Gull, T. R., and Benvenuti, P.: 1984, *Astrophys. J.* **278**, 124.

Hutchings, J. B.: 1980, *Astrophys. J.* **237**, 285.

Hutchings, J. B.: 1980a, *Astrophys. J.* **235**, 413.

Hutchings, J. B.: 1982, *Astrophys. J.* **255**, 70.

Hutchings, J. B. *et al.*: 1985, *Publ. Astron. Soc. Pacific* **97**, 418.

Leitherer, C. *et al.*: 1985, *Astron. Astrophys.* **153**, 168.

Lequeux, J., Maucherat-Joubert M., Deharveng, J. M., and Kunth, D.: 1981, *Astron. Astrophys.* **103**, 305.

Maeder, A.: 1983, *Astron. Astrophys.* **120**, 113.

Maran, S. P., Aller, L. H., Gull, T. R., and Stecher, T. P.: 1982, *Astrophys. J.* **253**, L43.

Massey, P. and Conti, P. S.: 1983, *Astrophys. J.* **273**, 576.

Massey, P. M. and Hutchings, J. B.: 1983, *Astrophys. J.* **275**, 578.

Massey, P. M., Hutchings, J. B., and Bianchi, L.: 1985, *Astron. J.* **90**, 2239.

Massey, P. M., Hutchings, J. B., and Bianchi, L.: 1986 (preprint).

Mathis, J. S., Chu Y-H., and Peterson, D. E.: 1985, *Astrophys. J.* **292**, 155.

Meaburn, J.: 1984, *Monthly Notices Roy. Astron. Soc.* **211**, 521.

Mead, J. M., Kondo, Y., and Boggess, A.: 1984, *IUE ESA Newsletter* **20**, 71.

Melnick, J.: 1985, *Astron. Astrophys.* **153**, 235.

Moffatt, A. F. J., Seggewiss, W., and Shara, M. M.: 1985, *Astrophys. J.* **295**, 109.

Nandy, K., Morgan, D. H., Willis, A. J., Gondhalekhar, P. M.: 1980, *Monthly Notices Roy. Astron. Soc.* **193**, 43.

Nandy, K. *et al.*: 1983, *Monthly Notices Roy. Astron. Soc.* **205**, 231.

Pagel, B. E. J.: 1985, *ESO Workshop on Production and CNO Elements Garching*, p. 155.

Peimbert, M., Pena, M., and Torres-Peimbert, S.: 1986, *Astron. Astrophys.* (in press).

Rosa, M., Joubert, M., and Benvenuti, P.: 1984, *Astron. Astrophys. Suppl.* **57**, 361.

Rosa, M. and Solf, J.: 1984, *Astron. Astrophys.* **130**, 29.

Sanz Fernandez de Cordoba, L. and Benvenuti, P.: 1984, *4th European IUE Conference ESA* **SP-218**, p. 73.

Shore, S. N. and Sanduleak, N.: 1984, *Astrophys. J. Suppl.* **55**, 1.

Smith, L. F. and Willis, A. J.: 1983, *Astron. Astrophys. Suppl.* **54**, 229.

Stahl, O. *et al.*: 1983, *Astron. Astrophys.* **127**, 49.

Stahl, O.,Wolf, B., and Zickgraf, F.-J.: 1986, 'Instabilities in Luminous Early Type Stars', in Lamers and de Loore (eds.), *Workshop in Honour of C. de Jager* (in press).

Tarenghi, M. *et al.*: 1981, *Astron. Astrophys. Suppl.* **43**, 353.

Thackeray, A. D.: 1974, *Monthly Notices Roy. Astron. Soc.* **168**, 221.

Thuan, T. X.: 1986, in D. Kunth and T. X. Thuan (eds.), *Star Forming Dwarf Galaxies and Related Objects* (in press).

Torres-Peimbert, S., Peimbert, M., and Daltabuilt, E.: 1980, *Astrophys. J.* **238**, 133.

van der Klis, M. *et al.*: 1982, *Astron. Astrophys.* **106**, 339.

Viallefond, F. and Thuan, T. X.: 1983, *Astrophys. J.* **269**, 444.

Viallefond, F. and Goss, W. M.: 1986, *Astron. Astrophys.* **154**, 357.

Walborn, N. R.: 1977, *Astrophys. J.* **215**, 53.

Weigelt, G. and Baier, G.: 1985, *Astron. Astrophys.* **150**, L18.

Wolf, B.: 1986, in C. de Loore, A. Willis, and P. G. Laskarides (eds.), 'Luminous Stars and Associations in Galaxies', *IAU Symp.* **116** (in press).

Wolf, B., Appenzeller, I., and Cassatella, A.: 1980, *Astron. Astrophys.* **88**, 15.

Wolf, B., Appenzeller, I., and Stahl, O.: 1981a, *Astron. Astrophys.* **103**, 94.

Wolf, B., Stahl, O., de Groot, M. J. H., and Sterken, C.: 1981b, *Astron. Astrophys.* **99**, 351.

Wolf, B. and Zickgraf, F.-J.: 1986, *Astron. Astrophys.* (in press).

Zickgraf, F.-J., Wolf, B., Stahl, O., Leitherer, C., and Klare, G.: 1985, *Astron. Astrophys.* **143**, 421.

Zickgraf, F.-J., Wolf, B., Stahl, O., Leitherer, C., and Appenzeller, I.: 1986, *Astron. Astrophys.* (in press = ESO Preprint No. 405).

STARBURST GALAXIES

DANIEL KUNTH

Institut d'Astrophysique de Paris, Paris, France

and

DANIEL WEEDMAN

The Pennsylvania State University, University Park, PA, U.S.A.

1. Introduction

Episodes of star formation that are too intense to continue over the life-times of galaxies are often termed starbursts. Such episodes have received increasing amounts of attention because they can be the dominant source of a galaxy's luminosity in many wavebands. Furthermore, nearby examples of starbursts have the potential of teaching us much about the formation of massive stars, and can show what newly forming galaxies might look like at early epochs of the universe.

Galaxies containing starbursts (equivalently, 'starburst galaxies') are conspicuous because of various radiation processes associated with the life and death of massive stars. In optical wavelengths, these galaxies are characterized by blue colors and strong emission lines, so they represent the majority of Markarian galaxies and emission line galaxies found in objective prism surveys. The massive, short lived stars also lead to significant energy release at X-ray, ultraviolet, infrared, and radio wavelengths. In the X-ray, accretion-powered radiation from binaries with compact, massive components contributes along with radiation from supernova remnants. In the infrared, dust heated by the ultraviolet and visible radiation absorbed from the massive stars produces conspicuous far infrared luminosity. At radio wavelengths, spectra are usually non-thermal, produced by the supernova remnants from the dying, massive stars. Much work is underway utilizing all of these wavebands to study the starburst phenomenon. In the present review, however, we concentrate on those results from the ultraviolet produced by the IUE.

There are four morphological categories of starburst galaxies which have been included in the observations that we reivew:

(a) Small, isolated, irregular galaxies containing starbursts throughout most of the visible galaxy. These have been defined as either 'extragalactic H II regions' or 'Blue Compact Galaxies = BCGs'; we use the latter terminology. Examples are NGC 4861 = Mkn 59 and I Zw 36 = Mkn 209 (Viallefond and Thuan, 1983). We also include in this category the nearby Magellanic Irregulars, such as NGC 4449 (Lequeux *et al.*, 1981) and NGC 4214 (Huchra *et al.*, 1983), which are close enough to be resolved into individual H II regions and bright stars.

(b) Larger, usually irregular, galaxies with widespread starbursts that seem to arise in most cases because of interactions between galaxies. We include the 'clumpy irregular galaxies' in this category, as well as the obvious examples of

Y. Kondo (ed.), Scientific Accomplishments of the IUE, pp. 623—635.
© 1987 *by D. Reidel Publishing Company.*

interacting starburst systems. The most luminous examples are NGC 3690 = Mkn 171 (Augarde and Lequeux, 1985) and NGC 6052 = Mkn 297 (Benvenuti *et al.*, 1982).

(c) Galaxies with the starburst restricted to the vicinity of the nucleus of an otherwise normal galaxy, such as NGC 7714 (Weedman *et al.*, 1981) and IC 5135 (Thuan 1984).

(d) Spiral galaxies with intense starbursts in the disk, either throughout the disk as in NGC 1068 (Weedman and Huenemoerder, 1985), or localized in the disk, so this category includes individual giant H II regions in nearby spiral galaxies, such as NGC 604 and NGC 5471 (Rosa, 1980).

At the present stage of research, such morphological classifications can be ambiguous, and the primary purpose of using them is to search for any differences that might be found in the physical parameters of starbursts within different morphologies. Eventually, that may yield information on how the star formation process depends on overall galaxy properties.

2. Ultraviolet Spectra of Starburst Regions

Many IUE spectra have been published, and a good way to tour the data is by consulting the 'Atlas of Ultraviolet Spectrograms' of giant H II regions in spirals, BCGs, and irregular galaxies published by Rosa *et al.* (1984), covering the IUE archives up to April 1982. Spectra published since then give more support to the general features revealed in the catalogue. These features are summarized below.

(1) In all examples the continua are flat or rising. Rosa *et al.* (1984) notice that no pronounced turn-over is observed unlike the UV spectra of ellipticals or the bulges of spiral galaxies. This simple observation tells immediately that the amounts of extinction must be moderate although not unimportant for further synthesis modeling. It further indicates that observers invariably pick out OB associations which stand out in UV light against the possible contribution of any stellar background of higher mass to light ratio. A rough estimate requires hundreds of early stars to match the UV total luminosity in objects such as NGC 604, a giant H II region of the spiral galaxy M33, up to many thousands in BCGs and clumpy irregular galaxies (Benvenuti *et al.*, 1979, 1982).

(2) The single most important result that an IUE observation of any blue galaxy can produce is proof that the luminous ultraviolet continuum arises from hot stars. This is particularly important for starburst nuclei, whose spectral signatures at many other wavelengths are similar for both activity powered by starbursts and that powered by non-thermal sources. Many prominent absorption features can be identified with lines of interstellar and stellar origin. They are observed in the short wavelength range when there are data with good signal to noise ratio. The most conspicuous lines are Si IV $\lambda1403$ and C IV $\lambda1549$ for which structures of the absorption lines characterize hot stars. The lines are broader than the spectral resolution and correspond to terminal wind velocities between 2000 and 4000 km $^{-1}$. These stellar features are prominent in NGC 604 (Rosa, 1980) and other H II regions in which Wolf—Rayet stars have been found from optical data (D'Odorico and Rosa, 1981). They are present but less conspicuous in Magellanic

irregular galaxies such as NGC 4214 and NGC 4670 (Huchra *et al.*, 1983) where main sequence O stars dominate the emission in the UV.

For an on-going starburst, with luminosity dominated by O and B stars, no absorption lines are readily visible in the optical portion of the spectrum, especially as the Balmer lines are often filled in by emission. Simply trying to detect these UV absorption lines justifies an IUE exposure, but in many cases the signal to noise has been adequate to use the lines for help in spectral synthesis modeling. Such spectral features are especially important for judging the kinds of stars present, because the overall spectral shape may be significantly affected by dust absorption. Comparison of the absorption features with those in the IUE Spectral Atlas for individual stars makes possible a judgment of what kinds of stars are most responsible for the UV luminosity. The most important discriminant is whether or not these absorption lines show a P Cygni profile (emission on the red side of absorption, with a broad absorption line blueshifted with respect to the systemic velocity). Such a profile is attributed to mass loss from massive O stars. If this profile were very conspicuous, it would be a sign of Wolf-Rayet stars or very massive, early supergiants. A typical profile in starburst regions of varying morphologies shows distinct but not dramatic emission, leading to the conclusion that main sequence stars earlier than O7 or supergiants earlier than B0 are present. The most probable contributors to the ultraviolet light are stars with initial masses near $20\ M_\odot$. Starburst regions without any sign of a P Cygni profile would be attributed to less massive, later type O or B stars. There are no obvious correlations between morphology and the form of these profiles, although our impression is that the more conspicuous P Cygni profiles are in BCGs and individual giant H II regions rather than in starburst nuclei or large interacting systems such as NGC 3690.

(3) All observers have noted the weakness of ultraviolet nebular emission lines in starburst galaxies, contrary to what occurs in the optical region. This was one of the biggest surprises to observers, until it was realized that the continuous spectra of the hot stars simply overwhelm the emission lines. When emission lines are seen, the most conspicuous are identified with nebular C III] $\lambda1909$, C IV $\lambda1549$, Si IV $\lambda1403$, and N V $\lambda1240$ lines. Occasionally the He II $\lambda1640$ line is seen in emission, but it is difficult to separate the stellar and nebular contribution. The Ly α line cannot be seen in most cases because of the strong geocoronal line as well as the presence of interstellar absorption. However, ultraviolet spectra have been obtained of some starburst galaxies with redshifts large enough to make the Ly α line well separated from the geocoronal. These observations have been used to discuss the prospects of detecting primeval galaxies via their emission lines and are reviewed in the last section.

If all young stellar systems observed with IUE are considered, one would tend to identify two spectral categories:

— Large irregular galaxies, interactive systems and starburst nuclei with conspicuous absorption features and steep or flat continua in the UV. Their properties are closely similar to those of giant H II regions such as NGC 604 or NGC 5471.

— BCGs with steeper UV spectra and very weak absorption lines. In most cases the nebular emission lines are weak (Huchra *et al.*, 1982), except in I Zw 36. Their metallicity is much lower than in the previous category.

Examples of such spectra are shown in Figures 1 and 2, which illustrate the typical absorption features seen for one category and the absence of such features in the other.

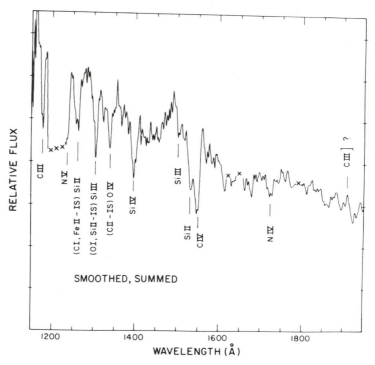

Fig. 1.   Composite IUE spectrum of two Magellanic Irregular galaxies, reproduced from Huchra *et al.* (1983), showing conspicuous absorption features.

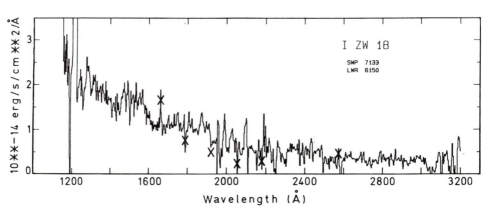

Fig. 2.   Spectrum of the extremely metal poor BCG I Zw 18, taken from the IUE atlas of Rosa *et al.* (1984), showing no strong absorption features.

Using the slope of the UV continuum along with absorption line equivalent widths and data from other wavelength bands, various synthesis modeling of the stars in the starburst can be undertaken. Differences in the ultraviolet spectral properties are likely to depend on metallicity, reddening effects and the present star formation rate. The aim of a spectral synthesis model is to sort out these parameters and ultimately determine the initial mass function and the possible presence of supermassive stars in starburst regions. There are many important objectives for spectral modeling. For example, it is necessary to know the total mass that cycles through starbursts in order to understand the impact on galaxy evolution. This requires knowing the IMF, the duration of the starburst, and the frequency with which starbursts occur in galaxies. All of these may depend in as yet unknown ways on the starburst morphology. Few general conclusions can be stated as yet, but the IUE spectra are crucial for such modeling efforts.

3. Dust and the Reddening Problem

Absorption by dust can be very significant in the UV, and accomodating this absorption produces a major uncertainty in any modeling. A review of the reddening correction treatment is given in Lequeux et al. (1981). They make a case for a burst model and compare the observed UV spectra with a set of models from which they compute the expected far UV emission at selected windows. Under these assumptions, the intrinsic far UV spectrum of extragalactic H II regions should be that of a zero-age cluster and depends very little on the IMF. The rule of thumb is that whatever the IMF, the flux at 1250 Å is about 3.2 times that at 1900 Å. In all cases they proceed by correcting for the extinction in our Galaxy and outside the star forming region using H I observations. The internal extinction is estimated from the $H\alpha/H\beta$ ratios corrected for these reddening effects to the value given by case B recombination theory. This procedure in practice has difficulties, in particular that the use of reddening parameters derived from the nebular gas is not necessarily relevant to the stellar component. Moreover the $H\alpha/H\beta$ may itself be affected by underlying stellar absorption which steepens the observed emission Balmer decrement. An interesting situation is found whereby a flat UV spectrum is observed together with no evidence of significant dust obscuration from the optical or radio data. This is clearly the case for NGC 2366 and suggests that the standard Galactic extinction law is not valid. The main conclusion reached by Lequeux et al. (1981) is that the far UV extinction curve rises at $\lambda < 1900$ Å as the metallicity decreases.

Another way to obtain reddening corrections is use of the 2200 Å absorption feature. This again requires knowledge of the appropriate extinction law, which may differ from object to object as it differs within the LMC from one place to another (Nandy et al., 1980). When such a procedure is applied using the Galactic extinction law (Seaton, 1979), a color excess no greater than 0.5 is required to remove the 2200 Å feature (Rosa et al., 1984; Gondhalekar, 1986). A third approach to the problem is to treat the reddening as a free parameter. This approach is followed by Thuan and collaborators (Thuan, 1986) using optimizing synthesis methods. For BCGs, the derived intrinsic extinction is always smaller than the one derived from the Balmer decrement and corrected for Galactic extinction. This again supports the view that dust does not in general strongly affect the ultraviolet component at least for the most low luminosity metal-poor compact galaxies (I Zw 18, Mkn 36, Mkn 209). All the studies agree with the conclusion reached by Israel and Koornneef (1979) that dust-scattered light contributes very little to the ultraviolet energy distribution.

It is to be expected that the ultraviolet continuum observed arises from those stars that happen to be obscured the least, so there could be large numbers of stars completely obscured while the few that remain dominate the spectrum which is seen. Until recently, there was no way for most starburst galaxies even to estimate how much total absorption of ultraviolet radiation occurs. Now, by comparing ultraviolet fluxes with the far infrared results from the Infrared Astronomy Satellite, it is at least possible to determine how much ultraviolet flux is completely unobserved, and to compare this with what is seen by IUE. Such comparisons have not yet appeared in the literature, although Weedman and Huenemoerder (1985), using non-IRAS data for infrared fluxes, found that less than 1% of the ultraviolet flux escapes the starburst disk of NGC 1068.

Such an estimate is made by considering the monochromatic ultraviolet luminosity which should be produced by a starburst, and then comparing with the bolometric luminosity produced by the same starburst. If there is extensive dust absorption and re-radiation in the infrared, the total infrared luminosity is approximately the same as the bolometric luminosity which has been absorbed. Consequently, the ultraviolet luminosity required for producing the infrared re-emission can be compared with the observed ultraviolet luminosity. The comparison of this reprocessed UV with that observed estimates the total obscuration in the UV. Fortunately, the results are relatively insensitive to the IMF of the starburst. To demonstrate this, we use the simplest possible starburst model — a model in which the starburst is observed in an on-going, equilibrium phase. That is, newly formed massive stars replace those that die, so the age of the starburst or the duration of star formation is not a necessary parameter, and the luminosity can be considered to arise from main sequence, unevolved stars. This approximation is particularly reasonable for cases in which we observe the averaged effects of several independent starburst regions over the disk of a galaxy.

The equations for dealing with such a model are given by Gehrz et al. (1983) where they are applied to NGC 3690. A useful advantage of this simple model compared to more sophisticated synthesis models discussed below is that the effects of changes in the slope of the IMF or in the upper mass limit can be easily seen. We show some examples in context of comparing the flux at 1285 Å with the total bolometric flux. These results utilize the relation between mass and monochromatic luminosity at 1285 Å for individual stars taken from the atmosphere models of Kurucz (1979), as tabulated in Weedman and Huenemoerder (1985). The entries in Table I show the ratio $L(1285)/L(\text{bol})$ for $L(1285)$ in ergs s^{-1} A^{-1} and $L(\text{bol})$ in ergs s^{-1} for starbursts with different IMF slope and upper mass limit. The important result is that, even for large ranges in these parameters, the ratio remains very nearly 10^{-3}.

A reasonable approximation to the bolometric flux from a starburst galaxy is given in the IRAS catalog (Lonsdale *et al.*, 1985) as FIR, which is the total flux from 40 to 120 μm; empirically, this seems to be the wavelength range over which most of the luminosity is emitted by thermal dust sources. This energy, at least, has to be accounted for by hot stars whose luminosity is absorbed by the dust. Using the results in Table I, we can estimate, therefore, that the obscured ultraviolet luminosity at 1285 Å is 10^{-3} FIR. We show in Table II some examples of how this obscured flux compares with the flux observed by IUE.

TABLE I
Ratio of 1285 Å luminosity to bolometric luminosity

IMF slope =	3.5		2.5		1.5	
upper mass =	25	50	25	50	25	50
ratio $\times 10^4$ =	9.3	8.9	9.3	8.3	9.1	7.5

TABLE II
Observed UV luminosity compared to obscured UV luminosity

Object	$f(1285) \times 10^{14}$ (spectrum shape)	FIR $\times 10^{10}$	Observed $f(1285)$ / Obscured $f(1285)$
Nuclear			
NGC 7714	4 (flat)	5	0.08
Mkn 54	2.2 (rising)	0.5	0.40
IC 5135	0.6 (flat)	8.7	0.01
Clumpy irregulars			
NGC 6052 = Mkn 297	1.3 (flat)	3.5	0.04
NGC 7673 = Mkn 325	2.6 (rising)	2.6	0.10
BCGs			
Mkn 209 = I Zw 36	3 (rising)	<0.5	>0.60
NGC 4861 = Mkn 59	15 (rising)	0.9	>1.00
Giant H II:			
NGC 5471	10 (rising)	0.9	>1.00

In general, the entries in Table II confirm the conclusion reached from the synthesis modeling: those starburst systems with flat UV spectra are more severely affected by dust absorption than systems with rising spectra. The BCGs and giant H II regions are generally much less contaminated by dust than other morphologies, and some nuclear starbursts or clumpy irregulars have far more luminosity obscured in the ultraviolet than emerges to the IUE observer.

4. Synthesis Modeling

After allowance is made for dust extinction, spectra can be compared with spectra generated by evolutionary synthesis models. These models are similar in their conceptual scheme to the ones initially developed by Tinsley (1972) and Searle, Sargent and Bagnuolo (1973). The star formation rate is generally assumed to be either exponential with an *e*-folding time parameter or an instantaneous burst, or a composite of both. The initial mass function is defined by the stellar mass spectrum at birth, which is

$$dN(m)/dm \propto m^{-x}$$

where $x = 2.35$ is the Salpeter value. The time evolution of the model is calculated using stellar evolutionary tracks. The model spectra are finally obtained by using stellar atmosphere models and/or available stellar spectra (Heck *et al.*, 1984; Wu *et al.*, 1981). External constraints may be added to produce a model with satisfactory energy distribution, such as the number of Lyman continuum photons required to match the observed Hβ luminosity, the amount of released synthesized oxygen, the age of the burst, the slope and the upper mass cut-off of the IMF. The population distribution is then deduced from the modeling process. Up to now this technique has been used by various groups but with few attempts to match the strengths of the stellar absorption features. These features have been used to infer the presence of hot stars only in a crude way. WR stars are revealed from the existence of P Cygni profiles but do not give a spectral signature as strong as that found in the optical for similar objects (see Kunth and Sargent, 1981 for the case of NGC 3125). In fact, models developed by Lequeux *et al.* (1981), Viallefond and Thuan (1984) and Huchra *et al.* (1983) have modeled the ultraviolet continuum energy distribution alone. From their model Huchra *et al.* (1983) conclude that the regions observed in very blue Magellanic irregular galaxies consist of young bursts of star formation superposed on an underlying old population. Stars with effective temperatures $30\,000\,\mathrm{K} < T_{\mathrm{eff}} < 35\,000\,\mathrm{K}$ and initial masses near $20\,M_\odot$ are dominant in the IUE range. This is because the star formation rate is high enough to maintain a well populated upper main sequence which dominates the UV light. No abnormal IMF is required to match the strength of the absorption features.

The galaxy sample studied by Lequeux *et al.* (1981) encompasses a larger range of metallicities from $Z_\odot/2$ to $Z_\odot/30$ for I Zw 18. A single-burst model is used rather than a continuous star formation model mostly because the metallicity enrichment would exceed the observed one if star formation lasts much longer than 10^7 yr. Since the far UV spectrum of a zero-age cluster is fairly insensitive to the assumed IMF, the slope of the IMF is poorly determined but is consistent with that of the solar neighborhood. There may be, however, some indications that the IMF is flatter ($x \sim 1.5$) at the high mass end. Such a trend is invoked by Viallefond and Thuan (1984) for I Zw 36. That such a flattening is due to the metallicity (Terlevich and Melnick 1983) is not confirmed by a recent study of the cluster CENA 13 in NGC 5128 that has a metallicity higher than solar (Rosa *et al.*, 1984). The age of the burst in the model of Lequeux *et al.* (1981) is constrained by the Lyman continuum flux required to match the observed Hα luminosity and is smaller than 10^6 yr. With the assumption of a single burst, the main information obtained from the slope of the far UV spectrum is that the extinction law varies from object to object and may be connected with its metallicity. It is not clear to us how much this conclusion might change if a model with more continuous star formation were used instead.

The most recent attempt to model the ultraviolet spectra of blue compact galaxies using evolutionary models is by Gondhalekar (1986). These models are also required to match the absorption lines. From comparing BCG spectra with spectra synthetized from stars of solar metallicity, Gondhalekar finds that the metallicities of the stellar atmospheres vary in BCGs from well below solar metallicity to well above. The upper mass limit of the IMF is found to vary from 60 $M_\odot$ to 1 $M_\odot$; it is not clear, however, how well this trend has been disentangled from aging models and extinction effects.

Thuan (1986) has given a thorough review of the technique called 'optimizing synthesis' used to derive the distribution of stars which best fits an observed energy distribution. The output stars are found in different evolutionary stages and are subject to some astrophysical constraints. Unlike an evolutionary model, linear or quadratic programming techniques allow an evaluation of the goodness of the fit and the uncertainties in the derived parameters. Such a technique has been used in the past mostly in the optical and is discussed in the early work of O'Connell (1976). Benvenuti (1983) applied this technique to the UV using only the few stellar spectra then available whereas now one can benefit from the recent IUE atlases. In building up such a synthesis model the completeness of the input stellar library is an even more critical factor than for an evolutionary approach. The problems encountered are that atlases contain only Galactic stars with solar neighborhood abundances and that these stars have to be dereddened.

While the first point has not been overcome, Thuan finds that by using low extinction Galactic stars dereddened with the standard galactic extinction law (Savage and Mathis, 1979), the agreement with theoretical spectra from Kurucz (1979) is satisfactory. Unfortunately, this only applies to spectral types later than O7. For earlier types, the stars generally belong to the Galactic plane and are more reddened. Thuan finds it necessary to use extinction laws different from the standard one.

The critical step in the optimizing synthesis procedure is to select the most prominent features in the continuum and in the lines which discriminate the best between different stellar types and luminosity classes. The main result is that the O I $\lambda 1304$ line index discriminates well between O and B stars, while the Si IV $\lambda 1394$ and C IV $\lambda 1550$ lines separate the main-sequence stars from the supergiants. The continuum slope as measured by the ratio between the fluxes at 1280 and 1710 Å can only distinguish between stars earlier and later than A0, a result also emphasized by Lequeux et al. (1981). All the models were constructed using intrinsic extinction as a free parameter and a Galactic extinction curve was adopted. The best fits reveal that, as suggested by Benvenuti (1983) and Barbieri and Kunth (1980), the extinction in the ultraviolet is much smaller than that deduced from the optical or the radio data (see also Israel and Kennicutt, 1980).

This may be a selection effect caused by the tendency to see those stars which are less obscured. A surprising result is that discontinuities — no stars between O7 to B0 and B2 to B7 — are obtained in the final stellar luminosity function. These are tentatively attributed to the bursting nature of the star formation in BCGs. No supergiants are required by the model fits, constraining the bursts to have ages $\leqslant 10^7$ yr.

Abundance effects can also be checked in principle using emission line observations. Over the $\lambda\lambda 1200-3200$ Å range obtained with the short and long wavelength cameras, few nebular lines are detected, and those most often measured are C IV $\lambda 1550$, [O III] $\lambda 1666$, Si III] $\lambda 1892$, C III] $\lambda 1909$ and a weak C II] line at 2325 Å. In most cases, the BCGs are too weak to exhibit anything but the C III] $\lambda 1909$ line. This line is sensitive to both the metallicity and the stellar effective temperature (requiring $T_{eff} > 40\,000$ K). Three objects have been analysed for their abundance using UV lines; these are NGC 2363 in the irregular galaxy NGC 2366 (Peimbert et al., 1986), Mkn 209 (Viallefond and Thuan, 1984) and Tololo 1924−416 (Bergvall, 1985). The nebular lines can be corrected for reddening with less difficulty than with the UV continuum mainly because the procedure can be checked using theoretical ratios such as [O III] $\lambda 1666$/[O III] $\lambda 5007$. It is also necessary to normalize the UV and visual line intensities to a common aperture. Using the parameters for the electron temperatures and densities deduced from the optical, the total C abundance is derived from models computed by Stasinska (1982) or Mathis (1982, 1985). In all cases the C/O is about half that in the Orion nebula. This difference can be understood by assuming that only young stars in metal poor objects have enriched the interstellar medium with O while at a later stage the intermediate mass stars will produce the rest of the observed C/O value.

5. Constraints on Detection of Primeval Galaxies

The IUE data provide an excellent data base for the empirical determination of UV spectra for use in predicting the appearance of similar galaxies at high redshift. Simply resolving an object as a galaxy at $z \gtrsim 1.5$ requires a high surface brightness in the ultraviolet (Weedman and Huenemoerder, 1985). Some theoretical models predict that primeval galaxies occur in a bright starburst phase and that the ultraviolet continuum and emission lines will be strong (Partridge and Peebles, 1967; Meier, 1976). Such galaxies should be conspicuous in Ly α if the simple recombination theory for the Ly α/Hβ ratio is appropriate. Therefore, the study of Ly α in star forming galaxies has been carried out with IUE to provide clues for identifying primeval galaxies at $z > 3$. A second interest has been to provide examples of the Ly α transfer problem in sites which differ from that of active nuclei.

Using the short wavelength camera and low resolution, 8 blue galaxies with a redshift large enough for the line to be separated from the geocoronal emission (i.e. $z \gtrsim 0.02$) have been observed to date (Meier and Terlevich, 1981; Hartmann et al., 1984; Deharveng et al., 1986, hereafter DKJ). The Ly α/Hβ result is shown in Figure 3 as taken from DKJ and indicates that an average Ly α is weaker than

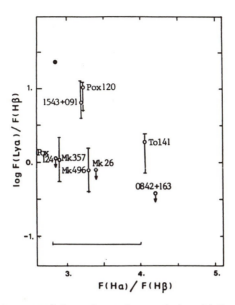

Fig. 3. Comparison of Hα and Hβ fluxes for starburst galaxies with Ly α fluxes determined from IUE data, taken from Deharveng *et al.* (1986). Filled circle: value for case B recombination theory.

expected from the theoretical ratio. The explanation put forward by Hartmann *et al.* (1984) is that relatively small amounts of dust combined with large neutral hydrogen densities can significantly reduce Ly α fluxes. Moreover the underlying stellar and interstellar absorptions are factors that can affect the Ly α line intensity. Figure 4 taken from DKJ shows that the trend reported by Meier and Terlevich on a small sample may still be valid but requires further interpretation. From the detected cases DKJ conclude that the surface density of galaxies which could be detected by their Ly α emission is very low at redshifts larger than 2.

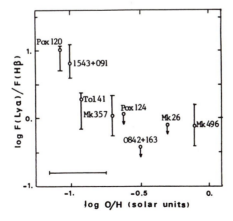

Fig. 4. Comparison of O/H abundance ratio with Ly α/Hβ ratio to test if dust reddening of Ly α correlates with metallicity, taken from Deharveng *et al.* (1986).

This situation arises mainly because fewer and fewer objects are bright enough to be detected against the background sky brightness which, in addition, increases toward the red. This situation is not encouraging for the detection of very distant galaxies by their Ly α line and is not very much improved by assuming evolution both in space density of blue galaxies and line luminosity unless galaxies occur in much brighter phases than the extrapolation from low redshift, gas-rich dwarfs suggests (Djorgovski *et al.*, 1985). Following up these conclusions suggested by observations with the IUE will be a prime objective for the next generation of space observatories, as the consequences are fundamental to our efforts to understand galaxy formation and evolution.

References

Augarde, R. and Lequeux, J.: 1985, *Astron. Astrophys.* **147**, 273.

Barbieri, C. and Kunth, D.: 1980, in M. Tarenghi and K. Kjär (eds.), *Proceedings of the ESO/ESA Workshop on Dwarf Galaxies*, p. 113.

Benvenuti, P.: 1983, in R. West (ed.), 'Highlights of Astronomy', *IAU Symp.* **6**, 631.

Benvenuti, P., Casini, C., and Heidmann, J.: 1979, *Nature* **282**, 272.

Benvenuti, P., Casini, C., and Heidmann, J.: 1982, *Monthly Notices Roy. Astron. Soc.* **198**, 825.

Bergvall, N.: 1985, *Astron. Astrophys.* **146**, 269.

Deharveng, J. M., Joubert, M., and Kunth, D.: 1986, in D. Kunth, T. X. Thuan, and J. T. T. Van (eds.), *First IAP Meeting on Star Forming Dwarf Galaxies and Related Objects*, editions Frontieres.

Djorgovski, S., Spinrad, H., McCarthy, P., and Strauss, M. A.: 1985, *Astrophys. J.* **299**, L1.

D'Odorico, S. and Rosa, M.: 1981, *Astrophys. J.* **248**, 1015.

Gehrz, R. D., Sramek, R. A., and Weedman, D. W.: 1983, *Astrophys. J.* **267**, 551.

Gondhalekar, P. M.: 1986, in D. Kunth, T. X. Thuan, and J. T. T. Van (eds.), *First IAP Meeting on Star Forming Dwarf Galaxies and Related Objects*, editions Frontieres.

Hartmann, L. W., Huchra, J. P., and Geller, M. J.: 1984, *Astrophys. J.* **287**, 487.

Heck, A., Egret, D., Jaschek, M., and Jaschek, C.: 1984, *IUE Low-dispersion Spectra Reference Atlas* − Part 1, Normal Stars. ESA SP-1052.

Huchra, J. P., Geller, M. J., Gallagher, J., Hunter, D., Hartmann, L., Fabbiano, G., and Aaronson, M.: 1983, *Astrophys. J.* **274**, 125.

Huchra, J., Geller, M., Gallagher, J., and Hunter, D.: 1982, in Y. Kondo (ed.), *Advances in Ultraviolet Astronomy: Four Years of IUE Research* (NASA).

Israel, F. P. and Koornneef, J.: 1979, *Astrophys. J.* **230**, 390.

Israel, F. P. and Kennicutt, R. C.: 1980, *Astrophys. J. Letters* **21**, 1.

Koornneef, J.: 1978, *Astron. Astrophys.* **64**, 179.

Kunth, D. and Sargent, W. L. W.: 1981, *Astron. Astrophys.* **101**, L5.

Kurucz, R.: 1979, *Astrophys. J. Suppl.* **40**, 1.

Lamb, S. A., Gallagher, J. S., Hjellming, M. S., and Hunter, D. A.: 1985, *Astrophys. J.* **291**, 63.

Lequeux, J., Maucherat-Joubert, M., Deharveng, J. and Kunth, D.: 1981, *Astron. Astrophys.* **103**, 305.

Lonsdale, C. J., Helou, G., Good, J. C., and Rice, W.: 1985, *Cataloged Galaxies and Quasars Observed in the IRAS Survey*, Pasadena, Jet Propulsion Laboratory.

Mathis, J. S.: 1982, *Astrophys. J.* **261**, 195.

Mathis, J. S.: 1985, *Astrophys. J.* **291**, 247.

Meier, D. L.: 1976, *Astrophys. J.* **207**, 343.

Meier, D. L. and Terlevich, R.: 1981, *Astrophys. J.* **246**, L109.

Nandy, K., Morgan, D. H., Willis, A. J., Wilson, R., Gondhalekar, P. M., and Houziaux, L.: 1980, *Nature* **283**, 725.

O'Connell, R. W.: 1976, *Astrophys. J.* **206**, 370.

Partridge, R. B. and Peebles, P. J.: 1967, *Astrophys. J.* **147**, 868.

Peimbert, M., Pena, M., and Torres-Peimbert, S.: 1986, preprint.

Rosa, M.: 1980, *Astron. Astrophys.* **85**, L21.

Rosa, M., Moellenhoff, C., and D'Odorico, S.: 1984, preprint.

Rosa, M., Joubert, M., and Benvenuti, P.: 1984, *Astron. Astrophys. Suppl.* **57**, 361.

Savage, B. D. and Mathis, J. S.: 1979, *Ann. Rev. Astrophys.* **17**, 73.

Searle, L. and Sargent, W. L. W.: 1972, *Astrophys. J.* **173**, 25.

Searle, L., Sargent, W. L. W., and Bagnuolo, W.: 1973, *Astrophys. J.* **179**, 427.

Seaton, M. J.: 1979, *Monthly Notices Roy. Astron. Soc.* **185**, 57.

Stasinska, G.: 1982, *Astron. Astrophys. Suppl.* **48**, 299.

Terlevich, R. and Melnick, J.: 1983, ESO preprint 264.

Tinsley, B. M.: 1972, *Astron. Astrophys.* **20**, 383.

Thuan, T. X.: 1984, *Astrophys. J.* **281**, 126.

Thuan, T. X.: 1986, in D. Kunth, T. X. Thuan, and J. T. T. Van (eds.), *First IAP Meeting on Star Forming Dwarf Galaxies and Related Objects,* editions Frontieres.

Viallefond, F. and Thuan, T. X.: 1983, *Astrophys. J.* **269**, 444.

Weedman, D. W. and Huenemoerder, D. P.: 1985, *Astrophys. J.* **291**, 72.

Weedman, D. W., Feldman, F. R., Balzano, V. A., Ramsey, L. W., Sramek, R. A., and Wu, C.-C.: 1981, *Astrophys. J.* **248**, 105.

Wu, C. C., Boggess, A., Holm, A., Schiffer, F., and Turnrose, B.: 1983, *IUE Ultraviolet Spectral Atlas,* Greenbelt, GSFC-NASA.

GLOBULAR CLUSTERS

VITTORIO CASTELLANI

Instituto Astronomico, Universita di Roma,
Rome, Italy

and

ANGELO CASSATELLA

Lab. Astrofisica Spaziale, Frascati, Rome, Italy

1. Globular Clusters as Pop. II Systems

A typical globular cluster (GC) is formed by 10^5-10^6 stars, gravitationally bound in a system with an approximately spherical shape.

Globular clusters are well known characteristic members of our galactic halo and, more generally, of the halo of spiral galaxies. Being also present in elliptical and irregular galaxies, they represent a rather diffuse component of the universe.

The GCs of our galaxy and, as far as we know, of spiral galaxies, are representative of Pop. II, i.e. of metal-poor stars born in the protohalo around 1.5×10^{10} yrs ago.

The occurrence of young GCs in the Large Magellanic Clouds (MCs) provides evidence for objects which are simultaneously young ($\sim 10^7-10^8$ yr) and metal-poor, a combination of characteristics not found in the Galaxy. Following recent discussions, we agree in extending the definition of Pop. II to all metal-poor stars, irrespective of their age. In this way the concept of population is directly connected with the 'genetic distance' of matter from the Big-Bang, independently of the time separating the matter from that event.

Within this framework, we will discuss in the following the invaluable contribution given by IUE to the study and knowledge of GCs as Pop. II systems scattered in the universe. A recent review on the general properties of Pop. II objects can be found in Castellani (1985).

2. UV Radiation from Globular Clusters

The vital importance of UV observations for the study of GCs is demonstrated by the HR diagram in Figure 1 from Welch and Code (1980), showing the magnitudes in three wavelength bands around 1550, 2460, and 5460 Å as a function of $B - V$ for the stars in a typical galactic globular cluster. The figure shows that the UV is largely dominated by horizontal branch (HB) stars at 1550 Å, while at 2460 Å a minor contribution to the flux is provided also by Turn-Off (TO) and sub giant stars. On the other hand, the flux in the visual is essentially due to red giants only. It is then clear from Figure 1 that UV observations of GCs are an essential complement to visual and infrared observations.

Y. Kondo (ed.), Scientific Accomplishments of the IUE, pp. 637–654.
© *1987 by D. Reidel Publishing Company.*

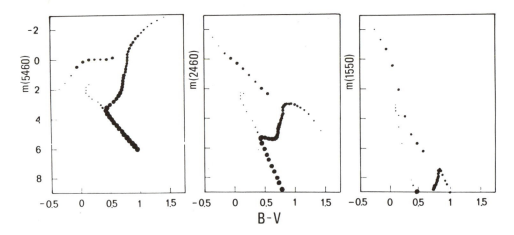

Fig. 1. Colour-magnitude diagrams for a model of the globular cluster M3 seen at different wavelengths: 5460, 2460, and 1550 Å (Welch and Code, 1980). The diagram shows, in particular, that HB stars are the major contributors to the UV, while they have a negligible influence on the visual flux (Welch and Code, 1980). The size of the points is proportional to the logarithm of the number of stars in each location.

Depending on the distribution in temperatures of HB stars, old GCs can appear very bright in the UV. This is shown in Figure 2, where the expected UV energy distribution of a typical galactic GC is reported as a function of the temperature distribution of the HB stars. The figure shows, as is largely confirmed by the

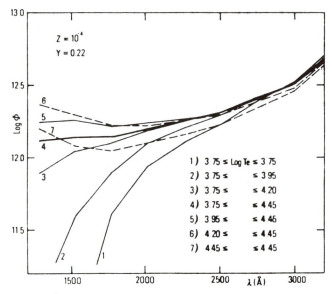

Fig. 2. The figure shows how large the changes of the UV energy distribution can be when the HB is uniformly populated with stars with effective temperatures varying within the limits indicated (Caloi *et al.*, 1983b).

observations, that the presence of a hard UV spectrum does not necessarily imply, by itself, the existence of a young stellar population, as sometimes reported.

Photometric measurements of UV fluxes from galactic GCs were obtained for the first time with the OAO—2 satellite (Welch and Code 1980) in a field of view of 78.5 square arc min. Further photometric investigations were performed by the ANS satellite, in five broad bands between 1550 and 3300 Å, and with a field of view of 2.5 square arc min. Considering the observed diameters of GCs, these observations are representative, in the large majority of cases, of the integrated properties of the clusters.

Based on observations with ANS, van Albada *et al.* (1979, 1981) showed that galactic GCs can be classified into four classes: extremely blue (EB), blue (B), intermediate (I) and Red (R), following the slope of the UV energy distribution. This classification is essentially related to the temperature distribution of HB stars.

IUE, due to its high sensitivity, has allowed us for the first time to attain not only several globular clusters, but also several of their member stars. Unfortunately, the IUE slit-integrated observations of GCs may be affected by sampling problems due to the small size of the 'large' $10'' \times 20''$ entrance aperture, problems which can be particularily severe especially for some galactic clusters. However, further important information, not obtainable before, was gained on the spatial structure of the sources, by using the spatial resolution capabilities of the instrument perpendicular to the direction of the dispersion (see e.g. Cassatella *et al.*, 1985). Using this capability it has been possible, in some cases, to detect the presence of bright field stars eventually contaminating the OAO—2 and ANS measurements, or to distinguish and to isolate the contribution of different sources falling in the aperture (Dupree *et al.*, 1979; Altamore *et al.*, 1981, 1983; Caloi *et al.*, 1983a).

Due to the characteristics of the sources, GCs are in general observed at low dispersion. However, a few observations at high resolution are also available.

The Logs of the IUE observations of the galactic clusters (complete to 1983) and of the LMC clusters are reported by Caloi *et al.* (1983a) and Cassatella *et al.* (1986), respectively.

3. Observations of Single Stars in Galactic Globular Clusters

IUE observations of single stars in galactic GCs present some difficulty due to crowding problems or to their faintness. The observations are usually made to obtain information on the spectral distribution and on the temperature of the hot stars. So far, the IUE observations of individual members of GCs were concentrated on hot HB stars, and on the objects known in the literature as UV-bright stars, which appear in some clusters (see Zinn *et al.*, 1972; and Harris *et al.*, 1983, for a more recent compilation) at about three magnitudes above the HB. We know from theory that the latter objects are stars which left the top of the giant branch, and are crossing the HR diagram at $\log L/L_{\odot} \sim 3$, before reaching the pre-white dwarf and, finally, the WD stage.

IUE data of ten UV-bright stars have been carefully discussed in a recent paper by de Boer (1985). One additional object, ROA 5701 in ω Cen, has been

observed by Cacciari *et al.* (1984). Figure 3 shows, for example, the spectrum of
the star ROB 162 in NGC 6397. These observations have nicely confirmed, on
the basis of the derived effective temperatures and luminosities, the nature of these
objects, and have indicated that UV-bright stars can play an important role, for
example, in ionizing the halo gas and in contributing to the UV emission from
elliptical galaxies (de Boer 1985).

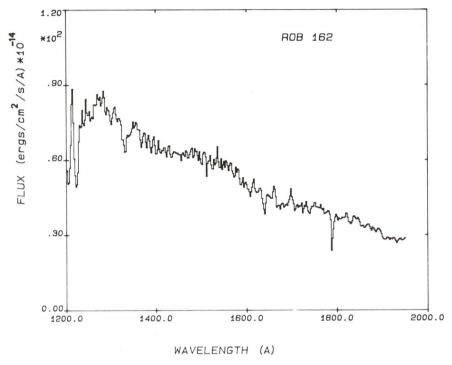

Fig. 3. IUE energy distribution of the UV-bright star ROB 162 in NGC 6397 (Caloi *et al.*, 1982).

The study of the stellar and interstellar features in the UV-bright stars has
provided important information on both the properties of the stellar atmospheres
and of the interstellar gas. For example, in the case of the planetary nebula
K 648 in M15, the strong C IV absorption indicated an extremely high carbon
abundance, $[C/H] = 0.05$, in agreement with theoretical predictions about carbon
dredge-up as produced by thermal pulses in the AGB progenitor (de Boer 1985).

A particularly delicate problem is the determination of the interstellar reddening towards globular clusters (e.g. de Boer, 1985). Any inaccuracy in the determination of the extinction law and on $E(B-V)$ have far reaching consequences
on the present knowledge of the evolutionary stage of the cluster members and
on important parameters like the ages of GCs. For example, we recall that the
puzzling non-canonical behaviour of the luminosity of RR Lyrae stars (the so-
called 'Sandage effect'; Sandage 1981, 1982) is entirely based on the adopted
values of $E(B-V)$. The reddening is of course best determined from the UV

when observing individual hot members of the GCs. Indeed, there is some evidence that discrepancies exist between colour excesses derived from the optical photometry and those from the ultraviolet. In the case of M15, for example, the IUE data are consistent with a colour excess which is considerably lower than previously evaluated from optical photometry (Cacciari *et al.*, 1984), a result which deserves further investigation.

The other class of individual objects which need UV observations are the hot HB stars. A reliable determination of the effective temperature and luminosity for these objects, impossible without UV data, is again of crucial importance. The location of hot HB stars in the theoretical HR diagram is governed by severe theoretical constraints and it is related to outstanding questions like the original abundance of helium, or the role played by stellar rotation in the evolution of GCs stars.

IUE observations of hot HB stars require a considerable amount of observing time, since data of reasonable quality can only be obtained with typically one shift per camera. We regret that, until now, a limited number of such stars have been observed.

The investigation on HB stars in GCs started when de Boer and Code (1981) first detected two hot stars near the center of M13. Additional objects were later detected in NGC 6752 and in NGC 6397 (see Caloi and Castellani 1983 for a detailed list). Figure 4 shows the spectra of three HB stars in the EB-type cluster NGC 6752, a cluster known for having a very hot tail of HB stars . The figure indicates that good quality UV data could be obtained in spite of the faintness of the objects, ranging from $V = 15.8$ to 17.5.

So far, eight HB stars with temperatures ranging from 10 000 to 20 000 K, and a O-subdwarf with $T_{eff} \sim 40\,000$ K have been observed in NGC 6752 (Heber *et al.*, 1986). As a result of these investigations, it has been possible to establish that the hottest HB stars in NGC 6752 represent the natural counterpart of field sdB and sdOB stars, so that the problem of a possible difference between the loose halo and the cluster halo seems to be vanishing. At the same time, the reliable effective temperatures obtained by comparing IUE data with atmospheric models provided rather accurate values for the surface gravity and the chemical composition for the above members of NGC 6752. Consequently, it has been possible to show that the helium abundance is anticorrelated with gravity, the former varying between 0.2% and 3% (Heber *et al.*, 1986). This result should be regarded as the first observational evidence for the efficiency of gravitational sedimentation, a process which inhibits a direct observation of the original helium abundance.

Extension of these investigations to a sufficient number of GCs would be highly desirable, as it will enable one to discriminate among some open working hypotheses on the evolutionary scenario for these objects, as well as to clarify some debated evidence from optical observations like, e.g., the suspected overluminosity of the HB stars in M5 (Buonanno *et al.*, 1981).

4. Integrated Properties of Galactic Globular Clusters

As mentioned in Section 2, due to the size of the large entrance aperture of the IUE spectrographs, only clusters sufficiently concentrated or distant provide,

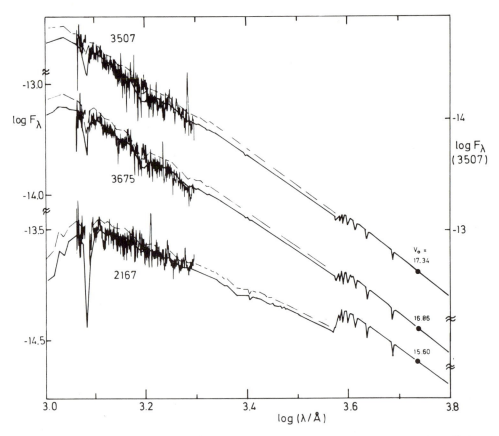

Fig. 4. Energy distribution of three HB stars in NGC 6752 compared to model predictions.
Top: star No. 3507; the models have $T_{eff} = 27\,500$ K (full drawn) and $T_{eff} = 30\,000$ K (dashed).
Middle: star No. 3675; the models have $T_{eff} = 25\,000$ K (full drawn) and $T_{eff} = 27\,500$ K (dashed).
Bottom: star No. 2167; the models have $T_{eff} = 15\,000$ K (full drawn) and $T_{eff} = 16\,000$ K (dashed).
The data are from Heber *et al.* (1986).

within the aperture, a sufficient number of stars which can be considered as
representative of the whole cluster population. Taking such a limitation into
account, one can say that nearly all the GCs suitable for IUE investigation have
been now observed in the Galaxy. The first study of the IUE slit-integrated energy
distribution of galactic GCs is due to Dupree *et al.* (1979). Figure 5 displays, for
example, the spectrum of the central region of the cluster NGC 6715 = M54. As
in all the other clusters studied, the strongest absorption feature in M54 is Mg II
2800 Å, mostly intrinsic in the cluster, as demonstrated by its absence in the
spectra of its UV-bright members.

Figure 6 displays the IUE energy distribution of four typical galactic GCs
belonging to different classes from R (NGC 104 = 47 Tuc) to EB (NGC 6093 =
M92). The progressive flux increase towards shorter wavelengths, from the top to
the bottom of the figure, is essentially due to the increasing contribution by hot

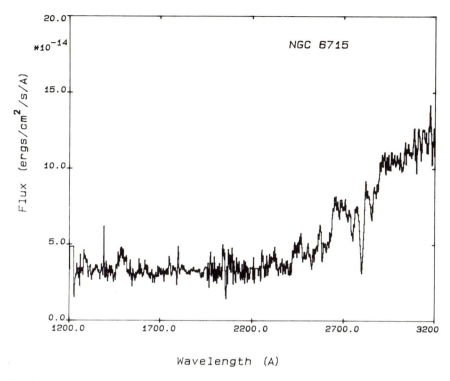

Fig. 5.   IUE slit-integrated spectrum of the nucleus of NGC 6715 = M54 (Caloi *et al.*, 1984).

HB stars. Note that even small variations of the HB population produce large effects on the UV energy distribution: M92, which has a smaller number of RR Lyrae stars than NGC 7078 (M15), has a harder spectrum than M15, in agreement with the idea that the whole HB is shifted towards higher temperatures. Figure 6 shows also how hazardous can be the belief in a 'typical UV emission from Pop. II objects'. Indeed, Pop. II systems can contribute very differently at the shortest wavelengths, the maximum emission corresponding, rather curiously, to a subset of intermediate metallicity (EB) clusters. This lack of correlation with metallicity is known as the 'second parameter problem'.

In this context IUE has provided new important insights on the problem of EB clusters. The galactic globular cluster NGC 6626 (= M62), known as an RR Lyrae rich, Oosterhoff I type globular cluster, was classified as EB according to the ANS photometry, whereas EB clusters are known for having only blue HB stars, without or with very few RR Lyrae variables. For this reason ANS measurements were suspected to be biased by hot field stars in the foreground. However, a study of the spatial distribution of the source using IUE data has shown no evidence for contamination by foreground sources (Caloi *et al.*, 1985). Consequently, it has been suggested that M62, in spite of its strong population of RR Lyrae stars, contains very hot HB stars, being then the only GC known with this peculiarity. Moreover, it has been found that the flux of M62 shortward of 2500 Å (after

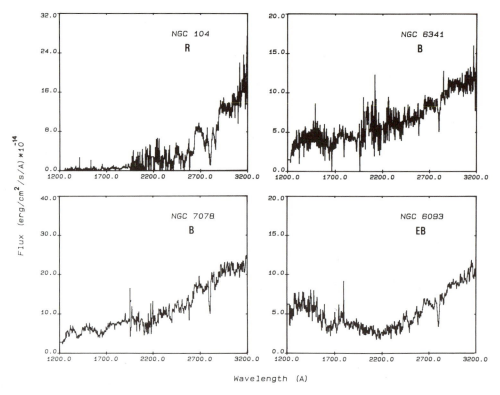

Fig. 6.   UV energy distribution of five galactic globular clusters with different ages and metallicities. An evolutionary spectral synthesis for these clusters has been presented by Nesci (1981).

correction for the reddening as given by van Albada, de Boer and Dickens, 1981) is the highest so far observed by IUE in any galactic GC. Recent optical observations at the 3.6 m ESO telescope at La Silla fully support the above scenario, confirming that M62 has the hottest HB of any galactic cluster, in spite of being the most metal rich cluster of the EB class (Caloi *et al.*, 1986). Hopefully, a breakthrough in the understanding of these facts will come about in the near future. Figure 7 shows the comparison between the UV spectrum of M62 with different assumed reddening values, and the spectrum of the nucleus of the galaxies M31 and NGC 4649.

Finally, one has to stress that in some cases there are some significant discrepancies between the shape of the IUE spectra in Figure 6 and their classification via ANS or OAO—2 data. However, one has to take into account that the three above sets of data correspond to very different samplings on the clusters and, in particular, that IUE is generally looking at the very centre of the clusters. At least part of the above discrepancies might be related to the presence of different populations across the clusters, although the problem is still under debate.

The sample of data in Figure 6 represent a milestone for detecting and studying the contribution of old Pop. II stars to UV fluxes in the galactic nucleus, in

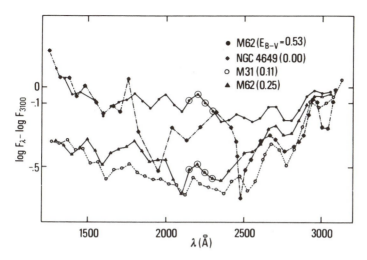

Fig. 7. The IUE spectrum of the central region of the galactic cluster M62 (dots: $E(B-V) = 0.53$; triangles $E(B-V) = 0.25$) is compared with IUE spectra of the nucleus of M31 (open circles, $E(B-V) = 0.11$) and from the nucleus of NGC 4649 (asterisks, no reddening correction). Circles indicate data for M62 affected by spikes. The data are from Caloi *et al.* (1985).

elliptical galaxies and in the halo of spiral galaxies. Furthermore, they can be used, at least in principle, to derive evolutionary constraints for these systems. For example, it has been earlier suggested (van Albada *et al.* (1981) that varying the cluster age should affect the UV spectra since the luminosity of TO stars, which contribute to the spectrum mainly longward of 1800 Å, has changed. Unfortunately, the UV energy distribution depends dramatically on the distribution in mass (and thus in colours) of HB stars which, from the theoretical point of view, is still a free parameter. In fact, although a good fit to the data in Figure 6 could be obtained by Nesci (1981) using the common assumption $Y = 0.25$, $t = 1.3 \times 10^{10}$ yr, the fit itself turns out to be strongly dependent on the temperature distribution of the HB stars, masking any possible effect of age. An interesting approach to the problem has been followed by Nesci (1983a, b): relying on a statistically significant HR diagram of a given GC, the number distribution of HB stars as a function of T can be derived together with their bolometric luminosities. The actual contribution of HB stars can then be computed and compared with IUE data to derive the age of the clusters. However, the method is possibly affected by significant uncertainties in the theoretical chain colour-temperature-UV spectra. Again, we need a more complete sample of spectra of hot HB stars in various clusters to set firm observational constraints on such a chain. Similar considerations hold for evolutionary parameters other than the age of the clusters, like stellar rotation and original helium content.

5. Globular Clusters in Other Galaxies

As quoted before, GCs are a very common component of the Universe. Outside our Galaxy, one finds GCs in the Magellanic Clouds and in the nearby dwarf spheroidal in Fornax and, further away, in Andromeda (M31) and in its small companions like M33 and NGC 185. Clusters in the halo of M31 do not seem to differ much, as far as their integrated properties are concerned, from the family of galactic globular clusters, though a trend for redder colours has been suggested. This occurrence is probably connected with the larger metallicity of M31.

However, striking differences exist between the halo clusters, both in Andromeda and in the Galaxy, and the GCs of the Magellanic Clouds. As a consequence of their age, the visual light from clusters in the halo of the spiral galaxies is dominated by red giants, so that they have integrated red $(B - V)$ colours. In contrast, in addition to the red clusters, also a considerable number of blue clusters are found in the MCs, as demonstrated by the bimodal distribution in the $B - V$ colour-frequency histogram in Figure 8 (van den Berg, 1981). The blue clusters of the MCs, as confirmed by the available HR diagrams, have their main sequences populated up to very early spectral types, indicating that their ages cannot, in some cases, exceed some million years.

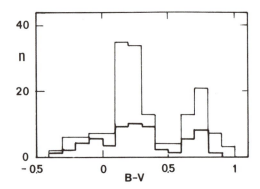

Fig. 8. Frequency distribution of colours in the LMC (upper histogram) and SMC (lower histogram), from van den Berg (1981).

We are then facing the intriguing problem of why GCs have stopped forming very early after the two above spiral galaxies have born, while the MCs are retaining the capability of forming large aggregates of stars. The fact that young globular clusters are present also in other irregular galaxies like NGC 55 (da Costa and Grahm, 1982) and M33 (Christian and Schommer, 1982) suggests in itself that the history of star formation is strongly linked to the dynamics of the parent galaxy. A likely scenario suggests that, while GCs could not form anymore in early gas-depleted haloes, they were prevented in forming in the disk at later stages due to the large effects of differential rotation (Iben and Renzini, 1983).

With IUE it has been possible to obtain for the first time UV spectra of GCs in external galaxies. So far, about 30 GCs have been observed in the LMC, about ten in the SMC, one in Fornax and one in M31.

Such observations are very important both for the old and the young GCs. As far as the old GCs are concerned, their study in the Galaxy and in the MCs is important to assess fundamental questions like whether there is a common scenario for their formation and development or whether the 'second parameter problem' is everywhere connected with the evolution of the population of GCs. We expect that these differences will give insight in the scenario of galactic evolution.

As for young globular clusters, simply recalling that most of the flux is emitted in the UV range, illustrates clearly the impact of IUE. The investigations are mainly directed to relate the UV integrated properties with the ages of the clusters and other structural parameters as the Initial Mass Function or the metallicity. These studies will not only contribute to clarify the evolutionary stage of the MCs, but will also give insight on the integrated properties of young Pop. II systems.

While the next section will be devoted to the rather advanced investigations on the MCs, we quote here that IUE observations of a GC in Fornax show important analogies with the galactic cluster M15 (Cacciari et al., 1986), a result further confirmed by a visual HR diagram (Buonanno et al., 1983).

An interesting problem has been raised by IUE observations of the X-ray cluster Bo 158 in M31 (Cacciari et al., 1982). It has been suggested that this well concentrated cluster has an UV excess typical of EB-galactic clusters, in spite of being moderately metal rich. If so, this would suggest that the second parameter problem exists also for the halo in M31, as in the Galaxy: this result clearly deserves further investigation.

6. Globular Clusters in the Magellanic Clouds

The large efficiency of the MCs in producing stellar clusters becomes evident when considering that about 1600 clusters are known in the LMC and about 120 in the SMC (Gascoigne, 1971), and that Hodge and Sexton (1966) estimated the total number of clusters in the LMC as great as 6000. Van den Bergh (1981) reported UBV photometry for 147 clusters in the LMC and 61 in the SMC, several of them globular.

Figure 9 shows the UV energy distribution of two young and one old globular clusters of the LMC. The UV spectra of the young clusters yield considerably better information on their ages than visual colours do. In fact, for ages of, say, some million years, only main sequence stars are present, and the most luminous stars are OB stars at the top of the main sequence (MS). As the age of the cluster increases, the luminosity of the stars at the top of the MS regularly decreases, but luminous giant stars begin to appear, as expected evidence for advanced phases of central and shell helium burning. The visual contribution by these giants overlaps the light from MS stars, in a way not easy to foresee since these advanced evolutionary stages of intermediate mass stars are still poorly known. However, such evolved giants do not contribute significantly to the far UV, so that at these short wavelengths most of the flux is due to MS stars. Consequently, the UV energy distribution of young clusters is expected to changed smoothly as a function of age.

Such a fortunate situation breaks down only for very old clusters, in which HB stars appear, and become the dominant sources in the UV. However, since there are no theoretical prescriptions about the amount of mass loss governing the

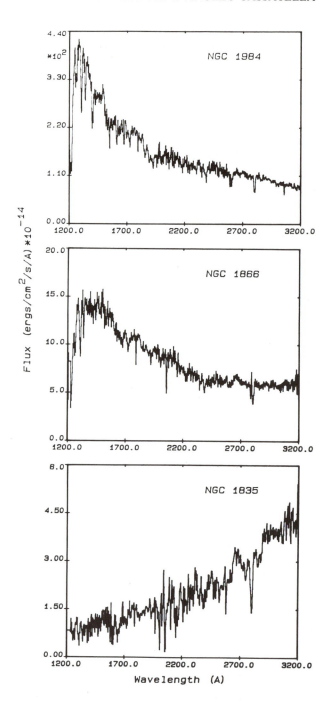

Fig. 9.   IUE observed energy distribution of three LMC globular clusters of different ages: 7×10^6, 8×10^7, and 10^{10} yr for NGC 1984, 1866 and 1835, respectively (the ages are from Hodge 1983 and references therein).

temperature of HB stars, we cannot predict, at present, the UV luminosity of old, HB-rich, globular clusters.

The first IUE observations of GCs in the MCs were reported by de Boer (1982), Cacciari *et al.* (1982) and Geyer and Cassatella (1983).

The importance of IUE spectra in the problem of the age (and metallicity) of LMC clusters has been stressed by Cohen, Rich and Persson (1984) who used IUE UV colours for ranking the LMC GCs in age, interpreting a previous classification based on visual colours in terms of a correlated variation of the ages and metallicities of the clusters. More recently, Barbaro and Olivi (1986) showed that current theoretical prescriptions can account fairly well for the UV energy distribution of clusters younger than about one billion years. These investigations provide a very promising opportunity for decoding, through GCs, the history of star formation in the Magellanic Clouds.

Van Albada *et al.* (1981) used a two-colour diagram, C(15—33) vs C(18—25), to study the properties of the galactic GCs from ANS data. As expected, galactic GCs arrange themselves along a sequence going from R to EB-type, through types I and B. It was suggested at that time that a separation existed in the quoted diagram between the loci of B and EB clusters. If so, it would have implied a discontinuity in some (unknown) evolutionary parameters between the above two classes. A deeper reexamination of this problem seems to rule out such a possibility, as demonstrated in Figure 10, in which the gap between the two above classes has practically disappeared (de Boer 1985).

It is of particular interest to compare the location of the galactic GCs with that of the LMC clusters in the UV two-colour diagram. This comparison is done in Figure 11 from Cassatella *et al.* (1986), which includes all LMC clusters so far observed with IUE for which sufficiently good quality spectra were obtained. A large number of young blue clusters is occupying the top-left corner of the figure, a location which is, of course, absolutely empty of galactic GCs. The sudden disappearence of clusters in the proximity of $C(18—25)_0 \gtrsim -0.6$ deserves particular attention, as it might support the results of Frogel and Blanco (1983) who found evidence for two discrete epochs of star formation in the LMC, the latter having started about 10^8 yr ago. Note in this respect that NGC 1866, whose age has been estimated 8×10^7 yr (Hodge, 1983), has C(18—25) ~ −0.6. However, other possibilities exist to explain the above discontinuity (see Renzini and Buzzoni, 1986).

As for the old clusters in Figure 11, only a few were observed in the LMC, due to the long exposure times needed. The figure shows that three of them overlap the location characterizing the galactic B clusters (like M15 or NGC 6397), one object (NGC 1987) appears in an intermediate position, and none shows EB characteristics. In the Galaxy, the B and EB clusters have roughly the same frequency. It would be important to complete the survey of the old GCs in the LMC to look for the counterparts of the EB galactic clusters. This would in fact provide information on the role of the "second parameter" in the LMC globular clusters.

Interesting results have been reported by Böhm-Vitense, Hodge and Proffit (1985) for the populous LMC cluster NGC 2100. These authors could obtain IUE

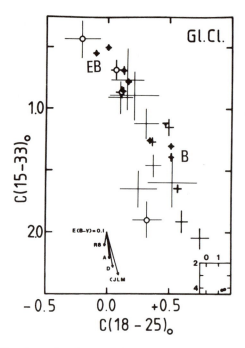

Fig. 10. Dereddened UV colours for the total light of galactic GCs (de Boer, 1985). The GCs exhibit a smooth distribution of their UV colours over the range 1500–3300 Å. This is in contrast with the results of Van Albada *et al.* (1981) who found a gap in a similar diagram, and suggested the existence of two distinct classes of GCs.

spectra of individual hot members of the cluster and compared them with spectra of galactic stars with similar effective temperature and luminosity to look for possible differences in mass loss. Interesting enough, they found no evidence for mass loss in the cluster members, so supporting earlier results by other authors (see e.g. Hutchings, 1982) that mass loss is comparatively smaller in the LMC. Through these observations it was also possible to find significant differences between the extinction law towards the cluster's center and towards its members. It would be important to extend such studies to other young clusters of the MCs.

Before concluding this section, we recall that, due to their brightness in the UV, it has been possible to obtain good IUE high resolution spectra of a few clusters in the LMC and SMC. Figure 12 shows, for example, a portion of the high resolution spectrum of the young LMC cluster NGC 2004. Such data will be important not only for a more detailed knowledge of the chemical composition and population of the clusters, but also for the study of the interstellar medium towards the LMC.

The above studies for the LMC are waiting to be extended to the clusters of the SMC. Data have been already collected for about ten clusters. Figure 13 reports, for example, the spectrum of SL 56. When the sample will be increased, it will be possible approach the problem of the evolution of GCs in the SMC, in comparison with similar results for the LMC and the galactic halo.

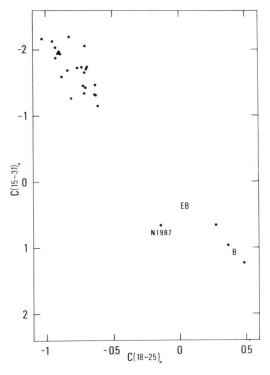

Fig. 11. UV two colour diagram for 28 globular clusters of the LMC (Cassatella *et al.*, 1986). The colours are corrected for galactic foreground and LMC reddening. The location of the extreme blue (*EB*) and blue (*B*) galactic cluster is indicated. The upper left part of the diagram is populated by young blue clusters, not found in the Galaxy.

7. Future Investigations

IUE has enabled us to obtain the first UV spectra not only of the GCs in the Galaxy, but also in the Magellanic Clouds, in M31 and in Fornax. In addition, individual member stars were observed both in the galactic clusters (HB and UV-bright stars) and in the LMC (generally hot stars slightly evolved from the upper main sequence). These observations have to be extended further with IUE, in order to obtain a more complete set of data to be used for systematic comparisons in the different galaxies.

Of course, IUE has also created the need for more data, especially of the fainter objects. Therefore, looking one step further in the future, we want to single out a few investigations which now appear necessary to get a deeper knowledge of Pop. II systems.

As for the integrated properties of GCs, we need to push the investigation at least as far as the globular cluster systems in M31, to have another sample of the behaviour of GCs in the halo of spiral galaxies. This research is open, in principle, to the capabilities of the Space Telescope, but unfortunately it is forbidden by the narrowness of the entrance slit in its instrumentation.

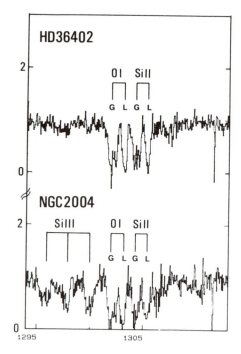

Fig. 12. A IUE high resolution spectrum of the young LMC globular cluster NGC 2004 (Geyer *et al.*, 1986) is compared with a spectrum of the LMC star HD36402. Both objects show the galactic and LMC interstellar absorption of O I 1302.17 and Si II 1304.37 Å. The interstellar lines in HD36402 were studied by Savage and de Boer (1981). NGC 2004 shows, in addition, intrinsic lines from Si III red shifted by about 300 km s^{-1}, in agreement with the cluster's radial velocity (Freeman *et al.* (1983). The Si III lines confirm spectroscopically the presence of a hot star population.

Moreover, we are waiting for an extensive exploration of the UV characteristics of hot Pop. II stars in GCs, which we know to contain important information on the history of the clusters and of their parent galaxies. In addition, we need UV spectral data in order to establish for Pop. II stars the reference spectral framework which has already been developed for Pop. I objects. A pioneering study in this direction is that of Cacciari (1985), who reported UV fluxes for 36 metal-poor field halo stars. These reference data need to be extended further as they are the basis for the interpretation of the UV properties of objects like, e.g., galactic nuclei and elliptical galaxies.

Beside spectrophotometry, also multicolour photometry in the UV domain is a valuable tool for the study of GCs: we have shown, in fact, that the IUE capabilities have often been fruitfully 'under-used' in order to simulate broad band photometry. In particular, a space experiment with an imaging capability in well-selected UV bands will speed up enormously the above investigations, allowing one to extend the knowledge of Pop. II systems far away from the Galaxy and the local group of galaxies; at least, we hope, up to the exceptional sample of globular clusters (and related galaxies) in the Virgo Cluster and in other nearby clusters of galaxies.

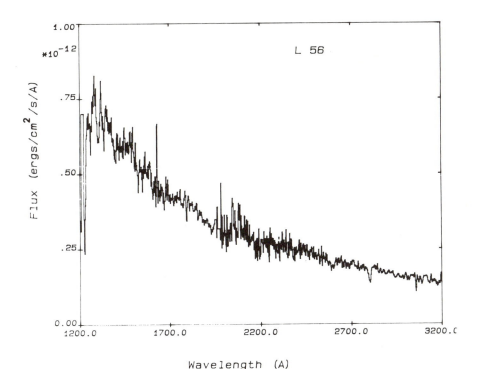

Fig. 13. Observed UV energy distribution of the young SMC cluster SL 56 (Geyer and Cassatella, 1986).

References

Altamore, A., Angeletti, L., Capuzzo Dolcetta, R., and Giannone, P.: 1981, *Astron. Astrophys.* **103**, 424.

Altamore, A., Angeletti, L., Capuzzo Dolcetta, R., and Giannone, P.: 1983, *Astron. Astrophys.* **118**, 332.

Barbaro, G. and Olivi, F. M.: 1986, 'Spectrophotometric Models of Galaxies', in C. Chiosi and A. Renzini (eds.), *Spectral Evolution of Galaxies*, D. Reidel Publ. Co., Dordrecht, Holland, pp. 283—307.

Böhm-Vitense, E., Hodge, P., Proffit, C.: 1985, *Astrophys. J.* **292**, 130.

Buonanno, R., Castellani, V., Corsi, C. E., and Fusi-Pecci, F.: 1981, *Astron. Astrophys.* **101**, 1.

Buonanno, R., Buscema, G., Corsi, C. E., Fusi-Pecci, F., and Iannicola, G.: 1983, *Astron. Astrophys. Suppl. Ser.* **51**, 83.

Cacciari, C.: 1985, *Astron. Astrophys. Suppl. Ser.* **61**, 407.

Cacciari, C., Cassatella, A., Bianchi, L., Fusi-Pecci, F., and Kron, R. G.: 1982, *Astrophys. J.* **261**, 77.

Cacciari, C. *et al.*: 1982, *Proc. Third Europ. IUE Conf.*, ESA SP-176, p. 519.

Cacciari, C., Caloi, V., Castellani, V., and Fusi-Pecci, F.: 1984, *Astron. Astrophys.* **139**, 285.

Cacciari, C., Clementini, G., Fusi Pecci, F., and Zinn, R. J.: 1986, (in prep.).

Caloi, V. and Castellani, V.: 1983, *Mem. Soc. Astron. It.* **54**, 419.

Caloi, V., Castellani, V., Panagia, N.: 1982, *Astron. Astrophys.* **107**, 145.

Caloi, V., Castellani, V., and Galluccio, V.: 1983a, *Mem. Soc. Astron. It.* **54**, 789.

Caloi, C., Castellani, V., Nesci, R., and Rossi, L.: 1983b, *Mem. Soc. Astron. It.* **54**, 763.

Caloi, V., Castellani, V., Galluccio, D., and Wamsteker, W.: 1984, *Astron. Astrophys.* **138**, 485.

Caloi, V., Castellani, V., and Tarenghi, M.: 1985, *Astron. Astrophys.* **145**, 286.

Caloi, V., Castellani, V., and Piccolo, F.: 1986, *Astron. Astrophys. Suppl.* **67**, 181.

Cassatella, A., Barbero, J., and Benvenuti, P.: 1985, *Astron. Astrophys.* **144**, 335.

Cassatella, A., Barbero, J., Geyer, E.: 1986, submitted to *Astrophys. J. Suppl. Ser.*

Cassatella, A. and Geyer, E.: 1986, (in prep.).

Castellani, V.: 1985, *Fund. of Cosmic Phys.* p. 317.

Christian, C. A. and Schommer, R. A.: 1982, *Astrophys. J. Suppl. Ser.* **49**, 405.

Cohen, J. G., Rich, R. M. and Persson, S. E.: 1984, *Astrophys. J.* **285**, 595.

da Costa, G. S. and Grahm, J. A.: 1982, *Astrophys. J.* **261**, 70.

de Boer, K. S.: 1982, in A. G. D. Philip and D. S. Hayes (eds.), 'Astrophysical Parameters for Globular Clusters', *IAU Coll.* **68**, 4.

de Boer, K. S.: 1985, *Astron. Astrophys.* **142**, 321.

de Boer, K. S. and Code, A. D.: 1981, *Astrophys. J.* **243**, L33.

Dupree, A. K., Hartmann, L., Black, J. H., Davis, R. J., Matilsky, T. A., Raymond, J. C., and Gursky, H.: 1979, *Astrophys. J.* **23**, L89.

Freeman, K. C., Iliiworth, G., and Oemler, Jr., A.: 1983, *Astrophys. J.* **272**, 488.

Frogel, J. A. and Blanco, V. M.: 1983, *Astrophys. J.* **274**, L57.

Gascoigne, S. C. B.: 1971, 'Cluster in the Magellanic Clouds', in A. B. Muller (ed.), *The Magellanic Clouds*, D. Reidel Publ. Co., Dordrecht, Holland, pp. 25—30.

Geyer E. H., Cassatella, A.: 1983, in S. van den Berg and K. S. de Boer (eds.), 'Structure and Evolution of the Magellanic Clouds', *IAU Symp.* **108**, 55.

Geyer, E. H., Cassatella, A., and Pettini, M.: 1986, (in prep.).

Harris, H. C., Nemec, J. M., and Hesser, J. E.: 1983, *Publ. Astron. Soc. Pacific* **95**, 25.

Heber, U., Kudritzki, R. P., Caloi, V., Castellani, V., Danziger, J., and Gilmozzi, R.: 1986, *Astron. Astrophys.* **162**, 171.

Hodge, P. W.: 1983, *Astrophys. J.* **264**, 470.

Hodge, P. W. and Sexton, J. A.: 1966, *Astrophys. J.* **71**, 363.

Hutchings, J. B.: 1982, *Astrophys. J.* **255**, 70.

Iben, I. Jr. and Renzini, A.: 1983, *Ann. Rev. Astron. Astrophys.* **21**, 271.

Nesci, R.: 1981, *Astron. Astrophys.* **99**, 120.

Nesci, R.: 1983a, *Astron. Astrophys.* **121**, 226.

Nesci, R.: 1983b, *Astron. Astrophys.* **121**, 325.

Renzini, A., Buzzoni, A.: 1986, 'Global Properties of Stellar Populations and the Spectral Evolution of Galaxies', in C. Chiosi and A. Renzini (eds.), *Spectral Evolution of Galaxies*, D. Reidel Publ. Co., Dordrecht, Holland, pp. 195—235.

Sandage, A.: 1981, *Astrophys. J.* **248**, 161.

Sandage, A.: 1982, *Astrophys. J.* **252**, 553.

Savage, B. D. and de Boer, K. S.: 1981, *Astrophys. J.* **243**, 460.

van Albada, T. S., de Boer, K. S., Dickens, R. J.: 1979, *Astron. Astrophys.* **75**, L11.

van Albada, T. S., de Boer, K. S., and Dickens, R. J.: 1981, *Monthly Notices Roy. Astron. Soc.* **195**, 591.

van Albada, T. S., Dickens, R. J., and Wevers, B. M. H. R.: 1981, *Monthly Notices Roy. Astron. Soc.* **196**, 833.

van den Bergh, S.: 1981, *Astron. Astrophys. Supl. Ser.* **46**, 79.

Welch, G. A. and Code, A. D.: 1980, *Astrophys. J.* **236**, 798.

Zinn, R. J., Newell, E. B., and Gibson, J. B.: 1972, *Astron. Astrophys.* **18**, 390.

ACTIVE GALACTIC NUCLEI

MATTHEW A. MALKAN

University of California, Los Angeles, CA, U.S.A.

DANIELLE ALLOIN

Observatoire de Paris-Meudon, France

and

STEVEN SHORE

New Mexico Institute of Technology, Socorro, NM, U.S.A.

1. Introduction

Well before the discovery of quasars, a special class of galaxies was known to harbor unusually powerful nuclei (Seyfert, 1943). Two primary defining characteristics of these active galactic nuclei (AGNs) are an intense point-like nuclear continuum source and strong, broad emission lines, neither of which could arise from ordinary stellar processes. The very broad emission lines, seen only in permitted transitions, have full-widths at half-maximum of a few thousand to more than ten thousand km s^{-1}. They are produced in the broad-line region (BLR), which is 0.1 pc or less in diameter, and contains many rapidly moving clouds with high electron densities ($\sim 10^{10}$ cm^{-3}). Forbidden lines with typical Doppler velocity widths of 500—1000 km s^{-1} are emitted from a surrounding ~ 100 pc narrow-line region (NLR), in which the characteristic electron densities range from 10^3 to 10^7 cm^{-3}.

The more recently discovered quasars, which show virtually the same emission line and continuum spectra as Seyfert 1 nuclei, are now known to be the high end of the AGN luminosity function. At the lowest nuclear luminosities, Seyfert 2 and low-ionization emission-line ('liner') galaxies lack a readily observable BLR, but have an NLR which appears to be photoionized by a nonstellar continuum. Evidently there is a continuous range of AGN luminosities spanning at least eight orders of magnitude.

Before IUE very little was known about any active galaxies at wavelengths below 3300 Å. Their ultraviolet spectra could only be inferred from optical observations of extremely luminous quasars with redshifts of 2 or more. IUE made it possible to investigate AGN properties over an enormous range of luminosities. The rest wavelength region between 1200 and 3200 Å is especially critical because it includes a large fraction of the total luminosity of many AGNs. In addition, at optical wavelengths the only strong permitted lines available are the hydrogen Balmer series. The ultraviolet contains most of the strong resonance lines from several abundant ions (H I, C IV, N V, Si IV, Mg II) which are the major coolants of the BLR. The same lines are also seen in absorption in some AGNs.

Y. Kondo (ed.), Scientific Accomplishments of the IUE, pp. 655—669.
© 1987 *by D. Reidel Publishing Company.*

2. The Ultraviolet Continua of AGNS

Spectrophotometric ground-based optical studies of the nearest, brightest Seyfert 1 galaxies indicated that their point-like nuclei were dominated by a nonstellar continuum source which resembled a power law in flux, with a slope near -1.1 (Malkan and Filippenko, 1983). Yet similar studies of high-redshift quasars had found much flatter slopes ($\alpha \simeq -0.6$) at ultraviolet rest wavelengths. The continuum shape was clarified when IUE and ground-based observations were combined. This doubling of the spectral coverage showed that there is a strong flattening in the continuum of Seyfert 1 nuclei and quasars (see the example of Markarian 335 in Figure 2). Sometimes called the 'blue bump', this continuum component rises with frequency in the visual, and peaks somewhere in the middle of the wavelength range of the SWP camera (Oke and Zimmermann, 1979; Oke and Goodrich, 1981). IUE observations detected the same excess flux at wavelengths below 2000 Å in several low-luminosity Seyfert 1 nuclei (Perola *et al.*, 1982; Clavel *et al.*, 1983; Clavel and Joly, 1984; Ulrich and Boisson, 1983). This ultraviolet component is so strong that it constitutes most of the total energy output of many AGNs.

Combined optical and ultraviolet spectrophotometry demonstrated that the continua of most Seyfert 1 nuclei and quasars are strikingly similar. For example, Figure 1 shows the ratio of fluxes at 4220 and 1460 Å as a function of luminosity. These spectrophotometric indices have no contamination from absorption lines, Fe II line or Balmer continuum emission. Much of the scatter at low luminosities is due to contaminating starlight at 4220 Å, which was not subtracted. The solid line shows the locus of mixtures of intrinsic quasar light (with $f_{\nu 4220}/f_{\nu 1460} = 1.80$) and a fixed amount of starlight (corresponding to $M_B = -18.4$ within the observing aperture). The bluest AGNs and quasars have optical/ultraviolet colors confined to a very narrow range of values, with little dependence on luminosity. This tightly limits the amount of reddening present in these objects (Malkan, 1984). The arrows in Figure 1 show the effect of a reddening of $E_{B-V} = 0.10$ mag. If many of the AGNs in the figure had larger reddenings, the optical/ultraviolet colors would be more widely distributed than shown. Since there is certainly no bias against the discovery of extremely blue AGNs, the ridge line in Figure 1 must represent the typical color of an intrinsically unreddened AGN. Evidently, by selecting the bluer AGNs, it is possible to find objects in which the effects of reddening are very small. IUE is producing growing evidence that dust and its associated reddening and thermal infrared emission is more abundant in the least luminous AGNs, while it seems unlikely that dust survives near the more luminous quasar nuclei.

The ultraviolet excess continuum component seems to be universally present in AGN energy distributions, although in some objects reddening makes it difficult to detect. This excess was first quantitatively described as a single-temperature blackbody by Malkan and Sargent (1982). More recent analysis by Edelson and Malkan (1986) confirms that in a wide range of AGNs, the best fitting blackbody temperature is 26 000 K. Figure 2 shows their model fits for the 0.1—100 μm spectrum of the Seyfert 1 galaxy Markarian 335. Independently analyzing the continuum of NGC 7469, Westin (1984) came to similar conclusions. The power

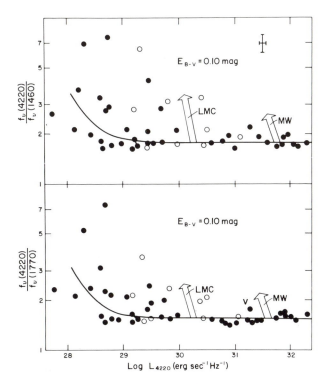

Fig. 1. The ratio of fluxes at 4220 and 1460 Å as a function of AGN luminosity. The solid line shows the locus of mixtures of intrinsic quasar light (with $f_{\nu 4200}/f_{\nu 1460} = 1.80$) and a fixed amount of starlight (corresponding to $M_B = -18.4$ within the observing aperture). The bluest AGNs and quasars have optical/ultraviolet colors confined to a very narrow range of values. This tightly limits the amount of reddening present in these objects (Malkan, 1984). The arrows show the effect of a reddening of $E_{B-V} = 0.10$ mag. for reddening laws in the Milky Way or the LMC.

law, which dominates the red to far-infrared continuum, has a sharp low-frequency turnover at 60 μm.

Analysis of the ultraviolet spectra of high-redshift quasars indicates that this ultraviolet component is in fact broader than a single-temperature Planck function (Malkan, 1983). The data are better fitted with simple geometrically thin, optically thick accretion disk models, as illustrated for the quasar PKS 0405—123 in Figure 3. The derived black hole masses and accretion rates imply that the most luminous quasars are accreting at nearly their Eddington limits. Less luminous AGNs must have lower mass accretion rates. Detailed disk model-fitting to their multi-wavelength continua will be required to determine whether they have much smaller black hole masses than quasars, or accretion rates far below the Eddington limit.

A few workers (e.g. Wills *et al.*, 1985; Netzer *et al.*, 1985) have proposed that this ultraviolet excess could be described by a relatively flat power law (with $-0.6 < \alpha < +0.3$). However, the sum of such a power-law component and the steeper infrared power law would predict more flux in the red (0.7—1.2 μm) than

MKN 335

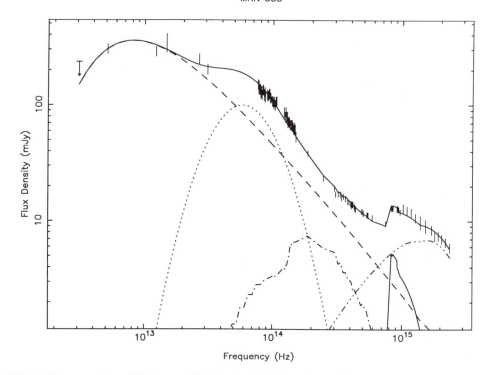

Fig. 2.   Decomposition of Edelson and Malkan's (1986) fit to the multi-wavelength continuum of the Seyfert 1 galaxy Markarian 335. The dashed line is a power law with slope −1.2; the dot-dash line is the flux from starlight; the triple-dot-dash line is a blackbody with temperature 26 000 K; the short solid curve in the ultraviolet is recombination radiation from the BLR; the dotted curve is a parabola centered at 5.8×10^{12} Hz with a FWHM of 3.3 octaves. Their sum is the long solid line, which fits the spectrophotometric data, shown by vertical error bars.

is actually observed (Malkan and Sargent, 1982). To reproduce the sharp inflection seen around 1 μm requires that one of the 'power law' components always has a turnover near that wavelength. No suggestion has been advanced for why the (presumably nonthermal) ultraviolet 'power law' should have a low-frequency cut-off in the red. It has been proposed that the near-infrared continuum could be predominantly thermal, and would therefore have a high-frequency cut-off beyond the Wien peak of the hottest dust grains present. This would require that all Seyfert 1 nuclei and quasars have significant thermal near-infrared emission from dust grains at least as hot as 1500 K, and that fluxes of the starlight in Seyfert 1 galaxies (e.g. by Malkan and Filippenko, 1983) were seriously underestimated.

Joint optical/ultraviolet studies of AGNs have made it possible, for the first time, to estimate the strength of additional features in the near-ultraviolet. For example, Long Wavelength IUE spectra with high signal-to-noise ratios have revealed low-contrast structure which is due to the blending of many Fe II emission

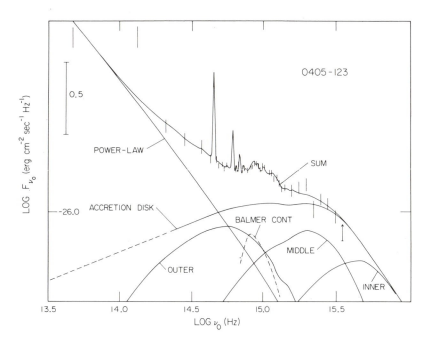

Fig. 3.   Decomposition of Malkan's (1983) fit to the multi-wavelength continuum of the quasar PKS 0405–123. The lower curves marked 'inner', 'middle', and 'outer' show the thermal emission from the accretion disk inside 20 gravitational radii, from 20 to 100 gravitational radii, and beyond 100 gravitational radii, respectively. The combination of those three curves gives the accretion disk spectrum, which when added to the power law and Balmer continuum components gives the fit to the data identified by 'sum'.

lines (Kollatschny and Fricke, 1983; Veron-Cetty *et al.*, 1983). The strength and density of these multiplets are so high between 2200 and 2700 Å that they may produce a pseudo-continuum. Individual low-contrast features are difficult to measure; accurate estimation of the ultraviolet Fe II line fluxes requires models which predict the relative strengths of hundreds or thousands of transitions, and knowledge of the shape of the underlying continuum. If it is a simple, flat power law (rather than a curved thermal spectrum, for example), the total Fe II emission in the ultraviolet can be as much as equal to the flux in Ly α (Wills *et al.*, 1985). Assuming a flat power-law continuum for the Seyfert 1 galaxies 3C 390.3 and Markarian 290, Netzer *et al.* (1985) inferred that their total Fe II flux (most of which is in the ultraviolet) is 0.6 and 0.9 times that of Ly α, respectively. Models of BLR clouds could hardly explain this larger value for the Fe II line flux, which is believed to arise in gas with very high densities and optical depths (Wills *et al.*, 1985).

3. Ultraviolet Emission Lines

The excellent correlation between the luminosities of the nuclear continuum and of the broad emission lines has long been taken as strong evidence that the dominant energy source for the lines is photoionization. The correlation is not perfectly linear: Baldwin (1977a) pointed out that the C IV 1550 Å line luminosity increases less than proportionally with ultraviolet continuum luminosity. IUE has now thoroughly investigated this decrease in C IV equivalent width with absolute luminosity, often called the 'Baldwin effect', over the full range of AGN luminosities. Data from Barr *et al.* (1983), Clavel and Joly (1984), Wu *et al.* (1983) and Malkan (1983) are plotted in Figure 4, which illustrates the decline of C IV equivalent widths, especially at ultraviolet continuum luminosities above $Lv \approx 10^{29.5}$ erg sec^{-1} Hz^{-1}.

Several interpretations of the effect shown in Figure 4 have been proposed. The less luminous AGNs could have BLRs with larger covering fractions (portions of the sky covered by BLR clouds, as seen from the central continuum source) (Wu *et al.*, 1983; Wampler *et al.*, 1984). Malkan and Sargent (1982) and Netzer (1986) speculated that the continuum measured at 1450 Å may not be directly proportional to the flux in the far ultraviolet which actually produces the C IV emission. Mushotzky and Ferland (1984) proposed that the ionization parameter, the ratio

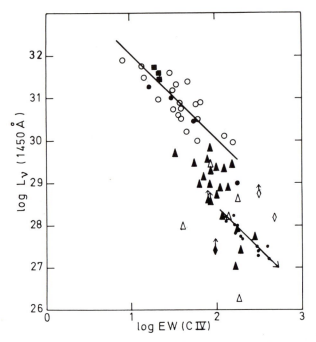

Fig. 4. The inverse correlation of rest frame C IV λ1550 equivalent width with luminosity at 1450 Å for Seyfert 1 nuclei and quasars, frequently known as the 'Baldwin effect'. Open circles are from Baldwin (1977b), filled squares are from Malkan (1983), filled circles and triangles are from Wu *et al* (1983), and open triangles are from Clavel and Joly (1984). The filled diamond represents a Seyfert 2 galaxy, and the open diamonds represent broad-line radiogalaxies (Wu *et al.*, 1983). The relation found for quasars, shown by the upper solid line, clearly does not extend to lower luminosities. The lower solid line is a fit to observations of a single variable Seyfert 1 nucleus, NGC 3783, from Barr *et al.* (1983).

of ionizing photons to electrons, is systematically higher in low-luminosity AGNs. A successful hypothesis should also explain the decreases in C IV equivalent width which are correlated with increases in the luminosity of an individual variable Seyfert 1 galaxy (e.g. NGC 3783, as described by Barr *et al.*, 1983).

Baldwin (1977b) first measured the Ly α/Hβ intensity ratio by piecing together composite spectra of quasars with large and small redshifts, and found it to range from 2 to 5. These values are an order of magnitude smaller than 40, the ratio predicted by recombination theory. IUE has now established that less luminous AGNs, measured at both optical and ultraviolet frequencies, have Ly α/Hβ $\approx$ 2—20, with an average value of 5 (Wu *et al.*, 1980; Oke and Goodrich, 1981; Lacy *et al.*, 1982). One proposed explanation is that enough dust is present in the BLR to redden the Ly α/Hβ ratio from the recombination value to its observed value. But reddening by itself cannot simultaneously explain the observed Lα/Hα/Hβ/Pα ratios — they imply that the Balmer line fluxes are enhanced, not that Lya is depressed. In several models, most of the Balmer line emission originates in the extended zones of optically thick BLR clouds, which are partly ionized by the soft X-rays in the continuum, while Ly α comes from the fully ionized front of these clouds (Kwan and Krolik, 1981; Canfield and Puetter, 1981). The Lyman edge optical depth may be as high as 10^{4-6}, and $\tau_{Ly\,\alpha}$ can exceed 10^8, building up a large population of hydrogen atoms in their first excited states. At high electron densities ($n_e > 10^9$ cm^{-3}), collisions thermalize trapped Ly α photons, and many new Balmer photons are created.

IUE's spectral resolution of ~6 Å in low-dispersion observing has facilitated accurate measurement of strong emission line profiles in well-exposed spectra. Ground-based studies of the profiles of ultraviolet emission lines in high-redshift quasars indicated that different ions (C III and C IV, for example) showed essentially identical velocity distributions (e.g. Richstone *et al.*, 1980). And in extremely bright AGNs, such as Fairall 9, IUE found that the C IV and Ly α profiles were similar and symmetric, as in quasars (Gregory *et al.*, 1982). Many AGNs have noticeably asymmetric line profiles, however, and they differ from one transition to another. For example, C IV is usually broader than Ly α, and shows significant excess emission in its red wing (Wu *et al.*, 1981). In 3C 382 the profile of Ly α has a relatively stronger central core than C IV (Tadhunter *et al.*, 1986).

4. Extended Structures

Although IUE has limited spatial resolution, it has studied and resolved a few types of extended structures in active galaxies.

Some Seyfert nuclei have been found in interacting galaxies. Kollatschny and Fricke (1984) observed Markarian 266, a pair of interacting galaxies 11 arc seconds apart. One shows the emission-line spectrum of a Seyfert 2 nucleus; the other, a brighter continuum source, is a liner. These examples raise a possibility that galaxy interaction can help to stimulate nuclear activity, although it is certainly insufficient in itself.

Snijders *et al.* (1982) found four regions of strong ultraviolet emission around the nucleus of the brightest Seyfert 2 galaxy, NGC 1068. All coincide with the bright inner spiral arms, and three could be OB associations with UV luminosities 10—20% that of the nucleus. The fourth region has nearly the same luminosity as the nucleus, and a harder spectrum. It shows P Cygni emission from C IV and N V, like those seen in NGC 7714, with outflow velocities of 2000 km s^{-1}.

Perola and Tarenghi (1980) used IUE to observe the nucleus and jet of M 87. The nuclear UV light is spatially resolved, and has a spectrum similar to that of other elliptical galaxies: it declines steeply from 3300 to 2000 Å, and then rises at shorter wavelengths. As in other ellipticals, this new short-wavelength component is presumed to be stellar, since it is spatially extended. It has been attributed to a population of evolved stars with effective temperatures of 30 000 K or higher (see the discussion of the ultraviolet spectra of old stellar populations, pp. 637—654). The ultraviolet flux from the spatially unresolved knots in the jet is significantly higher than an extrapolation of their optical spectra using $f_\nu \propto \nu^{-1.7}$. The only detectable spectral features in any of the spectra were Mg II 2800 Å absorption, presumed to be from interstellar gas or stars in the galaxy.

IUE has detected extended envelopes of Ly α emission around some radio-galaxies. In PKS 2158—380, extended C IV 1550 Å and He II 1640 Å emission is also detected (Fosbury et al., 1982). The high state of ionization is inconsistent with local photoionization by hot stars. More probably the gas is photoionized by the point-like ultraviolet source in the galactic nucleus, which has a power-law slope of -1.4.

Extended Ly α emission has also been detected in cooling flows around elliptical galaxies in the centers of rich clusters. For example, a short-wavelength spectrum of NGC 1275, the Seyfert galaxy in the Perseus cluster, shows extended redshifted Ly α emission (Fabian et al., 1984). Unlike the geocoronal Ly α line, the redshifted component does not completely fill IUE's observing aperture. The extended Ly α flux is 8.6×10^{-14} erg cm^{-2} sec^{-1}, with no detectable emission at any other wavelength. A second SWP exposure displaced by 10 arc sec, confirms the line emission. The Ly α/Hβ ratio is probably compatible with the expectations of radiative recombination. Since the upper limit for C IV emission is 1.4×10^{-14} erg cm^{-2} sec^{-1}, there is no evidence for the presence of shocks.

Extended Ly α emission is detected in another possible cooling flow in the cluster Abell 1795 (Norgaard et al., 1984). The Ly α/Hβ ratio may be close to the recombination value, depending on the adopted reddening. As in Perseus, there is no evidence that stars are forming in the flow. Unlike NGC 1275, the central galaxy in Abell 1795 is not active, arguing against any close link between cooling flows and nuclear activity.

5. Detailed Variability Studies of NGC 4151 and Other AGNs

IUE has been especially useful in the study of AGN variability. Less luminous AGNs appear to have the largest amplitudes of line and continuum variability, up to one or even two magnitudes, over timescales as short as a week. As the nearest and brightest Seyfert 1 galaxy, NGC 4151 is also the most extensively studied. Accordingly, it has been a prime target for IUE, and led to a new approach in extragalactic research. As early as 1978, European astronomers realized that an intensive variability study of a single object could yield critical insights into the nature of all AGNs. The requirement of a long series of coordinated observations dictated that many extragalactic observers collaborate rather than collect data individually as most had done before. This resulted in the most impressive database ever assembled on a single AGN (over 300 IUE spectra), and has produced discoveries that could not have come from a similar investment of IUE time in scattered, uncoordinated observations.

From 1978 to 1979 the sampling frequency of observations was modified to probe variability which IUE found to be more rapid than had been anticipated. The results of observations up to 1983 have been discussed by several groups (Penston et al., 1979; Penston et al., 1981; Ferland and Mushotzky, 1982; Perola et al., 1982; Stoner et al., 1984; Ulrich et al., 1984; Bromage et al., 1985; Veron et al., 1985; Ulrich et al., 1985).

In several Seyfert 1 galaxies the observed continuum variations are strongest at the shortest ultraviolet wavelengths, with amplitudes much larger than those seen at optical and infrared wavelengths (NGC 4151 — Perola et al., 1982; NGC 3783 — Barr, Willis and Wilson, 1983; NGC 4593 — Clavel et al., 1983; NGC 7469 — Westin, 1984; NGC 5548 and Fairall 9 — Wamsteker et al., 1984). This effect may be explained partly by the decreasing fraction of nonstellar light at shorter wavelengths when the nucleus fades.

The two-folding timescales (i.e. times for the continuum brightness to double) derived from adjacent observations range from 5 to 30 days in NGC 4151, implying a source size of less than 0.01 pc. Although the timescales for flux increases and decreases are generally similar, Boksenberg et al. (1978) tentatively suggested a rise time as short as 24 hr in their data. So far, no ultraviolet continuum variations above the limiting sensitivity of 5% have been detected on a timescale of 8 hr. In contrast, X-ray flare-like events in NGC 4151 have been reported with rise-times shorter than 12 hr and decay-times of 2 to 4 days. These imply that the X-ray flux originates from a smaller volume than the ultraviolet and optical, and possibly from a different emission mechanism.

Since even the nearest AGNs have BLRs no larger than tenths of a milliarc-second, their structure can be probed only by variability studies. Quasars seldom show any emission-line variability greater than 10%, even when their continua vary substantially. The variability of line emission seems to be strongest and most rapid in the least luminous AGNs, probably because their BLRs are smaller. For this reason, IUE observations of the variability of the strong $C IV$ 1550, $C III$] 1909, and $Mg II$ 2800 emission lines in NGC 4151 and other low-luminosity AGNs have provided key insights and have changed previous thinking about the structure and kinematics of the BLR.

IUE's detailed monitoring programs have been especially useful in revealing the connections between variability in the continuum, and the profiles of broad emission lines. When NGC 4151's ultraviolet continuum is faint, all the ultraviolet permitted emission lines are relatively narrow, with profiles similar to those of the Balmer lines. When the continuum is bright, however, the base of the $C IV$ line broadens, with emission wings extending up to 16 000 km s^{-1} (Figure 5). IUE observations of other AGNs also revealed that the full-widths at zero-intensity of some emission lines, especially C IV, increase when their continua brighten. The C IV lines of not only NGC 4151, but also NGC 3516 (Ulrich and Boisson, 1983), NGC 3783 (Barr et al., 1983b), NGC 4593 (Clavel, 1983), and NGC 5548 (Gregory et al., 1982) develop strong wings with velocities extending to a few 10^4 km s^{-1}. Since the Mg II and C III] emission lines always have FWHMs less than 6000—10 000 km s^{-1}, the high-velocity clouds must emit primarily in C IV, presumably because of their higher ionization and electron density.

IUE studies have found that the intensities of the broad emission lines vary in response to changes in the continuum, but often with smaller amplitude (Kollatschny and Fricke, 1981, Clavel et al., 1983; Clavel, 1983; Ulrich and Boisson, 1983; Barr et al., 1983a; Westin 1984). This may be because a brightening of the central source takes a significant time to light up the entire BLR. Continuum variations on timescales shorter than the light-crossing time of the BLR tend to be smoothed out. In four well studied cases, NGC 3783 (Barr et al., 1983), NGC 4151 (Perola et al., 1982), NGC 4593 (Clavel, 1983), and NGC 7469 (Westin, 1984), the C IV line responds to a change in the photoionizing continuum with a time lag of one to three weeks. Thus the region producing the broad C IV wings in these AGNs is believed to have a light-crossing time of a few light-weeks. The Mg II emission line is only loosely correlated with the continuum: in NGC 3783 the time lag may be four months, consistent with a much larger emitting volume. In NGC 4151, the lack of significant dependence of C III] 1909 Å intensity on continuum flux suggests a dimension for the outer parts of the BLR exceeding a light-year. These light-travel size estimates depend on the assumption that all line emissivities are linearly proportional to the far ultraviolet flux.

Only a few AGNs have been monitored by IUE intensively enough to allow systematic searches for lags between the line and continuum variability. Reliable detection of the light-echo time in the BLR requires fortuitous temporal sampling of the AGN. Yet even the limited data available strongly suggest that the BLR encompasses a continuous range of decreasing density, ionization, and velocity with increasing distance from the nucleus. For example, the variable broad wings seen in C IV but not C III emission demonstrate that the gas in the BLR has systematically higher velocities and ionization parameters closer to the center. Quasars are thought to have a relatively constant ionization parameter throughout the BLR, implying that $n_e \propto r^{-2}$. In contrast, densities in AGN BLRs seem to fall off less sharply with increasing radius. The IUE results suggest that the BLR gas merges smoothly into the NLR, which is known to show similar velocity and density stratification from optical studies (Pelat et al., 1981).

A remarkable feature apparently unique to NGC 4151 is the presence of two narrow emission lines at 1523 and 1599 Å (Ulrich et al., 1984). Since their strength is rapidly variable, they are associated with gas in the BLR. The wavelengths are not variable, and have not been identified with any plausible transitions at the same redshift as the other emission lines. Ulrich et al. (1984) proposed that these lines are satellites of C IV emission, at −6100 and +8500 km s^{-1}, suggestive of a high-velocity bipolar outflow, possibly a jet.

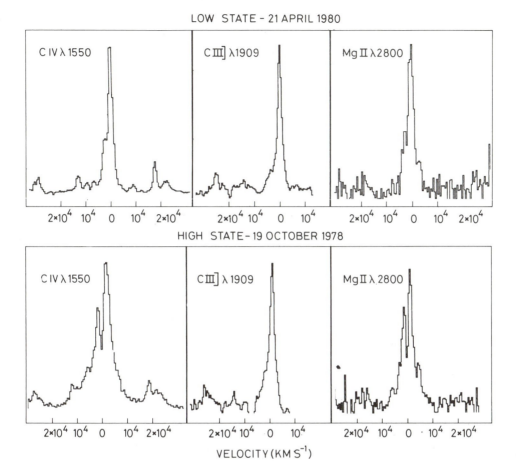

Fig. 5. Typical emission line profiles of C IV 1550 Å, C III] 1909 Å, and Mg II 2800 Å in NGC 4151 (from Ulrich *et al.*, 1984, reproduced with permission of Monthly Notices of the Royal Astronomical Society). All profiles are plotted on the same velocity scale, given in km s^{-1}. The upper spectrum was observed when the nucleus was faint. The lower spectrum, which shows the extremely broad wings of C IV, was obtained when the nucleus was bright.

6. Absorption Lines

IUE's access to ultraviolet wavelengths is particularly valuable in studying NGC 4151's rich absorption spectrum, which includes resonance lines of H I, C II, C IV, N V, O I, Si IV, and excited fine-structure lines of C II, O I, Si II (Bromage *et al.*, 1985). Strong absorption extends from +100 to −1000 km s^{-1} (with respect to the NLR radial velocity of +960 km s^{-1}). This is intrinsic to the active nucleus, perhaps similar to the outflow phenomenon seen in 'broad-absorption' line quasars. IUE also detected absorption lines from metastable levels of C III (at 1176 Å) and Si III (at 1297 Å), analagous to the He I and Balmer absorptions viewed optically. The line strengths indicate large populations in these metastable levels, which must be produced collisionally, at electron densities of at least $10^{8.5}$ cm^{-3}. The depth of the C IV absorpton in high-resolution spectra obtained by Penston *et al.* (1979) indicates that the absorbers cover not only the continuum source, but probably 80% of the BLR as well. The high ionization and electron density in the absorbing region imply that it is just on the outskirts of the BLR, 5 $\times$ 10^{17} cm from the nucleus, in an outflowing shell or a large number of small clouds. The large velocity spread of the absorbers does not change during continuum variations, indicating that the changes in absorption strength result primarily from changes in the column densities of absorbing species. An estimated lower limit to the hydrogen column density through the absorption-line region is $N(H) > 10^{21}$ cm^{-2}, which is compatible with 4×10^{22} to 2×10^{23} cm^{-2}, the range of values implied by photoelectric absorption in the soft X-ray spectrum.

NGC 3516 also shows C IV and N V absorption which is positively correlated with the continuum brightness (Boisson and Ulrich, 1983). In contrast, NGC 4593 appears to have absorption lines only from lowly-ionized species (O I, C II, Si II, and Mg II) (Clavel *et al.*, 1983).

7. Low-Luminosity Active Galaxies

Observing AGNs in the ultraviolet minimizes the contaminating effects of the host galaxy because the starlight is much redder than the nonstellar nucleus. Thus the ultraviolet is one of the best spectral regions for measuring low-luminosity active nuclei. Nonetheless, studying the low end of the luminosity function of AGNs, particularly the much fainter nonstellar light of Seyfert 2 nuclei and 'liners', has pushed IUE close to the limit of its sensitivity.

A few ultraviolet spectra of X-ray selected AGNs have been obtained (Bergeron *et al.*, 1981; Barr *et al.*, 1983). These studies have been hampered by the large amounts of dust obscuring the nuclei of these galaxies. This is not unexpected, since they tend to be edge-on spiral galaxies which may contain heavily reddened BLRs (Lawrence and Elvis, 1982). For example, NGC 5506 appears to be reddened by at least 0.2, more likely 0.4 magnitudes in E_{B-V} (Bergeron *et al.*, 1981). The ultraviolet spectrum of NGC 3081 is also significantly reddened.

Most detections of Seyfert 2 galaxies in the ultraviolet are quite noisy. At optical and near-infrared wavelengths their fluxes are accurately measured, but are heavily contaminated by starlight. And at longer infrared wavelengths the emission is dominated by thermal emission from warm dust grains enshrouding the nucleus (Rudy *et al.*, 1985). The reddening from this nuclear dust (typically corresponding to a magnitude or more of visual extinction) makes accurate measurement of the nonstellar flux extremely difficult. Some studies indicate that the nonstellar continuum of Seyfert 2 nuclei can be described by a power law with a slope similar to that in Seyfert 1 galaxies (Malkan and Oke, 1983; Ferland and Osterbrock, 1986). The Seyfert 2 galaxy NGC 4507 is relatively bright in the ultraviolet, and shows no clear evidence of any reddening above the foreground amount expected from our Galaxy (Durret and Bergeron, 1986). Its nonstellar nuclear continuum resembles a power law with a slope of 1.2. The relatively strong C IV emission lines indicate that the NLR gas is photoionized by a fairly energetic continuum. Allowing for even a little reddening, the ionizing ultraviolet continuum has enough power to account for all the energy seen in the emission lines. Nonetheless, the large thermal luminosity re-radiated in the far-infrared may imply that we do not have a direct view of the nuclear source.

Thuan (1984) has detected some weak unresolved ultraviolet emission lines in IC 5153, and possibly NGC 5153. Although these galaxies have moderately high-excitation optical emission lines, they are probably best explained by high rates of current star formation, and are discussed on 'Ultraviolet Properties of Starburst Galaxies', pp. 623–635.

8. Conclusions

IUE has greatly improved our understanding of all aspects of activity in the nuclei of galaxies. For a telescope of its size, this impact is truly unprecedented. IUE has unified the study of active galactic nuclei and quasars. It has helped to overcome the contaminating effects of reddening and starlight contributions in AGN spectra. IUE was the keystone of the first detailed analyses of the multi-wavelength energy distributions of quasars and AGNs. And it has probed the structure of the BLR, and its relation to the ionizing continuum.

As with most astronomical instruments, IUE is most valuable when combined with telescopes working at other wavelengths. The combination of observations from a variety of detectors is especially important in studying AGNs, since their energy is unusually widely distributed over the electromagnetic spectrum. IUE substantially narrowed the most critical wavelength gap in our coverage of the energy distributions of AGNs, which had been the seven octaves between the atmospheric UV cut-off and the soft X-rays. Three of these octaves may never become accessible unless we can find extragalactic lines-of-sight virtually devoid of H I gas.

The variability of most AGNs requires that multi-wavelength observations be made as nearly simultaneously as possible. This exemplifies an important new trend in observational astronomy: the critical need to coordinate observations over a wide range of wavelengths. IUE's valuable experience proves that such flexibility in scheduling space- and ground-based observatories leads to significant scientific advances. As discussed on pp. 759–769, IUE has also highlighted the value of archival research with a well organized, calibrated database. The important lessons learned from IUE should be incorporated into the planning of future astronomical observing facilities.

References

Baldwin, J. A.: 1977b, *Astrophys. J.* **214**, 679.
Baldwin, J. A.: 1977a, *Monthly Notices Roy. Astron. Soc.* **178**, 67.
Barr, P., Willis, A. J., and Wilson, R.: 1983a, *Monthly Notices Roy. Astron. Soc.* **202**, 453.
Barr, P., Willis, A. J., and Wilson, R.: 1983b, *Monthly Notices Roy. Astron. Soc.* **203**, 201.
Bergeron, J., Maccacaro, T., and Perola, G. C.: 1981, *Astron. Astrophys.* **97**, 94.
Bertola, F., Cappaccioli, M., Holm, A. V., and Oke, J. B.: 1980, *Astrophys. J.* **237**, L65.

Bromage, G. E., Boksenberg, A., Clavel, J., Elvius, A., Penston, M. V., Perola, G. C., Pettini, M., Snijders, M. A. J., Tanzi, E. G., and Ulrich, M. H., 1985, *Monthly Notices Roy. Astron. Soc.* **215**, 1.

Boksenberg, A. *et al.*: 1978, *Nature* **27**, 404.

Canfield, R. C. and Puetter, R. C.: 1981, *Astrophys. J.* **243**, 390.

Clavel, J.: 1983, *Monthly Notices Roy. Astron. Soc.* **204**, 189.

Clavel, J. and Joly, M.: 1984, *Astron. Astrophys.* **131**, 87.

Clavel, J., Joly, M., Collin-Souffrin, S., Bergeron, J., and Penston, M. V.: 1983, *Monthly Notices Roy. Astron. Soc.* **202**, 85.

Davidson, K. and Netzer, H.: 1979, *Rev. Mod. Phys.* **51**, 715.

Durret, F. and Bergeron, J.: 1986, *Astron. Astrophys.* **156**, 51.

Edelson, R. A. and Malkan, M. A.: 1986, *Astrophys, J.* **308**, 59.

Ferland, G. J. and Mushotzky, R. F.: 1982, *Astrophys. J.* **262**, 564.

Ferland, G. J. and Osterbrock, D. E.: 1986, *Astrophys. J.* **300**, 658.

Ferland, G. J., Rees, M. J., Longair, M. S., and Perryman, M. A. C.: 1979, *Monthly Notices Roy. Astron. Soc.* **187**, 65.

Fosbury, R. A. E., Boksenberg, A., Snijders, M. A. J., Danziger, J. J., Disney, M. J., Goss, W. M., Penston, M. V., Wamsteker, W., Wellington, K. J., and Wilson, A. S.: 1982, *Monthly Notices Roy. Astron. Soc.* **201**, 991.

Grandi, S. A.: 1982, *Astrophys. J.* **255**, 25.

Gregory, S., Ptak, R., and Stoner, R.: 1982, *Astrophys. J.* **261**, 30.

Hubbard, E. N. and Puetter, R. C.: 1985, *Astrophys. J.* **290**, 394.

Kollatschny, W. and Fricke, K. J.: 1981, *Astron. Astrophys.* **102**, 123.

Kollatschny, W. and Fricke, K. J.: 1983, *Astron. Astrophys.* **125**, 276.

Kollatschny, W. and Fricke, K. J.: 1984, *Astron. Astrophys.* **135**, 171.

Kwan, J. and Krolik, J.: 1981, *Astrophys. J.* **250**, 478.

Lacy, J. H., Soifer, B. T., Neugebauer, G., Matthews, K., Malkan, M., Becklin, E. E., Wu, C. C., Boggess, A., and Gull, T. R.: 1982, *Astrophys. J.* **256**, 75.

Lawrence, A. and Elvis, M., 1982, *Astrophys. J.* **256**, 410.

Malkan, M. A.: 1983, *Astrophys. J.* **268**, 582.

Malkan, M. A.: 1984, in W. Brinkmann and J. Trumper (eds.), *Proc. Garching Conf. on Ultraviolet and X-Ray Emission from Active Galactic Nuclei* (Garching: MPE), p. 121.

Malkan, M. A. and Filippenko, A. V.: 1983, *Astrophys. J.* **275**, 477.

Malkan, M. A. and Oke, J. B.: 1983, *Astrophys. J.* **265**, 92.

Malkan, M. A. and Sargent, W. L.: 1982, *Astrophys. J.* **254**, 22.

Mushotzky, R. and Ferland, G. J.: 1984, *Astrophys. J.* **278**, 558.

Netzer, H., Wamsteker, W., Wills, B. J., and Wills, D.: 1985, *Astrophys. J.* **292**, 143.

Neugebauer, G., Morton, D., Oke, J. B., Becklin, E. E., Daltabuit, E., Matthews, K., Persson, S. E., Smith, A. M., Soifer, B. T., Torres-Peimbert, S., and Wynn-Williams, C. G.: 1980, *Astrophys. J.* **238**, 502.

Oke, J. B. and Goodrich, R. W.: 1981, *Astrophys. J.* **243**, 445.

Oke, J. B. and Zimmerman, B.: 1979, *Astrophys. J.* **231**, L15.

Pelat, D., Alloin, D., and Fosbury, R. A. E.: 1981, *Monthly Notices Roy. Astron. Soc.* **195**, 787.

Penston, M. V., Clavel, J., Snijders, M. A., Boksenberg, A., and Fosbury, R. A. E.: 1979, *Monthly Notices Roy. Astron. Soc.* **189**, 45.

Penston, M. V., Boksenberg, A., Bromage, C. E., Clavel, J., Elvius, A., Gondhalekar, P. M., Jordan, C., Lind, J., Lindegren, L., Perola, G. C., Pettini, M., Snijders, M. A. J., Tanzi, E. J., Tarenghi, M., and Ulrich, M. H.: 1981, *Monthly Notices Roy. Astron. Soc.* **196**, 857.

Perola, G. C. and Tarenghi, M.: 1980, *Astrophys. J.* **240**, 447.

Perola, G. C., Boksenberg, A., Bromage, G. E., Clavel, J., Elvis, M., Elvius, A., Gondhalekar, P. M., Lind, J., Lloyd, C., Penston, M. V., Pettini, M., Snijders M. A. J., Tanzi, E. G., Tarhengi, M., Ulrich, M. H., and Warwick, R. S.: 1982, *Monthly Notices Roy. Astron. Soc.* **200**, 293.

Rudy, R. J., Cohen, R. D., and Puetter, R. C.: 1985, *Astrophys. J.* **288**, L29.

Richstone, D. O., Ratnatunga, K., and Schaeffer, J.: 1980, *Astrophys. J.* **240**, 1.

Seyfert, C. K.: 1943, *Astrophys. J.* **97**, 28.

Soifer, B. T., Neugebauer, G., Oke, J. B., and Matthews, K.: 1981, *Astrophys. J.* **243**, 369.

Stoner, R. and Ptak, R.: 1984, *Astrophys. J.* **280**, 516.

Stoner, R., Ptak, R., and Gregory, S.: 1984, *Astrophys. J.* **285**, 69.

Tadhunter, C. N., Perez, E., and Fosbury, R. A.: 1986, *Monthly Notices Roy. Astron. Soc.* **219**, 557.

Ulrich, M. H. and Boisson, C.: 1983, *Astrophys. J.* **267**, 515.

Ulrich, M. H. *et al.*: 1985, *Nature* **313**, 745.

Ulrich, M. H., Altamore, A., Boksenberg, A., Bromage, G. E., Clavel, J., Elvius, A., Penston, M. V., Perola, G. C., and Snijders, M. A. J.: 1985, *Nature* **313**, 747.

Ulrich, M. H., Boksenberg, A., Bromage, G. E., Clavel, J., Elvius, A., Penston, M. V., Perola, G. C., Pettini, M., Snijders, M. A. J., Tanzi, E. G., and Tarenghi, M.: 1984, *Monthly Notices Roy. Astron. Soc.* **206**, 221.

Veron, P., Veron-Cetty, M. P., and Tarenghi, M.: 1985, *Astron. Astrophys.* **150**, 317.

Veron-Cetty, M. P., Veron, P., and Tarenghi, M.: 1983, *Astron. Astrophys.* **119**, 69.

Wampler, E. J., Gaskell, C. M., Burke, W. L., and Baldwin, J. A.: 1984, *Astrophys. J.* **276**, 403.

Wamsteker, W. and Barr, P.: 1985, *Astrophys. J.* **292**, L45.

Westin, B. A. M.: 1984, *Astron. Astrophys.* **132**, 136.

Wills, B. J., Netzer, H., and Wills, D.: 1985, *Astrophys. J.* **288**, 94.

Wu, C. C., Boggess, A., and Gull, T. R.: 1980, *Astrophys. J.* **242**, 14.

Wu, C. C., Boggess, A., and Gull, T. R.: 1981, *Astrophys. J.* **247**, 449.

Wu, C. C., Boggess, A., and Gull, T. R.: 1983, *Astrophys. J.* **266**, 28.

QUASARS

P. M. GONDHALEKAR

Rutherford Appleton Laboratory, Chilton, England

and

G. FERLAND

Department of Astronomy, Ohio State University, Columbus, OH, U.S.A.

1. Introduction

The ultraviolet observations of quasars provide a direct measure of the ionizing non-thermal radiation in these objects and the fluxes of the resonance emission lines. These data provide information on the processes taking place close to the "central engine" which drives these luminous objects and also on the physical state of the gas which emits the line radiation and the dynamics of the region in which these lines are formed. The observations of quasars with the IUE thus provide an opportunity of not only studying these objects over a wide wavelength range, but also the possibility of comparing quasars at different epochs (i.e. study the evolution of quasar properties). Through observations of the absorption lines formed in the quasar spectra it is also possible to explore the intervening intergalactic space.

The first observation of a quasar at ultraviolet wavelengths was made by the Astronomical Netherlands Satellite (ANS). This satellite observed 3C273 in five wide wavelength bands between 1550 and 3300 Å (Wu, 1977). The second ultraviolet observation of a quasar was also that of 3C273, in this case a rocket spectrograph was used to obtain a spectrum of this quasar between 1200 and 1700 Å at a resolution of 10—15 Å (Davidsen *et al.*, 1977).

IUE has now been operational for eight years and up to 30 August, 1985, 151 quasars had been observed. In all 405 images have been obtained of which 99 have been obtained with the LWP camera, 152 with the LWR camera and 254 with the SWP camera. The distribution of observed quasars in redshift and V-magnitude are shown in Figure 1. About 20% of quasars observed with IUE have redshifts between 0.1 and 0.5 (in this chapter only objects with redshifts greater than 0.1 have been considered as quasars) although objects of redshifts up to 3.53 have been observed with IUE. This, of course, does not mean that all objects observed have been detected with IUE and indeed there has been no positive detection of a spectrum of a quasar of $z_e > 3.0$ even after an integration time of 15 hr. A number of quasars with $z_e = 2.0 \pm 0.25$ have been detected in the long wavelength region (2400 to 3000 Å) but only one quasar (PG 1115 + 08, $z_e = 1.72$) has been detected in the short wavelength region (1200 to 1950 Å) (Green *et al.*, 1980). The redshift 'depth' to which IUE is capable of observing is disussed in detail in Section 4.

About 50% of quasars observed with IUE are of visual magnitude between 15.5

Y. Kondo (ed.), Scientific Accomplishments of the IUE, pp. 671—683.
© 1987 *by D. Reidel Publishing Company.*

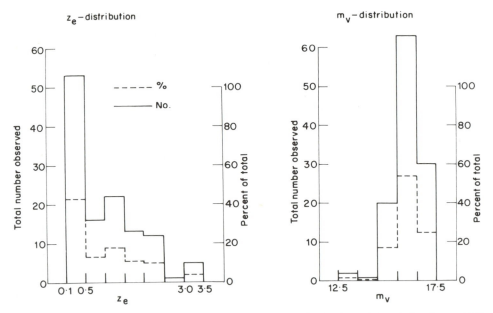

Fig. 1. The distribution of quasars observed with IUE. The distribution with emission line redshift z_e and visual magnitude m_v are shown. The full line shows the distribution by number of the quasars observed and the dashed line shows the percentage of total numbers observed. These data are only for quasars observed till 30 August, 1985.

and 16.5 (the redshifts and visual magnitudes of the quasars were obtained from the catalogues of Hewitt and Burbridge (1980) and Veron-Cetty and Veron (1984)). This is clearly a selection effect — not many quasars brighter than $15^m.5$ are known and quasars fainter than $16^m.5$ cannot be detected with IUE. Out of the 151 quasars observed with IUE, 14 are not listed in the catalogue of Hewitt and Burbridge (1980) or in the catalogue of Veron-Cetty and Veron (1984). This suggests that quasars observable with IUE are being discovered in new surveys and hopefully more such quasars will be discovered in future.

In the eight years since the launch of the IUE there has been a rapid progress in addressing questions concerning the nature of quasars and in particular their relationship to other types of galaxies. This review will attempt to concentrate more on the 'big picture', that is observations and interpretations which have played (in our opinion) fundamental roles in altering our understanding of these objects. Because of limited space this review cannot be a comprehensive survey of all the contributions made by IUE and we apologize to the authors whose work we have not been able to discuss here.

In Section 2 we discuss (briefly) the special problems encountered in observations of quasars with IUE and in the reduction of these data. The scientific achievements of IUE are discussed in Section 3 and the limitations of observing quasars with IUE are discussed in Section 4. In Section 5 observations in the near future which should lead to a better understanding of the quasar phenomenon are briefly discussed.

2. Observations and Data Reduction

A large majority of quasars cannot be detected even with the most sensitive mode of Fine Error Sensor (FES) and quasars are normally acquired by IUE by a blind off-set technique. Here the spacecraft is slewed to a bright star ($m_v < 12$) within 8 arcmin of a quasar. The quasar can then be acquired by making a measured slew from the bright star to the quasar. For accurately known positions of the bright star and the quasar, the quasar image can be centred in the large (10×20 arcsec) aperture of IUE.

The signal levels detected for most quasars are low and apart from 3C273 all quasars have been detected only in the low resolution mode. In this mode the average resolution of the SWP camera is about 5 Å (FWHM) and that of LWR(P) camera is about 6 Å (Casatella *et al.*, 1984). This mode of IUE is adequate for studying the gross structure of the most prominent emission lines in quasar spectra. The resolution is also adequate to investigate the large scale structure in the BAL objects but is not sufficient for investigations of the fine structure in the emission lines or narrow absorption lines like the Ly α forest.

Because the signal levels detected for most quasars are low, the integration time with IUE for sufficient signal-to-noise ratio, are long. The minimum integration time in most cases is ~6 hr. During the integration of a quasar spectrum the IUE cameras also integrate a large background signal, this background is due to Cerenkov radiation produced in the face-plate of the camera by the ambient electrons. The level of background signal can be different for each image as the signal level depends on the orientation and position of the spacecraft in its orbit and on the level of solar activity. The low signal levels embedded in high background pose special problems in extracting quasar spectra from IUE images. The IUE detectors are known to be non-linear at low signal levels (Holm, 1981; Settle *et al.*, 1981), however, this does not directly affect the detection of faint spectra if long integration times are involved. The particle background raises a faint spectrum out of the non-linear part of the Intensity Transfer Function (ITF). The extracted faint spectrum signal then depends critically on small gain changes, noise in the background and uncertainties in the background subtraction. Unfortunately an 'observatory consensus' on extraction of faint signals embedded in high background levels has not emerged and most astronomers have an individual approach to this problem.

The small gain changes in the IUE Scientific Instrument (SI) are the most difficult to quantify. These changes depend on the history of camera usage, state of the other subsystems on the IUE spacecraft during integration, the temperature of the camera and/or the head amplifier, the leakage of Earth's magnetic field into the IUE SI and the microchanges which may take place in the camera electronics during integration of an image. Taking these uncertainties into account it is reasonable to believe that the amplitude of faint signals (e.g. continuum) can only be measured to an accuracy of 10%—20%.

The noise in the background and the background signal level also depend on the history of camera usage and the preparation of a camera for a new image. An ideal situation would be to prepare a camera for a image and leave it in 'Stand-By' for 6—8 hr before integrating a new image. During this stand-by any trace of previous camera usage (e.g. residual images due to high exposures, negative images etc.) would die out of the camera phosphor and the camera target would stabilize. Unfortunately such ideal situation rarely occurs (IUE time is too precious for cameras to be left idle for 6—8 hr!) and a solution to background noise and subtraction has to be found in image analysis. Most astronomers prefer to re-extract a quasar spectrum from the line-by-line spectrum (LBLS) provided by IUE observatories. The LBLS file is adequate for extracting a quasar spectrum as long as high spatial resolution is not required (as in the analysis of spectra of the double quasar Q0957 + 561A, B). In order to prevent the noise or blemishes in the background affecting a spectrum it is normal to smooth the background in the LBLS file. This is done by folding a filter function through the background pixels and rejecting pixels with signals above (or below) a preset limit. A mean background level is usually obtained from both sides of a spectrum in order to account for any shading in the camera response.

The FWHM of the point spread function of IUE optics is about 3 arcsec or approximately 3 pixels at the camera face-plate. The extracted spectrum provided by the IUE observatories is usually obtained with a slit which is about 12 pixels long. A faint spectrum (signal levels $\sim 10^{-15}$ ergs cm^{-2} s^{-1} Å^{-1}) embedded in a high background will be degraded by the noise in the background if a slit of this length is used to extract the spectrum. The signal level and the signal-to-noise ratio in a spectrum are shown in Figure 2 as a function of slit length. The results shown in Figure 2 were obtained from a LBLS files created from a photometrically corrected IUE image, these LBLS files had higher spatial resolution than the LBLS files provided by IUE observatories and some differences are to be expected if similar analysis is done with the standard LBLS file. Also the results in Figure 2 are for two quasars selected at random and some differences are to be expected if this analysis is repeated for other quasar spectra with different signal and background levels. A spectrum extracted with a slit 6 pixels long appears to have most of the signal in a image and the best signal-to-noise ratio. However, with a slit of this length it is necessary to track a spectrum accurately, any deviation of the slit track from a spectrum can result in spurious features in a spectrum. It is also possible to extract a spectrum with a shaped (Gaussian or skewed Gaussian) slit. This makes tracking much more critical and the improvement in the quality of a spectrum does not justify the increase in computer time required to obtain a spectrum.

Because the integration times required to detect quasar spectra are long the probability of a cosmic ray 'hit' on the IUE cameras during integration of a spectrum is also high. If a cosmic ray event intersects a spectrum then data at that location are lost. If an event is in the background and is not smoothed out by filtering then an absorption feature can be injected into a spectrum and lead to loss of data. Thus quasar spectra being intrinsically weak and in high noise environment require considerable care in extraction and analysis. Because astronomers have an individual approach to extracting quasar spectra from IUE images, spectra extracted by two astronomers from the same image can sometimes differ signifi-

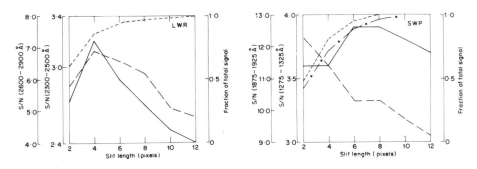

Fig. 2. The signal-to-noise ratio and the fraction of the total signal as a function of the slit length used to track a spectrum. Camera SWP; (——) S/N between 1275–1325 Å, (———) S/N between 1875–1925 Å, (—·—·—) fraction of total signal between 1275–1325 Å, (-------) fraction of total signal between 1875–1925 Å. Camera LWR; (——) S/N between 2300–2500 Å, (———) S/N between 2600–2900 Å, and (-------) fraction of total signal between 2600–2900 Å. These data are only appropriate for low signal level ($\sim 10^{-15}$ ergs cm^{-1} s^{-1} Å^{-1}) spectra embedded in high background.

cantly and lead to ambiguous interpretation. This is unfortunate and at present unavoidable.

3. Contributions of IUE to Our Present Understanding of Quasars

Almost since their discovery some twenty-five years ago quasars have been widely regarded as a phenomenon occurring in the nuclei of distant galaxies. Within the last ten years this view has been substantiated by detection of the host galaxies. A fundamental question is the nature of their energy source; these objects are the most luminous known in the Universe and optical variability studies reveal the central engine to be of a fairly small size (roughly the size of the solar system). Interpretations now centre ˙on accretion as a likely energy source, with the strongest argument in favour of it being that nobody can think of another viable energy source. These questions are largely theoretical in nature and beyond the scope of this review. (Several reviews of ideas concerning the central engine are to be found in the recent Minnesota lectures on Active Galactic Nuclei; see Jones, 1986).

Besides questions concerning the basic energy source of quasars, other important problems include the following: (a) What is the relationship (if any) between quasars and the nearby Seyfert galaxies? This is important because we can more easily study Seyfert galaxies (they have greater apparent brightnesses and are easily resolved spatially); it was not possible to compare emission lines of Seyfert galaxies and high red-shift quasars before IUE. (b) How can observed properties of quasars, such as the emission line spectrum or the continuum, be used to calibrate them as 'standard candles'? This question is important because quasars can be detected at large redshift; if it were possible to measure their distances reliably and directly, say from their emission line spectra, then both the Hubble constant and the deceleration parameter could be determined directly. Finally, (c) how can the chemical composition of the emission line regions be measured reliably? If it were possible to measure the composition of the emitting gas using the emission-line spectrum (something done all the time with other emission line objects such as H II regions and planetary nebulae) then the initial chemical evolution of galaxies at high redshift and large look-back time could be measured. All three of these questions have cosmological or galactic implications which transcend basic questions concerning the nature of quasars.

3.1. QUASARS AS ACTIVE GALACTIC NUCLEI

This section should rather be titled, active galactic nuclei (AGN) as quasars, since a major contribution of IUE has been to explore the ultraviolet spectrum of low-redshift Seyfert and radio galaxies. Such work allows these nearby objects, whose identification with the nuclei of otherwise fairly normal galaxies is certain, to be compared directly with high-redshift quasars, in which only the redshifted ultraviolet portion of the spectrum is observable from the ground and study of the host galaxy is difficult.

Perhaps the best way to compare properties of high redshift quasars and low redshift Seyfert and radio galaxies is in a statistical sense, as has been done by Wu *et al.* (1980, 1981, and 1983, see also Bergeron *et al.*, 1981, Clavel and Joly, 1984). They were able to compare IUE data on Seyfert galaxies with high redshift data from Baldwin (1977) and Osmer and Smith (1980). Although differences were noted, the most striking result is the overall similarity of the spectra of the two groups; the most prominent lines, Ly α, C IV $\lambda 1549$, C III] $\lambda 1909$ and Mg II $\lambda 2798$ are present with roughly the same relative intensities as in high redshift quasars (Baldwin and Netzer, 1978). The implications of this result are manyfold: the sample includes objects with luminosities distributed over a range of $\sim 10^5$, the fact that the relative intensities and equivalent widths are so similar (although not exactly, see below) is an indication that photoionization is the dominant process in the emission line clouds and these clouds are somehow able to accommodate the range in luminosity by forming at distances from the central object which keep the 'ionization parameter', U (the ratio of ionizing photon to free electron densities) constant. (This had been noted previously by Davidson, 1977). The physics governing these similarities is not yet clear, perhaps the process is basically a selection effect (and clouds with 'improper' values of U are not easily observed in the optical and ultraviolet), or clouds can only form and be stable in certain regions near the central engine (as argued by Ferland and Elitzur, 1984).

The second conclusion of this study is that there is a continuous range of spectra joining the quasars and Seyfert galaxies. This fact links the two groups in a manner which is basically irrefutable; the search is now on for an understanding of the physics governing the full range of the active galaxy phenomenon. The importance of this result cannot be understated; it is possible to set such constraints as central masses (from stellar kinematics), the spatial extent of the emission line regions (especially the narrow-line region), and to study such properties as the occurrence of companion galaxies or the morphology of the host system from observations of Seyfert galaxies. Largely because of these aspects, a great deal of the attention which was focused on quasars ten years ago is now centred on the more easily studied Seyfert galaxies.

3.2. QUASARS AS STANDARD CANDLES

One goal of AGN research is to be able to measure the intrinsic luminosity of distant quasars from observed properties such as the emission line spectrum. A first step was taken by Baldwin (1977), who found a possible anticorrelation between the continuum luminosity at $\lambda 1450$ Å and the equivalent width of C IV $\lambda 1549$ emission line. The correlation was strongest for flat radio spectrum sources. Baldwin *et al.* (1978) and Wampler *et al.* (1984) confirmed this anticorrelation for flat spectrum radio sources by observing complete samples of such quasars. By concentrating on complete sample of flat radio spectrum sources these authors were able to eliminate optical selection effects as the cause of the 'Baldwin effect'. Optically selected quasars are subject to strong selection effects but nonetheless they too show the Baldwin effect (Kiang *et al.*, 1983).

In attempting to use the Baldwin effect as a luminosity calibrator for establishing parameters in cosmological models the observational difficulty lies in observing low redshift objects. This is where the IUE has played a vital role. Wampler *et al.* (1984) observed a small but complete sample of low redshift flat radio spectrum quasars and showed that they too showed a Baldwin relationship identical with that of the high redshift samples. Comparison of the IUE and high redshift samples suggests a value of the deceleration parameter q_0 of the order of unity.

One does not expect the Baldwin effect to extend to lower luminosity objects since the C IV equivalent width would then exceed the maximum value possible for a quasar emission line with a reasonable spectral energy distribution (i.e. a covering factor of over 100% would be needed). Wu *et al.* (1983) and Wampler *et al.* (1984) found that this was indeed the case. The increase in equivalent width stopped with Seyfert galaxies and there was thus no correlation between the brightest and faintest Seyfert galaxies. They did however find a continuous progression of equivalent widths, with high luminosity Seyfert galaxies being similar to low luminosity quasars.

The anti-correlation between continuum luminosity and equivalent width of C IV is not understood at present. Wampler *et al.* (1984) suggest that higher luminosity quasars either have (a) softer UV continua, (b) anisotropic continuum emission, (c) systematically different cloud properties, or (d) lower covering factor. Only the first of these is probably ruled our so far. Mushotzky and Ferland (1984) have produced a model invoking a variable ionization parameter and covering factor, while Netzer (1986) has suggested that an accretion disk is producing anisotropic continuum emission. It is also still possible that selection effects are influencing the Baldwin effect. Optical variability must also play a part. Murdoch (1983) and Wamsteker and Colina (1986) suggest that variability of a luminosity-limited sample of objects will be a factor influencing the correlation. Until the Baldwin effect is understood, it cannot be considered a reliable luminosity indicator. This is unfortunate as it removes the most luminous objects in the universe as probes to obtain parameters fundamental to cosmology.

3.3. THE COMPOSITION OF QUASAR EMISSION-LINE REGIONS

The fact that high-redshift quasars observed with look-back times well in excess of 5 billion years and the nearby Seyfert galaxies have similar emission-line spectra

(Wu *et al.*, 1983) is surprising in the light of current thinking regarding the chemical evolution of galaxies. The implication is that there cannot be large differences in chemical compositions among the various classes. Model analysis of high-redshift quasars (Davidson, 1977; Gaskell *et al.*, 1981) suggested that their emission-line regions have roughly solar abundances and models of low-redshift objects continue to employ this mixture. The (roughly) similar abundances are a (presently) poorly understood clue to both the early chemical evolution of galaxies and the origins of the emission-line-region gas.

Although this area of study is still in its infancy, some things can be said about the chemical composition of clouds. Wills *et al.* (1985) studied the Fe II spectra of many objects and came to the conclusion that iron may be overabundant. Continued study of low redshift objects with the IUE confirm the general presence of strong Fe II emission, often as a difficult to observe pseudo-continuum (Netzer *et al.*, 1985). A better understanding of the composition of emission-line region clouds will await a more complete model of the origin of and physical conditions within the broad-line region clouds.

3.4. Steps to a Comprehensive Picture

Quasars can be said to be understood when physical models are capable of reproducing their observed properties. The emission-line regions have the potential of producing the greatest return in knowledge, at least for the questions outlined above and it is towards progress in this direction that we will concentrate here. Important steps towards modelling these regions have been made (see Collin-Souffrin *et al.*, 1986) but many fundamental questions remain unanswered.

IUE has addressed several fundamental questions concerning the nature of the emission-line regions. A first question is the form of the continuum at ionizing wavelengths; this has been addressed by, among others, Gondhalekar *et al.* (1986), Bechtold *et al.* (1984), Netzer (1985), Green *et al.* (1980), MacAlpine (1981, 1986), and Garilli and Tagliaferri (1986), Elvis and Fabbiano (1984). It is now generally agreed that quasars have a multi-component continuum in the 1000 Å to 10 μ region. In the near infrared and long wavelength visible the continuum can be fitted by a powerlaw. At shorter wavelengths (1000 Å $< \lambda <$ 5000 Å) the continuum flattens; the region around 3000 Å is believed to be due to Balmer and Fe II emission of gas associated with BLR. It is the nature of the continuum at shorter wavelengths that concerns us here. This region of continuum, observed principally by IUE, has been 'modelled' by Malkan and Sargent (1982) and Gondhalekar *et al.* (1986) and these authors show that a single temperature blackbody emission can adequately represent the short wavelength continuum in quasars. Malkan (1983) has shown that a better fit is obtained if an accretion disk spectrum is used instead of a blackbody. A word of caution is necessary here; in these 'models' a blackbody or an accretion disk spectrum is fitted to a quasar spectrum this does not mean that a disk has been discovered in quasars. What this does mean is that there may be a thermal component to quasar continuum but the origin of this component is uncertain at present.

Energy-budget arguments based on observations of the ultraviolet and optical emission lines suggest that the continuum must rise into the far ultraviolet (that is, have a energy slope smaller than unity). Direct observation of the far-UV continuum sometimes detects a steepening in slope, which has been interpreted as due to absorption by intervening systems. The growing consensus of opinion is that quasar radiation field may actually peak at energies near 50 eV.

A byproduct of these studies of the ultraviolet radiation field has been the reconnaissance for emission and absorption lines in the 600 to 912 Å region (Green *et al.*, 1980; Gondhalekar *et al.*, 1986; Bruchweiler *et al.*, 1986). The absence of Gunn-Peterson trough due to hydrogen in high redshift quasars assures us that if lines are formed shortwards of 912 Å then they will be detected. The only high ionization lines which have been detected are O VI $\lambda 1034$ Å and N V $\lambda 1240$ Å and maybe higher signal-to-noise data are required to detect weak high ionization lines which may be present. However, the present generation of photoionization models are unable to reproduce the high ionization lines which have been observed at short wavelengths (the so caleld 'O VI $\lambda 1034$ problem') and the most important result of these observations will be to constrain future models.

The most direct challenge to existing photoionization models is provided by the correlation of line and continuum variability in AGNs. In this area the most significant results have been provided by IUE. The strong broad emission lines, which are a signature of AGNs, are formed in BLR which contains numerous small, high density ($\sim 10^{10}$ cm^{-3}) clouds. The clouds are assumed to be in Keplerian motion under the gravitational influence of a central object. The clouds are ionized by ultraviolet radiation emitted at or near the central source and the estimate of the distance (R) between clouds and the source of ionizing radiation is based on this assumption. Computations to reproduce the observed BLR spectrum lead to the 'best' estimates of R and for a typical quasar $R \sim 1$ pc and for a typical Seyfert 1 galaxy $R \sim 0.1$ pc. Direct measurements of R are highly desirable and are just becoming available from comparison of continuum and line variability. Line-continuum variability has been studied in several low-redshift Seyfert 1 galaxies (Urich *et al.*, 1984; Peterson *et al.*, 1983, 1985; Gaskell and Sparks, 1986) and upper limits for delay between continuum and line variability in quasars have been obtained (Gondhalekar *et al.*, 1986). These are only a few such measurements of R but enough to shake our confidence in the photoionization models. In some objects no time lag is seen (Bregman *et al.*, 1986), indicating that R is small compared to the variation time scale.

4. Limitations of IUE Observations

Before the launch of IUE observations of quasars and in particular spectro-photometry of quasars with a 45 cm telescope would have been unthinkable. The impact of IUE on 'quasar science' proves that a 45 cm telescope in space equipped with an imaging detector is comparable in its scientific performance to considerably larger telescopes on ground. Removed from the interfering atmosphere of the Earth and with a well defined operating procedure and the consistent use of optimum calibration procedures and calibration data make IUE a perfect tool for investigating a number of astronomical phenomenon. For quasar astronomy IUE does have its limits. Of course, it would be nice to have a larger telescope, spectrographs with higher resolution etc., however, here only the limitations of IUE as we have at present are considered.

The principal limitations of IUE for faint object observations are the gain variability and the accumulation of particle background in long exposures. The gain variations (as discussed in Section 2) limit the accuracy of low signal levels to 10—20%. Thus variations in the continuum level in a quasar spectrum can only be believable above this limit. The stability of observations gained by a observatory in space and the consistent use of optimum calibration data is lost at low signal levels by microvariations of gain in the IUE cameras. This uncertainty in the continuum level is reflected in the uncertainty in the integrated line flux.

The integration of particle background limits the observations of faint spectra in two ways. Firstly, it is not possible to integrate faint spectra indefinitely as the particle background would saturate the IUE cameras. Secondly, the background noise defines the lowest signal level that can be detected. For exposure times of about six hours signal levels of 5×10^{-15} ergs cm^{-2} s^{-1} Å^{-1} can be detected in the SWP camera with a signal-to-noise ratio greater than 3 (in low dispersion) and this signal level is high enough for uncertainty due to gain variation to be unimportant. For LWR(P) camera the comparable signal level is 1×10^{-15} ergs cm^{-2} s^{-1} Å^{-1} and higher sensitivity (by a factor of 2) is possible between 2600 and 2900 Å particularly in the LWP camera. In both long wavelength cameras the sensitivity shortwards of 2400 Å is considerably lower.

In Figure 3 the 'redshift depth' to which IUE can observe is shown as a function of luminosity of a quasar at 2500 Å. To obtain the curve for SWP camera a detection of continuum signal at 1300 Å (the highest sensitivity region of SWP camera, Bohlin *et al.*, 1980) was assumed. The minimum detectable signal was assumed to be 5×10^{-15} ergs cm^{-2} s^{-2} Å^{-1}. At 1300 Å the continuum between 1182 and 289 Å (in the rest-frame of quasars) can be detected for quasars between redshifts of 0.1 and 3.5. The curve for LWR(P) camera was calculated for detection at 2800 Å and the minimum detectable signal was assumed to be 1×10^{-15} ergs cm^{-2} s^{-1} Å^{-1}. At 2800 Å continuum between 2545 and 622 Å will be detected for quasars between redshifts of 0.1 and 3.5. To calculate the luminosity at 2500 Å the slope of the continuum shortwards of 1216 Å was assumed to be 1.4 and longwards of 1216 Å a slope of 0.8 was assumed (O'Brien, 1986). The 2500 Å luminosities of quasars observed with IUE have been computed from available *V* colour (Wills and Lynds, 1978). It can be seen from Figure 3 that at low redshifts IUE is capable of observing quasars fainter than those detected at present. However, at high redshifts IUE is clearly being 'used to its limits' and not surprisingly detection of high redshift quasars with IUE has proved to be difficult.

A number of assumptions have been implicitly made in calculating the curves in Figure 3: the quasar continuum is assumed not to be reddened — intrinsically or extrinsically, the intergalactic medium is assumed to be transparent to EUV radiation. The slopes in the UV/EUV regions may not be true for all quasars. Figure 3 clearly indicates that IUE is capable of detecting some of the high redshift quasars known at present. The fact that quasars of redshifts greater than 1.75 have not been detected in the short wavelength region suggests that either the

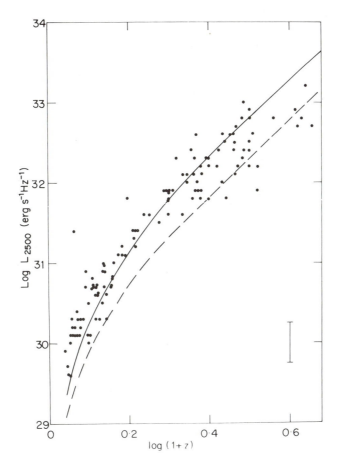

Fig. 3. The sensitivity of IUE. The 'redshift depth' to which IUE can observe is given as a function of luminosity at 2500 Å. Full line for detection at 1300 Å and dashed line for detection at 2800 Å. The error bar shows the uncertainty in the luminosity at 2500 Å. The dots are 2500 Å luminosities of quasars observed with IUE, these luminosities are obtained from the V-magnitudes of observed quasars.

simple assumptions about continuum slope made above are not true or the intergalactic medium (IGM) is not transparent to EUV radiation along all (but one) slightlines of $z_e > 1.75$ observed with IUE. There is some reason to believe that the second possibility is more likely than the first. If this is true then the distribution of optically thick hydrogen clouds (halos) in the IGM is considerably more extensive at lower redshifts ($z \sim 1.75$) than that suggested by observations of halos at higher redshifts. This has implications for evolution of galactic halos and needs to be investigated further.

5. The Future

Attempts at understanding the emission-line regions of quasars are necessarily handicapped by the simplicity of existing models. The present generation of models characterize the emitting clouds as single density, constant pressure 'blobs' located a fixed distance from the central object. The lack of high quality, high spatial and spectral resolution data, showing the true nature of the inner regions, makes further progress entirely theoretical. In particular, the lack of knowledge of the kinematics of gas in the BLR is a major uncertainty in these models. Arguably, the most important accomplishment of IUE has been that it has shown that the spectra of quasars and Seyfert galaxies are basically similar, and this opens the way to high spatial and spectral resolution studies of nearby brighter Seyfert and radio galaxies (from the ground and with the Hubble Space Telescope). These studies should shed light on the inner workings of the more-difficult-to-observe high redshift objects.

The continuum and line variability studies have provided the most significant challenge to our understanding of BLR. Clearly these studies should be continued and extended to more luminous objects. Long term observations of a few well chosen objects may provide us with more information of BLR than the 'single-shot' observations of many objects attempted at present. The variability studies also need to be extended to other regions of electromagnetic spectrum. Simultaneous observations from gamma ray to radio are not technically impossible but new generations of spacecraft are required to make these observations. The benefits of such synoptic multifrequency studies are enormous and these, coupled with fast timing studies, may be our only hope of understanding the true nature of quasar continuum and eventually the central engine.

Acknowledgements

We would like to acknowledge useful comments by G. M. Gaskell.

References

Baldwin, J. A.: 1977, *Astrophys. J.* **214**, 679.

Baldwin, J. A. and Netzer, H.: 1978, *Astrophys. J.* **226**, 1.

Baldwin, J. A., Burke, W. L., Gaskell, G. M., and Wampler, E. J.: 1978, *Nature* **273**, 431.

Bohlin, R. C., Holm, A. V., Savage, B. D., Snijders, M. A. J., and Sparks, W. M.: 1980, *Astron. Astrophys.* **85**, 1.

Bechtold, J., Green, R. F., Weymann, R. J., Schmidt, M., Estabrook, F. B., Sherman, R. D., Wahlquist, H. D., and Heckman, T. M.: 1984, *Astrophys. J.* **281**, 76.

Bergeron, J., Maccacaro, T., and Perola, C.: 1981, *Astron. Astrophys.* **97**, 94.

Bregman, J. N., Glassgold, A., Huggins, P., and Kinney, A.: 1986, *Astrophys. J.* **301**, 698.

Casatella, A., Barbero, J., and Benvenuti, P.: 1985, *Astron. Astrophys.* **144**, 335.

Clavel, J. and Joly, M.: 1984, *Astron. Astrophys.* **131**, 87.

Collin-Souffrin, S., Dumont, S.: 1986, *Astron. Astrophys.* **166**, 13.

Davidsen, A. F., Hartig, G. F., and Fastie, W. G.: 1977, *Nature* **269**, 203.

Davidson, K.: 1977, *Astrophys. J.* **218**, 20.

Elvis, M. and Fabbiano, G.: 1984, *Astrophys. J.* **280**, 91.

Ferland, G. J. and Elitzur, M.: 1984, *Astrophys. J. Letters* **285**, L11.

Garilli, G. and Tagliaferri, G.: 1986, *Astrophys. J.* **301**, 703.

Gaskell, G. M., Shields, G. A., and Wampler, E. J.: 1981, *Astrophys. J.* **249**, 443.

Gaskell, G. M. and Sparke, L. S.: 1986, *Astrophys. J.* **305**, 175.

Gondhalekar, P. M., O'Brien, P., and Wilson, R.: 1986, *Monthly Notices Roy. Astron. Soc.* **222**, 71.

Green, R. F., Pier, J., Schmidt, M., Estabrook, F., Lane, A., and Wahlquist, H. D.: 1980, *Astrophys. J.* **239**, 483.

Hewitt, A. and Burbidge, G.: 1980, *Astrophys. J. Suppl.* **43**, 57.

Holm, A. V.: 1981, *NASA IUE Newsletter*, No. 15.

Jones, T. W.: 1986, *Publ. Astron. Soc. Pacific* **98**, 129.

Kiang, T., Chang, F. M., and Zhou, Y. Y.: 1983, *Monthly Notices Roy. Astron. Soc.* **203**, 25.

MacAlpine, G. M.: 1981, *Astrophys. J.* **251**, 465.

MacAlpine, G. M.: 1986, *Publ. Astron. Soc. Pacific* **98**, 134.

Malkan, M. A. and Sargent, W. L. W.: 1982, *Astrophys. J.* **254**, 22.

Malkan, M. A.: 1983, *Astrophys. J.* **268**, 582.

Murdoch, M. S.: 1983, *Monthly Notices Roy. Astron. Soc.* **202**, 987.

Mushotzky, R. F. and Ferland, G. J.: 1984, *Astrophys. J.* **278**, 558.

Netzer, H.: 1985, *Astrophys. J.* **289**, 451.

Netzer, H.: 1986, *Monthly Notices Roy. Astron. Soc.* **216**, 63.

Netzer, H., Wamsteker, W., Wills, B. J., and Wills, D.: 1985, *Astrophys. J.* **292**, 143.

O'Brien, P.: 1986, Ph.D. Thesis, University of London.

Osmer, P. S. and Smith, M. G.: 1980, *Astrophys. J. Supplement* **42**, 333.

Peterson, B. M., Foltz, B., Crenshaw, D., Meyers, K., and Byard, P.: 1983, *Astron. J.* **88**, 926.

Peterson, B. M., Meyers, K., Capriotti, E., Foltz, C., Wilkes, B., and Miller, H.: 1985, *Astrophys. J.* **292**.

Settle, J., Shuttleworth, T., and Sandford, M. C. W.: 1981, *NASA IUE Newsletter*, No. 15.

Ulrich, M. H. *et al.*: 1984, *Monthly Notices Roy. Astron. Soc.* **202**, 221.

Véron-Cetty, M-P. and Véron, P.: 1984, ESO SP No. 1.

Wampler, E. J., Gaskell, C. M., Burke, W. C., and Baldwin, J. A.: 1984, *Astrophys. J.* **276**, 403.

Wamstekar, W. and Colina, L.: 1986, *Astrophys. J.* **311**, 617.

Wills, D. and Lynds, R.: 1978, *Astrophys. J. Supplement* **36**, 317.

Wills, B. J., Netzer, H., and Wills, D.: 1985, *Astrophys. J.* **288**, 94.

Wu, C-C.: 1977, *Astrophys. J. Letters* **217**, L117.

Wu, C-C., Boggess, A., and Gull, T. R.: 1980, *Astrophys. J.* **242**, 14.

Wu, C-C., Boggess, A., and Gull, T. R.: 1981, *Astrophys. J.* **247**, 449.

Wu, C-C., Boggess, A., and Gull, T. R.: 1983, *Astrophys. J.* **266**, 28.

BLAZARS

JOEL N. BREGMAN

National Radio Astronomy Observatory[1], Charlottesville, VA, U.S.A.

LAURA MARASCHI

Dipartimento de Fisica, Universita Degli Studi di Milano, Italy

and

C. MEGAN URRY

Physics Department, Massachusetts Institute of Technology, Cambridge, MA, U.S.A.

1. Introduction

Early in the study of quasars and active galactic nuclei it was discovered that the optical brightness of some sources varied by a magnitude or more. Further studies in the past two decades revealed that there is a class of objects in which variability is accompanied by high polarization and flat radio spectra. These characteristics are found in sources with emission lines (Optically Violently Variable quasars) and without emission lines (BL Lacertae objects). Collectively, these sources are referred to as blazars and have been the subject of a great deal of study (review by Angel and Stockman, 1980).

The properties of these objects make them ideally suited for the study of continuum emission. The variability, which is detectable in some sources with observations separated by only hours or days, carries important information about the size of the emitting region. The polarization of the optical continuum has been measured to be above 30% in several objects, indicating an ordered emitting region that may be suitable for simple modeling. Models that have been developed to understand the blazar phenomenon interpret the radio, infrared, and optical emission as incoherent synchrotron radiation. It is suggested, but not proven, that the X-ray emission is caused by another mechanism, such as inverse Compton radiation (Jones *et al.*, 1974). Observational evidence that bears on the connection of the X-ray emission with that at lower frequencies is a major issue.

In view of observations at other wavebands and theoretical models, the ultraviolet region is of critical importance. It is where the continuum is expected to change from being dominated by synchrotron radiation to being dominated by inverse Compton radiation. Synchrotron loss processes are more severe in the ultraviolet than in the optical-infrared regions and evidence of such losses contains independent information about magnetic fields and energy densities. The ultraviolet region is also the highest frequency at which these sources could be observed regularly for several years. Not a single blazar had been studied in the

[1] National Radio Astronomy Observatory is operated by Associated Universities, Inc., under contract with the National Science Foundation.

Y. Kondo (ed.), Scientific Accomplishments of the IUE, pp. 685–702.
© 1987 *by D. Reidel Publishing Company.*

ultraviolet prior to the launch of the International Ultraviolet Explorer. Here we summarize a variety of vigorous programs carried out during the past eight years on the IUE that are aimed at developing a deeper understanding of the continuum properties in this volatile class of objects.

2. Ultraviolet Spectra of the Blazar Class

The blazar class has been discussed most recently by Angel and Stockman (1980), who listed all known objects that met the criteria of variability, polarization, and flux properties. Sources meeting their selection criteria that have been observed by the IUE are summarized in Table I. In keeping with these criteria, NGC 1275 has been included, although it is considered in more detail elsewhere in this volume. Four additional blazars that were discovered since 1980 have been included: H 0323 + 022, 0716 + 71, H 1218 + 304, and PKS 2005—489. The data in Table I are based upon information either published or brought to our knowledge before Feburary 1986 and includes the results of a systematic search of the data archives up to December 1983, which were partially published in Maraschi (1984), Treves *et al.* (1986), and Ghisellini *et al.* (1986).

In Table I, the coordinate designation and names of the objects are given in columns 1, 2, the redshift is listed in column 3 as is an associated letter that indicates how the redshift was determined. Redshifts derived from an underlying galaxy is designated by 'g', while 'w' and 's' indicate the presence of weak or strong emission lines; redshifts determined from absorption lines are given as lower limits. The flux densities at 2500 and 1500 Å (columns 4, 5) have been dereddened with the value of A_v (column 7) according to the ultraviolet extinction law of Seaton (1980); minimum and maximum fluxes are listed for sources in which variability was detected. The spectral index α ($F_v \propto v^{-\alpha}$; column 6) was usually determined by a power law fit to the combined long and short wavelength spectra. This procedure ignores the possibility of curvature of the spectrum and of rapid variability (shorter than a day). However, in our experience, it gives a more reliable estimate of the spectral shape than individual fits to each wavelength range. The resulting errors, ignoring extinction uncertainties, are, for reasonably exposed spectra, of the order of 10% in α. Sources with faint spectra or lacking exposures with both cameras have poorly determined slopes ($\pm$ 50%). References and comments are reported in column 8. In cases of repeated observation the largert and smallest reported slopes are given.

2.1. ULTRAVIOLET SPECTRAL SLOPE OF BLAZARS AND COMPARISON WITH OTHER CLASSES OF OBJECTS

An advantage of using the IUE to determine continuum slopes is that the ultraviolet waveband is practically free from contamination by starlight from the underlying galaxy. However, an important selection effect should be mentioned about the observation of blazars. Because these sources are among the faintest objects observed with the IUE, it is likely that blazars known to have an especially steep optical spectrum were not chosen for study with the IUE (for fear of nondetections). Consequently, many sources that might have especially steep

TABLE I

Ultraviolet data for blazars

Object	Name	z	$F_{\nu(2500)}$	$F_{\nu(1500)}$	α	A_v	Reference — Comments
0215+015		>1.686	0.7	—	—	—	Blades et al., 1985; mup
0219+428	3C 66A	0.44 w	1.3	0.57	1.6	0.15	Maccagni et al., 1983
							Worrall et al., 1984c[m]
0235+164	AO	>0.852	0.1	—	—	—	Snijders et al., 1982
0316+41	NGC 1275	0.0172	3.8	1.9	1.6	0.6	Maraschi et al., 1984
0323+022	H	~0.13 g	0.3	0.1	1.4	—	Feigelson et al., 1986[m]
0521−365	PKS	0.055 w	0.36	0.16	1.5	—	Danziger et al., 1983
			0.25	0.1	1.5	—	Ulrich et al., 1984
0537−441	PKS	0.894 w	1−2	—	1.5	0.2	Maraschi et al., 1985[m]
0548−322	H	0.069 g		0.3	0.8	—	Urry et al., 1982; mup
0716+71		—	0.17	0.1	1	—	Frichi et al., 1981
							Maraschi et al., 1984
0735+178	PKS	>0.424 s	0.8	0.3−1.5	1.8−1.5	0.15	Bregman et al., 1981, 1984[m]
0736+017		0.191 s	0.63	0.35	1	0.3	
0754+10	OI 90.4	—	1.5	0.7	1.5	—	Worrall et al., 1984c[m]
			1	0.35	2.1	—	
0829+046	OJ 049	—	1.7	0.6	2.1	—	Maraschi et al., 1984; mup
0851+202	OJ 287	0.306 w	0.8	0.35	1.8	0.1	Worrall et al., 1982[m]
			5	2.8	1.2	—	Maraschi et al., 1986
0912+297	OK 222	—	1	0.5	1.5		Worrall et al., 1986[m]
1101+384	Mkn 421	0.0308 g	3.6	2.2	1.2		Ulrich et al., 1984; mup
			12.1	8.7	0.75		
1133+704	Mkn 180	0.046	0.61	0.42	0.7	—	Mufson et al., 1984[m]
1156+295	TON 599	0.729 s	0.75	0.3	1.7		Glassgold et al., 1983[m]
			4.5	1.7			
1215+303	ON 325		—	0.6	1		Worrall et al., 1984b[m]
1219+305	2A	~0.13 g	0.6	0.4	0.64		Urry et al., 1984
1219+285	ON 231	0.102 g	0.7	0.3	1.7		Worrall et al., 1986[m]
1308+32	B2	0.996 s	0.5	—	0.7		Maraschi et al., 1984; mup
1418+54	OQ 530	—	0.53	0.21	1.8		Worrall et al., 1984b[m]

Table I (Continued)

Object	Name	z	$F_{\nu(2500)}$	$F_{\nu(1500)}$	α	A_v	Reference — Comments
1514 − 24	APLib	0.049 w	1.9	—	2	0.2	
1641 + 399	3C 345	0.585 s	0.9–1.5	0.4–0.75	1.3–1.8		Bregman et al., 1986b[m] Snijders et al., 1979
1652 + 398	Mk 501	0.0337 g	1.8–2.6	0.9–1.3	0.7–1.1		Kondo et al., 1981[m] Mufson et al., 1984[m]
1727 + 502	IZw 187	0.0554 g	0.32	0.2	1		Bregman et al., 1982a[m]
1807 + 698	3C 371	0.05 w	1.7	0.7	1.7	0.13	Worrall et al., 1984a[m]
1845 + 797	3C 390.3	0.056 s	0.7	0.4	1.2	0.3	Ferland et al., 1979 Oke and Goodrich, 1981
2005 − 489	PKS	—	6.3	3.5	1.4		Wall et al., 1986
2155 − 304	PKS	0.118 g	4.5–11	8.7–1.7	1.2–0.7		Maraschi et al., 1986; mup
2200 + 420	BLLAC	0.069 w	1.5	—	3	1	Bregman et al., 1982b[m]
2223 + 056	3C 446	1.404 s	0.1–0.7	—	3	0.15	Bregman et al., 1986a, c[m] Brown et al., 1986 Garilli and Tagliaferri, 1986

m   denotes multifrequency spectrum.
mup indicates that multifrequency data exist but are unpublished.

ultraviolet spectrum were never observed, so the distribution of spectral slopes reported in Figure 1 is probably deficient for $\alpha > 2$. Therefore, $\langle \alpha \rangle = 1.4$ is less than what would have been determined if a complete sample of blazars could have been observed. Nevertheless, the mean slope is still substantially greater than that found for quasars by Richstone and Schmidt (1980), who derived a mean spectral index of 0.6 at 2500 Å in the rest frame. Much of the dispersion (and the flatness) in the spectral indices of quasars found by Richstone and Schmidt (1980) probably arises from the presence of a blue bump component and it has been claimed that the underlying power law index is remarkably constant at about 1.1 (Malkan and Sargent, 1982; Elvis, 1985). The blue bump component is far less important in blazars (and frequently not present; see Section 3), so the considerable dispersion of the spectral index in blazars probably carries direct information about the underlying electron population or about the spectral shift of a spectrum for which Doppler boosting is important.

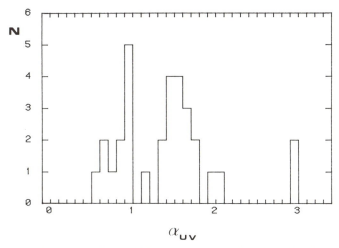

Fig. 1.   Distribution of spectral indices of blazars observed in the ultraviolet. In cases of observed spectral variability, the arithmetic mean between maximum and minimum values has been used. The mean slope, 1.4 is greater than the value of 0.6 found for normal quasars.

An interesting point concerns the X-ray selected blazars. We define a sample of sources appearing above the completeness limit of the X-ray catalogues of Piccinotti *et al.* (1982) and Wood *et al.* (1984), which cover a large fraction of the sky. Six sources in Table I satisfy this criterion: H 0323 + 022, PKS 0548−322, Mkn 421, H 1218 + 304, Mkn 501, and PKS 2155−304. Ultraviolet spectral indices less than 1.0 have been seen in only six objects, four of which are in this subsample. In fact, the mean ultraviolet spectral index of this subsample is 0.9, which is significantly flatter than the average for the whole group. The association of the strong X-ray emission with the flat ultraviolet spectrum is suggestive of a connection between the two spectral domains (e.g., a common emission mechanism). A deeper examination of this connection for the entire sample is given below (Section 3).

2.2. Ultraviolet variability

Several of the 33 blazars listed in Table I have been observed repeatedly and in at least 11 cases, the flux density was found to vary by more than 50% and in some cases, greater than an order of magnitude (1156 + 295, Mkn 421, 3C 446, OJ 287). However, it is impossible to determine quantitatively the frequency with which ultraviolet flares occur because many of the objects were chosen for study based upon their degree of optical variability just prior to IUE observations. The shortest spacing between IUE observations is generally a day or longer; few observations address the issue of variation during a single day (the exposure times themselves are 2—8 hr). Given these limitations, the timescale of variability for ultraviolet emission appears to be approximately the same as for optical variation. Changes on a timescale of a day are rare but have been observed while variability with a timescale of a week or greater is more common.

Variation in the ultraviolet spectral slope has also been detected but is rarely, if even large. Even when the intensity variation is great, the spectral index varies by less than 0.5. No strong correlation exists between the intensity level and spectral slope, although weak trends were found for a few cases in which the spectrum hardens as the source brightens (Mkn 421, Ulrich *et al.*, 1984; OJ 287, Maraschi *et al.*, 1986; Figure 2). This flux — spectrum correlation, which is barely evident in blazars, is well defined in Seyfert galaxies (Perola *et al.*, 1982; Barr *et al.*, 1983).

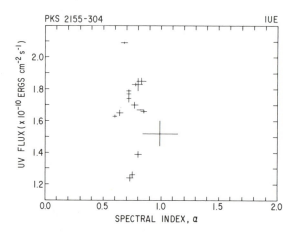

Fig. 2.   The distribution of spectral index and flux for PKS 2155—305 during 1978—81 (Urry *et al.*, 1986). No apparent trend is evident. Slopes determined after 1982 (not shown) tend to be steeper and fainter than during 1978—82 (Maraschi *et al.*, 1986).

2.3. Ultraviolet emission line characteristics

Emission lines provide an unambiguous measurement of the velocity of thermal gas for the inner few parsecs and they contain important information about the ionizing continuum. Since the ionizing continuum of BL Lacertae objects is indistinguishable from that of violently variable quasars, both classes of sources have sufficient ionizing radiation to create emission lines (provided that the emission line gas sees the same ionizing continuum as we). It has been suggested that BL Lacertae objects might have emission lines, but the line ratios would be considerably different from quasars and that only the highest ionization lines, such as O VI $\lambda 1035$ Å, would be visible (Krolik *et al.*, 1978). No ultraviolet emission lines have been detected in any of the BL Lacertae objects observed, with upper limits to the equivalent widths of about 10A.

Several objects on Angel and Stockman's list with known optical emission lines were observed with the IUE. The source with the largest equivalent width broad lines is 3C 390.3, which is one of the only blazars that is a broad line radio galaxy. However, nearly all other blazars may be characterized as having smaller equivalent widths for the broad emission lines than normal quasars. This result occurs partly because many of these objects were chosen for observation when the continuum was bright, so that the broad lines, which do not vary in most sources, appear washed out and have small equivalent widths. Of greater importance is the steepness of the continuum in reducing the equivalent width of the lines. Rather than compare equivalent widths, it may be more significant to calculate the covering factor of the broad line clouds, which is the fraction of ionizing photons that should be absorbed in order to account for the Lα line strength. The covering factors for the violently variable quasars PKS 0521—36, 3C 466, and 3C 345 are similar to the ordinary quasars observed by Richstone and Schmidt (1980), by Kinney *et al.* (1985), and by Bregman *et al.* (1986a, b). Consequently, the strengths of the broad lines in these objects are about 'normal' for the number of ionizing photons available.

The presence of narrow lines depends upon the host galaxy and whether or not the object is nearby. Weak narrow emission lines are seen in 3C 371, which is a low redshift N galaxy with interesting absorption lines but no broad emission lines (Miller, 1975; Worrall *et al.*, 1984).

Of special interest is 3C 446, the only high luminosity OVV in which broad emission line variability has been seen, both in the optical and the ultraviolet (Stephens and Miller, 1985; Bergman *et al* 1986b). The line and continuum fluxes usually change together, thereby maintaining constant equivalent width. The time-scale for variability of Lα is two months (rest frame). Because the distance of the broad line region from the ionizing source as deduced from ionization and luminosity considerations is about a light decade, the broad emission line gas was inferred to be distributed asymmetrically about the central source. Much of the broad line gas must lie close to our line of sight on the near side of the quasar.

3. The Connection of the Ultraviolet Spectra to Spectra at Other Frequencies

3.1. SINGLE EPOCH MULTIFREQUENCY SPECTRA

Because of the rapid variability in the flux density of these objects in all observable wavebands, the connection between the ultraviolet emission and that in other spectral regions requires simultaneous observations (i.e., taken over a time period less than the shortest variability timescale, which is a day to a few weeks, depending upon the particular source). Although the ability to obtain high quality optical, near infrared and radio data existed prior to the launch of the IUE, sensitive X-ray, far infrared and submillimeter measurements became available only during the lifetime of the IUE. Despite the difficulties of assembling large observing teams and coordinating many observatories, some of them space-based, several independent programs were undertaken to obtain single epoch spectra spanning almost ten decades in frequency. Two dozen BL Lacs and OVVs were observed in these programs, many of them more than once (Table I). Additional data from ground-based observations alone, spanning a smaller frequency range, have also been obtained for other sources. Compilations and discussions of multifrequency data for samples of blazars are given by Cruz Gonzales and Huchra (1984) and Ghisellini et al. (1986).

The multifrequency spectra of blazars, which are listed in Table I, has some striking similarities. The radio spectra are flat or rising in the 10^8—10^{10} Hz region, probably due to a superposition of components that become optically thick at different frequencies. In the 1 mm—60 μm region, the spectrum steepens and may be characterized by a power law of index 0.6 to 1.4 in the infrared and near infrared regions. Between the infrared and the ultraviolet regions, spectral steepening occurs (mean change in slope of 0.5), Ghisellini et al. (1986), the magnitude of which varies considerably in the current sample. For example, 0235 + 164, BL Lacertae, and 3C 446 have spectra that steepen exponentially between the infrared and ultraviolet regions. This dramatic spectral steepening, which occurs at a few microns in the class of objects known as red quasars, is probably a common phenomenon that can occur in the infrared through X-ray range (Rieke et al., 1982). This spectral behavior has been interpreted as resulting from an abrupt cutoff in the underlying electron spectrum (Beichman et al., 1981; Bregman et al., 1981) associated with the difficulties of accelerating particles above a specific energy. In these sources the spectral steepening is determined by the high frequency tail of the single electron emission function rather than by the detailed shape in the cutoff of the electron spectrum.

Less dramatic spectral softening that occurs over decades of frequency space may reflect changes in the actual shape of the underlying electron spectrum, complex mass motions or nonhomogeneous emitting regions.

The connection between the ultraviolet and X-ray emission is sometimes investigated by determining whether the ultraviolet continuum extrapolates smoothly to the X-ray data. In doing so, we bear in mind that most X-ray observations were not simultaneous with the ultraviolet measurements. If this extrapolation is carried out assuming that the continuum does not steepen between the ultraviolet and X-ray region, the extrapolation passes at least a factor of two below the X-ray flux in about 40% of the sources in Table I. This calcualtion was also performed by permitting spectral curvature between the ultraviolet and X-ray region, where the adopted spectral curvature is determined by the difference in the infrared and ultraviolet slopes (i.e., constant curvature for the infrared through X-ray region as defined in a log F_ν log ν plot). The extrapolation falls at least a factor of two below the X-ray flux in about 65% of the objects. In these cases, the X-ray flux is probably produced by a different mechanism than the ultraviolet emission or it arises in a different spatial location. Of the sources for which an extrapolation of the ultraviolet continuum passes through or above the X-ray flux, most are either X-ray or optically selected objects, even though these are a minority of the entire sample (PKS 0548—322, Mkn 421, Mkn 180, 1218 + 304, Ap Lib, Mkn 501, IZw—187, 2155—304). These are the 'X-ray strong blazars' discussed by Ledden and O'Dell (1985) or the 'radio weak blazars' discussed by Maraschi *et al.* (1986). Ultraviolet and multifrequency data suggest that for these sources, the X-ray emission may be a high frequency extension of the synchrotron emission that produces the ultraviolet, optical, and infrared emission. Studies of the X-ray spectra reveal that PKS 2155—304, 0548—322, Mrk 501, Mrk 421, and 1218 + 304 have steep soft X-ray spectral slopes that flatten at higher energies (Urry and Mushotzky, 1982; Urry, 1984, 1986; Singh and Garmire, 1985, Morini *et al.*, 1986; Makino *et al.*, 1986, Brodie *et al.*, 1986). This may be understood as steepening of the synchrotron emission in the soft X-ray region and the dominance of another emission process, such as inverse Compton radiation at higher energies.

The relation in the continuum properties of blazars is in contrast to the situation for ordinary quasars, where the ultraviolet spectrum is considerably harder and a similar extrapolation passes well above the X-ray emission. In ordinary quasars, the blue bump makes an important contribution to the ultraviolet continuum. For the BL Lacertae objects, no blue bump has ever been unambiguously identified (the blue bump reported in IZw—187 by Bregman *et al.*, 1982a was not confirmed in subsequent observations). In some of the violently variable quasars, such as 3C 345, a blue bump is seen, but it is fairly weak and does not lead to the hard ultraviolet spectrum seen in ordinary quasars (Bregman *et al.*, 1986a). In other objects, such as 3C 390.3 where the broad line strength is considerable, the blue bump appears to be stronger (Oke *et al.*, 1984). This suggests that the presence and strength of a blue bump is directly related to the strength of the broad emission lines.

For nearly all of the radio selected blazars, the primary contribution to the observed power generally comes from the submillimeter through optical region (assuming either isotropic emission or similar beaming of all components), with the contributions from the radio, ultraviolet, and X-ray regions being at least several times smaller (Figure 3). In contrast, most of the X-ray and optically selected blazars have their peak power per logarithmic bandwidth in the ultraviolet and X-ray region (Figure 3).

3.2. VARIABILITY IN OTHER WAVEBANDS COMPARED TO ULTRAVIOLET ACTIVITY

A comparison of the variation in the flux density at ultraviolet wavelengths with that at other wavebands is a particularly powerful method of probing the physical connection between the spectral regions. Unfortunately, this requires a considerable effort by a number of observers and has been carried out for only a fraction of the blazar sample (Table I).

Flux variations in the optical and infrared regions are generally correlated with those in the ultraviolet, but the variation in the two regions is not always identical.

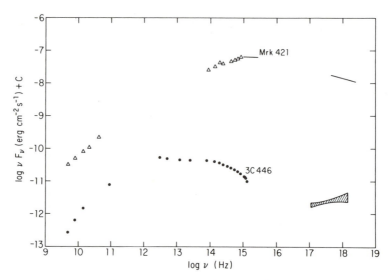

Fig. 3. The power per logarithmic bandwidth for Mrk 421 (Makino *et al.*, 1986) and 3C 446 (Bregman *et al.*, 1986c). For 3C 446, most of the observed power emerges in the far infrared through optical region while for Mrk 421, most of the observed power emerges near the ultraviolet region. For Mrk 421, the bright soft X-ray emission may be a continuation of the synchrotron emission that is responsible for the radio, infrared, optical, and ultraviolet continuum. The lack of a smooth connection between the ultraviolet and X-ray continuum in 3C 446 suggests that the X-ray flux is produced by a mechanism other than the synchrotron process, probably by the inverse Compton mechanism.

Flux density variations in which the shape of the spectrum is preserved in infrared-optical through ultraviolet regions have been seen in 3C 371, 3C 66A, 0735 + 178, 3C 446, and OJ 287 (Worrall *et al.*, 1984a, b; Bregman *et al.*, 1984, 1986c). This was also the case for 1156 + 295 (Glassgold *et al.*, 1983), in which a sequence of observations was made near and immediately after the peak intensity of a particularly violent outburst (Figure 4). No pronounced spectral variation occurred as the source dimmed, although some separate infrared data showed modest but temporary steepening. In some of the sources, spectral variation has occasionally been seen (3C 371, 0735 + 178, 3C 345; Worrall *et al.*, 1984; Bregman *et al.*, 1984, 1986a). The sense of the variation is that the ultraviolet emission often (but not always) shows greater change than optical or infrared emission. In the X-ray bright sources, the X-ray variation is often greater and more rapid than at lower frequencies (see below). Despite these few cases, the data are too sparse to determine in general whether the spectrum of a source softens or hardens during brightness variations.

   The connection of the ultraviolet variability to that in the radio and submillimeter wavelengths is less direct. Radio variability is clearly not directly connected to the ultraviolet brightness since there are several examples in which variation is seen in only one waveband (usually the ultraviolet) or variations in both wavebands occur but in the opposite sense (3C 371, 0735 + 178, 3C 446; Worrall *et al.*, 1984, Bregman *et al.*, 1984, 1986c). It is tempting to expect that variability in

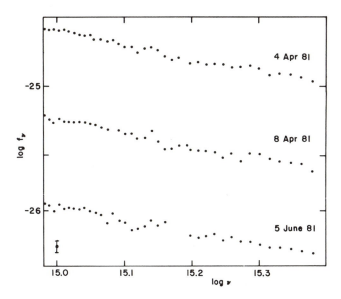

Fig. 4. The ultraviolet continuum of $1156 + 295$ decreased rapidly during April 1981 following a dramatic flare in which it reached $B = 13$ mag. The spectral slope of 1.7 was preserved as the source dimmed.

the ultraviolet is connected to that in the radio through a common emission process (synchrotron radiation), but that this connection is obscured by time delays, energy losses, or particle acceleration. A connection between optical and radio outbursts has been found from long term monitoring data for OJ 287 (Pomphrey *et al.*, 1976, Balonek, 1982), AO $0235 + 164$ (Balonek and Dent, 1980), 3C 345 (Balonek, 1982, Bregman *et al.*, 1986a), and is suspected for several other blazars (Balonek, 1982). This decade old finding for OJ 287 was recently reaffirmed during the 1983 outburst in which the source was highly variable from the radio through the ultraviolet band at the same time (Holmes *et al.*, 1984; Moles *et al.*, 1984; Aller *et al.*, 1985; Maraschi *et al.*, 1986).

A comparison between the ultraviolet and X-ray variability displays an unusually broad range of behavior. The ultraviolet emission from $0735 + 178$ increased by nearly a factor of four in observations separated by a year, but simultaneous X-ray observations failed to show any increase (Bregman *et al.*, 1984; the systematic uncertainties in the absolute flux of Einstein IPC measurements are about 20%). The subsequent decrease in the ultraviolet brightness over a year was also unaccompanied by any change in the X-ray emission. Only modest radio variability occurred during this time, which is consistent with the limits on the X-ray variation and led to the suggestion that the soft X-ray emission and the radio emission may be linked through inverse Compton scattering of radio photons (the difference between the optical-ultraviolet and the X-ray emission is supported by at least one multifrequency spectrum in which an extrapolation of the ultraviolet continuum passes below the X-ray datum).

In several other sources, the ultraviolet and X-ray emission tends to be more closely related. In Mkn 421, a modest decrease in the both the ultraviolet and X-ray flux were seen in widely separated observations (1978 and 1984; Pounds, 1985). Makino *et al.* (1986) have recently completed an intensive study of Mkn 421 and find that the ultraviolet, optical, and infrared fluxes decreased by about 20—30% in five weeks while the radio flux remained unchanged (to within 7%) and the X-ray flux decreased by a factor of about two. A similar behavior was found for PKS 2155—304 (Morini *et al.*, 1985, 1986). In both cases, the change in the hard X-ray band was greater than in the soft X-ray band, which could arise from an inhomogeneous emitting region or from distinct components. In contrast is the behavior of 3C 446, where the ultraviolet, optical, and X-ray emission decreased together while slower, less dramatic dimming occurred at radio frequencies (Figure 5; Bregman *et al.*, 1986c). Similar, but less well documented

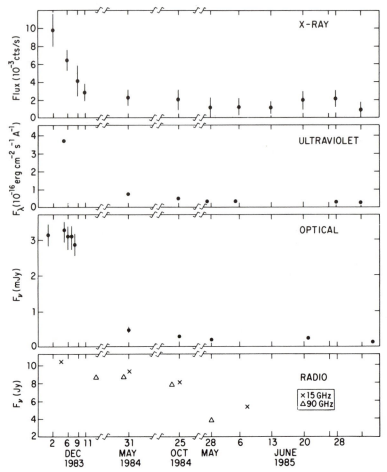

Fig. 5. Flux density variations at X-ray, ultraviolet, optical, and radio frequencies for 3C 446 between December 1983 and June 1985. Rapid variations in the X-ray flux during December 1983 are not seen in the optical waveband. The ultraviolet and optical flux variations were identical and are similar to the long term variation of the X-ray emission. Variability at radio frequencies is slower and of lesser magnitude than at higher frequencies.

variability is seen in OJ 287 (Pollock *et al.*, 1985) and PKS 0537—441 (Tanzi *et al.*, 1986), but an especially perplexing source is 1156 + 295 (McHardy *et al.*, 1986). In 1156 + 295, a well resolved X-ray outburst occurred between two optical outbursts that were separated by a few months; it is ambiguous which optical outburst (if any) the X-ray outburst is associated with.

The wide variety of X-ray variation when compared to the emission at ultraviolet and other wavelengths suggests that there may be more than one X-ray emission process or more than one X-ray emission region. The X-ray emission (particularly in bright sources) may be explained as an extension of the ultraviolet synchrotron emission (e.g. Mkn 421). This may be the case for optically and X-ray selected blazars in general (Ledden and O'Dell, 1985). However, this picture cannot explain the behavior of 3C 446, in which an extrapolation of the steep ultraviolet spectrum passes orders of magnitude below the X-ray emission. In cases such as this, the X-ray emission is likely to arise from the inverse Compton process in which the scattering occurs in the infrared-ultraviolet emission region. The likely explanation for the X-ray emission in 0735 + 178 is that it arises from inverse Compton scattering that occurs in the radio emission region (Bregman *et al.*, 1984).

4. Theoretical Considerations

Synchrotron emission is the dominant process in the radio through ultraviolet region, while the X-ray emission probably arises from either the synchrotron or the inverse Compton process. These processes have been combined into a synchrotron self-Compton model, in which relativistic plasma produces emission at radio through ultraviolet frequencies via the synchrotron process while some of these photons are inverse Compton scattered by the plasma to X-ray frequencies (e.g. Jones *et al.*, 1974). Through VLBI constraints and radio flux variability, it was recognized that the emitting plasma might have relativistic bulk motion, with Lorentz factors of a few. The interpretation of flat radio spectra as a superposition of partly opaque components (Condon and Dressel, 1973; Marscher, 1977) and VLBI studies that show the source size changing with frequency (e.g. Cotton *et al.*, 1980) have led to models with non-homogeneous plasma, often in a jet-like geometry (Blandford and Konigl, 1979; Marscher, 1980; Konigl, 1981; Reynolds, 1982; Ghisellini *et al.*, 1985; Hutter and Mufson, 1986). In these jet models, the electron density and magnetic field decrease with distance alone the jet. The geometry and velocity structure of the jet is either assumed or calculated from some adopted hydromagnetic model. Although there are too few observational diagnostics to make a unique model selection, a range of parameters can be usefully calculated.

The application of jet models to the data leads to the following typical values for the physical conditions in the plasma responsible for the far infrared through ultraviolet emission: magnetic fields that of 10^{-2}—10^2 G, particle densities of 10^2—10^6 cm^{-3}, and sizes of light days to light months. Calculations of the Doppler boosting factor depend upon assumptions about how variability timescales translate into physical sizes and whether the timescale for electrons to lose their radiation contains information about the source size. Estimates for the boosting

factor range from unity in several cases (no relativistic bulk motion required) to
> 10, but most sources have values of 2—5. The suggestion has also been put
forward that X-ray selected objects are viewed at large angles while radio selected
ones are viewed at small angles (Maraschi *et al.*, 1986). It is interesting that in a
few sources, the energy density in particles, magnetic fields, and photons are
comparable in the far infrared through ultraviolet emitting region.

5. Summary and Conclusions

The 33 blazars that have been observed with the IUE since launch have permitted
investigators to carry out detailed line and continuum studies. The observations
indicate that the lack of broad emission lines in BL Lacertae objects is not due to
the absence of ionizing photons. Either sufficient gas is not present, or the ionizing
flux we observe is not incident upon the gas clouds. The broad emission lines in
violently variable quasars have small equivalent widths when compared to normal
quasars, but the line fluxes are nearly normal for the number of ionizing photons
(the convering factor is nearly normal). In 3C 446, variation of $L\alpha$ in two months
implies that considerable broad emission line gas lies along our line of sight and on
the near side of 3C 446.

The ultraviolet continuum emission from blazars is nearly always steeper than
the optical and infrared continuum. Ultraviolet slopes range from 0.5 to 3, with a
mean of 1.4, which is significantly steeper than the mean slope of ordinary quasars,
0.6. In several sources, the ultraviolet flux density decreased by more than an
order of magnitude in observations separated by months or more while variation
of a factor of four was seen in observations of $1156 + 295$ that were separated by
four days. Only modest ultraviolet slope variation has even been observed
($\Delta \alpha < 0.5$), and there is little, if any correlation between spectral slope variation
and flux density. The observed variability seems to be similar to, if not identical to,
variation at optical and infrared wavelengths. This is not generally true when the
comparison is made with X-ray variations. Changes in the X-ray flux are often
faster and more dramatic than in the ultraviolet waveband. Such changes are
sometimes well correlated with ultraviolet and optical variation, but there are cases
where the ultraviolet emission varies while the X-ray emission does not and cases
where the X-ray emission variation is unaccompanied by similar behavior at
ultraviolet frequencies. This suggests that in some sources the X-ray emission may
be produced by a different process or in a different spatial region than the
ultraviolet emission.

Multifrequency spectra reveal that the nearly universal flat radio spectrum
steepens in the millimeter or submillimeter region and there is a smooth connection between the infrared, optical, and ultraviolet regions. Along with the variability data, this suggests that the ultraviolet radiation, like the infrared and optical
emission, is produced by the synchrotron process. In one extreme type of source
(BL Lac and 3C 446), the optical and ultraviolet continua steepen extremely
rapidly with increasing frequency so that they resemble Red Quasars. Because an
extrapolation of the ultraviolet continuum to higher frequencies passes orders of
magnitude below the X-ray flux, the X-ray emission may be produced by the

inverse Compton mechanism. Although spectral steepening is less dramatic in most blazars, a similar separation between the ultraviolet and X-ray emission exists for about half or more of the sources. At the other extreme, objects such as Mkn 421 and 2155—304 have hard ultraviolet spectra, modest spectral steepening, and an extrapolation of the ultraviolet continuum would connect smoothly to the soft X-ray flux; synchrotron emission probably produces the X-ray emission. A high proportion of sources with this behavior were either X-ray or optically selected.

It is encouraging that multifrequency data provide enough information for use in non-homogeneous synchrotron-self-Compton models. Although there is some latitude in the choice of models, typical physical properties of the plasma emitting the far infrared through ultraviolet emission is: 10^{-2}—10^2 G, densities of 10^2—10^6 cm^{-3}, sizes of light days to light months, and Doppler boosting factors between unity (no relativistic bulk motion) and 10.

Although tremendous progress in our understanding of blazars has been achieved, neither the complex connection between the ultraviolet and the X-ray region nor the detailed structure of the emitting plasma is fully understood. One powerful tool in unraveling these issues is through multiwavelength variability, and a few such studies have already added considerably to our understanding (Section 3.2). Improved model predictions and a more ambitious program of multifrequency variability with better time coverage and for more sources may permit the emitting plasma to be mapped spatially and provide us with a deeper understanding of the structure of the emitting plasma.

Acknowledgements

L. Maraschi is grateful to E. G. Tanzi and A. Treves for a long standing collaboration on the IUE observing program. J. Bregman would like to acknowledge A. E. Glassgold, P. J. Huggins, and A. L. Kinney, with whom IUE observations were planned and obtained.

References

Aller, H. D., Aller, M. F., Latimer, G. E., and Hodge, P. E.: 1985, *J. Sup. Ser.* **59**, 513.
Angel, J. R. P. and Stockman, H. S.: 1980, *Ann. Rev. Astron.* **8**, 321.
Balonek, T. J.: 1982, Ph.D. Thesis, Univ. Mass..
Balonek, T. J. and Dent, W. A.: 1980, *Astrophys J. Letters* **240**, L3.
Barr, P., Willis, A. J., and Wilson, R.: 1983, *Monthly Notices Roy. Astron. Soc.* **202**, 453.
Beichman, C. A., Neugebauer, G., Soifer, B. T., Wooten, H. A., Roellig, T., and Harvey, P. M.: 1981, *Nature* **293**, 711.
Blades, J. C., Hunstead, R. W., Murdoch, H. S., Pettini, M.: 1985, *ESA SP* **288**, 580.
Blandford, R. D. and Konigl, A.: 1979, *Astrophys. J.* **232**, 34.
Bregman, J. N., Lebofsky, M. J., Aller, M. F., Rieke, G. H., Aller, H. D., Hodge, P. E., Glassgold, A. E., and Huggins, P. J.: 1981, *Nature* **293**, 714.
Bregman, J. N., Glassgold, A. E., Huggins, P. J., Pollock, J. T., Pica, A. J., Smith, A. G., Webb, J. R., Ku, W. H.-M., Rudy, R. J., LeVan, P. D., Williams, P. M., Brand, P. W. J. L., Neugebauer, G., Balonek, T. J., Dent, W. A., Aller, H. D., Aller, M. F., and Hodge, P. E.: 1982a, *Astrophys. J.* **253**, 19.
Bregman, J. N. *et al.*: 1982b. *ESA-SP* **176**, 589.

Bregman, J. N., Glassgold, A. E., Huggins, P. J., Aller, H. D., Aller, M. F., Hodge, P. E., Rieke, G. H., Lebofsky, M. J., Pollock, J. T., Pica, A. J., Leacock, R. J., Smith, A. G., Webb, J., Balonek, T. J., Dent, W. A., O'Dea, C. P., Ku, W. H.-M., Schwartz, D. A., Miller, J. S., Rudy, R. J., and LeVan, P. D.: 1984, *Astrophys. J.* **276**, 454.

Bregman, J. N., Glassgold, A. E., Huggins, P. J., Neugebauer, G., Soifer, B. T., Matthews, K., and Elias, J., Webb, J., Pollock, J. T., Pica, A. J., Leacock, R. J., Smith, A. G., Aller, H. D., Aller, M. F., Hodge, P. E., Dent, W. A., Balonek, T. J., Barvanis, R. E., Roellig, T. P. L., Wisniewski, W. Z., Rieke, G. H., Lebofsky, M. J., Wills, B. J., Wills, D., Ku, W. H.-M., Bregman, J. D., Witteborn, F. C., Lester, D. F., Impey, C. D., and Hackwell, J. A.: 1986a, *Astrophys. J.* **301**, 708.

Bregman, J. N., Glassgold, A. E., Huggins, P. J., and Kinney, A. L.: 1986b, *Astrophys. J.* **301**. 698.

Bregman, J. N. *et al.*: 1986c, (in prep.).

Brodie, J., Boyer, S., and Tennant, A.: 1986, *Astrophys. J.* (in press.)

Brown, L. M. J., Robson, E. I., Gear, W. K., Crosthwaite, R. P., McHardy, I. M., Hanson, C. G., Geldzahler, B. J., and Webb, J. R.: 1986, *Monthly Notics Roy. Astron. Soc.* (in press).

Condon, J. J. and Dressel, L. L.: 1973, *Astrophys. Letters* **15**, 203.

Cotton, W. D., Wittels, J. J., Shapiro, I. I., Marcaide, J., Owen, F. N., Spangler, S. R., Rius, A. Angulo, C., Clark, T. A., and Knight, C. A.: 1980, *Astrophys. J. Letters* **238**, L123.

Cruz Gonzales, I. and Huchra, J. P.: 1984, *Astron. J.* **89**, 411.

Donziger, I. J., Bergeron, J., Fosbury, R. A. E., Maraschi, L., Tanzi, E. G., and Treves, A.: 1983, *Monthly Notices Roy. Astron. Soc.* **203**, 565.

Elvis, M.: 1986, in M. Sitho (ed.), in *Continuum Emission in Active Salactic Nuclei*, KPNO, pp. 25—28.

Elvis, M., Green, R. F., Bechtold, J., Schmidt, M., Neugebauer, G., Soifer, B. T., Matthews, K., and Fabbiano, G.: 1986, *Astrophys. J.* Nov. 1, (in press).

Feigclson, E. D. *et al.*: *Astrophys. J.* **302**, 337.

Ferland, G. J., Rees, M. J., Longair, M. S., and Perryman, M. A. C.: 1979, *Monthly Notices Roy. Astron. Soc.* **187**, 65.

Fricke, K. J., Kollatschny, W., and Schleicher, H.: 1981, *Astron. Astrophys.* **100**, 1.

Garilli, B. and Tagliaferri, G.: 1986, *Astrophys. J.* **301**, 703.

Ghisellini, G., Maraschi, L., and Treves, A.: 1985, *Astron. Astrophys.* **146**, 204.

Ghisellini, G., Maraschi, L., Tanzi, E. G., and Treves, A.: 1986, *Astrophys. J.* (in press).

Holmes, P. A., Brand, P. W. J. L., Impey, C. D., Williams, P. M., Smith, P., Elston, R., Balonek, T., Zeilik, M., Burns, J., Heckert, P., Barvainis, R., Kenny, J., Schmidt, G., and Puschell, J.: 1984, *Monthly Notices Roy. Astron. Soc.* **211**, 497.

Glassgold, A. E., Bregman, J. N., Hüggins, P. J., Kinney, A. L., Pica, A. J., Pollock, J. T., Leacock, R. J., Smith, A. G., Webb, J. R., Wisniewski, W. Z., Jeske, N., Spinrad, H., Henry, R. B. C., Miller, J. S., Impey, C., Neugebauer, G., Aller, M. F., Aller, H. D., Hodge, P. E., Balonek, T. J., Dent, W. A., and O'Dea, C. P.: 1983, *Astrophys. J.* **274**, 101.

Holmes, P. A., Brand, P. W. J. L., Impey, C. D., Williams, P. M., Smith, P., Elston, R., Balonek, T., Zeilik, M., Burns, J., Heckert, P., Barvainis, R., Kenny, J., Schmidt, G., and Puschell, J.: 1984, *Monthly Notices Roy. Astron. Soc.* **211**, 497.

Hutter, D. J. and Mufson, S. L.: 1986, *Astrophys. J.* (in press).

Jones, T. W., O'Dell, S. L., and Stein, W. A.: 1974, *Astrophys. J.* **188**, 353.

Kinney, A. L., Huggins, P. J., Bregman, J. N., and Glassgold, A. E.: 1985, *Astrophys. J.* **291**, 128.

Kondo, Y. *et al.*: 1981, *Astrophys. J.* **243**, 690.

Konigl, A.: 1981, *Astrophys. J.* **243**, 700.

Krolik, J. H., McKee, C. F., and Tarter, C. B.: 1978, 'Pittsburgh Conference On BL Lac Objects', A. M. Wolfe (ed.), Univ. Pittsburg, p. 277.

Ledden, J. E. and O'Dell, S. L.: 1985, *Astrophys. J.* **298**, 630.

Maccagni, D., Maraschi, L., Tanzi, E. G., Tarenghi, M., and Chiappetti, L.: 1983, *Astrophys. J.* **273**, 75.

Makino, F., Tanaka, Y., Matsuoka, M., Koyama, K., Inoue, H., Makishima, K., Hoshi, R., Hayakawa, S., Kondo, Y., Urry, C. M., Mufson, S. L., Hackney, K. R., Hackney, R. L., Kikuchi, S., Mikami, Y., Wisniewski, W. Z., Hiromoto, N., Nishida, M., Burnell, J., Brand, P., Williams, P. M., Smith, M. G., Takahara, F., Inoue, M., Tsuboi, M., Tabara, H., Kato, T., Aller, M. F., Aller, H. D.: 1986, *Astrophys. J.* (submitted).

Malkan, M. A. and Sargent, W. L.: 1982, *Astrophys. J.* **254**, 22.

Maraschi, L., Tanzi, E. G., and Treves, A.: 1984, *Adv. Space Res.* **3**, 167.

Maraschi, L., Schwartz, D. A., Tanzi, E. G., and Treves, A.: 1985, *Astrophys. J.* **294**, 615.

Maraschi, L., Ghisellini, G., Tanzi, E. G., and Treves, A.: 1986, *Astrophys. J.* (in press).

Maraschi, L., Tagliaferri, G., Tanzi, E., G., and Treves, A.: 1986, *Astrophy. J.* (in press).

Marscher, A. P.: 1977, *Astrophys. J.* **216**, 244.

Marscher, A. P.: 1980, *Astrophys. J.* **235**, 386.

McHardy, I.: 1985, *Space Sci. Rev.* **40**, 559.

McHardy, I. *et al.*: 1986, (in prep.).

Miller, J. S.: 1975, *Astrophys. J. Letters* **200**, L55.

Moles, M., Garcia-Pelayo, J., and Masegosa, J.: 1984, *Monthly Notices Roy. Astron. Soc.* **211**, 621.

Morini, M., Maccagni, D., Maraschi, L., Molteni, D., Tanzi, E. G., and Treves, A.: 1985, *Space Sci. Rev.* **40**, 601.

Morini, M., Chiappetti, L., Maccagni, D., Maraschi, L., Molteni, D., Tanzi, E. G., Treves, A., and Wolter, A.: 1986, *Astrophys. J. Letters* (in press).

Mufson, S. L., Hutter, D. J., Hackney, K. R., Hackney, R. L., Urry, C. M., Mushotzky, R. F., Kondo, Y., Wisniewksi, W. Z., Aller, H. D., Aller, M. F., and Hodge, P. E.: 1984, *Astrophys. J.* **285**, 571.

Oke, J. B. and Goodrich, W. W.: 1981, *Astrophys. J.* **243**, 445.

Oke, J. B., Shields, G. A., and Korycansky, D. G.: 1984, *Astrophys. J.* **277**, 64.

Perola, G. C., Boksenberg, A., Bromage, G. E., Clavel, J., Elvis, M., Elvius, A., Gondhalekar, P. M., Lind, J., Lloyd, C., Penston, M. V., Pettini, M., Snijders, M. A. J., Tanzi, E. G., Tarenghi, M., Ulrich, M. H., and Warwick, R. S.: 1982, *Monthly Notices Roy. Astron. Soc.* **200**, 293.

Piccinotti, G., Mushotzky, R. F., Boldt, E. A., Holt, S. S., Marshall, F. E., Serlemitsos, P. J., and Shafer R. A.: 1982, *Astrophys. J.* **253**, 485.

Pollock, A. M. T., Brand, P. W. J. L., Bregman, J. N., and Robson, E. I.: 1985, *Space Sci. Rev.* **40**, 607.

Pomphrey, R. B., Smith, A. G., Leacock, R. J., Olsson, C. N., Scott, R. L., Pollock, J. T., Edwards, P. L., and Dent, W. A.: 1976, *Astron. J.* **81**, 489.

Pounds, K. A.: 1985, in Y. Tanaka and W. H. G. Lewin (eds.), *Galactic and Extragalactic Compact X-Ray Sources*, Tokyo, Institute of Space and Astronautical Science, p. 261.

Reynolds, S. P.: 1982, *Astrophys. J.* **256**, 38.

Richstone, D. O. and Schmidt, M.: 1980, *Astrophys. J.* **235**, 361.

Rieke, G. H., Lebofsky, M. J., and Wisniewski, W. Z.: 1982, *Astrophys. J.* **263**, 73.

Seaton, M. J.: 1980, *Monthly Notices Roy. Astron. Soc.* **187**, 73.

Singh, K. P. and Garmire, G. P.: 1985, *Astrophys. J.* **297**, 199.

Snijders, M. A. J. *et al.*: 1979, *Monthly Notices Roy. Astron. Soc.* **189**, 873.

Snijders, M. A. J., Boksenberg, A., Penston, M. V., and Sargent, W. L. W.: 1982, *Monthly Notices Roy. Astron. Soc.* **201**, 801.

Stephens, S. A. and Miller, J. S.: 1984, *Bull. AAS,* **16**, 1007.

Tanzi, E. G., Barr, P., Bouchet, P., Chiappetti, L., Cristiani, S., Falomo, R., Giommi, P., Maraschi, L., and Treves, A.: 1986, submitted to *Astrophys. J.*

Treves, A., Ghisellini, G., Maraschi, L., and Tanzi, E. G.: 1986, in *Structure and Evolution of Active Galactic Nuclei*, Trieste, April 10—13, 1985, (in press).

Ulrich, M. H., Hackney, K. R. H., Hackney, R. L., and Kondo, Y.: 1984, *Astrophys. J.* **276**, 466.

Urry, C. M.: 1984, Ph.D. Thesis, The Johns Hopkins University.

Urry, C. M. and Mushotzky, R. P.: 1982, *Astrophys. J.* **253**, 38.

Urry, C. M., Mushotzky, R. F., Kondo, Y., Hackney, K. R. H., and Hackney, R. L.: 1982, *Astrophys. J.* **261**., 12.

Urry, C. M., Mushotzky, R. P., and Holdt, S. S.: 1986, *Astrophys. J.* **305**, 369.

Wall, J. V., Danziger, J. J., Pettini, M., Warwick, R. S., Wamsteker, W.: 1986, *Monthly Notices Roy. Astron. Soc.*, (in press).

Wood, K. S. *et al.*: 1984, *Astrophys. J. Supp. Ser.* **56**, 507.

Worrall, D. M. *et al.*: 1982, *Astrophys. J.* **261**, 403.

Worrall, D. M. *et al.*: 1984, *Astrophys. J.* **178**, 521.

Worrall, D. M., Puschell, J. J., Bruhweiler, F. C., Miller, H. R., Rudy, R. J., Ku, W. H.-M., Aller, M. F., Aller, H. D., Hodge, P. E., Matthews, K., Neugebauer, G., Soifer, B. T., Webb, J. R., Pica, A. J., Pollock, J. T., Smith, A. G., and Leacock, R. J.: 1984a, *Astrophys. J.* **278**, 521.

Worrall, D. M., Puschell, J. J., Bruhweiler, F. C., Sitko, M. L., Stein, W. A., Aller, M. F., Aller, H. D., Hodge, P. E., Rugy, R. J., Miller, H. R., Wisniewski, W. Z., Cordova, F. A., and Mason, K. O.: 1984b, *Astrophys. J.* **284**, 512.

Worrall, D. M., Puschell, J. J., Rodriguez-Espinosa, J. M., Bruhweiler, F. C., Miller, H. R., Aller, M. F., and Aller, H. D.: 1984c, *Astrophys. J.* **286**, 711.

Worrall, D. M. *et al.*: 1986, *Astrophys. J.* (in press).

QUASAR ABSORPTION LINES AND GALAXY HALOS

JACQUELINE BERGERON

Institut d'Astrophysique de Paris, Centre National de la Recherche Scientifique, Paris, France

BLAIR SAVAGE

Washburn Observatory, University of Wisconsin, Madison, WI, U.S.A.

and

RICHARD F. GREEN

Kitt Peak National Observatory, National Optical Astronomy Observatories, Tucson, A2, U.S.A.

1. Introduction

The study of quasar absorption lines addresses the questions of the evolution of the halo environments of galaxies at early epochs and the development and change of the intergalactic medium. Ultraviolet observations of individual absorbers with IUE are still limited, because of the faintness of the quasar probes, but nevertheless can provide decisive answers to some of the basic questions. In addition, observations of Lyman limit discontinuities have provided an independent survey of metal-line absorbing systems at low column density, while extensive observations of distant stellar and extragalactic probes have given great insight into the structure of the Milky Way halo, as a basis for comparison with the quasar absorption systems.

We will restrict the discussion of quasar spectra to systems with narrow absorption features, which are now commonly assumed to arise in intervening material (see, e.g., the review of Weymann *et al.*, 1981). These systems comprise two populations: (1) the Ly α forest, which typically has low neutral hydrogen column density and no or extremely weak absorption from heavier elements; and (2) the metal-rich systems, which are often identified by their C IV or Mg II absorption; C IV is typically much stronger than C II. The Ly α forest absorbers could be primordial intergalactic or intracluster clouds, or extremely extended halos around galaxies. The metal-line systems can be subdivided into displaced and associated systems, the latter with similar absorption and emission redshifts ($z_{abs} \sim z_{em}$). The displaced systems are often tentatively identified with extended disks or halos of galaxies, as first suggested by Bahcall and Spitzer (1969). The associated systems are often of higher ionization than the displaced systems and often show a complex velocity structure. They appear preferentially associated with steep-spectrum radio sources (Foltz *et al.*, 1986).

For metal-rich systems already identified by absorption lines in the optical, study of their Ly α and C IV lines and Lyman limits provides important information on their ionization and H I column densities. Their properties can be compared with those of high latitude gas in our Galaxy and in the Magellanic Clouds. Further, associated galaxies could be detected by direct imaging and the

Y. Kondo (ed.), Scientific Accomplishments of the IUE, pp. 703—725.
© 1987 *by D. Reidel Publishing Company.*

stellar and gaseous content inferred from spectroscopic follow-up. There are several basic scientific questions that cannot be resolved by ground-based observations alone, for which IUE results have a direct bearing. We summarize them here by topic before discussing them in more detail:

1.1. LYMAN α FOREST

Is there strong number density and column density evolution in the Ly α forest clouds? Do the results support the picture of Ikeuchi and Ostriker (1986) of a hot intergalactic medium with clouds forming from Rayleigh—Taylor instabilities? Do they instead require a dark matter scenario, such as that of Rees (1986)?

1.2. NEARBY GALAXIES

For some quasars projected near galaxies, Ca II and/or Na I absorption from the galaxies have been detected in the quasar spectra. Do their UV resonance lines show a link to Mg II absorption systems for $z < 1$? How do their Ca II/Mg II equivalent width ratios compare to results from Ca II and Mg II absorption line surveys?

1.3. DISPLACED ABSORPTION SYSTEMS

For $z < 1$, only low ionization species are detectable in absorption in the optical range. Thus, the total range of ionization in these systems can be determined only by UV observations. The H I column density must be determined from the UV for $z < 1.6$. At high redshift, low excitation systems with strong C II or Mg II absorption but very weak C IV are extremely rare. Are they more frequent at lower z; is there a cosmological evolution of the ionization level? Some Mg II systems would then have properties similar to those of clouds within galactic disks. Do the Lyman limit absorbers show the same cross-section as C IV or Mg II systems? High redshift Mg II or Fe II absorbers always show a corresponding Lyman limit discontinuity. Is there any evolution in these Lyman limit systems?

1.4. $z_a \sim z_e$ SYSTEMS

For some low redshift associated systems, are there clear cases for which the absorption originates in dilute gas of low velocity dispersion? What is their ionization level? Are they mainly found for steep-spectrum radio sources as at higher z? Are these systems related to the finding by Yee and Green (1987) that a substantial fraction of powerful radio quasars are situated in clusters of galaxies of Abell richness class 1 or greater for $z > 0.6$?

2. IUE Observations of Quasar Absorption Systems

2.1. THE LYMAN α FOREST

Individual absorption features of low column density can at present be studied only in the optical range, because IUE sensitivity is not sufficient to detect absorption lines of modest strength (observed equivalent width < 5 Å) in faint quasar probes. Exceptions are the brightest Seyfert galaxies and quasars or the case when absorption lines fall near the peaks of strong emission features. Therefore, discovery of low-redshift absorption systems is not easy with IUE if a strong continuum discontinuity is not present.

Thus, it has not been possible to perform a survey with IUE of Ly α forest systems. Not only does the cosmological evolution of these absorbers remain to be studied for $0 < z < 2$, but the existence of a low-redshift Ly α forest population is not yet established. Only two systems at intermediate redshift could at first glance be considered candidates. For both, a Lyman limit discontinuity has been observed with IUE, at $z = 0.713$ in PG $1718 + 481$ (Bechtold et al., 1984) and $z = 1.218$ in PG $1247 + 268$ (Green et al., 1980). An upper limit of 50 mÅ was placed on the Mg II doublet components observed in the optical; considering that the system showed only unity optical depth at the Lyman limit, the N(Mg II)/N (H I) ratio is not inconsistent with that of other Lyman limit systems. On the other hand, the system could also be of high excitation with no Mg II absorption and a C IV doublet too weak to be detected with the LWP camera, as would most C IV absorption lines if they are similar to their higher redshift counterparts (see, e.g., Young et al., 1982). The second Lyman limit system is most probably associated with a weak C IV doublet reported at a $z = 1.223$ (see Figure 1 and Table 2 in Young et al., 1982); a 5 Å error in the reported wavelength of the Ly α absorption observed with IUE would be sufficient to resolve the discrepancy. Discovery of a genuine very low redshift Ly α forest system would be of prime interest for detection of the absorbing object by direct imaging. A possible case is the Ly α systems at $z = 0.289$ in PKS $2128-12$ (Bergeron and Kunth, 1983). Two other full-shift SWP images have been obtained to confirm this feature (Bergeron, in prep.), but observations with HST will be required to detect an associated weak C IV doublet.

An indirect test of the absorption line density in the Ly α forest is to measure the depression of the continuum to the violet of the Ly α emission line relative to an extrapolation from the long wavelength side. Oke and Korycansky (1982) devised an index to measure the depression for their optical data on high redshift objects. Two wavelength regions are considered: the region between Ly α and Ly β emitted, in which Ly α will be the primary absorber, and the region between Ly β and the Lyman limit, in which higher order Lyman series members also contribute to the absorption. If the spectrum of equivalent widths is constant with redshift, then the total absorption from Ly α forest systems is proportional to the number of absorbers per unit redshift. Bechtold et al. (1984) used this fact and a sample of quasars observed with IUE in the redshift range 1 to 2 to combine with the Oke and Korycansky (1982) high redshift sample. Bechtold et al. (1984) found that for $dN/dz \propto (1 + z)^\gamma$, then $\gamma = 1.29 \pm 0.21$ for z between 1.0 and 3.5, consistent with the non-evolving value of 1.0 for $q_0 = 0$. Murdoch et al. (1986) summarize some nine investigations of the number density evolution of Ly α forest absorbers based on counting individual absorption lines. They show that the contradictory results previously obtained could be explained by an inverse effet in dN/dz present within individual QSOs. Their own analysis yields $\gamma = 2.17 \pm 0.36$ for redshifts between 1.50 and 3.78, i.e., a strongly evolving population of absorbers. The continuum method is more uncertain than the sample statistical error suggests; traces of reddening or intrinsic curvature would mimic stronger absorption in the lower redshift objects. The counting of individual absorbers in the next generation of high signal to noise quasar UV spectra will be required for a definitive answer at lower redshifts.

2.2. NEARBY GALAXIES

Two quasar-galaxy pairs have been observed with IUE. For the pair 3C 232–NGC 3067, Ca II from the galaxy has been detected in the quasar spectrum by Boksenberg and Sargent (1978) with w(Ca II K, H) = 0.43, 0.26 Å. We have analyzed the UV observations of Snijders (1980) and obtained w(Mg II D) $\simeq$ 6 Å and w(C IV D) < 1.5 Å, thus yielding a ratio of about 9 for the Mg II to Ca II doublet ratio. For the pair PKS 1327–206–ESO 1327–2041, a strong Na I absorption doublet is present at the redshift of the galaxy with w(Na I D) = 2.2 Å (Kunth and Bergeron, 1984). There are also a weak Ca II doublet, with w(Ca II K, H) = 0.55, 0.25 Å and an Fe II 2600 absorption line ($w \simeq$ 4 Å), which fell on the Ly α emission from the quasar, the only region of the ultraviolet with sufficient signal-to-noise ratio to detect weak absorption (Bergeron et al., 1986). Since in higher redshift systems observed in the optical, the Fe II 2600 absorption line is at most as strong as Mg II 2796 (see e.g., Boissé and Bergeron, 1987). the Mg II to Ca II doublet ratio should be equal to at least five.

A high Mg II to Ca II ratio is also observed for the $z = 0.8365$ system in Q0002–422, with w(Mg II D)/w(Ca II D) ~ 10 (Boissé and Bergeron, 1985). For high latitude extragalactic probes of halo gas, the absorption lines most often studied are those detectable in the SWP camera and in the optical. The equivalent width line ratio of Si II 1260 to Ca II K is measured to be 6.2 and 7.5 in the directions of Mk 509 and F 9 (York et al., 1982). For a line of sight to the LMC probing halo gas, Savage and Jeske (1981) find a ratio of one component of the Mg II doublet to the Si II 1260 feature of about 2 in equivalent width.

These results suggest that line ratios between singly ionized elements can be similar in high latitude gas in the Galaxy, regions at moderately large radial distances in nearby galaxies, and some higher redshift Mg II systems. At present we cannot infer the ionization level in the systems associated with quasar-galaxy pairs. This value is needed to compare to the ionization state of higher redshift absorption line systems, which have been suggested by Wolfe (1983) to be more highly ionized than the high latitude gas in the Galaxy. More extensive discussion of the physical state of the Milky Way halo and comparison to quasar absorption systems will be presented later.

2.3. INTERVENING ABSORBERS

2.3.1. *Lyman Limit Systems*

The one absorption feature from an intervening system that can be incontrovertibly identified in low signal to noise detections of faint quasars is the Lyman limit discontinuity. These systems are almost invariably metal containing, so that absorption lines from other species can be identified with optical spectroscopy. An example is shown in Figure 1 for the quasar PG 1634 + 706, with two closely

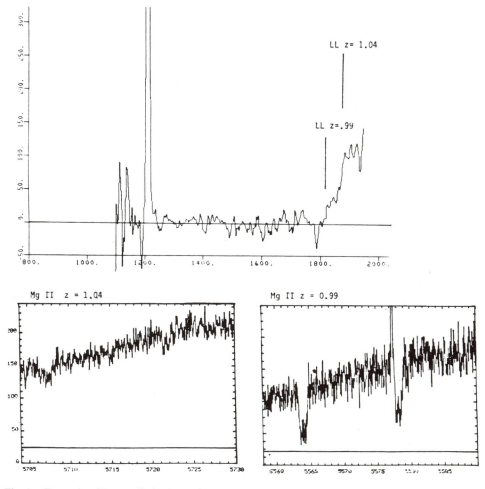

Fig. 1. Example of Lyman limit absorption systems. The upper panel shows the IUE SWP spectrum of the quasar PG 1634 + 706, containing two Lyman limit discontinuities, as marked. The lower panels show the corresponding Mg II 2800 doublet absorption, as detected by the MMT echelle spectrograph at 10 km s^{-1} resolution.

spaced Lyman limit absorptions in the SWP spectrum, and their corresponding Mg II doublets observed in the optical. Since unity optical depth at the Lyman limit corresponds to a neutral hydrogen column density of 1.6×10^{17} cm^{-2}, the survey of Bechtold *et al.* (1984) for Lyman limit absorptions was sensitive to lower column density systems than either 21 cm surveys or absorption line surveys for C IV or Mg II with moderate limiting equivalent widths.

Statistical analysis of Lyman limit systems requires the application of survival analysis techniques, because only the highest redshift optically thick system is detectable. This method was first used by Tytler (1982) to analyze a sample of high redshift objects with optical data, and seven lower redshift objects observed with IUE. Bechtold *et al.* (1984) expanded this sample to 35 Lyman limit systems along 72 lines of sight by including their own new IUE data and that of Tytler (1982), Green *et al.* (1980), Snijders (1980), and Elvis and Fabbiano (1982). Figure 2 shows the inverse of the mean free path to absorption for Lyman limit systems as a function of redshift. In the figure, $A(z)$ represents the comoving cross section of the cloud to absorption and $n(z)$ represents the number density of absorbers. The solid line is that expected for a constant cross-section and no evolution of the number density of absorbers, and is seen to be consistent with the data at a slope of 3. The normalization of the curve yields 1.3 absorbers per unit redshift locally, corresponding to 7.5 optically thick Lyman limit systems on a typical line of sight to redshift 3.5 (for $q_0 = 0.5$).

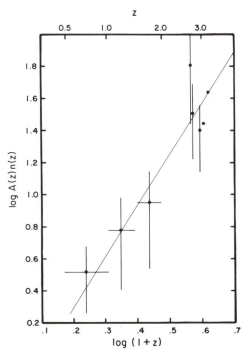

Fig. 2. Inverse of the mean free path to Lyman limit absorption, $A(z)n(z)$, as a function of redshift. The solid line is the relation expected for no evolution with epoch; i.e., a constant cross-sectional area and constant comoving volume density of absorbers.

As discussed below, the Lyman limit absorbers are metal-containing systems. Since they also show a constant comoving volume density, it is reasonable to associate them with galaxies. One can use the mean free path and the luminosity function of galaxies to derive the cross-section to absorption. Bechtold *et al.* (1984) parameterized the cross section by the effective radius of a galaxy with the characteristic luminosity at the knee in the luminosity function. That radius takes on values between 6 and 10 Holmberg radii of an L^* galaxy, R_H, which is about 16 kpc. By comparison, Weymann *et al.* (1979) and Young *et al.* (1982) find mean free paths for C IV and Mg II absorption corresponding to 2—6 R_H, with the Mg II radius smaller than that for C IV. Roughly similar size estimates have been obtained by other observers (for a review, see Wolfe, 1986). The interpretation of this difference in terms of a physical model for the distribution of absorbing gas is complicated because of varying conditions of ionization and density. The indication is, however, that the Lyman limit absorbers sample a lower total column density, and may represent conditions in the outer halo regions of distant galaxies.

An interesting question is what the observed curvature in the spectral energy distribution of high redshift quasars tells us about intervening absorbing systems. For quasars near redshift 2, there is an abrupt change in slope near the Ly α emission line. Bechtold *et al.* (1984) find an observed change in the power-law spectral index of up to 5 units; the result correlates strongly with redshift, but not with the intrinsic luminosity of the quasar. This suggests that the curvature is induced by intervening material, although Eastman *et al.* (1983) constructed a model for absorption by cloudlets ejected at relativistic velocities from the quasar itself. Bechtold *et al.* (1984) postulated that the effect was caused by the cumulative absorptions from Lyman continuum edges with optical depths around 0.1. The derived mean free path predicts 3.7 optically thick Lyman limit systems in these objects; the exponential distribution of column densities found by Atwood *et al.* (1985) then suggests that an additional 14 Lyman limit systems with optical depth around 0.1 might be found on the same sight line. Since they are found preferentially at the higher redshifts, the Lyman line plus continuum absorption could account for the curvature.

An alternative to atomic absorption is dust reddening. We would require an amount of dust associated with the optically thick absorbers sufficient to induce a decrease in flux of 0.2 in log f_ν between 1500 and 1000 Å. The most optimistic case would be a metal-poor SMC-type extinction curve, proportional to $1/\lambda$ (e.g., Lequeux *et al.* 1984). Since the dust to gas ratio in that environment is inferred to be some 17 times lower than Galactic, one would need $\sim 1 \times 10^{22}$ cm^{-2} column density of total hydrogen per absorbing system to produce the observed inflection in the energy distribution. For optical depths at the Lyman limit less than 10, this result implies a hydrogen neutral fraction of less than 1 in 10^4; although this condition is not impossible, modeling of some of the 'mixed ionization' systems by Bechtold *et al.* (1987) suggests neutral fractions in the range of 0.01 to 0.1.

2.3.2. *Properties of Intervening Systems*

At $z < 1$, only systems showing low excitation absorption lines have been studied, since no high excitation displaced system (C IV absorption detected but no Mg II) has been discovered with IUE. One should remain aware that, if there is no cosmological evolution in the ionization level of the metal-rich systems, the low excitation absorbers should constitute only a small fraction of the low z systems.

An important property is a possible link between the ionization level and the opacity in the Lyman continuum τ_{LL}. For most low excitation systems observed with IUE, τ_{LL} is larger than unity. Indeed a discontinuity at the Lyman limit of low z Mg II systems, as well as a strong Ly α absorption line have been detected at $z = 0.420$ in PKS 0735 + 178 (Bregman *et al.* 1981), $z = 0.432$ in PKS 2128 − 12 (Bergeron and Kunth, 1983), $z = 0.49$ in PKS 0454 − 22 (Kinney *et al.*, 1985) and $z \sim 0.52$ in the BL Lac object $0235 + 164$ (Snijders *et al.*, 1982). The observed Lyman jump is usually not very sharp and sometimes redshifted from the expected limit for the Mg II system. This probably arises mainly from absorption by high members of the Lyman series but also from instrumental effects, and shifts as large as 20 Å can be observed. In higher redshift low excitation systems with $w(\text{C II } 1334) \sim w(\text{C IV } 1548)$, a jump at the Lyman edge has also been found at $z = 1.840$ in Q $1101 - 264$ (Boksenberg and Snijders, 1981) and $z = 1.807$ in B2 $1225 + 317$ (Snijders *et al.*, 1981). In PKS $0237 - 233$ Bechtold *et al.* (1984) have detected a discontinuity at $\lambda 2345$ which they attribute to a C IV system at $z = 1.560$. Optical observations of Young *et al.* (1982) and Boissé and Bergeron (1985) show that this system, in fact at $z = 1.596$, is of high excitation with no associated Mg II absorption. However there is a strong low excitation absorber at $z = 1.672$ and the jump observed can belong to this system since the UV data have been heavily binned. Large opacities in the Lyman continuum of low excitation absorbers are also found for the higher redshift Lyman limit systems studied by Tytler (1982). The detailed optical observations in the C II—C IV region of absorbers with strong Lyman discontinuities are available in five cases and strong lines of singly ionized elements are always present (two of the cases are among those mentioned above). However, at least at low z, not all Mg II systems have large H I column densities. One clear counterexample is the $z = 0.852$ weak Mg II system in the BL Lac object $0235 + 164$ for which no large jump (of a factor of 2) is observed, implying $N(\text{HI}) \leqslant 1 \times 10^{17}$ cm^{-2} (Snijders *et al.*, 1982). Another such case has been found for the $z_a \sim z_e$ absorbers (see next section).

Low excitation systems alone cannot account entirely for all the observed Lyman discontinuities. The observed density of optically thick Lyman limits implies that all Ly α lines with $w_r(\text{Ly } \alpha) > 1.7$ Å show corresponding Lyman discontinuities (Bechtold *et al.*, 1984), whereas for the Mg II/Fe II absorbers $w_r(\text{Ly } \alpha)$ is larger than 2 to 3 Å. Systems which could have optically thick Lyman edges are cases of intermediate ionization level with weak C II absorption but no Mg II or Fe II, as indeed observed for the $z = 1.649$ system in PKS $0215 + 015$ (Blades *et al.*, 1985; Bergeron and D'Odorico, 1986). However most of the high excitation systems (C IV absorption detected but no C II or Mg II) should have optically thin Lyman limits. From the equivalent width spectrum for Ly α-metal absorbers (Sargent *et al.*, 1980), one can derive the fraction of metal-rich systems with $\tau_{LL} < 1$, using the statistical analysis of Bechtold *et al.* (1984). Taking a conservative lower limit for $w_r(\text{Ly } \alpha$-metal) of 0.6 Å, one find about twice as many metal-rich systems with $0.6 < w_r(\text{Ly } \alpha) < 1.7$ Å than with $w_r(\text{Ly } \alpha) > 1.7$ Å. Thus most of the high excitation systems should have optically thin Lyman edges. This can be directly verified for a few sytems. In the sample observed by Bechtold *et al.* (1984) there is a moderately strong C IV system at $z = 1.98$ in TON1530 ($w_r(\text{Ly } \alpha) = 1.2$ Å) with no discontinuity at the Lyman limit and this is also the case for 3 C IV systems in B2 $1225 + 317$ with $N(\text{H I}) < 3.5 \times 10^{16}$ cm^{-2} and $w_r(\text{Ly } \alpha) = 0.35$ to 1.2 Å (Sargent *et al.*, 1980; Snijders *et al.*, 1981).

Little information has been published on the C IV/Mg II ionic ratio in low redshift systems. For $0.29 < z < 0.55$ ($\lambda\lambda 2000—2400$), the sensitivity of the IUE LW cameras is too low to detect C IV counterpart of known Mg II systems even in fairly bright QSOs as PKS $2128 - 12$ (Bergeron and Kunth, 1983). An attempt was made for Mg II systems at higher redshift, $0.6 < z < 1$, which led to C IV detection for two systems. In PKS $2145 + 06$, the C IV and Si IV doublets at $z = 0.7897$, shown in Figure 3, are strong with observed equivalent widths of 3.2 and 2.7 Å respectively leading to a $w(\text{C IV D})/w(\text{Mg II D})$ ratio of 1.7, and the C IV doublet has also been detected in PKS $0454 + 039$ at $z = 0.8594$ with $w_{\text{obs}}(\text{C IV D}) = 3$ Å and $w(\text{C IV D})/w(\text{Mg II D}) = 0.6$ (Bergeron, unpublished). Equivalent width ratios $w(\text{C IV D})/w(\text{Mg II D}) \sim 1$ together with strong individual lines are frequently found for higher redshift Mg II systems (see e.g., Boissé and Bergeron, 1985). Since most of the Mg II absorbers have large H I column densities ($> 4 \times 10^{17}$ cm^{-2}), the above property is only a consequence of large ionic column densities, $N(\text{C IV})$ or $N(\text{Mg II}) > 10^{14}$ cm^{-2}, and moderate velocity dispersions $b \sim 5$ to 30 km s^{-1}. This should also applies to the two $z \sim 0.8$ systems mentioned above.

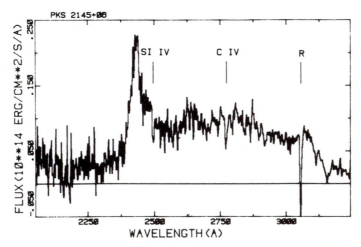

Fig. 3. The $z_a = 0.7897$ C IV and Si IV absorption doublets in the spectrum of PKS $2145 + 06$. The strong emission line is Ly α at $z_e = 1.000$ and a reseau mark falls on the C IV emission doublet. Two full-shift images, LWP 4922 and LWP 4944, have been combined to obtain this spectrum, and in both the C IV and Si IV absorptions are present.

Even a rough estimate of the heavy element abundances can seldom be inferred for low z absorbers. In PKS $2128 - 12$, the H I column density, derived from the Lyman limit discontinuity and Ly α (damping limit) is in the range $4 \times 10^{17} - 8 \times 10^{18}$ cm^{-2} (Bergeron and Kunth, 1983). The velocity dispersion in the absorbing cloud is small, $b \sim 8$ km s^{-1}, Mg is not predominantly in a neutral form, $N(\text{Mg I}) = 1.9 \times 10^{12}$ cm^{-2} and $N(\text{Mg II}) = 1.7 \times 10^{15}$ cm^{-2} (Tytler et al., 1987), but since C IV is not detected the ionization level of the absorber and the ionic ratio Mg II/Mg are unknown. The relative abundance $N(\text{Mg II})/N(\text{H I})$ ($\geq 2.1 \times 10^{-4}$) is at least one order of magnitude larger than the Mg cosmic abundance. This suggests that Mg II is in an ionized zone, H II /H I > 1, and that the abundances of the cloud are probably not much smaller than the cosmic values.

The gas density n can only be directly derived from the observations if fine structure lines are detected. However in one particular case, it is possible to get an idea of n. In the gravitational lens $0957 + 561$ A, B the system of low excitation at $z = 1.391$ has a strong Ly α absorption, and the H I column densities obtained in the damping limit are 1×10^{20} and 3×10^{19} cm^{-2} for components A and B respectively (Gondhalekar and Wilson, 1982). The strength of the metallic absorption lines are very similar in the two QSO images, and this leads to a lower limit on the characteristic size of the absorbing cloud of about 1 kpc (Weymann et al., 1979; Young et al., 1981). This implies an upper limit for n(H I) of a few 10^{-2} cm^{-3}.

There are two objects for which IUE spectra have played a significant role in the detailed interpretation of the absorption spectra. One is the (temporarily) bright BL Lac object $0215 + 015$, analyzed by Blades et al. (1985); the other is the (relatively) bright high-redshift quasar B2 $1225 + 317$, studied by Bechtold et al. (1987). The first case produced an IUE spectrum of sufficient quality that many individual absorption features could be identified and measured. Blades et al. discuss the system at $z = 1.345$, which shows lines of C IV and Si IV as well as singly ionized C, Mg, Al, Si, and Fe. The C IV and Si IV have higher column densities than observed for Galactic disk lines of sight, and have a similar ratio to those found by Pettini and West (1982) for halo clouds in the Galaxy. Their conclusion is that this system traverses both the disk and halo of an intervening galaxy. The low-ionization species have a depletion pattern that is much more uniform than that induced by dust grains, and suggests that the clouds might be genuinely underabundant with respect to solar. Two systems at $z = 1.549$ and 1.649 were studied in an earlier paper by Pettini et al. (1983), and were resolved into seven and nine velocity components spread over 300 and 900 km s^{-1}, respectively. The $z = 1.549$ system seems to show N V and O VI, but no Si II, classifying it as a high-ionization system. The authors interpreted the two systems as arising from intervening clusters of galaxies.

Bechtold et al. (1987) modeled the $z = 1.795$ absorption system in the quasar B2 $1225 + 317$. It was shown by York et al. (1984) that the system could be resolved into at least 12 velocity components, spread over 500 km s^{-1}. The higher velocity components are of high ionization, while the stronger components are 'mixed'; the N(C IV)/N(Si IV) ratio varies strongly from component to component. The relative column densities of Si II, Mg II, Zn II, Fe II, and H I were found to be consistent with the local ISM values. A photoionization model was used to investigate the diffuse radiation field longward of the Lyman limit with the Mg I/Mg II C I/C II, and Si II/Si IV ratios as a function of the total neutral hydrogen column, along with the density estimate from the limit on C II/C II*. The specific intensity at 1000 Å was in the range $-19.9 < \log J_v < -19.1$ ergs s^{-1} cm^{-2} Hz^{-1}, consistent with the near-ultraviolet radiation field in the halo of a spiral galaxy.

2.2. $z_a \sim z_e$ SYSTEMS

There are a few low redshift systems at less than 5000 km s^{-1} from the QSO velocity. The two lowest redshift ones are probably of intrinsic origin. In the bright QSO PG $1351 + 64$ ($z_e = 0.088$), extensively observed with IUE, the C IV and

N V absorption features, blueshifted by 1800 km s^{-1}, are comparable to those observed in NGC 4151 (Treves *et al.*, 1985) and variations of these lines should be search for to check for a possible association with material in the nucleus of PG 1351 + 64. Strong and resolved C IV and N V absorptions have also been detected in 3C 371 (z_e = 0.052) and their large velocity dispersion also suggests an intrinsic origin (Worrall *et al.*, 1984). For these two low z absorbers, observations of the Mg II region are not reported.

In 3C 232 there is a very weak Mg II —Fe II system at z = 0.5134 (Boksenberg and Sargent, 1978). The lines are narrow as usually observed for displaced systems. In the UV, Ly α absorption is not detected and there is no discontinuity at the Lyman limit (see Figure 1 in Bruhweiler *et al.*, 1986). At higher redshift $z_a \sim z_e$ systems are mostly of high excitation without drop at the Lyman edge, as in PG 1115 + 080, z_a = 1.731 (Green *et al.*, 1980) or PKS 0237—233, z_a = 2.2019 (Bechtold *et al.*, 1984).

The z_a = z_e = 0.200 system in PKS 2135—14, discovered by Bergeron and Kunth (1983) show absorption lines (Ly α, C IV and N V) unresolved at the resolution of IUE and there is no clear evidence of line variation. In the optical range the Mg II doublet is not detected thus the velocity dispersion of the absorber could not be accurately determined (Bergeron and Boissé, 1986). The QSO has a gas-rich companion galaxy also with a redshift of 0.200 (Stockton, 1982). Only higher resolution UV observations will allow to determine whether the absorption lines arise indeed in low density gas of small velocity dispersion, associated with either the nebulosity underlying the QSO or the companion galaxy, or in the nuclear regions of the QSO.

The origin of the z = 0.401 systems in PKS 1912 — 54 is less ambiguous although the emission and absorption redshifts are equal within 40 km s^{-1}. The Mg II and Fe II lines are resolved and have not been observed to vary; the velocity dispersion, b = 51 km s^{-1}, although large is within range observed for displaced systems (Bergeron and Boissé, 1986). Strong Ly α, C IV and N V lines are detecteds in the UV as well as a discontinuity at the Lyman edge (Bergeron and Kunth, 1983). The H I column density is in the range $4 \times 10^{17} - 2.6 \times 10^{18}$ cm^{-2} and those of Mg I, Mg II, and Fe II are 1.1×10^{12}, 9.3×10^{13}, and 6.0×10^{13} cm^{-2}. The spread in ionization degrees from Mg I to N V is among the largest known in absorption line systems.

The three cases for which the $z_a \sim z_e$ absorption systems are most probably associated with low density gas at large distances (> 10 kpc) from the active nucleus are observed in radio-loud QSOs, two of which having steep radio spectra. Another $z_a \sim z_e$ system is known at $z < 1$, in B2 1011 + 280, which is also a steep radio spectrum QSO (Peterson and Strittmatter, 1978). At large redshift Foltz *et al.* (1986) have also found that $z_a \sim z_e$ absorptions arise preferentially in radio-loud QSOs and comment on the possibility that radio properties correlate with environment.

3. Milky Way Halo Gas and Quasar Absorption Lines

Bahcall and Spitzer (1969) originally suggested that *some* of the narrow quasar

absorption line systems may arise in the gaseous halos and/or extended disks of
intervening galaxies. The confirmation via the IUE satellite that our own galaxy
possesses a gaseous halo has lent support to the intervening halo hypothesis. It is
therefore of interest to compare the absorption produced by our own halo and the
absorption found in quasar systems. In the following discussion we first summarize
the current state of understanding of the nature and distribution of gas in the
Milky Way halo and then indicate the possible relationship between Milky Way
halo gas and the gas sampled in selected quasar absorption line systems.

3.1. THE PHYSICAL NATURE AND DISTRIBUTION OF MILKY WAY HALO GAS

The first studies with IUE of Milky Way halo gas involved measures of the
interstellar absorption lines in the spectra of hot stars situated in the Large and
Small Magellanic Clouds (Savage and deBoer, 1979, 1981). Other studies utilized
Milky Way halo stars as background sources (Pettini and West, 1982; deBoer and
Savage, 1983, 1984; Savage and Massa, 1985, 1986). Extragalactic supernovae,
the brightest quasars and Seyfert galaxies have also been used as probes (Pettini
et al., 1984; York et al., 1983, 1984). The interest in the study of Milky Way halo
gas also extends to ground based optical astronomy, to 21 cm radio astronomy,
and to X-ray astronomy. On the theoretical side, numerous papers discussing the
ionization, support, and kinematics of the gas have also appeared. The results of
these and other studies of Milky Way halo gas have been reviewed extensively
(Jenkins, 1981; York, 1982; deBoer, 1983, 1985; Blades, 1984; and Savage,
1984, 1986). An important source of information about halo gas is found in the
Proceedings of the National Radio Astronomical Observatory Conference on the
Gaseous Halo of Galaxies (Bregman and Lockman (eds.), 1986).

  Figure 4 shows IUE absorption line spectra toward HD 36402 in the LMC and
HD 5980 in the SMC. The spectra were produced by averaging five individual
IUE high dispersion SWP spectra obtained with the star in different positions of
the large entrance aperture to reduce fixed pattern noise. Other examples of
absorption toward the LMC and SMC can be found in Savage and deBoer (1979,
1981), deBoer and Savage (1980), and Fitzpatrick and Savage (1983). The strong
absorption extending away from zero LSR velocity in weakly and highly ionized
features originates in Milky Way disk and halo gas. The primary evidence that the
gas detected has a large z extent follows from the fact that foreground stars
($|z| < 0.5$ kpc) in the direction of the LMC and SMC have very weak lines of
such species as C IV and Si IV while the LMC and SMC stars have strong
absorption in these ions extending smoothly away from LSR velocities near
0 km s^{-1}. At high velocities (250 km s^{-1} for the LMC and 150 km s^{-1} for the
SMC) absorption associated with the Magellanic Clouds is apparent.

  A subject of considerable controversy in recent years has been the possible
origin of absorption at intermediate velocities. The absorption is apparent in most
LMC star spectra near 60 and 120 km s^{-1}. If this absorption is associated with the
Milky Way and if the gas in the halo corotates with the gas in the disk, then the
absorbing clouds would have z distances of -7 and -15 kpc, respectively (Savage
and deBoer, 1981). However, there is evidence that the corotation assumption is
not valid for halo gas at $|z| > 1$ kpc. In fact, the rotational motion of high z gas
seems to lag behind the rotation of the disk gas (deBoer and Savage 1983; Kaelble
et al., 1985; Savage and Massa, 1987). For halo gas rotating more slowly than disk

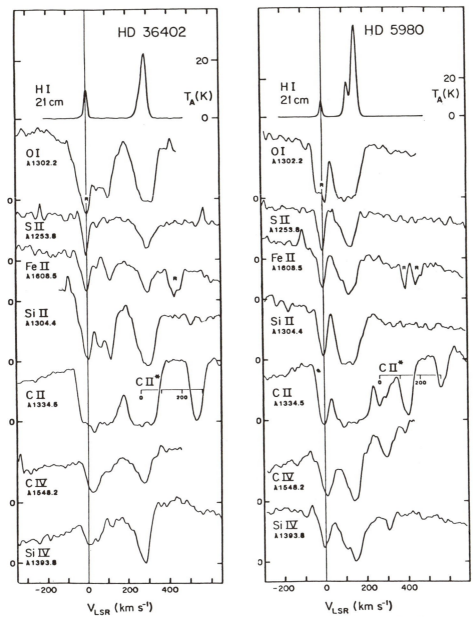

Fig. 4. Sample UV line profiles toward HD 36402 in the LMC and HD 5980 in the SMC. Intensity is plotted versus LSR velocity. The various lines plotted include low and high stages of ionization. These data were obtained by averaging five SWP spectra of each star. To reduce the effects of fixed pattern noise in the IUE detector, the stars were placed at different positions in the large entrance aperture. The 'R' identifies detector registration marks. Absorption extending away from zero LSR velocity is produced in the Milky Way disk and halo gas. The LMC disk and halo gas absorbs near 270 km s^{-1} and SMC disk and halo gas absorbs near 150 km s^{-1}. Absorption at intermediate velocities may be Milky Way and/or Magellanic Cloud in origin (see the text). Near zero velocity the Milky Way produces strong absorption in the lines of Si IV and C IV. This absorption occurs in a highly ionized gaseous layer having a scale height of about 3 kpc. At the top of the two figures, H I 21 cm emission line data are plotted for the two lines of sight to illustrate the extreme sensitivity of UV absorption line measurements compared to the 21 cm measurements.

gas, the absorption line velocities for gas toward the LMC would imply clouds closer to the disk. However, Songaila *et al.* (1986) have suggested that the intermediate velocity features seen toward the LMC are more likely associated with the LMC than with the Milky Way.

A reasonable way to settle the issue would be to obtain high quality 21 cm emission line data and evaluate whether or not the intermediate velocity gas 'connects' in velocity with the Milky Way disk. Unfortunately, the emission in question is very weak and careful attention must be paid to the ratio antenna side lobe contamination. Near the longitude of the LMC (1 = 270°) strong 21 cm emission associated with the Milky Way having a velocity of +60 km s^{-1} can be seen extending to latitudes of about −15° (Goniadzky and Jech, 1970). Given the extreme sensitivity of the far ultraviolet absorption lines to small amounts of gas (see the 21 cm profiles in Figure 4), it is not too unreasonable to expect to see in absorption the same gas at the latitude of the LMC (b = −33°). Part of the reason for the extension of the Galactic emission to these negative latitudes in this direction is the warp of the Milky Way to negative b for the directions 1 = 240 to 300°. For the opposite side of the Galaxy (1 = 60 to 120°) the warp of the distant parts of the Milky Way extends to positive latitudes and in the velocity range −80 to −180 km s^{-1}, this distant gas of the Milky Way can be easily recognized in the Bell Laboratories 21 cm H I survey data of Stark *et al.* (1986) to extend to latitudes of about +25 to +30° at limiting column density of about 5 × 10^{18} atoms cm^{-3} (Danly, private communication). However, it is known that the warp in the direction 1 = 260 to 300° is less pronounced that in the direction 1 = 60 to 120° (Henderson *et al.*, 1982).

The technique of examining the spatial extent of the 21 cm emitting gas to infer its possible origin has been applied to the IUE absorption line data for the SMC line of sight (Fitzpatrick, 1985). This sight line shows a complex velocity structure with sheets of absorbing and emitting gas at heliocentric velocities of 100 and 200 km s^{-1} which are clearly associated with the SMC.

Because of the uncertainties of the motions of halo gas, it is unreliable at this time to use the observed velocities of the gas to estimate its distance. A good procedure for estimating the z extent of halo gas is to obtain measures of absorption toward stars at different z distances. With such data one can investigate the stratification of the gas away from the Galactic plane by examining at what z distance the measured column densities projected onto the z axis ($N \sin |b|$) no longer increase. Measures of this type were obtained as part of the important halo gas survey of Pettini and West (1982). In that work, it was concluded that the density distribution of Si IV and C IV peaked at a z distance of about 2 to 3 kpc. Savage and Massa (1985, 1987) have recently completed work on a new survey of Si IV, C IV and N V in Galactic halo and disk gas. They report absorption line measurements toward 40 B stars at a wide variety of z distances. When these data are combined with the earlier Pettini and West (1982) data, a somewhat different picture for the z distribution of the highly ionized gas emerges. Figures 5a and b shows N(ion) $|\sin b|$ plotted against $|z|$ for Si IV and C IV from Savage and Massa (1987) for a group of approximately 50 stars observed by IUE. Figure 5c shows the expected behavior of N(ion) $|\sin b|$ versus $|z|$ if the gas has a simple exponential distribution with scale heights of 0.3, 1.0, 3.0, and 10 kpc. The data are roughly consistent with a scale height of 3 kpc. However, there is some evidence for an enhancement in the density over the simple exponential curve for z near 1 to 2 kpc. The work of Savage and Massa (1985, 1987) included sight lines extending over great distances (7 to 9 kpc) through the low halo ($|z| < 2$ kpc). Such measurements permitted the clear detection of N V absorption in halo gas for 10 lines of sight. These data also revealed that the character of halo gas in the inner Galaxy (galactocentric distances between 4 and 6 kpc) is not substantially different from that of halo gas in the vicinity of the sun.

Although most of the attention has focused on the highly ionized gas in the halo as traced by Si IV, C IV, and N V, the neutral and weakly ionized halo gas is actually about 10 times more abundant (Savage and deBoer, 1981). In fact, if we were to borrow terminology from the quasar absorption line literature, the absorption would be referred to as 'mixed ionization' with lines such as C II $\lambda 1335$, Si II $\lambda 1260$ and Mg II $\lambda 2800$ being about three times stronger than C IV $\lambda 1548$ (see Savage and Jeske, 1981).

The z extent and motion of neutral halo gas in the inner Galaxy has been studied through its 21 cm emission by Lockman (1984). He concludes there is emission from corotating H I in the inner Galaxy to a distance away from the plane of about 1 kpc. This high z gas is not present within about 3 kpc from the Galactic center. Over the region 4 to 8 kpc from the Galactic center, the extended (halo) component of H I has a scale height of about 0.5 kpc and a density of 0.01 atoms cm^{-3} at $|z| = 0.75$ kpc. Lockman's technique for separating emission from the halo gas and closer-by disk gas does not work for halo gas rotating more slowly than disk gas. Therefore, more gas may exist in the inner halo of the Galaxy than the numbers above suggest. An analysis of ultraviolet absorption line data for halo stars in the inner Galaxy suggests a breakdown of the corotation assumption for the highly ionized halo gas (Savage and Massa, 1987). It will be of great interest to try to interrelate ultraviolet absorption line data and 21 cm H I emission line data for similar lines of sight, in order to understand better the kinematics of the low and highly ionized gas.

Another way of evaluating the z extent of the neutral halo gas is by inter-comparing H I Ly α absorption data for halo stars at known z distances and H I 21 cm emission data for the same direction. Larger column densities from the 21 cm measurements would be expected if gas exists in the halo beyond the star. From such a program involving 25 stars, Lockman et al. (1986) have shown for high-latitude lines of sight that 5×10^{19} atoms cm^{-2} or 15% of the H I emitting gas exists in the halo with $|z| > 1$ kpc. A two-component model with half the H I mass in clouds having a Gaussian vertical distribution with $\sigma_z = 0.135$ kpc and half the H I in an extended distribution with a 0.5 kpc exponential scale height fits the data well. When evaluating this result, the difficulties of disentangling the interstellar absorption from the stellar Ly α profiles must be kept in mind. A similarly extended component of H I was also apparent in the H I Ly α survey with the Copernicus satellite (Bohlin et al., 1978).

The weakly ionized halo gas has been particularly important for studies of the kinematics of halo gas. In addition to the information it provides about the

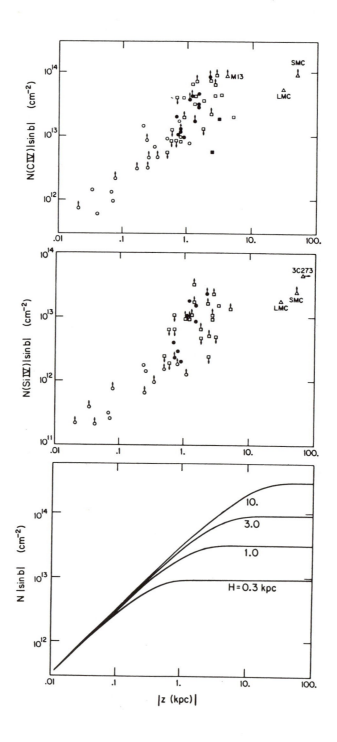

corotation issue discussed above, there now is strong evidence for infall of the weakly ionized gas toward the Galactic plane with velocities in the z direction of up to 100 km s^{-1} (deBoer and Savage, 1984; Danly, 1986a). However, there are strong differences between the north and south Galactic polar regions with the north Galactic zone showing significantly more gas moving at large negative velocities (Danly, 1986a). This gas is visible in the 21 cm emission line profiles for the Galactic polar regions presented by Kulkarni and Fich (1985). These data reveal that toward the North Galactic Pole ($75 < b < 90°$) half the H I is falling toward the plane with v_z between -20 and -100 km s^{-1}. These observed motions are in direct conflict with the summary by York (1982) that "no peculiar velocities are seen comparable to free fall". Toward the North Galactic Pole ($b > 60°$), a better summary based on the neutral gas observations would be: "Substantial portions of the northern sky are falling with velocities of up to -100 km s^{-1}".

Blades and Morton (1983) and York (1982) have argued, based on optical Ca II measurements that the velocity structure of the absorption seen along extended paths through the Milky Way halo is usually confined to 1 or 2 components over a velocity range of less than 50 km s^{-1} centered on zero LSR velocity. D'Odorico et al. (1985) disagree with this claim and point out that multiple Ca II profiles are in fact almost invariably seen when the sight lines through the halo are in directions where Galactic rotation produces a spread of velocities and when spectra with adequate signal to noise and resolution are obtained. Since the Ca II doublet is not a sensitive tracer of the low column density halo gas, it is important to look at the existing ultraviolet data. Danly (1986a, b) has shown that the strong ultraviolet lines of Si II, C II, and Mg II exhibit full widths at half intensity ranging from about 100 to 140 km s^{-1} toward distant stars at the North Galactic Pole and from 70 to 110 km s^{-1} toward distant stars at the South Galactic Pole. The observations of weaker ultraviolet lines toward these same stars often reveals the multicomponent absorbing nature of this gas. Thus, the multicomponent absorbing nature of the medium even exists in those directions relatively unaffected by Galactic rotation. For a number of the sight lines observed by Savage and Massa (1987) toward the inner Galaxy, the strong ultraviolet lines of Si II, C II, and Mg II sometimes attain full widths at half intensity as large as 200 km s^{-1}. *To summarize,*

Fig. 5. In the upper two figures, N(C IV)$|\sin b|$ and N(Si IV)$|\sin b|$ are plotted against $|z(\text{kpc})|$ for approximately 60 stars located at various distances away from the Galactic plane. The figures are from Savage and Massa (1987) and include data for approximately 30 B stars from their survey plotted as open and closed circles. The results from the Pettini and West (1982) halo gas survey are shown as squares. Extragalactic measurements from Savage and deBoer (1981) and York et al. (1983) are shown as triangles. The lower figure illustrates the expected behavior of N(ion)$|\sin b|$ versus $|z|$ if the gas has a simple exponential density distribution with scale heights of 0.3, 1.0, 3.0, and 10 kpc. The lower figure is most easily compared to the data by making a transparency of the figure and placing it over the data points. The Si IV and C IV data points are very roughly described by an exponential distribution with a scale height of about 3 kpc. However, these estimates are strongly influenced by the few extragalactic data points (LMC, SMC, and 3C 273). The Si IV and C IV measurements suggest the presence of an abrupt (factor of 2 to 3) increase in N(ion)$|\sin b|$ over the prediction of the simple exponential curve near $|z| \simeq 1$ kpc. This increase could be the result of the photoproduction of Si IV and C IV by the extragalactic EUV radiation (see the theoretical curves of Chevalier and Fransson, 1984).

when looking through 3 to 5 kpc of halo gas in the Milky Way, it is common to record multicomponent absorption extending over a velocity range of 100 to 150 km s⁻¹. Furthermore, a sight line extending from pole to pole at the sun's position in the Milky Way would produce multicomponent absorption having a full width at half intensity in the strong ultraviolet lines of more than 150 km s⁻¹ and extending from less than −90 to more than +60 km s⁻¹.

The study of elemental abundances in halo gas with the IUE satellite has been hampered by the low signal to noise of typical data and by the low resolution of the IUE spectrograph for interstellar studies. For reviews of the abundance question, see Jenkins (1983) and Pettini (1985). For the intermediate velocity gas toward the LMC (which may or may not be associated with the Milky Way) Savage and deBoer (1981) performed a curve of growth study and obtained column density estimates or limits for neutral O and singly ionized Mg, Al, Si, and Fe. When these results are combined with high sensitivity 21 cm radio data of H I emission at similar velocities from McGee *et al.* (1983), the data indicate that the composition is within a factor of three of solar. While these results are very uncertain, it is important to note that in the interstellar medium of the Galactic disk, which contains dust, elements such as Al and Fe generally have very low gas phase abundances because these elements are mostly incorporated into the dust. The most detailed halo gas abundance study based on ultraviolet data has been for the high velocity gas toward the bright halo star HD 93521 observed by the Copernicus satellite (Caldwell, 1979). Although this star is in the low halo ($|z| \simeq$ 0.7 kpc), the elemental abundance results are informative. In the clouds at $v_{\mathrm{LSR}} < -10$ km s⁻¹, the abundances of Fe, Si, Ar, and P are found to be nearly solar. This result for HD 93521 is also confirmed for Ti in the important ground-based Ti II measurements of Albert (1983). For a group of 19 halo stars, Albert (1983) also finds that the Ti gas phase abundance increases strongly with distance above the plane. The available information therefore suggests that the gas in the low and perhaps distant halo has abundances approaching those found in the sun. This result is in marked contrast to the depleted gas phase abundances usually found for disk gas.

A theory for the origin of the gaseous Galactic halo must explain the support and ionization of the gas. Two competing models for the support of the gas are the galactic fountain model (Shapiro and Field, 1976; Bregman, 1980; also see Chevalier and Oegerle, 1979; Habe and Ikeuchi, 1980) and the cosmic ray supported halo models of Chevalier and Fransson (1984). For the ionization of the gas, the possibilities for the production of the highly ionized species (Si IV, C IV, and N V) include electron collisional ionization in 0.8 to 3 × 10⁵ K gas and photoionization by hot white dwarf stars (Dupree and Raymond, 1983), by normal Population I stars, and by the extragalactic EUV background (Bregman and Harrington, 1986). However, the various calculations have difficulty producing the observed amount of N V (Savage and Massa, 1985, 1987). N V is an important ion since among those ions accessible to the IUE, it requires the greatest amount of energy for its production (77 eV). Most hot stars containing He have strong edges at 54 eV. Therefore, the only stellar sources that might be capable of converting N IV into N V are the very hot hydrogen white dwarfs and subdwarfs.

Savage and Massa (1987) concluded that the explanation for the support and ionization of halo gas will probably require a blending of the ideas from the galactic fountain models and the photoionized halo models. In the galactic fountain model, hot (10^6 K) disk gas which is buoyant (thermal scale height $\simeq$ 7 kpc) flows outward away from the galactic disk, cools, and returns to the plane as falling clouds. If the disk region supplying the gas is over pressured by a recent supernova explosion, the process is probably best described as a 'fountain'. However, if the process relies more on the buoyancy of the gas, the description 'galactic convection' seems more appropriate. In either case, the result is the input of gas into the low halo region. If this fountain or convective flow actually occurs, the origin of the absorption lines of Si IV, C IV, and N V might be in thermally ionized cooling gas associated with the flow. However, some of the highly ionized atoms (e.g., Si IV and C IV) may also be produced by the photoionization of cool (10^4 K) fountain gas. An interesting aspect of the fountain flow is that the flow naturally produces a highly inhomogeneous gaseous region with a wide range of ionization and a complex multicomponent velocity structure.

3.2. ABSORPTION BY THE MILKY WAY HALO VERSUS ABSORPTION FOUND IN SELECTED QUASAR ABSORPTION LINE SYSTEMS

As discussed in Section 2, unity cross-section for absorption can be from two to ten Holmberg radii. If some quasar absorption lines are indeed produced in the halos and/or extended disks of intervening galaxies, then the region of the halo and/or disk being sampled is probably a region very far from the center of the intervening galaxy. This result reveals a major problem in relating observations of Milky Way halo gas to observations of halos and/or disks as possibly revealed by quasar absorption lines. The Milky Way halo gas measurements obtained by the IUE spacecraft mostly refer to halo gas above and below the Galactic disk at galactocentric distances ranging from about 4 to about 15 kpc, while the quasar lines are more likely to refer to halo and/or disk gas at galactocentric distances many times larger.

A number of factors could change the character of halo and disk gas with galactocentric distance. Some of these might include:

(1) Changes in the vigor of the galactic fountain flow with galactocentric distance.

(2) Changes in the gravitational acceleration in the z direction with galactocentric distance.

(3) Galaxy radial abundance gradients which will influence the detectability of halo gas observed via metal absorption lines.

(4) Changes in the relative importance of photoionization to collisional ionization with galactocentric distance which will modify the ionization state of the halo and disk gas.

(5) Changes in the physical density and EUV optical depth of the halo gas and associated disk gas with galactocentric radius which will modify the ionization state of the gas.

(6) Changes in the kinematical properties of halo gas with galactocentric distance will almost certainly occur because of changes in the vigor of the fountain and changes in the gravitational acceleration.

Because of the very different nature of the sampling of Milky Way halo gas and the halo and/or disk gas possibly seen in selected quasar systems, one can certainly question the basic validity of a comparison. Ignoring this very major problem, it is interesting that some quasar absorption line systems do bear a strong resemblance to the absorption produced by Milky Way halo gas (Savage and Jeske, 1981). However, the quasar absorption line phenomenon is observed to extend over such a large range of parameter space that it is not surprising to find absorption line systems resembling Milky Way halo absorption. The type of mixed ion systems discussed by Savage and Jeske (1981) which are very similar to Milky Way halo absorption are not the most common type of quasar absorption system observed.

Three absorption line systems were discussed above which have total velocity widths of 300 to 900 km s^{-1}. Although it is possible that these sight lines traverse particularly rich clusters of galaxies, another interpretation has been offered by York et al. (1986). They suggest that a line of sight passing through a star-forming Magellanic type irregular system in the halo of an L^* galaxy could produce absorption with the complexity, mixed ionization and velocity spread observed. As examples, lower limits to emission-line widths in Hα are 350 and 250 km s^{-1} for NGC 604 and II Zw 40 (Gallagher and Hunter, 1983). Active star formation in such systems would produce H II regions and many supernovae, with accompanying shocks and expanding remnants. Chevalier and Clegg (1985) present a supernova wind model for M 82 that accelerates typical interstellar clouds to velocities of up to 400 km s^{-1}. A full sight line could then show a velocity spread of over 800 km s^{-1}. Gas temperatures could range from that of cold clouds to over 10^5 K. Although the free fall time for such satellite galaxies is short, the pressure produced by active star formation can keep the total system extended for times comparable to a Hubble time. For some complex systems, the quasar absorption lines may be a tracer of star formation history.

IUE has opened the door to understanding the quasar absorption line systems, especially by providing a solid basis for comparison with the Milky Way halo. The definitive test of whether the extended gaseous halos and/or disks of distant galaxies produce quasar absorption lines will have to await the launch of the Hubble Space Telescope. At that time, quasar and Seyfert galaxy probes situated off the edges of known foreground galaxies can be observed in relatively short times. From these data it will be possible to determine the extent and ionization characteristics of the gaseous halos around galaxies of a variety of types and sizes. From that point, we can step out in redshift to begin to interpret the information on the evolutionary history of galactic halos and the intergalactic medium contained in quasar absorption spectra.

Acknowledgements

The work of B.D.S. was supported by NASA grant NAG5-186. That of R.F.G. was supported in part by NASA grant NAG5-38.

References

Albert, C. E.: 1983, *Astrophys. J.* **272**, 509.
Atwood, B., Baldwin, J. A., and Carswell, R. F.: 1985, *Astrophys. J.* **292**, 58.
Bahcall, J. N. and Spitzer, L.: 1969, *Astrophys. J. Letters* **156**, L63.
Bechtold, J., Green, R. F., Weymann, R. J., Schmidt, M., Estabrook, F. B., Sherman, R. D., Wahlquist, H. D., and Heckman, T. M.: 1984, *Astrophys. J.* **281**, 76.
Bechtold, J., Green, R. F., and York, D. G.: 1987, *Astrophys. J.* **312**, 50.
Bergeron, J. and Kunth, D.: 1983, *Monthly Notices Roy. Astron. Soc.* **205**, 1053.
Bergeron, J. and D'Odorico, S.: 1986, *Monthly Notices Roy. Astron. Soc.* **220**, 883.
Bergeron, J. and Boissé, P.: 1986, *Astron, Astrophys.* **168**, 6.
Bergeron, J., Kunth, D., and D'Odorico, S.: 1987, *Astron. Astrophys.* (in press).
Blades, C.: 1984, in *Proceedings of the Fourth European IUE Conference*, Paris, European Space Agency, ESA SP-218, p. 161.
Blades, J. C. and Morton, D. C.: 1983, *Monthly Notices Roy. Astron. Soc.* **204**, 317.
Blades, J. C., Hunstead, R. W., Murdoch, H. S., and Pettini, M.: 1985, *Astrophys. J.* **288**, 580.
Bohlin, R. C., Savage, B. D., and Drake, J. F.: 1978, *Astrophys. J.* **224**, 132.
Boissé, P. and Bergeron, J.: 1985, *Astron. Astrophys.* **145**, 59.
Boksenberg, A. and Sargent, W. L. W.: 1978, *Astrophys. J.* **220**, 42.
Boksenberg, A. and Snijders, M. A. J.: 1981, *Monthly Notices Roy. Astron. Soc.* **194**, 353.
Bregman, J. N.: 1980, *Astrophys. J.* **236**, 577.
Bregman, J. N., Glassgold, A. E., and Huggins, P. J.: 1981, *Astrophys. J.* **249**, 13.
Bregman, J. N. and Harrington, P. J.: 1986, *Astrophys. J.* **309**, 833.
Bruhweiler, F. C., Kafatos, M., and Sofia, U. J.: 1986, *Astrophys. J. Letters* **303**, L31.
Caldwell, J. A. R.: 1979, Ph.D. Thesis, Princeton University.
Chevalier, R. A. and Clegg, A. W.: 1985, *Nature* **317**, 44.
Chevalier, R. A. and Fransson, C.: 1984, *Astrophys. J. Letters* **279**, L43.
Chevalier, R. A. and Oegerle, W. R.: 1979, *Astrophys. J.* **227**, 398.
Danly, L.: 1986a, in J. N. Bregman and F. J. Lockman (eds.), *Proceedings of the NRAO Conference on Gaseous Galactic Halos*, (Greenbank: NRAO SP), (in press).
Danly, L.: 1986b, in preparation.
D'Odorico, S., Pettini, M., and Ponz, D.: 1985, *Astrophys. J.* **299**, 852.
deBoer, K. S.: 1983, in H. van Woerden, W. B. Burton, and R. J. Allen (eds.), 'The Milky Way Galaxy', *IAU Symp.* **106**, 415.
deBoer, K. S.: 1985, *Mitt. der Astron. Gesell.* **63**, 21.
deBoer, K. S. and Savage, B. D.: 1980, *Astrophys. J.* **238**, 86.
deBoer, K. S. and Savage, B. D.: 1983, *Astrophys. J.* **265**, 210.
deBoer, K. S. and Savage, B. D.: 1984, *Astron. Astrophys. Letters* **136**, L7.
Dupree, A. K. and Raymond, J. C.: 1983, *Astrophys. J. Letters* **275**, L71.
Eastman, R. G., MacAlpine, G. M., and Richstone, D. O.: 1983, *Astrophys. J.* **275**, 53.
Elvis, M. and Fabbiano, G.: 1982, in Y. Kondo, J. M. Mead, and R. D. Chapman (eds.), *Advances in Ultraviolet Astronomy: Four Years of IUE Research*, NASA CP-2238, p. 205.
Fitzpatrick, E. L.: 1985, *Astrophys. J. Suppl.* **59**, 77.
Fitzpatrick, E. L. and Savage, B. D.: 1983, *Astrophys. J.* **267**, 93.
Foltz, C. B., Weymann, R. J., Peterson, B. M., Sun, L., Malkan, M., and Chaffee, Jr., F.: 1986, *Astrophys. J.* **307**, 504.
Fransson, C. and Chevalier, R. A.: 1985, *Astrophys. J.* **296**, 35.
Gallagher, J. S. and Hunter, D. A.: 1983, *Astrophys. J.* **274**, 141.
Gondhalekar, P. M. and Wilson, R.: 1982, *Nature* **296**, 415.
Goniadzky, D. and Jech, A.: 1970, in W. Becker and G. Contopoulos (eds.), The Spiral Structure of Our Galaxy', *IAU Symp.* **38**, 157.
Green, R. F., Pier, J. R., Schmidt, M., Estabrook, F. B., Lane, A. L., and Wahlquist, H. D.: 1980, *Astrophys. J.* **239**, 483.
Habe, A. and Ikeuchi, S.: 1980, *Prog. Theor. Phys.* **64**, 1995.
Hartquist, T. W., Pettini, M., and Tallant, A.: 1984, *Astrophys. J.* **276**, 519.

Henderson, A. P., Jackson, P. D., and Kerr, F. J.: 1982, *Astrophys. J.* **263**, 116.

Ikeuchi, S. and Ostriker, J. P.: 1986, *Astrophys. J.* **301**, 522.

Jenkins, E. B.: 1981, in R. D. Chapman (ed.), *The Universe at Ultraviolet Wavelengths*, Greenbelt: NASA CP-2171, p. 541.

Jenkins, E. B.: 1983, in W. H. L. Shuter (ed.), 'Changes in Interstellar Atomic Abundances from the Galactic Plane to the Halo', *Kinematics, Dynamics and Structure of the Milky Way*, D. Reidel Publ. Co., Dordrecht, Holland, pp. 21—30.

Kaelble, A., deBoer, K. S., and Grewing, M.: 1985, *Astron. Astrophys.* **143**, 408.

Kinney, A. L., Huggins, P. J., Bregman, J. N., and Glassgold, A. E.: 1985, *Astrophys. J.* **291**, 128.

Kulkarni, S. R. and Fitch, M.: 1985, *Astrophys. J.* **289**, 792.

Kunth, D. and Bergeron, J.: 1984, *Monthly Notices Roy. Astron. Soc.* **210**, 873.

Lequeux, J., Maurice, E., Prévot, L., Prévot-Burnichon, M.-L., and Rocca-Volmerange, B.: 1984, in S. van den Bergh and K. S. deBoer (eds.), 'Structure and Evolution of the Magellanic Clouds', *IAU Symp.* **77**, 405.

Lockman, F. J.: 1984, *Astrophys. J.* **283**, 90.

Lockman, F. J., Hobbs, L. M., and Shull, M.: 1986, preprint.

McGee, R. X., Newton, L. M., and Morton, D. C.: 1983, *Monthly Notices Roy. Astron. Soc.* **205**, 1191.

Murdoch, H. S., Hunstead, R. W., Pettini, M., and Blades, J. C., 1986, *Astrophys. J.* **309**, 19.

Oke, J. B. and Korycansky, D. G.: 1982, *Astrophys. J.* **255**, 11.

Peterson, B. M. and Strittmatter, P. A.: 1978, *Astrophys. J.* **226**, 21.

Pettini, M.: 1985, in *ESO Workshop on Production and Distribution of C, N, and O Elements*, Garching: European Southern Observatory Pub.

Pettini, M. and West, K. A.: 1982, *Astrophys. J.* **260**, 561.

Pettini, M., Benvenuti, P., Blades, J. C., Boggess, A., Boksenberg, A., Grewing, M., Holm, A., King, D. L., Panagia, N., Penston, N. V., Savage, B. D., Wamsteker, W., and Wu C-C.: 1984, *Monthly Notices Roy. Astron. Soc.* **199**, 409.

Pettini, M., Hunstead, R. W., Murdoch, H. S., and Blades, J. C.: 1983, *Astrophys. J.* **273**, 436.

Rees, M.: 1986, *Monthly Notices Roy. Astron. Soc.* **218**, 21P.

Sargent, W. L. W., Young, P. J., Boksenberg, A., and Tytler, D.: 1980, *Astrophys. J. Suppl.* **42**, 41.

Savage, B. D.: 1984, in J. M. Meade, R. D. Chapman, and Y. Kondo (eds.), *Future of UV Astronomy Based on Six Years of IUE Research*, Greenbelt: NASA CP-2349, p. 3.

Savage, B. D.: 1986, in J. N. Bregman and F. J. Lockman (eds.), *Proceedings of the NRAO Conference on Gaseous Galactic Halos*, Greenbank: NRAO SP (in press).

Savage, B. D. and deBoer, K. S.: 1979, *Astrophys. J. Letters* **230**, L77.

Savage, B. D. and deBoer, K. S.: 1981, *Astrophys. J.* **243**, 460.

Savage, B. D. and Jeske, N. A.: 1981, *Astrophys. J.* **244**, 768.

Savage, B. D. and Massa, D.: 1985, *Astrophys. J. Letters* **295**, L9.

Savage, B. D. and Massa, D.: 1987, *Astrophys. J.* **314**, 380.

Shapiro, P. R. and Field, G. B.: 1976, *Astrophys. J.* **205**, 762.

Snijders, M. A. J.: 1980, in *Second European IUE Conference*, Tübingen, ESA SP-157, p. lxxi.

Snijders, M. A. J., Pettini, M., and Boksenberg, A.: 1981, *Astrophys. J.* **245**, 386.

Snijders, M. A. J., Boksenberg, A., Penston, M. V., and Sargent, W. L. W.: 1982, *Monthly Notices Roy. Astron. Soc.* **201**, 801.

Songaila, A., Blades, J. C., Hu, E. M., and Cowie, L. L.: 1986, *Astrophys. J.* **303**, 198.

Stark, A. A., Bally, J., Linke, R. A., and Heiles, C.: 1986 (in prep.).

Stockton, A.: 1982, *Astrophys. J.* **257**, 33.

Treves, A., Drew, J., Falomo, R., Maraschi, L., Tanzi, E. G., and Wilson, R.: 1985, *Monthly Notices Roy. Astron. Soc.* **216**, 529.

Tytler, D.: 1982, *Nature* **298**, 427.

Tytler, D., Boksenberg, A., Sargent, W. L. W., Young, P., and Kunth, D.: 1987, *Astrophys. J.* (in press).

Weymann, R. J., Chaffee, F. H., Jr., Davis, M., Carleton, N. P., Walsh, D., and Carswell, R. F.: 1979, *Astrophys. J. Letters* **233**, L43.

Weymann, R. J., Carswell, R. F., and Smith, M. G.: 1981, *Ann. Rev. Astron. Astrophys.* **19**, 41.

Weymann, R. J., Williams, R. E., Peterson, B. M., and Turnshek, D. A.: 1979, *Astrophys. J.* **234**, 33.

Wolfe, A. M.: 1983, *Astrophys. J. Letters* **268**, L1.

Wolfe, A. M.: 1986, in J. N. Bregman and F. J. Lockman (eds.), *Proceedings of the NRAO Conference on Gaseous Galactic Halos*, (Greenbank: NRAO SP), (in press).

Worrall, D. M., Puschell, J. J., Bruhweiler, F. C., Miller, H. R., Rudy, R. J., Ku, W. H.-M., Aller, M. F., Aller, H. D., Hodge, P. E., Matthews, K., Neugebauer, G., Soifer, B. T., Webb, J. R., Pica, A. J., Pollock, J. T., Smith, A. G., and Leacock, R. J.: 1984, *Astrophys. J.* **278**, 521.

Yee, H. K. C. and Green, R. F.: 1987, *Astrophys. J.* (in press).

York, D. G.: 1982, *Ann. Rev. Astron. Astrophys.* **20**, 221.

York, D. G., Blades, J. C., Cowie, L. L., Morton, D. C., Songaila, A., and Wu, C.-C.: 1982, *Astrophys. J.* **255**, 467.

York, D. G., Wu, C. C., Ratcliff, S., Blades, J. C., Cowie, L. L., and Morton, D. C.: 1983, *Astrophys. J.* **274**, 136.

York, D. G., Green, R. F., Bechtold, J., and Chaffee Jr., F. H.: 1984, *Astrophys. J. Letters* **280**, L1.

York, D. G., Ratcliff, S., Blades, J. C., Cowie, L. L., Morton, D. C., and Wu, C. C.: 1984, *Astrophys. J.* **276**, 92.

York, D. G., Dopita, M., Green, R. F., and Bechtold, J.: 1986, *Astrophys, J.* **311**, 610.

Young, P. J., Sargent, W. L. W., Boksenberg, A., and Oke, J. B.: 1981, *Astrophys. J.* **249**, 415.

Young, P. J., Sargent, W. L. W., and Boksenberg, A.: 1982, *Astrophys. J. Suppl.* **48**, 455.

PART VI

HOW TO USE IUE DATA

Edited by J. L. LINSKY

HOW TO USE IUE DATA

A. W. HARRIS

Rutherford Appleton Laboratory, Chilton, U.K.

and

G. SONNEBORN[1]

Astronomy Programs, Computer Sciences Corporation, Beltsville, MA, U.S.A.

Introduction

The very large body of data in the IUE data-banks will continue to be of enormous value to astronomers for many years beyond the termination of the satellite's active life. However, the effective use of this valuable data archive requires that the user have some knowledge of the instrumentation and the techniques employed in data acquisition and processing. This article provides a brief overview for prospective users of IUE data who may not be very familiar with the IUE instrumentation or standard 'Spectral Image Processing System' (IUESIPS). In particular, we alert potential users to problems that may arise in the analysis and interpretation of their IUE data and direct them to sources of more detailed technical information, generally published in the IUE Newsletters.

The reader who is totally unfamiliar with IUE should read the first two articles of this book and Boggess *et al.* (1978) initially for a general introduction to the satellite, its instrumentation and the philosophy behind the mode of operation.

The detectors and their data characteristics are described in Section 1. Section 2 discusses the Fine-Error Sensor, IUE aperture orientation, some special observing techniques, and the science image header. Section 3 summarizes limitations on IUE data quality imposed by the spacecraft hardware (signal-to-noise ratio, camera sensitivity, spectral and spatial resolution, and spectrograph scattered light). The last section describes the standard image processing steps applied by IUESIPS (geometric correction, photometric correction with the intensity transfer function, spectral extraction, photometric calibration, and wavelength calibration) and IUESIPS data products.

1. The Detectors

The long- and short-wavelength spectrographs of IUE each have one primary and one redundant camera, which integrate the spectrograph image in SEC vidicon detectors. Exposures are controlled by the on-board computer, which gives reliable exposure length control in units of 0.4096 sec. At the conclusion of the exposure, the image, a 768 × 768 pixel array, is scanned and transmitted to the ground. The video signal from the potassium chloride SEC target is digitized in

---

[1] Staff member of the International Ultraviolet Explorer Observatory, at the Laboratory of Astronomy and Solar Physics, NASA Goddard Space Flight Center.

Y. Kondo (ed.), Scientific Accomplishments of the IUE, pp. 729—749.
© 1987 *by D. Reidel Publishing Company.*

to one of 256 discrete levels (0 to 255 Data Numbers, or DN) by an eight-bit analog-to-digital converter. An 'optimum' exposure level is approximately 200 DN, due to non-linear camera response at higher DN levels. All exposure-level information is lost above 255 DN. Overexposures are measured with respect to 'optimum' levels.

All four television cameras are of identical construction and, since they are only sensitive to visible light, are preceeded by ultraviolet-to-visible converters (UVCs). For a detailed description of the cameras and technical details see Boggess *et al.* (1978) and Coleman *et al.* (1977). The performances and operational histories of the cameras are quite different and each is affected by certain anomalies and idiosyncracies of which the user of IUE data should be aware.

1.1. SHORT-WAVE PRIME (SWP)

This is the most used and least problematic camera. Its role as the standard short-wavelength camera has remained unchanged throughout the history of the project to date. The SWP camera provides complete wavelength coverage from 1150 to 1975 Å in low dispersion. In high dispersion, complete coverage is available from 1145 to 1930 Å, with partial coverage between 1930 and 2198 Å. Since mid-1979 its sensitivity has decreased by less than 1% per year at any wavelength (Sonneborn and Garhart, 1986), although a rather rapid decrease of about 6% was observed at 1550 and 1850 Å during the first year after launch (Schiffer, 1982). To date, the SWP camera has obtained 51% of all IUE images.

1.2. SHORT-WAVE REDUNDANT (SWR)

The SWR camera has suffered from intermittent failure of its read-out section since launch and has never been made available for Guest Observer use. Some images were taken during the commissioning period, however.

1.3. LONG-WAVE PRIME (LWP)

Until 16 October, 1983 this was, despite its name, the back-up long-wavelength camera due to the frequent scan control malfunction of its read-out section. However, on that date it was promoted to standard (default) long-wavelength camera due to the 'flare' problem which has developed in the LWR camera (see below). The LWP camera provides complete wavelength coverage from 1910 to 3300 Å in low dispersion. In high dispersion, complete coverage is available from 1845 to 2980 Å, with partial coverage between 2980 and 3230 Å. Due to the improved behavior of the camera with increased usage and implementation of a 'bad scan detection' software patch, the scan control malfunction no longer presents a problem. No change of sensitivity with time has been apparent to date (Sonneborn and Garhart, 1986). In terms of both sensitivity and signal-to-noise ratio (S/N) the LWP is the better camera longward of 2500 Å (Barylak, 1982; Blades and Cassatella, 1982). While serving as the long-wavelength back-up camera, the LWP camera was made available in 1981 to Guest Observers having programs for which these characteristics were particularly important.

1.4. LONG-WAVE REDUNDANT (LWR)

This was the standard long-wavelength camera for the first 5.5 yr of Guest

Observer operations. The LWR camera provides complete wavelength coverage from 1860 to 3300 Å in low dispersion. In high dispersion, complete coverage is available from 1845 to 3105 Å, with partial coverage between 3105 and 3230 Å. The flare discharge in the camera's UVC, which produces a bright patch near the lower rim of an image, has been contaminating long exposure images since April 1983. The serious and steadily worsening impact of this anomaly was recognised in September 1983, and prompted the switch to the LWP as default long-wavelength camera in October 1983 (Harris, 1984, 1985a). Tests have shown, however, that with a 10% reduction in the UVC operating voltage, flare-free images can be obtained routinely with essentially no change in the photometric properties of the camera, apart from a 27% drop in sensitivity which is almost uniform in wavelength (Harris, 1985b; Imhoff, 1985b).

Due to its higher signal-to-noise ratio shortward of 2500 Å, the LWR is the more attractive long-wavelength camera for certain programs, such as studies of the 2200 Å interstellar absorption feature. Since November 1985, the camera has been available for Guest Observer use but only at the reduced UVC voltage of 4.5 kV. The LWR camera continues to exhibit a significant sensitivity decrease with time, independent of UVC voltage and level of use (Sonneborn and Garhart, 1986). This is worst around 2300 Å, where the decrease has been steady at a rate of -3.5% per year over most of the project lifetime (Clavel et al., 1986; Holm, 1985). Finally, users of LWR data should be aware that images may be contaminated by a narrow band of microphonic noise (up to 100 DN) cutting horizontally across the image.*

1.5. DATA-QUALITY LIMITATIONS

The quality of IUE data is limited by several phenomena and artifacts which are common to all cameras. The most important of these from the user's point of view are:

1.5.1. *Reseau Marks*

These marks appear on images as small dark dots, 2 to 3 pixels square, in a regular array. They provide a reference grid to enable correction during processing for the geometric distortion of the image by the camera electron optics (Turnrose and Thompson, 1984; see also Section 4 below). A spectral order falling on a reseau gives rise to a very narrow 'absorption' feature in the processed spectrum. Reseaux are flagged in extracted spectral files and plots; on photowrites their regular spacing usually faciliates their separation from real absorption features.

1.5.2. *Image Blemishes*

There are a number of permanent 'hot-pixels' on the targets of all cameras. These

---

* All three operational cameras have some periodic noise interference present in their images. In the SWP and LWP cameras this mircophonic noise is generally of very low amplitude (1—3 DN) compared to random background noise. In the LWR camera, however, the noise has different characteristics, being localized in a small number of image lines, well-modelled by an exponentially damped sinusoid. Features due to this anomaly have been flagged in standard IUESIPS processed spectra since 28 September/10 November, 1981 (low-/high-dispersion spectra respectively) at GSFC and 11 March, 1982 at VILSPA.

appear as bright spots on the images and produce false narrow 'emission' features where they coincide with spectral orders. Locations of permanent bright spots in SWP and LWR images are listed by pixel position and approximate wavelength by Ponz (1980a, b) and by Turnrose and Thompson (1984). Random bright spots also appear in IUE images due to radioactive decay of atoms in the UVC phosphor; their number increases with exposure time. Bright spots have been flagged by IUESIPS since 19 October, 1982 at VILSPA and 19 November 1982 at GSFC. The flagging routine checks for unusually large brightness values (DN > surrounding background + 90) confined to a very limited area. The routine is not infallible, however, and may flag genuine bright emission lines (see Grady and Imhoff, 1985b). Other permanent blemishes in LWR and LWP images are listed by Settle et al. (1981) and by Turnrose and Thompson (1984). Hackney et al. (1984) give the wavelength positions of some permanent SWP features which only show up in long exposures (> 2 hr). Many random features are caused by cosmic-ray hits in the UVCs. These can leave streaks or comet-like patterns on an image, which will give rise to spurious spectral features when they intersect an order or contaminate the adjacent background. See Imhoff and Grady (1985) and Grady and Imhoff (1985b) for examples of how camera artifacts and cosmic ray hits can lead to misinterpretation of IUE data.

1.5.3. *Phosphorescence*

After the end of an exposure, the UVC continues to emit some light with an intensity proportional to the exciting exposure level and a power-law decay with time (Coleman et al., 1977). A uniform low-level afterglow results each time a camera is prepared for an exposure since the PREP sequence involves heavy illumination by tungsten flood lamps. In particular, following an exposure where parts of the spectrum are overexposed by five times or more, a special 'over-exposed' camera PREP is normally performed. This includes an eight-times optimal exposure by the flood lamps to erase the effects of the overexposed spectrum. A long exposure that immediately follows this PREP sequence will have a significantly increased background level due to this afterglow. A more serious problem occurs when there is phosphorescence after an overexposed spectral image or even a series of many optimum exposures. The potential user of IUE data should be aware that the data bank contains many examples of long-exposure images contaminated by residual 'ghost' spectral orders originating in this way which produce spurious features in the processed data. Indeed, a weak long-exposure spectrum may be due entirely to a preceeding overexposure in the same dispersion mode! In order to check whether a particular image is affected by a preceeding overexposure, the user should examine the merged log of IUE observations ordered by date of observation. A detailed discussion of the problem of UVC phosphorescence is given by Snijders (1983).

1.5.4. *Radiation-Induced Background*

This builds up uniformly over the image area during an exposure and is the result of Cerenkov photons produced in the camera faceplate by high-energy electrons from the outer Van Allen Belt. Characteristics and trends of this background source are discussed by Walter and Imhoff (1983), Broude and Imhoff (1984) and Imhoff (1985a).

1.5.5. *Microphonics and Data Corruption*

The LWR camera microphonic noise previously described is the worst case of this type of noise in the three operational cameras. However, the sensitivity of the cameras to mechanical activity in the spacecraft has produced serious microphonics, up to 50 DN, on a small number of images which were read during spacecraft maneuvers. In the cases of the SWP and LWP cameras, significant microphonics are rare and not flagged by IUESIPS; their presence on an image can normally be verified by referring to the photowrite and comments recorded in the handwritten log by the Telescope Operator and Resident Astronomer.

A very small fraction of IUE images has been affected by the loss of data during transmission from the satellite to the ground. It is readily apparent when a large portion of the spectrum is missing. However, a momentary loss of signal results in the loss of data in units of telemetry minor frames (one minor frame contains 96 consecutive pixels). Portions of the spectrum so affected are not flagged by IUESIPS, and can give the appearance of narrow spectral features in the extracted data. The loss of any portion of the image is readily apparent from inspection of the photowrite. There is also a section of the science header archived with the image which contains a data quality flag for each minor frame of camera telemetry (see Turnrose and Thompson, 1984, Section 9).

2. The FES and Normal Data Acquisition

IUE target acquisition and telescope pointing control during exposures, rely on the Fine-Error Sensor (FES) in the Scientific Instrument. The actual spacecraft pointing control and slewing is performed by the on-board computer (OBC). The FES provides the only means of mapping the telescope field of view and measuring stellar positions and brightnesses. FES star tracking data is also used by the OBC for automatic offset guiding during exposures. The most important aspects of the FES operations with regard to the quality of IUE spectral data are the accuracy with which a target can be centered in the spectrograph apertures and the pointing precision of the tracking, or offset-guiding mode. The large apertures are approximately 10×20 arcsec ovals; the small apertures are 3 arcsec diameter circles (see Panek, 1982 and Turnrose and Thompson, 1984 for accurate dimensions). Sonneborn *et al.* (1987) give a thorough discussion of IUE observing techniques.

The FES 'prime' mode is used to measure the position of an object's center of light and its brightness. The FES positional resolution is about 0.26 arc sec, so that an object brighter than about $m_v = 13.5$ can be centered in an aperture with an error of less than 0.5 arc sec. The photometric calibration of the FES brightness measurements is a function of time, $B - V$, and FES tracking mode. The reader is referred to Imhoff and Wasatonic (1986) for the most recent discussion of the FES photometric calibration and references to earlier work. While the FES was not designed to function as a photometer, it has often been used as such. The FES signal (counts) measured at the time of a given observation are recorded on the observing 'scripts' at GSFC and VILSPA and have been entered in the Merged Log of IUE Observations.

2.1. IUE APERTURE ORIENTATION

Occasionally it is important to know the astronomical orientation of the spectrograph apertures, for example during observations of extended sources and close double stars. The equations for the spacecraft roll angle and the position angles of the apertures are described in detail by Sonneborn et al. (1987). Here we give a brief summary because the orientation, especially with respect to the spectrum, is not always obvious. As a visual aid in finding the orientation of the apertures, the formats of the FES field and the cameras are depicted in Figures 1 and 2.

Since the FES (and scientific instruments) are fixed within the spacecraft, the FES coordinate system (labelled X and Y) is defined relative to the inertial axes of the spacecraft. These coordinates are used to specify the locations of the apertures, guide stars, etc. in the telescope field of view. The aperture and spectrum orientation in the plane of the sky is a function of the *spacecraft roll angle*. The spacecraft roll is defined to be the angle, in the plane of the sky, between astronomical north and a reference vector pointing to the Sun. This angle, measured 0° to 360° eastward from this vector, changes as a function of the day of the year for a given right ascension and declination because the spacecraft orientation must keep the solar arrays optimally facing the Sun.

Angles in the FES field of view are defined with respect to a *reference vector* which points to the Sun and is parallel to the pitch direction. When the spacecraft roll is 0° the reference vector points north. Due to the odd number of reflections in the optical path to the FES, the FES image is normally displayed in a left-handed coordinate system. The reference vector for FES system No. 2 is rotated approximately 28.31° in the FES coordinate system from the negative Y axis toward the negative X axis. FES No. 2 has been the system in use since 1978.

Let the FES *position angle* in the plane of the sky be defined such that 0° points north and 90° east. The position angle is related to the spacecraft roll by:

$$\text{position angle} = \text{aperture orientation angle} - \text{spacecraft roll angle},$$

where the orientation angles of the apertures are defined with respect to the reference vector and increase counterclockwise from the reference vector. The aperture orientation angles are given in Table I. The spacecraft roll angle for a given observation may be found on the NASA observing scripts, the VILSPA observing log, and in the image header (lines 84 and 85) archived with the image (see Turnrose and Thompson 1984, Section 9).

The orientations of the apertures relative to the FES coordinate system are shown in Figure 1. The orientation of the field as projected on the camera faceplates may also be determined. The north direction in the FES image is rotated counterclockwise from the reference vector by an amount equal to the spacecraft roll angle (see Figure 1). Figure 2 shows the placement of the large and small apertures with respect to low dispersion LWP, LWR and SWP spectra. (The apertures have been enlarged for clarity). While the orientation of the FES

TABLE I

IUE aperture plate orientation angles

Aperture	Orientation angle
LWLA[a] major axis	73 ± 3 deg
SWLA major axis	73 ± 2
LWLA to LWSA	238
SWLA to SWSA	242
LWLA to SWLA	275
LWSA to SWSA	279

[a] LW and SW refer to the long- and short- wavelength spectrographs; LA and SA refer to the large and small apertures, respectively.

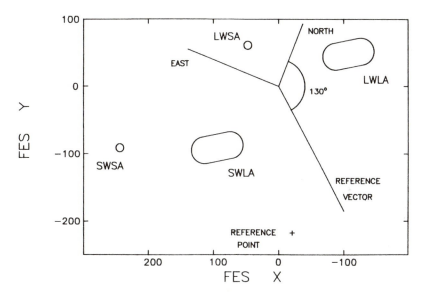

Fig. 1. The central portion of the IUE aperture plate as seen by FES No. 2. The locations of the apertures and the reference point are shown. The direction of the reference vector is also indicated. The North and East directions are shown for a spacecraft roll of 130 deg. The spatial scale is approximately 0.26 arc sec per FES unit.

coordinate system is different for each camera, the dispersion direction is approximately aligned with the FES Y axis in low dispersion and with the FES X axis in high dispersion.

2.2. SPECIAL OBSERVING TECHNIQUES

2.2.1. *Moving Targets*

IUE has successfully observed all types of solar system objects (Earth, Moon,

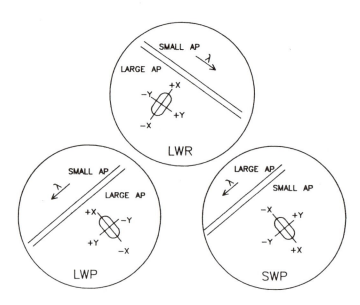

Fig. 2. Projection of the apertures onto the camera faceplates. The locations of large and small aperture low-dispersion spectra are shown for the LWP, LWR, and SWP cameras. The projection of the large aperture and the FES No. 2 X and Y directions are displayed relative to the apertures and spectra for each camera. The size of the large aperture is shown greatly enlarged for clarity.

comets, asterioids, Venus, Mars, Jupiter, Saturn, Jovian and Saturnian satellites, the Io torus, Uranus, and Neptune. Solar system objects, with the exception of the outermost planets, move across the sky fast enough to require motion compensation by the spacecraft's attitude control system. This involves trimming the gyros to the drift rate of the target, so that the telescope tracks at the same rate as the target motion. During an exposure, the telescope pointing is normally held by gyro control, although occasionally a planetary satellite has been used for offset guiding. Exposures longer than 10 to 15 min are usually performed in multiple segments. Between segments the target's center of light is repositioned to eliminate accumulated gyro drift errors. Under the best circumstances the gyro drift can be kept to less than 0.001 arc sec/sec. Even so, the telescope pointing would change one arc sec in 16 min at this drift rate.

The Earth, Moon, Jupiter, Saturn, and Venus are so bright that the Fine-Error Sensor is saturated when viewing them. Therefore, a technique has been developed for using the FES measurement of scattered light in the telescope to put an object as bright as Venus into an aperture. The large angular size of the Moon and the Earth require that they be observed by blind offset techniques under gyro control. The coordinates of solar system objects with a geocentric distance less than one AU (the Moon and some comets and asterioids) should be corrected for the parallax shift due to IUE's orbit. The Galilean satellites of Jupiter can be tracked to within two arc min of the planet.

Due to the complexity of individual observations, users of solar system spectra from the archives are strongly advised to contact the original investigators to obtain details of specific observations. The documentation in the IUE image header for individual spectra is frequently inadequate to completely describe the observations in cases such as these.

2.2.2. *Blind Offset Acquisitions*

The acquisition of targets invisible (in visual light, $m_v > 14.5$) to the FES is one of the most difficult type of observations routinely carried out by the IUE spacecraft. Such objects are acquired by blind-offset techniques, where the telescope is maneuvered a short distance to the invisible target from a nearby star. Blind-offset maneuvers performed with the three-gyro control system had errors less than 1 arc sec for slews less than 10 arc min, and errors less than 2.5 arc sec for slews less than 30 arc min. Similar maneuvers performed on the two-gyro/Fine Sun Sensor control system (after 17 August 1985, Sonneborn, 1985a) have errors less than 2 arc sec for slews less than 15 arc min (Sonneborn 1985b). The indication that a blind-offset acquisition was performed for a particular spectral image will be found on the GSFC observing script or the VILSPA observing log. The GSFC observing scripts usually contain the name and coordinates of the offset star used for the blind offset.

2.2.3. *Trailed and Multiple Exposures*

The *S/N* of low-dispersion spectra may be improved by trailing the target along the long axis of the large aperture. Such trailed spectra, if optimally exposed, have *S/N* values nearly twice that of untrailed spectra (see Section 3.1 for actual *S/N* values of untrailed spectra).

Trailed exposures may also be performed in high dispersion by trailing the star along the short axis of the large aperture. However, this reduces the width of the interorder background required to properly extract the echelle orders. Since the crowding of the orders is greatest at shorter wavelengths, extracted spectra from high dispersion trailed exposures should be analyzed only when the spectral features of interest are at relatively long wavelengths. The improvement in *S/N* per spectral resolution element is about 40% for high dispersion spectra which are optimally exposed.

The relationships between trailed (t_T) and point-source (t_{ps}) exposure times and trailing rates are given below.

low dispersion
$$t_T = 3.7\, t_{ps}$$
$$\text{trail rate} = 20.0/t_T \text{ arc sec/sec}$$

high dispersion
$$t_T = 1.85\, t_{ps}$$
$$\text{trail rate} = 10.0/t_T \text{ arc sec/sec.}$$

For trailed exposures longer than about 10 min, a sequence of point-source exposures with the target displaced a fixed amount along the large aperture is usually performed instead. This will appear as a 'pseudo-trailed' exposure. Usually

3 or 4 such spectra are placed within the aperture. The pseudotrail has a similar factor of 2 improvement in *S/N*.

For special purposes, such as time-critical observations, it is often desirable to place two separate exposures within the large aperture at low dispersion. Since the standard image processing system does not extract the individual spectra, the analysis of time-resolved observations requires that the spectra be extracted by the user from the spatially resolved spectrum file (see Section 4.3). Even when the maximum practical separation of the two exposures in the large aperture (10—12 arc sec) is used, there is still a measurable amount of point spread function overlap between the two spectra (Panek, 1982). The amount of separation is not sufficient to eliminate overlap of the two spectra. Good results have been obtained by fitting Gaussian profiles to cuts across the spectra (Panek and Holm, 1984).

2.3. THE SCIENCE IMAGE HEADER

Associated with each image is a set of 100 72-byte header, or label, records. These header records are generated automatically by the IUE operations ground system software during image acquisition and readout. The header contains spacecraft scientific and engineering data, an event 'round-robin', information provided by the observer about the object, and comments about the observation. The time-tagged event 'round-robin' section documents the sequence of procedures used to slew the spacecraft, acquire the target, start the exposure, and read the image. The label is appended in a sequential manner by IUESIPS to record significant processing parameters. Turnrose and Thompson (1984, Section 9) describe the image header contents. During the first year or so of the IUE mission, the ground system science header software was not fully operational. Consequently, very early IUE images have incomplete headers.

Occasionally a problem occurs with archiving an image during real-time operations and the spectral data must be recovered from the history tapes which record the spacecraft telemetry stream. Unfortunately, since the science header is generated by the ground system and is not contained in the telemetry stream, the image header cannot be recreated in its entirety from the history tapes. Images recovered and archived in this manner have incomplete information in the header and are usually identifiable by the phrase 'history replay' or 'history tape image recovery' in the Telescope Operator comments (the first five to seven lines of the header). Other errors in the header information can occur, so it is important to check critical data (such as exposure time) against other records and the IUE merged log.

3. Data Accuracy: Limitations Imposed by the Hardware

In addition to the various camera anomalies discussed in Section 1, there are other factors governed primarily by the instrumentation which constrain the quality of IUE data.

3.1. SIGNAL-TO-NOISE RATIO

The signal-to-noise ratio (*S/N*) of IUE data is camera dependent and varies greatly with exposure level and wavelength. Up to a signal level of about 200 DN, the *S/N* increases roughly linearly with exposure level. Beyond this the camera target begins to saturate and the extra exposure time required per DN increase in signal leads to a disproportionate increase in noise level. For optimum (untrailed) exposures typical *S/N* levels of low-dispersion spectra are in the range 15—30 for the SWP and 10—25 for the LWR and LWP cameras (Harris *et al.*, 1984; Settle *et al.*, 1981). The noise in IUE data is due to read-out noise, 'fixed-pattern' noise, and cosmic ray events. For a detailed comparison of the *S/N* characteristics of the two long-wavelength cameras in high dispersion see Barylak (1982). The *S/N* can be enhanced by averaging several spectra. However, due to small-scale spatial variations in camera gain and imperfect photometric correction by the ITF (see Section 4), some 'fixed-pattern' noise is generally apparent in IUE data. Since adding spectra will not reduce this non-random noise component, the improvement in *S/N* obtained by co-adding spectra is therefore not as great as would be expected if the noise were purely statistical. Examples of improvements in *S/N* obtained by co-adding IUE spectra are typically a factor of 1.4 for 3 spectra (Harris *et al.*, 1984) and factors of 1.6 and 1.7 for 5 and 6 spectra respectively (York and Jura, 1982). The latter reference, and also Adelman and Leckrone (1985), Joseph (1985), and Welty *et al.* (1986), include detailed discussions of fixed-pattern noise in high resolution spectra. A study by West and Shuttleworth (1981), based on high resolution data, indicates that there is little to be gained by averaging more than 5 spectra, while Clarke (1981a) finds that the *S/N* continues to improve significantly with up to 8 co-added low dispersion spectra (see also Adelman and Leckrone, 1985). Since the fixed-pattern structure on an image is spatially variable on a small scale, one should co-add spectra taken with the target source located at different positions in the large aperture (or at different heliocentric radial velocities) so that spectral features do not always fall on the same pixels. Furthermore, since the spectral image position with respect to the camera target is temperature- and time-dependent, it is advisable to average spectra obtained at different epochs.

3.2. SENSITIVITY

The sensitivity of the IUE cameras is determined by the quantum efficiency of the UVC photocathode, which is highly wavelength-dependent, and the photoelectron sensitivity of the remaining system which is dependent on position in the image. The sensitivity of the SWP camera increases with wavelength, apart from a dip near 1500 Å. The LWR and LWP cameras both have sensitivities which peak near 2800 Å. The performance of the SWP camera is superior to that of the long-wavelength cameras in the 1850—2100 Å overlap region. These detector characteristics are reflected in the minimum flux detectability curves for the whole instrument plotted in terms of absolute flux in Figure 3. The range of magnitudes of various continuum sources plotted for comparison highlights the very broad ($\sim 10^6$) dynamic range of IUE. The large variation in sensitivity as a function of wavelength implies that optimum exposure of one region of a spectral image normally requires significant over- or under-exposure of another. Consequently, two or three images with different exposure times may be required for optimum

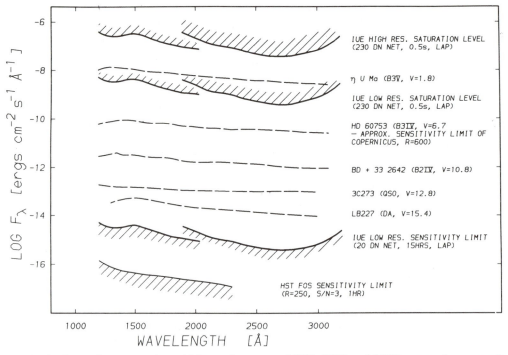

Fig. 3. Dynamic range and sensitivity performance of IUE (SWP and LWR cameras) compared with fluxes of typical continuum sources. Approximate minimum flux detectability limits of *Copernicus* and the HST Faint Object Spectrograph are also shown. The latter is based on the performance of the blue digicon/G160L grating combination given by Ford (1985).

coverage of a particular spectrum. Exposure time is quantized by the on-board computer in units of 0.4096 sec, one unit representing the shortest possible exposure time. The upper limits of IUE's dynamic range shown in Figure 3 are the flux levels which would just give about the maximum *gross* raw signal of 255 DN in each dispersion mode in this time. Excessive exposure gives rise to camera target saturation. Data points thus affected are flagged in IUESIPS extracted spectra files and plots (Turnrose and Thompson, 1984), and users are urged to be aware of these flags when analyzing their data since localized saturation could be mistaken for a broad absorption feature in a calibrated spectrum. A spurious feature arising in this way is illustrated in Figure 4 and discussed by McLachlan and Nandy (1984) and Savage and Sitko (1984). In addition, the maximum exposure level of an image is normally indicated in the comments field of the IUE cumulative merged log of observations.

The sensitivities of the IUE cameras also vary with time (see Section 1) and head-amplifier temperature (THDA). Sensitivity monitoring of the IUE instruments forms an integral part of the Three-Agency routine calibration program and periodic analyses are published in the Three-Agency meeting reports and IUE Newsletters (e.g. Sonneborn and Garhart, 1986).

When targets are acquired by blind-offset or observed through the small

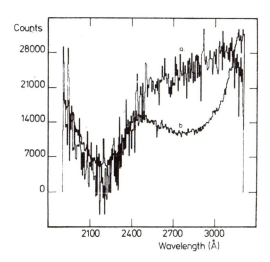

Fig. 4. A spurious 'absorption' feature is the result of overexposure around the 2800 Å sensitivity peak of the LWR camera in a large aperture spectrum (b) of HD 14250. A correctly exposed small aperture spectrum (a) of the star from the same image is plotted for comparison (from McLachlan and Nandy, 1984).

aperture, IUE data users should be aware that small errors in target positioning with respect to the apertures can lead to significant differences in apparent absolute flux level between spectra of the same object. The transmission of the small aperture is typically about 50% that of the large aperture, but variations in measured flux ratios of 10% about this figure are common. Similarly, effective exposure times of trailed spectra are not precisely known due to slight maneuvering errors as the target is trailed through the large aperture.

3.3. Resolution

The spectral and spatial resolutions achievable with IUE data are limited by the size of the point spread function (PSF) in the camera image plane. In general this varies with wavelength and is dependent on camera used, dispersion mode and telescope focussing conditions. For the SWP camera in high dispersion, Evans and Imhoff (1985) find that the spectral resolution is practically independent of the generation of IUESIPS software used (see Section 4). At 1300 Å they obtain a FWHM value of about 0.1 Å, corresponding to a resolution of $R = 13\,000$. This decreases to $R = 10\,500$ at 2000 Å. For the LWR camera in high dispersion, using the new software (see Table II below for exact dates), Evans and Imhoff (1985) find that R *increases* with wavelength from $R = 13\,300$ at 2000 Å to $R = 16\,300$ at 3100 Å. The same trend is found using the old software (which has a lower sampling frequency), but with R tending to be slightly lower throughout the range. For both cameras the difference in resolution between the large and small apertures with point sources is small (Penston, 1979).

In low dispersion Cassatella *et al.* (1985) find for the SWP camera that R varies between 270 and 330, with the maximum value obtained at 1300—1400 Å.

Similarly, for the LWR camera they find that R passes through a maximum of about 420 around 2400 Å. Perpendicular to the dispersion direction Cassatella *et al.* (1985) find that the PSF is strongly dependent on wavelength and telescope focussing conditions, being smallest for negative values of the focus step. Knowledge of the cross-dispersion PSF profile is important for extracting time-resolved or spatial information from large aperture trailed, multiple or extended source spectra. Cassatella *et al.* (1985) find that whereas for LWR and LWP spectra good fits are obtained with Gaussian profiles, the PSF for SWP spectra is asymmetric and better represented by a 'skewed Gaussian' function. For a detailed review of IUE resolution studies see Turnrose and Thompson (1984).

3.4. SCATTERED LIGHT

It is well known that the short wavelength region of low dispersion SWP spectra can be contaminated by longer wavelength light scattered by the grating (for example, see Clarke, 1981b; Crivellari and Praderie, 1982). The amount of scattered light is normally negligible, but it can be a serious problem in observations of late-type stars with UV flux distributions that rise very steeply to longer wavelengths, requiring long exposure times to register the short wavelength flux. The effect of the scattered light is to produce an apparent plateau of emission shortward of the steep spectral slope (see Figure 5). Since the contamination is confined to the low dispersion order it is difficult to distinguish it from the stellar

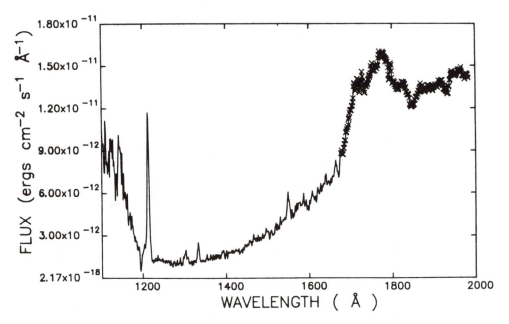

Fig. 5. A spurious far-ultraviolet continuum below 1600 Å results from longer wavelength light being scattered by the short-wavelength grating. The exposure (SWP 22548) is one of Procyon (HD 61421), spectral type F5 IV. In particular, note the very steep false 'continuum' below Lyman alpha. The 'X's' indicate those portions of the spectrum which are saturated (DN = 255) in the raw image.

spectrum. Basri *et al.* (1985) have made a quantitative study of the problem using a computer model of the grating scattering function. They conclude that serious contamination by scattered light can occur below 1800 Å for solar-type stars and at longer wavelengths for later spectral types. A further effect of grating scattering is to reduce the apparent contrast between narrow emission lines and the continuum, since some light is scattered from an emission line into the adjacent continuum. However, Feldman (1986) and Basri (1985) discuss a modification to the near-field scattering profile of Basri *et al.* (1985) which was found to be necessary after comparison of the Lyman alpha profile observed in cometary spectra with the predicted profile. The observations indicated that much less light is scattered from emission line cores than originally estimated.

4. IUE Image Processing

The IUE Spectral Image Processing System (IUESIPS) has been designed to provide the Guest Observer with spectral data reduced from the raw images in an accurate and standard manner and as free of instrumental effects as possible. Image processing steps necessary to produce calibrated IUE spectra are complex and require detailed knowledge of the IUE scientific instrument's performance and calibration. The two ground stations use essentially identical software.

The IUESIPS software has evolved during the course of the IUE mission. The most significant changes in the processing algorithms are listed in Table II with their implementation dates at GSFC and VILSPA. A detailed description of the 'new' software is given by Turnrose and Thompson (1984); the 'old' software is described by Turnrose *et al.* (1981). The detailed chronology of the IUESIPS configurations is described by Gass and Thompson (1985), Thompson (1984), and Turnrose *et al.* (1984).

The principal image processing steps have remained qualitatively the same during the mission. These steps include implicit (in the 'new' software) or explicit (in the 'old' software) correction for geometrical distortion of the camera system by reference to the reseau grid superposed on each image; photometric correction of the image, by means of pixel-by-pixel intensity transfer functions (ITFs), which correct for sensitivity variation across the SEC vidicon detector and its non-linear response; wavelength calibration of the spectral orders with analytical dispersion relations determined from on-board platinum-neon lamp spectral images; extraction of spectral flux as a function of wavelength from the photometrically corrected image; and reduction of the extracted spectral flux to an absolute flux system by means of absolute calibrations obtained from observations of photometric standard stars (low dispersion only).

These steps are briefly summarized below. Grady and Imhoff (1985a) give a more complete description, and the reader should refer to Turnrose and Thompson (1984) for a detailed discussion of the image processing procedures.

4.1. GEOMETRIC CORRECTION

Geometric correction of IUE images is required because the 'pixels' which are read out are read-beam locations on the camera target, not physical pixels. Since

TABLE II

Summary of most significant IUESIPS reduction software changes pertinent to archive data users

---

Low Dispersion

Correct SWP ITF error	7 July, 1979 (GSFC)
	7 August, 1979 (VILSPA)

— Removed photometric error at 20% exposure level of ITF

Implementation of 'new' software	4 November, 1980 (GSFC)
	10 March, 1981 (VILSPA)

— Doubled spectral extraction frequency and apparent spectral resolution by reducing the effective slit width by a factor of 2.
— Geometric resampling handled differently.
— Increased point-to-point noise (factor of $\sqrt{2}$).
— Better background handling.
— Basic photometry unchanged.

Extended line-by-line file	1 October, 1985 (GSFC/VILSPA)

— Increased spatial resolution, perpendicular to dispersion.

High Dispersion

Time/temperature corrected geometric and wavelength calibrations	19 May 1981 (GSFC)
	11 March, 1982 (VILSPA)

— Reduced residual internal wavelength errors ($1\sigma \leqslant 2$—3 km s^{-1}).

Improved spectral registration at crowded orders	28 August, 1981 (GSFC)
	11 March, 1982 (VILSPA)

— Better background placement, hence better net fluxes

Implementation of 'new' software	10 November, 1981 — LWR, SWP (GSFC)
	7 January, 1982 — LWP (GSFC)
	11 March, 1982 (VILSPA)

— Doubled spectral extraction frequency and apparent spectral resolution by reducing the effective slit width by a factor of 2.
— Explicit geometric resampling eliminated.
— Increased (but more realistic) point-to-point noise.
  (~ factor of 2 unfiltered, $\sqrt{2}$ when filtered).
— Further improved background placement, and better handling.
— Better photometry (increased net fluxes at short wavelengths, due to lower background; better stability).

---

the pointing of the read-out beam varies with temperature and local exposure levels, the raw data are mapped to a geometrically-correct domain using a reseau grid on the camera faceplate before the other corrections are applied. The details of the mapping procedure depend on whether processing was performed with the 'old' or 'new' software (see Table II for implementation dates). In the 'old' software, the raw data were geometrically corrected. In the 'new' software, the calibration data are mapped onto the raw image in a geometrically correct manner.

4.2. INTENSITY TRANSFER FUNCTIONS

The image is transformed from the initial array of 8-bit raw signal data numbers

(DN) to an array of 16-bit *linearized* flux numbers (FN) by means of an 'intensity transfer function' (ITF). This is necessary since the signal stored on the potassium chloride camera target is a non-linear function of exposure, especially at high exposure levels. Furthermore, the form of the response varies from one pixel to another. Hence an ITF actually consists of a set of FN/DN transformation curves, one for each pixel in the image.

The ITFs are derived from on-board UV-floodlamp flat-field images of varying exposure level. An ITF is constructed by averaging several (usually 4 or 5 images in early ITFs) individual flat-field images for each of 11 or 12 discrete exposure levels. The linearization function is then derived for each pixel in the raw image by interpolating in the geometrically corrected ITF domain. The accuracy with which the reseau grids of two images can be registered, typically about 0.2 pixels, fundamentally limits the photometric accuracy of IUE data. This registration uncertainty affects the ITF in the geometric correction of the individual flat-field images, the generation of average flat-field images at each intensity level in the ITF, and the mapping of the final ITF dataset to match the format of the raw spectral image. Analysis of ITF images has shown that while the ITF does a good job of removing the large scale sensitivity variations in the IUE cameras, the geometrical smoothing inherent in the generation of the ITF means that there is little compensation for pixel-to-pixel sensitivity variations. Methods of compensating for pixel-to-pixel noise in IUE spectra have been discussed by York and Jura (1982), Adelman and Leckrone (1985), and Welty *et al.* (1986).

4.3. SPECTRAL EXTRACTION

Once the wavelength-calibration dispersion relations are known, the positions of the order(s) in the geometrically corrected reference frame are determined. The wavelength scale must be registered with the spectrum, which may be offset either accidentally or intentionally as part of the observing program. To extract each order, the dispersion relations must be mapped back into the photometrically corrected raw-image space. For low-dispersion images the mapping is done for each pixel. In high dispersion, due to the sheer volume of data, the mapping is done for selected pixels, and the position of the order between these points is determined by bi-linear interpolation between the mapped pixels.

Low-dispersion spatially-resolved extracted spectra are generated by passing a numerical slit along the dispersion direction, and calculating the extracted flux values every $\sqrt{2}/2$ pixels along and approximately perpendicular to the dispersion direction. In the old software, the slit was $\sqrt{2}$ pixels wide. Extracting data from the image in this way is equivalent to forming an appropriately weighted average of the four surrounding pixels for each pixel in the appropriate portion of the photometrically corrected image. For large aperture observations the size of the extraction slit in the spatial dimension is controlled by the extraction type, such as point source, trailed, or extended source extraction as selected by the observer (unless the image has been reprocessed). The extraction of fluxes in the spatial direction is along lines of constant wavelength. For images processed prior to 1 October, 1985, 110 lines are extracted in the spatial direction and averaged in pairs to produce the 55 line spatially-resolved image file (see Munoz Peiro, 1985 for details). Images processed after this date omit the pair-wise averaging and have spatially resolved files which have 110 lines. This is referred to as the 'extended line-by-line' extraction. Slit-integrated spectra are formed by adding 9 (point source) or 15 (extended source) lines of the spatially resolved data.

The high-dispersion spectra are formed by passing a numerical slit along the measured and interpolated positions of the orders. The extraction slit width varies as a function of order number across the high dispersion image. The background (inter-order) spectra are extracted by passing a slit one pixel square in area halfway between adjacent orders, and averaging the inter-order spectra on each side of the order of interest. The net spectrum is then the difference between the gross spectrum and the interpolated background spectrum. There is no spatially resolved information in the extracted high dispersion spectra.

4.4. PHOTOMETRIC CALIBRATION

The philosophy behind the photometric calibration of IUE is described in detail by Bohlin *et al.* (1980). The IUESIPS extraction procedure produces spectra in units of time-integrated net FN. Conversion to absolute flux units requires multiplication by S_λ^{-1}/t, where S_λ^{-1} is the 'inverse sensitivity function' in units of 10^{-14} erg cm^{-2} Å^{-1} FN^{-1}, and t is the exposure time in seconds. In practice, there is no officially adopted S_λ^{-1} for high dispersion and in this case IUESIPS final output files and plots are left in 'ripple-corrected' net FN units (for details of the echelle blaze, or 'ripple', correction see Ahmad, 1981; and Ake, 1981, 1982).

However, Cassatella *et al.* (1981, 1982, 1983) have described a method for obtaining absolute photometric calibration in high dispersion and presented resulting tables of calibration factors. For low dispersion IUESIPS uses officially adopted S_λ^{-1} curves for each camera to give absolutely calibrated time-integrated fluxes in units of 10^{-14} erg cm^{-2} Å^{-1} (division by the exposure time is left to the user). The curves currently used by IUESIPS are those of Bohlin and Holm (1980, the so-called 'May 1980' calibration — see also Holm *et al.*, 1982) for the SWP and LWR cameras and of Cassatella and Harris (1983) for the LWP camera. The adoption of the SWP and LWR calibrations coincided with the implementation of the new software (see Section 4) on 4 November 1980 at GSFC and 10 March 1981 at VILSPA. For details of earlier calibrations see Turnrose and Harvel (1982). The LWP absolute calibration was installed in IUESIPS on 11 October, 1983 at VILSPA and 19 October, 1983 at GSFC. Prior to this there was no official calibration for the LWP. All IUESIPS absolute calibrations are valid for 'point' (untrailed) spectra taken in the large aperture.

The fundamental standard for the IUE absolute calibration is the flux for Eta UMa (HD 120315, $m_v = 1.8$) adopted by Bohlin *et al.* (1980). This star was well observed by earlier UV space experiments, such as TD—1 and OAO—2. The S_λ^{-1} curves were derived using several secondary standards, also observed by the earlier UV experiments, whose spectra were reduced to the Bohlin *et al.* (1980) Eta UMa flux scale using correction factors and then ratioed to IUE FN spectra of the same stars. The reliability of the absolute calibration so produced is clearly limited by the uncertainty in the adopted flux for Eta UMa, which is considered to be of the order of $\pm 10\%$. However, the relative accuracy within a particular spectrum over regions of $\leqslant 100$ Å should be better than this (Bohlin *et al.*, 1980).

The sensitivity decrease of the cameras with time since the absolute calibrations were derived must be taken into account when very accurate absolute fluxes are required. In the case of the LWR camera, for which the sensitivity decrease is largest, a correction algorithm is available (Clavel *et al.*, 1986). Work on deriving a similar method of correcting SWP data is in progress.

New ITFs and absolute calibrations are now being developed but have not as yet been installed in IUESIPS at either ground station.

4.5. WAVELENGTH CALIBRATION

Frequent high and low dispersion wavelength exposures to the on-board platinum-neon calibration lamp, through the small aperture, are made to monitor the wavelength calibration of IUE spectra. These images are used to check the validity of mean analytic dispersion relations which are used in IUESIPS data processing to convert from image position to wavelength. Results from this monitoring lead to occasional updates to the mean dispersion constants. The position of the spectral format on an image shifts slightly with time and camera head-amplifier temperature (THDA), and for each image appropriate corrections are applied to the mean dispersion relations to compensate. Additional corrections are applied after spectral extraction to reduce the wavelength scale to a heliocentric frame of reference and, for wavelengths greater than 2000 Å, to convert from vacuum to air wavelengths. The wavelength assignments in high dispersion are considered to be self-consistent to ± 3 km s^{-1} (1σ), although errors in positioning a target in the (large) aperture can give rise to substantial absolute errors: a pointing error of 1 arc sec in the dispersion direction corresponds to about 5 km s^{-1} in high dispersion. For detailed discussions of the IUE wavelength calibration and corrections see Thompson *et al.* (1982) and Turnrose and Thompson (1984).

4.6. IUESIPS DATA PRODUCTS

IUESIPS provides several data products to the Guest Observer for each IUE image. The primary data source is magnetic tape data files. Guest Observers are also provided with photowrites of each image and, optionally, CalComp plots of the gross and net spectra. There are magnetic tape files for the raw and photometrically corrected versions of the image. Low-dispersion spectra have both a line-by-line spectrum file and a file containing the net extracted, absolutely-calibrated spectrum. If both large and small apertures were exposed on the same image, there are separate line-by-line and extracted spectrum files for each aperture. In addition to the raw and photometrically corrected image files, there is a single extracted spectrum file for high-dispersion images. The specifics of the tape formats for the various file types may be found in Turnrose and Thompson (1984).

References

Adelman, S. J. and Leckrone, D. S.: 1985, *NASA IUE Newsl.*, No. 28, 35.
Ahmad, I. A.: 1981 *NASA IUE Newsl.*, No. 14, 129.
Ake, T. B.: 1981, *NASA IUE Newsl.*, No. 15, 60.

Ake, T. B.: 1982, *NASA IUE Newsl.*, No. 19, 37.

Barylak, M.: 1982, *ESA IUE Newsl.*, No. 15, 31 (*NASA IUE Newsl.*, No. 21, 55).

Basri, G., Clarke, J. T., and Haisch, B. M.: 1985, *Astron. Astrophys.* **144**, 161.

Basri, G.: 1985, *NASA IUE Newsl.*, No. 28, 58.

Blades, J. C. and Cassatella, A.: 1982, *ESA IUE Newsl.*, No. 15, 38 (*NASA IUE Newsl.*, No. 21, 62).

Boggess, A. *et al.*: 1978, *Nature* **275**, 377.

Bohlin, R. C. and Holm, A. V.: 1980, *NASA IUE Newsl.*, No. 10, 37 (*ESA IUE Newsl.*, No. 11, 18).

Bohlin, R. C., Holm, A. V., Savage, B. D., Snijders, M. A. J., and Sparks, W. M. 1980, *Astron. Astrophys.* **85**, 1.

Broude, S. M. and Imhoff, C. L.: 1984, *NASA IUE Newsl.*, No. 24, 127.

Cassatella, A., Barbero, J., and Benvenuti, P.: 1985, *Astron. Astrophys.* **144**, 335.

Cassatella, A., and Harris, A. W.: 1983, *ESA IUE Newsl.*, No. 17, 12 (*NASA IUE Newsl.*, No. 23, 21).

Cassatella, A., Ponz, D., and Selvelli, P. L.: 1981, *ESA IUE Newsl.*, No. 10, 31 (*NASA IUE Newsl.*, No. 14, 170).

Cassatella, A., Ponz, D., and Selvelli, P. L.: 1982, *ESA IUE Newsl.*, No. 15, 43.

Cassatella, A., Ponz, D., and Selvelli, P. L.: 1983, Resport presented at the March IUE 3 Agency Meeting, U.K.

Clarke, J. T.: 1981a, *NASA IUE Newsl.*, No. 14, 149.

Clarke, J. T.: 1981b, *NASA IUE Newsl.*, No. 14, 143.

Clavel, J., Gilmozzi, R., and Prieto, A.: 1986, *ESA IUE Newsl.*, No. 26, 65 (*NASA IUE Newsl.*, No. 31, 83).

Coleman, C. I. *et al.*: 1977, 'IUE Camera Users Guide', IUE Technical Note No. 31, Appleton Lab. and UCL.

Crivellari, L. and Praderie, F.: 1982, *Astron. Astrophys.* **107**, 75.

Evans, N. R. and Imhoff, C. L.: 1985, *NASA IUE Newsl.*, No. 28, 77.

Feldman, P. D.: 1986, *Astron. Astrophys.* **159**, 342.

Ford, H. C.: 1985, 'Faint Object Spectrograph Instrument Handbook', Space Tel. Sci. Inst.

Gass, J. and Thompson, R. W.: 1985, *NASA IUE Newsl.*, No. 28, 102.

Grady, C. A. and Imhoff, C. L.: 1985a, *NASA IUE Newsl.*, No. 28, 86.

Grady, C. A. and Imhoff, C. L.: 1985b, *NASA IUE Newsl.*, No. 28, 140.

Hackney, R. L., Hackney, K. R. H., and Kondo, Y.: 1982, in Kondo, Y. *et al.* (eds.), 'Advances in Ultraviolet Astronomy; Four Years of IUE Research', NASA CP-2238, p. 335.

Harris, A. W.: 1984, *ESA IUE Newsl.*, No. 20, 64.

Harris, A. W.: 1985a, *ESA IUE Newsl.*, No. 22, 19.

Harris, A. W.: 1985b, *ESA IUE Newsl.*, No. 24, 17 (*NASA IUE Newsl.*, No. 28, 22).

Harris, A. W. *et al.*: 1984, in Mead, J. M. *et al.* (eds.), 'Future of Ultraviolet Astronomy Based on Six Years of IUE Research', NASA CP-2349, p. 516.

Holm, A. V.: 1985, *NASA IUE Newsl.*, No. 26, 11.

Holm, A. V., Bohlin, R. C., Cassatella, A., Ponz, D. P., and Schiffer, F. H.: 1982, *Astron. Astrophys.* **112**, 341.

Imhoff, C. L.: 1985a, *NASA IUE Newsl.*, No. 27, 9.

Imhoff, C. L.: 1985b, *NASA IUE Newsl.*, No. 28, 10 (*ESA IUE Newsl.*, No. 25, 45).

Imhoff, C. L. and Grady, C. A.: 1985, *NASA IUE Newsl.*, No. 26, 66.

Imhoff, C. L. and Wasatonic, R. P.: 1986, *NASA IUE Newsl.*, No. 29, 45.

Joseph, C.: 1985, Ph.D. dissertation, University of Colorado.

McLachlan, A. and Nandy, K.: 1984, *The Observatory*, **104**, 29.

Munoz Peiro, J. R.: 1985, *ESA IUE Newsl.*, No. 23, 58 (*NASA IUE Newsl.*, No. 27, 27).

Panek, R. J.: 1982, *NASA IUE Newsl.*, No. 18, 68.

Panek, R. J. and Holm, A. V.: 1984, *Astrophys. J.* **277**, 700.

Penston, M.: 1979, Report Presented at the March IUE 3-Agency Meeting.

Ponz, J. D.: 1980a, *ESA IUE Newsl.*, No. 8, 12.

Ponz, J. D.: 1980b, in Weiss, W. W. *et al.* (eds.), 'IUE Data Reduction', Austrian Solar and Space Agency, p. 75.

Savage, B. D. and Sitko, M. L.: 1984, *Astrophys. Space Sci.* **100**, 427.

Schiffer, F. H.: 1982, *NASA IUE Newsl.*, No. 18, 64.

Settle, J., Shuttleworth, T., and Sandford, M. C. W.: 1981, *NASA IUE Newsl.*, No. 15, 97.

Snijders, M. A. J.: 1983, *ESA IUE Newsl.*, No. 16, 10 (*NASA IUE Newsl.*, No. 23, 56).

Sonneborn, G.: 1985a, *NASA IUE Newsl.*, No. 28, 147.

Sonneborn, G.: 1985b, *NASA IUE Newsl.*, No. 28, 154.

Sonneborn, G. and Garhart, M. P.: 1986, *NASA IUE Newsl.*, No. 31, 29.

Sonneborn, G., Oliversen, N. A., Imhoff, C. L., Pitts, R. E., and Holm, A. V.: 1987, 'IUE Observing Guide', *NASA IUE Newsl.*, No. 32, 1.

Thompson, R. W.: 1984, *NASA IUE Newsl.*, No. 25, 1.

Thompson, R. W., Turnrose, B. E., and Bohlin, R. C.: 1982, *Astron. Astrophys.* **107**, 11.

Turnrose, B. E. and Harvel, C. A.: 1982, *NASA IUE Newsl.*, No. 16 (*ESA IUE Newsl.*, No. 14).

Turnrose, B. E., Harvel, C. A., and Stone, D. F.: 1981, 'IUE Image Processing Information Manual, Version 1.1', CSC/TM-81/6268.

Turnrose, B. E. and Thompson, R. W.: 1984, 'IUE Image Processing Information Manual Version 2.0', CSC/TM-84/6058.

Turnrose, B. E., Thompson, R. W., and Gass, J.: 1984, *NASA IUE Newsl.*, No. 25, 40.

Walter, S. O. and Imhoff, C. L.: 1983, *NASA IUE Newsl.*, No. 20, 9.

Welty, D. E., York, D. G., and Hobbs, L. M.: 1986, *Publications of the Astronomical Society of the Pacific* **98**, 857.

West, K. and Shuttleworth, T.: 1981, *ESA IUE Newsl.*, No. 12, 27 (*NASA IUE Newsl.*, No. 19, 58).

York, D. G. and Jura, M.: 1982, *Astrophys. J.* **254**, 88.

PART VII

THE FUTURE

Edited by A. BOGGESS

THE ROLE OF IUE IN THE HUBBLE SPACE TELESCOPE ERA

F. DUCCIO MACCHETTO

Space Telescope Science Institute, Baltimore, MD, U.S.A.

and

RICHARD C. HENRY

Department of Physics and Astronomy, The Johns Hopkins University, Baltimore, MD, U.S.A.

1. Introduction

When will Space Telescope be launched? Also, when will the IUE satellite die? If there *is* a time of overlap of these two, what will be the the role of IUE, given the extraordinary capabilities of the Hubble Space Telescope (HST)?

The first two questions do not have certain answers, but the third question definitely does. If a brand new IUE were launched at the same time as HST, it would provide a perfect complement to HST.

The analog is ground-based astronomy, where if there were only the 5-m Hale telescope, we would have a very distorted program in ground-based observational astronomy, indeed. But in fact, the Hale telescope is complemented by a host of smaller telescopes around the world; these various telescopes are used to study those problems that are scaled to the abilities of the telescope in question. The time of the largest telescope need not be wasted on problems that can be successfully attacked using smaller telescopes.

While the HST has capabilities in both the visible and the ultraviolet parts of the electromagnetic spectrum, the same complement of ground-based telescopes that is available to support the Hale telescope, will also support the HST *in the visible*. It is at *ultraviolet* wavelengths that special depth is needed to support HST, and it is IUE alone that provides that depth.

In this paper, we examine the character of IUE as a support telescope for HST, calling out a few specific examples as illustration. We have of course used examples where we are thoroughly familiar with the issues from our own research; many other examples will occur to the reader.

We also, in the next section, remark on how IUE and HST operate in a similar manner; there is flexible access to IUE to *permit* the efficient use of IUE as support for HST. Of course this is no accident: the planned method for utilizing HST itself, is largely modeled on IUE's own pioneering management scheme.

Following our discussion of examples, from three scientific fields, of IUE/ST complementarity, we provide a specific comparison of IUE and ST capabilities for ultraviolet spectroscopy, and finally we give more detail on the rationale for IUE operation in the era of HST.

2. IUE as Precursor to HST

IUE was developed as the first true space observatory. Previous astronomical

Y. Kondo (ed.), Scientific Accomplishments of the IUE, pp. 753—758.
© 1987 *by D. Reidel Publishing Company.*

satellites had been built with the concept of Principal Investigators (PIs) defining and building the on board scientific instrumentation. The PIs and their co-investigators were also responsible for defining a scientific observing program, gathering the data, and carrying out its scientific analysis. While most PIs showed generosity and understanding in sharing some of the data with colleagues outside their teams, this was done on an ad-hoc basis and with little if any organization, scientific review, and so on.

The philosophy behind IUE was diametrically opposite. An international team of scientists under the overall guidance of a NASA-appointed Project Scientist, was responsible for the design and for overseeing the implementation of the scientific instrument, the spacecraft, and the ground system. The latter included the relatively novel concept of a systematic approach to data reduction software, to be developed as part of the observatory to be applied in a standard way to all the data. This ensured that all astronomers, regardless of the resources of their institution or their software skills, had equal opportunity toward understanding and making scientific use of IUE data.

The other major difference between IUE and previous astronomical satellites was that no person or team had guaranteed or privileged access to the data. Observing time was divided among the international partners, with 2/3 for NASA, 1/6 for ESA and 1/6 for the U.K. Separate selection committees in each organization assigned time to astronomers from their respective areas. In every case however, proposals were selected solely on the basis of scientific merit.

The Hubble Space Telescope has adopted the IUE philosophy in many areas. The first one is that of a true space observatory. The HST will indeed be a major observatory open to the wide international community. While, unlike IUE, the concept of having instruments built by PI teams has been used, these teams get only a fraction of the total observing time. Most of the time will be assigned to astronomers from all over the world selected through a strict peer-review process. In the case of HST the funding agencies, NASA and ESA, have agreed that proposals originating in both areas should be reviewed by the same international allocation committee thereby further strengthening international collaboration.

A second area where HST has followed and considerably expanded the IUE philosophy is that of data analysis. Because of the increased complexity of the HST instruments and their number when compared to IUE, it was decided that common data reduction, calibration and analysis tools should be developed for use by the HST observers.

In addition to the greater sophistication of the calibration reduction software, the Space Telescope Science Institute is developing scientific analysis tools that will allow astronomers to carry out much more research-oriented data analysis than is possible with the standard IUE software. This trend to make greater use of common data analysis tools was started with IUE of observers sharing data reduction programs, and is becoming a standard feature in astronomical research.

One other area which was pioneered by IUE is that of data archives and their utilization. The next paper (pp. 759—769) describes how IUE archives are organized and are being utilized. In the HST, the same basic concept of open access to the data archive, following an initial period of proprietary data, was adopted. However, this concept has now been refined and extended. By using modern communication techniques it will be possible to remotely access by computer the catalogue information and determine whether a particular observation set exists or is available. In addition the Space Telescope Science Institute will make available to a number of centers, notably the European Coordinating Facility and major observatories in the US, a copy of the archived data in the form of optical discs. This will enormously enhance the research opportunities on the archived data and will make possible the pursuit of statistical studies on an extensive and homogeneous data set.

The final lesson that IUE has taught, and HST has adopted, is that of fostering international cooperation. This is evident in all areas of the two projects: engineering and technical management of the project during the development phase; construction and testing in the many industries involved in the hardware in the USA and Europe; scientific guidance of the project and of the Science Institute by an international team of astronomers; and scientific exploitation by the world wide astronomical community. Not only is the access to these observatories open to individual scientists from every country, but IUE has spurred international collaborations to carry out studies which would otherwise not have been possible. Examples of these are described in the next section.

3. Science: IUE and HST

Access to space has opened up entirely new fields of scientific investigation. Study of the chromospheres of late-type stars will form our first scientific example. A rocket experiment in 1970 by Warren Moos and his collaborators revealed that chromospheric hydrogen Lyman alpha radiation from nearby stars does succeed in penetrating the local interstellar medium, and can be detected. They, and others, used the *Copernicus* satellite to obtain the first high-resolution profiles of this emission, and they also detected interstellar deuterium Lyman alpha, the study of which is important for cosmology.

The IUE made this field explode. The reason was that unlike Copernicus, which was a slow scanning spectrometer, IUE uses panoramic detectors which capture an enormous spectral range simultaneously. Jeff Linsky and others have exploited this to greatly increase our knowledge and understanding of stellar chromospheres.

The HST will build on this work in a powerful way, both in the study of stellar chromospheres, and of the interstellar medium (using chromospheric emissions as light sources for absorption studies). The HST will have spectral resolution comparable to *Copernicus*; will record the spectrum of one entire line simultaneously; and if we are lucky, and IUE survives, can be complemented by a *simultaneous* IUE spectrum of the other important chromospheric emissions. This is particularly important with binary stars, where phase effects appear.

The long observing time required to obtain a good IUE spectrum of faint cool stars has led to collaboration (to obtain joint ESA/NASA observing shifts) between American and European observers.

Our second scientific example concerns NGC 4151. The monitoring of the variability of NGC 4151 over many years was only possible because of the

reproducible behavior of IUE over that period of time and because of the
dedication of a number of astronomers from several of the ESA countries. The
exciting results obtained so far give us a much greater insight into the physics of
the central energy source of this Seyfert galaxy. However, continued monitoring
over many more years will be needed, for example to establish during the next
non-quiescent phase the true temporal correlations between variations in the
ultraviolet continuum and variations in the lines, still a matter of some scientific
debate. This will require continued observations by IUE in the foreseeable future.
In addition, high resolution spatial imaging and spectroscopic observations with
HST will give us greater insight into the physical structure and conditions at one
point in time. With HST there is chance of physically resolving the narrow line
region in this galaxy and therefore correlate temporal variations observed with
IUE with actual physical sizes. On the other hand, because of the expected huge
oversubscription factor, HST is unlikely to be able to spend as much time as IUE
carrying out a monitoring program over the required long time periods. IUE will
remain an essential complement to HST for this program.

Our final example concerns extensive international collaboration in the study of
supernovae. This program was started under the aegis of the 'target of opportunity'
program in ESA. An international team of scientists joined forces to carry out this
program whose aim is the study of extragalactic supernovae bright enough for IUE
to observe over as long a time interval as feasible.

Prior to the occurence of the first supernovae a well-defined set of observing
sequences was defined. These could later be tuned in real time thanks to the great
scheduling and observing flexibility of IUE, something that HST in its low orbit
unfortunately, does *not* have! The program was quickly expanded to include
observations by other astronomers at ground-based optical, infrared and radio
telescopes, and on other occasions X-ray and γ-ray satellites such as Einstein,
Exosat and Cos-B.

After the first two or three supernova events, the IUE operation became so
smooth and efficient that in many cases IUE observations were obtained before
ground based observatories could begin their work. This quick turn around,
typically 24 hr or less from the time of notification, has allowed the observers to
'catch' the supernovae even during their brightening phase before maximum.
Supernovae were then followed, with timely observations spread over days, weeks,
and months, as long as the supernova was bright enough for IUE to observe. This
program has already provided a new understanding of the nature of supernova
progenitors and of the physics of the ejected envelope after the supernova explosion. This will allow, for example, supernovae of Type I to be calibrated as
standard candles to determine the distance of far-away galaxies. The latter program
is clearly a HST program and will greatly benefit from the precursor work done by
IUE. However, the study of supernovae in nearby galaxies will continue to require
extensive IUE observations, because of IUE scheduling flexibility and of the
important requirement to continue observations during many weeks and months.
HST could of course be used in a similar mode in studying supernovae in distant
galaxies and will provide the opportunity to carry out a detailed comparison of SN
light curves to establish whether or not differences exist which may be related to

the evolutionary state of the parent galaxy. It is obvious that without the precursor work done with IUE and its continuation during the foreseeable lifetime of IUE, supernova study in all its aspects and implication would be greatly hampered.

4. IUE Comparison with HST

The IUE compares most favorably with the HST when spectroscopy of extended sources is considered. Figure 1, which is part of a diagram prepared by William P. Blair for the Hopkins Ultraviolet Telescope project, gives an indication of the comparison in the far ultraviolet. The sensitivity levels shown for the four instruments represent a signal-to-noise of 5 Å^{-1} in thirty minutes exposure. Background effects are included. "Voyager" is the ultraviolet spectrometer aboard the Voyager spacecraft, now travelling beyond Uranus, while 'HUT-1st' is the first-order spectrometer on the Hopkins Ultraviolet Telescope, one of three instruments making up the ASTRO observatory, which was scheduled to fly on the next Space Shuttle mission following the Challenger tragedy.

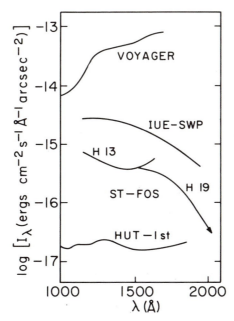

Fig. 1.   A comparison, for 1′ extended sources, of the sensitivity of IUE and HST, as well as two other ultraviolet instruments. (We thank William P. Blair for permission to use this figure).

Blair's calculation presents $\log I_\lambda$ versus λ, where I_λ is the source *intensity* (ergs cm^{-2} s^{-1} Å^{-1} arcsec^{-2}) required to yield S/N = 5 Å^{-1} in 30 min. For IUE, the large aperture was assumed used. For HUT the 18″ × 120″ aperture was assumed. For FOS, a 1″ × 1″ slit was assumed.

It is apparent that for this special case, there is less than an order of magnitude difference in sensitivity between HST and IUE. This is not to denigrate HST, it is to demonstrate that IUE is a fitting complement, especially in view of the fact that IUE obtains its entire spectrum simultaneously.

The IUE is, and if it survives it will remain, a powerful ultraviolet observatory to complement the HST.

5. Rationale: IUE in the Era of HST

The Hubble Space Telescope will be overbooked by a large factor.

Observing proposals for HST will be declined, when there is any possibility of successfully attacking the problem in question from the ground or at other wavelengths. Only proposals that need some of the special abilities of HST can be accepted.

If large ground-based telescopes, with their light-gathering power in some cases much greater than that of HST, and with their highly varied complement of focal-plane instruments (including sophisticated CCD detectors, fruit of HST development), can make significant progress on a problem, this will free up HST observing time for other important projects which can *only* be addressed by HST.

The point is, that every non-HST observing capability that we have, that can attack a problem that would otherwise have to be done by HST, provides to science an equivalent amount of observing time on the billion-dollar HST.

There are many ground based facilities, optical and radio, that can make this profound 'equivalent financial' contribution (by saving HST time). But a large chunk of HST's capability is in the ultraviolet, and apart from ASTRO (a planned 9-day Shuttle mission), there is *no* planned low-cost mission that can free up HST observing time. . . . except for IUE, or the launch of a replacement. Running IUE (very low cost), or launching a replacement (not inordinately expensive) *buys observing time on the billion-dollar HST.*

This is the fundamental rationale for maintaining or expanding IUE capability in the era of the Hubble Space Telescope.

6. Conclusion

IUE was both a technical and a scientific precursor to the HST, and still has observing capabilities which in limited areas are comparable to those of HST. Any observing facility which can relieve HST of observing tasks, buys for science that precious commodity, HST observing time. IUE is unique in that it is the only *ultraviolet* observing facility which is available to relieve HST in a major area of its function, ultraviolet spectroscopy. For this reason, we conclude that maintainance of IUE capability is an extremely high priority, not only for ultraviolet spectroscopy, but for all of observational astronomy.

MINING THE IUE ARCHIVE

DAVID GIARETTA

Space Telescope Science Institute, Baltimore, MD, U.S.A.

and

Rutherford Appleton Laboratory, Chilton, Didcot, Oxfordshire, England

JAYLEE M. MEAD

Laboratory for Astronomy and Solar Physics, Goddard Space Flight Center, Greenbelt, MD, U.S.A.

and

PIERO BENVENUTI

Space-Telescope — European Coordinating Facility, European Southern Observatory, Garching bei Munchen, F.R.G.

1. Introduction

All the data obtained with IUE are archived by each of the three agencies; at the time of writing, this exceeds 55 000 images. The data for which the six months embargo on release has expired are available to the whole astronomical community. This paper describes the IUE archives and the uses to which they have been put, and attempts to throw some light on possible future developments.

2. History

The primary NASA archive is the National Space Science Data Center (NSSDC) located at the NASA Goddard Space Flight Center. It contains the complete set of raw and processed data. Soon after launch, West designed a 'Merged Observing Log' and put it in computerized form. The observatory made a 'browse library' at Goddard, consisting of photographic reconstructions, known as 'photowrites'.

In 1981, NASA established two regional data analysis facilities (RDAF), one at Goddard and the other at the University of Colorado, and provided the means of reviewing the observing log. Lindler converted the log to an on-line relational database and developed software to allow astronomers to select the observations of particular interest to them. In 1982, Heap, Sullivan, and Wade copied the spectral data files at the NSSDC and installed them on Goddard's mass-storage device, thus providing on-line access to all IUE spectra. A tape copy of these data, known as the Condensed Data Archives, was also sent to the University of Colorado. The RDAFs then retrieved data for users of the two facilities.

In 1985, Heap, Sullivan, and Bohlin developed the 'Goddard Catalogue of Ultraviolet Fluxes' and installed it as an on-line database at the two RDAFs. This catalog gives the ultraviolet flux registered in 30 passbands for all objects observed at low-dispersion. In April 1986, the Goddard RDAF announced the capability to supply astronomers worldwide with IUE spectra from the Condensed Data

Y. Kondo (ed.), Scientific Accomplishments of the IUE, pp. 759—769.
© 1987 *by D. Reidel Publishing Company.*

Archives (CDA). In addition, Sullivan *et al.* (1986) developed software that allows astronomers to access the CDA directly from remote sites and to plot or print out the spectra.

The ESA archive is maintained at the ground station at VILSPA. There are currently differences in detailed tape formats between the NSSDC and VILSPA, but otherwise the data should be identical. The VILSPA archive serves the astronomical community of the ESA nations (apart from the UK) and many others worldwide.

A copy of the IUE archive is kept by SERC in the World Data Center Cl at Rutherford Appleton Laboratory (RAL). The UK astronomical community is served by this archive, with additional support from the IUE Support Office at RAL. The format of this data is entirely compatible with the GSFC archive.

The development of the archives is documented in various newsletters distributed by ESA, NASA and SERC and in the minutes of the biannual 3-Agency meetings. Copies of these and other related information may be found at the IUE observatories at GSFC and VILSPA, and at the IUE Support Office, RAL.

3. What is the IUE Archive

After eight years of successful operation the IUE archive is recognized as one of the most durable scientific outputs of the IUE project. Its value is not only due to its size but particularly to its relative homogeneity which is based on a well understood and monitored calibration of the data.

Before launch, the NASA IUE Project was given no responsibility regarding the archive other than to deposit data promptly at the NSSDC. ESA and SERC also did little planning for archival data, expecting essentially to receive whatever NASA sent them. After the archiving was begun, it was recognized that the only way the archive would reach the full potential that was expected of it would be for the project to actively establish requirements for and monitor such things as accounting procedures, distribution policies and standards, quality standards for data products, and the contents and formats of catalogs and other ancillary products. There were several reasons for this, but in essence only the project was in a position to be fully aware of the contents of the archive and how it had to perform in order to meet the overall mission success criteria. After it was realized what had to be done, a concerted effort was made to design and implement an archive service that would meet the needs.

Looking now at the future from the IUE experience, it is quite clear that the archives of other space observatories (and possibly of major ground-based facilities) will also become a standard tool for astrophysical research. It is important therefore to critically analyze the IUE archive with the aim of pointing out deficiencies and inconveniences which can still be taken into account in the planning and implementation of similar future databases.

We present such an analysis in the following sections, along with a brief description of the different components of the IUE archive.

3.1. IUE DATA

The data for each image is archived as a set usually consisting of a raw image, a processed image and one or more extracted spectra. Most images contain a simple spectrum of one object. Some have multiple spectra of the same object, for example, low dispersion spectra in each of the apertures. A few even have several spectra of different objects within the same image. In certain cases the Fine Error Sensor (FES) image associated with a spectral image has also been archived if it is thought to be useful for interpretation of the data. This shows a 16 arc minute square area of the sky usually centered on the target. All the flight calibration images — both flat fields and wavelength calibrations — may be found in the archive.

The IUE processing software has been continually evolving; the instruments have been recalibrated several times; as a result, the archive is not a completely homogeneously processed data set.

3.2. IUE MERGED OBSERVING LOG

The index to the data is the IUE Merged Observing Log. This is made at the two ground stations and is produced every two months, ordered in three ways: right ascension, object class and date. The data which appears in the log consists of astronomical information about the target, data on the instrument configuration, and processing information about the calibration and algorithms used.

As noted in the preface to the Merged Log, certain data fields are closely controlled and monitored by the observatory staff. Extra care should be taken when using data in the remaining fields since they are obtained from guest observers and are not closely monitored. In particular, the following comments on some of the fields should be taken into account:

Object ID

Supposedly, it identifies the target of the observation. Unfortunately, due to the lack of a clear naming policy and to the lack of discipline on the astronomers' part, the content of this field is inconsistent (even in its format) and therefore practically useless, in particular for any automatic search.

Target's Coordinates

This field contains the coordinates of the object as they were provided by the observer. This field, together with the object ID, should allow a proper identification of the target. However, errors of a few arc seconds may be expected. In the case of an extended object (e.g., a galaxy), these coordinates usually refer to its center, irrespective of the position which was actually observed. Information on the exact pointing during the exposure has to be found elsewhere.

Object Class

This two-digit code identifies the astrophysical class to which the object belongs. Although this classification scheme is very simple and occasionally ambiguous, it is essential for any systematic approach to the IUE Log. It should be noted that this scheme was not defined before the launch of IUE, but was gradually designed and improved according to its use.

The Merged Log is available on microfiche distributed periodically by each agency, often timed to be useful for proposal preparation. Each agency also makes the log available for on-line queries, as mentioned earlier.

Some data is omitted from the Merged Log but is available in hand-written form in the Observatories' exposure log sheets. Other useful data is kept in the form of photowrite images at GSFC which may be browsed through in order to perform a very quick evaluation of image quality and spurious features before requesting retrieval. Copies of the photowrite images can be obtained from the NSSDC. A total of 4340 photowrite images had been supplied from the NSSDC archive through 31 August 1985. Many astronomers consider the photowrites to be a very useful tool.

3.3. OBJECT DISTRIBUTION

It is informative to examine the distribution of types of objects observed with IUE, and their positions on the sky. Figure 1 is a histogram of observations according to the IUE Merged Log object classes, roughly organized as follows: solar system, Wolf—Rayet and O stars, B stars, A stars, F—M stars, variables and symbiotics, miscellaneous objects, nebulae, and extragalactic objects, each group having 10 subdivisions, as described in the Merged Log.

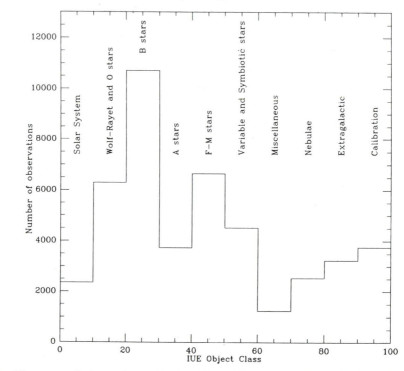

Fig. 1. Histogram of observations with IUE Merged Log object classes. This and the next two figures use data up to the beginning of 1985.

Figure 2 shows the positions of all the observations made with IUE, as given in the current Merged Log, in a whole sky equal area projection with equatorial 1950 coordinates. There are several features worthy of note, the main ones of which are the widespread and fairly uniform distribution over the sky, together with the concentrations in the galactic plane and the Large Magellanic Cloud. A histogram of the Merged Log observations by magnitude is shown in Figure 3.

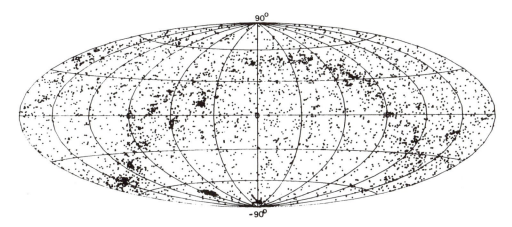

Fig. 2. Positions of all observations made with IUE, in a whole sky equal area projection with equatorial 1950 coordinates.

4. Use of the Archive Data

Images of a particular object or of a class of objects can be requested for analysis or re-analysis. The archival user may have a different research interest from the original observer and thus pursue a new interpretation for the data. For example, the original observer may have used the object as a simple light source in order to investigate the medium between Earth and the object; another user may be interested in the object itself. Sometimes archival data may supplement data obtained about other objects or about the same object in additional wavelength ranges or from different instruments. Yet again there may be fortuitously repeated observations of the same object spread over time — although this was guarded against later in the mission — and these may be combined to examine variability. Often the data has been used in the preparation of other observing programs, both those using IUE and also other instruments. Many of the earlier chapters in this book describe scientific papers based on extensive use of the IUE archives.

Associated with the archive itself is the necessity of knowing of related work using the same or similar data. Observations with IUE have resulted in the publication of more than a thousand papers in referred journals describing studies obtained using data from this satellite.

The *Bibliographical Index of Objects Observed by IUE 1978—85* (Mead *et al.*, 1986) is the result of a search of 1134 of these papers in order to record the names of the astronomical objects discussed, along with each reference. This

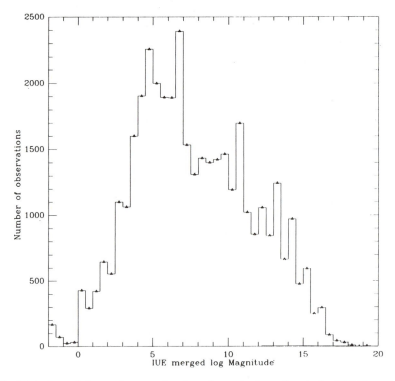

Fig. 3.   Histogram of observations as a function of magnitude as given in the IUE Merged Log.

bibliographical index enables a user to tell where to find published papers describing IUE observations for the objects of interest.

4.1. SURVEYS AND ATLASES

Microfiche copies of the IUE low dispersion archive through to the end of 1984 were produced and widely distributed by Giaretta and RAL. Several other surveys and atlases using the IUE archives have been published, and others are in progress. The purpose of these tools is to facilitate additional research using IUE data.

The *IUE Ultraviolet Spectral Atlas* (Wu *et al.*, 1983) provides low dispersion trailed IUE spectra for a representative set of spectral type standard stars with a reasonably good coverage of the Hertzsprung—Russell diagram.

The *Atlas of High Resolution IUE Spectra of Late-Type Stars* (*2500—3230 Å*) (Wing *et al.*, 1983) presents the ultraviolet spectra of 13 late-type stars in the high-resolution (0.3 Å) mode.

The *IUE Low-Dispersion Spectra Reference Atlas — Part 1. Normal Stars* (Heck *et al.*, 1984) provides a set of representative IUE low-dispersion spectra for the purpose of constructing a classification scheme in the UV, thus producing a sequence of standard stars in which the features transit smoothly from one standard to the next, staying as nearly as possible within the general MK system.

The *International Ultraviolet Explorer Atlas of O-Type Spectra from 1200—1900 Å* (Walborn *et al.*, 1985) investigates the existence of systematic trends in the ultraviolet line spectra of the O stars, including the prominent stellar wind features, and the degree to which they correlate with the optical spectral classifications.

5. How the Archive is Operated

A common theme between the operation of the archives run by the three agencies is one of continual development as new resources become available. The developments are driven by, and almost certainly also help to fuel, increased usage of the archives. There are interesting lessons to be learned by studying the operation of the archives and their developments in more detail.

5.1. HOW THE DATA IS REQUESTED

The basic service which an archive provides is to retrieve requested data and send it to the requestor. This is distinct from other systems which provide some quick look capability. The main difference is simply quantity — the archives can provide copies of the images as well as the extracted spectra in volumes measured in megabytes. Quick look facilities typically allow only a small amount of data for each spectrum.

Taking such things as accuracy of reproduction for granted, the service may be judged by two criteria. The first and probably most important is the time it takes between a request and final receipt of the data. The second is the convenience of use.

An analysis of the operation shows that the four key areas where avoidable delays occur are those where some human agent has to intervene. This may be illustrated by the original method of request used by all agencies; this method remains a fall-back system for a user with no other access to the system. First, the astronomer would mail a written list of images to the data center. Second, someone had to find the archive location of each item and then initiate a tape copying job on a computer. Third, computer resources being limited, a request would generally be broken into several parts to be performed in sequence. SERC and NASA both use large mainframe central computing facilities which tend to be lightly loaded at night. But since the retrieval was tied to the working hours of the archives' personnel, little advantage could be taken of this fact. Fourth, the retrieved data had to be transferred to the astronomer.

Each of these areas is being addressed by the agencies in their different ways. Electronic mail is replacing paper requests; the computer itself can automatically perform the cross referencing of image identifier with archive location. Improvements in the third area require the requestor to play an active part in the retrieval. Since astronomers are not tied to normal business hours, great improvements may be made relatively painlessly by making it possible for an astronomer to make the request directly to the computer at off-peak hours. Since the data is usually measured in tens of megabytes per request, the return medium at the moment is magnetic tape. This usually means that the fourth delay remains governed by the mail service. It should be mentioned that in all cases the service is essentially free.

An example is the system used by the SERC in which the request is made via Starlink and the Joint Academic network (Giaretta, 1984). The requestor is allocated a tape to use and then at his or her leisure runs a program on a local node which takes camera and image numbers and remotely submits a batch job to a central mainframe at RAL. When the copying job is performed, a short summary listing is sent over the network back to the user where it can be inspected and further copying jobs submitted. A running estimate is kept of the length of output tape used so that efficient use may be made of the tape. When the tape is full, or the user has all the data required, a network message may be sent to RAL and shortly thereafter the tape is dispatched. A similar mechanism exists at the Goddard mainframe computer, as described earlier.

The effect of these developments may be illustrated by noting that when the archives started operating, the turn around time could be several weeks, and there were large backlogs. Currently the turn around time is generally determined by the postal service taking the data tape to the astronomer. There are now essentially no backlogs of requests. Moreover, searches of astronomical databases cross-referenced to the IUE yield virtually direct requesting possible with no transcription necessary.

5.2. WHAT SORT OF THINGS ARE RETRIEVED

Together with the developments of the retrieval process there have been changes in the type of data requested. The three forms of the data are: raw images, corrected images and extracted spectra. At the beginning of operations, extracted spectra were of primary concern to astronomers. However, drawbacks or restrictions were perceived in the IUE project processing. Several groups developed new, interactive ways of overcoming various of these and the software packages then tended to become available to the wider community, for example, through the RDAFs in the USA and Starlink in the UK. These packages work on the corrected images and hence attention for retrieval tended to switch to this type of data. This trend will almost certainly be continued to allow the user to fully reprocess raw data unless the whole archive is reprocessed in a uniform way. Many studies involve intercomparison of spectra for subtle effects which may be obscured by differing processing methods.

5.3. HOW THE DATA IS DISTRIBUTED

When a large volume of data is retrieved for a user, the usual medium for distribution is magnetic tape, generally 1600 bpi unblocked, although other densities are possible. It should be noted that the format is not FITS but rather VICAR tape format. A full description is given in the handbooks produced by the observatories.

A large subset of the data bank is available in two other forms; both are for quick looks at the data, prior to detailed analysis from the image data. First plots on microfiche of all low dispersion spectra up to late 1984 have been produced by SERC, and these have been circulated widely. No processing was performed on the data, and the spectra are shown as produced by IUESIPS. GSFC may produce another set of similar fiche in the future. The second subset consisting of all extracted spectra, high and low dispersion, is stored on a mass storage device at GSFC and is available essentially on line. Users may log in remotely via a modem and display plots, data listings and catalog information for spectra they select. This data has been additionally processed beyond the IUESIPS product by the application of a range of well accepted corrections in order to give a more homogeneous set of spectra. Even so the data is specifically for quick look evaluation. For further details see Sullivan *et al.*, 1986.

5.4. THE QUANTITY OF DATA DISTRIBUTED

Figure 4 displays the cumulative plots of IUE image retrieval from the various

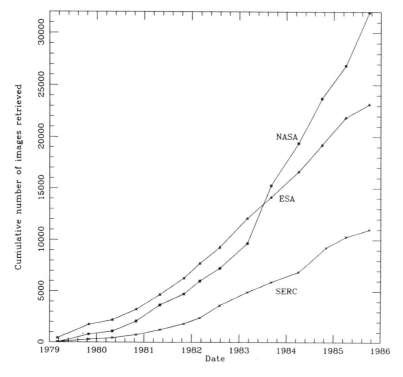

Fig. 4. Cumulative plot of the number of images which have been retrieved from the IUE archives. Data taken from the minutes of the biannual 3-Agency Reports.

archives. This plot shows that interest in the data continues to grow in all the astronomical communities served. The rate of retrieval exceeds the rate at which new data is being added to the archive by a large factor, the latter being about 6000 images per year. The total number of image retrievals equals approximately the total number of images in the archive. Note that the sudden rise in NASA retrievals after 1983 occurs from inclusion of RDAF statistics at this point after the capability was established for initiating image retrieval directly through the RDAFs.

6. Related Facilities

In parallel with the growth of the archives, facilities have been developed to assist the astronomical community to take full advantage of the data stored there. Such specialized organizations are necessary as repositories of expertise about the instrument characteristics, and also as providers of the resources required for the large scale, long term production of specialized analysis software which is inevitably needed. The RDAFs set up by NASA and the IUE Special Interest Group within the SERC Starlink Project have performed these functions admirably and have without doubt contributed greatly to the large volume of IUE related papers.

Astronomers visiting the RDAFs can request archival data in advance and have it waiting on-line at their arrival. In addition there are trained assistants to work with visitors by helping them to get started on the computer, if necessary, and advising them in the choice of software and other resource tools. The availability of such assistance has been of great value to guest astronomers in helping them handle their data analysis on their own as soon as possible and thus accomplishing the maximum work in the minimum time. Similar, but more limited facilities, are available at VILSPA.

7. How Will this Change in the Future

One can foresee that in the near future the main limitations or difficulties with the archives will be remedied. These include the non-uniformity of the processing of the data and the bulky form of hundreds of tapes. Already plans are being made for the reprocessing of at least part and possibly all the archive. New technologies such as optical disks will enable the physical size to be reduced dramatically. In this time-frame it is possible that individual images may be available essentially on-line to be transmitted across computer networks. However, it is difficult to foresee the transmission of large numbers of images this way. Staggered requesting may get around this, but magnetic tape and the postal system may well continue to be the main form of data transmission for some time.

On past experience we should expect access to the data to be made yet more convenient. Indeed as new technologies become available one may anticipate further copies of the archive to be distributed, although perhaps only after termination of satellite operations when the archive has stopped growing. These developments should be considered especially in the context of European networks such as Starlink.

Of course, the eventual ending of IUE satellite operations will have large consequences in terms of the funding of the project by all three agencies. However, plans are being made to at least allow a reasonable time for reprocessing operations. Since the main archives are permanent data repositories, the IUE data will be safely maintained regardless of the fate of the satellite.

An event which will also have the greatest consequence for the IUE archive is the system being set up for the dissemination of Space Telescope data. In principle, the IUE archive will be a fraction the eventual size of the ST archive and may be accommodated therein. There are already plans at least to keep a copy of the IUE archive at the Space Telescope Science Institute (STScI) at Baltimore, and also at the ST-European Coordination Facility (ST-ECF) at Munich, although it is not clear whether or not IUE data will be distributed from these places.

As a final note it would appear from the statistics on usage, and on results from questionnaires sent out to the astronomical community by the IUE project, that interest in, and use of, the IUE archive will flourish for many years to come.

It would be impossible to recognize everyone who has worked on the setting up, maintenance, and retrieval of IUE archival data; however, a few names stand out especially for their contributions in this area: W. Warren Jr. of the NSSDC, D. de Pablo of VILSPA, and K. Doidge of WDC, RAL.

Acknowledgement

The authors would like to thank A. Boggess and S. Heap for valuable comments on this paper.

References

Giaretta, D. L.: 1984, *IUEDEARCH*, Starlink User Note 58.
Heap, S. R., Sullivan, E. C., and Wade, C.: 1982, *Bull. Amer. Astron. Soc.* **14**, 919.
Heck, A., Egret, D., Jaschek, M., and Jaschek, C.: 1984, *IUE Low-Dispersion Spectra Reference Atlas — Part 1. Normal Stars*, ESA SP-1052.
Mead, J. M., Brotzman, L. E., and Kondo, Y.: 1986, Bibliographical Index of Objects Observed by IUE 1978—85, *NASA IUE Newsl.*, (Spec. Ed.) (in press).
Sullivan, E., Bohlin, R., Heap, S., and Mead, J. M.: 1986, Direct Access to the IUE Spectral Archive, *NASA IUE Newsl.*, (in press).
Walborn, N. R., Nichols-Bohlin, J., and Panek, R. J.: 1985, *International Ultraviolet Explorer Atlas of O-Type Spectra from 1200 to 1900 Å*, NASA Ref. Publ. 1155.
Wing, R. F., Carpenter, K. G., and Wahlgren, G. M.: 1983, *Atlas of IUE Spectra of Late-Type Stars 2500—3200 Å*, Spec. Publ. 1, Perkins Observatory.
Wu, C.-C., Ake, T. B., Boggess, A., Bohlin, R. C., Imhoff, C. L., Holm, A. V., Levay, Z. G., Panek, R. J., Schiffer, F. H., and Turnrose, B. E.: 1983, The IUE Ultraviolet Spectral Atlas, *NASA IUE Newsl.*, No. 22 (Spec. Ed.).

NAME INDEX

SUBJECT INDEX

LINE INDEX

STAR INDEX